Arctic Mineral Resources

Arctic Mineral Resources

Science and Technology

Special Issue Editor

Sergey V. Krivovichev

MDPI • Basel • Beijing • Wuhan • Barcelona • Belgrade

MDPI

Special Issue Editor
Sergey V. Krivovichev
Saint-Petersburg State University
Russia

Editorial Office
MDPI
St. Alban-Anlage 66
4052 Basel, Switzerland

This is a reprint of articles from the Special Issue published online in the open access journal *Minerals* (ISSN 2075-163X) from 2018 to 2019 (available at: https://www.mdpi.com/journal/minerals/special_issues/Arctic_mineral)

For citation purposes, cite each article independently as indicated on the article page online and as indicated below:

LastName, A.A.; LastName, B.B.; LastName, C.C. Article Title. *Journal Name* **Year**, *Article Number*, Page Range.

ISBN 978-3-03897-824-4 (Pbk)
ISBN 978-3-03897-825-1 (PDF)

Cover image courtesy of Sergey V. Krivovichev.

Contents

About the Special Issue Editor

Sergey V. Krivovichev, was born in 1972 in Leningrad and received his PhD and Doctor of Sciences degrees from St. Petersburg State University. He is currently a Head of the Federal Kola Science Centre of the Russian Academy of Sciences (Apatity, Russia) and Professor of Crystallography at St. Petersburg State University. During his career, he served as the President of the International Mineralogical Association, Chief Editor of the European Journal of Mineralogy, and Associate Editor of several other mineralogical journals, including Minerals. His research interests include Arctic geology and mineralogy, the structural and topological diversity of minerals and inorganic compounds (with special emphasis on uranium, copper, and lead oxocompounds), structural complexity, and mineral evolution. He is a Corresponding Member of the Russian Academy of Sciences and a Foreign Member of the Turin Academy of Sciences.

Preface to "Arctic Mineral Resources: Science and Technology"

The Arctic zone of the Earth is a major source of mineral and other natural resources for the future development of science and technology. It contains a large supply of strategic mineral deposits, including rare earths, copper, phosphorus, niobium, platinum-group elements, and other critical metals. The continued melting of the sea ice due to climate change makes these resources more accessible than ever before. However, the mineral exploration in the Arctic has always been a challenge due to the climatic restrictions, remote location, and vulnerability of Arctic ecosystems. This book covers a broad range of topics related to the problem of Arctic mineral resources, including geological, geochemical, and mineralogical aspects of their occurrence and formation; chemical technologies; and environmental and economic problems related to mineral exploration. The contributions can be tentatively classified into four major types: geodynamics and metallogeny, mineralogy and petrology, mineralogy and crystallography, and mining and chemical technologies associated with the exploration of mineral deposits and the use of raw materials for manufacturing new products. The book can be of interest for all those interested in Arctic issues and especially in Arctic mineral resources and associated problems of mineralogy, geology, geochemistry, and technology.

Sergey V. Krivovichev
Special Issue Editor

Editorial

Editorial for Special Issue "Arctic Mineral Resources: Science and Technology"

Sergey V. Krivovichev [1,2]

1   Kola Science Center, Russian Academy of Sciences, Fersmana str. 14, 184209 Apatity, Russia; krivovichev@admksc.apatity.ru or s.krivovichev@spbu.ru
2   Department of Crystallography, Institute of Earth Sciences, St. Petersburg State University, University Emb. 7/9, 199034 St. Petersburg, Russia

Received: 6 March 2019; Accepted: 14 March 2019; Published: 22 March 2019

check for updates

The Arctic zone of the Earth is a major source of mineral and other natural resources for the future development of science and technology. It contains a large supply of strategic mineral deposits, including rare earths, copper, phosphorus, niobium, platinum-group elements, and other critical metals. The continuing melting of the sea ice due to climate change makes these resources more accessible than ever before. However, mineral exploration in the Arctic has always been a challenge, due to the climatic restrictions, remote location and vulnerability of Arctic ecosystems. This Special Issue covers a broad range of topics related to the problem of Arctic mineral resources, including geological, geochemical and mineralogical aspects of their occurrence and formation, chemical technologies, environmental and economic problems of mineral exploration. The contributions can be broadly classified into four major types: geodynamics and metallogeny, mineralogy and petrology, mineralogy and crystallography, and mining and chemical technologies associated with exploration of mineral deposits and the use of raw materials for manufacturing new products.

The issue opens up with a review on the relations between geodynamics and oil and gas potential of the Yenisei-Khatanga Basin (Polar Siberia, central part of the Russian Arctic) provided by Vernikovsky et al. [1]. The main oil and gas generating deposits of the basin are Jurassic and Lower Cretaceous mudstones, and the latter contain 90% of known hydrocarbon reserves. The article by Kozlov et al. [2] is devoted to the Western part of the Russian Arctic and examines the interplay between geodynamic evolution and metallogeny of the region, which is extremely rich in unique and rare mineral deposits. The distribution of the latter is investigated in several subsequent papers. Kalinin et al. [3] reports on gold prospects in the region with special focus on regional metallogeny and distribution of mineralization, whereas the article by Kompanchenko et al. [4] focuses on vanadium mineralization. Uranium mineralization and the age of its formation is the subject of the paper by Kaulina et al. [5]. The region is also famous for its platinum-group-element (PGE) deposits, which are investigated by Bayanova et al. [6], who provides a comprehensive overview on the mechanisms of their formation and time evolution associated with long-lived plume activity. In a companion paper, Groshev and Karykowski [7] report on the results of the study of the main anorthosite layer of the West-Pana intrusion (Kola peninsula), which hosts the PGE mineralization. The Cu-Ni productive suite of the Pechenga structure on the Russian-Norway border is considered by Zhamaletdinov [8]. The idea of the study initiated back in Soviet time was to estimate the prospects for discovery of Cu-Ni deposits in northern Norway by means of single electric pulses generated by the 80 MW magneto-hydrodynamic (MHD) generator "Khibiny". The conclusion of the study is that the most promising potential for Cu-Ni deposits Pil'gujarvi formation of the Northern wing of the Pechenga structure is rather quickly wedged out in Norway, while the conductive horizons of the Southern part of Pechenga, which have a weak prospect for Cu-Ni ores, continue into Norway nearly without a loss of power and integral electrical conductivity.

The understanding of the structure, origin and evolution of mineral deposits is impossible without the use of comprehensive information on the geochemical and petrochemical features of their rocks and minerals. Thus, the second part of the issue contains a collection of articles on mineralogical and petrological characterization of mineral deposits and its use in petrogenetic models. Khibiny (Kola peninsula, Russia) is one of the world largest peralkaline intrusions that hosts giant apatite deposits, which are mined by the "PhosAgro" company. Kogarko [9] reports on the results of detailed chemical studies of apatite in the Khibiny apatite-nepheline deposits, which prompted her to conclude that the main mechanism for their formation was the gravitational settling of large nepheline crystals in the lower part of the magma chamber, while small apatite crystals were concentrated in its upper part. The formation mechanisms of zeolite deposits in carbonatites and their geochemical and mineralogical constraints are examined by Zozulya et al. [10] using the Breivikbotn deposit (Northern Norway) as an example. Three following papers of the issue [11–13] are devoted to 3-D mineralogical mapping of the Kovdor phoscorite-carbonatite complex (Kola peninsula, Russia). On the basis of detailed mineralogical studies (including crystal-structure analysis) of forsterite [11], sulfides [12] and pyrochlore-supergroup minerals [13], the authors construct a 3-D model of the deposit and try to decipher major features of its genesis and evolution. Zircon macrocrysts from the Drybones Bay kimberlite pipe (Northwest Territories, Canada) are the subject of a detailed study of Reguir et al. [14], who, on the basis of a high-resolution trace element and geochronological study, provide strong evidence that the macrocrysts are the products of interaction between a shallow (<100 km) mantle source and transient kimberlitic melt.

Descriptive mineralogy and crystal chemistry provide an essential fundamental basis for our understanding of mineral matter and basic characteristics of their formation and behavior under changing chemical and physical parameters. The third part of the issue includes three papers on rare minerals from Kola peninsula. The review of Lyalina et al. [15] concentrates on beryllium mineralization. Twenty-eight different mineral species of Be have been found on the Kola peninsula, and their chemistry, crystallography and geological and mineralogical occurrences are reviewed in detail. The paper by Selivanova et al. [16] reports on the compositional and textural variations in hainite-(Y) and batievaite-(Y), two rare-earth-containing minerals from the Sakharjok Massif, Keivy alkaline province (Eastern part of the Kola peninsula). Zolotarev et al. [17] provides the first crystal-structure data on shkatulkalite, a rare mineral from the Lovozero massif that was discovered in 1996, but could not be structurally characterized until now.

The last part of the issue is devoted to technological problems associated with the exploration of mineral deposits and the use of the raw materials. Chanturiya et al. [18] review advanced techniques of saponite recovery from diamond processing plant water and the areas of saponite application, including modified saponite-based products. The Arkhangelsk-area diamond deposits are considered as an example. Masloboev et al. [19] provide a timely and comprehensive analysis of hydrometallurgical processing of low-grade sulfide ore and mine waste in the Arctic regions. The paper presents perspectives and challenges for heap leaching of sulfide and mixed ores from the Udokan (Russia) and Talvivaara (Finland) deposits, as well as technogenic waste dumps such as the Allarechensky deposit dumps (Russia). Antigorite is an important constituent of mining wastes and can be used in production of alkaline-activated binders, which is the subject of the report by Kalinkina et al. [20]. Finally, Gerasimova et al. [21] show how titanite ores such as those found in the Khibiny deposit can be used for industrial production of nanostructured and microporous titanosilicates that can be used as sorbents for the extraction of radionuclides from liquid radioactive wastes.

In total, the issue contains three reviews and eighteen full research articles on various problems associated with investigation of Arctic mineral resources. As Arctic issues become more and more relevant, we hope that the issue provides a useful overview of a broad range of topics related to the geological and technological exploration of the Arctic.

Author Contributions: S.V.K. wrote this editorial.

Conflicts of Interest: The author declares no conflict of interest.

References

1. Vernikovsky, V.; Shemin, G.; Deev, E.; Metelkin, D.; Matushkin, N.; Pervukhina, N. Geodynamics and Oil and Gas Potential of the Yenisei-Khatanga Basin (Polar Siberia). *Minerals* **2018**, *8*, 510. [CrossRef]
2. Kozlov, N.; Sorokhtin, N.; Martynov, E. Geodynamic Evolution and Metallogeny of Archaean Structural and Compositional Complexes in the Northwestern Russian Arctic. *Minerals* **2018**, *8*, 573. [CrossRef]
3. Kalinin, A.; Kazanov, O.; Bezrukov, V.; Prokofiev, V. Gold Prospects in the Western Segment of the Russian Arctic: Regional Metallogeny and Distribution of Mineralization. *Minerals* **2019**, *9*, 137. [CrossRef]
4. Kompanchenko, A.; Voloshin, A.; Balagansky, V. Vanadium Mineralization in the Kola Region, Fennoscandian Shield. *Minerals* **2018**, *8*, 474. [CrossRef]
5. Kaulina, T.; Kalinin, A.; Il'chenko, V.; Gannibal, M.; Avedisyan, A.; Elizarov, D.; Nerovich, L.; Nitkina, E. Age and Formation Conditions of U Mineralization in the Litsa Area and the Salla-Kuolajarvi Zone (Kola Region, Russia). *Minerals* **2018**, *8*, 563. [CrossRef]
6. Bayanova, T.; Korchagin, A.; Mitrofanov, A.; Serov, P.; Ekimova, N.; Nitkina, E.; Kamensky, I.; Elizarov, D.; Huber, M. Long-Lived Mantle Plume and Polyphase Evolution of Palaeoproterozoic PGE Intrusions in the Fennoscandian Shield. *Minerals* **2019**, *9*, 59. [CrossRef]
7. Groshev, N.; Karykowski, B. The Main Anorthosite Layer of the West-Pana Intrusion, Kola Region: Geology and U-Pb Age Dating. *Minerals* **2019**, *9*, 71. [CrossRef]
8. Zhamaletdinov, A. Study of the Cu-Ni Productive Suite of the Pechenga Structure on the Russian-Norway Border Zone with the Use of MHD Installation "Khibiny". *Minerals* **2019**, *9*, 96. [CrossRef]
9. Kogarko, L. Chemical Composition and Petrogenetic Implications of Apatite in the Khibiny Apatite-Nepheline Deposits (Kola Peninsula). *Minerals* **2018**, *8*, 532. [CrossRef]
10. Zozulya, D.; Kullerud, K.; Ravna, E.; Savchenko, Y.; Selivanova, E.; Timofeeva, M. Mineralogical and Geochemical Constraints on Magma Evolution and Late-Stage Crystallization History of the Breivikbotn Silicocarbonatite, Seiland Igneous Province in Northern Norway: Prerequisites for Zeolite Deposits in Carbonatite Complexes. *Minerals* **2018**, *8*, 537. [CrossRef]
11. Mikhailova, J.; Ivanyuk, G.; Kalashnikov, A.; Pakhomovsky, Y.; Bazai, A.; Panikorovskii, T.; Yakovenchuk, V.; Konopleva, N.; Goryainov, P. Three-D Mineralogical Mapping of the Kovdor Phoscorite–Carbonatite Complex, NW Russia: I. Forsterite. *Minerals* **2018**, *8*, 260. [CrossRef]
12. Ivanyuk, G.; Pakhomovsky, Y.; Panikorovskii, T.; Mikhailova, J.; Kalashnikov, A.; Bazai, A.; Yakovenchuk, V.; Konopleva, N.; Goryainov, P. Three-D Mineralogical Mapping of the Kovdor Phoscorite-Carbonatite Complex, NW Russia: II. Sulfides. *Minerals* **2018**, *8*, 292. [CrossRef]
13. Ivanyuk, G.; Konopleva, N.; Yakovenchuk, V.; Pakhomovsky, Y.; Panikorovskii, T.; Kalashnikov, A.; Bocharov, V.; Bazai, A.; Mikhailova, J.; Goryainov, P. Three-D Mineralogical Mapping of the Kovdor Phoscorite-Carbonatite Complex, NW Russia: III. Pyrochlore Supergroup Minerals. *Minerals* **2018**, *8*, 277. [CrossRef]
14. Reguir, E.; Chakhmouradian, A.; Elliott, B.; Sheng, A.; Yang, P. Zircon Macrocrysts from the Drybones Bay Kimberlite Pipe (Northwest Territories, Canada): A High-Resolution Trace Element and Geochronological Study. *Minerals* **2018**, *8*, 481. [CrossRef]
15. Lyalina, L.; Selivanova, E.; Zozulya, D.; Ivanyuk, G. Beryllium Mineralogy of the Kola Peninsula, Russia—A Review. *Minerals* **2019**, *9*, 12. [CrossRef]
16. Selivanova, E.; Lyalina, L.; Savchenko, Y. Compositional and Textural Variations in Hainite-(Y) and Batievaite-(Y), Two Rinkite-Group Minerals from the Sakharjok Massif, Keivy Alkaline Province, NW Russia. *Minerals* **2018**, *8*, 458. [CrossRef]
17. Zolotarev, A.; Selivanova, E.; Krivovichev, S.; Savchenko, Y.; Panikorovskii, T.; Lyalina, L.; Pautov, L.; Yakovenchuk, V. Shkatulkalite, a Rare Mineral from the Lovozero Massif, Kola Peninsula: A Re-Investigation. *Minerals* **2018**, *8*, 303. [CrossRef]
18. Chanturiya, V.; Minenko, V.; Makarov, D.; Suvorova, O.; Selivanova, E. Advanced Techniques of Saponite Recovery from Diamond Processing Plant Water and Areas of Saponite Application. *Minerals* **2018**, *8*, 549. [CrossRef]

Minerals **2019**, *9*, 192

19. Masloboev, V.; Seleznev, S.; Svetlov, A.; Makarov, D. Hydrometallurgical Processing of Low-Grade Sulfide Ore and Mine Waste in the Arctic Regions: Perspectives and Challenges. *Minerals* **2018**, *8*, 436. [CrossRef]
20. Kalinkina, E.; Gurevich, B.; Kalinkin, A. Alkali-Activated Binder Based on Milled Antigorite. *Minerals* **2018**, *8*, 503. [CrossRef]
21. Gerasimova, L.; Nikolaev, A.; Maslova, M.; Shchukina, E.; Samburov, G.; Yakovenchuk, V.; Ivanyuk, G. Titanite Ores of the Khibiny Apatite-Nepheline-Deposits: Selective Mining, Processing and Application for Titanosilicate Synthesis. *Minerals* **2018**, *8*, 446. [CrossRef]

minerals

MDPI

Review

Geodynamics and Oil and Gas Potential of the Yenisei-Khatanga Basin (Polar Siberia)

Valery Vernikovsky [1,2,*], Georgy Shemin [1,2], Evgeny Deev [1,2], Dmitry Metelkin [1,2], Nikolay Matushkin [1,2] and Natalia Pervukhina [1]

[1] A.A. Trofimuk Institute of Petroleum Geology and Geophysics, Siberian Branch of the Russian Academy of Sciences, prosp. Akad. Koptyuga, 3, Novosibirsk 630090, Russia; SheminGG@ipgg.sbras.ru (G.S.); deevev@ngs.ru (E.D.); metelkindv@ipgg.sbras.ru (D.M.); matushkinny@ipgg.sbras.ru (N.M.); PervuhinaNV@ipgg.sbras.ru (N.P.)

[2] Department of Geology and Geophysics, Novosibirsk State University, Pirogova st. 1, Novosibirsk 630090, Russia

* Correspondence: VernikovskyVA@ipgg.sbras.ru; Tel.: +7-383-3636720

Received: 12 October 2018; Accepted: 3 November 2018; Published: 6 November 2018

check for updates

Abstract: The geodynamic development of the north–western (Arctic) margin of the Siberian craton is comprehensively analyzed for the first time based on our database as well as on the analysis of published material, from Precambrian-Paleozoic and Mesozoic folded structures to the formation of the Mesozoic–Cenozoic Yenisei-Khatanga sedimentary basin. We identify the main stages of the region's tectonic evolution related to collision and accretion processes, mainly subduction and rifting. It is demonstrated that the prototype of the Yenisei-Khatanga basin was a wide late Paleozoic foreland basin that extended from Southern Taimyr to the Tunguska syneclise and deepened towards Taimyr. The formation of the Yenisei-Khatanga basin, as well as of the West-Siberian basin, was due to continental rifting in the Permian–Triassic. The study describes the main oil and gas generating deposits of the basin, which are mainly Jurassic and Lower Cretaceous mudstones. It is shown that the Lower Cretaceous deposits contain 90% of known hydrocarbon reserves. These are mostly stacked reservoirs with gas, gas condensate and condensate with rims. The study also presents data on oil and gas reservoirs, plays and seals in the Triassic, Jurassic and Cretaceous complexes.

Keywords: Yenisei-Khatanga basin; Siberian craton; Arctic; oil; gas; petroleum potential; geodynamics

1. Introduction

The structures of the north–western margin of the Siberian craton have attracted the attention of geologists for over a hundred years [1–15] and many others. The earliest and most essential questions on the study area are: (a) How did the Taimyr-Severnaya Zemlya folded area form—a key structure of the Artic, and is it a part of the Siberian craton; (b) what are the sequence and stages of its formation, how is the formation of this folded area related to the formation of the Yenisei-Khatanga basin; (c) what serves as the basement for the latter. V.E. Khain wrote in his book "Tectonics of continents and oceans" (2001) [3]: *"There is no doubt that the basin and the rift that generated it were emplaced within the Siberian craton, and that its territory has the same platform cover, since it also is present in Southern Taimyr. At the same time, it is supposed that in the zone of the Mesozoic basin there was a Riphean rift, and the Permian-Early Triassic rifting was preceded by arched uplift"* [3] (p. 228). The significant remoteness of these structures, harsh climate conditions, minuscule amounts of geostructural, geochronological, geophysical investigations, and even less drilling works prevented researchers from building a noncontradictory model for the formation of this segment of the Earth's lithosphere. Nonetheless, the interest in this region began to rise in the last decades. This was due to the need to estimate the oil and gas potential of this territory and to the discovery of the first seeps and deposits of petroleum, even though the Yenisei-Khatanga basin definitely does not have the same resource potential as its huge neighbor—the West Siberian basin [16].

In this paper we present our point of view on the geodynamic history of the north–western margin of the Siberian craton and the formation of the Yenisei-Khatanga basin. Our reconstructions generalize our own results obtained during many years of studies of the various geological structures of the Taimyr Peninsula and the Severnaya Zemlya Archipelago, as well as a body of modern geological, geophysical, geodynamic, paleomagnetic and petro-geochemical data. This includes materials of seismic surveys and drilling data in the Yenisei-Khatanga basin.

The Yenisei-Khatanga basin is located in the north–western part of the margin of the Siberian craton (Figures 1 and 2). It extends for 900 km from the Gydan Bay of the Kara Sea in the west to the Khatanga Bay of the Laptev Sea in the east. From the north, the basin is bordered by the Byrranga Mountains of the Taimyr Peninsula, and from the south, by the Putorana Plateau. The maximum width of the basin reaches 450 km. The area of the basin is approximately 300,000 km^2.

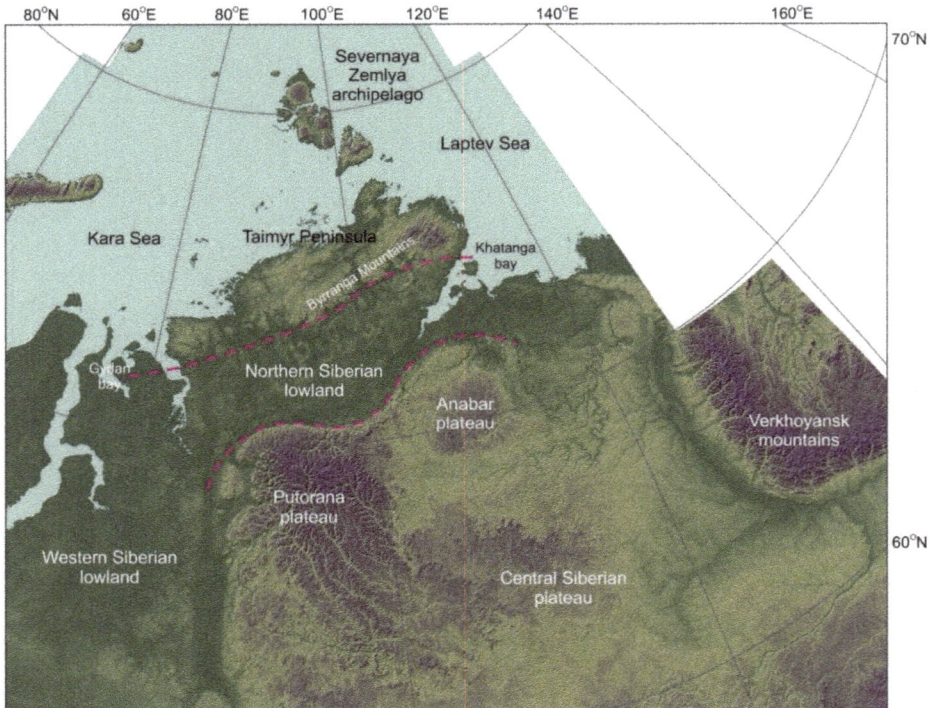

Figure 1. Geographic position of the Yenisei-Khatanga Basin (magenta line).

The Yenisei-Khatanga basin is a Permian–Cenozoic sedimentary basin dividing the northern margin of the Siberian craton and the Taimyr-Severnaya Zemlya fold-and-thrust belt (Figures 2 and 3). The basin was formed above a continental rift of late Permian–Early Triassic age and filled with a 12 km thick sequence of terrigenous Triassic, Jurassic, Cretaceous and Cenozoic deposits. The northern and southern boundaries of the basin are traced along the wedging-out of its deposits [2,3]. In the western part the basin widens and transitions into the West Siberian sedimentary basin, with which it has a similar geological history. The margin between them is conventionally defined where the sub-latitudinal strikes of structures typical of the Yenisei-Khatanga basin change into sub-longitudinal ones that are typical of the West Siberian plate. In the east, the Yenisei-Khatanga basin is overlapped by the Laptev Sea [3].

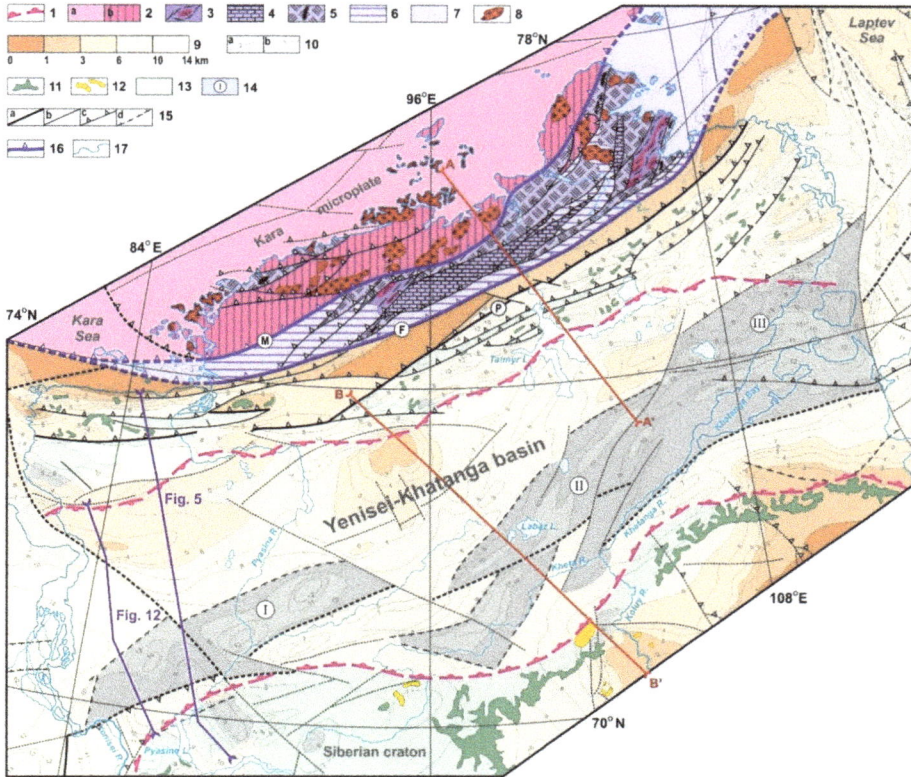

Figure 2. Tectonic map of the north–western part of the Siberian craton, the Taimyr folded area and the Yenisei-Khatanga basin. Composed using References [5,6] with additions by the authors. 1—boundary of the Yenisei-Khatanga basin along outcrops of Early Jurassic and younger rocks; 2—Kara microcontinent (a) including metamorphosed and deformed flyschoid passive margin sediments (NP–Є) of the North Taimyr Zone (b); 3–7—the Central Taimyr accretionary belt: 3—gneiss complexes of cratonic terranes with 940–850 Ma granitoids; 4—terranes of carbonate shelves (MP(?)–NP); 5—island arc terranes, ophiolites and plagiogranites (750–630 Ma), and accretionary wedge complexes; 6—molasses (NP), carbonate, black shales, shelf and hemipelagic sediments (NP₃–Є); 7—offshore part of the belt; 8—syncollisional (306–300 Ma) and postcollisional (264–258 Ma) granites; 9—depths of the Siberian craton crystalline basement and of the deformed, mostly Paleozoic rocks within the Yenisei-Khatanga basin and the West Siberian basin; 10—isohypses (depth in km) of the seismic basement roof: (a)—established, (b)—inferred; 11–13—Siberian traps (P₂-T₁): 11—dolerites and their differentiates; 12—alkaline ultramafic rocks and carbonatites; 13—flood basalts and associating volcanogenic-sedimentary deposits; 14—zones of continental rifting and range formation: I—Tanama-Malokhet, II—Rassokhin and III—Balakhnin from [17]; 15—large faults (a), other faults (b), thrusts and upthrows (c), normal faults (d); 16—sutures with thrust kinematics; 17—coastline and water bodies. Letters indicate named sutures and thrusts: M—Main Taimyr, F—Pyasina-Faddey, P—Pogranichny. Lines AA′ and BB′ show locations of geological-geophysical sections in Figure 3.

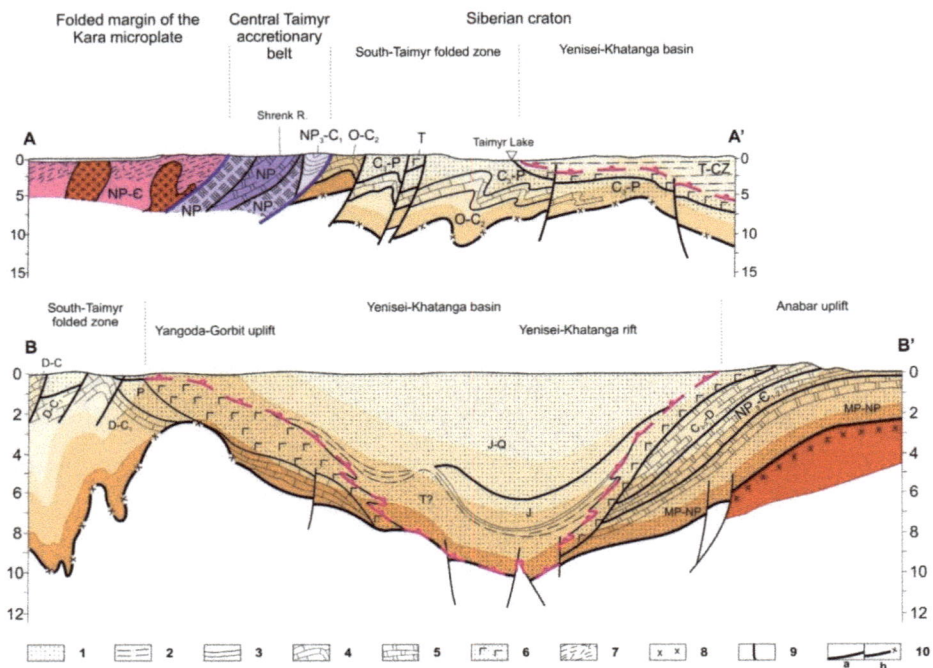

Figure 3. Geological-geophysical sections along lines AA′ and BB′. 1—sands and sandstones; 2—sandy-argillaceous sediments; 3—mudstones, and shales; 4—limestones; 5—dolomites; 6—basalts, tuffs, terrigenous rocks (P_2–T_1); 7—folded structure of the crystalline basement Kara microcontinent (shown figuratively); 8—gneiss-granite and granulite-mafic complexes of the crystalline basement; 9—faults; 10—roof of the seismic basement: (a) Established, (b) inferred from indirect data and single observations. Remaining symbols in caption to Figure 2. Vertical scale is in km.

The main information on the structure of the Mesozoic sedimentary infill of the Yenisei-Khatanga basin comes from seismic surveys and drilling. However, the focus areas of exploration activity including seismic lines and wells, are concentrated in the western part of the basin, which is directly adjacent to the West Siberian petroleum basin (Figure 4). Some wells have penetrated the full thickness of the Mesozoic deposits, including the volcanogenic-sedimentary Lower Triassic.

Seismic data, materials from well drilling, and geological analogies with surrounding structures show that the basement of the basin is composed of deformed siliceous-carbonate, carbonate, salt-bearing, terrigenous and terrigenous-volcanic complexes of a wide range of ages from the Mesoproterozoic to the Early Triassic [17–20]. The thickness of these units is significantly greater than the thickness of the sedimentary infill of the Yenisei-Khatanga basin itself [21,22]. The formation of these complexes is directly linked to the evolution of the northern margin of the Siberian craton: The change from a passive continental margin setting to an active one in the Neoproterozoic, followed by the formation of the Taimyr-Severnaya Zemlya fold-and-thrust belt in the late Paleozoic, and of the Yenisei-Khatanga rift system in the Permian–Triassic. Significant structural alterations at the Jurassic–Cretaceous boundary, the Paleogene and the Neogene, are also manifested within the Yenisei-Khatanga basin itself.

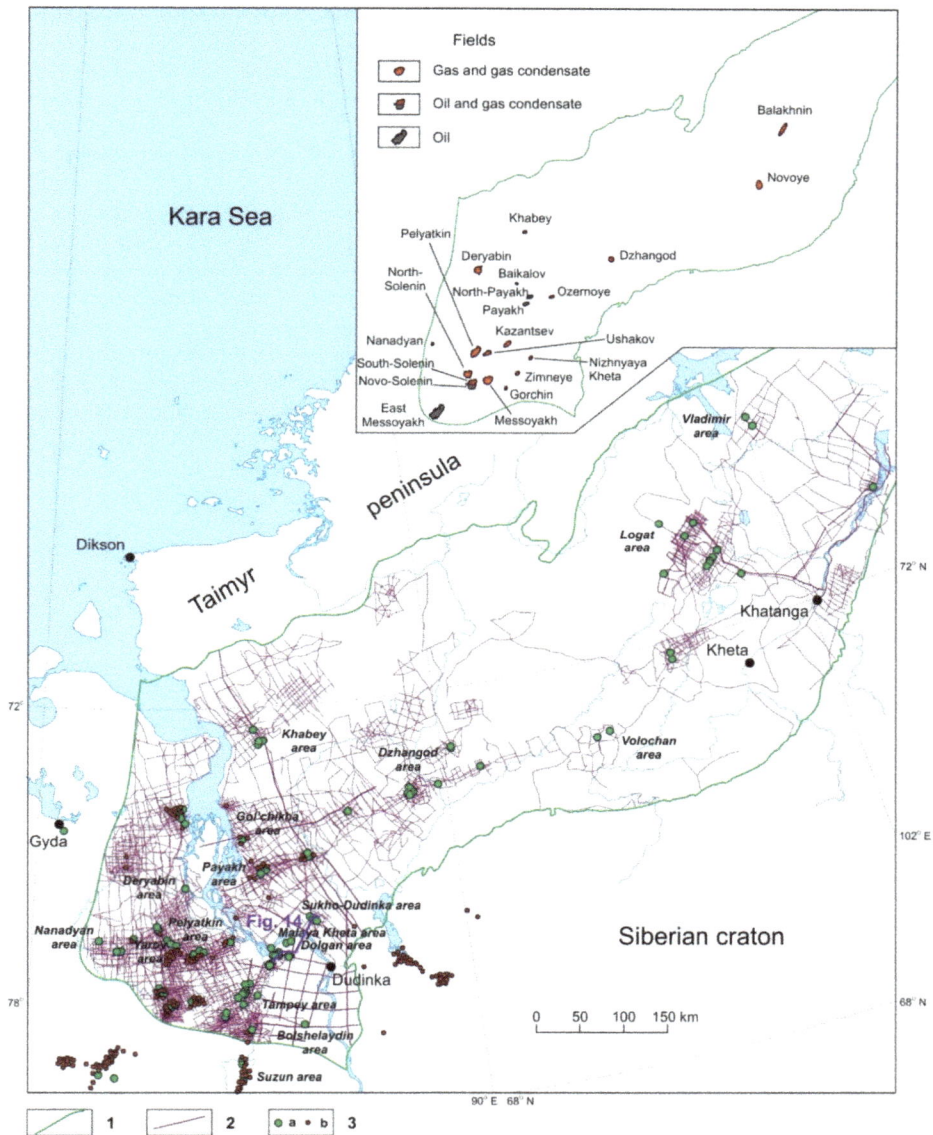

Figure 4. State of exploration of the Yenisei-Khatanga Basin by seismic surveys and deep drilling. 1—boundaries of the Yenisei-Khatanga basin; 2—seismic lines; 3—deep wells that penetrated: Pre-Jurassic and Jurassic deposits (a), Cretaceous deposits (b). Inset shows the location of oil and gas deposits.

2. Tectonic Evolution of the north–western Margin of the Siberian Craton and the Formation of the Basement of the Yenisei-Khatanga Basin in the Mesoproterozoic—Early Triassic

In the north the Yenisei-Khatanga basin borders to the southern margin of the Taimyr fold belt, and in the south to the northern part of the Siberian craton. Therefore, understanding the geological evolution from the Mesoproterozoic to the Triassic of these adjacent areas that form the basement for the Yenisei-Khatanga basin is highly significant [1–3,5,7].

According to results of deep seismic sounding, the Moho boundary depth under the Yenisei-Khatanga basin varies from 27–33 km in the central part to 36–38 km under the North-Siberian monocline and to 40–43 km under Taimyr [6,23,24]. The crystalline basement is supposedly composed of metamorphic rocks of Archean–Paleoproterozoic age. The depth of the basement surface varies from 6–7 to 12 km in different estimates [5,21].

To demonstrate the tectonic evolution of the north–western margin of the Siberian craton we propose our own version of paleotectonic reconstructions (Figure 5) [25]. The reconstructions that span the entire time interval from the end of the Mesoproterozoic to the Early Triassic are based on geological, geochronological, and paleomagnetic data for the territory of the Siberian craton and adjacent structures. These data are analyzed in detail in References [25–29]. These results take into account published global paleotectonic reconstructions [30–33], however, the central place here is occupied by the Siberian craton and the territory of the Yenisei-Khatanga basin. Therefore, some tectonic events in the margin of the craton that are not related to the evolution of the basement of the Yenisei-Khatanga basin are not reflected in our reconstructions. The quantity and quality of paleomagnetic information on the craton and the terranes of its boundaries is not the same for the time slices presented in the figure. Some are well characterized, while others, only in general traits. Nonetheless, the general trends of tectonic development of the region are reconstructed quite clearly. Together with additional geological indicators, this fully compensates the lack of some paleomagnetic data.

According to available geological data and paleogeographic (paleogeodynamic) reconstructions (Figure 5a), the north–western margin of the Siberian craton was a passive continental margin in the Mesoproterozoic–early Neoproterozoic [5,6,34–36], as were its western and eastern margins [37–39]. The composition of Mesoproterozoic deposits, judging from coeval formations of the Siberian craton, was siliceous-carbonate, terrigenous and carbonate, and their thickness reached 2–3 km [17]. Recent common depth point data indicate that the thickness of this sequence in the northern margin of the Siberian craton is up to 12 km (Figure 6).

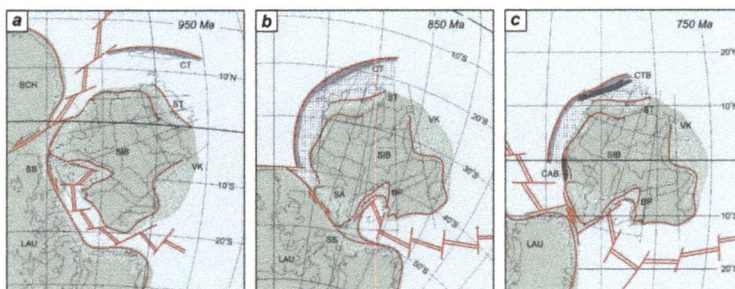

Figure 5. *Cont.*

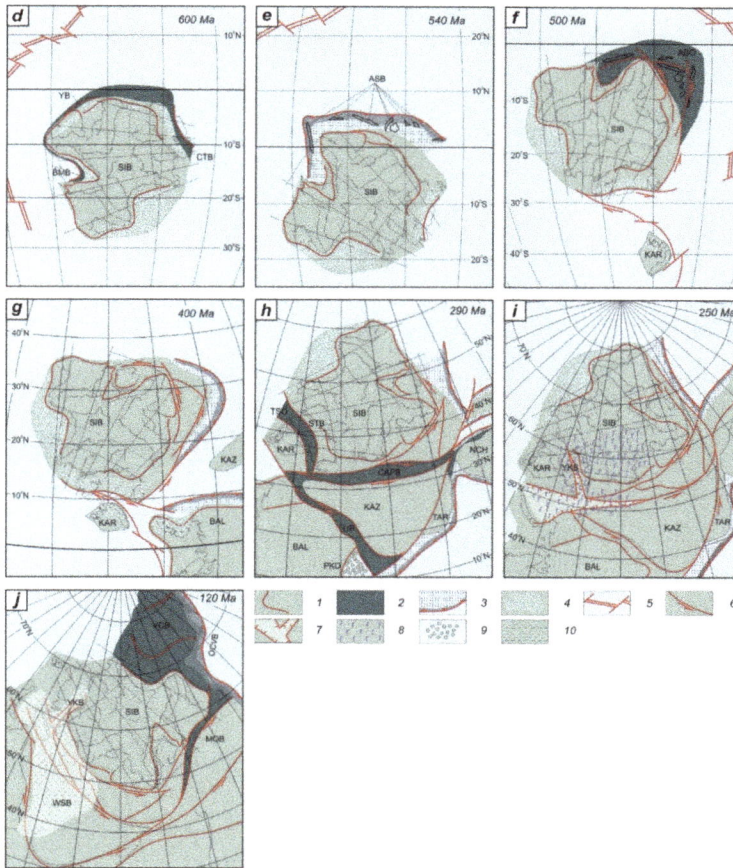

Figure 5. Paleotectonic reconstructions for the Siberian paleocontinent from the end of the Mesoproterozoic to the Early Cretaceous (adapted from Reference [25]). 1—continental masses and the most important contours; 2—accretionary structures and orogenic belts of corresponding ages; 3—subduction systems, including volcanic belts and back-arc basins; 4—marginal basins, shelves of passive continental margins; 5—inferred location of spreading zones; 6—general strike of transform/strike-slip zones with their strike-slip kinematics; 7—schematic area of thinning of the continental crust in the West Siberian graben-rift system; 8—schematic location of flood basalts of the Siberian trap late Permian—Early Triassic formation; 9—remnant basins with sub-oceanic crust; 10—schematic distribution of Mesozoic-Cenozoic deposits of the West Siberian sedimentation basin. Letter abbreviations: Continental blocks: SIB—Siberian, BAL—Baltica (East European), KAR—Kara, KAZ –Kazakhstan, LAU—Laurentia (North America), NCH—North China, TAR—Tarim, SCH—South China; passive continental margin basins, marginal seas: VK—Verkhoyan, BP—Baikal-Patom, SA—Sayan, SS—South-Siberian (hypothetical), ST—South Taimyr; orogenic belt: ABO—Altai-Baikal orogen, BMB—Baikal-Muya belt, VCB—Verkhoyan-Chukotka belt, MOB—Mongolian-Okhotsk belt, YB—Yenisei belt, TSO—Taimyr-Severnaya Zemlya orogen, UB—Ural belt, CAB—Central Angara belt, CAPB—Central Asian late Paleozoic belt, CAT—Central Angara terrane, CTB—Central Taimyr belt; island arc terranes, active continental margin fragments and volcanic-plutonic belts: ASB—Altai-Sayan-Baikal, CT—Central Taimyr, OCVB—Okhotsk-Chukotka volcanic-plutonic belt; other structures: PKD—Peri Caspian depression, WSP—West Siberian plate, STB—South Taimyr foreland basin, YKB—Yenisei-Khatanga basin.

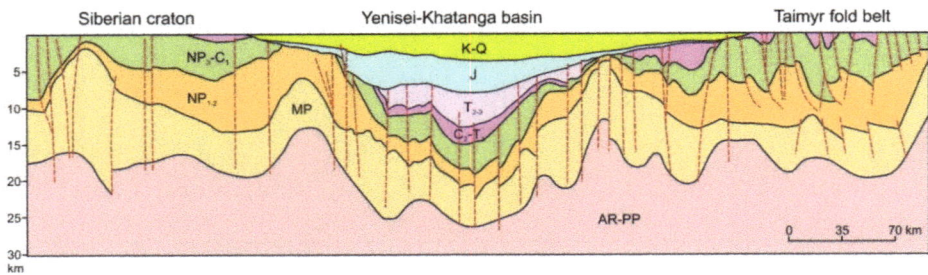

Figure 6. Seismic-geological section along the common depth point traverse line Norilsk–Kara Sea, modified after Reference [22]. The location of the line is shown on Figure 2.

However, as early as the early Neoproterozoic, an island arc system started forming near the north–western margin of Siberia, which means the passive continental margin transformed into an active one (Figure 5a–c) [25,28,29,40–43]. Ophiolite and island arcs continued forming here until the end of the Neoproterozoic in the Chelyuskin and Stanovoy belt (750–730 Ma) and 660 Ma in the Ust-Taimyr belt. Then, during the Ediacaran they accreted and were included into the Siberian craton as the composite Central Taimyr accretionary terrane [44] (Figures 2 and 5d). The age of this accretion event is substantiated by isotope-geochemical data (Sm–Nd, Rb–Sr, Ar–Ar methods) for garnet amphibolites (606–570 Ma) occurring in the floor of the allochthon in the juncture zone of the Central and Southern Taimyr units [5,45]. This development stage of the active continental margin ended with rifting on the margin of the continent accompanied by the outpouring of thin trachybasalt and basalt flows, often with pillow and rope jointing, as well as rhyolites and felsites. In addition to these effusive rocks, the section includes corresponding tuffs. Volcanics contain tuff conglomerates, gravelstones, sandstones often with red coloration. Gabbro dikes have been identified intruding the underlying carbonate rocks. Many paleovolcanos have been identified in the axial part of the Central Taimyr zone, in the north–eastern end of Cape Chelyuskin, and Cape Khariton Laptev [4,46,47]. The development of the active continental margin ended in deformation of Meso-Neoproterozoic deposits, uplift of the region, and a nondepositional hiatus [39,48].

From the Ediacaran to the early Carboniferous the 5–6 km thick sedimentary sequence of the platform cover accumulated on the north–western passive continental margin of the Siberian craton (Figure 5e–g). It is composed of mainly carbonate, carbonate-terrigenous and salt bearing formations (Figures 3 and 6). Periodic transitions from shallow marine sedimentation to a continental one caused hiatuses in deposition and washing out of lower deposits, which is indicated by unconformities on various stratigraphic levels. Passive margin complexes are traced in the basement of the Yenisei-Khatanga basin further to the north in the structure of the Southern and Central Taimyr zones [1,17,19–21,49]. Marine limestones, dolomites, marls, mudstones of the upper Cambrian, Ordovician, and Silurian have been penetrated on the Yenisei-Khatanga basin by the Tochino well (570–2002 m interval) in the south-western part of the basin [18].

At the same time with the accumulation of shallow marine carbonate and carbonate-terrigenous deposits, in the late Cambrian–Ordovician a deep water (over 750 m) elongated basin began forming between the Central and Southern Taimyr [1,3]. The axis of this SW–NE oriented basin was located southerly of the suture between the Central Taimyr accretionary terrane and the continent, in the frontal part of the large Pyasina-Faddey thrust. This means that the basin was formed in the front of the Neoproterozoic fold belt that was the result of accretion-type orogeny. In this basin, black graptolite shales accumulated with interlayers of dark bituminous limestones containing pteropod fauna, with a total thickness of 1.3–2.0 km. To the south of the basin, on the shelf, limestones and dolomites were deposited with abundant fauna of brachiopods, gastropods, bryozoans, corals and other shallow water benthic organisms [46].

During the Silurian and Devonian, the region south of the southern margin of the Taimyr fold belt was occupied by the formation of a Lower–Middle Devonian evaporite salt bearing formation, which was overlain by a carbonate (dolomites, limestones) shallow marine formation accumulated in the north. The position of the deep-water basin did not change, and they were filled with a Silurian black shale formation, and a Devonian argillaceous-carbonate formation. The deep-water basin existed for over 100 million years, and the subsidence ended by the beginning of the Carboniferous.

At the Mississippian–Pennsylvanian boundary there was a transition to mainly shore-marine and continental sedimentation conditions on the Taimyr shelf of the Siberian craton. The reasons for the structural transformation and change from the carbonate sedimentation to a shore-marine and continental one was the change in tectonic setting on the Taimyr margin of the Siberian craton. Zonenshain et al. [7] explained this by the closing of the South-Anyui Ocean that was located between Siberia and the hypothetical Arctida paleocontinent. Pogrebitsky [1] connected the cause of the structural transformation and sedimentation mode change to the Taimyr orogeny, defining the early Permian as its most intense stage. Modern geostructural, petrological, geochronological and paleomagnetic data prove conclusively, that this change in sedimentation was related to the collision of the Siberian continent, including the accretionary structures of the Central Taimyr belt, with the Kara microcontinent [3,5,6,26–28,50,51]. This collision resulted in the formation of a large orogen. It is reflected in the formation of regional metamorphism zonation in the crust of southern margin of the Kara microcontinent (Figure 5h) (from the amphibolite to the greenschist facies) that is intruded by syncollisional calc-alkaline granitoids. Their oldest ages correspond to the late Carboniferous–early Permian (306–275 Ma) [41,51,52]. In the late Permian (264–258 Ma) postcollisional granite plutons were emplaced in the Central and Northern Taimyr zones [41] forming wide contact aureoles. It is noted that from the end of the Permian period vertical displacements were supplemented with dextral strike-slips and associating thrusts [1,53], which is confirmed by our own paleomagnetic research [27] (Figure 5g,h).

Transform zones with large displacements, along which the Kara plate "slid" during the entire early Paleozoic, finally led to the collision of the Kara and Siberian continents, and to the subsequent formation of the imbricate structure of the Taimyr-Severnaya Zemlya folded area, in accordance with the proposed model (Figure 7).

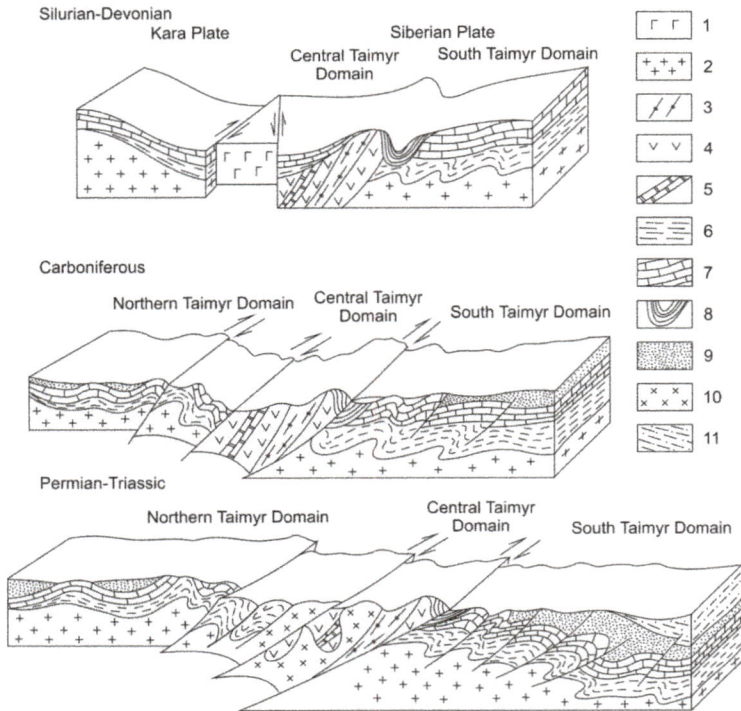

Figure 7. Model of transformation of the Taimyr margin of Siberia during interaction with the Kara microcontinent (Reference [54] with alterations). Composed based of sections of the Taimyr fold-and-thrust system [6], not to scale. 1—oceanic complexes; 2—early Precambrian complexes of the Kara microcontinent and Siberian craton crystalline basement; 3–5—late Precambrian complexes of the Central Taimyr accretionary zone: 3—gneiss, 4—volcanogenic-sedimentary, 5—carbonate; 6—Neoproterozoic-Cambrian flyschoid deposits of the Kara and Siberian continental margins; 7—Paleozoic mainly carbonate deposits of the shelf (Ordovician-Silurian on the Kara microcontinent and Ordovician-early Carboniferous on the Taimyr margin of the Siberian continent); 8—Ediacaran—early Carboniferous argillaceous-carbonate and black shales deposits of the Pyasina-Faddey deep-water basin; 9—late Paleozoic mainly terrigenous deposits (Devonian and Carboniferous–Permian for the Kara part and late Carboniferous–Permian of Southern Taimyr); 10—late Paleozoic (300–260 Ma) collisional granitoids; 11—Triassic terrigenous-volcanogenic deposits, including the Trap complex (base horizons of the Mesozoic-Cenozoic Yenisei-Khatanga basin).

The growth of the orogen caused a tremendous source for terrigenous material, filling the foreland basins in the frontal part (Southern Taimyr). As the orogen developed and compression progressed, the deep strike-slips that caused the collision of continental masses transformed into thrusts [55]. The most important among them is the Main Taimyr thrust, which can be regarded as the main suture of the late Paleozoic orogen. In the front of the orogen, in the South Taimyr zone, there were many upthrow-thrusts forming in concordance with the compression. Their offset gradually disappeared towards the Siberian continent, which conditioned the formation of the significantly large South Taimyr foreland basin [2,3,5].

As a result, a considerable amount of clastic material and coal bearing formations with a thickness of 500–1000 m were accumulated in the South Taimyr foreland basin during the entire middle Carboniferous and Permian in continental environments [2,19,21]. These deposits have been penetrated in the west of the Yenisei-Khatanga basin by the Volochan-1 and 2 wells (Figure 4) and their age has been determined as Late Wordian–Capitanian [18]. By the end of the Permian, the northern flank of the South Taimyr basin became displaced towards its axis to 150 km [1], which confirms our proposed model (Figure 7) and indicates that the strong horizontal compression in dextral strike-slip settings continued until, and including, the Triassic.

The similarity of the upper Paleozoic section of Southern Taimyr with the section of the northern part of the Siberian craton, with a great role of marine facies and with typical deposits thicknesses leads us to assume that both regions belonged to the same wide basin that deepened towards Taimyr and from the west to the east [2]. The South Taimyr foreland basin and its configuration became the prototype for the Yenisei-Khatanga basin. The latter is believed to have emplaced along with the West Siberian basin due to continental rifting at the Permian–Triassic boundary, which turned out to be coeval with the spectacular manifestations of the Siberian superplume (Figure 5i) [56,57]. The products of trap magmatism spread over a huge area on the territory of the Siberian craton, on Southern Taimyr, in Western Siberia, including the present-day arctic territories of the South Kara basin [21,57–65].

The position of rift structures in the basement of the Yenisei-Khatanga basin and in the Southern Taimyr zone is marked by intense linear magnetic and gravity anomalies (Figure 8). The axial graben of the Yenisei-Khatanga riftogenic structure was controlled by the east–north–east striking Malokhet-Rassokhin-Balakhnin fault system (Figure 2), whose strike-slip kinematics is related to the oblique collision formation mechanism of the Taimyr orogen formation mentioned above (Figure 6). The Earth's crust thickness in the axial graben is reduced to 25–28 km, unlike 35–40 km on its flanks [19,20,53]. Considered together with similar structures of Western Siberia, this rift forms a kind of triple junction [66,67] (Figure 5j). Geophysical data indicate that the total thickness of effusive-sedimentary late Permian–Early Triassic series of the rifting stage reaches 2–3 km [53]. They have been penetrated by wells in the western part of the Yenisei-Khatanga basin on the Malaya Kheta, Dolgan, Sukho-Dudinka and Volochan areas [18] (Figure 4).

In summary, the basement of the Yenisei-Khatanga basin is dominated by variously deformed complexes, mainly of continental margin and intracontinental origin, that have direct connections to the evolution of the late Paleozoic Taimyr-Severnaya Zemlya orogen. Only the overlapping Late Triassic and younger sedimentary deposits should be compared with the infilling stage of the Yenisei-Khatanga basin—a large rift depression that appeared first due to the collision between the Siberian craton and the Kara plate, and then the subsequent rifting, due to the activity of the Siberian superplume [57], with simultaneous strike-slip motions that affected large areas [68].

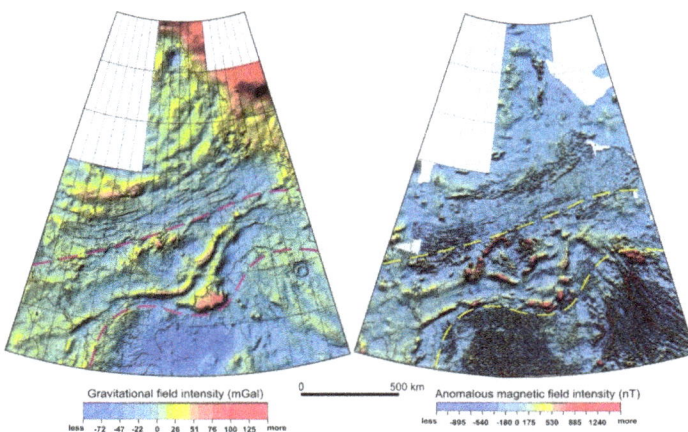

Figure 8. Maps of the gravitational field and anomalous magnetic field anomalies for the Severnaya Zemlya—Taimyr Peninsula region from Reference [69] with our simplifications. Magenta and yellow dashed lines show the approximate boundaries of the Yenisei-Khatanga basin.

3. Structure and Development History of the Yenisei-Khatanga Basin

3.1. Main Structural Elements

The views on the inner structure of the Yenisei-Khatanga basin are mainly based on seismic data, supported by well data, most of which were obtained in the ranges of the axial area of the basin. On Figures 9 and 10 are structural maps, on which isohypses show the base of the Jurassic deposits, the top of the Middle Jurassic, as well as swells and depressions of various scale.

In the flanks of the basin its monoclinal sedimentary section dips sharply towards the axial part (Figure 10). The northern flank (the Taimyr monocline) is less steep (2°–6°). This structure is complicated by the Taimyr and Yangoda-Gorbit basement highs. The southern flank (the North Siberian monocline) is narrow and steep (up to 8°–15°) in the north–west and widens and flattens in the east. In its south–western part there is the small Agapa basement high. The width of the monoclines varies from 20 to 150 km. The reduced thicknesses of the Jurassic–Cretaceous deposits on the crest is only 2–3 km. The monoclines are connected to the central zone of the basin by a system of faults, which provide a stepwise subsidence of the basement surface towards the axial part of the basin [19–21,53].

In the central zone of the basin there is a combination of large contrasting positive and negative structures. The first noticeable one is the chain of three echelon-like basement ranges (Tanama-Malokhet, Rassokhin, Balakhnin) (Figures 2 and 10). They cross the basin diagonally from the southern flank in the west to the northern one in the east. The Tanama-Malokhet range is 300 km long and 60–140 km wide, the Rassokhin range is 570 × 45–60 km and the Balakhnin range is 260 × 40–60 km.

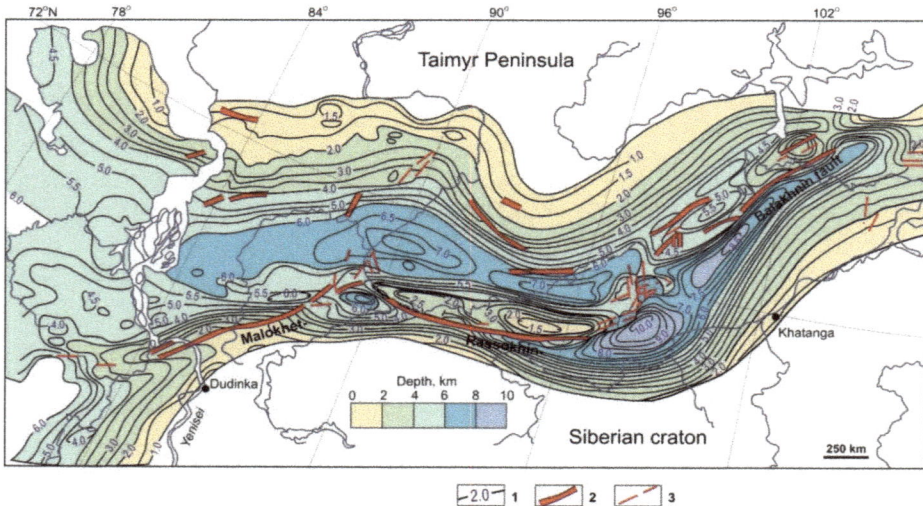

Figure 9. Relief map of the base of the Yenisei-Khatanga basin Jurassic infill (composed by V.Yu. Bogatova, V.V. Grebenyuk, T.A. Divina) from Reference [70]. 1—stratoisohypses (depth in km) for the base of the Jurassic—erosional truncation intervals of the Triassic formations in the eastern basin (seismic reflector V) and of the unconformable bedding of the Zimnii Formation on the Triassic rocks in the western half (seismic reflector III); 2—fault zones according to seismic data; 3—faults according to geological and geophysical data.

To the north and south of the chain of basement ranges the most subsided areas of the Yenisei-Khatanga basin are located. In the southern subsided part, several en-echelon depressions are located: Dolgan and Dudyptin depressions, the Boganid-Zhdanikhin trough, in the northern part—the

Noskov depression and the Central Taimyr trough (Figure 10). The greatest depths of the base of Jurassic are observed in the Boganid-Zhdanikhin (8.5–10 km) and the Central Taimyr (7–7.5 km) trough.

Abrupt differences in depths (up to 5.5 km for the base of the Jurassic deposits) of the structural surfaces from the most depressed parts of the Yenisei-Khatanga basin to the roofs of the ranges undoubtedly indicate that the positive and negative structures are bordered by faults. The Tanama-Malokhet, Rassokhin and Balakhnin ranges form a system of large linear anticlines along the northern flank of the Malokhet-Rassokhin-Balakhnin fault [71]. In several cases, geophysical data indicate the thrusting of the ranges from north to south [72]. The fault is clearly reflected on the map of the gravitational field (Figure 8) by an intense positive anomaly (up to 70 mG) 10–50 km wide that can be traced for over 1500 km [69]. Deep seismic sounding indicates a tectonic step in the Moho discontinuity in the fault zone, reaching 4–7 km depending on estimates [24].

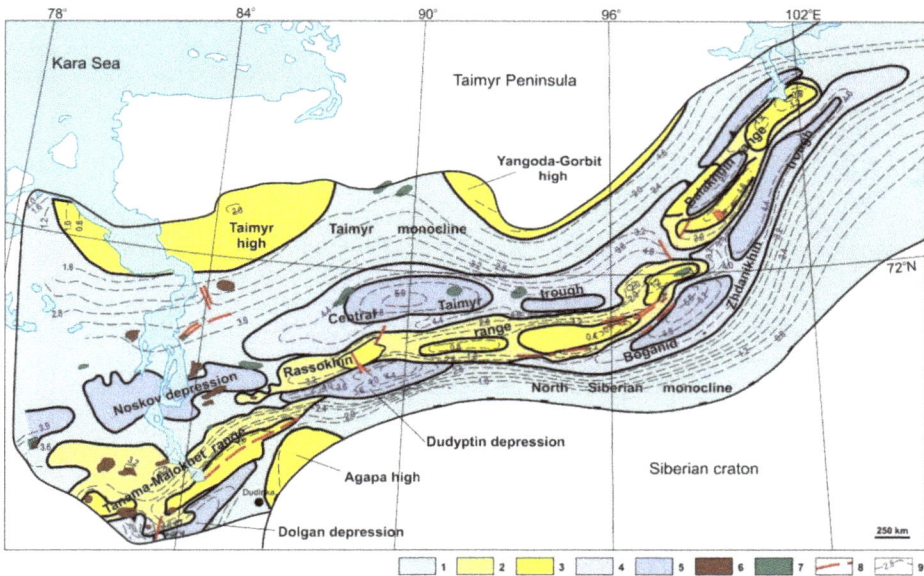

Figure 10. A fragment of the structural-tectonic map of the oil and gas bearing provinces of the Siberian craton from Reference [73] with our alterations. 1—Yenisei-Khatanga basin; 2—basement ranges; 3—basement high/swell; 4—first order negative structures; 5—second order negative structures; 6—hydrocarbon fields; 7—prepared structures; 8—faults; 9—stratoisohpyses for the top of the Middle Jurassic.

The consistent subsidence of the Yenisei-Khatanga basin during the Triassic and the Jurassic changed to an abrupt structural transformation at the Jurassic–Cretaceous boundary [17]. During this time there was an intensive growth of the Malokhet-Messoyakh range that had been developing in the Triassic–Jurassic, the Rassokhin range, that was formed in the Middle–Late Jurassic, and the newly formed Balakhnin range. Due to the intense growth of the range chain, the adjacent troughs structures also developed. Due to the intense uplift along the roofs of the ranges, some of the deposits were eroded. In this area, different levels of the Upper Jurassic–Valanginian deposits were eroded, and on some fields, the washing out also affected the Middle Jurassic deposits. The Upper Jurassic and Lower Cretaceous deposits were accumulated with reduced thicknesses [21,71,74].

To determine the reasons for the tectonic forces and the formation of the inverted structure of the basin in the Late Jurassic–Early Cretaceous, one must consider the tectonic events that took place at the same time on the adjacent territories. Regarding time, this transformation can be correlated with the late Cimmerian deformations in the Verkhoyansk fold-and-thrust belt. It is precisely in the Late Jurassic that the Kolyma-Omolon microcontinent collided with the margin of the Siberian craton [3]. This compressional deformation from the south–east affected the Yenisei-Khatanga basin, especially its eastern part, where they led to a partial inversion and the formation of a system of elongated ranges [6]. The intense compression from the east also manifested in deformations of the Jurassic deposits of Cape Tsvetkov on the Taimyr Peninsula, which in recent years has been progressively considered a result of the late Cimmerian Verkhoyansk folding [69,75].

Thus, the combination of large rift-related extension in the north–east of the region with compression strain in the south–east led to the transformation of the Yenisei-Khatanga basin's structure in the Late Jurassic–Early Cretaceous time. Beginning with the Late Cretaceous, due to the levelling of the Taimyr Mountains, various uplifts on the Siberian craton became the main sources of material for the Yenisei-Khatanga basin. In the Paleogene–Neogene there was a general uplift of the Siberian craton, Taimyr, the Kara plate, and the Yenisei-Khatanga basin [6]. An erosion region formed here, from which material was transported in the sedimentation basins of the Kara and Laptev seas. During this time, weak inversion motions took place in the axial zone of the Yenisei-Khatanga basin (the Rassokhin and Balakhnin ranges), as well as active growth of salt domes near Khatanga bay [53]. All these tectonic processes on the shelf and the continental margin reflected the largest event of this stage—the opening of the Eurasian basin and the appearance of the Arctic mid-oceanic ridge–Gakkel Ridge.

3.2. Sedimentary Infill of the Basin and Sedimentation History

The base of the sedimentary infill of the Yenisei-Khatanga basin is formed by Triassic deposits (Figure 11). In the western part of the basin the terrigenous Middle–Late Triassic Tampey Group (Series) is determined [76]. Its sections have been penetrated on the Zimnii, Gol'chikha, Vladimir, and Nizhnyaya Kheta fields, and some others (Figure 4). The base of the Tampey Group is associated with the "A" reflector (Figures 11 and 12), and the top of the Group—with the "Ia (III)" reflector [77–79]. The Lower–Middle Triassic lagoonal, deltaic, shore-, shallow-, and deep-marine conglomerates, sandstones, siltstones and mudstones of the Pronchishev, Angardam, and Tuorakhayata groups are concentrated in the eastern part of the basin [80]. The thickest Triassic deposits are predicted in the eastern part of the basin. Here, on the Vladimir field, a 1380 m section of Triassic deposits has been penetrated [81].

Jurassic deposits within the Yenisei-Khatanga basin have a complete stratigraphic succession and occur everywhere. Their thickness decreases, and the lower horizons wedge out in the flank areas of the basin, as well as in individual large uplifts of its inner part.

The Lower–Middle Jurassic deposits are divided into nine formations (Figures 11 and 13). Their cyclic alternation reflects eustatic variations of the sea level in the sedimentation basin [82]. Those formations that accumulated during marine transgressions (Levin, Kiterbyut, Laidin, and Leontiev formations) are composed mainly of mudstones, sometimes bituminous mudstones with marine fauna. They form clearly traceable reflectors on seismic profiles: The "T" group ("Tlv", "Tkt", "Tld", "Tln") (Figures 11 and 12). The formations composed mainly of sandstones and mudstones (Zimnii, Sharapov, Nadoyakh, Vym, and Malyshev formations) accumulated in shallow-marine settings in stages of regression. Close to the paleo-coasts, their composition includes interbeds and lenses of coals and pebble conglomerates. The thickness of the Lower–Middle Jurassic deposits can reach 4500 m [81].

In Callovian–Lower Berriasian times, the transgression reached its peak. During this stage, mudstones, bituminous mudstones with siltstones and sandstones interlayers of the Gol'chikha Formation with a thickness of up to 950 m were accumulating in the basin (Figure 11). In the south–west of the basin the Upper Jurassic deposits reach 1000 m in thickness and have a three-part structure. The middle part is taken by the sandstones and siltstones of the Sigovoe Formation (Oxfordian–Kimmeridgian), which are overlapped by the Tochino Formation (Callovian) mudstones and themselves overlay the Yanov Stan Formation (Upper Kimmeridgian–Lower Berriasian) mudstones [82]. The top of the Yanov Stan formation corresponds to the "IIa" reflector (Figures 11 and 12), the top of the Gol'chikha Formation–to the "B" reflector, the Tochino Formation–to the "Ttc" reflector [81,83]. In some individual large uplifts in the Yenisei-Khatanga basin, the deposits of the upper and partially the Middle Jurassic are eroded.

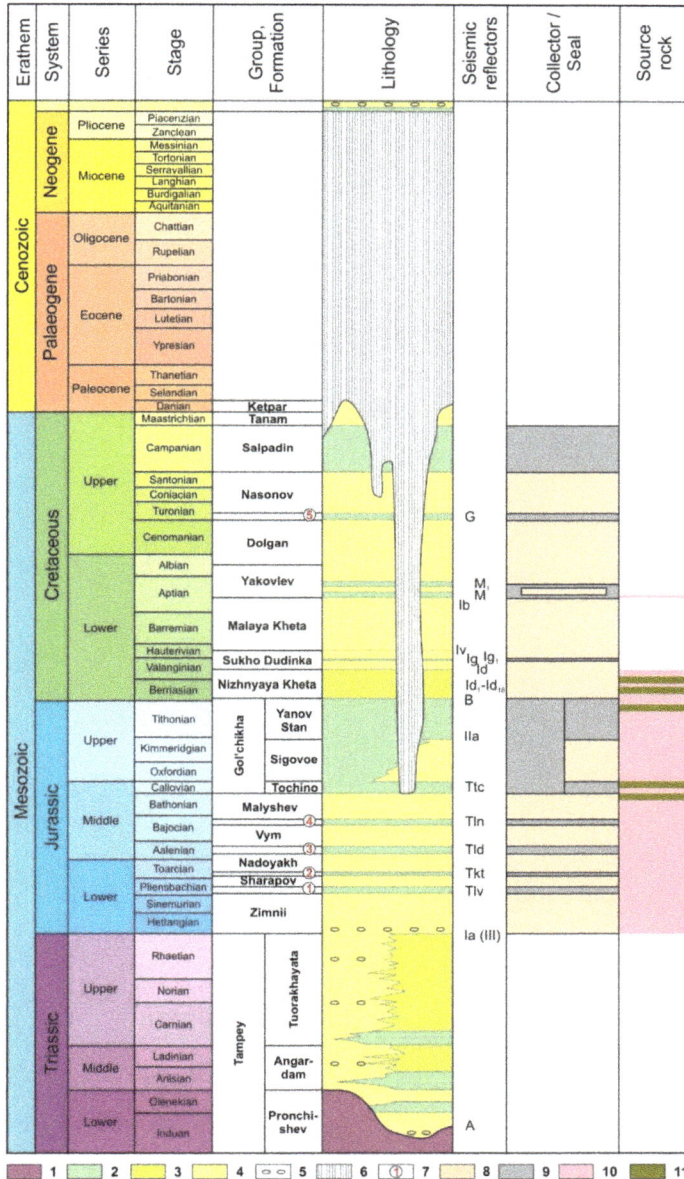

Figure 11. Lithostratigraphic scale of the Yenisei-Khatanga basin's sedimentary fill. 1—heterogeneous basement (Pz–T$_1$); 2—mudstone and clay; 3—interbedding of mudstone, sandstone, and siltstone; 4—mostly sandstone and siltstone with interbeds of mudstone and clays; 5—conglomerate, pebble; 6—scouring and lack of deposits; 7—formations: ①—Levin, ②—Kyterbyut, ③—Laidin, ④—Leontiev, ⑤—Dorozhkovo; 8—plays; 9—seals; 10—main oil and gas generating sediments; 11—main oil generation sediments.

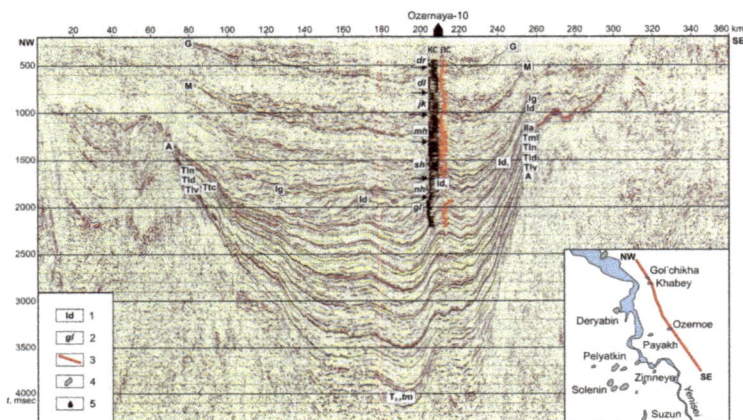

Figure 12. Stratified common depth point seismic section along the regional 1-1 line [81]. 1—seismic reflectors (their stratigraphic correlation is shown in Figure 11); 2—formations: gl—Gol'chikha, nh—Nizhnyaya Kheta, sh—Sukho Dudinka, mh—Malaya Kheta, jk—Yakovlev, dl—Dolgan, dr—Dorozhkovo, T$_{2-3}$tm—Tampey; 3—seismic section line; 4—hydrocarbon fields; 5—well. The location of the line is shown in Figure 2.

Figure 13. Correlation scheme of Lower and Middle Jurassic deposits in the south-western part of the Yenisei-Khatanga basin between the Sukho Dudinka and Nizhnyaya Kheta fields by G.G. Shemin and N.N. Kostagacheva. Location of the scheme line is shown in Figure 4. 1—sandstone, 2—silty sandstone, 3—interbedding of sandstone and siltstone, 4—sandy siltstone, 5—siltstone, 6—sandy-clayey siltstone, 7—clayey siltstone, 8—silty mudstone and clayey siltstone, 9—silty mudstone, 10—mudstones, 11—coal bearing rocks; 12—piece of the lithological column: a—section intervals built using core material and well log data, b—section intervals built using well log data; 13—hiatuses; 14—facies changes; 15–18—boundaries: 15—formations, 16—subformations, 17—cyclically structured layers, 18—layers; 19–21—biostratigraphic determinations of age: 19—ammonites, 20—bivalves, 21—foraminifera. NGL—neutron gamma-ray logging, GL—gamma-ray logging, SP—self-potential logging, RL—resistivity logging.

Cretaceous deposits are also located in the entire area of the Yenisei-Khatanga basin. Seismic data indicate that their thickness exceeds 3000 m with maximum thicknesses observed in the western part of the basin and gradually decrease to the east.

On seismic sections in the western part of the Yenisei-Khatanga basin the Berriassian–Hauterivian deposits (Nizhnyaya Kheta, Sukho Dudinka formations and their stratigraphic analogues) have a clinoform structure (Figures 11 and 12) [81,84]. The Nizhnyaya Kheta Formation (Berriassian–Lower Valanginian) is mainly composed of mudstones with beds of sandstones in the base and the top. Its thickness can reach 600–1360 m [74]. The top of the formation corresponds to the "Id" reflector. The Sukho Dudinka Formation (upper Valanginian–lower Hauterivian, thickness: 200–800 m) is composed of sandstones and mudstones with siltstones interbeds in the lower and upper part. The middle of the formation is dominated by mudstones with sandstones interbeds. Its lower part corresponds to the "Ig" reflector, its top is correlated with the "Iv" reflector [84]. The clinoform complex itself is framed on the seismic sections by reflectors "IIa" in the top of the Yanov Stan Formation and "Ig$_1$" within the Sukho Dudinka Formation. Its inner structure is underlined by reflectors "Ig", "Id", and "Id$_{1-18}$". The occurrence area of the clinoform complex exceeds 74,000 km^2. In total, 18 clinoforms can be identified on the CDPM sections. The clinoforms dip to the north–west and become younger in the same direction (Figure 12). Reflectors are generated by mudstone layers that formed during brief transgressions. The spatial distribution of the clinoforms reflects the asymmetry of the sedimentary basin, whose northern part was more intensely subsided. The main input of terrigenous material in this sedimentation basin was from the south and the south–east from the uplifted dry land of the Siberian craton. Discharge of material from the Taimyr-Severnaya Zemlya orogen was significantly less, and along its margin one poorly defined clinoform is developed, directed in the opposite direction. On the left bank of the Yenisei River, the thickness of individual clinoforms exceeds 450 m. Their slope angles can vary from 1–2 to 4–6 degrees. Along paleo-slopes, gravity flows transported clastic material into the deep-water part of the basin forming deep-water turbidite fans and mass-transport derived sediment bodies. Towards the east–north–east the amount of clinoforms decreases, they become less steep, and the reflector pattern becomes more chaotic. To the east of the 90 degrees meridian, the clinoforms disappear from the seismic sections. Sedimentation took place here in conditions of more active input of terrigenous material into the sedimentary basin, which is reflected in increasing thicknesses of the Berriassian–Hauterivian deposits and in their increasing lithological homogeneity [84].

The Lower Cretaceous section is overlain by the Malaya Kheta Formation (Upper Hauterivian–Lower Aptian, thickness: 100–500 m) composed of sandstones with rare interbeds of siltstones, mudstones, conglomerates and coals [74]. The deposits formed in alluvial plain, near-shore, and coastal flood plain environments. The upper part of the formation corresponds to the "Ib" reflector (Figures 11 and 12). It is conformably overlain by the Yakovlev Formation (thickness: 150–580 m) composed of mudstones and siltstones with interbeds of sandstones and coals. The age of the formation is between the Middle Aptian and the Middle Albian. The coal-mudstones layers of the lower part of the formation are correlated with the "M" and "M$_1$" reflectors. From the end of the Albian to the Cenomanian the sandstones, siltstones, and mudstones of the Dolgan Formation (thickness: 160–780 m) accumulated in shallow-nearshore and lagoonal-continental environment. The Early Turonian marine transgression was the time of accumulation of the Dorozhkovo Formation (thickness: 40–130 m) mudstones with siltstones and sometime sandstones interbeds. The top of the mudstones corresponds to the "G" reflector (Figures 11 and 12).

During the Middle Turonian–Maastrichtian the Yenisei-Khatanga basin continued to experience a steady, compensated subsidence. The sedimentation environment alternated between shallow-marine, shore-marine and continental environments. In the Middle Turonian–Santonian, the Nasonov Formation (thickness: 300–600 m) accumulated. It is composed of rhythmically alternating mudstones, siltstones, and sandstones layers with phosphorite horizons at the base of the layers. The Nasonov Formation is overlain by the Salpadin Formation (Campanian, 170 m thick), which is composed mainly of clays, opoka-like clays, argillaceous siltstones with rare interbeds of fine-grained sands that accumulated in shallow-marine conditions. The base of the formation is marked by a phosphorite horizon. The lower part of the formation has iron ore peas and oolites. The Cretaceous section ends with the Tanam Formation (Maastrichtian, up to 140 m thick) composed of sands and silts with interbeds of silty clays and a phosphorite horizon at the base [74].

The Upper Turonian–Maastrichtian deposits, as well as the Paleogene–Neogene formation are eroded for a significant part of the basin [74]. Only the Ketpar Formation was preserved (Danian, thickness up to 300 m), composed of sands and silts with clay interbeds. In the lower part of the formation there are gravels with clasts of bauxites.

In the Pleistocene the territory of the Yenisei-Khatanga basin was periodically occupied by marine transgressions. Regressions were accompanied by increased glacial activity. As a result, a cover of Quaternary deposits formed with thicknesses of 250–300 m. They are composed of marine clays and sands, moraine diamictons, fluvio-glacial and alluvial sands and pebblestones, lacustrine-glacier sands, silts and clays, subaerial clay loams. They overlie the washed-out surface of the Cretaceous, sometimes Jurassic and Paleogenic deposits.

4. Petroleum Geology

4.1. Discovered Hydrocarbon Fields

In the Yenisei-Khatanga basin 21 gas, gas condensate and oil fields have been discovered (Figure 4). The most oil perspective is the western part of the basin that is intensively studied by seismic surveys and drilling. Nearly all known fields are concentrated here. In the eastern part, only the Balakhnin and Novoye fields are located.

They are mainly multiplay fields (Figure 14). The depth of productive reservoirs varies from 750 to 3550 m. Mainly, the reservoirs belong to the Lower Cretaceous plays. Along with roof pools; there can often be lithologically and tectonically screened reservoirs.

The preservation of hydrocarbons accumulations in the plays was affected by widespread tectonic motions in the Yenisei-Khatanga basin in the Late Jurassic–Early Cretaceous and at the neotectonic stage. Tectonic motions favored the growth of swells and a general uplift of the territory, which caused the erosion of deposits, the local destruction of the seals, and the development of faults. This led to the migration of hydrocarbons to the upper stratigraphic levels, a significant seepage into the atmosphere and a decrease of the total quantity of hydrocarbon resources [17]. A favorable factor for the preservation of the hydrocarbons became the formation of the 700 m thick permafrost zone at the latest stage, which constitutes an additional seal. The second consequence of the appearance of the permafrost rocks was the formation of gas-hydrate accumulations.

4.2. Estimated Hydrocarbon Resources

The estimated mean volumes of undiscovered resources for the Yenisei-Khatanga Basin Province are approximately 25 billion boe: 5.6 billion barrels of crude oil, 100 trillion cubic feet of natural gas, and 2.7 billion barrels of natural gas liquids [85,86]. The total initial resources of hydrocarbons in the Jurassic–Cretaceous infill of the Yenisei-Khatanga basin for early 2009 were estimated [87] as 84.6′billion boe: gas—88%, oil—7%, condensate—5%. The Lower Cretaceous clinoform deposits can contain: Gas—134 trillion cubic feet, oil—12 billion boe (2.4 billion boe recoverable), condensate—2.7 billion boe (1.35′billion boe recoverable) [81]. Hydrocarbon resources in the basement of the basin and the Triassic deposits have not yet been estimated.

4.3. Hydrocarbon Systems and Plays

4.3.1. Source Rocks

Geochemical data only allow us to say that the hydrocarbons that formed in the basement of the basin did not migrate into the Mesozoic–Cenozoic sedimentary infill. Moreover, the oil and gas potential of the pre-middle Carboniferous complexes was already depleted by the time the Yenisei-Khatanga basin was formed [17].

The Triassic mudstones with thicknesses from tens to hundreds of meters could also have oil generation properties. According to various sources the organic carbon (C_{org}) concentrations here vary from 0.1–0.4 to 2–3% [74,80]. These marine sediments contain significant amounts of Type-II Organic Matter (OM).

The main oil and gas generating deposits of the Yenisei-Khatanga basin are the mudstones of the Jurassic and the Lower Cretaceous (Figure 11). Among them, there are no identified deposits distinguished by elevated contents of OM. The rocks contain OM of mixed type (Type-III and Type-II). The OM accumulated both in diffuse and concentrated (coals) form [17,88]. In general, the Jurassic–Cretaceous deposits section is weakly differentiated in terms of C_{org} content. The Middle Jurassic–Barremian mudstones have C_{org} concentrations of 0.34–1.6% [17,74], and in some cases 3% [88]. The Jurassic–Lower Cretaceous deposits are usually considered as gas generating [17]. However, some levels in the Yanov Stan, Gol'chikha, Malyshevka, and Nizhnyaya Kheta formations could have been oil generating [88]. Gases actively migrated from the deep parts of the basin in the upper horizons, and the oils accumulated mostly in the generating complex.

4.3.2. Plays and Seals

The Triassic complex. The oil and gas potential of the Triassic deposits of the Yenisei-Khatanga basin is scarcely studied. The structure of the Triassic section in the central and eastern parts of the basin allow a tentative identification of three to four sandy-silty plays with thicknesses of up to 300 m (Figure 11), overlapped by seals composed of marine mudstones with thicknesses ranging from 30 to 250 m. The porosity of the sandstones varies from 2% to 28%, the permeability from 0 to 138 mD [80]. Borehole sampling of the sandy layers of the Tampey Group yielded reservoir water with a flow rate of 38.4 m^3/day [81]. Possible perspectives of the Triassic complex are indicated by oil and gas seepage within the Anabar-Khatanga saddle [89].

The Jurassic complex. The Jurassic deposits of the Yenisei-Khatanga basin show an alternating pattern of mainly sandy-silty and mudstone formations forming plays and seals. Reservoir rocks have been identified in the Zimnii, Sharapov, Nadoyakh, Vym, Malyshev and Sigovoye formations (Figure 14, Table 1). Mudstones of the Levin, Kiterbyut, Laidin, Leontiev, Gol'chikha, Tochino, and Yanov Stan formations are seals (Figure 11, Table 1). The plays are wedging-out near the margins of the basin, and in the upper parts of the ranges, and are mudded-off in the eastern part of the basin. Mudstones of the Levin (thickness 50–600 m), Kiterbyut (20–60 m), Laidin (20–160 m), Leontiev (45–475 m), Gol'chikha (140–1000 m), Tochino (25–200 m) and Yanov Stan (50–400 m) formations are seals (Figure 14).

The Cretaceous complex. The Cretaceous deposits are the main oil and gas bearing complex of the Yenisei-Khatanga basin. Just the Lower Cretaceous deposits include up to 90% of the known reserves of hydrocarbons. In the Cretaceous section, the Nizhnyaya Kheta, Sukho Dudinka, Malaya Kheta, lower Yakovlev, upper Yakovlev-Dolgan, and Nasonov plays are defined (Figure 11, Table 1). These are mostly stacked reservoirs with gas, gas condensate and condensate with rims [17,74,81,84,87]. The seals are represented by mudstones of the Dorozhkovo (40–130 m), Yakovlev (two layers with a thickness of tens of meters), and Salpadin (60–170 m) formations. In the Berriasian–Hauterivian section the regional and zonal seals are formed by argillaceous layers with thicknesses of the first tens of meters that accumulated in periods of regional transgressions and high sea levels. Among them the most widespread is the Pelyatin layer with a thickness of 50–140 m. Additional sealing of the sandy strata in the Nizhnyaya Kheta-Sukho Dudinka play is provided by argillaceous layers formed in distal and slope parts of clinoforms [17,74,81,84].

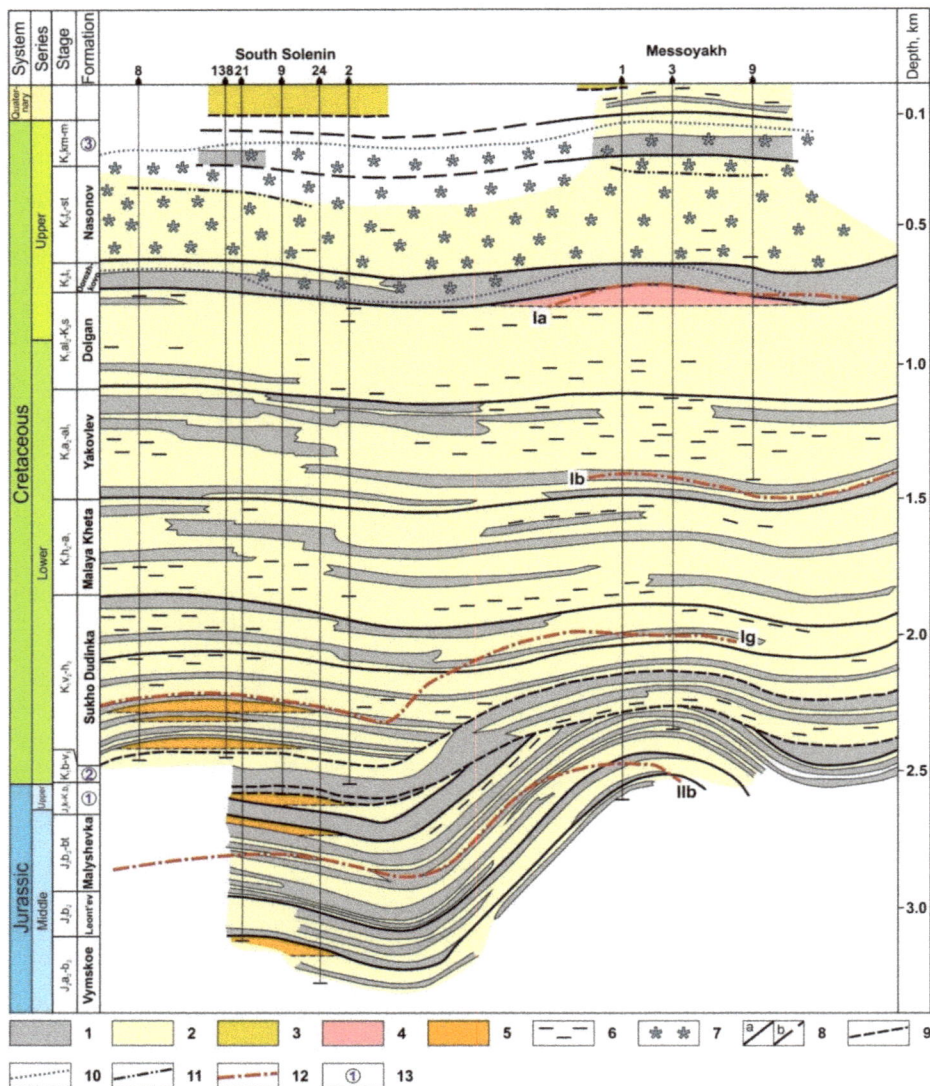

Figure 14. Geological section across the South Solenin and Messoyakh fields from Reference [90] with our alterations. 1—mudstone and clay; 2—sandstone and sand; 3—Quaternary sand, clay, pebble; 4—gas deposit, 5—gas condensate deposit; 6—coal; 7—hydrates formation zone; 8—stratigraphic boundaries: a—revealed, b—inferred; 9—unconformities; 10—boundaries of hydrates formation zone; 11—permafrost floor; 12—formations: ①—Tochino, Sigovoe, Yanov Stan, ②—Nizhnyaya Kheta, ③—Salpadin, Tanam. The location of the fields is shown in Figure 4.

Table 1. Play characteristics in the Jurassic–Cretaceous sedimentary infill of the Yenisei-Khatanga basin. Compiled using References [17,74,81,84,87,91].

Play	Age	Lithology	h (m)	Φ (%)	k (mD)	Hydrocarbons
Zimnyaya	Hettangian–beginning of Late Pliensbachian	Interlayering of sand-, silt-, and mudstones, beds of conglomerates. Individual sand and conglomerate layers: 3–30 m.	200–1600	3.5–24	0.43–38	Gas influx on the Zimneye field (yield of up to 72 thousand m^3/day) and the Tampey (yield of 2–2.5 thousand m^3/day) exploration area
Sharapovo	End of Late Pliensbachian	Alternation of mudstones and siltstones with beds of sandstones. The thickest sandstone layer: 20–200 m.	20–300	2.59–20.25	0.1–35.9	
Nadoyakh	End of Early Toarcian–beginning of Aalenian	Alternation of mudstones and siltstones with beds of sandstones. Two thick sandy layers 60–110 and 30–110 m.	185–300	2.59–20.25	0.1–35.9	Noncommercial gas seeps on the Malaya Kheta (up to 2.7 thousand m^3/day) exploration area
Vymskoe	End of Early Toarcian–beginning of Aalenian	Rhythms of mudstones, siltstones and sandstones. Thickness of sandy and silty layers reaches 20–30 m.	60–450	1.2–28	0.5–75	Gas play on the Balakhnin field, influx of gas on the South Solenin and Messoyakh fields, the Malaya Kheta and Yarov exploration areas (up to 4–4.5 thousand m^3/day). Oil seeps on the South Solenin field.
Malyshevka	End of Aalenian–beginning of Early Bajocian	Interlayering of sandstones, siltstones and mudstones	140–1000	4.5–33.4	80–550	Gas play of the Khabey field. Gas and oil seeps on the Zimneye, North Solenin, South Solenin, Messoyakh, Ozernoye, Dzhangod and Novoye field
Sigovoe	Oxfordian–Kimmeridgian	Sandstones and siltstones with mudstones interbeds	50–400	2.3–25	≤81.3	Gas play of the Nizhnyaya Kheta field. Gas condensate and oil seeps of the North Solenin field
Nizhnyaya Kheta–Sukho Dudinka	Berriassian–Lower Hauterivian	Rhythms of sandy and muddy layers. Echelon-, and lens-like sandy-silty bodies at the base of clinoform slopes.	950–2150	14–20	50–600	Oil plays on the Novo-Solenin, Baikalov, Payakh and North Payakh fields; gas and gas condensate—on the South Solenin, North Solenin, Deryabin, Baikalov, Zimneye, Ushakov, Khabey, Pelyatkin and other fields
Malaya Kheta	Upper Hauterivian–Lower Aptian	Sandstones with rare interbeds of siltstones, mudstones, conglomerates and coals. Sandstone layers up to 20–30 m thick.	100–500	15–27	150–571	Gas plays on the Pelyatkin, Khabey, South Solenin, North Solenin and Ozernoye fields
Lower Yakovlev	Upper Aptian	Interlayering of siltstones, sandstones, and mudstones. Sandy layers from several meters up to 30 m thick.	50–200	16–22	6–20	Gas and gas-oil plays, noncommercial gas influx on the South Solenin, North Solenin, and Baikalov fields
Upper Yakovlev-Dolgan	Albian-Cenomanian	Interlayering of silty-sandy and clay layers and beds of uneven thickness. Sandy layers reach tens to hundreds of meters in thickness.	250–1100	22–30	10–1000	Gas plays on the Pelyatkin and Messoyakh fields

Table 1. *Cont.*

Play	Age	Lithology	h (m)	Φ (%)	k (mD)	Hydrocarbons
Nasonovskaya	Middle Turonian–Santonian	Rhythms of mudstones, siltstones and sandstones with phosphorite horizons at the base of each rhythm. Up to 10 sandy layers 5–40 m thick.	250–500	av. 14	≤15	Gas play on the Kazantsev deposit (partially in gas-hydrate state)

Note: h—thickness, Φ—effective porosity, k—permeability.

5. Conclusions

Studies of the structure and tectonic evolution of the north–western margin of the Siberian craton and the arctic structures adjacent to the Kara Sea and the West-Siberian plate indicate that the basement of the Yenisei-Khatanga basin, the basin itself and the rift that generated it were all formed within the Siberian craton. This review presents our opinion on the geodynamic evolution of the north–western (Arctic) margin of the Siberian craton from the Precambrian–Paleozoic and Mesozoic folded structures that formed due to accretion and collisional processes to the formation of the rift-related late Paleozoic–Cenozoic Yenisei-Khatanga sedimentary basin. Geological, geophysical, paleomagnetic and petro-geochemical data obtained in recent years conclusively confirm the sequence of tectonic events previously proposed [5,44]. Moreover, new paleomagnetic data allowed us to create more accurate paleogeodynamic reconstructions for the Siberian craton, the Kara microcontinent and their interactions in the Paleozoic and Mesozoic [25,27,28]. We were also able to show that the predecessor of the Yenisei-Khatanga was a wide late Paleozoic basin that occupied the space between Southern Taimyr and the Siberian craton and deepened towards Taimyr. The formation of the Yenisei-Khatanga basin itself, like the formation of the West-Siberian basin, is related to continental rifting at the Permian–Triassic boundary. The formation of inversion structures in the basin in the Late Jurassic–Early Cretaceous [75] remain debatable. The time of this inversion can be correlated with late Cimmerian deformations in the Verkhoyansk fold belt [3]. This compression deformation affected the Jurassic deposits on Cape Tsvetkov on the Taimyr Peninsula [69,75], as well as the Yenisei-Khatanga basin, especially in its eastern part, where it led to partial inversion and the formation of a system of elongated ranges [6,72].

The Yenisei-Khatanga petroleum bearing area is a part of the West-Siberian petroleum province. The sedimentary cover is also represented by mostly sandy-silty Mesozoic deposits. The main oil and gas potential of the region are related to the Lower Cretaceous clinoform complex and the Jurassic deposits. The Mesozoic deposits of the region contain a huge amount of dispersed organic matter and have favorable conditions for catagenic maturation, mostly for gas generation. The productive part of the section of the sedimentary infill is significantly mudded-off and characterized by significant areal changes in composition. Most of the area has seals with high sealing properties. Only in areas where they partially or completely wedge-out due to stratigraphic breaks their sealing properties become decreased or are low.

Author Contributions: Conceptualization, V.V., E.D., and N.M.; Methodology, G.S. and D.M.; Formal Analysis, G.S. and E.D.; Resources, G.S.; Data Curation, N.P.; Writing—Original Draft Preparation, V.V., E.D., G.S., D.M., and N.M.; Writing—Review & Editing, V.V., E.D., N.M.; Visualization, N.M. and E.D.; Supervision, V.V.; Funding Acquisition, V.V. and D.M.

Funding: This work was financially supported by grants from the RSF (project No. 14-37-00030), the RFBR (projects 16-05-00523, 18-05-70035), and the Integrated research program of the SB RAS (project No. II.1.28.1).

Acknowledgments: The authors express deep gratitude to the reviewers for their comprehensive and useful suggestions, which allowed us to significantly improve this paper.

Conflicts of Interest: The authors declare no conflict of interest. The funders had no role in the design of the study; in the collection, analyses, or interpretation of data; in the writing of the manuscript, and in the decision to publish the results.

References

1. Pogrebitsky, Yu.E. *Paleotectonic Analysis of the Taimyr Fold System*; Nedra: Leningrad, Russia, 1971; 248p. (In Russian)

2. Khain, V.E. *Regional Geotectonics. Pre-Alpian Asia and Australia*; Nedra: Moscow, Russia, 1979; 356p. (In Russian)

3. Khain, V.E. *Tectonics of Continents and Oceans (Year 2000)*; Scientific Word: Moscow, Russia, 2001; 606p. (In Russian)

4. Zabiyaka, A.I.; Zabiyaka, I.D.; Vernikovsky, V.A.; Serdyuk, S.S.; Zlobin, M.M. *Geological Structure and Tectonic Evolution of Northeastern Taimyr*; Nauka: Novosibirsk, Russia, 1986; 144p. (In Russian)

5. Vernikovsky, V.A. *Geodynamic Evolution of Taimyr Folded Area*; Publishing House SB RAS: Novosibirsk, Russia, 1996; 202p. (In Russian)

6. Bogdanov, N.A.; Khain, V.E.; Rosen, O.M.; Shipilov, E.V.; Vernikovsky, V.A.; Drachev, S.L.; Kostyuchenko, S.H.; Kuzmichev, A.B.; Sekretov, S.B. *Tectonic Map of the Kara and Laptev Seas and North Siberia (Scale 1:2,500,000)*; Explanatory Note; Institute of the Lithosphere of Marginal and Inner Seas RAS: Moscow, Russia, 1998; 127p. (In Russian)

7. Zonenshain, L.P.; Kuzmin, M.I.; Natapov, L.M. *Geology of the USSR: A Plate Tectonic Synthesis*; Page, B.M., Ed.; Geodynamic Series, 21; American Geophysical Union: Washington, DC, USA, 1990; 242p.

8. Pease, V.; Vernikovsky, V. The tectono-magmatic evolution of the taimyr peninsula: Further constraints from new ion-microprobe data. *Polarforschung* **2000**, *68*, 171–178.

9. Walderhaug, H.J.; Eide, E.A.; Scott, R.A.; Inger, S.; Golionko, E.G. Palaeomagnetism and 40Ar/39Ar geochronology from the South Taimyr igneous complex, Arctic Russia: A Middle–Late Triassic magmatic pulse after Siberian flood-basalt volcanism. *Geophys. J. Int.* **2005**, *163*, 501–517. [CrossRef]

10. Drachev, S.S.; Malyshev, N.A.; Nikishin, A.M. Tectonic history and petroleum geology of the Russian Arctic Shelves: An overview. In *Petroleum Geology: From Mature Basins to New Frontiers—Proceedings of the 7th Petroleum Geology Conference*; Vining, B.A., Pickering, S.C., Eds.; Geological Society: London, UK, 2010; pp. 591–619.

11. Nikishin, V.A.; Malysgev, N.A.; Nikishin, A.M.; Obmetko, V.V. Late Permian-Triassic rift system of the South Kara sedimentary basin. *Vestnik Moskovskogo Univ. Seriya 4.* **2011**, *6*, 3–9. (In Russian) [CrossRef]

12. Cherepanova, Y.; Artemieva, I.M.; Thybo, H.; Chemia, Z. Crustal structure of the Siberian craton and the West Siberian basin: An appraisal of existing seismic data. *Tectonophysics* **2013**, *609*, 154–183. [CrossRef]

13. Zhang, X.; Omma, J.; Pease, V.; Scott, R. Provenance of Late Paleozoic-Mesozoic Sandstones, Taimyr Peninsula, the Arctic. *Geosciences* **2013**, *3*, 502–527. [CrossRef]

14. Klitzke, P.; Faleide, J.I.; Scheck-Wenderoth, M.; Sippel, J. A lithosphere-scale structural model of the Barents Sea and Kara Sea region. *Solid Earth* **2015**, *6*, 153–172. [CrossRef]

15. Drachev, S.S. Fold belts and sedimentary basins of the Eurasian Arctic. *Arktos* **2016**, *2*, 21. [CrossRef]

16. Gautier, D.L.; Bird, K.J.; Charpentier, R.R.; Grantz, A.; Houseknecht, D.W.; Klett, T.R.; Moore, T.E.; Pitman, J.K.; Schenk, C.J.; Schuenemeyer, J.H.; et al. Oil and gas resourse potential north of the Arctic Circle. *Geol. Soc. Lond. Mem.* **2011**, *35*, 131–144. [CrossRef]

17. Kontorovich, A.E.; Grebenyuk, V.V.; Kuznetsov, L.L.; Kulikov, D.P.; Khmelevsky, V.B.; Azarnov, A.N.; Nakaryakov, V.D.; Polyakova, I.D.; Sibgatullin, V.G.; Soboleva, E.I.; et al. *The Yenisey-Khatanga Basin*; UIGGM SB RAS: Novosibirsk, Russia, 1994; 71p. (In Russian)

18. Bochkarev, V.S.; Braduchan, Yu.V.; Voronin, A.S.; Generalov, P.P.; Kovrigina, E.K.; Kulakhmetov, N.Kh.; Mitusheva, V.S.; Stavitsky, B.P.; Faybusovich, Ya.E.; Shemraeva, S.V. *State Geological Map of the Russian Federation. Scale 1:1,000,000 (New Series). Sheet R-43-(45)—Gydan-Dudinka. Explanatory Note*; VSEGEI: St. Petersburg, Russia, 2000; 187p. (In Russian)

19. Kovrigina, E.K. *State Geological Map of the Russian Federation. Scale 1:1,000,000 (New Series). Sheet R-(45)-47—Norilsk. Explanatory Note*; VSEGEI: St. Petersburg, Russia, 2000; 479p. (In Russian)

20. Pogrebitsky, Yu.E.; Lopatin, B.G. *State Geological Map of the Russian Federation. Scale 1:1,000,000 (New Series). Sheet S-44-46—Ust-Tareya*; Explanatory Note; VSEGEI: St. Petersburg, Russia, 2000; 251p. (In Russian)

21. Talvirsky, D.B. *Tectonics of the Yenisei-Khatanga Oil and Gas Bearing Region and Adjacent Territories from Geophysical Data*; Nedra: Moscow, Russia, 1976; 168p. (In Russian)

22. Kazais, V.I. Innovation decision of regional structure task in difficult areas of Arctic (Taimyr sector). *Oil Gas Geol.* **2012**, *1*, 74–91. (In Russian)

23. Egorkin, A.V.; Kostiuchenko, S.L. Inhomogeneity of the upper mantle structure of Siberia. In *1993 CCSS Workshop Proceedings Volume*; Mooney, W.D., Ed.; United States Geological Survey: Menlo Park, CA, USA, 1995; pp. 122–132.

24. Dolmatova, I.V.; Peshkova, I.N. Model of a rift destruction of the northern palaeomargin of the Siberian continent (Yenisei-Khatanga basin). *Geol. Geofizika Razrabotka Neftyanykh Mestorozhdeniy* **2001**, *7*, 30–33. (In Russian)

25. Metelkin, D.V.; Vernikovsky, V.A.; Kazansky, A.Yu. Tectonic evolution of the Siberian paleocontinent from the Neoproterozoic to the Late Mesozoic: Paleomagnetic record and reconstructions. *Russ. Geol. Geophys.* **2012**, *53*, 675–688. [CrossRef]

26. Metelkin, D.V.; Kazansky, A.Yu.; Vernikovsky, V.A.; Gee, D.; Torsvik, T. First paleomagnetic data on the Early Paleozoic rocks from the Severnaya Zemlya archipelago and their geodynamic interpretation. *Russ. Geol. Geophys.* **2000**, *41*, 1767–1772.

27. Metelkin, D.V.; Vernikovsky, V.A.; Kazansky, A.Yu.; Bogolepova, O.K.; Gubanov, A.P. Paleozoic history of the Kara microcontinent and its relation to Siberia and Baltica: Paleomagnetism, paleogeography and tectonics. *Tectonophysics* **2005**, *398*, 225–243. [CrossRef]

28. Metelkin, D.V.; Vernikovsky, V.A.; Matushkin, N.Yu. Arctida between Rodinia and Pangea. *Prec. Res.* **2015**, *259*, 114–129. [CrossRef]

29. Vernikovsky, V.A.; Dobretsov, N.L.; Metelkin, D.V.; Matushkin, N.Yu.; Koulakov, I.Yu. Concerning tectonics and the tectonic evolution of the Arctic. *Russ. Geol. Geophys.* **2013**, *54*, 838–858. [CrossRef]

30. Golonka, J.; Krobicki, M.; Pajak, J.; Van Giang, N.; Zuchiewicz, W. *Global PlateTectonics and Paleogeography of Southeast Asia*; AGN University of Science and Technology: Krakov, Poland, 2006; 128p.

31. Cocks, L.R.M.; Torsvik, T.H. Siberia, the wandering northern terrane, and its changing geography through the Paleozoic. *Earth Sci. Rev.* **2007**, *82*, 29–74. [CrossRef]

32. Lawver, L.A.; Ganagan, L.M.; Norton, I. Paleogeographic and tectonic evolution of the Arctic region during the Paleozoic. In *Arctic Petroleum Geology*; Spencer, A.M., Embry, A.F., Gautier, D.L., Stoupakova, A.V., Sørensen, K., Eds.; The Geological Society: London, UK, 2011; Volume 35, pp. 61–77.

33. Torsvik, T.H.; Cocks, L.R.M. *Earth History and Palaeogeography*; Cambridge University Press: Cambridge, UK, 2017; 317p.

34. Pisarevsky, S.A.; Natapov, L.M. Siberia and Rodinia. *Tectonophysics* **2003**, *375*, 221–245. [CrossRef]

35. Vernikovsky, V.A.; Vernikovskaya, A.E.; Pease, V.L.; Gee, D.G. Neoproterozoic orogeny along the margins of Siberia. *Geol. Soc. Lond. Mem.* **2004**, *30*, 233–247. [CrossRef]

36. Vernikovsky, V.A.; Metelkin, D.V.; Vernikovskaya, A.E.; Matushkin, N.Yu.; Kazansky, A.Yu.; Kadilnikov, P.I.; Romanova, I.V.; Wingate, M.T.D.; Larionov, A.N.; Rodionov, N.V. Neoproterozoic tectonic structure of the Yenisei Ridge and formation of the western margin of the Siberian craton based on new geological, paleomagnetic, and geochronological data. *Russ. Geol. Geophys.* **2016**, *57*, 47–68. [CrossRef]

37. Semikhatov, M.A.; Ovchinnikova, G.V.; Gorokhov, I.M.; Kuznetsov, A.B.; Vasil'eva, I.M.; Gorokhovsky, B.M.; Podkovyrov, V.V. Isotopic age of Middle-Upper Riphean boundary: Pb-Pb geochronology of Lakhanda Group carbonates, Eastern Siberia. *Dokl. Earth Sci.* **2000**, *372*, 216–221.

38. Petrov, P.Y.; Semikhatov, M.A. Sequence organization and growth patterns of late Mesoproterozoic stromatolite reefs: An example from the Burovaya Formation, Turukhansk Uplift, Siberia. *Prec. Res.* **2001**, *111*, 257–281. [CrossRef]

39. Frolov, S.V.; Akhmanov, G.G.; Bakay, E.A.; Lubnina, N.V.; Korobova, N.I.; Karnyushina, E.E.; Kozlova, E.V. Meso-Neoproterozoic petroleum systems of the Eastern Siberian sedimentary basins. *Prec. Res.* **2014**, *259*, 95–113. [CrossRef]

40. Khain, V.E.; Gusev, G.S.; Khain, E.V.; Vernikovsky, V.A.; Volobuyev, M.I. Circum-Siberian Neoproterozoic Ophiolite Belt. *Ofioliti* **1997**, *22*, 195–200.

41. Vernikovsky, V.A.; Sal'nikova, E.B.; Kotov, A.B.; Ponomarchuk, V.A.; Kovach, V.P.; Travin, A.V. Age of post-collisional granitoids, North Taimyr: U-Pb, Sm-Nd, Rb-Sr and Ar-Ar data. *Doklady RAN* **1998**, *363*, 375–378. (In Russian)

42. Vernikovsky, V.A.; Vernikovskaya, A.E.; Sal'nikova, E.B.; Berezhnaya, N.G.; Larionov, A.N.; Kotov, A.B.; Kovach, V.P.; Vernikovskaya, I.V.; Matushkin, N.Yu.; Yasenev, A.M. Late Riphean alkaline magmatism in the western margin of the Siberian Craton: A result of continental rifting or accretionary events? *Dokl. Earth Sci.* **2008**, *419*, 226–230. [CrossRef]

43. Proskurnin, V.F.; Vernikovsky, V.A.; Metelkin, D.V.; Petrushkov, B.S.; Vernikovskaya, A.E.; Gavrish, A.V.; Bagaeva, A.A.; Matushkin, N.Yu.; Vinogradova, N.P.; Larionov, A.N. Rhyolite–granite association in the Central Taimyr zone: Evidence of accretionary-collisional events in the Neoproterozoic. *Russ. Geol. Geoph.* **2014**, *55*, 18–32. [CrossRef]

44. Vernikovsky, V.A.; Vernikovskaya, A.E. Central Taimyr accretionary belt (Arctic Asia): Meso-Neoproterozoic tectonic evolution and Rodinia break up. *Prec. Res.* **2001**, *110*, 127–141. [CrossRef]

45. Vernikovskii, V.A.; Kotov, A.B.; Ponomarchuk, V.A.; Sal'nikova, E.B.; Kovach, V.P.; Travin, A.V. Late Riphean-Vendian tectonic event responsible for the formation of North Taimyr (based on Sm-Nd, Rb-Sr, K-Ar isotopic dating of garnet amphibolites from the Stanovoy Ophiolitic Belt). *Doklady Akad. Nauk* **1997**, *352*, 218–221.

46. Bezzubtsev, V.V.; Zalyalyaev, P.Sh.; Sakovich, A.B. *Geological Map of Mountainous Taimyr, Scale 1:500,000*; Explanatory Note; Krasnoyarskgeologiya: Krasnoyarsk, Russia, 1986; 177p. (In Russian)

47. Lopatin, V.M.; Natapov, L.M.; Uflyand, A.K.; Ushakov, A.N.; Chernov, D.V. Geodynamic nature of the Riphean volcanic belt of Taimyr. In *Paleovolcanism of the Altai-Sayan Folded Area and the Siberian Platform*; Nauka: Novosibirsk, Russia, 1991; pp. 58–62.

48. Nikishin, A.M.; Sobornov, K.O.; Prokopiev, A.V.; Frolov, S.V. Tectonic evolution of the Siberian Platform during the Vendian and Phanerozoic. *Mosc. Univ. Geol. Bull.* **2010**, *65*, 1–16. [CrossRef]

49. Deviatov, V.P.; Savchenko, V.I. New data for hydrocarbon resources reevaluation of Anabar-Khatanga oil-and-gas bearing area. *Oil Gas Geol.* **2012**, *1*, 55–61. (In Russian)

50. Vernikovsky, V.A. Metamorphic formations and geodynamics of the Northern Taimyr. *Russ. Geol. Geophys.* **1992**, *33*, 42–49.

51. Vernikovsky, V.A.; Neimark, L.A.; Ponomarchuk, V.A.; Vernikovskaya, A.E. Geochemistry and age of collisional granitoides and metamorphites of the Kara microcontinent (Northern Taimyr). *Russ. Geol. Geophys.* **1995**, *36*, 46–60.

52. Vernikovsky, V.A.; Kovach, V.P.; Kotov, A.B.; Vernikovskaya, A.E.; Sal'nikova, E.B. Sources of granitoids and the evolutionary stages of the continental crust in the Taimyr folded area. *Geochem. Int.* **1999**, *37*, 493–502.

53. Pogrebitsky, Yu.E.; Shanurenko, N.K. *State Geological Map of the Russian Federation. Scale 1:1,000,000. Sheet S-47-49—Lake Taimyr. Explanatory Note*; VSEGEI: St. Petersburg, Russia, 1998; 231p. (In Russian)

54. Metelkin, D.V. *Evolution of the Structures of Central Asia and the Role of Strike-Slip Tectonics According to Paleomagnetic Data*; IPGG SB RAS: Novosibirsk, Russia, 2012; 460p. (In Russian)

55. Inger, S.; Scott, R.A.; Golionko, B.G. Tectonic evolution of the Taimyr Peninsula, northern Russia: Implications for Arctic continental assembly. *J. Geol. Soc.* **1999**, *6*, 1069–1072. [CrossRef]

56. Dobretsov, N.L. Permian-Triassic magmatism and sedimentation in Eurasia as result of superplume. *Doklady Akad. Nauk* **1997**, *354*, 220–223.

57. Dobretsov, N.L.; Vernikovsky, V.A. Mantle plumes and their geological manifestations. *Int. Geol. Rev.* **2001**, *43*, 771–787. [CrossRef]

58. Renne, P.R.; Basu, A.R. Rapid eruption of the Siberian Traps flood basalts at the Permo–Triassic Boundary. *Science* **1991**, *253*, 176–179. [CrossRef] [PubMed]

59. Campbell, I.H.; Czamanske, G.K.; Fedorenko, V.A.; Hill, R.I.; Stepanov, V. Synchronism of the Siberian Traps and the Permian–Triassic Boundary. *Science* **1992**, *258*, 1760–1763. [CrossRef] [PubMed]

60. Deev, E.V.; Votakh, O.A.; Belyaev, S.Yu.; Zinov'ev, S.V.; Levchuk, M.A. Tectonic map of the basement of the Mid-Ob' plate complex (West Siberia). *Russ. Geol. Geophys.* **2001**, *42*, 968–978.

61. Saunders, A.D.; England, R.W.; Reichow, M.K.; White, R.V. A mantle plume origin for the Siberian Traps. Uplift and extension in the West Siberian Basin. *Lithos* **2005**, *79*, 407–424. [CrossRef]

62. Dobretsov, N.L.; Kirdyashkin, A.A.; Kirdyashkin, A.G.; Vernikovsky, V.A.; Gladkov, I.N. Modelling of thermochemical plumes and implications for the origin of the Siberian Traps. *Lithos* **2008**, *100*, 66–92. [CrossRef]

63. Dobretsov, N.L.; Vernikovsky, V.A.; Karyakin, Yu.V.; Korago, E.A.; Simonov, V.A. Mesozoic–Cenozoic volcanism and geodynamic events in the Central and Eastern Arctic. *Russ. Geol. Geophys.* **2013**, *54*, 874–887. [CrossRef]

64. Reichow, M.K.; Saunders, A.D.; White, R.V.; Pringle, M.S.; Al'Mukhamedov, A.I.; Medvedev, A.Ya.; Kirda, N.P. ^{40}Ar/^{39}Ar dates from the West Siberian basin: Siberian flood basalt province doubled. *Science* **2002**, *296*, 1846–1849. [CrossRef] [PubMed]

65. Vernikovsky, V.A.; Pease, V.; Vernikovskaya, A.E.; Romanov, A.P.; Gee, D.G.; Travin, A.V. First report of early Triassic A-type granite and syenite intrusions from Taimyr: Product of the northern Eurasian superplume? *Lithos* **2003**, *66*, 23–36. [CrossRef]

66. Aplonov, S.V. An aborted Triassic ocean in West Siberia. *Tectonics* **1988**, *7*, 1103–1122. [CrossRef]

67. Aplonov, S.V. The tectonic evolution of West Siberia: An attempt at a geophysical analysis. *Tectonophysics* **1995**, *245*, 61–84. [CrossRef]

68. Metelkin, D.V.; Vernikovsky, V.A.; Kazansky, A.Yu.; Wingate, M.T.D. Late Mesozoic tectonics of Central Asia based on paleomagnetic evidence. *Gondwana Res.* **2010**, *18*, 400–419. [CrossRef]

69. Proskurnin, V.F.; Simonov, O.N.; Sobolev, N.N.; Tuganova, E.V.; Uklein, V.N. Tectonic zoning of the north of Central Siberia (Taimyr AD). In *Natural Resources of Taimyr*; Ministry of Natural Resources: Dudinka, Russia, 2003; Volume 1, pp. 178–209.

70. Staroseltsev, V.S. Lena-Yenisei cascade of Mesozoic depressions in the northern margin of the Siberian platform in connection with petroleum potential. *Geol. Miner. Resour. Siberia* **2013**, *1*, 46–53. (In Russian)

71. Baldin, V.A.; Kunin, K.N.; Kunin, N.Ya. New considerations on the structure and origin of the diagonal megabank system in the Yenisei-Khatanga basin. *Oil Gas Geol.* **1997**, *3*, 26–34. (In Russian)

72. Sobornov, K.; Afanasenkov, A.; Gogonenkov, G. Strike-Slip Faulting in the Northern Part of the West Siberian Basin and Enisey-Khatanga Trough: Structural Expression, Development and Implication for Petroleum Exploration. 2015. Available online: http://www.searchanddiscovery.com/pdfz/documents/2015/10784sobornov/ndx_sobornov.pdf.html (accessed on 31 October 2018).

73. Structural-Tectonic Map of the Siberian Platform by, V.S. Staroseltsev. Available online: http://www.sniiggims.ru/maps/strtect.html (accessed on 5 October 2018).

74. Glagolev, P.L.; Mazanov, V.F.; Mikhailova, M.P. *Geology and Oil and Gas Potential of the Yenisei-Khatanga Basin*; IGiRGI: Moscow, Russia, 1994; 118p. (In Russian)

75. Khudoley, A.K.; Verzhbitsky, V.; Zastrozhnov, D.; O'Sullivan, P.B.; Ershovaa, V.; Proskurninc, V.F.; Tuchkova, M.I.; Rogovf, M.; Kyserg, T.K.; Malysheva, S.V.; et al. Late Paleozoic—Mesozoic tectonic evolution of the Eastern Taimyr-Severnaya Zemlya Fold and Thrust Belt and adjoining Yenisey-Khatanga Depression. *J. Geodyn.* **2018**, *119*, 221–241. [CrossRef]

76. Bochkarev, V.S.; Brekhuntsov, A.M.; Deshchenya, N.P. The Paleozoic and Triassic evolution of West Siberia (data of comprehensive studies). *Russ. Geol. Geophys.* **2003**, *44*, 115–140.

77. Deyev, E.V.; Zinoviev, S.V. Morphotectonics of the basement surface of the West-Siberian basin within Nadym-Tazovsk interfluves. *Geotecton. Metall.* **2003**, *27*, 11–23.

78. Deev, E.V.; Zinoviev, S.V.; Chikov, B.M. Structural model of Pre-Cretaceous complexes of the West-Siberian basin (using the example of Nadym-Tazovsk interfluve). *Lithosphere* **2004**, *2*, 61–80. (In Russian)

79. Kontorovich, V.A.; Ayunova, D.V.; Gubin, I.A.; Ershov, S.V.; Kalinin, A.Yu.; Kalinina, L.M.; Kanakov, M.S.; Solov'ev, M.V.; Surikova, E.S.; Shestakova, N.I. Seismic stratigraphy, formation history and gas potential of the Nadym-Pur interfluve area (West Siberia). *Russ. Geol. Geophys.* **2016**, *57*, 1248–1258. [CrossRef]

80. Kazakov, A.M.; Konstantinov, A.G.; Kurushin, N.I.; Mogutcheva, N.K.; Sobolev, E.S.; Fradkina, A.F.; Yadryonkin, A.V.; Devyatov, V.P.; Smirnov, L.V. *Stratigraphy of Oil and Gas Basins of Siberia. Triassic System*; Publishing House SB RAS: Novosibirsk, Russia, 2002; 322p. (In Russian)

81. Isayev, A.V.; Devyatov, V.P.; Karpukhin, S.M.; Krinin, V.A. Oil and gas prospects of the Yenisei-Khatanga regional trough. *Oil Gas Geol.* **2010**, *4*, 13–23. (In Russian). Available online: https://drive.google.com/file/d/0B_bCwpRhYduYdHAwdWRrT05fM1k/view (accessed on 3 October 2018).

82. Shurygin, B.N.; Nikitenko, B.L.; Devyatov, V.P.; Ilyina, V.I.; Meledina, S.V.; Gaideburova, E.A.; Dzyuba, O.S.; Kazakov, A.M.; Mogucheva, N.K. *Stratigraphy of Oil and Gas Basins of Siberia. Jurassic System*; Publishing House SB RAS: Novosibirsk, Russia, 2000; 480p. (In Russian)

83. Kontorovich, V.A. The tectonic framework and petroleum prospects of the western Yenisei-Khatanga regional trough. *Russ. Geol. Geophys.* **2011**, *52*, 804–824. [CrossRef]

84. Isayev, A.V.; Krinin, V.A.; Filiptsov, Yu.A.; Karpukhin, S.M.; Sklyarov, V.R. Potential oil-and-gas bearing objects in the clinoform complexes of the Yenisey-Khatanga regional trough: Results of seismic-geological modelling. *Geol. Miner. Resour. Siberia* **2011**, *2*, 74–82. (In Russian)
85. Bird, K.J.; Charpentier, R.R.; Gautier, D.L.; Houseknecht, D.W.; Klett, T.R.; Pitman, J.K.; Moore, T.E.; Schenk, C.J.; Tennyson, M.E.; Wandrey, C.J. *Circum-Arctic Resource Appraisal: Estimates of Undiscovered Oil and Gas North of the Arctic Circle. U.S. Geological Survey Fact Sheet FS-2008-3049*; U.S. Geological Survey: Reston, VA, USA, 2008; 4p.
86. Klett, T.; Pitman, J. Assessment of undiscovered petroleum resources of the Yenisey–Khatanga Basin. *First Break* **2010**, *28*, 107–110. Available online: http://fb.eage.org/publication/content?id=38390 (accessed on 6 November 2018).
87. Afanasenkov, A.P. VNIGNI Perspectives of Studying the Oil and Gas Potential of Individual Regions of Russia on the Example of the Gydan-Khatanga and Anabar-Lena Oil and Gas Perspective Zones. 2014. Available online: www.vnigni.ru/downloads/011_Afanasenkov.pdf (accessed on 6 November 2018).
88. Kim, N.S.; Rodchenko, A.P. Organic geochemistry and petroleum potential of Jurassic and Cretaceous deposits of the Yenisei–Khatanga regional trough. *Russ. Geol. Geoph.* **2013**, *54*, 966–979. [CrossRef]
89. Pronkin, A.; Savchenko, V.; Shumsky, B. Hydrocarbon potential of the Khatanga gulf and adjoining land. *Offsore [Russia]* **2013**, *1*, 18–23. (In Russian)
90. Afanasenkov, A.P. Historical and Genetic Conditions for the Oil and Gas Potential of the Jurassic Deposits of the Yenisei-Khatanga Oil and Gas Bearing Region. Ph.D. Thesis, Moscow State University, Moscow, Russia, 15 May 1987.
91. Shemin, G.G. *Regional Oil and Gas Reservoirs of Jurassic Deposits in the Northern West Siberian Province*; Publishing House SB RAS: Novosibirsk, Russia, 2014; 362p. (In Russian)

minerals

MDPI

Article

Geodynamic Evolution and Metallogeny of Archaean Structural and Compositional Complexes in the Northwestern Russian Arctic

Nikolay E. Kozlov [1],*, Nikolay O. Sorokhtin [2] and Eugeny V. Martynov [1]

[1] Geological Institute of the Kola Science Centre RAS, 14 Fersman Street, Apatity 184209, Russia;
    mart@geoksc.apatity.ru

[2] P.P. Shirshov Institute of Oceanology RAS, 36 Nakhimovsky Prospect, Moscow 117997, Russia;
    nsorokhtin@mail.ru

* Correspondence: kozlov@geoksc.apatity.ru; Tel.: +7-81555-79-656

Received: 19 October 2018; Accepted: 3 December 2018; Published: 6 December 2018

✓ check for updates

Abstract: This paper highlights the geodynamic evolution of the early Precambrian rock associations in the northwestern part of the Russian Arctic where the rocks are exposed in the Kola region (northeastern Baltic Shield). The evolution is shown to predetermine the metallogenic potential of the area. It is emphasized that the Earth's evolution is a non-linear process. Thus, we cannot draw direct analogies with Phanerozoic time or purely apply the principle of actualism, which is still widely used by experts in Precambrian geology to study the premetamorphic history of ancient deposits. In both cases, the principles should be adjusted. This article provides a novel technique for reconstructing geodynamic regimes of protolith formation in the early Precambrian. The technique identifies changing trends in geodynamic regimes during the formation of the Archean structural and compositional complexes in the Kola region. These trends fit into the earlier suggested general scheme of their formation, thus enhancing its reliability. The metallogeny of the ore areas is specified. The results of the geodynamic reconstructions explain most of the location patterns of minerals within the Kola region. Thus, the authors consider the metallogenic forecast based on geodynamic reconstructions to be a promising trend for further research.

Keywords: evolution of the composition; basic rocks; Precambrian; search of trend differences; geodynamic evolution; metallogeny

---

1. Introduction

The basis for studying the geodynamic evolution of Early Precambrian rock assemblages in the northwestern Russian Arctic that crop out within the Kola region (northeastern Baltic Shield) is provided by data on their compositional variations with time. The authors used geological–structural and isotope–geochemical methods combined with geophysical investigations to outline stages of the most ancient structures in the region [1–5]. Notably, the results of geochronological analyses, however accurate, still provide ambiguous interpretations of ages. Thus, the age of the Patchemvarek metagabbro massif (2935 ± 6 Ma) is considered to be the beginning of the Kolmozero-Voronya Greenstone Belt formation [4], or the age of the "best-preserved fragments of the ancient protolith in pregranulitic and pregranitic associations of the Murmansk microcontinent" [2]. The time of the protolith formation in basic rocks of the Lapland (Lapland-Kolvitsa) Granulite Belt is also in doubt. Some authors [1,2] believe the protoliths to be Paleoproterozoic, while others [6–8] have argued that at least a few of the rocks in the belt section are Archean. Certain supporters of the first concept consider the protoliths to be Paleoproterozoic with the proviso that mafic granulates should not be older than the Paleoproterozoic gabbro-anorthosite massifs [2].

At the same time, it is clear that the volcanic evolution can only be explored by ranking the studied rock assemblages according to the time of the formation of their protoliths. Understanding that the future of this issue belongs to geochronology, we assign some certain role in this to the methods relied on applying the data on the rock composition that allow spotting trends in compositional changes in rocks and their associations over time, as shown in previous studies ([7–11] etc.). It is also important that these methods concurrently describe such trends using statistically evaluated indicators. Though there are many evolution models, data on the composition of metamorphites composing early Precambrian domains have rarely been considered. To close the gap, we have worked for a long time to address the above issue. In this respect, the first detailed survey [12] and one of the latest papers dedicated to this issue should be noted [13].

2. Geological Setting

A detailed description of the Archean geology of the northeastern Baltic Shield has been provided [10]. Structural and compositional complexes originated here in the early Mesoarchean era (probably in the Paleoarchean era) and formed by the end of the Mesoarchean era to the beginning of the Neoarchean era. This period marked the genesis of two major segments (lithospheric plates) of the continental crust, i.e., the Belomorian (Belomorsky mobile belt) and Kola granulite-gneiss areas, which show unique patterns of crustal evolution. The later stage covers the time span from the Mesoarchean and Neoarchean border up to date and indicates the time when the united Archean Karelian-Kola lithospheric plate existed. As a result, a united collision structure formed within the northeastern Baltic Shield through collided basic-ultrabasic domains and areas of the continental crust [10,14–16].

The Kola Granulite-Gneiss Area (GGnA) occupies the northeastern part of the Baltic Shield (Figure 1). It borders the Belomorskaya mobile area in the south and southwest and is overlapped by the Russian Plate cover in the east and southeast. In the north and northeast, the Kola GGnA is constrained by the Karpinsky fault in the Barents Sea. In general, it shows a typical mosaic structure of the continental crust and consists of structurally different domains (Figure 1). There are six in the region: The Murmansk, Kola-Norwegian, Keivy, Eastern Kola, Chapoma, Umba, and Tersky domains. Most of these formations are divided by narrow linear belts of the greenstone type, i.e., the Kolmozero-Voronya (Titovsko-Kolmozersky), Sergozero-Strelnensky, and Lapland granulite belts. There are a few points of contact where two closely jointed crustal domains are marked by specific structural and compositional parageneses [10,16].

At present, despite the extensive and detailed investigation of the Archean structural and compositional complexes of the Kola GGnA, reconstructing its earlier stages is still a challenge, and datings of its generation are obscure [8–11,13,17–21]. We can clearly define structural-metamorphic complexes that show the upper Archean collision of the continental crust, which is well-observed in the eastern part of the Baltic Shield. Recent detailed research of the most ancient complexes in the Kola-Norwegian domain yield no evidence of relic parageneses at the earliest stages of the continental crust generation.

Generally, the Kola-Norwegian domain has an imbricated thrust structure that formed in the late Archean era and at the Rebolian folding phase. However, dome-shaped folds of the same age still occur in the southwestern part of the domain, which indicates that intensive granitization prevails in this area (the Priimandrovsky district). The Archean complexes are represented by repeatedly and heterogeneously metamorphosed formations from amphibolite facies of the high-temperature stage to low-gradient granulite facies.

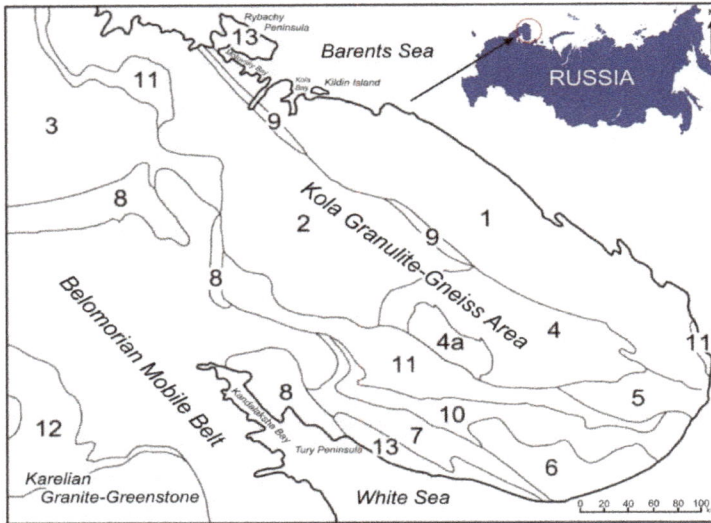

Figure 1. Zoning of the early Precambrian geostructural elements of the Earth's crust in the northeastern Baltic Shield (Modified from [10]). Crustal domains: 1—Murmansk, 2—Kola-Norwegian, 3—Lotta, 4—Keyvy, 4a—Verkhneponoysky, 5—Eastern Kola, 6—Chapoma, 7—Tersky. Greenstone and granulite belts (8–10—Archaean; 11, 12—Neoproterozoic): 8—Lapland-Kolvitsa; 9—Titovka-Kolmozero (Kolmozero-Voronya); 10—Sergozersky–Strelninsky; 11—Pechenga-Imandra-Varzuga-Ustponoysky; 12—Northern Karelian; 13—Riphean rift and continental marginal sediments.

The Belomorian Mobile Belt (MB) is a separate crustal block that occurs between two major tectonic structures of the Baltic Shield, i.e., the Karelian granite-greenstone area (GGrA) and the Kola granulite-gneiss area (GGnA) (Figure 1) [10]. The width of the junction zone with the latter varies from 2–3 to 15–20 km and is well-detected according to geophysical data. In the magnetic field, the border is marked by chains of linear suture-oriented anomalies. The suture zone has linear-type structures. Here, margins of the Kola block thrust over the Belomorian block. Most of the contact is marked by the Lapland-Kolvitsa Granulite Belt. It comprises basic metamorphic rocks altered both in amphibolite and in granulite facies, including the eclogite subfacies of granulite facies [8,18,22].

3. Materials and Methods

The order in which the Archean complexes (rock assemblages) of the northeastern Kola region formed has been previously outlined during the analysis of the composition of the metabasites [13]. Since the results of these studies provide the basis for further reconstructions, we deem it necessary to briefly point out the applied methods and present the obtained results.

The behavior of the chemical rock compositions was described using the method of modeling the trend of variations in chemical compositions of rocks that compose the studied rock assemblages as compared to the partial order entered by the researcher towards the combination of these assemblages. The algorithm applied to search for this trend has been mainly highlighted in Russian scientific journals and will be briefly described here. It shall also be noted that all the algorithms applied in this research are published in international scientific sources. Suppose $Z = \{Z_i\}$ represents a set of n-dimensional random variables, and partial order '$\rightarrow$' is given on set $Z*Z$. If c is an n-dimensional vector of unit length, the scalar product (c, Z_i) is a one-dimensional random variable. This random variable may be described by its mathematical expectation, $M(c, Z_i)$. Each of these random variables (Z_iZ) are represented by a sample of n-dimensional vectors $(V_i = \{v_{ij}\})$ in Euclidian space, R^n (i.e., the sample V_i describes the chemical compositions of rocks composing the corresponding rock assemblage). To compare the mathematical expectations of the random variables (c, Z_i and c, Z_j), the Puri–Sen–Tamura rank-order test for the equality of means is proposed to be used [23,24]. The choice of the specified statistical criterion is defined by both its stability against the violation of the normality (or even unimodality) condition for the random variable distribution and against the availability of anomalous observations in the samples.

For the above criterion to be used, it is necessary to estimate the means (median Me(c, Z_i) is chosen as the estimator) and calculate the Puri-Sen-Tamura statistics $\Lambda((c, Z_i),(c, Z_j))$ for all pairs of random variables (Z_i, Z_j). The statistical modeling (where the set of variables is specified by ratio '→') involves the search for the n-dimensional vector c of unit length, for which, at a chosen significance level, δ, the following conditions are satisfied Equations (1) and (2):

$$Me(c, Z_i) < Me(\mathbf{c}, Z_j) \tag{1}$$

$$\Lambda((c, Z_i),(\mathbf{c}, Z_j)) > \chi^2(\delta) \tag{2}$$

where $\chi^2(\delta)$ denotes a value of χ^2-inverse distribution for the significance level δ of all pairs $<Z_i,Z_j>$ so that $Z_i{\to}Z_j$.

The trend modeling task lies in the approximation of the given partial order by linear function P: $Z{\to}R$, where R is a set of real numbers. The approximation quality is estimated by the value of the function: $J(P) = max_c min_{(i,j)}\Lambda(\{(c,v_{ik})\}_k,\{(c,v_{jl})\}_l)$, where on the set of pairs $\{(Z_i, Z_j)\}$ so that $Z_i{\to}Z_j$.

Thus, the modeling of the trend in differences of chemical compositions of rocks that compose the studied rock assemblages as compared to the partial order entered by the researcher towards the combinations of these assemblages reduced itself to a classical functional J(P) optimization task on the given set in view of the entered constraints. The optimal solution was searched for using the known Nelder–Mead algorithm [25].

Vector c, which is further referred to as the partial-order factor, describes the common trend in the behavior of the chemical compositions in relation to the given partial order.

As an illustration of the above, we shall provide the following hypothetic example (Figure 2). In this case, the coordinate axes convey no specific geological meaning. It is only valuable here that the space is two-dimensional, and thus, it is enough to preset two arbitrary axes for introducing a coordinate system in this space. Suppose that some combination of rock assemblages {A, B, C, S, H, D, E, K, M} is given for the two-dimensional feature space. Each of the assemblages is represented by a figurative point (for example, a means point). The partial order (for example, 'younger' or '→') in this combination is displayed as a graph, namely B is younger than S, C is younger than D, D is younger than E and H is younger than K. It is required to find the linear function P implemented as axis F specified by vector c in the feature space so that functional J(P) takes on a minimum value in the ratio (B→S, C→D, D→E, H→K).

To find the partial order that describes the compositional changes in the Archean metabasites from younger to older formations, the following conditions were formulated based on the geological setting and geochronological data [2,8,10,21]:

1. The Titovka-Kolmozero (Kolmozero-Voronya) Belt is the youngest when compared to the Murmansk and Kola-Norwegian domains, since its rock assemblages formed at the pre-metamorphic stage as a result of interaction of the domains at their border;

2. The Lapland-Kolvitsa Granulite Belt, regardless of its age, is younger than the Lotta domain and Belomorian Mobile Belt, due to the interaction of which it was emplaced as a volcanogenic-sedimentary complex;

3. According to available data, rocks of ancient complexes in Karelia, Canada, and Greenland formed at an earlier stage of the Earth's evolution as compared to the Archean metamorphites of the Kola region.

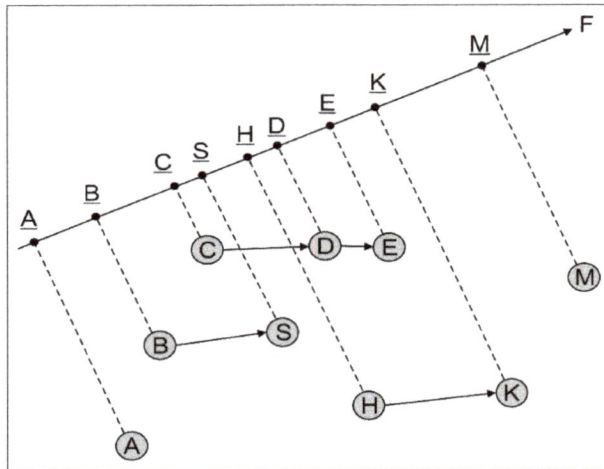

Figure 2. Partial order relation presented as a graph: B precedes (→) S; C→D; D→E; H→K. Location of A and M, as well as B and S regarding C, D and E, as well as H and K on axis F is not introduced initially [13]. Axis F describes the location of rock assemblages towards to each other with time.

It shall be noted that the ambiguous interpretation of the protolith age of the Lapland-Kolvitsa granulite belt, which was mentioned earlier in the introduction, does not affect the second condition since it is satisfied when accepting both the Archean and Paleoproterozoic time of emplacement for this structure.

The formulated conditions define the following partial order:

1. The Murmansk Domain precedes (→) the Titovka-Kolmozero Belt, and the Kola-Norwegian Domain precedes (→) the Titovka-Kolmozero Belt;
2. The Lotta Domain precedes (→) the Lapland-Kolvitsa granulite belt, and the Belomorian Mobile Belt precedes (→) the Lapland-Kolvitsa granulite belt;
3. Each of the oldest rock assemblages in Karelia, Canada, and Greenland precede (→) each of the Archean rock assemblages in the Kola region.

As a result, we found the partial order factor (F_1) which describes the trend in the compositional changes of the metabasites in the studied rock complexes. Alongside, the above formulated conditions have to be concurrently fulfilled with the differences in rocks for each task to be significant at the 5% significance level (Figure 3). The metamorphites of the Archean complexes are clearly seen to group on it and form sets that we code-named groups A, B and C.

Notably, searching for factor F_1 was based on data on the composition of rock assemblages where relative age can be defined quite accurately based on geological data. We did not introduce the locations of complexes listed in different conditions of the task as well as other structures on the factor. As far as we remember, the location of most Archean complexes on factor F_1 does not contradict the sequence of formation during the northeastern Baltic Shield evolution. Its pattern has been suggested by a number of authors on the basis of complex rock study compared to geophysical and geochronological data [2,8,20].

Since any petrogeochemical reconstructions are probabilistic, the only way to increase their reliability is to introduce new data that would enlarge the study area. As long as these reconstructions do not clash with the already obtained results, we seem to take a step that approaches high probability, but never reaches it. In this research, such a step was taken in studying the evolution of the geodynamic regimes of metavolcanite formation in the Archean era in the Kola region.

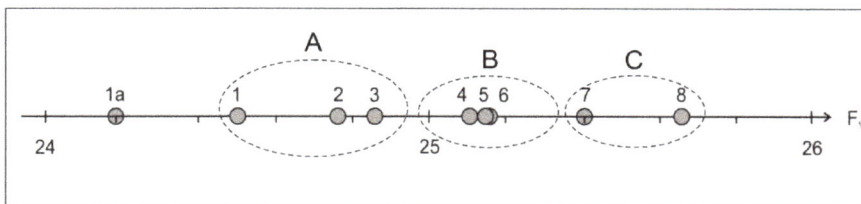

Figure 3. Location of compositions of Archaean metabasites in the Kola region compared to formations in Karelia, Canada, Greenland, on factor $F_1 = 0.34SiO_2 - 0.46TiO_2 + 0.25Al_2O_3 + 0.19\Sigma Fe + 0.19MnO - 0.07MgO + 0.38CaO - 0.24Na_2O - 0.57K_2O$ [13]: 1a—ancient formations in Karelia, Canada, Greenland, 1–8—domains of the Kola region: 1—Keyvy, 2—Lotta (Allarechka and Notozero complexes), 3—Chapoma, Tersky and partly Umba, 4—Kola-Norwegian, 5—Murmansk, 6—Belomorian Mobile Belt; 7–8—Archaean belts of the Kola region: 7—Lapland-Kolvitsa, 8—Titovka-Kolmozero. A, B, C—enlarged groups to be described further.

In this study, data on the geology and mineral composition of the early Precambrian complexes were collated and compared. As with the previous task, basic rocks were studied to make various reconstructions. Protoliths of their metamorphosed analogues are best recognized by their original features. Furthermore, they are widespread within most of the early Precambrian structures of the region.

Previously, we defined regular and statistically significant differences in the petrochemical characteristics of the Precambrian metabasalts and Phanerozoic basalts (Figure 4), which provide evidence that processes of the Earth's evolution were non-linear [9]. The nonlinearity of the evolutional processes is attested by the following fact established in this research. As a result of studying the mutual arrangement of reference images (Phanerozoic basalts) and rock assemblages of Precambrian metabasalts in the feature space, it turned out that the shifting of their petrochemical features could not be statistically reliably described using one linear function. This meant that at least two linear functions needed to be applied to describe such a shift, as shown in Figure 4.

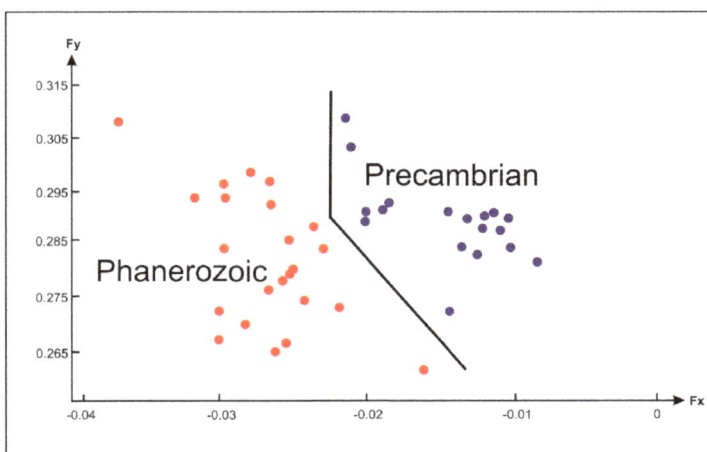

Figure 4. Location of median values of compositions of basic rocks from different Phanerozoic (red dots) and Precambrian (blue dots) rock assemblages [9] on diagram $Fx - Fy$, where: $Fx = -0.1SiO_2 - 0.82TiO_2 - 0.11Al_2O_3 + 0.34FeO + 0.06MgO + 0.03CaO - 0.28Na_2O + 0.32K_2O$; $Fy = 0.45SiO_2 - 0.28TiO_2 + 0.21Al_2O_3 + 0.22FeO - 0.06MgO + 0.24CaO - 0.27Na_2O - 0.75K_2O$.

In addition, the conventionally applied methods for the reconstruction of geodynamic regimes are often based on reference Phanerozoic rock assemblages that are represented by statistically invaluable samples. This particularly concerns reconstruction methods that apply various diagrams [26–36]. This is the reason why we were quite careful when using such methods for modeling geodynamic regimes. Thus, the principle of direct analogies with the Phanerozoic cannot be applied for the reconstruction of the Precambrian geodynamic settings without making necessary adjustments, i.e., considering the shift of rock images in the feature space when moving from the Precambrian to the Phanerozoic time and the representativity of the applied reference Phanerozoic rock assemblages.

In this study, we compared Precambrian and Phanerozoic rock complexes that were genetically linked to certain regimes on the basis of the idea of the specific nature of the Precambrian period in the Earth's evolution, and on some commonness in the geodynamic settings manifested throughout the whole geological history, i.e., homolog rows of geodynamic regimes [7]. This comparison was performed on more than 1100 bulk rock chemical analyses of Precambrian metamorphites and about 1100 bulk rock chemical analyses of Phanerozoic magmatic rocks.

We analyzed data on the abundance of certain varieties of original (premetamorphic) rock associations in the study areas. This work was based on reconstructions of the original features of rocks according to the petrogeochemical characteristics of compositions. For this, the method described in Reference [37] was applied. Like other techniques elaborated for these purposes, this method provides no solution for this task (for felsic metamorphites, first of all). This was solved in two ways. When points of rock composition occurred within the field covering sedimentary and magmatic formations on the reconstruction diagrams, such rocks were all referred to the first group in one comparison variant. In another, they were referred to the second group. Next, patterns of reconstruction variants were compared. We assumed all intermediate decisions to be within these extreme decisions. Once patterns of both variants coincided, the final conclusion on a spreading character was made.

As stated above, systematic differences in the composition of the Precambrian rocks and their Phanerozoic homologs constrain the application of numerous diagram methods to reconstruct early Precambrian geodynamic regimes. The authors have worked on this issue for a long time [7,9]. To obtain more accurate results, we elaborated on a method that describes the mentioned non-linearity even more efficiently than previously used techniques. Instead of 10 petrogenic elements (SiO_2, TiO_2, Al_2O_3, FeO, Fe_2O_3, MnO, MgO, CaO, Na_2O, and K_2O), we used eight parameters of the chemical composition, turning the FeO, Fe_2O_3, an MnO parameters into a new one, provisionally named $\sum FeO$. We described the shifting of the Precambrian rock assemblages regarding the Phanerozoic standards in 8D feature space (the Precambrian rock assemblages and Phanerozoic standards shown by multiple figurative points) in terms of a quadric surface dividing them. The criterion to determine this surface as dividing was the statistical significance of difference between multiple figurative points of each rock assemblage with a multitude of projections of these points on the surface. Multiple figurative points of the Precambrian and Phanerozoic rock assemblages were arranged on either side of this surface (Figure 5). Furthermore, we used an optimal dividing surface for which the minimal proximity of the whole rock assemblage totality to the surface was maximal. Here, we applied a quadratic decision rule as follows Equation (3):

$$(z - Y)S_2^{-1}(z - Y)' - (z - X)S_1^{-1}(z - X)' \geq 2ln\frac{|S_1|^{1/2}}{|S_2|^{1/2}} \tag{3}$$

where S_1 and S_2 are the estimates of covariance matrixes, and X and Y are averages of the Precambrian and Phanerozoic rock assemblages.

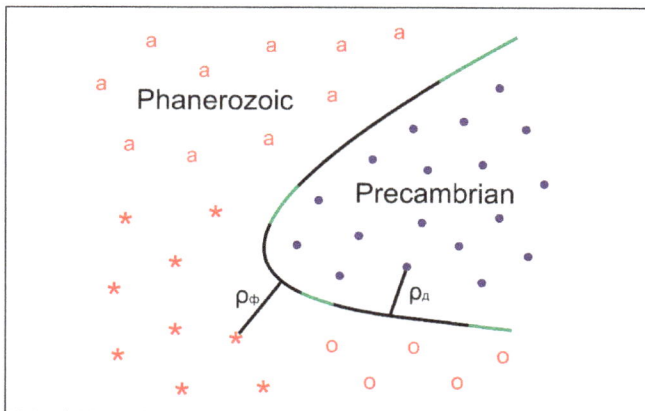

Figure 5. Schematic reconstruction of settings based on a quadric surface: a, *, o for the Phanerozoic rock assemblages formed in different geodynamic settings, • for the Precambrian rock assemblages; ρ_ϕ for the proximity of a certain Phanerozoic rock assemblage represented by a set of figurative points in a space of chemical rock composition which has parameters similar to the simulated quadratic surface in relation to chosen metrics, and $\rho_\text{д}$ for the proximity of a certain Precambrian rock assemblage to the surface respectively.

To determine the geodynamic settings of the studied Precambrian rock assemblage, we used multitude figurative points for the Precambrian rock assemblages and Phanerozoic standards on the designed dividing surface. Thus, we used different proximity degrees for the projection points of figurative points that indicated the chemical compositions of the Precambrian and Phanerozoic rock assemblages. It also allowed us to verify the obtained results with statistical criteria.

Since the distribution of figurative points of the rock assemblages does not comply with any of the known classic criteria, we applied nonparametric ones. In this case, we used the Puri-Sen-Tamura criterion [23,24] that is stable to a disrupted normal (and even unimodal) distribution of studied random variables and to anomalous observations in samplings. As noted, there are a number of proximity degrees. Choosing one of them is a challenge since there are no science-based arguments in one's favor. Therefore, the authors applied the following technique. First, the regime of the studied Precambrian rock assemblage origin was reconstructed using different proximity degrees. Next, the obtained reconstructions were compared. Once decisions for most of the applied proximity degrees coincided, the decision was considered accurate.

The study of many Precambrian rock assemblages showed that ancient complexes that formed in different geodynamic settings differed less than their Phanerozoic homologs. The reason for this is still unclear. However, we may state that when Precambrian complexes are geodynamically reconstructed, it is more accurate to consider changing trends in regimes than their complete analogues.

4. Results

Table 1 and Figure 6 provide reconstructed geodynamic regimes of protolith formation in Precambrian complexes of the Kola region. Since sampling in the Chapoma, Tersky and partly in the Umba domains was not representative enough (group 3 in Figure 3), it was not considered in the current research. It is impossible to refer geodynamic settings of protolith formation in any complex to a certain type because of the reasons provided in the previous section. As shown by a previous study (Figure 3), there is a clear pattern of interchanging regimes in geological evolution. Petrogeochemical characteristics of these regimes make them similar to traps that are closer to continental reefs. At the final stage of the formation of the Archean complexes, they become similar to young arcs. In the geological time span, such an interchange of regimes is possible within a small area.

Thus, a geodynamic evolution with varying plume and subduction magmatism ratios in one region was described for younger complexes in the central and eastern Arctic [38].

Table 1. Reconstructed formation settings of some Precambrian complexes.

Groups, Figure 3	NoNo, Figure 3	Structures	TRAP	CR	MA	YA	DA	MOR
A	1	Keyvy Domain	**3.611**	4.021	4.022	4.075	4.174	5.014
	2	Lotta Domain	**2.667 ***	3.198	3.430	3.042	3.864	4.002
B	4	Kola-Norwegian Domain	2.448	**2.278**	3.265	2.308	2.791	2.781
	5	Murmansk Domain	2.549	**2.361**	3.145	2.371	3.074	3.144
	6	Belomorian Mobile Belt	2.549	**2.270**	3.050	2.349	2.833	2.775
C	7	Lapland-Kolvitsa Belt	2.535	2.135	2.982	**2.117**	2.675	2.435
	8	Titovsko-Kolmozersky Belt	3.555	3.284	4.045	**3.170**	3.855	3.275

* "distance" of Precambrian samplings to the respective standard (Phanerozoic) groups provisionally called "a proximity coefficient". The lower values of the provided coefficients (bold type), the closer compared rock assemblages are. Underlined figures indicate values of a minimal difference, if they insignificantly differ from the next-largest value with the significance level of $\alpha = 0.05$. Bold indicated values of a minimal difference. TRAP—traps, CR—continental rifts, MA, YA, DA—mature, young and developed arcs, respectively, MOR—mid-ocean ridges.

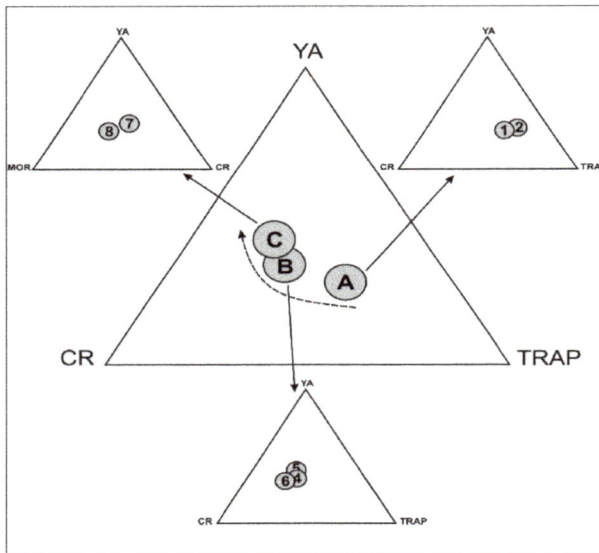

Figure 6. Location of composition points of the Archaean complexes in the Kola region on diagrams showing relative proximity to certain geodynamic regimes. Notations are specified in Table 1.

Many researchers believe that metabasites of ancient structures are close to rocks of trap formations that are genetically confined to plumes. Metabasites of younger structures are considered similar to the regimes typical of plate-tectonic processes [39]. This concept is rather interesting and matches the well-known concept that plate-tectonic processes substituted the plume-tectonic ones at early stages of the Earth's evolution.

Therefore, we may assume that disseminated rifting developed in the ancient crust during the early stages of formation of supracrustal complexes in the Kola region. The crust has features of the continental one. Relics of it have been possibly found in zircon cores in recent studies [40–42]. Thus, we suggest a more complicated multistage genesis of the continental crust within the Keivy domain when compared to other domains in the region. This was previously supposed in Reference [10] on the basis of regional geological-geophysical data [10]. The Lotta domain metabasites are also the most similar to trap formations.

These data should be considered jointly with similar and different features in the mineralogical composition of rock associations of Archean domains in the Kola region. We made a tentative conclusion that its central part has a line composed of metamorphites of the Kola–Norwegian and Keivy domains. This line clearly differs from rocks of the Murmansk and Belomorian domains and has features of the Keivy domain [10]. This conclusion was verified based on a set of fact data obtained in the recent decade. More than 2100 bulk rock chemical analyses of metamorphic rocks of different compositions were made to verify this. The sampling range of metamorphites from the Chapoma, Tersky and part of the Umba domains was considerably extended when felsic to intermediate rocks were included as study areas.

In this study, the Keivy, Lotta, Chapoma and Tersky (also part of the Umba) domains were defined as similar (Figure 7). The Keivy domain is still the closest to the Kola–Norwegian domain. Notably, substances of the Lotta domain were more similar to those of the Chapoma, Tersky and part of the Umba domains. The similar composition of metabasites in these areas was noted earlier in Reference [10], when the evolution of Archean formations was studied. Thus, it proves the suggestion set by Ivanov et al. on the possible allocation of supracrustal formations of the mentioned domains as a single zone [17].

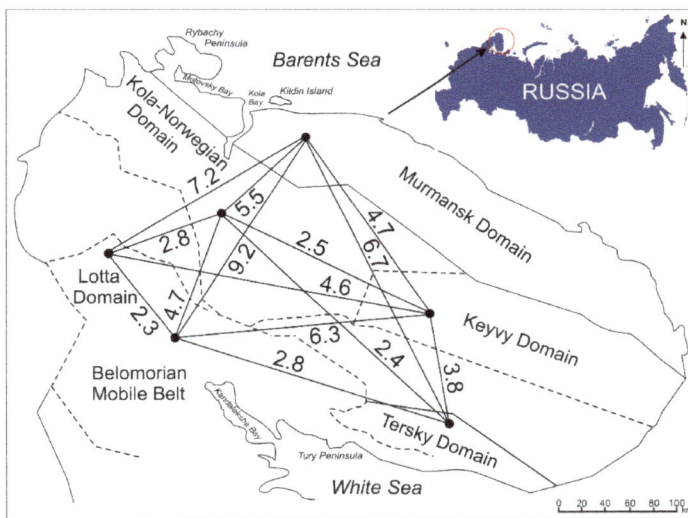

Figure 7. Compared mineralogical compositions of different Archaean structures in the Kola region. Structures are numbered as in Figure 3. Numbers near lines indicate a proximity coefficient of respective rock assemblages. The lower the coefficient of difference, the more the compared rock assemblages are similar.

Notably, the mineral composition of the Murmansk Domain metamorphites was significantly different from other structures in the region. In general, ratios of conditional proximity values indicating compositions of rock associations in different domains (Figure 7) were close to the data we obtained earlier ([10], Figure 3), but specified them greatly. Previously, we linked a minimal value of this coefficient to the fact that rock associations of the Murmansk domain belonged to the so called granite–greenstone areas (GGrA). Their ancient age of origin was also considered. As for rock associations from other domains, we remembered them belonging to the intermediate-type or granulite–gneiss areas (GGnA) [21].

With the new data, we may presume the coefficients of difference obtained for the studied complexes, just like previously described estimates of proximity to rock associations in Canada, Greenland and Northern Karelia, to reflect not so much the time of the protolith formation, as the geodynamics of processes within these structures, from the first stages of the protovolcanism and protosedimentation to further transformations at different levels of the Archean crust, which was suggested by Sorokhtin ([10], Chapter 5). The fact that there is a certain zoning within the Baltic Shield, i.e., GGrA rim GGnA and intermediate-type areas in the southwest and northeast, means this area deserves further research.

The ratio consistently changes within the studied domains, which are arranged according to the suggested chronological order of the formation of their protoliths, varieties, reconstructed metavolcanic (metavolcanites and metatuffites) and metasedimentary rocks (Figure 8). We believe this ratio to reflect the intensity of volcanism at certain stages of the region development. The Archean complexes of the northeastern Baltic Shield provide evidence of pulsating volcanic activity, as mentioned earlier [12].

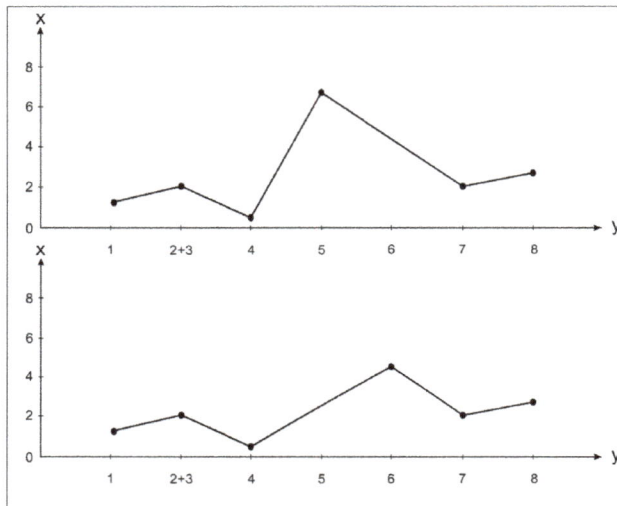

Figure 8. Varying ratios of metavolcanites and metasedimentary rocks distribution (axis x) in the studied complexes (axis y). Groups are numbered as in Figure 3. Constructed in two graphs, for the Murmansk Domain (group 5) and the Belomorian Mobile Belt (group 6).

5. Discussion

In the late Archean era, the geodynamic evolution of the northeastern Baltic Shield was associated with highly intensive structural metamorphism of the continental-crustal matter. Some new matter was added and border structures (collision sutures) were formed. These were marked by greenstone and granulite belts. As a result, a single collision complex occurred and united certain domains of the continental crust. These domains are rimmed by greenstone belts where the protooceanic lithosphere was taken in. Each of the crustal domains composing the collision zone had its own evolution that was dampened by further imposed events.

The obtained reconstructions are considered promising for the metallogenic assessment of the studied areas. As noted, the Paleoproterozoic PGE mineralization of the Kola region is likely to be linked with its Archean history [19,21]. When geodynamic reconstructions are compared to map data on deposits and ore occurrences in the Kola region [5], trends in the distribution of minerals and their association with the most ancient complexes in the region can be traced (Figure 9).

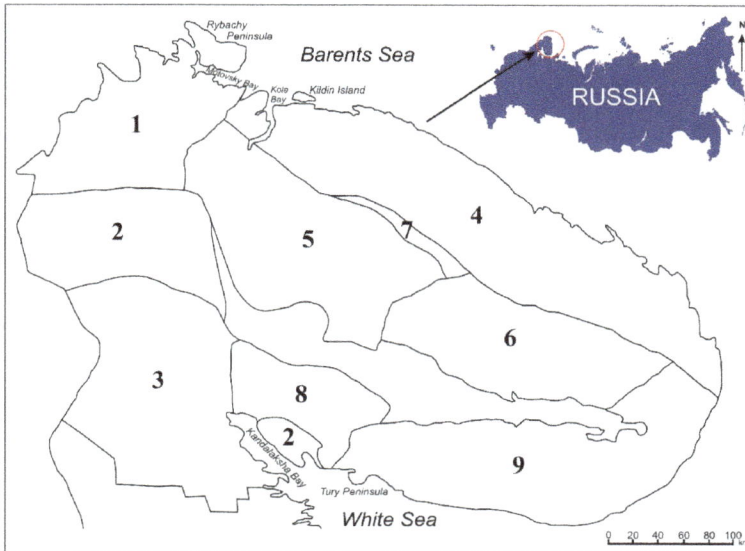

Figure 9. Metallogenic areas in the Murmansk region (Modified from [5]): 1—Pechenga-Allarechka area (nickel, copper, cobalt, sulphur, gold, silver, platinum, palladium, rhodium, ruthenium, iridium, selenium, tellurium, mica and ceramic pegmatites); 2—Lapland ore area (copper, nickel, vanadium, titanium, iron, manganese, molybdenum, graphite and gold, chromium, construction and decorative materials); 3—Belomorian ore area (deposits of muscovite and ceramic raw materials, sulfide (Cu-Ni) and Pt-bearing mineralization, raw materials for the aluminium industry); 4—Murmansk ore area (construction and decorative granite, ferruginous quartzites and ore occurrences of the thorium-uranium and uranium type); 5—Olenegorsk ore area (major deposits of ferruginous quartzites, apatite-silicate ores, apatite-magnetite ores of the Kovdor type, apatite-carbonate ores, tantalum-niobium ores, apatite and perovskite-titanium-magnetite ores, construction materials); 6—Keyvy ore area (ore occurrences of niobium, tantalum, zirconium, yttrium, thorium, uranium, tin, wolfram, vanadium and cobalt, major deposits of kyanite, abrasive garnet and amazonite, quartzites, high-grade vein quartz, muscovite, sillimanite, etc.); 7—Kolmozero-Voronya ore area (deposits and occurrences of gold, silver, molybdenum, lead, zinc, copper, nickel, iron, lithium, berillum, tantalum, niobium, caesium); 8—Kandalaksha ore area (titanium-magnetite and perovskite-titanium-magnetite ores with rare earths, phosphate raw materials, copper mineralization); 9—Tersky ore area (major occurrences of muscovite and ceramic pegmatites, decorative red sandstones, occurrences of molybdenite, carbonatites with apatite and rare metal mineralization, barite, amethyst, gold occurrences, diamond finds).

Previously, we linked increased concentrations of molybdenum and gold with the early Precambrian proto-island arc formations of the Lapland (Lapland-Kolvitsa) Granulite Belt [20]. Analyses of the metallogenic specification of the collision zones [43] suggest increased concentrations of titanium, manganese, vanadium, chromium and probably copper, nickel, and iron, which relate to the early Precambrian geodynamics of the formation of protoliths within the belt.

The authors recognize that increased concentrations of these elements in, e.g., the Lapland-Kolvitsa Belt stem from its geological history including its multistage metamorphism. Dehydration and anatexis of the protooceanic crust in collision zones are likely to follow a complicated multistage pattern. Spatial-temporal changes in metamorphic transformations mean that rock associations of the subducting protooceanic lithospheric plate were subjected to progressive metamorphism in the contact zone with the thrusting continent. They consistently passed all stages of transformation, from the lowest to the highest ones. A mineralized gas-saturated fluid formed in these conditions and was transported up the faults. When the fluid cooled, it provided retrograde contact-metasomatic changes in surrounding rocks. Numerous ultrabasic protrusions and protoophiolites of the Lapland-Kolvitsa Granulite Belt were also subjected to retrograde metamorphism as they passed the peak of changes. All chemical reactions in the collision zones and the plate underthrust zones were irreversible and happened with the heat uptake or release in certain oxidizing-reducing conditions. The geological time played an important role in the above processes, balancing all physical-chemical parameters of the fold system.

Thus, increased concentrations of some elements can be related to the early geodynamic history of the Lapland-Kolvitsa Belt. The authors consider it to be the basic hypothesis of possible ore sources. This should be applied on a case-by-case basis, considering all rock transformations in certain ore occurrences. Similarly, increased concentrations of gold, silver, molybdenum, lead, zinc, beryllium, tantalum, niobium and possibly copper, nickel and iron in the Titovka-Kolmozero (Kolmozero-Voronya) suture zone can be linked to the geodynamics of its formation.

Iron ores could occur in the Kola-Norwegian domain since it is geodynamically close to continental rift areas that typically contain this element. In the southwestern part of the domain, there is a thick ferruginous volcanic-sedimentary complex stretching northwestwards along its margin. Though the complex was intensively and repeatedly subjected to later structural metamorphism, a trough-like mode of its occurrence can be detected. It contains ferruginous quartzites, metapelites, bipyroxene crystal schists, and amphibolites. Carbonate-bearing rocks are rare. Notably, bottom parts of the sections are mainly composed of amphibolites and are superimposed by biotite gneisses. Aluminous gneisses and ferruginous quartzites are typical for the top parts of the sections. We believe this sedimentation pattern to be evidence of the cratonization of the Earth's crust.

The fact that thick volcanic–sedimentary sequences of an iron-ore accretion prism formed at the edge of the Kola-Norwegian domain in the Neoarchean era requires study of the preconditions and patterns of ferruginous sedimentation at the Earth's surface. According to recent works, there are three main stages of iron ore sedimentation in the Earth's history [44]. The first stage is the most ancient and occurred 3.8–3.6 Ga ago. The second stage was 2.9–2.6 Ga ago and the third one was 2.3–1.7 Ga ago. The studied complex refers to the second stage of the iron ore sedimentation of mantle silicates. In the Proterozoic and Phanerozoic, it was subjected to the barodiffusive separation of the mantle material. The switch in mechanisms at the turn of the Archean and Proterozoic made the Archean geology basically different from that of other epochs and also predetermined the specific distribution of iron in the convective Earth's mantle.

We suggest that the Lotta domain copper–nickel occurrences within the Allarechka area formed partly because the domain is similar to trap formations with the same metallogeny. Both of these have similar geodynamic features of protolith formation. As for the Keivy domain, basic rocks that are similar to magmatic trap formations are minor and are likely to have no significant impact on the metallogeny of the area. Further processing of supracrustal formations greatly affected the domain metallogeny when weathering crusts formed in a steady period [12], followed by the regional metasomatism [45].

6. Conclusions

The obtained data were based on both new and previous research and prove the suggestion [10,11] that the oldest core of the Kola protocontinent formed in the northeast of the region and accreted further west and southwestwards. These data suggest the following development model of rock association in the Kola region.

There were two main stages of the continental crust evolution in the southeastern Baltic Shield. The first stage marked the origin of the continental crust of the Kola GGnA and Karelian GGrA. According to our data, the Kola GGnA originated when its core formed and is composed of basic rocks of the Keivy domain bottom, where the material was preserved as basic rocks of the Patchervtundra and Lebyazhinskaya suits. The core was accreted in the southern and southwestern direction by formations that constituted the belt (intracratonic activation zone). This belt comprises supracrustal rocks of the Lotta, Chapoma, Tersky and part of the Umba domains.

In the west, complexes of the Kola-Norwegian domain developed at the next stage. Supracrustal complexes of the Belomorian domain formed in the further converging of the Kola GGrA and Karelian GGnA until they produced the single Archean Karelian-Kola lithospheric plate. Original volcanogenic formations of the Murmansk Domain could not occur at the first stage, but at the same time as the rock associations in the Belomorian domain. Since the Murmansk domain rocks have a peculiar composition, it is possible that before the Rebolian, the domain developed separately from the rest of the area and only joined other domains at its early stages.

Geological-structural and geophysical data [10] suggest that the Keivy microcontinent sunk to some depth by surrounding continental massifs that overthrusted it at a certain stage of its development. Orogenic belts formed around this structure later in the Neoarchean era at the Rebolian phase of folding. It complies with the idea [9,10] that since the Archean rock associations in the southwestern part of the region had been gradually accreting the older (Mesoarchean, probably Paleoproterozoic) core located in the north to northeast. At that time, the crust of the eastern Kola region was thicker and older and determined the occurrence of deeper magmatic sources here. In this, the rock associations of the Kola–Norwegian domain are likely to be the main removal area for the Keivy structure sediments, which made compositions of their rock associations similar. Thus, in the Neoarchean era, the Keivy domain was a median massif. Since the Keivy domain refers to structures that are often confined to hydrocarbon deposits in the Phanerozoic eon, it is easier to reason the occurrence of "methane" graphite in the Keivy crystalline schists [46,47].

Furthermore, the Lapland Belt of the proto-island arc type formed as the Belomorian domain interacted with domains to its northeast. The Titovka-Kolmozero (Kolmozero-Voronya) suture zone occurred, when the Kola-Norwegian and Murmansk domains interacted. This zone has features of both island arc and rift formations. Metabasites show these features as they are more akin to volcanites of young arcs in the Lapland-Kolvitsa Belt, while there is a shift towards mid-ocean ridge (MOR) volcanites in the Titovka-Kolmozero suture zone.

This paper provides data on the geodynamic evolution of the Archean structural and compositional complexes in the Kola region. Compared to the information that the specific volcanism in the early Precambrian era could predetermine authentic compositions of products at other stages of magmatism, we may consider the metallogenic forecast based on geodynamic reconstructions as promising for further research.

Author Contributions: N.E.K. conceived the research, designed experiments and prepared the considerable part of the manuscript. N.O.S. provided the metallogenic analysis and wrote the part on the regional metallogeny jointly with N.E.K., E.V.M. elaborated methods of geodynamic reconstructions and applied them to the current research. E.V.M. prepared the part on mathematical processing of data. All authors discussed the manuscript.

Funding: The research leading to this paper was funded by the Geological Institute of the Kola Science Centre RAS, public contract No. 0231-2015-0007.

Acknowledgments: The authors express their deepest gratitude to T.S. Marchuk for her highly skilled and qualified processing of the geochemical material and editing of the current paper. We also appreciate unknown reviewers for constructive comments and corrections.

Conflicts of Interest: The authors declare no conflict of interests.

References

1. Daly, J.S.; Balagansky, V.V.; Timmerman, M.J.; Whitehouse, M.J. The Lapland-Kola orogen: Palaeoproterozoic collision and accretion of the northern Fennoscandian lithosphere. *Geol. Soc. Lond. Mem.* **2006**, *32*, 579–598. [CrossRef]
2. *Deep Structure, Evolution and Minerals of the Early Precambrian Basement of the East-European Platform: Interpretation of Materials after Profile 1-EB, Profiles 4B and TATSEIS*; Geokart: Moscow, Russia, 2010; p. 408. Available online: https://www.twirpx.com/file/836342/ (accessed on 12 October 2018). (In Russian)
3. Mitrofanov, F.P. Modern problems and some solutions of the Precambrian geology of cratons. *Litosfera* **2001**, *1*, 5–14.
4. Mitrofanov, F.P.; Bayanova, T.B. Geochronology of rocks and processes in Archaean domains of the Kola province of the Baltic Shield. *Mineral. J.* **2004**, *26*, 33–39.
5. Pozhilenko, V.I.; Gavrilenko, B.V.; Zhirov, D.V.; Zhabin, S.V. *Geology of Ore Areas of the Murmansk Region*; KSC RAS: Apatity, Russia, 2002; 359p. Available online: http://www.geokniga.org/bookfiles/geokniga-geologiya-rudnyh-rayonov-murmanskoy-oblasti-2002.pdf (accessed on 12 October 2018). (In Russian)
6. Korikovsky, S.P.; Kotov, A.B.; Sal'nikova, E.B.; Aranovich, L.Y.; Korpechkov, D.I.; Yakovleva, S.Z.; Tolmacheva, E.V.; Anisimova, I.V. The age of the protolith of metamorphic rocks in the southeastern part of the Lapland granulite belt, southern Kola Peninsula: Correlation with the Belomorian mobile belt in the context of the problem of Archean eclogites. *Petrology* **2014**, *22*, 91–108. [CrossRef]
7. Kozlov, N.E. Mineral Composition of Metamorphic Complexes of High-Pressure Granulitic Belts and Issue of Their Protoliths Formation (on Example of the Lapland Granulites). Ph.D. Thesis, Institute of Precambrian Geology and Geochronology RAS, St Petersburg, Russia, November 1995. Available online: https://dlib.rsl.ru/viewer/01000012868#?page=1 (accessed on 12 October 2018). (In Russian)
8. Kozlov, N.E.; Avedisyan, A.A.; Balashov, Y.A.; Ivanov, A.A.; Kamienskaya, A.D.; Muhamedova, I.W. Some new aspects of geology, deep structure, geochemistry and geochronology of the Lapland Granulite Belt, Baltic Shield. Geology of the eastern Finnmark—Western Kola Peninsula region. *Geol. Surv. Nor. Bull.* **1995**, *7*, 157–166.
9. Kozlov, N.E.; Martynov, E.V.; Predovskii, A.A. Petrochemical reconstruction of the primary nature of metamorphic rocks and geodynamic settings of their protolites (new approaches and constraints). *Russ. Geol. Geophys.* **1999**, *40*, 1217–1225.
10. Kozlov, N.E.; Sorokhtin, N.O.; Glaznev, V.N.; Kozlova, N.E.; Ivanov, A.A.; Kudryashov, N.M.; Martynov, E.V.; Tyuremnov, V.A.; Matyushkin, A.V.; Osipenko, L.G. *The Archaean Geology of the Baltic Shield*; Nauka: St Petersburg, Russia, 2006; 329p, ISBN 5-02-025095-3. (In Russian)
11. Kozlov, N.E.; Sorokhtin, N.O.; Martynov, E.V.; Kozlova, N.E. Basic rocks of the Lapland granulite belt and compositional heterogeneity of the Early Precambrian mantle. *Geochem. Int.* **2010**, *48*, 505–509. [CrossRef]
12. Predovsky, A.A.; Melezhik, V.A.; Bolotov, V.I.; Fedotov, G.A.; Basalaev, A.A.; Kozlov, N.E.; Ivanov, A.A.; Zhangurov, A.A.; Skufjin, P.K.; Lyubtsov, V.V. *Volcanism and Sedimentation Genesis in the Precambrian in the NE Baltic Shield*; Nauka: Leningrad, Russia, 1987; p. 185. (In Russian)
13. Kozlov, N.E.; Martynov, E.V.; Sorokhtin, N.O.; Marchuk, T.S. Evolution of the mineral composition of the early Precambrian metabasites in the Kola region. *Proc. MSTU Vestn.* **2014**, *17*, 304–313. (In Russian)
14. Bridgwater, D.; Marker, M.; Mengel, F. *The Eastern Extension of the Early Proterozoic Torngat Orogenic Zone across the Atlantic*; LITHOPROBE Report 27; Memorial University of New Foundland: St. John's, NL, Canada, 1992; pp. 76–91.
15. Gorbatschev, R.; Bogdanova, S. Frontiers in the Baltic Shield. *Precambrian Res.* **1993**, *64*, 3–21. [CrossRef]
16. Mitrofanov, F.P. (Ed.) *Geology of the Kola Peninsula (Baltic Shield) + Geological Map of the N.-E. Baltic Shield*; KSC RAS: Apatity, Russia, 1995; 145p. Available online: http://b-ok.org/book/3149477/ff5165 (accessed on 12 October 2018).
17. Ivanov, A.A.; Kozlov, N.E.; Martynov, E.V. The most ancient Archaean collisional zone in the Notheastern Baltic Shield. In Proceedings of the International Conference of Tectonic & Metallogenegy of Early/Mid Precambrian Orogenic belts, Montreal, QC, Canada, 28 August–1 September 1995.
18. Kozlov, N.E.; Ivanov, A.A. Composition of metamorphic rocks and some aspects of evolution of the Lapland Granulite Belt on the Kola Peninsula, USSR. *Geol. Surv. Nor. Bull.* **1991**, *421*, 19–32.

19. Kozlov, N.E.; Ivanov, A.A.; Martynov, E.V.; Sorokhtin, N.O.; Kozlova, N.E. Comparison of Archaean complexes of the Murmansk domain with other Archaean complexes of the Baltic Shield, Greenland, and Canada. *Dokl. Earth Sci.* **2005**, *403*, 685–687.

20. Kozlov, N.E.; Ivanov, A.A.; Nerovich, L.I. *Granulites of the Baltic Shield Proto-Island Arc Formations at the Early Pre-Cambrian Stage of the Earth Crust Evolutions*; Preprint; USSA Academy of Sciences, Geological Institute of KSC: Apatity, Russia, 1990; p. 16.

21. Kozlov, N.E.; Martynov, E.V.; Sorokhtin, N.O.; Ivanov, A.A.; Kudryashov, N.M.; Kozlova, N.E. Metamorphic rocks of the Murmansk domain (Kola Peninsula) as compared with the oldest rock associations of the northeastern Baltic shield, Canada, and Greenland. *Geochem. Int.* **2008**, *46*, 608–613. [CrossRef]

22. Belyaev, O.A.; Kozlov, N.E. *Geology, Geochemistry and Metamorphism of the Lapland Granulite Belt and Adjacent Areas in the Vuotso Area, Northern Finland*; Report of investigation/Geological Survey of Finland; Geologian Tutkimuskeskus: Espoo, Finland, 1997; ISBN 951-690-656-7. Available online: http://tupa.gtk.fi/julkaisu/tutkimusraportti/tr_138.pdf (accessed on 12 October 2018).

23. Puri, M.L.; Sen, P.K. *Nonparametric Methods in Multivariate Analysis*; John Wiley & Sons: New York, NY, USA, 1971; p. 677.

24. Tamura, R. Multivariate Nonparametric Several-Sample Tests. *Ann. Math. Stat.* **1966**, *37*, 611–618. [CrossRef]

25. Nelder, J.A.; Mead, R. A Simplex Method for Function Minimization. *Comput. J.* **1965**, *7*, 308–313. [CrossRef]

26. Pearce, J.A. Statistical Analysis of Major Element Patterns in Basalts. *J. Petrol.* **1976**, *17*, 15–43. [CrossRef]

27. Pearce, J.A.; Harris, N.B.W.; Tindle, A.G. Trace Element Discrimination Diagrams for the Tectonic Interpretation of Granitic Rocks. *J. Petrol.* **1984**, *25*, 956–983. [CrossRef]

28. Kerr, A.C.; White, R.V.; Saunders, A.D. LIP Reading: Recognizing Oceanic Plateaux in the Geological Record. *J. Petrol.* **2000**, *41*, 1041–1056. [CrossRef]

29. Makrygina, V.A. Geochemistry of metamorphism and possibilities for the reconstruction of ancient lithogenesis features. *Russ. Geol. Geophys.* **1986**, *6*, 89–99.

30. Dobretsov, N.L. Petrochemical features of oceanic and early-geosynclinal basalts. *Russ. Geol. Geophys.* **1975**, *2*, 11–25.

31. McLaughlin, R.J.W. Geochemical changes due to weathering under varying climatic conditions. *Geochim. Cosmochim. Acta* **1955**, *8*, 109–130. [CrossRef]

32. Bhatia, M.R. Plate Tectonics and Geochemical Composition of Sandstones. *J. Geol.* **1983**, *91*, 611–627. [CrossRef]

33. Grachev, A.F. *Rift Zones of the Earth*; Nedra: Moscow, Russia, 1987; p. 285.

34. Mullen, E.D. MnO/TiO2/P2O5: A minor element discriminant for basaltic rocks of oceanic environments and its implications for petrogenesis. *Earth Planet. Sci. Lett.* **1983**, *62*, 53–62. [CrossRef]

35. Piskunov, B.N. *Geological and Petrological Specificity of the Island-Arc Volcanism*; Nauka: Moscow, Russia, 1987; p. 237.

36. Taylor, S.R.; McLennan, S.M. *The Continental Crust: Its Composition and Evolution*; Blackwell: Oxford, UK, 1985; p. 312. Available online: https://docplayer.ru/52865795-S-r-teylor-s-m-mak-lennan-kontinentalnaya-kora-ee-sostav-i-evolyuciya-izdatelstvo-mir.html (accessed on 12 October 2018).

37. Predovsky, A.A. *Reconstruction of Conditions of Sedimentation Genesis and Volcanism in the Early Precambrian*; Nauka: Leningrad, Russia, 1980; p. 152. (In Russian)

38. Dobretsov, N.L.; Vernikovsky, V.A.; Karyakin, Y.V.; Korago, E.A.; Simonov, V.A. Mesozoic–Cenozoic volcanism and geodynamic events in the Central and Eastern Arctic. *Russ. Geol. Geophys.* **2013**, *54*, 874–887. [CrossRef]

39. Kuzmin, M.I.; Yarmolyuk, V.V. Plate tectonics and mantle plumes as a basis of deep-seated Earth's tectonic activity for the last 2 Ga. *Russ. Geol. Geophys.* **2016**, *57*, 8–21. [CrossRef]

40. Bayanova, T.B.; Kunakkuzin, E.L.; Serov, P.A.; Fedotov, D.A.; Borisenko, E.S.; Elizarov, D.V.; Larionov, A.V. Precise U-Pb (ID-TIMS) and SHRIMP-II ages on single zircon and Nd-Sr signatures from Achaean TTG and high aluminum gneisses on the Fennoscandian Shield. In Proceedings of the 32nd Nordic Geological Winter Meeting, Helsinki, Finland, 13–15 January 2016. Available online: https://docplayer.net/44703126-Federation-of-finnish-learned-societies.html (accessed on 17 October 2018).

41. Bridgwater, D.; Scott, D.; Marker, M.; Balagansky, V.; Bushmin, S.; Alexejev, N. LAM-ICP-MS 207Pb/206Pb ages from detrital zircons and the provenance of sediments in the Lapland-Kola belt. In Proceedings of the SVEKALAPKO Workshop, Lammi, Finland, 28–30 November 1996; Oulu University: Oulu, Finland, 1996.

42. Myskova, T.A.; Glebovitsky, V.A.; Mil'kevich, R.I.; Shuleshko, I.K.; Berezhnaya, N.G.; Lepekhina, E.N.; Matukov, D.I.; Antonov, A.V.; Sergeev, S.A. Findings of the oldest (3600 Ma) zircons in gneisses of the Kola Group, Central Kola Block, Baltic Shield: Evidence from U-Pb (SHRIMP-II) data. *Dokl. Earth Sci.* **2005**, *402*, 547–550.

43. Sorokhtin, N.O.; Lobkovsky, L.I.; Kozlov, N.E. Metallogeny of subduction zones. *Proc. MSTU* **2017**, *20*, 111–128. (In Russian) [CrossRef]

44. Sorokhtin, O.G.; Chilingarian, G.V.; Sorokhtin, N.O. *Evolution of Earth and Its Climate: Birth, Life and Death of Earth*; Elsevier Science Ltd.: Amsterdam, The Netherlands, 2011; ISBN 9780444537584.

45. Kozlov, N.E.; Fomina, E.N.; Martynov, E.V.; Sorokhtin, N.O.; Marchuk, T.S. On reasons of specific composition of the Keivy domain rocks (the Kola Peninsula). *Proc. MSTU* **2017**, *20*, 83–94. [CrossRef]

46. Bushmin, S.A.; Glebovitskii, V.A.; Prasolov, E.M.; Lokhov, K.I.; Vapnik, E.A.; Savva, E.V.; Shcheglova, T.P. Origin and composition of fluid responsible for metasomatic processes in shear zones of the Bolshie Keivy tectonic nappe, baltic shield: Carbon isotope composition of graphite. *Dokl. Earth Sci.* **2011**, *438*, 701–704. [CrossRef]

47. Fomina, E.N.; Kozlov, E.N.; Lokhova, O.V.; Lokhov, K.I. Graphite as an indicator of contact influence of Western Keivy alkaline granite intrusion, the Kola Peninsula. *Proc. MSTU* **2017**, *20*, 129–139. [CrossRef]

minerals

MDPI

Article

Gold Prospects in the Western Segment of the Russian Arctic: Regional Metallogeny and Distribution of Mineralization

Arkady A. Kalinin [1,*], Oleg V. Kazanov [2], Vladimir I. Bezrukov [3] and Vsevolod Yu. Prokofiev [4]

[1] Geological Institute, Kola Science Center, Russian Academy of Sciences, Apatity 184209, Russia
[2] N.M. Fedorovsky All-Russian Research Institute of Mineral Raw Materials, Moscow 119017, Russia; okazanov@gmail.com
[3] A.P. Karpinsky Russian Geological Research Institute, S-Petersburg 199106, Russia; vladimir_bezrukov@vsegei.ru
[4] Institute of Geology of Ore Deposits, Petrography, Mineralogy, and Geochemistry, Russian Academy of Sciences, Moscow 119017, Russia; vpr2004@rabler.ru
* Correspondence: kalinin@geoksc.apatity.ru; Tel.: +7-921-663-68-36

Received: 18 December 2018; Accepted: 21 February 2019; Published: 26 February 2019

check for updates

Abstract: Location of the deposits and occurrences of gold mineralization in metamorphic complexes of the Kola region is controlled by tectonic zones at the regional scale at the boundaries of major segments of the Fennoscandian Shield. Three zones are the most important: (1) the system of Neoarchean greenstone belts Kolmozero–Voron'ya–Ura-guba along the southern boundary of the Murmansk craton; (2) the suture, delineating the core of the Lapland–Kola orogeny in the north; and (3) the series of overthrusts and faults at the eastern flank of the Salla–Kuolajarvi belt. Gold deposits and occurrences are located within greenstone belts of Neoarchean and Paleoproterozoic age, and hosted by rocks of different primary compositions (mafic metavolcanics, diorite porphyry, and metasedimentary terrigenous rocks). The grade of metamorphism varies from greenschist to upper amphibolite facies, but the mineralized rocks are mainly lower amphibolite metamorphosed, close to the transition from greenschist to amphibolite facies. Gold deposits and occurrences in the northeastern part of the Fennoscandian Shield formed during two periods: the Neoarchean 2.7–2.6 Ga and the Paleoproterozoic 1.9–1.7 Ga. According to paleo-geodynamic reconstructions, these were the periods of collisional and accretionary orogeny in the region. Those Archean greenstone belts, which were reworked in the Paleoproterozoic (e.g., Strel'na and Tiksheozero belts), can contain gold deposits of Paleoproterozoic age.

Keywords: greenstone belt; gold; Kola Peninsula; Northern Karelia; rock alteration

1. Introduction

The western segment of the Russian Arctic (Kola region, for short below), covering the Murmansk region and Russian Northern Karelia, is not considered a gold district in Russia. Annual gold production in the Kola region is 0.1 t [1]: gold, together with platinum group elements (PGEs), are byproducts of nickel–copper metallurgy, extracted by the Kola Mining-Metallurgical Company from the Pechenga nickel–copper ores (Figure 1), which contain ~0.1 ppm Au. Thus, gold is the third-most important component (after Pt and Pd) in low-sulfide PGE ore from layered mafic–ultramafic massifs in the region—the Fyodorovo–Pansky intrusion, and the Vuruchuaivench and Monchetundrovsky massifs (Figure 1). The gold content in regular low-sulfide ores varies from 0.08 ppm (the Fyodorova Tundra deposit) [2] to 0.29 ppm (the East Chuarvy deposit) [3]. Gold as an associated component of

Ni–Cu and PGE ores currently makes up more than 99% of the gold resources in the western segment of the Russian Arctic.

The distribution of gold resources in Northern Finland (Finnish Lapland), in the area adjacent to the Murmansk region and Russian Northern Karelia, differs significantly: gold in Ni–Cu (the Kevitsa deposit) and in the PGE low-sulfide deposits (the Ahmavaara, Kontijärvi, Siika-Kämä reef, etc.) only makes up 15% of the gold resources in Finnish Lapland, and the other 85% relates to gold deposits in metamorphic complexes within greenstone belts [4,5]. Presently in Northern Finland, there is one world class deposit (Suurikuusikko) with indicated and inferred resources of 260 t and an annual production of 5.5–6 t Au [6], as well as a number of medium and small deposits, and dozens of prospective gold occurrences (Figure 1) [7].

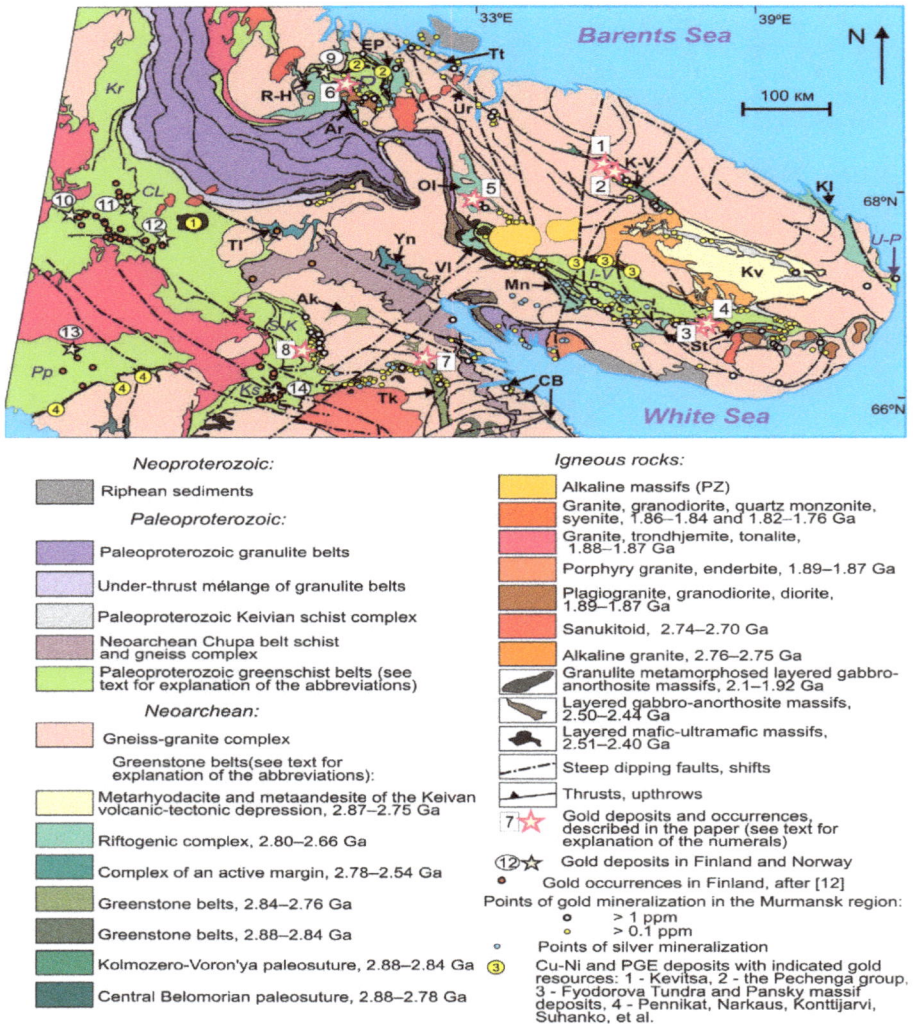

Figure 1. Schematic tectonic–geological map (modified from Reference [8]), showing the locations of greenstone belts and gold mineralization in the northeastern part of the Fennoscandian Shield.

Considering the similarity of lithological and geological structures of Northern Finland and the Kola region, the described difference in the distribution of gold resources (and deposits) can be explained, first of all, with different levels of exploration for gold, which are much lower in the Russian territory. This indicates a high probability of discovery of gold deposits in metamorphic complexes in the western segment of the Russian Arctic in future.

Data on certain gold occurrences in the Kola region can be found in the Russian geological literature, but no systematic information on gold mineralization in the Murmansk region and Northern Karelia has previously been published. For example, only two deposits (the Mayskoe quartz vein and Pellapahk porphyry) were noted in the paper by Krister Sundblad [9] devoted to gold deposits in Northern Europe. More gold deposits and occurrences were mentioned in Reference [10], published in 2012, those were the Mayskoe, Vorgovy, Olennoe, Oleninskoe, Nyal'm-1, and -2, Pellapahk; but the deposits and occurrences were not described in Reference [10] in detail.

During geological investigations carried out in the Kola region in the beginning of the 21st century, geological structures' prospective for gold mineralization were defined, new gold occurrences were found, and new data on the above-named deposits and occurrences obtained. Brief information about deposits and occurrences, studied during the last two decades, is provided in this paper.

The following terms are used throughout this study for the description of gold mineralization:

- A deposit—a concentration of minerals on the surface or underground with delineated mineralized bodies and indicated or inferred resources; quantity, quality, and bedding of the ore meet the exploitation requirements;
- An occurrence—mineralized rocks with gold content over 1 ppm for more than 1-m thickness according to core or trench sampling, with an undefined scale of mineralization and no contoured mineralized bodies;
- Points of gold mineralization—small occurrences with gold content more than 0.1 ppm for thickness over 1 m in core or trench samples, or more than 1 ppm in hand samples.

The following abbreviations are used below in the figures with the images of minerals:

Asp—arsenopyrite, Cc—calcite, Cp—chalcopyrite, Di—diopside, Gr—garnet, Grs—gersdorfite, Hb—hornblende, Lo—löllingite, Mr—marcasite, Pl—plagioclase, Po—pyrrhotite, Py—pyrite, Qu—quartz, Tit—titanite, Uy—Uytenbogaardtite, and Zo—zoisite.

2. Materials and Methods

Data on gold occurrences and points of mineralization (Figure 1) were collected in the Fund of Geological Information on Mineral Resources in the Murmansk Region from the open access reports on the results of geological mapping and exploration, carried out in the Kola region by state geological surveys and mining companies, beginning in the 1930s and continuing up to the present. Data for gold deposits and occurrences in the northern part of Finland (Figure 1) were taken from the database FINGOLD [7], which contains information on occurrences of gold mineralization with >1 ppm Au for intervals over 1 m in trench and/or core samples.

The authors of the present paper took part in fieldwork connected with the exploration for gold during different time periods from the 1970s to 2017, and visited all the described gold occurrences and deposits, and collected the necessary materials to study the geology, petrography, geochemistry, and mineralogy of the deposits. The only exception was the Olenegorsk group of banded iron formations deposits (BIFs), where information on this gold mineralization was based on published data. Schematic geological maps of gold deposits and occurrences given in this paper were compiled by the authors themselves, if there are no references cited in the figure captions.

The following abbreviations are used in Figure 1: Paleoproterozoic greenschist belts (letters in italics), Polmak–Pasvik–Pechenga–Imandra-Varzuga belt system: P—Pechenga belt, I-V—Imandra-Varzuga belt, U-P—Ust'-Ponoy belt; Lapland—Karelia belt system: Kr—Karasjok, CL—Central Lapland belt, S-K—Salla-Kuolajarvi belt, Ks—Kuusamo belt; Pp—Perapohja belt.

Neoarchean greenstone belts (letters in bold): K-V—Kolmozero–Voron'ya belt, CB—Central Belomorian belt, Tk—Tiksheozero belt, Ak—Alakurti, Yn—Yona belt, Tl—Tulppio–Kareka Tundra belt, St—Strel'na belt, Mn—Munozero, Vl—Vochelambina, Ol—Olenegorsky belt, R-H—Runijoki-Hihnajarvi, Ar—Allarechka belt, EP—East-Pechenga belt (NE frame of the Pechenga belt), Tt—Titovka, Ur—Ura-guba, Kl—Kachalovsky, Kv—Keivan volcanic-tectonic depression. Gold deposits and occurrences: 1—Oleninskoe, 2—Nyal'm, 3—Sergozero, 4—Vorgovy, 5—Olenegorsk BIF deposits, 6—Porojarv, 7—iKichany, 8—Mayskoe, 9—Gjeddevanett, 10—Saattopora, 11—Suurikuusikko, 12—Pahtavaara, 13—Rompas and Rajapalot, 14—Juomasuo.

Investigations of wallrock alteration, metasomatic zonation, and determination of pre-ore, gold-related, and post-ore mineral assemblages in altered rocks were based on the study of rocks in the outcrops and in drillcore, on examination of mineral relations in thin and polished sections, as well as on the results of assays of primary and altered rocks.

The samples were assayed for major (rock-forming) elements in the chemical laboratory of the Geological Institute, Kola Science Centre, Russian Academy of Sciences, with flame atomic absorption spectrometry (FAAS). Data on minor elements, which determined the geochemical characteristics of the deposits, were obtained by ICP-MS in IRGIREDMET (Irkutsk) and in the Institute of Geology and Geochemistry of the Ural Branch of the Russian Academy of Sciences, Ekaterinburg.

Gold content in the samples from the Sergozerskoe, Vorgovy, and Kichany occurrences was determined in IRGIREDMET with ICP-MS after fire assay pre-concentration of precious metals; the data on gold content in the samples from the Oleninskoe deposit and Porojarvi occurrence were obtained in the chemical laboratory of the Geological Institute, Kola Science Centre RAS by FAAS with pre-concentration of PGE, gold, and silver with p-alkylaniline and oil sulfides. Data on gold grades in other deposits were taken from published papers and/or public releases by mining companies.

Mineral composition of the ores was studied in polished sections with reflected light microscope Axioplan 2 Imaging (Karl Zeiss, Jena, Germany) and with the electron microscope LEO-1450 (Karl Zeiss, Jena, Germany) in the Geological Institute of the Kola Science Center. Preliminary estimation of the composition of mineral species was done with the energy-dispersive system Bruker XFlash-5010. Microprobe analysis (MS-46, CAMECA, France, 22 kV; 30–40 nA, standards (analytical lines): $Fe_{10}S_{11}$ (FeKα, SKα), Bi_2Se_3 (BiMα, SeKα), $LiNd(MoO_4)_2$ (MoLα), Co (CoKα), Ni (NiKα), Pd (PdLα), Ag (AgLα), Te (TeLα), Au (AuLα)) was performed for grains larger than 20 μm. Identification of rare mineral phases was verified with X-ray analysis in the Geological Institute of the Kola Science Center.

Fluid inclusions were studied with a Linkam THMSG-600 freezing/heating stage (Linkam Scientific, Epsom, UK) equipped with an Olympus BX51 optical microscope, video camera, and computer at the Institute of Geology of Ore Deposits, Mineralogy, Geochemistry, and Petrography, Russian Academy of Sciences, Moscow, Russia. The composition of salts in fluid inclusions was estimated from the eutectic temperature [11], and the salinity was estimated from the final melting temperatures of ice according to the experimental data of the $NaCl–H_2O$ system [12]. The salinity of the aqueous solution in the $CO_2–H_2O$ inclusions was estimated from the melting temperature CO_2 hydrates [13]. The CO_2 and CH_4 concentrations were estimated from volumetric ratios of phases and the densities of CO_2 and CH_4 in gas phase. Pressure was determined for immiscible fluids from the intersection of the isochore and isotherm. The salinity and pressure of the fluid was calculated with the FLINCOR program [14].

3. Geological Setting

The area under consideration in this study (shown with the blue rectangle in Figure 2A) occupies the northwestern part of the Fennoscandian Shield, including the Murmansk craton, Kola Province, northern part of the Belomorian mobile belt, and a small part of the Karelian craton (Figure 2A) [15].

The Murmansk craton is composed of diverse granite gneisses and granitoids, which contain some xenoliths of Archean supracrustal rocks.

The Kola Province is the Archean tectonic collage of the Kola–Norwegian, Keivy, Sosnovka, and Kolmozero–Voron'ya terranes. The terranes consist of the Archean greenstone, schist, paragneiss, granulite, and granite-gneiss complexes that underwent structural deformation and metamorphism in the Archean and Paleoproterozoic, except the Kola–Norwegian Terrane, which almost escaped the Paleoproterozoic process.

The Belomorian Mobile belt is made up largely of Meso- and Neoarchean granite gneisses, greenstone rocks, and paragneiss complexes. The province is distinguished by intense repeated deformations and high- and moderate-pressure metamorphic events that occurred in both the Neoarchean and Paleoproterozoic.

The Kola and Belomorian provinces evolved in structural deformation and metamorphism in the Paleoproterozoic, being elements of the Paleoproterozoic Lapland–Kola collisional Svecofennian (1.95–1.75 Ga) orogen (Figure 2A). The Pechenga–Imandra–Varzuga and Lapland–Kola collisional sutures are localized in the central part of this orogen (Figure 2B).

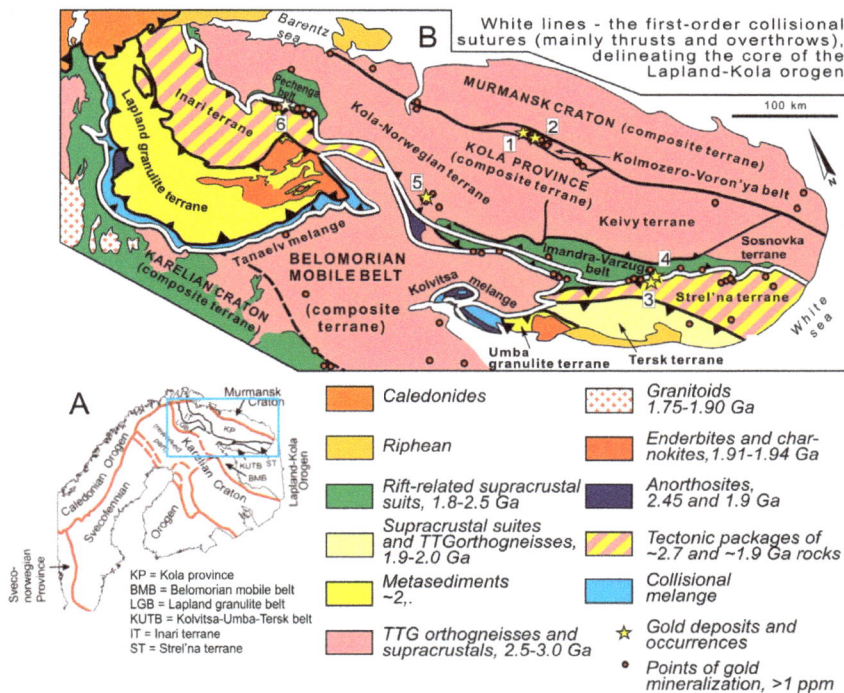

Figure 2. Schematic tectonic map of the Fennoscandian Shield (**A**) and of the Kola region and adjacent regions of Northern Fennoscandia (**B**) [15].

The Karelian craton is a classic Neoarchaean granite-greenstone province, containing a few remnants of the Mesoarchean crust. It is cut by a series of Palaeoproterozoic rifts and related layered mafic intrusions.

The Paleoproterozoic volcanic-sedimentary rocks of 1.8–2.5 Ga age form two rift-related systems of greenschist belts. The Polmak–Pasvik–Pechenga–Imandra–Varzuga belt system stretches southwest from Norway across the Kola Peninsula to the White Sea for about 500 km along the southern boundary of the Kola Province. The largest parts of this system are the Pechenga and Imandra–Varzuga belts (Figure 1). The Lapland–Karelian belt system can be traced for more than 1000 km along the northern boundary of the Karelian craton from Norway across Finnish Lapland and Russian

Karelia. In the northern part of the Fennoscandian Shield the system consists of the Karasjok, Lapland, Salla–Kuolajarvi, and Kuusamo belts (Figure 1).

4. Brief Information on Gold Deposits in the Western Segment of the Russian Arctic

In the Murmansk region and Northern Karelia there are four minor gold deposits with indicated and inferred resources (not complying with NI 43-101 and other international and national Codes). Two of them—the Oleninskoe (10 t Au) and Nyal'm (7.5 t)—are located in the Kolmozero–Voron'ya greenstone belt, the Sergozerskoe (13 t Au)—in the Strel'na belt, and the Mayskoe (0.122 t Au) in the Salla–Kuolajarvi belt (Figures 1 and 2). The latter is the only gold deposit ever developed in the region (51 kg mined between 1998–2000) [16]. Gold occurrences are known in the Strel'na belt (Vorgovy), in the Pechenga belt (Porojarvi), and in the Tiksheozero belt (Kichany). Gold mineralization in the iron-producing BIF deposits Olenegorskoe and Kirovogorskoe have a statute of gold occurrences as well.

4.1. Gold Deposits in the Kolmozero–Voron'ya Greenstone Belt

Two small gold deposits (Oleninskoe and Nyal'm) and dozens of occurrences and points of mineralization were found in the northwestern part of the Neoarchean greenstone belt Kolmozero–Voron'ya, which separates two major blocks of the Fennoscandian Shield—the Murmansk craton and the Kola–Norwegian terrane (Figures 1 and 2). Both deposits are located in the axial part of the belt and relate to the stratigraphic sequence of the Oleny Ridge amphibolite.

The Oleninskoe deposit (#1 in Figure 1) is located at the northwestern thinning of the Oleny Ridge amphibolite strata in a shear zone of northwest strike. The amphibolite and high-alumina metasedimentary schist host numerous granodiorite quartz porphyry dykes 0.1–6.0-m thick with the bedding concordant to the wallrocks (Figure 3). The latest rocks in the deposit area are granite-pegmatite veins, which cut all rocks, including those mineralized.

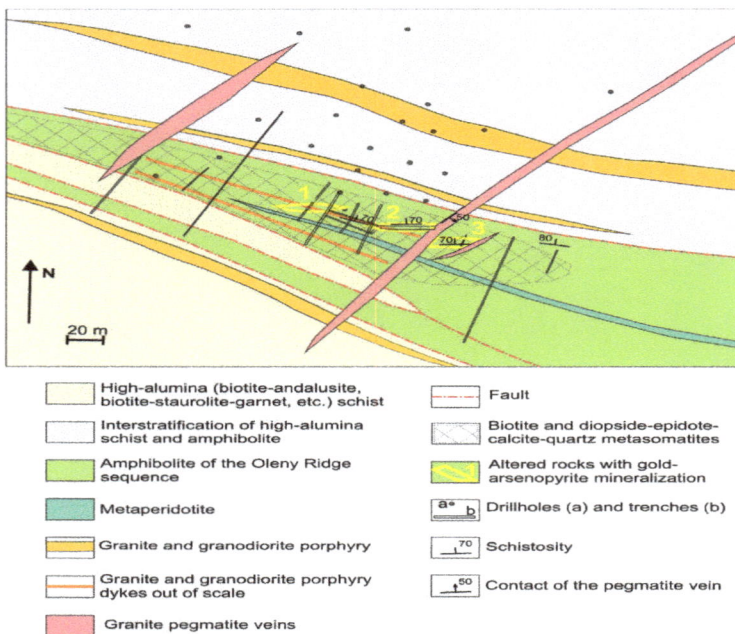

Figure 3. Schematic geological map of the Oleninskoe gold deposit.

The rocks are intensely altered (except pegmatite), the alteration zone is ~50-m thick and traced along the strike for 200–250 m. Pre-ore alteration processes, covering the whole zone of alteration, were biotitization (potassium metasomatism) and formation of diopside–zoisite–carbonate, diopside–zoisite–garnet mineral assemblages (calcium metasomatism) in the amphibolite (Figure 4, Table 1).

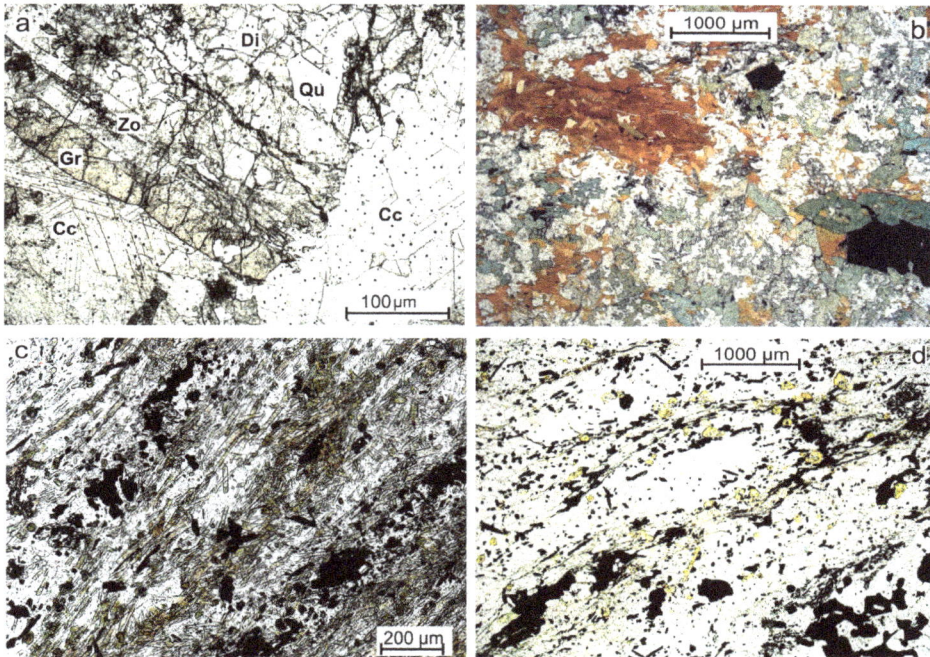

Figure 4. Altered rocks from the Oleninskoe deposit. Thin sections photos, transparent light, one polarizer: (**a**) diopside–zoisite–calcite–garnet metasomatite; (**b**) biotitizated hornblendite; and (**c**) and (**d**) quartz–tourmaline metasomatic rock: (**c**) after granite porphyry, with arsenopyrite dissemination and (**d**) after hornblende amphibolite, with pyrrhotite–arsenopyrite mineralization.

Gold-related alteration was the formation of quartz–muscovite–oligoclase, quartz–tourmaline, and quartz metasomatic rocks after both amphibolite and quartz porphyry (Table 2). Quartz-rich metasomatic rocks form an echelon-like series of lenticular bodies, cutting general schistosity in the host rocks at an acute angle of 10–15° (Figure 3), and control distribution of the gold–arsenopyrite mineralization. The lenses are up to 3.5-m thick (1.5 m on the average) with the length up to 50 m.

Arsenopyrite, pyrrhotite, and ilmenite are the most abundant ore minerals, which present in all altered rocks. Minor sulfides are chalcopyrite, sphalerite, and pentlandite. Quartz–tourmaline and quartz metasomatic rocks in lens 2 contain rich Pb–Ag–Sb mineralization (galena, freibergite, dyscrasite, boulangerite, semseyite, diaphorite, pyrargyrite, and other Sb-sulfosalts of Pb, Ag, and Cu, etc.—totalling more than 40 mineral phases [17]). Ilmenite in quartz metasomatic rocks is partly replaced by rutile along the grain boundaries.

Table 1. Chemical composition of the mineralized rocks of the Oleninskoe deposit, mas.%.

Sample #	AK-713	AK-708	AK-716	AK-715	AK-703	AK-704a	AK-706	VP-10	AK-710	114-79	114b-79	AK-707	AK-705	AK-24
SiO_2	47.20	51.19	49.94	47.12	49.24	46.90	71.56	69.53	64.22	76.52	69.11	62.38	89.12	82.27
TiO_2	0.75	1.21	0.67	0.98	1.37	1.17	0.34	0.23	0.97	0.44	0.68	0.28	0.05	0.11
Al_2O_3	12.20	11.61	17.00	15.39	12.89	12.31	14.69	16.20	11.91	8.22	10.64	2.71	1.62	2.11
Fe_2O_3	0.84	1.76	0.00	3.39	1.68	2.60	0.00	0.48	1.62	1.51	4.90	7.99	0.00	5.31
FeO	10.64	10.81	4.40	8.30	11.35	10.38	2.38	1.36	5.18	4.14	1.78	6.39	3.01	2.47
MnO	0.23	0.21	0.11	0.27	0.22	0.25	0.03	0.03	0.02	0.04	0.03	0.02	0.05	0.03
MgO	11.20	5.37	1.84	5.18	6.49	5.82	1.18	1.18	0.19	1.41	1.22	0.43	1.17	0.86
CaO	12.33	14.31	19.50	12.93	10.71	12.32	2.50	1.74	0.39	2.35	2.83	0.65	2.02	1.14
Na_2O	1.68	1.14	1.31	1.44	1.13	0.98	4.89	6.28	2.28	2.14	3.24	0.49	0.11	0.44
K_2O	0.10	0.21	0.07	1.04	1.17	1.33	1.01	1.56	1.94	0.60	0.47	0.26	0.07	0.19
H_2O^-	0.08	0.04	0.16	0.03	0.14	0.39	0.11	0.00	0.44	0.29	0.24	0.67	0.00	0.22
H_2O^+	2.42	1.06	1.39	1.14	1.57	1.69	0.73	0.83	2.60	1.15	0.67	3.86	0.28	0.28
S	0.07	0.60	0.05	1.37	1.08	2.46	0.07	0.01	1.15	1.95	2.26	3.57	0.04	2.20
P_2O_5	0.05	0.10	0.05	0.06	0.08	0.11	0.07	0.06	0.42	0.01	0.00	0.46	0.03	0.13
CO_2	<0.10	<0.10	3.32	0.11	<0.10	<0.10	<0.10	0.00	<0.10	0.06	0.00	<0.10	<0.10	0.02
F	0.019	0.016	0.014	0.01	0.098	0.057	0.017	0.000	0.014	0.019	0.000	0.010	0.011	0.018
Cl	0.009	0.009	0.011	0.01	0.009	0.009	0.011	n.a.	0.013	n.a.	n.a.	0.010	0.007	n.a.
Total	99.85	99.66	99.85	98.81 *	99.34	98.84 *	99.62	99.51	93.38 *	100.86	98.08 *	90.19 *	97.60 *	97.80 *
Ag (ppm)	0.26	1.46	0.46	1.68	2.3	18.05	0.8	n.a.	20.72	n.a.	n.a.	96.56	0.086	n.a.
Au (ppm)	0.016	0.009	<0.004	1.04	0.49	1.27	<0.004	n.a.	3.29	n.a.	n.a.	3.16	<0.004	n.a.
Pd (ppm)	<0.004	<0.004	<0.004	n.a.	<0.004	<0.004	<0.004	n.a.	<0.004	n.a.	n.a.	0.019	<0.004	n.a.
Cu	0.01	0.012	<0.005	0.015	0.016	0.026	<0.005	n.a.	0.018	n.a.	n.a.	0.029	<0.01	n.a.
Ni	0.028	0.009	0.009	0.019	0.008	0.01	0.005	n.a.	0.006	n.a.	n.a.	0.013	0.014	n.a.
Co	<0.005	<0.005	<0.005	<0.005	<0.005	<0.005	<0.005	n.a.	<0.005	n.a.	n.a.	<0.005	<0.01	n.a.

* The deficit appears due to high content of arsenopyrite mineralization, the samples were not assayed for As; n.a. = not assayed; AK-713, AK-708—amphibolite; AK-716—diopside-epidote-calcite–garnet metasomatite (skarnoid) after amphibolite; AK-715, 703, 704a—biotitized amphibolite; VP-10, AK-706—granite porphyry (dyke); AK-710, 114-79, 114b-79—quartz-sericite–tourmaline metasomatite; AK-707, AK-705, AK-24—quartz metasomatite.

Four types of minerals of Au–Ag series were recognized: 1—electrum 25–32 mas.% Au in association with arsenopyrite, löllingite, and pyrrhotite (Figure 5a); 2—electrum 33–47 mas.% Au in intergrowths with galena, dyscrasite, and sulfosalts; 3—gold (78–95 mas.% Au) in quartz; 4—native silver (<7 mas.% Au) in the crust of weathering of the ore (Table 3). Types 1 and 3 are widespread in the deposit, and 2 and 4 relate only to the quartz and quartz-tourmaline rocks rich in Pb–Ag–Sb.

Gold and silver content are not stable even within one and the same grain of electrum: some grains are zonal with the outer parts enriched in Ag (Figure 6a,b), other grains are cross-cut by Au-rich "veinlets" (Figure 6c).

Table 2. General characteristics of gold deposits in the Kolmozero–Voron'ya greenstone belt.

Deposit		Oleninskoe	Nyal'm
Tectonic structure		Intersection of NW shear zone and a sub-meridional fault	Tectonized contacts of porphyry intrusion
Host rocks		Amphibolite, quartz porphyry dykes (~2.83 Ga)	Gabbrodiorite–diorite–granodiorite porphyry (~2.83 Ga)
Regional metamorphism		(1) Neoarchean: 2.6–2.9 Ga, lower amphibolite T~ 600 °C, P = 3–4 kbar; (2) Paleoproterozoic: 1.8–1.9 Ga, lower amphibolite T~ 530 °C, P = 5.0–5.5 kbar.	
Alteration, assemblages of new formed minerals	-pre-ore	K–Ca: Biotite, diopside-epidote-carbonate	K–Ca–(Si) Beresite (quartz–carbonate–muscovite), biotite–epidote
	-gold-related	Si–B–(K) Quartz-tourmaline-muscovite, quartz-tourmaline, quartz	CO_2–Si Stockwork of quartz–carbonate veinlets
	-post-ore	Near-pegmatite: biotite and holmquistite	Not defined
Mineralized bodies		Mineralized lenses up to 3.5 × 50 m. The best sections: 83.8 ppm for 2.2 m in drillhole #BF-28 and 88.6 ppm for 1.5 m in BF-45 [18]; average grade 7.6 ppm Au	Mineralized stockwork up to 15-m thick with weighted average 1.2 ppm Au. The best intersections are 0.93 ppm Au for 32.8 m and 1.77 ppm for 18.4 m in two intervals in drillhole #BF-107 [18]
Texture of mineralization, sulfide content (%)		Disseminated, veinlet-disseminated, 1–15	Disseminated, 1–3
Main ore minerals		Arsenopyrite, pyrrhotite, ilmenite	Pyrrhotite
Minor ore minerals		Chalcopyrite, sphalerite, galena, freibergite, sheelite, pyrite, Sb-sulfosalts, löllingite, dyscrasite, electrum, uytenbogaardtite, aurostibite, petzite, chlorargyrite.	Chalcopyrite, pyrite, cobaltite, ilmenite, arsenopyrite
Geochemical association		Au–As–Ag (Sb–Cu–Pb–Zn–B–W). Au/Ag <0.2	Au–As. Au/Ag >1.
Age of mineralized rocks		Neoarchean—later than 2.83 Ga, but before 2.45 Ga	Neoarchean(?)–later than 2.83 Ga
Gold characteristics, fineness		4 types of Au–Ag alloys: 1) electrum (25–32 mas.% Au) with arsenopyrite; 2) electrum (33–47% Au) with galena and Sb-sulfosalts; 3) gold (78–95% Au) in quartz; 4) silver (<7% Au) in the weathered rocks.	Grains up to 1.0 mm in carbonate–quartz veinlets. The gold fineness is 890–940.

Other gold minerals in the Oleninskoe are gold-bearing dyscrasite (grains up to 1 mm) (Figure 6), aurostibite, petzite, calaverite (inclusions <10 μm in electrum of the 2nd type), and uytenbogaardtite, which forms rims around grains of electrum (Figure 5b).

The gold content in dyscrasite varies from 0 to 17 wt.% (Table 4), some grains are zonal with the outer parts enriched in S and poor in Au (Figure 6c).

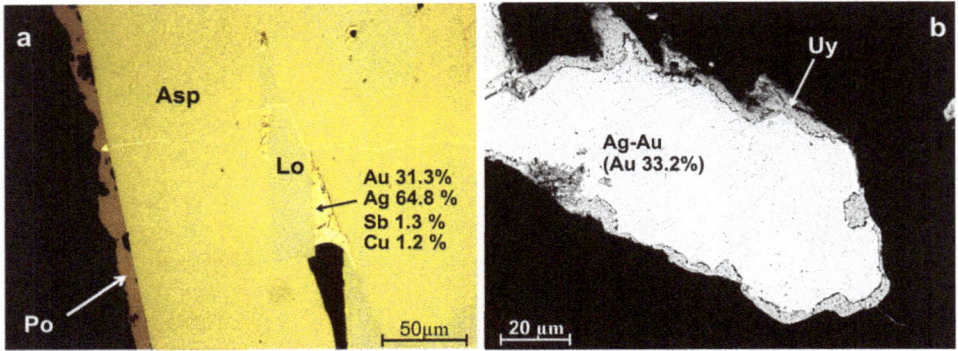

Figure 5. (**a**) Electrum of the 1st type at the boundary arsenopyrite (Asp)–löllingite (Lo) and in fissures in arsenopyrite, reflected light, one polarizer; (**b**) back scattered electron image of uytenbogardtiite, replacing electrum.

Table 3. Electron microprobe data in wt.% for native gold and silver from the Oleninskoe and Nyal'm (sample # 62.8–107) deposits.

Sample #	Type	Fe	Ni	Cu	Pb	Ag	Au	Sb	As	S	Total
090	1	1.20	n.a.	0.03	<0.01	72.35	25.52	0.05	1.61	0.14	100.91
9014-2	1	0.22	0.05	1.21	n.a.	64.77	31.34	1.27	0.53	0.27	99.66
AK-707-1	1	0.12	n.a.	n.a.	0.88	77.49	20.87	n.a.	n.a.	0.59	99.94
700	2	0.01	n.a.	<0.01	<0.01	61.79	33.63	1.78	<0.01	<0.01	97.20
700	2	0.01	n.a.	0.01	<0.01	55.13	37.29	2.41	<0.01	<0.01	94.86
711-8-23(C)	2	<0.01	n.a.	<0.01	n.a.	32.14	62.84	0.29	<0.01	n.a.	95.28
711-8-23(R)	2	<0.01	n.a.	<0.01	n.a.	51.68	42.46	0.34	<0.01	n.a.	94.48
711-2-1(C)	2	<0.01	n.a.	<0.01	n.a.	52.22	41.14	1.81	<0.01	n.a.	95.17
711-2-1(R)	2	<0.01	n.a.	<0.01	n.a.	37.07	58.10	0.65	<0.01	n.a.	95.81
AK-711A-1	2	0.14	n.a.	n.a.	n.a.	39.61	59.17	n.a.	n.a.	1.04	99.96
700	2	0.03	n.a.	0.02	<0.01	39.35	54.99	0.71	<0.01	<0.01	95.09
284	2	0.01	<0.01	<0.01	n.a.	34.58	61.16	0.49	0.08	<0.01	96.32
284	2	0.02	<0.01	<0.01	n.a.	24.16	73.20	0.16	0.08	<0.01	97.62
K-13C-29/1	3	n.a.	n.a.	n.a.	n.a.	21.80	77.87	n.a.	n.a.	n.a.	99.67
K-505.7	3	n.a.	n.a.	0.05	n.a.	12.77	86.65	n.a.	n.a.	n.a.	99.47
K-027-12/1	3	n.a.	n.a.	n.a.	n.a.	12.54	87.18	n.a.	n.a.	n.a.	99.72
BF-2029	3	0.05	n.a.	0.11	n.a.	12.07	87.83	n.a.	n.a.	0.14	100.20
K-027-12	3	n.a.	n.a.	n.a.	n.a.	11.85	87.63	n.a.	n.a.	n.a.	99.48
9015	4	0.41	n.a.	n.a.	n.a.	89.27	7.15	3.16	n.a.	n.a.	99.99
107-62.8		<0.01	n.a.	1.35	n.a.	8.73	89.50	n.a.	n.a.	n.a.	99.57
Formulae coefficients (calculated for Σ Metals = 1)											
090	1	0.025		0.001	0.000	0.790	0.153	0.000	0.025	0.005	
9014-2 3a	1	0.005	0.001	0.024		0.742	0.197	0.013	0.009	0.010	
AK-707-1	1	0.002			0.005	0.846	0.125			0.022	
700	2	0.000		0.000	0.000	0.755	0.225	0.019	0.000	0.000	
700	2	0.000		0.000	0.000	0.709	0.263	0.028	0.000	0.000	
711-8-23	2	0.000		0.000		0.687	0.309	0.004	0.000		
711-8-23'	2	0.000		0.000		0.481	0.515	0.004	0.000		
711-2-1'	2	0.000		0.000		0.684	0.295	0.021	0.000		
711-2-1	2	0.000		0.000		0.534	0.458	0.008	0.000		
AK-711A-1	2	0.003				0.523	0.428			0.046	
700	2	0.001		0.001	0.000	0.560	0.429	0.009	0.000	0.000	
284	2	0.000	0.000	0.000		0.504	0.488	0.006	0.002	0.000	
284	2	0.001	0.000	0.000		0.374	0.621	0.002	0.002	0.000	
K13C-29/1	3					0.338	0.662				
K-505.7	3			0.001		0.212	0.787				
K027-12/1	3					0.208	0.792				
BF-2029	3	0.002		0.003		0.200	0.796			0.004	
K-027-12	3					0.198	0.802				
9015-1 4-1a	4	0.008				0.922	0.040	0.029			
107-62.8		0.000		0.038		0.145	0.817				

Notes: n.a. = not analyzed; R = rim, and C = center of one and the same grain.

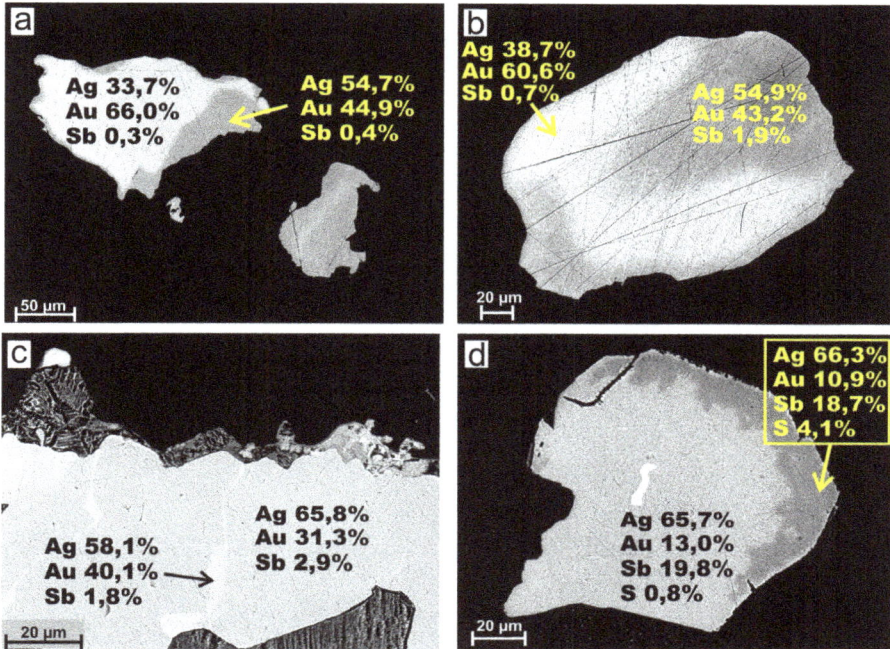

Figure 6. (**a–c**) Back scattered electron images of heterogenic electrum grains with electron microprobe data: (**a,b**) zonality in the electrum grains with the outer parts, enriched in Ag and Sb; (**c**) microveinlets of electrum, enriched in gold, in electrum (the second type); (**d**) zonality in dyscrasite with an inclusion of electrum.

Table 4. Electron microprobe data in wt.% for dyscrasite and uytenbogaardtiie.

Mineral	Sample #	Fe	Cu	Zn	Pb	Ag	Au	Sb	As	S	Total
	700	<0.01	0.02	0.01	<0.01	76.25	0.08	23.05	0.02	<0.01	99.43
	700	0.03	0.04	0.03	<0.01	63.64	12.22	17.93	0.06	<0.01	98.95
	284-3	0.01	0.02	0.01	0.07	69.84	10.46	19.00	<0.01	0.01	99.41
Dyscrasite	136-1	0.01	<0.01	0.01	1.26	76.96	0.44	20.81	<0.01	0.04	99.54
	700	<0.01	<0.01	n.a.	n.a.	73.78	2.05	23.73	n.a.	n.a.	99.57
	700	<0.01	n.a.	n.a.	n.a.	73.82	3.04	23.48	n.a.	n.a.	100.33
	700	<0.01	n.a.	n.a.	n.a.	71.37	7.45	21.29	n.a.	n.a.	100.13
Uytenbogaardtite	AK-711-4	0.15	n.a.	n.a.	n.a.	59.86	26.88	n.a.	<0.01	11.15	98.04
Formulae coefficients (calculated for Sb + As + S = 1 in dyscrasite, S = 2 in uytenbogaardtite)											
	700	0.000	0.001	0.000	0.000	3.728	0.002	0.999	0.001	0.000	
	700	0.004	0.004	0.003	0.000	3.985	0.419	0.994	0.005	0.000	
	284-3	0.001	0.002	0.001	0.002	4.144	0.340	0.999	0.000	0.001	
Dyscrasite	136-1	0.001	0.000	0.001	0.035	4.142	0.013	0.992	0.000	0.008	
	700	0.000	0.000			3.509	0.053	1.000			
	700	0.000				3.549	0.080	1.000			
	700	0.000				3.784	0.216	1.000			
Uytenbogaardtite	AK-711-4	0.016				3.191	0.785		0.000	2.000	

Note: n.a. = not analyzed.

Geochemical association of metals Au–As–Ag (Sb–Cu–Pb–Zn–B–W) reflects complicated character of the mineralization. The deposit is rich in silver, and Au/Ag ratio is less than 0.2.

Fluid inclusions in diopside from diopside–zoisite–carbonate matasomatic rock have high temperatures of homogenization 341–351 °C and low salinity 1.1 wt.% NaCl-equivalent [19] (Table 5). Temperature of homogenization of fluid inclusions in quartz from quartz metasomatite with gold–arsenopyrite mineralization is lower, 120–160 °C, the fluid has H_2O–CO_2 composition with minor CH_4, the salinity is high 13.4–13.5 wt.% NaCl-equivalent, the main cations are Na, Ca, and Mg [19].

Mineralization in the Oleninsloe formed in the Neoarchean between 2.83 Ga (the age of quartz porphyry dykes [20]) and 2.45 Ga (the age of the pegmatite veins, cutting the mineralized rocks [21]).

Table 5. Microthermometric data for individual fluid inclusions in rock-forming minerals from gold deposits in the Kola region.

Gold Deposit/Occurrence	Host Mineral	Temperature of Homogenization, °C	Fluid Composition, mole/kg	Salinity, wt.% NaCl-equivalent; Salt Composition	Pressure, Bar
Oleninskoe [19]	Diopside	341–351	No data	1.1	No data
	Quartz with gold mineralization	120–160	H_2O–CO_2–CH_4	13.4–13.5 Na, Ca, Mg	No data
Nyal'm (this study)	Quartz from veinlets	251–334	H_2O–CO_2–CH_4 CO_2 4.3–5.8 CH_4 0.5–0.7	2.5–5.6	1130–3290
Porojarvi [22]	Quartz from metasomatic quartzite	242–336	H_2O–CO_2–CH_4 CO_2 4.3–7.8 CH_4 0.6–0.9	0.4–3.8 Na, Ca, Mg, K (Au 4.6 ppm)	1200–2580
	Quartz from quartz-albite- carbonate metasomatite	342–370	H_2O–CO_2–CH_4 CO_2 5.3–8.8 CH_4 0.6–1.1	0.6–1.0 Na, Ca, Mg, K (Au 202 ppm)	2130–4550
	Quartz from propylite	280–320	H_2O–NaCl	25–30	No data
Mayskoe [23]	Vein quartz	320–350	No data	1.7–7.7	No data
	Granulated quartz from the ore shoots	190–270	CO_2–CH_4 CH_4 16–70%	22–26 Ca	940–500
	Quartz with gold in inclusions	140–198	No data	22–28 Ca	No data

Small gold deposit Nyal'm (#2 in Figure 1) is located 18 km SE from the Oleninskoe, at the SE thinning of the same stratum of the Oleny Ridge amphibolite. Geological setting, mineralogical, petrographical, and geochemical characteristics of the Nyal'm deposit differ significantly from those of the Oleninskoe (Table 2).

Mineralization at the Nyal'm is spatially confined to a stockwork of quartz–carbonate veinlets at tectonized contacts of a minor intrusion of diorite porphyry, hosted by a amphibolite and carbonaceous schist (Figure 7). The intrusion is differentiated: rock composition varies from gabbrodiorite in the center to granodiorite porphyry in the marginal part.

A network system of mineralized quartz–carbonate veins and veinlets forms two linear stockworks up to 15-m thick, orientated parallel to elongation of the intrusion. Concentration of veinlets in the stockworks is normally less than 10 vol.%, but in two vein zones (3.0- and 0.8-m thick) it increases to 80 vol.%.

Content of ore minerals is less than 3%, and pyrrhotite, ilmenite, and rutile prevail. Some pyrrhotite grains contain intergrowths of chalcopyrite, pyrite, flame-like pentlandite, and sporadically, sphalerite. Arsenopyrite and cobaltite were noted less frequently.

Visible native gold concentrates mainly in the central parts of quartz–carbonate veinlets. Gold grains are <0.7 mm in size, idiomorphic, isolated in quartz, intergrowths of gold with sulfides were not found. The gold fineness is high, from 870 to 920 [24]. The weighted average is 1.2 ppm Au in the stockworks, and 11.6 and 12.8 ppm in two "vein zones". Inferred resources of the deposit are estimated at 7.5 t Au (down to 100 m depth) [16].

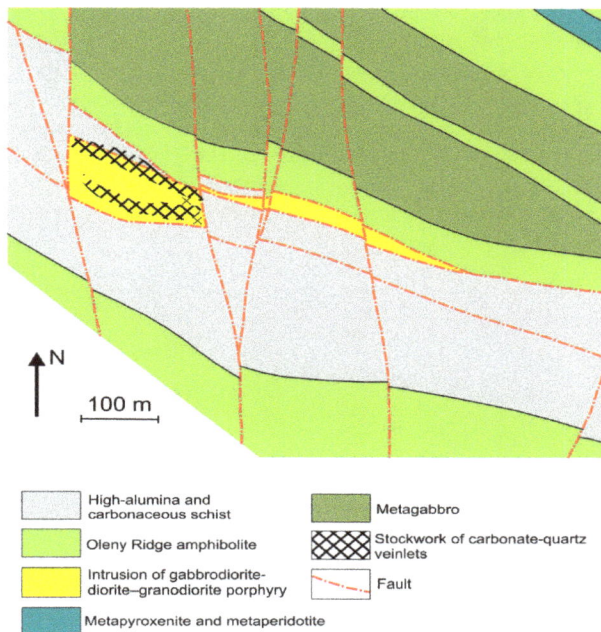

Figure 7. Schematic geological map of the Nyal'm-1 gold deposit area.

Temperature of homogenization of the primary fluid inclusions in quartz varies from 251 to 334 °C (Table 5). Composition of the fluid is H_2O–CO_2, with minor CH_4. Salinity is low, 2.5–5.6 wt.% NaCl -equivalent; Na, Mg, and Fe chlorides prevail. Estimated fluid pressure is 1.13–3.29 kbar.

The Mo–Cu porphyry deposit Pellapahk, mentioned in References [9,10] as a gold-bearing one, is located in the Kolmozero–Voron'ya belt, 2-km west of Oleninskoe. Mineralized stockwork in the Pellapahk deposit is hosted by altered granodiorite—granite porphyry, which underwent acidic leaching and turned into muscovite–kyanite, muscovite–andalusite, and muscovite metasomatic quartzite [25]. The mineralized stockwork occupies area 1500 × 350–600 m. Content of sulfide mineralization in the stockwork is 1–2 vol.%, sporadically up to 10 vol.%. Sulfide concentrate from the deposit contains 2.5 ppm Au [16] (the calculated gold content in the rock is 0.08 ppm), and 200 ppm Ag, hypothetic resources (P_3 according to the Russian classification of resources) of the deposit are estimated at 10 t Au [16].

According to information obtained during the last decade [25], the inferred ore resources of the deposit are estimated at 152 million tons, with an average Mo content of 0.028 mas.%, and Cu 0.154%. The resources of molybdenum and its grades are much lower than the average for the porphyry deposits, and this information significantly decreases interest in exploration of the Pellapahk deposit as a source of Mo, Cu, and, consequently, Au.

4.2. Gold Deposits and Occurrences in the Strel'na Greenstone Belt

The Strel'na greenstone belt is the easternmost structure of the Neoarchean belt system at the southern flank of the Imandra–Varzuga greenschist belt (Figure 1). It is a part of the Srel'na terrane, made mainly of the Neoarchean rocks intensely reworked in the Paleoproterozoic. The boundary between the Strel'na and Imandra–Varzuga belts is of tectonic character—an overthrust of the Neoarchean rocks (the Strel'na greenstone belt) over the Paleoproterozic metavolcanics (the Imandra Varzuga belt) (Figure 2).

The Strel'na belt hosts small gold deposit Sergozero, gold occurrence Vorgovy, and a number of points of gold mineralization (Figures 1 and 2).

The Sergozero deposit (#3 in Figure 1) is located 1.5–2-km south from the mentioned regional overthrust fault zone, at a second order shear zone, parallel to the main fault. Geological structure of the deposit is defined by the strata of amphibolite and amphibole schist with a composition corresponding to komatiite, komatiitic, and tholeiitic basalts, within a series of fine-grained muscovite–biotite and biotite schists (Figure 8). The rocks are greenschist—lower amphibolite metamorphosed (Table 6). Supracrustal rocks are intruded with dykes and minor intrusions of diorite porphyry (1875 ± 2 Ma, U-Pb, zircon [26]) and granite (~1.83 Ga?).

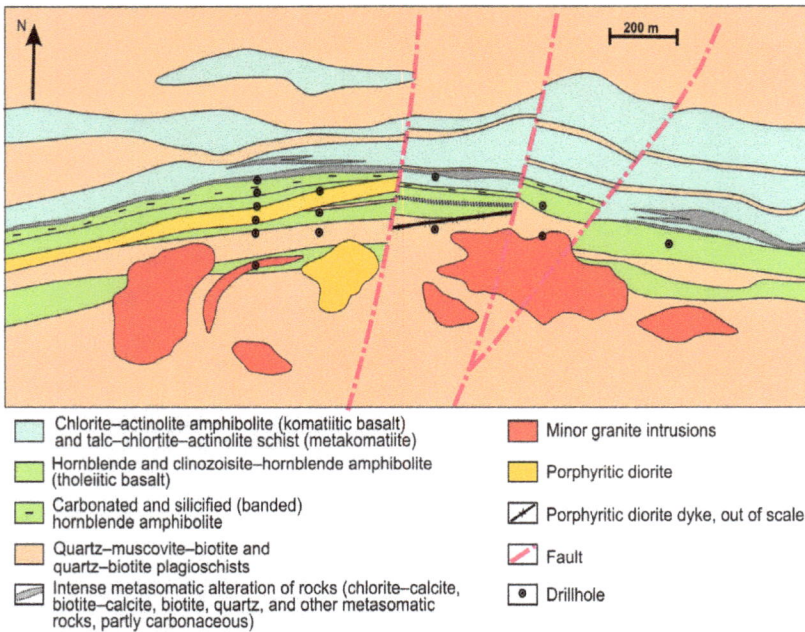

Figure 8. Schematic geological map of the Sergozero gold deposit.

All volcanic–sedimentary rocks in the deposit and diorite porphyry in the dykes are schistose, foliated, cut by hydrothermal veinlets, in some parts brecciated and altered. The prevalent alteration processes are chloritization (up to 60 vol.% of newly formed chlorite), biotitization, graphitization, silicification, and carbonatization. The total thickness of the zone of alteration varies from 60 to 120 m. This tract comprises the entire sequence of the hornblende amphibolite, the upper part of the chlorite–actinolite rocks, and the porphyry diorite dykes, which cut the amphibolites.

All rocks in the deposit, except granite, contain sulfide mineralization. The list of ore minerals includes pyrrhotite, arsenopyrite, pyrite, chalcopyrite, gersdorfite–cobaltite, sphalerite, pentlandite, galena, molybdenite, scheelite, native gold, and, less frequently, ulmannite and hessite. The mineralization is disseminated, veinlet-disseminated, nested; content of sulfides reaches 10 vol.%.

The main mineralized zone follows the contact between hornblende and actinolite–chlorite amphibolites. High gold grades (up to 5.7 ppm) relate to the biotite–calcite metasomatic rocks after hornblende amphibolite, chlorite–actinolite amphibolite, and diorite porphyry (Figure 9).

Table 6. General characteristics of the gold deposits in the Strel'na greenstone belt.

Deposit/occurrence		Sergozero	Vorgovy
Tectonic structure		Second-order faults, accompanying the regional overthrust of the Neoarchean Strel'na belt onto Paleoproterozoic rocks of the Imandra–Varzuga belt	Regional overthrust of the Neoarchean Strel'na belt onto Paleoproterozoic rocks of the Imandra–Varzuga belt
Host rocks		Hornblendite (tholeiitic basalt), chlorite–actinolite amphibolite (komatiite, komatiitic basalt), diorite porphyry dykes	Biotite gneisses and schists with interbeds of mafic-ultramafic metavolcanics
Regional metamorphism		Paleoproterozoic 1.8–1.9 Ga, greenschist—lower amphibolite	
Alteration (assemblages of new formed minerals)	-pre-ore	Chlorite	Quartz-chlorite-sericite and quartz-sericite
	-gold-related	$K–CO_2–Si$ Quartz–calcite–biotite	$Si–CO_2–(K)$ Stockwork of quartz—carbonate and quartz veinlets
	-post-ore	Quartz–calcite	Not defined
Mineralized bodies		Lenses up to 14-m thick and up to 150-m long. The best section is 5.3 ppm Au for 3.3 m in a drillhole by the Central Kola Expedition (CKE); average gold grade 3.1 ppm	Not contoured, the best section is 4 ppm for 1 m in a trench by the CKE
Texture of mineralization, sulfide content (vol.%)		Disseminated, 1–3	Nested, nested-disseminated, up to 0.1–5
Main ore minerals		Arsenopyrite, pyrrhotite, gersdorfite	Arsenopyrite, pyrrhotite
Minor ore minerals		Ilmenite, chalcopyrite, pyrite, pentlandite, cobaltite, sphalerite, galena, sheelite	Pyrite, sphalerite, galena, chalcopyrite, pentlandite, cobaltite, marcasite, mackinawite
Geochemical association, ratio Au/Ag		Au, As, Ni Au/Ag = 2–6	Au, As Au/Ag > 1
Age of mineralized rocks		Paleoproterozoic - later than 1875 ± 2 Ga (age of the diorite porphyry dyke); Rb–Sr age of quartz–calcite–biotite gold-related mineral association 1739 ± 86 Ma	Paleoproterozoic (?)
Gold characteristics, fineness in ‰		Fine-grained (<0.1 mm) gold with fineness 760–960 in association with arsenopyrite and gersdorfite	Spongy, porous, lumpy or flattened grains 0.1–0.25 mm, the gold fineness is 890–940.

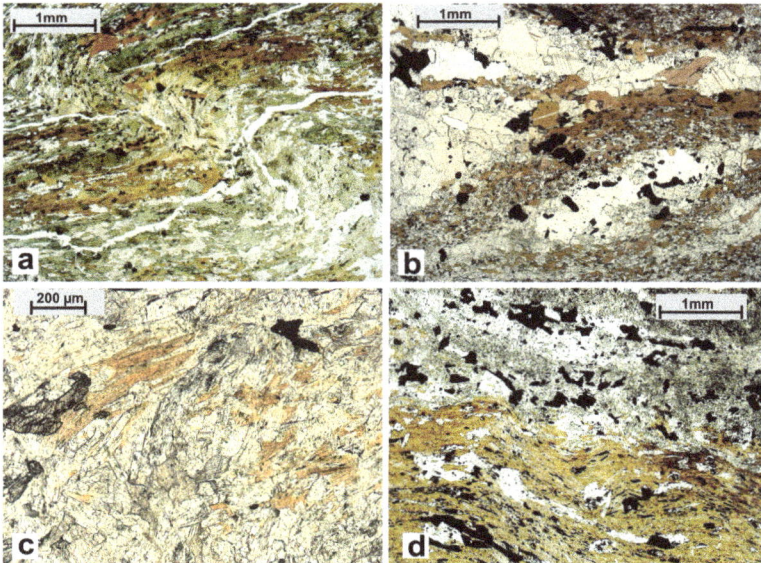

Figure 9. Altered rocks from the Sergozerskoe gold occurrence, thin sections photos, transparent light, one polarizer. (**a**) Biotitizated folded hornblende amphibolite, (**b–d**) calcite–biotite metasomatite: (**b**) after hornblende amphibolite, (**c**) after actinolite-chlorite amphibolite, (**d**) after diorite porphyry.

Native gold associates with arsenopyrite and gersdorfite. Gold grains more often are located at the boundary between sulfides and vein minerals (Figure 10). The prevalent gold grain dimension is 10–40 μm, maximum 0.1 mm, the grains are idiomorphic or lumpy. The fineness is high (957‰ on the average for 16 samples), medium (792‰ for 5 samples), and rarely low (672‰).

The best section is 2.13 ppm for 13.7 m, including 5.3 ppm for 3.3 m in drillhole #SRG-14. Inferred resources of the deposit are estimated at 13 t Au.

Mineralization in the Sergozero deposit formed in the Paleoproterozoic: age of diorite porphyry dyke, which is the youngest gold-bearing rock, is 1875 ± 2 Ma (U–Pb, zircon) [26]. Age of alteration processes, estimated with the Rb–Sr method for calcite–biotite metasomatic rocks is 1739 ± 86 Ma [26]. Thus, gold mineralization was superimposed on the Neoarchean amphibolites during the Paleoproterozoic tectonic events.

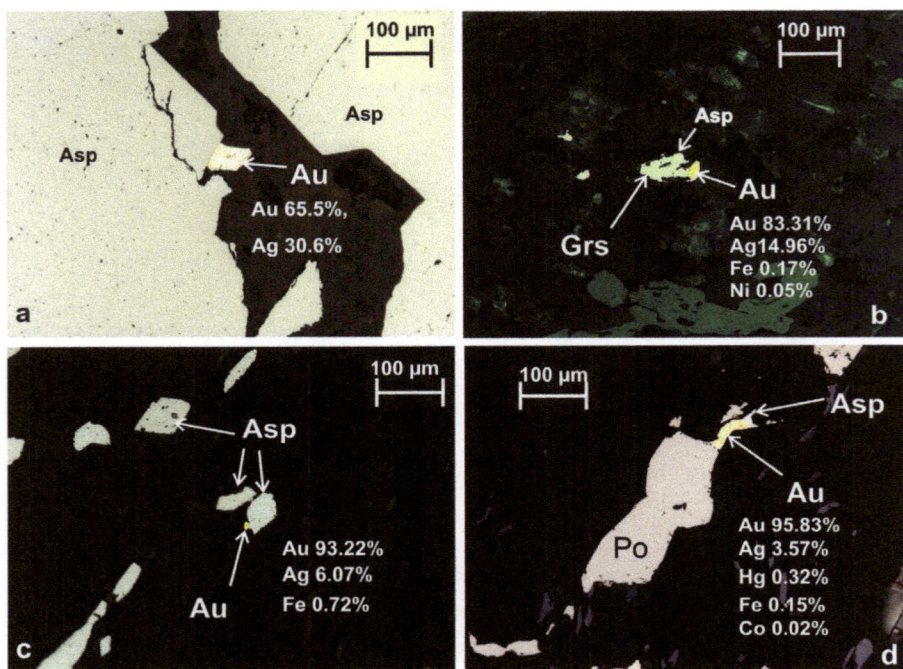

Figure 10. Gold with arsenopyrite and pyrrhotite, polished section photos, reflected light, one polarizer. (**a**) electrum and quarts in a fracture in arsenopyrite, (**b**) gold at the boundary of gersdorfite and silicate minerals, (**c**) gold at the boundary of arsenopyrite and silicate minerals, (**d**) gold with pyrrhotite and arsenopyrite.

The Vorgovy occurrence (#4 in Figure 1) is located 2-km north from the Sergozerskoe, at the zone of tectonic contact of the Neoarchean and Paleoproterozoic rocks (Figure 11): The Strel'na belt sequence of biotite schists and gneisses with interbeds of mafic and ultramafic metavolcanics is upthrown over the metavolcanics of the Imandra Varzuga belt at an angle of 60°. The rocks near the reverse fault zone are intensely deformed (folds with a limb span from a few centimeters to tens of meters), and metasomatically altered—transformed to quartz–sericite–chlorite and quartz–sericite schists, partly graphitized. Thickness of the zone of foliation and alteration is approximately 150 m (Figure 11).

Mineralization in the Vorgovy occurrence relates to a stockwork system of quartz and carbonate–quartz veins and veinlets hosted by chlorite–sericite–quartz metasomatic rocks after biotite gneiss. The stockwork zone is 100–150 m thick, extends for more than 1500 m, and dips southwest at an angle of 60°. Productive sulfide mineralization in veinlets includes arsenopyrite, pyrite, sphalerite, galena, and native gold. Despite the fact that superimposed veinlets and sulfide mineralization occur within a vertical interval of more than 100 m, the increased gold concentration up to 4 ppm, agreed with increase in arsenic content, have been found only in a 1-m interval immediately at the contact of sericite–chlorite–quartz schist and sericite quartzite.

Gold at the Vorgovy is spongy, porous, lumpy or flattened in shape. Predominant dimensions of gold grains are 0.1–0.25 mm. Fineness of gold is 890–940‰; Ag, Cu, and Fe occur as impurities in gold [24].

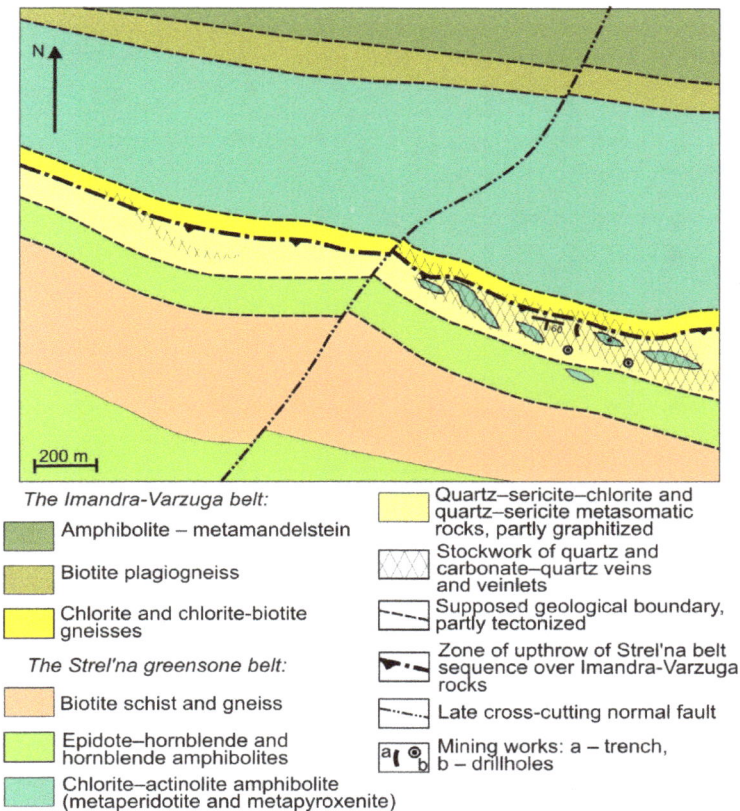

Figure 11. Schematic geological map of the Vorgovy gold occurrence.

4.3. Gold Occurrences in the Tiksheozero Greenstone Belt

The Tiksheozero belt is a part of greenstone belt system at the southwestern flank of the Belomorian mobile belt (Figure 1). The rocks in the belt were amphibolite metamorphosed twice–in the Neoarchean (T ~665–700 °C, P = 7–11 kbar), and in the Paleoproterozoic (T ~575–630 °C, P = 5.0–6.5 kbar) [27].

The Kichany occurrence (#7 in Figure 1) is located in the central part of the belt, in the Kichany synform (Figure 12) in a stratum of amphibolite (tholeiitic metabasalt) with interbeds of plagiogneiss–metatuffite. A series of thrusts at an angle of 30–40° splits the stratum to thin slabs, in the outcrops the thrusts reveal in intense jointing and schistosity of rocks [28].

The thrusts control zones of intense rock alteration. One of these zones, 50-m thick, was traced in the amphibolite along the shore of the lake Verhnie Kichany for more than 2.5 km (Figure 12). The main alteration processes in the amphibolite are calcium metasomatism with formation of diopside–epidote–calcite–Ca-rich almandine, or diopside–scapolite–epidote (±titanite) mineral assemblages, and silicification (quartz metasomatite with minor biotite) (Tables 7 and 8).

Arsenopyrite–pyrrhotite sulfide mineralization (with chalcopyrite, sphalerite, pyrite, marcasite, ilmenite, rutile, and native gold) was found in the zone of silicification at the boundary of garnet and feldspar amphibolites (Figure 13). Thickness of the mineralized zone is 1.2–1.5 m, and the length is more than 30 m. The weighted average gold content in the silicified amphibolite is 4.4 ppm for 1.3 m, including 0.3 m with 17.4 ppm Au.

Table 7. General characteristics of gold occurrences in the Tiksheozero and Olenegorsk greenstone belts.

Deposit/Occurrence		Kichany	Olenegorsk BIF Deposits [29–31]
Greenstone belt		Tiksheozero, Kichany structure (AR)	Olenegorsk (AR)
Tectonic structure		A series of thrusts in the central part of the Kichany synform	
Host rocks		Hornblendite and garnet amphibolite (tholeiitic basalts), interlayered with biotite gneiss	Magnetite–sulfide quartzite
Regional metamorphism		1) Neoarchean 2.6–2.9 Ga, upper amphibolite T~ 665–700 °C, P = 7–11 kbar; 2) Paleoproterozoic ~1.8 Ga, amphibolite T~ 575–630 °C, P = 5.0–6.5 kbar	Neoarchean ~2750 Ga, amphibolite
Alteration, assemblages of new formed minerals	-pre-ore	Ca Diopside–epidote–calcite–garnet, diopside–scapolite	Ca–Si Amphibole–epidote–garnet–calcite–quartz; diopside–magnetite
	-gold-related	Si–K Quartz, quartz–biotite	Ca–Si Hydrothermal veins of quartz, calcite, diopside–calcite, hornblende–almandine, quartz–calcite–almandine composition
	-post-ore	Not defined	Not defined
Mineralized bodies		The best section 4.4 ppm for 1.3 m in an outcrop	Average gold grade 1.36 ppm for 18 m in magnetite–sulfide quartzite, maximum 10.6 ppm
Texture of mineralization, sulfide content (vol.%)		Disseminated, 1–5	Disseminated, 1–5
Main ore minerals		Arsenopyrite, pyrrhotite, pyrite	Pyrrhotite, pyrite, chalcopyrite
Minor ore minerals		Chalcopyrite, marcasite, molybdenite, sphalerite, rutile, ilmenite, gold	Galena, sphalerite, molybdenite, electrum, sulfides and tellurides of Ag, Bi, and Au
Geochemical association, ratio Au/Ag		Au–As Au/Ag = 10	No data, Au/Ag = 0.4
Age of mineralized rocks		Paleoproterozoic (?)–1739 ± 15 Ma (U–Pb, titanite with inclusions of gold)	Neoarchean 2645–2675 Ma (U–Pb, zircon from hydrothermal veins)
Gold characteristics, fineness in ‰		Fine (<0.1 mm) hexahedral and irregular gold grains in association with arsenopyrite, marcasite, inclusions in quartz and in silicate minerals (oligoclase, titanite, hornblende), rare in pyrrhotite and chalcopyrite. Gold fineness 890–910.	Gold grains from non-magnetic heavy concentrate are of tabular, flaky, hooked, or of regular form, 0.2–0.5 mm. Gold fineness from 490 to 940.

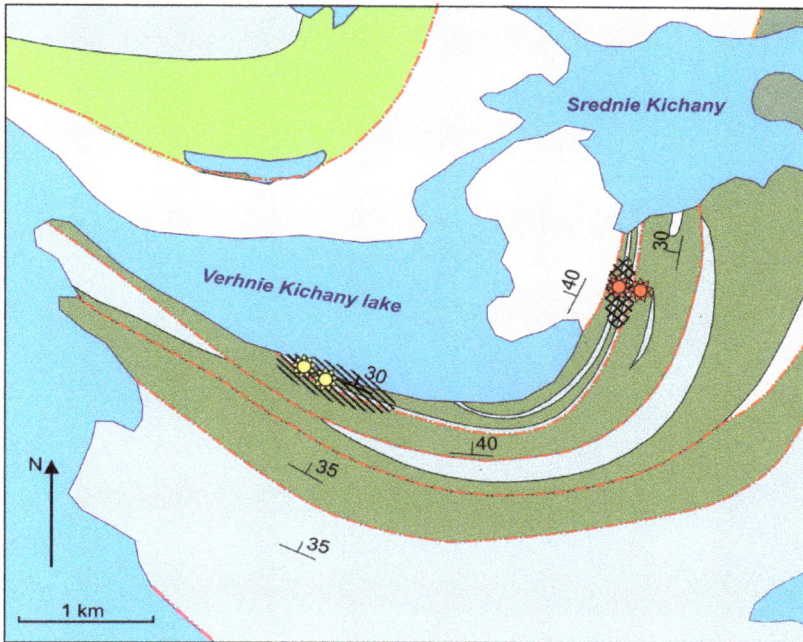

Figure 12. Schematic geological map of the Kichany gold occurrence.

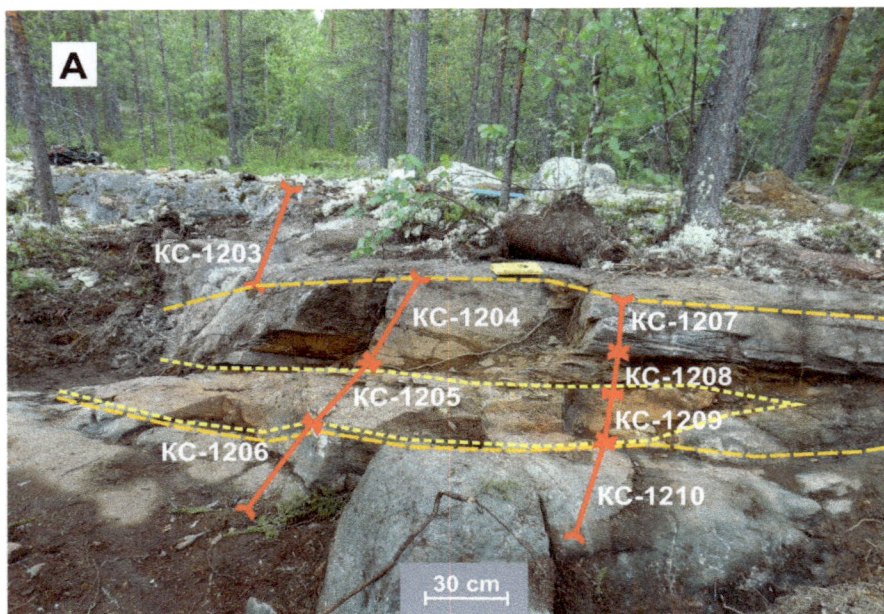

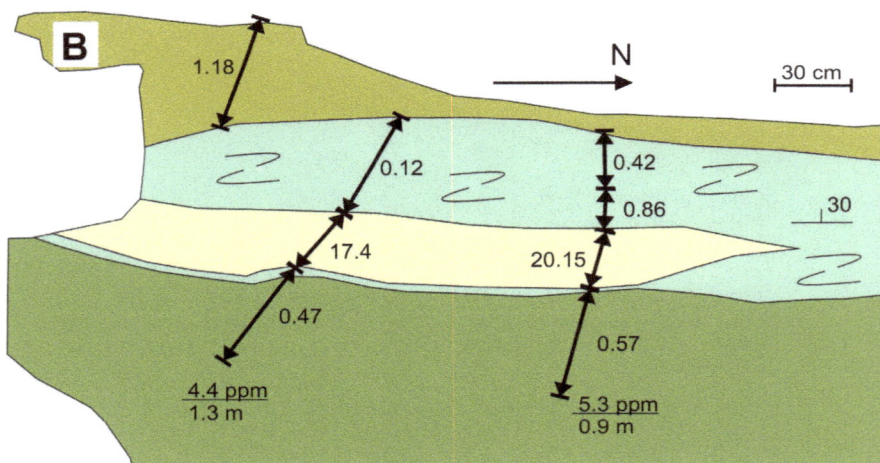

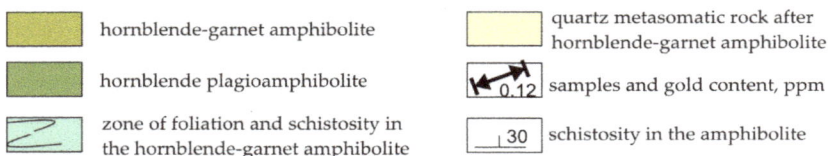

Figure 13. The Kichany gold occurrence. (**A**) Photo of the outcrop, with location of trench samples; (**B**) documentation of the same outcrop, with the results of assaying for gold.

Gold forms inclusions in silicate minerals (hornblende, oligoclase, titanite) in quartz, then it was noted at the boundary of sulfides (arsenopyrite and marcasite) with quartz, and less frequently, in inclusions in pyrrhotite and chalcopyrite (Figure 14).

Gold is fine-grained, mainly less than 0.1 mm. Grains are xenomorphic lumpy, or idiomorphic hexahedral. Some grains are spongy due to numerous quartz inclusions. Gold fineness is high (890–913), it makes alloys with silver; other impurities (Fe, Cu, Zn, Hg, As, Sb) were not detected.

Age of titanite (U–Pb method) from the silicified amphibolite with gold–arsenopyrite mineralization, which contains inclusions of gold (Figure 14c), is 1739 ± 15 Ma [28]. Mineralization either formed or was re-mobilized in the Paleoproterozoic during the Svecofennian metamorphic event.

Table 8. Chemical composition of mineralized rocks of the Kichany gold occurrence, the Tiksheozero belt, mas.%.

Sample #	KC-1203	KC-1204	KC-1104A	KC-1205	KC-1104	KC-1206	KC-1210	KC-1054
SiO_2	53.51	52.91	63.28	64.93	84.14	47.29	49.06	40.49
TiO_2	1.44	1.73	0.81	1.13	0.27	0.65	0.65	1.85
Al_2O_3	13.76	14.27	8.74	8.63	4.07	10.99	10.99	17.21
Fe_2O_3	0.89	2.3	4.53	0.86	0.84	1.23	1.09	5.56
FeO	9.99	9.65	7.09	8.67	4.64	9.23	9.52	7.08
MnO	0.16	0.18	0.06	0.086	0.04	0.23	0.22	0.19
MgO	4.79	4.44	1.96	3.71	1.05	12.96	10.4	3.46
CaO	7.98	7.57	3.72	4.56	1.67	11.32	11.79	18.81
Na_2O	2.69	2.96	1.98	1.58	0.92	1.4	1.55	0.96
K_2O	0.79	0.85	0.47	0.47	0.25	0.56	0.57	0.21
H_2O^-	0.44	0.14	0.26	0.47	0.14	0.16	0.18	0.27
H_2O^+	1.84	1.86	1.35	2.07	0.69	2.56	2.36	2.55
S	0.90	0.29	3.01	1.92	1.07	0.03	0.30	0.23
CO_2	0.53	0.43	<0.10	<0.10	<0.10	0.23	0.22	0.73
P_2O_5	0.06	0.07	n.a.	0.03	n.a.	0.02	0.02	0.15
F	0.013	0.013	0.009	0.009	0.005	0.015	0.013	0.016
Cl	0.027	0.033	0.005	0.015	0.008	0.012	0.017	0.017
As	n.a.	n.a.	2.86	n.a.	0.053	n.a.	n.a.	n.a.
Total	99.82	99.71	100.14	99.15	99.86	98.90	98.96	99.78
Au (ppm)	1.18	0.12	31.22	17.4	2.76	0.47	1.44	n.a.

Notes: n.a. = not analysed; KC-1203, KC-1204—hornblende-garnet amphibolite; KC-1104A, KC-1205—silicified hornblende-garnet amphibolite; KC-1104—quartz metasomatite after amphibolite; KC-1206, KC-1210—hornblende plagioamphibolite; KC-1054—diopside-epidote-calcite-garnet metasomatite (skarnoid) after amphibolite.

4.4. Gold in BIF Deposits in the Olenegorsk Greenstone Belt

Rocks of the banded iron formations are known in all Precambrian shields in greenstone belts of Archean and Paleoproterozoic age (3.0–2.3 Ga), including in Canada, Australia, South America, South Africa, in the Ukranian Shield and Voronezh Crystalline Massifs in Europe, etc., and the Fennoscandian Shield is no exception. Globally, some gold-bearing BIF deposits contain up to 10–16 ppm in BIF deposits in the São Francisco Craton (Brazil), up to 8–14 ppm in the Yilgarn Craton, and up to 10 ppm in Slave and Superior Provinces in North America. Gold mineralization is intrinsic to the BIF deposits rich in sulfides—to magnetite–sulfide, silicate–sulfide, carbonate–sulfide, and other facies of ferriginous quartzite [29].

In the Kola region, only two of nine iron-producing BIF deposits in the Olenegorsk belt were studied for gold—the Olenegorskoe (more details following) (#5 in Figure 1) and the Kirivogorskoe [29–31]. High gold grades in the Olenegorskoe deposit were found in magnetite–sulfide quartzite in the area with numerous pegmatite, pegmatite granite, and metagabbrodiorite veins. Thickness of the zone, enriched in gold, is 18–30 m, average gold grade is 1.36 ppm (for 18 m thickness), but in some samples, gold content reaches 5.5 ppm [29].

Aside from that, gold mineralization was found in the areas of intense calcium metasomatism in both the Olenegorskoe and Kiriovogorskoe deposits in skarnoids and hydrothermal metasomatic veins. Skarnoids (mineral assemblages, similar to skarn, but not connected with carbonate rocks and granitic magmatism) in the Olenegorskoe deposit develop at the boundary of magnetite–sulfide quartzite and high-alumina biotite gneiss, these metasomatic rocks consist of ferropargasite, Ca-rich almandine, epidote, calcite, quartz, and sulfides in different proportion [31]. In the Kirovogorkoe deposit skarnoids are of diopside–magnetite mineral composition.

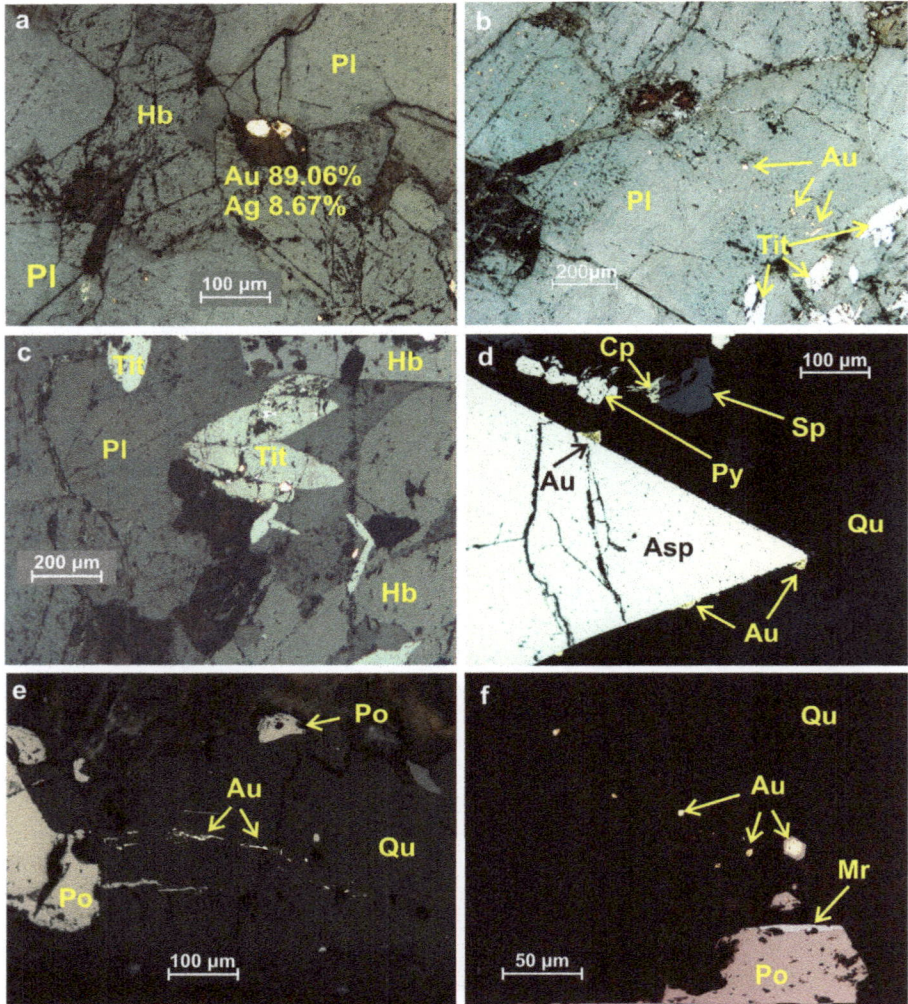

Figure 14. Gold in the Kichany occurrence, polished section photo, reflected light, one polarizer. (**a**) gold at the boundary plagioclase–hornblende, with the results of microprobe analysis, (**b**) a series of gold inclusions in plagioclase, (**c**) gold in titanite, (**d**) gold at the boundary of arsenopyrite metacrystal and quartz, (**e**) gold in fissures in quartz, and (**f**) gold in quartz, hexahedral gold grain is overgrown by pyrrhotite.

Hydrothermal metasomatic in 50–80 cm thick veins were found in the magnetite–sulfide quartzite, in the high-alumina biotite gneiss, and in the skarnoids. Quartz veins are the most common, in addition calcite, diopside–calcite, quartz–biotite–almandine, and hornblende–almandine veins were defined [32].

The skarnoids and hydrothermal veins contain sulfide mineralization (pyrrhotite, pyrite, chalcopyrite, less frequently galena, sphalerite, molybdenite) with electrum ($Au_{0.39-0.56}Ag_{0.33-0.51}Fe_{0.03-0.18}-Ag_{0.70}Au_{0.25}Fe_{0.05}$), native copper and bismuth, and Bi-tellurides. Electrum forms flakes flattened or wire-like grains in intergrowths with quartz, magnetite, hornblende, sulfides.

Age of zircon from the mineralized hydrothermal veins falls into interval 2645–2675 Ma (U–Pb method) [33], which is ~35 Ma younger than the zircon from the host biotite gneiss [33]; probably, sulfide mineralization with gold formed at the Neoarchean stage of regional metamorphism.

4.5. South Pechenga Zone of the Pechenga Greenschist Belt

The Pechenga belt is the best studied part of the Polmak–Pasvik–Pechenga–Imandra–Varzuga Paleoproterozoic belt system due to Pechenga group of Ni–Cu deposits in its northern part (Figure 1). Similar to other belts of the system, the Pechenga belt consists of two zones of metamorphosed sedimentary and volcanic sequences—the northern and southern [34]. The northern zones in the belts are characterized by weakly broken monoclinal sequences dipping at 20–60°, while in the southern zones the rocks are highly folded and faulted as a result of collision in the Late Paleoproterozoic [34]. Most of the Pechenga belt is a monocline formed by subconcordant sedimentary and volcanic sequences bounded by thrust faults plunging to the SSW. The oldest sequences are located in the north (footwall), and the youngest on the southern side.

Volcanic and tuffaceous sedimentary rocks of the Southern Pechenga structure, dated at 1.93–1.86 Ga, are localized within a long, but comparatively narrow tectono-stratigraphic area. The rocks are intensely folded and faulted and form a system of isoclinal folds overturned to the souteast, complicated by cross breaks and thrusts concordant to the general strike of the zone. The rocks are greenschist—lower amphibolite metamorphosed [35].

Points of gold mineralization (>0.1 g/t Au) were identified in different rocks in the belt [36,37]: in andesitic related to the Por'itash igneous complex; in volcanogenic massive sulfide pyrite–pyrrhotite ores; in tectonized and altered carbonaceous sequences near the meridional faults; in chlorite–carbonate altered ultramafics (picrite, peridotite, and serpentinite); in carbonate–quartz veins; and in metasomatic quartzite. But only in metasomatic quartzite the gold content reaches 1–5 ppm (#6 in Figure 1).

Location of metasomatic quartzite is governed by the intersection of northwestern-trending faults and thrusts (along the strike of the metasedimentary horizons) and later northeastern-trending faults (Figure 15). Metasomatic quartzite occurs as lenticular bodies with individual lenses less than 9-m thick, traced for a few tens of meters (up to 200 m). Along strike, lenses can be replaced by zones of carbonate–quartz veining. Lenses of quartzite are often folded together with the host rocks and cut through by later faults.

The mineralogy of quartzite is relatively simple: quartz makes 80–98 vol.% of the rock, dolomite, and rare calcite 1–5 vol.%; interstices between quartz grains are filled with amphibole (actinolite or grunerite), chamosite, oligoclase-albite, magnetite, sulfide minerals, and less frequently, carbonaceous matter. Amphibole- (the most abundant), chlorite–amphibole-, chlorite-, and magnetite-bearing metasomatic quartzites are distinguished by the dominant Fe–Mg mineral.

Sulfide-oxide disseminated mineralization (up to 3 vol.%) was observed in quartzite from all locations except for the Porojarvinskoe. The mineral assemblage is composed by sulfides pyrrhotite, chalcopyrite, pyrite, arsenopyrite, sphalerite, and molybdenite (the last two minerals are sporadic), and by oxides (magnetite, ilmenite, rutile); native gold was also identified (Table 9).

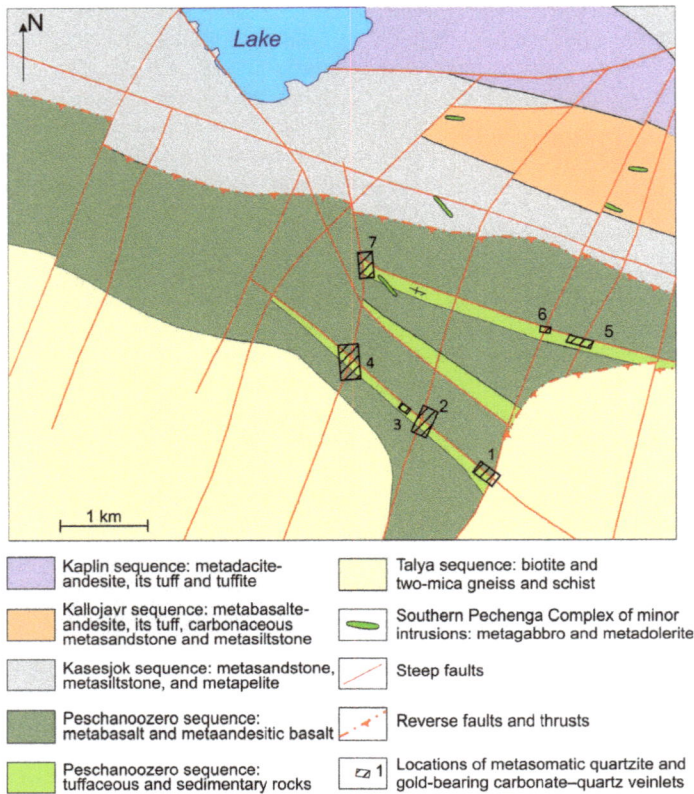

Figure 15. Schematic geological map of the Porojarvi area in the South Pechenga structure. Numerals are locations: (1) Timofeevsky, (2) Anomal'ny, (3) Zagadka, (4) Kontaktovy, (5) Svetlanovsky, (6) Istok, and (7) Porojarvinskoe.

Gold grades above 1 ppm were measured in amphibole quartzite with sulfide mineralization (Timofeevsky, Kontactovy), and without it (Porojarvinskoe). The best section in the Timofeevsky is 1.2 ppm for 3.2 m, and 3.3 ppm Au for 2.0 m in the Porojarvinskoe. Gold grains are fine (<0.2 mm), of irregular form.

Our study of fluid inclusions in quartz from mineralized rocks showed temperature of homogenization in the range 242–336 °C, the composition of the fluid was H_2O–CO_2 with minor methane, and low fluid salinity 0.4–3.8 wt.% NaCl-equivalent, Na, Ca, Mg, and K chlorides. The estimated fluid pressure is 1.20–2.58 kbar (Table 5).

Abundances of the main cations, measured by LA-ICP-MS [22], are the following (ppm): Na (983–4001), K (1055–6075), Mg (59–294), Ca (442–3176). It is interesting to note content of Ag (11.0–34) and Au (4.6–202) is relatively high in the fluid.

Isotope Sm–Nd study of quartz–albite–carbonate metasomatic rock with gold mineralization from the Zagadka location had a defined age of 1888 ± 22 Ma [38]. This age is consistent within the uncertainty limits, with the Rb–Sr dating of the volcanic rocks of the Bragino (1865 ± 58 Ma) and Kaplin (1855 ± 54 Ma) sequences [40], which likely corresponds to the age of metamorphism.

Table 9. General characteristics of gold occurrences in the Pechenga and Salla–Kuolajarvi greenstone belts.

Deposit/Occurrence		Porojarvi	Mayskoe
Greenstone belt		Pechenga, South Pechenga structure (PR)	Salla–Kuolajarvi (PR)
Tectonic structure		Intersection of northwestern and northeastern fault systems	Quartz veins along dolerite and gabbropyroxenite dykes filling northeastern faults
Host rocks		Biotite and muscovite-biotite schists, amphibole–chlorite and chlorite plagioschists, amphibolite.	Metabasalt, metaandesite, and their tuffs
Regional metamorphism		Paleoproterozoic ~1.9 Ga, upper greenschist—lower amphibolite	Paleoproterozoic ~1.9 Ga, upper greenschist
Alteration, assemblages of new formed minerals	-pre-ore	Chlorite-carbonate (propylitization)	Regional propylitization, Mg-metasomatism (Mg-amphibole–biotite–quartz, chlorite)
	-gold-related	Si (CO₂) Quartz–cummingtonite and quartz–cummingtonite–chlorite, quartz–albite–carbonate, stockwork of quartz–carbonate veinlets	K–Ba–Si Quartz–carbonate–feldspar (K-Ba)
	-post-ore	Not defined	Not defined
Mineralized bodies		The best cross section is 3.3 ppm Au for 2.0 m in a trench by CKE	Ore shoots in quartz veins:vein #1: ore shoot length 110 m, Au 13.1 ppm;vein #40: ore shoot length 80 m, Au 10.5 ppm
Texture of mineralization, sulfide content (vol.%)		Disseminated and nested-disseminated in quartzite and in carbonate-quartz veinlets, 0–2	Streaky and disseminated in quartz veins and vein exocontacts, 1–2
Main ore minerals		Arsenopyrite, pyrrhotite	Chalcopyrite, pyrite, sphalerite
Minor ore minerals		Magnetite, chalcopyrite, rutile, pyrite.	Galena, magnetite, cobaltite, marcasite, bornite, gold, sheelite, cobaltpentlandite, Bi-Pb tellurides
Geochemical association, ratio Au/Ag		Au–As Au/Ag = 4–10	Ag-Cu-Pb-Au (Zn, Co, Mo, Te, W)
Age of mineralized rocks		Sm–Nd isochrone age of quartz–albite–carbonate rock with gold mineralization 1888 ± 22 Ma [38]	Propylitization 1770 ± 9 Ma (Rb–Sr isochrone) (9610 ± 30 Ma [23]), Gold from ore shoots (Re–Os isochrone) 397 ± 15 Ma [39]
Gold characteristics, fineness in ‰		Fine (<0.05 mm) gold inclusions in arsenopyrite and in quartz. Gold fineness 790–800	Native gold in fissures in quartz, in aggregates with chalcopyrite and galena. Gold fineness 900–940

4.6. The Salla–Kuolajarvi Belt

The Salla–Kuolajarvi belt is a member of the Lapland–Karelian Paleoproterozoic greenschist belt system (Figure 1). Similar to the Pechenga and Imandra–Varzuga belts, and the Salla–Kuolajarvi belt is of asymmetric structure. The eastern zone (the Russian part) has a simple cross-section with rock sequences dipping west at an angle of 10° (in the center of the belt) to 70° (in the flank). The western part of the belt, in Finland, is represented by a set of granitized tectonic slabs (blocks), each block has its own structure [41].

Volcanic-sedimentary rocks in the Russian part formed in Jatulian (2.3–2.1 Ga) and Ludikovian (2.1–1.92 Ga), older rocks (Sumian–Sariolian, 2.5–2.3 Ga) are known only in the southeastern part of the belt [41].

The small deposit of Mayskoe (#8 in Figure 1), the only gold deposit ever mined in the region, is located in the central part of the Salla–Kuolajarvi belt. Quartz veins in the deposit are hosted by metavolcanics of the Apajarvi Formation (basalt, andesite, and mafic tuffs), intruded by dolerite dykes and ultramafic sills (Figure 16). The dykes are controlled by northeast trending faults and form two sub-parallel bodies, dipping nothwest at an angle of 60–80°. The faults reveal in 5–10 m thick zones of schistosity, foliation, and jointing. The same northeast trending faults control zones of pre-vein metasomatic alteration and location of two quartz veins (veins #1 and #40), i.e., the faults play the role of ore-hosting tectonic structures.

The wallrocks underwent processes of propylitization, Mg–metasomatism (Mg–amphibole–biotite–quartz assemblage and chloritization), and Si–K–Ba metasomatism (formation of quartz–K–Ba–feldspar–carbonates assemblage) [23,42]. Propylitic alteration is of regional character; but Mg and Si–K–Ba-metasomatism occur only within the vein-hosting fault zones. Quartz veins contain numerous xenoliths of intensely altered host metabasalt and metapyroxenite.

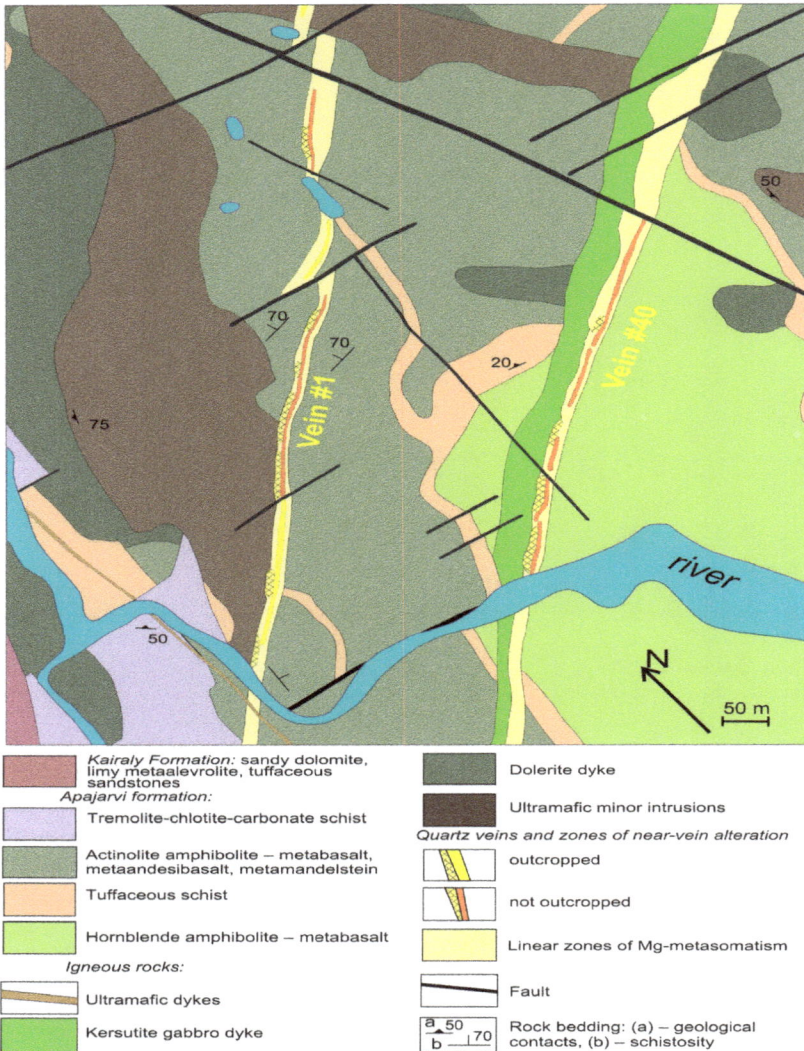

Figure 16. Schematic geological map of the Mayskoe deposit (modelled from Reference [23]).

Nested and veinlet sulfide mineralization is located mainly close to the vein selvages, or at the contact with xenoliths. Ore mineral assemblage includes chalcopyrite, pyrrhotite, sphalerite, galena, pyrite, cobaltite, native gold, sporadically Pb and Bi tellurides and selenides [43,44], and oxide minerals are rutile and scheelite. The metal association can be described as Ag–Cu–Pb–Au (Zn–Co–Te–W).

Gold in the veins is distributed unevenly. Ore shoots are localized in those parts of the veins, which are of complicated morphology, with xenoliths, intensely jointed, and with granular quartz. Ore shoots are 20–30 m in vertical section (80–110 m long on the surface), northeastern declines are at an angle of 15–20°. The gold grade is 13.1 ppm in the ore shoot of vein #1, and 10.5 ppm Au in vein #40.

Visible gold was noted in fractures in quartz, often in intergrowths with galena and sphalerite. Grains of gold are of irregular flattened form, rounded, hexahedral. The gold fineness is high (900–960). The study of fluid inclusions are shown in the following (Table 5) [23]:

- The temperature of homogenization of fluid inclusions in quartz from propylite is 350–430 °C. The fluids have very high salinity (up to 25–35 wt.% NaCl-equivalent), and its composition changed in time from NaCl to MgCl.
- The temperature of homogenization of fluid inclusions in non-mineralized quartz is 300–350 °C, the composition of the fluid is H_2O–NaCl, the salinity is moderate, 3–7 wt.% NaCl-equivalent
- In granular quartz with gold mineralization the temperature of homogenization of fluid inclusions is 190–270 °C, the fluid composition is CO_2–CH_4 with very high methane content from 16–25 to 62–70 mas.%, increasing with the approach to the chalcopyrite grain; the salinity of the fluid is high with 22–28 wt.% $CaCl_2$-equivalent, $CaCl_2$ dominates. The pressure is estimated at 0.5–0.94 kbar.

Isotope dating showed multi-stage history of mineralization development. Pre-ore propylites formed 1770 ± 9 Ma (Rb–Sr, minerals and rock) [45]. Age of Mg–metasomatism is 1610 ± 30 Ma (Rb–Sr, minerals and rock) [39]. K–Ar age of feldspar in Si–K–Ba metasomatite was estimated 1380 ± 40 Ma [23]. Finally, Re–Os age of gold is 397 ± 15 Ma [39].

5. Distribution and Timing of Formation of Gold Deposits in the Northeastern Part of the FenNoscandian Shield

Nearly all gold deposits, occurrences, and points of mineralization (75 of 80 points of mineralization with >1 ppm gold), known in metamorphic complexes in the Kola region and shown in Figure 1, are located within the Neoarchean and Paleoproterozoic greenstone belts. The distribution of gold occurrences is uneven; not all greenstone belts contain gold mineralization and the most prolific are the Kolmozero–Voron'ya and Strel'na belts, then the Olenegorsk belt and South Pechenga. According to genetic grouping of the belts [8], shown in Figure 1, the Kolmozero–Voron'ya belt is a paleosuture, other aforementioned belts are rift-related.

The locations of the deposits and occurrences are controlled by regional tectonic zones at the boundaries of major segments of the Fennoscandian Shield. One of these zones is the system of Neoarchean greenstone belts Kolmozero–Voron'ya–Ura-guba along the Murmansk craton—Kola-Norwegian terrane boundary (Figures 1 and 2). This tectonic structure includes the deposits Oleninskoe and Nyal'm and more than 20 occurrences and points of gold mineralization with >1 ppm gold.

Another important tectonic structure is a suture, delineating the core of the Lapland–Kola orogen in the north (Figure 2), composed of a series of overthrusts. This zone includes the Sergozerskoe deposit and gold occurrence Vorgovy in the Strel'na belt, Porojarvi occurrence in the South Pechenga structure, and more than 20 points of gold mineralization.

The third tectonic structure to be mentioned here is the series of overthrusts and faults at the eastern flank of the Salla–Kuolajarvi belt, traced farther southeast along the northern boundary of the Lapland–Karelian system of the Paleoproterozoic belts (Figure 1).

The primary composition of rocks, hosting gold mineralization in the deposits and occurrences in the region, is variable:

- Mafic-ultramafic metavolcanics play a principal role in the Oleninskoe, Sergozerskoe, Mayskoe deposits, and in the Kichany gold occurrence (Tiksheozero belt).
- Metasedimentary terrigenous rocks host gold mineralization in BIF deposits of the Olenegorsk group, in the Porojarvi occurrence (South Pechenga), and in the Vorgovy occurrence (the Strel'na belt).
- Diorite porphyry intrusion hosts mineralization in the Nyal'm deposit. Except the Nyal'm, diorite porphyry dykes in mafic volcanics contain gold mineralization in the Oleninskoe and

Sergozerskoe deposits, but it is unclear if these dykes play any role in mobilization and/or deposition of gold.

The metamorphic grade of mineralized rocks in the studied deposits varies from greenschist (in the central part of the Salla–Kuolajarvi belt) to upper amphibolite facies (the Tiksheozero and Olenegorsk belts), but mainly they are metamorphosed at lower amphibolite grade, close to the transition from greenschist to amphibolite facies (the Kolmozero–Voron'ya, Strel'na, South Pechenga, eastern flank of the Salla–Kuolajarvi belt) (Figure 17).

Under the conditions of transition from greenschist to amphibolite facies, the modal mineralogy of rocks and bulk-rock volatile content change, and consequently, favorable conditions for gold mobilization and migration emerge [46]. In the Paleoproterozoic belts with zonal metamorphism, the majority of occurrences and points of mineralization are located in the tracts of greenschist—amphibolite transition in the marginal part of the belts (occurrences in the South Pechenga structure and at the eastern flank of the Salla–Kuolajarvi belt) (Figure 17).

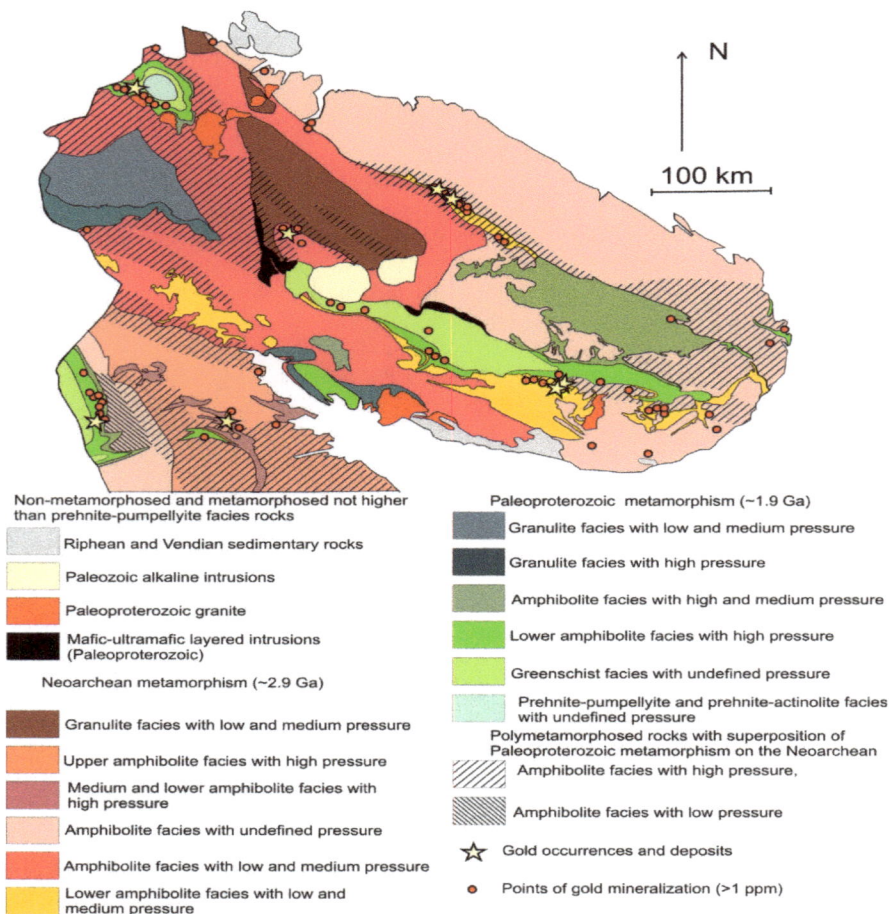

Figure 17. Distribution of gold deposits and occurrences in the schematic map of mineral facies of metamorphic rocks in the northeastern part of the Fennoscandian Shield (Modified from Reference [47]).

The Kolmozero–Voron'ya and Strel'na belts, two structures, which are the richest in gold mineralization, are low amphibolite high-gradient (high temperature, low pressure) metamorphosed. Moderate-to-high geothermal gradients are known to be favorable for the formation of gold deposits in metamorphic rocks [46].

The wallrocks were intensely altered in all studied deposits. Calcium rich minerals formed during pre-ore alteration in amphibolite metamorphosed complexes, the mineral assemblages include Ca–amphiboles (actinoliote, pargasite) and pyroxene (diopside), epidote-zoisite, Ca-rich almandine, calcite, and other carbonates (Oleninskoe, Kichany, Olenegorsk BIF deposits). Propylitization was the most common pre-ore alteration in the greenschist metamorphosed complexes, where mineral assemblage of chlorite, carbonate, and quartz formed. Gold-related alteration in the studied deposits was of a Si–CO$_2$–K character, with development of quartz, carbonate, biotite and/or sericite, and sulphide mineral assemblages.

Three groups of fluids inclusions were defined in quartz from the studied gold deposits (Table 5):

- Fluids of H$_2$O–CO$_2$ composition with low (0.4–7.7 wt.% NaCl-equivalent) salinity (Nyal'm, Porojarvi, vein quartz in Mayskoe deposit).
- Fluids of H$_2$O–CO$_2$ composition with high (13–35 wt.% NaCl-equivalent) salinity (Oleninskoe, quartz from propylite in Mayskoe deposit).
- Fluids of CH$_4$–CO$_2$ composition with high salinity 22–28 wt.% NaCl-equivalent (quartz with gold in Mayskoe deposit).

Fluids of H$_2$O–CO$_2$ composition with low salinity are favorable for formation of gold-only deposits in metamorphic complexes, because these fluids are able to mobilize Au from metamorphosed rocks and transport it in a form of Au–S complexes, but the content of base metals in these fluids is low [46,48]. Earlier we noted very high Au concentrations up to 202 ppm in fluids in the Porojarvi occurrence [39].

Fluids with high salinity do not form during regional metamorphism of the rocks, and they are probably connected with some magmatic source. In the Oleninskoe deposit, this magmatic source may be that one, which produced quartz porphyry dykes. In the Mayskoe, the presence of a large granite body at depth is likely due to a big negative gravity anomaly in the area of the deposit.

Fluids with high salinity can transport gold in chloride complexes, and these fluids are enriched in base metals as well [46,48,49]. Additional studies are needed to understand the origin of the fluids with extremely high methane and high salinity.

The main sulfide minerals in all gold deposits and occurrences in the region are arsenopyrite and pyrrhotite, with the only exception of the Mayskoe, where chalcopyrite prevails, and arsenic minerals are cobaltite and gersdorfite. Pyrite is not common in the deposits probably due to relatively high levels of metamorphism, when pyrrhotite formed instead of pyrite [50].

The geochemical association of metals (Au–As), and consequently, mineral composition of the ores are relatively simple in the deposits, which formed from H$_2$O–CO$_2$ fluids with low salinity and low concentration of base metals (Nyal'm, Porojarvi, probably Vorgovy, Sergozerskoe, Kichany). In the deposits with high fluid salinity (the Oleninskoe and Mayskoe) the metal association includes Au–Ag–As–Cu–Pb–Zn, and different Ag and base metal sulfides and Pb, Ag, and Bi tellurides play an important role in the mineral composition of the ore.

Gold deposits and occurrences in greenstone belts in the northeastern part of the Fennoscandian Shield (including Finnish Lapland) formed during two main periods, one in the Neoarchean, and another one in the Paleoproterozoic, both periods related to global collision events (Table 10), which coincided with the main peaks of gold deposit formation worldwide [51].

During the period 2.72–2.64 Ga, when a few Archean microcontinents consolidated and formed the supercontinent Kenorland, gold deposits Oleninskoe and Nyal'm formed in the Kolmozero–Voron'ya belt in between the Murmansk and Inari–Kola microcontinents under conditions of "continent–continent" collision. At the same period of time, gold mineralization formed in BIF deposits in the Olenegorsk belt.

In the Paleoproterozoic, two stages of formation of gold deposits were defined; these stages are connected with consolidation of separated fragments of the Kenorland (with accreted greenstone belts) into new supercontinent Columbia. "Arc–continent" collision occurred 1.86–1.93 Ga in the Pechenga–Imandra–Varzuga and Central Lapland rift-related greenschist belts. At this stage, gold deposits and occurrences formed along Sirkka and Kiistola shear zones in the Central Lapland belt in Finland (including Suurukuusikko deposit, the biggest in the Europe), and in the Porojarvi area in the Southern Pechenga.

At the final phase of Columbia supercontinent consolidation, 1.87–1.70 Ga, intracontinental collision events resulted in development of the Lapland–Kola–Belomorian orogeny. During this stage gold deposits and occurrences formed not only in Paleoproterozoic belts (in Kuusamo and Peräpohja schist belts in Finland), but in reworked Neoarchean greenstone belts (Strel'na and Tiksheozero greenstone belts).

The Mayskoe deposit has a long and complicated history of development. The stage 1.87–1.70 Ga is revealed only in pre-ore regional propylitization, and other metasomatic processes are dated to be Middle Proterozoic (1.6–1.4 Ga). The Re–Os age of gold 397 ± 15 Ma [39] corresponds to the time of the Paleozoic tectonic activization and formation of alkaline ultramafic intrusions in the Kola region, but this dating needs to be verified, and may not actually be the time of formation of the deposit, but rather the time of gold recrystallization.

Table 10. The main stages of formation of gold and gold-bearing deposits and occurrences in the northeastern part of the Fennoscandian Shield (based on data from References [8,32,52]).

Geodynamic Conditions	Time Interval, Ga	Geological Structures (Greenstone Belts, Intrusions)	Metallogenic Events	Deposits and Occurrences
Formation of core of the Murmansk, Inari–Kola, Khetolambina microcontinents	>2.93			
Formation of ancient island arc systems and their accretion to microcontinents	2.88–2.77	Kolmozero–Voron'ya belt (paleosuture), Central Belomorian paleosuture, Tiksheozero belt system	Deposition of volcanic–sedimenary rocks, hosting epigenetic gold mineralization and gold-bearing quartz conglomerates	
Formation of epigenetic rift greenstone belts	2.80–2.66	Strel'na, Ura-Guba, Allarechka, East Pansky, Olenegorsk, etc. greensone belts	Deposition of volcanic–sedimenary rocks, hosting epigenetic gold mineralization	
Consolidation of microcontinents into Kenorland supercontinent ("continent-continent" collision)	2.72–2.64	Kolmozero–Voron'ya belt	Epigenetic gold deposits in metamorphic complexes, Mo-Cu porphyry deposits	Oleninskoe, Nyal'm-1 (Au), Pellapahk (Mo–Cu)
		Olenegorsk belt		Epigenetic Au mineralization in BIF deposits
Rifting (mantle plume—initial stage of Kenorland break)	2.53–2.42	Fyodorovo-Pansky massif	Layered mafic–ultramafic intrusions (PGE, Ni–Cu deposits)	Fyodorova Tundra, Kievey, N.Kamennik, E.Chuarvy, Konttijarvi, Ahmavaara, Pennikat
"Dormant tectonic" period (sedimentation in intra-continental basins)	2.3–2.11	Salla–Kuolajarvi belt	Deposition of volcanic-sedimenary rocks, hosting epigenetic gold mineralization	
Rifting and re-activization of intra-continental basins, formation of intra-continental rifts of Red Sea type (mantle plume, break of Kenorland supercontinent)	2.11–1.92	Pechenga, Imandra–Varzuga	Layered mafic- ultramafic intrusions; intrusions of gabbro-verlite formation	Pechenga deposits (Cu, Ni + PGE, Au) (Zhdanovske, Zapolyarnoe, etc.)
		Salla–Kuolajarvi, Pechenga, Imandra–Varzuga belts	Deposition of volcanic –sedimenary rocks, hosting epigenetic gold mineralization	
Formation of Columbia supercontinent; accretionary orogenes of "igneous arc—continent" type	1.93–1.86	S.Pechenga zone	Epigenetic gold deposits	Occurrences in the Porojarvi area
Intracontinental collisions of "continent-continent" type; formation of Lapland–Kola orogen	1.87–1.7	Strel'na belt Tiksheozero belt	Epigenetic gold deposits	Sergozero, Vorgovy Kichany
Tectonic-magmatic activization in Mesoproterozoic and Paleozoic	<1.7	Salla-Kuolajarvi belt	Quartz-vein gold deposits	Mayskoe

The main prospects of the Kola region for new gold deposits are connected with the three abovementioned regional sutures. We can define four areas which have potential to become real goldfields in the case of successful exploration: the northwestern part of the Kolmozero–Voron'ya belt, the South Pechenga structure, the Strel'na belt at the northern flank of the Srel'na terrane, and the eastern flank of the Salla–Koulajarvi belt (Figures 1 and 2). The prospectivity is based on favorable geological conditions in terms of structures and due to numerous gold occurrences and minor deposits.

6. Conclusions and Implications

Gold deposits and occurrences in metamorphic complexes of the Kola region are located within the Neoarchean and Paleoproterozoic greenstone belts. The richest are the Kolmozero–Voron'ya and Strel'na belts, then the Olenegorsk belt and South Pechenga.

The location of the majority of deposits and occurrences is controlled by tectonic zones of regional scale at the boundaries of major segments of the Fennoscandian Shield: (1) the system of Neoarchean greenstone belts Kolmozero–Voron'ya–Ura-guba along the Murmansk craton—Kola-Norwegian terrane boundary, (2) a suture, delineating the core of the Lapland–Kola orogen in the north, and (3) the series of overthrusts and faults at the eastern flank of the Salla–Kuolajarvi belt, traced farther southeast along the northern boundary of the Lapland–Karelian system of Paleoproterozoic belts. The abovenamed suture zones are considered the most prospective for gold exploration in this region.

Gold deposits and occurrences in the northeastern part of the Fennoscandian Shield formed during two periods: in the Neoarchean at 2.7–2.6 Ga, and in the Paleoproterozoic at 1.9–1.7 Ga. According to paleo-geodynamic reconstructions these were the periods of collisional and accretionary orogeny in the region. The Archean greenstone belts, reworked in Paleoproterozoic (e.g., Strel'na and Tiksheozero belts), can contain gold deposits of Paleoproterozoic age.

The metamorphic grade of rocks in the studied deposits varies from greenschist to upper amphibolite facies, but the mineralized rocks are mainly metamorphosed at lower amphibolite grade, close to the transition from greenschist to amphibolite facies (deposits in the Kolmozero–Voron'ya, Strel'na, South Pechenga belts, and at the eastern flank of the Salla–Kuolajarvi belt). This level of metamorphism is considered as the most favorable for formation of gold deposits in metamorphic complexes.

Our study of fluid inclusions in quartz showed that fluids are mainly of H_2O–CO_2 composition, but salinity differs from one deposit to another. Fluids with low salinity are believed to be of metamorphic origin (Nyal'm deposit, Porojarvi gold occurrence), and fluids with high salinity (Mayskoe, Oleninskoe) are probably related to an undefined magmatic source.

Author Contributions: Methodology: A.A.K.; conceptualization: A.A.K., O.V.K., V.I.B.; archive data collecting: V.I.B.; study of fluid inclusions: V.Yu.P., original draft writing: A.A.K., O.V.K., V.I.B.; editing and review: A.A.K.

Funding: The work was carried out under Project 0226-2019-0053 of the Russian Academy of Sciences.

Acknowledgments: The authors thank Yevgeny Savchenko and Yekaterina Selivanova (GI KSC RAS) for microprobe and X-ray study of minerals, and Tatiana Kaulina and Dmitry Zozulya for helpful discussion of the results. The authors are grateful to the reviewers for their comments, which helped to improve the manuscript.

Conflicts of Interest: The authors declare no conflict of interest.

References

1. State Report 'On Statute and Exploitation of Mineral Resources of Russian Federation in 2015'. Available online: http://www.mnr.gov.ru/docs/gosudarstvennye_doklady/o_sostoyanii_i_ispolzovanii_mineralno_syrevykh_resursov_rossiyskoy_federatsii/ (accessed on 16 September 2018). (In Russian)
2. Subbotin, V.V.; Gabov, D.A.; Korchagin, A.U.; Savchenko, Y.E. Gold and Silver in the Composition of PGE Ores of the Fedorov-Pana Layered Intrusive Complex. *Herald Kola Sci. Centre RAS* **2017**, *1*, 53–65. (In Russian)

3. Ward, M.; Kalinin, A.; McLaughlin, D.; Voytekhovich, V. Kola Mining Geological Company Ltd (KMGC)—Prospecting for PGE in the East Pansky layered massif. In *The Neighborhood Cooperation and Eperience Exchange of Geological Prospecting and Survey of PGE Deposits in the Northern Fennoscandia*; KNTs RAN: Apatity, Russia, 2008; pp. 52–55.

4. Eilu, P.; Rasilainen, K.; Halkoaho, T.; Huovinen, I.; Karkkainen, N.; Kontoniemi, O.; Lepisto, K.; Niiranen, T.; Sorjonen-Ward, P. *Quantitative Assessment of Undiscovered Resources in Orogenic Gold Deposits in Finland*; Report of Investigation 216; Geological Survey of Finland: Espoo, Finland, 2015; 318p, ISBN 978-952-217-331-7.

5. Rasilainen, K.; Eilu, P.; Halkoaho, T. Assessment of undiscovered metal resources in Finland. *Ore Geol. Rev.* **2017**, *86*, 896–923. [CrossRef]

6. Suurikuusikko Mineral Deposit Report. Geological Survey of Finland. Available online: http://tupa.gtk.fi/karttasovellus/mdae/raportti/386_Suurikuusikko.pdf (accessed on 22 June 2018).

7. Eilu, P.; Pankka, H. *Fingold—A Public Database on Gold Deposits in Finland*; Geological Survey of Finland: Espoo, Finland, 2013; Available online: http://en.gtk.fi/informationservices/palvelukuvaukset/fingold.html (accessed on 21 March 2016).

8. Mints, M.V.; Suleimanov, A.K.; Babayants, P.S. *Deep Structure, Evolution, and Minerals in the Early Precabrian Basement of the East European Platform: Interpretation of Materials on Referent Profile 1-EV, Profiles 4V and TATSEIS*; GEOKART GEOS: Moscow, Russia, 2010; ISBN 978-5-89118-531-9. (In Russian)

9. Sundblad, K. Metallogeny of Gold in the Precambrian of Northern Europe. *Econ. Geol.* **2003**, *98*, 1271–1290. [CrossRef]

10. Eilu, P. (Ed.) *Mineral Deposits and Metallogeny of Fennoscandia*; Special Paper; Geological Survey of Finland: Espoo, Finland, 2012; Volume 53, 401p, ISBN 978-952-217-175-7.

11. Borisenko, A.S. Cryometric study of salt composition of the gas–liquid inclusions in minerals. *Geol. Geofiz.* **1977**, *8*, 16–27.

12. Bodnar, R.J.; Vityk, M.O. Interpretation of microterhmometric data for H_2O-NaCl fluid inclusions. In *Fluid Inclusionsin Minerals: Methods and Applications*; Benedetto DeVivo & Maria Luce Frezzotti, Pontignano: Siena, Italy, 1994; pp. 117–130.

13. Collins, P.L.P. Gas hydrates in CO_2-bearing fluid inclusions and the use of freezing data for estimation of salinity. *Econ. Geol.* **1979**, *74*, 1435–1444. [CrossRef]

14. Brown, P. FLINCOR: A computer program for the reduction and investigation of fluid inclusion data. *Am. Mineral.* **1989**, *74*, 1390–1393.

15. Daly, J.S.; Balagansky, V.V.; Timmerman, M.J.; Whitehouse, M.J. The Lapland-Kola orogen: Palaeoproterozoic collision and accretion of the northern Fennoscandian lithosphere. *Geol. Soc. Lond. Mem.* **2006**, *32*, 579–598. [CrossRef]

16. Korovkin, V.A.; Turyleva, L.V.; Rudenko, D.G.; Zhuravlev, V.A.; Klyuchnikova, G.N. *Underground of the North-West of Russian Federation*; VSEGEI: Sankt-Peterburg, Russia, 2003; 754p. (In Russian)

17. Kalinin, A.A.; Savchenko, Y.E.; Selivanova, E.A. Minerals of precious metals in the Oleninskoe gold deposit (Kola peninsula). *Zapiski RMO* **2017**, *146*, 43–58. (In Russian)

18. Ovoca Gold plc Operational Update 13/09/2007. Available online: http://www.ovocagold.com/upload/20070913_ayax__kola_&_goltsovoye_update.pdf (accessed on 2 April 2017).

19. Volkov, A.V.; Novikov, I.A. The Oleninskoe gold sulfide deposit (Kola peninsula, Russia). *Geol. Ore Deposits* **2002**, *44*, 361–372.

20. Kudryashov, N.M.; Kalinin, A.A.; Lyalina, L.M.; Serov, P.A.; Elizarov, D.V. Geochronological and isotope geochemical characteristics of rocks, hosting gold occurrences in the Archean greenstone belt Kolmozero-Voron'ya (Kola region). *Lithosphere* **2015**, *6*, 83–100. (In Russian)

21. Kudryashov, N.M.; Lyalina, L.M.; Apanasevich, E.A. Age of rare metal pegmatites from the Vasin Myl'k deposit (Kola region): Evidence from U-Pb geochronology of microlite. *Dokl. AN* **2015**, *461*, 321–325. [CrossRef]

22. Prokofiev, V.Y.; Kalinin, A.A.; Lobanov, K.V.; Banks, A.A.; Borovikov, D.A.; Chicherov, M.V. Composition of Fluids Responsible for Gold Mineralization in the Pechenga Structure of the Pechenga–Imandra–Varzuga Greenstone Belt, Kola Peninsula, Russia. *Geol. Ore Depos.* **2018**, *60*, 277–299. [CrossRef]

23. Safonov, Yu.G.; Volkov, A.V.; Vol'fson, A.A.; Genkin, A.D.; Krylova, T.L.; Chugaev, A.V. The Maysk quartz gold deposit (Northern Karelia): Geological, mineralogical, and geochemical studies and some genetic problems. *Geol. Ore Depos.* **2003**, *45*, 375–394.

24. Gavrilenko, B.V. Minerageny of Precious Metals and Diamonds in the North-Eastern Part of the Baltic Shield. Ph.D. Thesis, Moscow University, Moscow, Russia, 2003.

25. Kalinin, A.A.; Galkin, N.N. The Precambrian Copper-Molybdenum porphyry deposit Pellapahk (Kolmozero-Voron'ya greenstone belt). *Vestnik KNTs RAN* **2012**, *44*, 80–92. (In Russian)

26. Kalinin, A.A.; Kazanov, O.V.; Kudryashov, N.M.; Bakaev, G.F.; Petrov, S.V.; Elizarov, D.V.; Lyalina, L.M. New promising gold objects in the Strelna greenstone belt, Kola Peninsula. *Geol. Ore Depos.* **2017**, *59*, 453–481. [CrossRef]

27. Bibikova, E.V.; Bogdanova, S.V.; Glebovitsky, V.A.; Claesson, S.; Skiöld, T. Evolution of the Belomorian belt: NORDSIM U-Pb zircon dating of the Chupa paragneisses, magmatism, and metamorphic stages. *Petrology* **2004**, *12*, 195–210.

28. Kalinin, A.A.; Astaf'ev, B.Y.; Voinova, O.A.; Bayanova, T.B.; Khiller, V.V. Geological structure and prospects for minerals of the Tiksheozero greenstone belt. *Lithosphere* **2017**, *17*, 102–126. (In Russian) [CrossRef]

29. Starostin, V.I.; Pelymskiy, G.A.; Leonenko, E.I.; Sakiya, D.R. Gold in BIF deposits of the East European platform. *Izvestiya RAEN* **2005**, *14*, 27–42. (In Russian)

30. Golikov, N.N.; Goryainov, P.M.; Ivanyuk, G.Y.; Pahomovskiy, Ya.A.; Yakovenchuk, V.N. Auriferous iron formations of the Olenegorsk deposit (Kola Peninsula, Russia). *Geol. Ore Depos.* **1999**, *41*, 144–151.

31. Bazay, A.V.; Ivanyuk, G.Y. Native metals in the rocks of BIF in the Kola Peninsula. *Zapiski RMO* **2008**, *5*, 34–47. (In Russian)

32. Ivanyuk, G.Y.; Bazay, A.V.; Pakhomovskiy, Y.A.; Yakovenchuk, V.N.; Goryainov, P.M. Low temperature hydrothermal veins in the rocks of the Archean ferriferous formation. *Zapiski VMO* **2001**, *3*, 16–28. (In Russian)

33. Goryainov, P.M.; Ivanyuk, G.Y.; Bayanova, T.B.; Bazay, A.V.; Astaf'ev, B.Y.; Voinova, O.A. Composition, genesis, and age of rare-earth and precious metal mineralization in the rocks of banded iron formation in the Kola Peninsula. *Proc. Fersman Sci. Sess.* **2012**, *9*, 235–238. (In Russian)

34. Melezhik, V.A.; Sturt, B.A. General geology and evolutionary of the early Proterozoic Polmak-Pasvik-Pechenga-Imandra/Varzuga-Ust'Ponoy Greenstone Belt in the northeastern Baltic Shield. *Earth Sci. Rev.* **1994**, *36*, 205–241. [CrossRef]

35. Duk, G.G. *Structural Metamorphic Evolution of the Rocks in the Pechenga Complex*; Nauka: Moscow-Leningrad, Russia, 1977; 104p. (In Russian)

36. Ahmedov, A.M.; Voron'yaeva, L.V.; Pavlov, V.A.; Krupenik, V.A.; Kuznetsov, V.A.; Sveshnikova, K.Yu. Gold mineralization in the Southern-Pechenga structural zone (Kola Peninsula): Types of occurrences and prospects for finding economic gold grades. *Region. Geol. Metall.* **2004**, *20*, 139–151. (In Russian)

37. Voron'yaeva, L.V. Geology and Gold Mineralization in the Southern Pechenga Structural Zone. Ph.D. Thesis, VSEGEI, Sankt-Peterburg, Russia, 2008. (In Russian)

38. Kalinin, A.A.; Bayanova, T.B.; Lyalina, L.M.; Serov, P.A.; Elizarov, D.V. Occurrences of gold mineralization in the Southern Pechenga structural zone: New isotope geochronological data. In *Geology and Geochronology of the Rock-Forming and Ore Processes in Crystalline Shields, Proceedings of All-Russian (with International Participation) Conference, Apatity, Russia, 8–12 July 2013*; K & M: Apatity, Russia, 2013; pp. 69–71. (In Russian)

39. Bushmin, S.A.; Belyatskiy, B.V.; Krymskiy, R.Sh.; Glebovitskiy, V.A.; Buyko, A.K.; Savva, E.V.; Sergeev, S.A. Isochron Re-Os age of gold from Mayskoe gold-quartz vein deposit (Northern Karelia, Baltic Shield). *Dokl. AN* **2013**, *448*, 54–57. [CrossRef]

40. Balashov, Y.A. Geochronology of the Early Proterozoic rocks of the Pechenga-Varzuga structure in the Kola Peninsula. *Petrology* **1995**, *4*, 3–25. (In Russian)

41. Kulikov, V.S.; Kulikova, V.V. The Kuolajarvi synclinorium: A new view on the geological structure and combined cross section. In *Trudy Karel'skogo Nauchnogo Tsentra RAN*; Russian Academy of Sciences: Petrozavodsk, Russia, 2014; pp. 28–38. (In Russian)

42. Vol'fson, A.A.; Rusinov, V.L.; Krylova, T.L.; Chugaev, A.V. Metasomatic alterations in Precambrian metabasites of the Salla-Kuolajarvi graben near the Mayskoe gold field, Northern Karelia. *Petrology* **2005**, *13*, 161–186.

43. Gavrilenko, B.V.; Rezhenova, S.A. Ore minerals in gold-bearing quartz vein zones. In *Mineral'nye ParaGeNezisy Metamorficheskih i Metasomaticheskih Porod (Mineral Paragenesis in Metamorphic and Metasomatic rocks)*; Kola Filial AN SSSR: Apatity, Russia, 1987; pp. 58–67. (In Russian)

44. Kalinin, A.A.; Karpov, S.M.; Kalachyova, A.B.; Savchenko, Y.E. New data on mineralogy of the Mayskoe gold-quartz deposit (Northern Karelia). In *Trudy Fersmanovskoy Nauchnoy Sessii GI KNTs RAN (Proceedings of the Fersman Scientific Session)*; Kola Scientific Center of RAS: Apatity, Russia, 2018; pp. 172–175. (In Russian)

45. Vol'fson, A.A.; Chugaev, A.V.; Safonov, Y.G.; Rusinov, V.L.; Krylova, T.L. Results of Rb-Sr, Rb-Rb study of gold field Mayskoe, Northern Karelia. In *Isotope Dating of Geological Processes: New Methods and Results: Abstracts of Russian Conference, Moscow, Russia, 15–17 November 2000*; GEOS: Moscow, Russia, 2000; pp. 93–96. (In Russian)

46. Phillips, G.N.; Powell, R. Formation of gold deposits: A metamorphic devolatilization model. *J. Metamorph. Geol.* **2010**, *28*, 689–718. [CrossRef]

47. Belyaev, O.A.; Bushmin, S.A.; Voinov, A.S.; Volodichev, O.I.; Glebovitsky, V.A. *Map of Mineral Facies of Metamorphic Rocks in the Eastern Baltic Shield in the Scale 1:1 500 000*; Glebovitskiy, V.A., Ed.; VSEGEI, IPGGRAS: Sankt-Peterburg, Russia, 1991.

48. Zhu, Y.; An, F.; Tan, J. Geochemistry of hydrothermal gold deposits: A review. *Geosci. Front.* **2011**, *2*, 367–374. [CrossRef]

49. Pokrovski, G.S.; Akinfiev, N.N.; Borisova, A.Y.; Zotov, A.V.; Kouzmanov, K. Gold speciation and transport in geological fluids: Insights from experiments and physical-chemical modeling. *Geol. Soc. Lond. Spec. Publ.* **2014**, *402*, 9–70. [CrossRef]

50. Toulmin, P.; Barton, P.B. A thermodynamic study of pyrite and pyrrhotite. *Geochim. Cosmochim. Acta* **1964**, *28*, 641–671. [CrossRef]

51. Goldfarb, R.J.; Groves, D.I.; Gardoll, S. Orogenic Gold and geologic time: A global synthesis. *Ore Geol. Rev.* **2001**, *18*, 1–75. [CrossRef]

52. Rasilainen, K.; Eilu, P.; Halkoaho, T.; Iljina, M.; Karinen, T. Quantitative mineral resource assessment of undiscovered PGE resources in Finland. *Ore Geol. Rev.* **2010**, *38*, 270–287. [CrossRef]

minerals

MDPI

Article

Vanadium Mineralization in the Kola Region, Fennoscandian Shield

Alena A. Kompanchenko *, Anatoly V. Voloshin and Victor V. Balagansky

Geological Institute of the Federal Research Centre Kola Science Centre of the Russian Academy of Sciences, 14 Fersman Street, Apatity 184209, Russia; anatolyvoloshin@yandex.ru (A.V.V.); balagan@geoksc.apatity.ru (V.V.B.)
* Correspondence: komp-alena@yandex.ru; Tel.: +7-921-0488782

Received: 24 September 2018; Accepted: 18 October 2018; Published: 23 October 2018

check for
updates

Abstract: In the northern Fennoscandian Shield, vanadium mineralization occurs in the Paleoproterozoic Pechenga–Imandra-Varzuga (PIV) riftogenic structure. It is localized in sulfide ores hosted by sheared basic and ultrabasic metavolcanics in the Pyrrhotite Ravine and Bragino areas and was formed at the latest stages of the Lapland–Kola orogeny 1.90–1.86 Ga ago. An additional formation of vanadium minerals was derived from contact metamorphism and metasomatism produced by the Devonian Khibiny alkaline massif in the Pyrrhotite Ravine area. Vanadium forms its own rare minerals (karelianite, coulsonite, kyzylkumite, goldmanite, mukhinite, etc.), as well as occurring as an isomorphic admixture in rutile, ilmenite, crichtonite group, micas, chlorites, and other minerals. Vanadium is inferred to have originated from two sources: (1) basic and ultrabasic volcanics initially enriched in vanadium; and (2) metasomatizing fluids that circulated along shear zones. The crystallization of vanadium and vanadium-bearing minerals was accompanied by chromium and scandium mineralization. Vanadium mineralization in Paleoproterozoic formations throughout the world is briefly considered. The simultaneous development of vanadium, chromium and scandium mineralizations is a unique feature of the Kola sulfide ores. In other regions, sulfide ores contain only two of these three mineralizations produced by one ore-forming process.

Keywords: vanadium mineralization; mineralogy; Paleoproterozoic; Kola region; Arctic zone; Fennoscandian Shield

1. Introduction

Vanadium is a fairly widespread element. It is present in all major types of rocks of the Earth's crust, in meteorites, spectra of stars and the Sun. The vanadium content in the Earth's crust is estimated at 1.6×10^{-2} mass % and in oceans 3.0×10^{-7} mass % [1]. Vanadium in the Earth's crust can create compounds or be present in them in the form of V^{3+}, V^{4+} and V^{5+}.

Vanadium and vanadium-bearing minerals are formed in different genetic settings. In minerals of magmatic rocks vanadium is found mainly in the trivalent form as an isomorphic admixture. Typical Fe-Ti-V ores are related to mafic-ultramafic rocks (for example, the Kolvitsa massif in the Kola region [2], the Bushveld layered intrusion in South Africa [3,4], and the Pan-Xi intrusion in China [5,6]). In these ores, vanadium is present predominantly in oxide minerals as an isomorphic admixture. In the Buena Vista Hills Fe-V deposit in Nevada (USA), submicroscopic exsolution lamellae of coulsonite FeV_2O_4 were revealed in magnetite [7]. Rare vanadium minerals were discovered in rocks initially enriched with vanadium and metamorphosed up to amphibolite facies. For example, tanzanite, a blue vanadium-bearing variety of zoisite, is an extremely rare mineral and occurs in the only deposit of tanzanite in the world in Tanzania [8]. Natalyite $NaVSi_2O_6$ [9], magnesiocoulsonite MgV_2O_4 [10], oxyvanite $V^{3+}_2 V^{4+} O_5$ [11] were found among other vanadium minerals in the Sludyanka complex in the southern Baikal area (Russia).

A wide variety of vanadium minerals has been established in metamorphosed massive sulfide ores, described so far only in supracrustal units of several Paleoproterozoic riftogenic structures of the ancient shields of the world. Vanadium mineralization in sulfide ores was reported in the Kola region in Russia [12–14], the Outokumpu and Vihanti deposits in Finland [15,16], the Sätra deposit in Sweden [17], and the Rampura Agucha deposit in India [18]. The Outokumpu and Sätra deposits are the type of localities for karelianite V_2O_3 and vuorelainenite MnV_2O_4, respectively [15,17]. All these massive sulfide ore deposits and occurrences were formed c. 1.9 Ga ago and were metamorphosed up to amphibolite or even granulite facies. This work gives a comprehensive description of the vanadium mineralization in the Paleoproterozoic Pechenga–Imandra-Varzuga (PIV) riftogenic structure in the northern Fennoscandian Shield. It bases on the authors' data on the Pyrrhotite Ravine deposit and the Bragino occurrence of sulfide ores and literature data on a vanadium mineralization in xenoliths of Paleoproterozoic supracrustal rocks in the Devonian Khibiny alkaline massif. This work also contains a brief comparative analysis of vanadium mineralization in the Paleoproterozoic supracrustal units of the ancient shields of the world.

2. Geological Setting

The Polmak–Pasvik–Pechenga–Imandra-Varzuga riftogenic belt (further, Pechenga–Varzuga) is located in a transitional zone between the orogenic core and north-eastern foreland of the Paleoproterozoic Lapland–Kola collisional orogen [19,20] (Figure 1). This belt is a detailed geological record of the Paleoproterozoic in the Fennoscandian shield [21–25]. It includes an almost complete Paleoproterozoic stratigraphy and all major types of Paleoproterozoic plutonic rocks in northern Fennoscandia. Its largest and well studied Pechenga and Imandra-Varzuga structures are located in the Kola region (Figure 1) and are described in detail in [22].

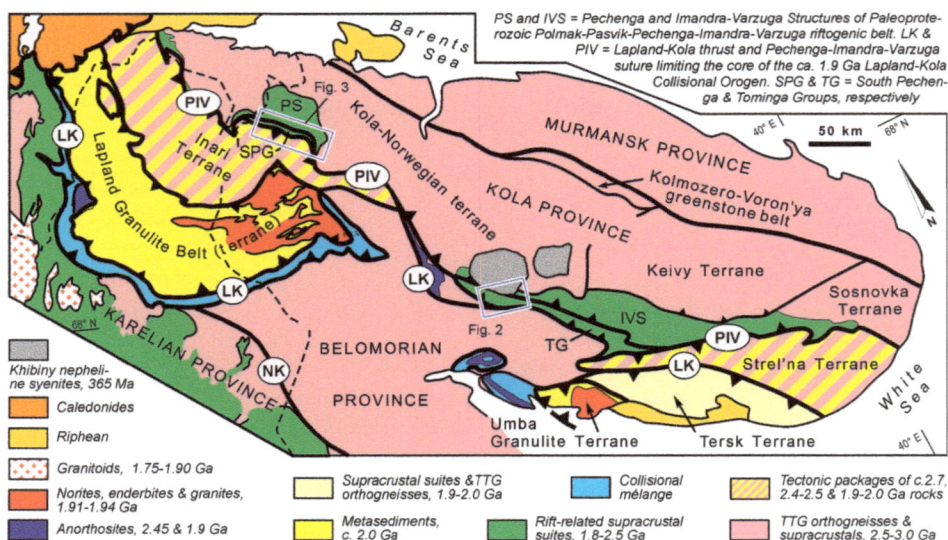

Figure 1. Tectonic map of the northern Fennoscandian Shield (modified from [19,20]).

2.1. Imandra-Varzuga Structure

This structure consists of wide northeastern and narrow southwestern structural zones [22,26] (Figure 1). The former extends along the entire Imandra-Varzuga structure, whereas the latter is

84

exposed only in its central and western parts. The north-eastern zone is appoximately a SW-dipping monocline. In the south-western zone rocks dip steeply to the south-west and north-east and are locally intensely tectonized and folded. Rocks are thrust over Archean basement gneisses in the western portion of the structure and are tectonically overlain by these gneisses in its central portion. Both the zones were metamorphosed under conditions of greenschist to amphibolite facies [27].

The Imandra-Varzuga stratigraphy consists of three groups (from bottom to top): Strelna, Varzuga, and Tominga [22,26]. The Strelna and Varzuga groups are dominated by basic metavolcanics (hereafter "meta" is omitted) and occur only in the north-eastern zone. The Strelna Group was more likely deposited from 2.53 Ga to 2.41 Ga or even to 2.36 Ga [24]. The deposition of the Varzuga Group took place appoximately during the Lomagundi-Jatuli ^{13}C excursion that occurred 2.28–2.05 Ga ago [23].

The Tominga Group is present only in the south-western zone (Figure 2). Apart from basic volcanics, it contains intermediate and felsic volcanic rocks. All these volcanics are alternated with sediments including graphite-bearing "black" schists. Five lithic units are established which may correspond to formations [26] but an accepted stratigraphy is lacking. The depositional environment of the Tominga Group differed considerably from that of the Strelna and Varzuga groups [26]. The Tominga supracrustal section appears to be tectonostratigraphical [23]. A Sm-Nd model age of a rhyodacite is 2.02 Ga [20], which suggests its origination from a middle Paleoproterozoic juvenile source. Sediments of the lower part of the Tominga Group are intruded by 1.94 Ga old granites and subvolcanic trachidacites of the upper part are dated at 1.91 Ga [28]. Felsic volcanics have an imprecise Rr-Sr isochron age of 1.87 ± 0.07 Ga [29].

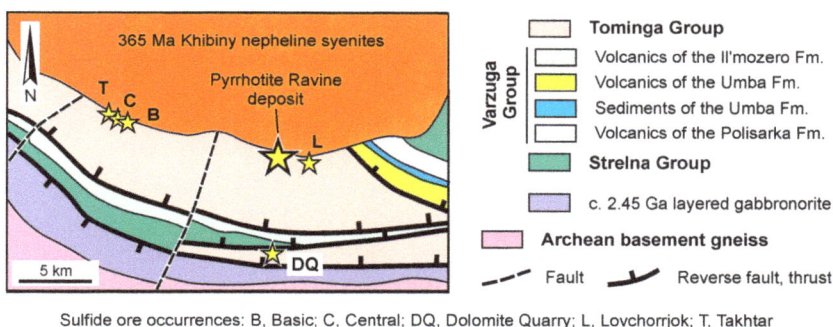

Figure 2. Geological map of the western Imandra-Varzuga structure (modified from [23]).

2.2. Pechenga Structure

The stratigraphy and architecture of the Pechenga structure is principally close to that of the Imandta-Varzuga structure but a counterpart of the Strelna Group which is lowermost is lacking in Pechenga. The Pechenga structure also consists of two zones, the wide Northern zone and the narrow Southern zone [23,30].

In the Northern zone rocks are folded into an open syncline (Figure 1) whose hinge line plunges at intermediate angle southwestwards in surface and gentler at depth. These were metamorphosed under lower greenschist/intermediate amphibolite facies conditions [27]. The stratigraphy of this zone is well studied on the surface and is supported by data on the Kola supedeep borehole [25,31]. It consists of rocks of the Pechenga [32] or North Pechenga Group [23] and is consistent with that of the Varzuga Group [32]. This congruence is supported by isotopic ages and data on stable isotopes of carbon (Lomagundi-Jatuli event [23,33–35]). A badly constrained Rb-Sr isochron age of 2.32 Ga was obtained for basic volcanics of the basal part of the the North Pechenga Group [30,36]. The deposition of basic volcanics and felsic tuffs of the uppermost part of this group and the emplacement of related intrusive rocks happened c. 1.98–1.99 Ga ago [23,33].

In the Southern zone rocks occur steeply or vertically. Reverse faults and thrusts limit this zone and dip deeply southwards and gentler at depth [24,30]. The south-western margin of the Southern zone is folded due to thrusting on it of the Kaskeljavr diorite and Shuoni sodic granite massifs (Figure 3). The South Pechenga rocks were metamorphosed under upper greenschist/intermediate amphibolite facies conditions [27]. These are often cataclized, mylonitized and metasomatically changed [37].

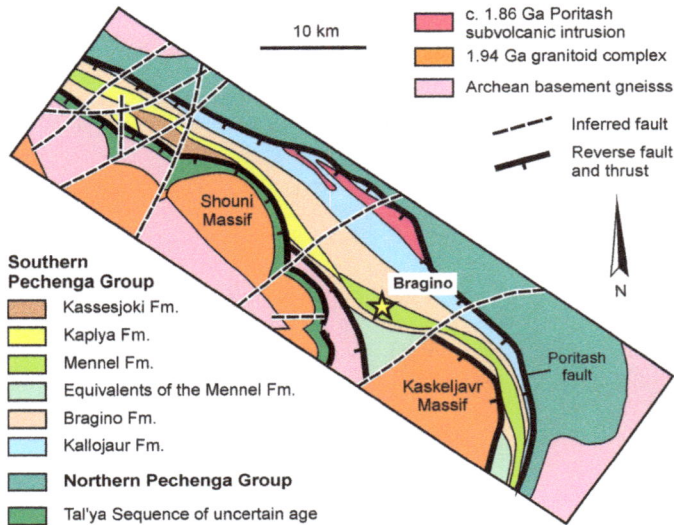

Figure 3. Geological map of the South Pechenga structural zone composed of the Southern Pechenga Group (modified from [23]).

A conventional stratigraphy of the Southern zone is lacking. According to [23,30], the stratigraphical section consists (from bottom to top) of the Kallojaur, Bragino, Mennel, Kaplya, and Kassesjoki formations of the South Pechenga Group [23,32] or the Ansemjoki and Porojärvi Groups [30]. A characteristic feature of the South Pechenga Group are rhythmically layered graphite-bearing "black" schists and a large amount of intermediate and felsic volcanics. There are imprecise Rb-Sr isochron whole-rock ages of 1.87 ± 0.06 Ga and 1.86 ± 0.05 Ga [36] and an imprecise Sm-Nd age of 1.89 ± 0.04 Ga [38]. It is possible that the Kaskeljavr diorites and the Shouni sodic granites dated at 1.94–1.95 Ga [39,40] are related to the South Pechenga intemediate and felsic volcanics (see a review in [23]). If it is the case, the South Pechenga Group is a counterpart of the Tominga Group.

3. Methods

The study of ore minerals in reflected light was conducted in optical microscope Axioplane. Minerals were identified based on their chemical compositions, Raman spectra, X-ray diffraction, and Electron Back Scattered Diffraction (EBSD). Chemical analyses of minerals were carried out by means of a Cameca MS-46 electron probe microanalyzer (Geological Institute, Kola Science Centre of the RAS, Apatity, Russia). Element distribution, morphology and intraphase heterogeneity also was determined using a LEO-1450 scanning electron microscope (SEM) (Carl Zeiss, Oberkochen, Germany) equipped with a Bruker XFlash-5010 Nano GmbH (Bruker, Bremen, Germany) energy-dispersive spectrometer (EDS) and a SEM Hitachi S-3400N (Hitachi, Tokyo, Japan) equipped with a EDS Oxford X-Max 20 (Oxford Instruments, Abington, UK). X-ray powder diffraction patterns were obtained on an URS 55b diffractometer. Raman spectra of minerals were measured by using an Almega XR ThermoScientific spectrometer (Thermo Fisher Scientific, Waltham, MA, USA) equipped with a confocal microscope Olympus BX51 with a $100\times$ objective (532 nm) and a Horiba Jobin Ivon LABRam HR800 (HORIBA Scientific, Kyoto, Japan) spectrometer equipped with a confocal microscope Olympus BX41 with a $50\times$ objective (488, 514 nm). An EBSD analysis was conducted on a SEM Hitachi S-3400N (Hitachi, Tokyo, Japan) equipped with an Oxford HKL Nordlys Nano detector (Oxford Instruments, Abington, UK). The chemical compositions of the mentioned minerals is available from the first author on request.

4. Vanadium Mineralization in the Kola Region

Basic volcanics of the Tominga and South Pechenga groups host massive sulfide ores. The Pyrrhotite Ravine deposit and the Basic, Central, Lovchorrjok and Takhtar occurrences of these ores are situated in the Imandra-Varzuga structure and the Bragino occurrence in the Pechenga structure (Figures 2 and 3). There are four textural types of sulfide ores: massive, banded, interspersed, and breccia-like. All ore types bear a vanadium mineralization [12,14]. The sulfide ores in both study areas were metamorphosed under epidote-amphibolite up to amphibolite facies during the Paleoproterozoic Lapland–Kola collisional orogeny. Their common feature is the high concentration of pyrrhotite in all studied ores (up to 95%). Another common feature is that all ores contain almost no galena.

Vanadium minerals always have an isomorphic admixture of chromium. The Cr content is markedly higher in oxide minerals than in silicates. The Cr content in vanadium minerals of the Bragino ores that are hosted by the ultrabasic volcanics is much higher than in vanadium minerals of Pyrrhotite Ravine ores hosted by basic volcanics.

4.1. Pyrrhotite Ravine Sulfide Ore Deposit

In the Imandra-Varzuga structure the vanadium mineralization was studied in the Pyrrhotite Ravine sulfide ore deposit (Figure 2). This deposit is situated in hornfels after basalts and picritobasalts of the Tominga Group. The hornfels were formed by the contact metamorphism resulting from the Khibiny alkaline massif [41], and this metamorphism affected sulfide ores. These volcanics were originally metamorphosed under biotite-chlorite-actinolite subfacies of greenschist facies [26]. Compared to other basic volcanics of the western Imandra-Varzuga structure they are characterized by elevated concentrations of V and Ti [42]. Abnormally high V contents were established in the Tominga graphite-bearing schists (the so-called black schists) but the origins of this anomaly and V containers have remained unclear [43].

The main ore mineral is pyrrhotite. Sphalerite, chalcopyrite, pyrite and molybdenite are the most common sulfides. The most typical accessory minerals are titanite, rutile, ilmenite, etc. Rare and single findings of argentopentlandite, breithauptite, pentlandite, brannerite, complex oxides of U-Pb composition, etc. are described. A noble metal mineralization is represented by gold, silver and their tellurides [13].

The vanadium mineralization in the Pyrrhotite Ravine sulfide ores are represented by four vanadium minerals (karelianite, coulsonite, goldmanite $Ca_3V_2(SiO_4)_3$ and mukhinite $Ca_2(Al_2V)[Si_2O_7][SiO_4]O(OH))$ and a large number of vanadium-bearing minerals (rutile, ilmenite, crichtonite group, micas, chlorites and etc.) (Table 1). Karelianite and coulsonite are usually found as relicts and highly corroded crystals surrounded by overgrowths of mukhinite, goldmanite and a mineral phase conditionally named "vanadomukhinite" (Figure 4a,b). "Vanadomukhinite" is a potentially new mineral of the epidote group which contains up to 22 mass % of V_2O_3 versus 10–12 mass % of V_2O_3 in mukhinite [13].

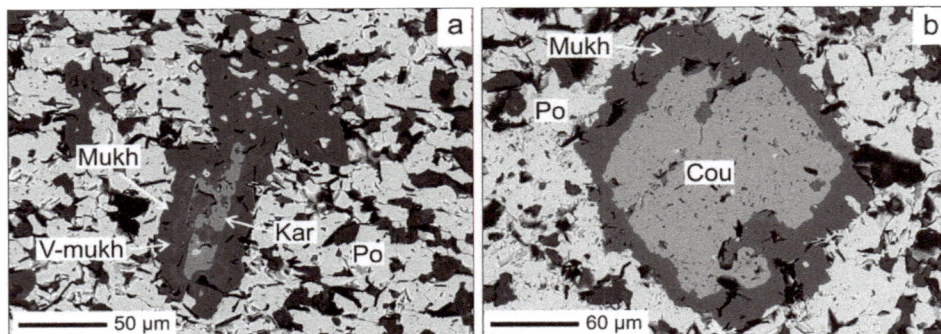

Figure 4. Vanadium mineralization in Pyrrhotite Ravine sulfide ore: karelianite (**a**) and coulsonite (**b**) with rim of vanadium silicates. Back-scattered electron (BSE) images. Cou, coulsonite; Kar, karelianite; Mukh, mukhinite; Po, pyrrhotite; V-mukh, vanadomukhinite. Mineral abbreviations hereafter on [44] apart from introduced by the authors for rare minerals.

Table 1. Major V-Cr-Ti-Sc-bearing minerals in some sulfide ore deposits in the Kola Region and throughout the world.

Mineral/Area	PR	Br	Vih	Otk	Sät	RA	Ctl	DC	Bgj
Oxides									
Berdesinskiite	−	−	+	−	−	−	+	−	−
Burydite	−	+	−	−	−	−	−	−	−
Chromite	−	+	−	+	−	−	+	−	+
Coulsonite	+	+	+	−	−	+	+	−	−
Crichtonite	+	+	−	−	−	−	−	+	−
Davidite-(Ce)	+	+	−	−	−	−	−	−	+
Davidite-(La)	+	+	−	−	−	−	−	−	+
Eskolaite	−	−	−	+	−	+	−	−	−
Ilmenite	+	+	−	−	−	−	−	+	−
Karelianite	+	−	+	+	+	+	−	−	−
Kyzylkumite	−	+	+	−	−	−	−	−	−
Lindsleyite	−	+	−	−	−	−	−	−	−
Loveringite	+	−	−	−	−	−	−	−	+
Nolanite	−	+	+	+	−	−	−	−	−
Rutile	+	+	+	−	−	+	−	+	−
Senaite	−	+	−	−	−	−	−	−	−
Shcherbinaite	−	−	+	−	−	−	−	−	−
Schreyerite	−	+	+	−	−	+	−	−	−
Tivanite	−	+	+	−	−	−	−	−	−
Silicates									
Aegirine	−	−	−	−	−	−	−	+	−
"Braginoite"	−	+	−	−	−	−	−	−	−
Chamosite	+	+	−	−	−	−	−	+	+
Clinochlore	+	+	−	+	−	−	−	−	−
Diopside	+	−	−	+	−	−	−	−	−
Fuchsite	−	−	−	+	−	−	−	−	−
Goldmanite	+	−	−	−	−	−	+	−	−
Jervisite	−	+	−	−	−	−	−	+	−
Mukhinite	+	−	−	−	−	−	−	−	−
Muscovite	+	+	−	−	−	−	−	−	−
Natalyite	−	−	−	−	−	−	−	+	−
Phlogopite	+	+	−	−	−	−	−	−	−

Table 1. *Cont.*

Mineral/Area	PR	Br	Vih	Otk	Sät	RA	Ctl	DC	Bgj
				Silicates					
Roscoelite	−	+	−	−	−	−	−	−	−
Thortveitite	−	+	−	−	−	−	−	+	+
Uvarovite	−	−	−	+	−	−	−	−	−
"Vanadomukhinite"	+	−	−	−	−	−	−	−	−
"Vanadoallanite-(Ce)"	−	+	−	−	−	−	−	−	−
"Vanadoallanite-(Nd)"	−	+	−	−	−	−	−	−	−

Note: Br, Bragino (this study); Bgj, Biggejavri [45]; Ctl, Catalonia [46]; DC, Deadhorse Creek [47]; Otk, Outokumpu [15]; PR, Pyrrhotite Ravine ([13,14] this study); RA, Rampura Agucha [18]; Sät, Sätra [17]; Vih, Vihanty [16]; "+", mineral established; "−", no data; quotes indicate minerals that have no exact analogs among known minerals and their names are conditional.

In the Pyrrhotite Ravine deposit, the vanadium mineralization was formed in two stages [13]. At the first stage, primary oxides (karelianite, coulsonite) were crystallized as a result of vanadium redistribution during a regional metamorphism of V-rich protoliths of basic volcanics. At the second stage, calcium-vanadium silicates were developed in skarns originated from a contact metamorphism of sulfide ores and calcium introduction into them due to an influence of the giant Devonian Khibiny alkaline intrusion.

4.2. Bragino Sulfide Ore Occurrence

The Bragino sulfide ore occurrence is located in picrobasalts of the Mennel Formation occurring among rocks of the Bragino Formation (Figure 3). Apart from basic volcanics, the Bragino Formation contains graphite-bearing tuffogenic sandstones and alevrolites with subordinate sulfide-graphite-bearing schists. The rocks experienced intensive metamorphic and deformational reworking and their greater part was changed in blastomylonites, mylonites and cataclasites in shear zones developed during the Paleoproterozoic Lapland–Kola collision. The Mennel basic volcanics are depleted by K, Sr, Rb, Ta, Zr, Hf, Ti, light rare-earth elements, and heavy rare-earth elements, are enriched by Ba, Th, Nb, and belong to low-alkali and Fe-Mg-enriched rocks of the normal-type mid-ocean ridge basalts [48].

In Bragino, similar to the Pyrrhotite Ravine, the main sulfide is pyrrhotite which can build up to 95% of ores. A massive coarse-grained pyrite ore in which pyrite forms euhedral crystals was found in pyrrhotite ores. Sphalerite, chalcopyrite and marcasite are established among the most common sulfides. Molybdenite, arsenopyrite, cobaltite, gold, and tellurides of Au, Ag, Bi, and Pd are also revealed. Non metallic minerals include quartz, albite, siderite, calcite, apatite, monazite, xenotime, micas, and chlorites [12]. The V_2O_5 content is less than 0.025 mass % in the sulfide ores.

The vanadium mineralization is represented by oxides and silicates (Table 1). All vanadium oxides may be divided into Fe-V and Ti-V groups. Among the Fe-V oxides coulsonite and nolanite $(V^{3+}, Fe^{3+}, Fe^{2+})_{10}O_{14}(OH)_2$ are revealed. Coulsonite was found in all types of the sulfide ores and is the most common vanadium mineral. It is characterized by a changeable composition due to the isomorphic substitution between V^{3+}, Cr^{3+} and Fe^{3+}. Chromite $(FeCr_2O_4)$ relicts were established in some crystals of coulsonite [12]. Coulsonite crystalls have variable size and morphology and the largest octahedral crystals are 300–500 μm across (Figure 5a). Coulsonite occurs in close intergrowths with V-bearing ilmenite (Figure 5b) and V-W-bearing rutile (Figure 5c). Nolanite was found exclusively in pyrite ores in which it is located in microcracks in coulsonite (Figure 5d).

The Ti-V oxides are represented by extremely rare minerals kyzylkumite $Ti_2VO_5(OH)$, tivanite $TiVO_3(OH)$ and byrudite $(Be,\square)(V, Ti)_3O_9$. Kyzylkumite, like nolanite, was discovered exclusively in massive pyrite ores in which it forms individual crystals in pyrite, intergrows with V-W-bearing rutile (Figure 5e) and is associated with Sc-V-bearing crichtonite group minerals (up to 23 mass % of V_2O_3 and 3 mass % of Sc_2O_3). Tivanite and byrudite were established in relics of quartz-albite veins (see below). These minerals usually occur together and form thin exolution lamellae in V-W-rutile. Byrudite sometimes forms individual crystals in association with other vanadium and vanadium-bearing minerals (Figure 5f). Tivanite and anorther mineral, tentatively classified as byrudite, were determined from the presence of TiO_2 and V_2O_3 as the main components in their chemical compositions and from crystal lattices obtained by an X-ray spectral analysis and an EBSD method, respectively. However, the byrudite chemical composition is significantly different from that of byrudite from the Byrud mine in Norway [49] and needs further study.

Intergrowths of vanadium-bearing allanite subgroup minerals (up to 12 mass % of V_2O_3) were found in association with thortvetite $Sc_2Si_2O_7$, V-bearing muscovite and monazite. According to their chemical composition, these minerals may be neodymium and cerium analogs of vanadoallanite-(La) $CaLa(VAlFe^{2+})[Si_2O_7][SiO_4]O(OH)$, and are conditionally named "vanadoallanite-(Nd)" and "vanadoallanite-(Ce)". In addition, a V-Al silicate was discovered which forms both single grains and rims around roscoelite $KV^{3+}_2(Si_3Al)O_{10}(OH)_2$. Its chemical composition has no analog among vanadium-bearing and vanadium minerals, and its Raman spectrum is close to that of the chlorite group. We think that it may be a new vanadium-bearing chlorite phase, conditionally named "braginoite". This and the allanite subgroup minerals also need further study.

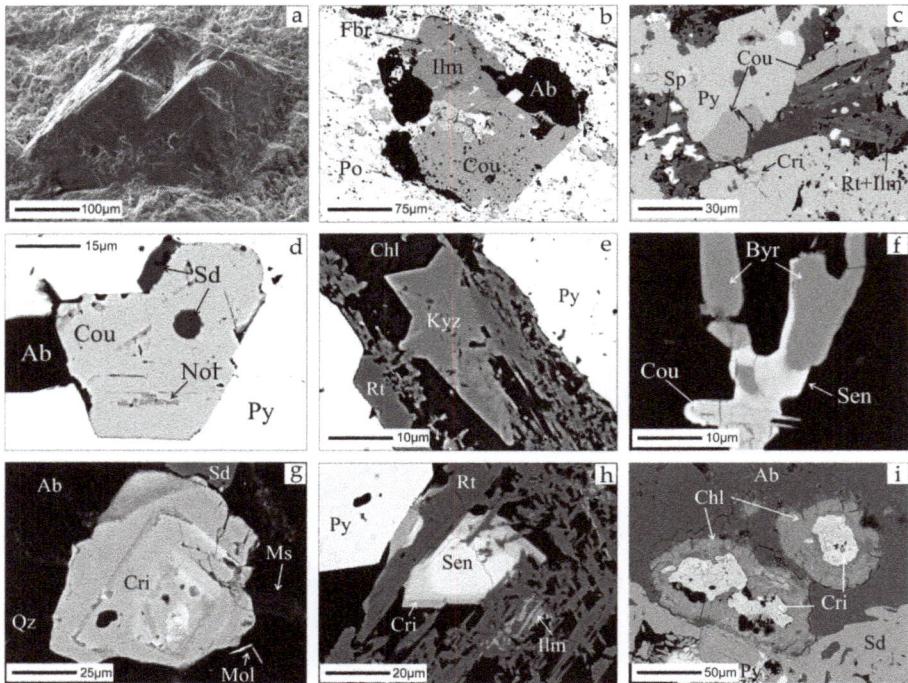

Figure 5. Vanadium and vanadium-bearing minerals in the Bragino massive sulfide ores: (**a–d**) Fe-V oxides; (**e,f**) Ti-V oxides; (**g,h**) Sc-V-bearing crichtonite group minerals (**a–i**), secondary electron and back-scattered electron images, respectively; Ab, albite; Byr, byrudite; Chl, chlorite; Cou, coulsonite; Cri, cricthonite group minerals; Fbr, ferberite; Ilm, ilmenite; Kyz, kyzylkumite; Ms, muscovite Mol, molybdenite; Po, pyrrhotite; Py, pyrite; Qz, quartz; Rt, rutile; Sd, siderite; Sen, senaite; Sp, sphalerite).

As an isomorphic admixture, vanadium is detected in ilmenite and rutile (up to 3 mass % and 11 mass % of V_2O_3, respectively), the mica group (roscoelite, muscovite with 3 mass % of V_2O_3, and phlogopite with 5 mass % of V_2O_3), the chlorite group (up to 12 mass % of V_2O_3 in chamosite and up to 3 mass % of V_2O_3 in clinochlore), and titanite (up to 5 mass % of VO_2).

The Bragino pyrite and pyrrhotite ores contain numerous elongated relics of quartz-albite veins dimensions of which vary from 1×2 to 10×50 mm (Figure 6a–c). The relics have zigzag-like contacts and their tiny fragments are observed in the sulfide ores along these contacts. This suggests their dissolution and replacement during the ore formation. The relics are often situated close to each other, their orientation is irregular but some of them form chains which appear to trace former veins.

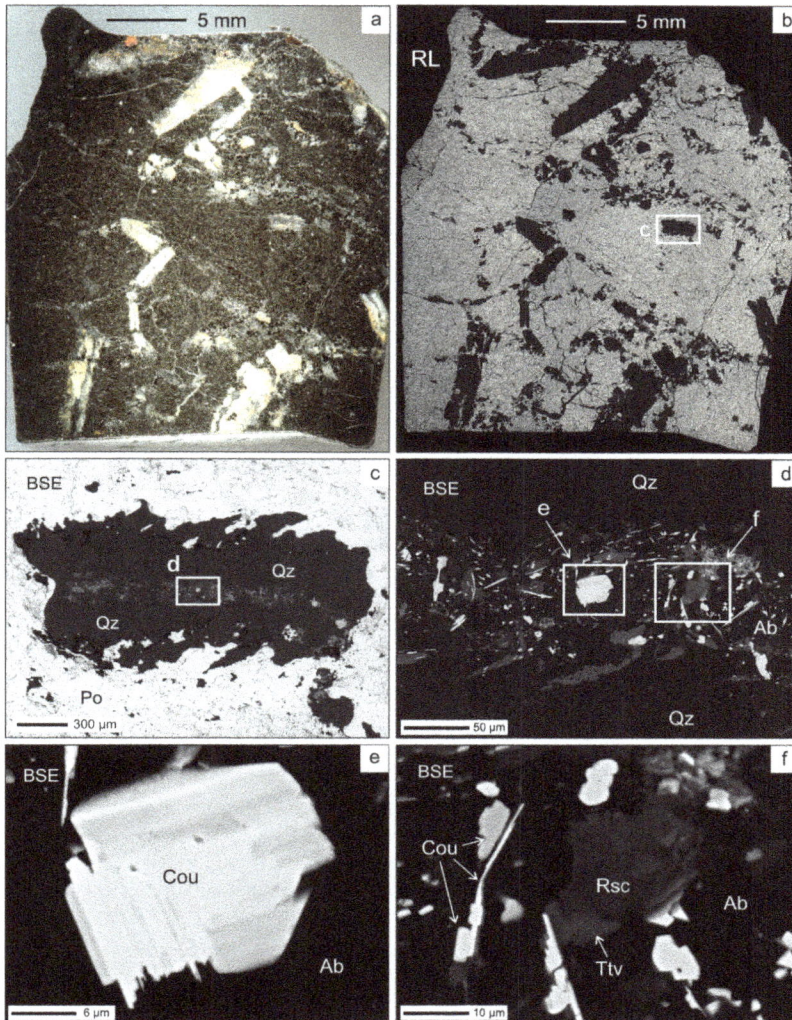

Figure 6. Appearance of a massive sulfide ore containing relics of quartz-albite veins in polished sample 16-K7-4-2 in ordinary (**a**) and reflected (**b**) light. (**c**) Zonal and banded structure of an elongated relic. (**d–f**) Internal structure (**d**) and mineralogy of the axial albite band (**f**). BSE and RL, back-scattered electron and reflected light images, respectively; Ab, albite; Cou, coulsonite; Qz, quartz; Rsc, roscoelite; Ttv, thortveitite.

The relics have a zonal and banded structure. Their axial band is composed of albite and two marginal bands are built up by quartz (Figure 6d). The albite axial band hosts a Cr-Sc-V-bearing mineralization (Figure 6e,f) which is represented by Fe-V oxides (coulsonite and its high-Ti variety, containing up to 18 mass % of TiO_2, and nolanite), Ti-V oxides (tivanite, byrudite), as well roscoelite, V-bearing muscovite, V-bearing rutile and Sc-V-bearing crichtonite group minerals. It is notable that the scandium mineral thortveitite which contains up to 4 mass % V_2O_5 occurs in the albite axial bands (Figure 6f). The Cr-Sc-V-bearing mineralization suggests that scandium and vanadium were brought by hydrothermal fluids.

4.3. Vanadium Mineralization not Related to Sulfide Ores

In addition to massive sulfide ores, vanadium and vanadium-bearing minerals are also found in rocks not related to sulfide ores but somehow connected with supracrustal units of the western part of the Imandra-Varzuga structure. A large xenolith of volcanogenic rocks in the foyaites of the Devonian Khibiny alkaline massif exposed on the Mt. Kaskasnyunchorr contains vuorelainenite, a member of the spinel group, which is associated with a V-bearing crichtonite and karelianite and often forms thin rims around the inclusions of rutile in pyrrhotite [50,51]. Vuorelainenite and karelianite are characterized by the high Cr_2O_3 content (up to 29 and 12 mass %, respectively). A small admixture of vanadium (less than 1 mass % of V_2O_3) is established in ilmenite and hematite. The V content in the crichtonite group minerals reaches 13.80 mass % of V_2O_3, but it usually is less than 1 mass %. Scandium in the minerals of this group is not established [52].

Minerals of the natalyite-aegirine series and vanadates of rare-earth elements are found in alkaline metasomatites, similar in their mineral composition to fenites, on the contact of basic volcanics with dolomites in the Dolomite Quarry which is located in the south from the Pyrrhotite Ravine deposit [53] (Figure 2). As an admixture vanadium is also detected in magnesioriebeckite (up to 1.3 mass % of V_2O_3). The V_2O_5 content in one analyzed sample of metasomatites is 0.41%.

5. Vanadium Mineralization Throughout the World

5.1. Vihanti, Finland

The Lampinsaari ore complex is located in intensely metamorphosed Paleoproterozoic mica gneisses with amphibolite intercalations in the Vihanti–Pyhäsalmi Zn-ore belt in the Svecofennides which adjoin the Archean Karelian Province [54]. The complex consists predominantly of acid metavolcanics, dolomites, skarns, and cordierite gneisses which host ore bodies. The general structure of the Lampinsaari ore complex was formed by an overthrust. Phosphorite bands and a phosphatic metatuff contain uraninite and uranium-bearing apatite, respectively. A "lead model age" from galena is 1.94 Ga, a Pb-Pb isochron age for quartz porphyry is 1.92 Ga, syntectonic plutons is 1.90 and 1.89 Ga (U-Pb, zircon) [54]. In supracrustal rocks of the Vihanti zinc ore deposit the V content usually does not exceed 0.03 mass % of V_2O_5. The highest amount of vanadium was determined in a graphite tuff (black schist) where the V_2O_5 concentration reaches 0.17 mass % [55]. Ore minerals are sphalerite (zinc) and pyrite-pyrrhotite (massive sulfide) [54]. A vanadium mineralization is located in the massive sulfides and is represented by simple oxides (karelianite and shcherbinaite V_2O_5), Fe-V oxides (coulsonite and nolanite), and Ti-V oxides (schreyerite $V_2Ti_3O_9$, kyzylkumite, berdesinskiite V_2TiO_5, and tivanite) (Table 1) [16]. A chromium admixture is characteristic of these minerals. There is no data about vanadium or vanadium-bearing silicates in Vihanti.

5.2. Outokumpu, Finland

Ore bodies of the Outokumpu deposit are located in Paleoproterozoic supracrustal rocks which include black schists, quartzites, dolomites, and chrome-bearing skarns, as well as in serpentinites that are connected with the c. 1.95 Ga Outokumpu ophiolite complex [56,57]. The V_2O_5 content in micaceous schists is 0.03 mass %, in argillaceous black schists 0.11 mass %, in siliceous schists

0.02 mass % and in serpentinites 0.01 mass % [56]. The Outokumpu ore bodies occur in sheared quartzites near their contacts with serpentinites and are broken down by faults into several blocks. The main ore minerals are pyrrhotite (23.2%), pyrite (21%), chalcopyrite (11%), and sphalerite (1.7%) [56]. Cobaltpentlandite, cubanite, mackinawite, magnetite, and stannite are also established [56]. Vanadium minerals are represented by karelianite which was described here in the first time in the world, vourelainenite, and nolanite [15]. The skarns are usually chromium-bearing and contain rare chromium minerals such as eskolaite Cr_2O_3, uvarovite, chrome diopside, chrome tremolite, and some others [56,58].

5.3. Sätra, Sweden

Pyrite-pyrrhotite ores of the Sätra deposit are located in the Doverstorp ore field which is typical of the Ammeberg-Tunaberg metallogenic belt in the southern Paleoproterozoic Bergslagen Province [17,59]. The supracrustal section in the Bergslagen Province consists of mainly 1.90–1.89 Ga felsic volcanics with subordinate sediments, including marble and basic volcanics that were accumulated in an island-arc environment and then were deformed and metamorphosed under conditions of predominantly upper greenschist to upper amphibolite facies [60]. The main mineral is pyrrhotite, in a smaller amount pyrite, chalcopyrite, and sphalerite are present. Accessory minerals are represented by galena, alabandite, gudmundite, freibergite, pyrophanite, rutile, spessartite, and some others. In the Sätra pyrite-pyrrhotite ores, vourelainenite (up to 31 mass % of Cr_2O_3) and schreyerite also were described, with the former having been discovered for the first time in the world [17]. Vanadium as an isomorphic admixture was found in rutile (up to 1.7 mass % of V_2O_3) and manganochromite (up to 33 mass % of V_2O_3) [17].

5.4. Rampura Agucha, India

The Rampura Agucha Zn-Pb-(Ag) stratiform deposit is located in supracrustal rocks of the Bhilwara belt at a contact with an Archean gneiss complex [18,61]. The Bhilwara belt is a thick pile of mainly metasedimentary rocks deposited during an intracratonic rifting of an Archean basement c. 2.0 Ga ago. A Pb model age of the Rampura Agucha deposit is 1.8 ± 0.04 Ga [62]. The deposit is located in sillimanite- and graphite-bearing mica schists, garnet-biotite-sillimanite gneisses, and amphibolites. A V mineralization predominantly occurs in graphite-sillimanite-mica schists which are enriched by vanadium and nickel [63]. The main vanadium minerals are karelianite, coulsonite, and schreyerite. Vanadium-bearing minerals are represented by eskolaite, chromite, and rutile. Like their host rocks, the Rampura Agucha ores were metamorphosed under upper amphibolite-facies conditions [63,64]. The source of vanadium and chromium may be an organic matter in black shales. Vanadium and chromium were released due to metamorphism and formed their oxides [18].

5.5. Biggejavri, Norway

The Biggejavri U-Sc-REE occurrence is located in the Paleoproterozoic Kautokeino volcano-sedimentary belt in Finnmark [45]. The supracrustal succession of this belt is dominated by basic volcanics and includes sandstones, and quartzites deposited in a rift basin. Supracrustals were highly deformed under upper greenschist/amphibolites-facies conditions during the Svecofennian orogeny 1.9–1.7 Ga ago. Specific features of deformation are numerous shear zones, including thrust-related, with the Bidjovagge gold mineralization being controlled by a shear zone [65]. Uranium, scandium and rare-earth-bearing mineralization is situated in albite felsites in the lower units of the belt. The origin of the albite felsites is currently under study. The albite felsites are slightly radioactive, medium- to fine-grained, and consist of more than 90% albite. These rocks also contain calcite, quartz, muscovite, chromite, rutile and Cr-Sc-V-bearing crichtonite group minerals (1–8 mass % of V_2O_3, 7–12 mass % of Cr_2O_3, 0.2–0.6 mass % of Sc_2O_3). Accessories are represented by monazite, thortveitite, brannerite, Cr-rich chlorite and others. The maximum V content in the albite felsites is 1100 ppm (the average 909 ppm) whereas the maximum Sc content is 136 ppm [45].

5.6. Deadhorse Creek, Canada

The Proterozoic Deadhorse Creek volcanoclastic breccia complex is located in Archean metasedimentary and metavolcanic rocks of the Schreiber–White River greenstone belt adjacent to the 1.11 Ga Coldwell alkaline complex [47]. This complex contains a metasomatically-altered breccia, a U-Be-Zr-rich mineralized zone, and a Zr-Y-Th-rich carbonate vein. A U-Pb zircon age of the U-Be mineralization is 1129 ± 6 Ma. The main minerals of the mineralized zone are represented by albite, potassium feldspar, quartz, calcite, and phenakite. Accessories include minerals of the aegirine-jervisite and aegirine-natalyite series, allanite, Ca-Mn silicates, Nb-V-bearing rutile [66], pyrite, thorite, thortveitite, xenotime, Sc-Nb-V-bearing crichtonite group minerals, and others [47]. The formation of this unusual mineralization occurred under the influence of CO_2-bearing and Cr-Nb-V-Ti-enriched alkaline fluids whose source remains enigmatic [47].

6. Discussion

6.1. Structural Position and Age of the Sulfide Ores in Pechenga–Varzuga

Based on a plate-tectonic model suggested in [25], the structural position of the studied sulfide ores has been determined. The model consists of Paleoproterozoic rifting that resulted from lithospheric thinning above a mantle plume head, oceanic separation, subduction, collision and post-orogenic relaxation. All these processes led to the development of the Paleoproterozoic Lapland–Kola orogen, a Himalayan-scale, high-pressure, collisional belt traceable across the Atlantic [67,68]. Rifting was occurring approximately from 2.53 Ga to 2.0 Ga (see reviews in [23,24]). The continental break-up and opening of the Lapland–Kola Red Sea type ocean happened south-west of the Pechenga–Imandra-Varzuga suture (PIV in Figure 1) c. 2.05 Ga [20] or c. 1.99 Ga ago [22]. These dates constrain a change of tectonic setting from a continental rifting to a passive continental margin [33]. The formation of subduction-related supracrustals (the Tominga and South Pechenga groups) and plutonic rocks was occurring from 1.97 Ga to 1.91 Ga [20,28,39,40,68]. The oceanic crust was completely subducted, which explains why basalts that are free of continental admixture are lacking in the upper part of the North Pechenga Group [33]. In the Kola region the subduction occurred 1.93–1.91 Ga ago [20] and in Finnish lapland 1.90–1.88 Ga ago [68]. Within the orogenic core and the footwall of the Lapland–Kola orogen (Belomorian Province), the collision resulted in a regional set of shear zones which strongly affected margins of all tectonic units [19,20]. It is this shearing that resulted in a strong tectonic, metamorphic and metasomatic reworking of the Tominga and South Pechenga groups.

Sulfide ores of the Pyrrhotite Ravine deposit and the Bragino occurrence are situated at a distance of 200 km from each other in the Pechenga and Imandra-Varzuga structures, parts of the Paleoproterozoic Pechenga–Imandra-Varzuga riftogenic belt (Figures 1–3). Both the study areas are located in basalts that were sheared and metasomatically changed under conditions of low-grade metamorphism. The basalts belong to the uppermost parts of the stratigraphical section of these two structures deposited in an island-arc environment 1.90–1.96 Ga ago. The basalts were deformed in shear zones that belong to the aforementioned regional set of collisional shear zones. Shear zones are good conduits for metasomatizing fluids and can control the localization of ore mineralization [69,70]. Two stages of collision were distinguished: frontal and transpressional collisions occurred at the first and second stages, respectively [25]. These stages are separated by the intrusion of a 1916 ± 10 Ma quartz diorite in island-arc-related supracrustal rocks [71] occurring in the Tersk Terrane south of the Imandra-Varzuga structure (Figure 1).

Vanadium and vanadium-bearing minerals are characteristic of gold deposits located in Archean greenstone throughout the world and these deposits are often related to late collisional shear zones, the so-called orogenic gold [69,72–75]. In the Tersk Terrane transpressional shear zones contain sulfide ore occurrences some of which are gold-bearing and were formed during the latest events of the transpressional stage [71,76]. The massive appearance of the sulfide ores suggests that their recrystallization took place during terminational events of the transpressional stage. The formation of the sulfide ores seems to have been ended by 1.86–1.87 Ga when Rb-Sr isotope systems were closed in rocks of the Bragino and Kaplya formations in the South Pechenga structural zone [36] and by 1.88–1.90 Ga, the time of closure of the Ar-Ar isotope system in hornblende in the Tersk terrane [77]. Therefore, the metamorphic recrystallization of the sulfide ores occurred in the Pyrrhotite Ravine and Bragino areas 1.86–1.90 Ga ago. This time interval is consistent with the main period (1.92–1.87 Ga) of formation of a syn- and post-orogenic gold and sulfide (mainly pyrrhotite) mineralization related to shear zones in the Central Lapland Greenstone Belt in northern Finland [78].

6.2. Comparative Analysis of Vanadium Mineralization in Kola and Other Regions

Comparison of the mineral compositions of the Pyrrhotite Ravine and Bragino sulfide ores demonstrates that these are principally similar. The only difference is that the Pyrrhotite Ravine deposit contains calcium silicates (goldmanite, mukhinite, and others, Table 1) which are spatially and genetically related to oxides and are lacking in the Bragino sulfide ores. This difference reflects, however, a contact metamorphism and metasomatism (a calcium input) that were produced by the giant Devonian Khibiny alkaline intrusion (Figure 2) and resulted in the formation of skarns. A similar in some extent example is reported from Silurian Sedex deposits in Spain [46]. In these deposits V- and Cr-rich mineral associations are the result of low-grade regional and subsequent medium- and high-grade contact metamorphism of pelitic rocks. Both of these metamorphic events were, however, isochemical. The protoliths contained abundant organic matter and were originally enriched in V and Cr.

Another principal similarity of the Pyrrhotite Ravine and Bragino sulfide ores is the scandium mineralization which associates with the vanadium minerals and is lacking in sulfide ores in other regions of the world (Table 1). It suggests that the metamorphic recrystallization of the Pyrrhotite Ravine and Bragino sulfide ores and the simultaneous development of the vanadium and scandium mineralization were produced from a mineral-forming process in the same environment. The identity of structural and tectonic positions of the Pyrrhotite Ravine and Bragino sulfide ores favors this suggestion.

A vanadium mineralization similar to that in the Pyrrhotite Ravine and Bragino sulfide ores is described in the Vihanti and Outokumpu deposits in Finland [15,16], the Sätra deposit in Sweden [17] and the Rampura Agucha deposit in India [18]. These deposits are also located in Paleoproterozoic sequences and were metamorphosed at low-grade and medium-grade conditions during Paleoproterozoic orogenies. Sulfide ores of the Lemarchant deposit in Canada [79] and the Waterloo deposit in Australia [80] bear vanadium sulfides sulvanite Cu_3VS_4 and colusite $Cu_{12}VAs_3S_{16}$ which have not been found in the Pyrrhotite Ravine and Bragino sulfide ores. In contrast to the Kola sulfide ores, the Lemarchant sulfide ores preserved an early exhalative/epithermal-type mineralization formed at temperature of 150–250 °C, whereas the Waterloo sulfide ores are located in rocks metamorphosed predominantly under lower greenschist-facies conditions.

A vanadium mineralization that is not linked with the formation of sulfide ores is reported from the Biggejavri deposit in Norway [45] and the Deadhorse Creek deposit in Canada [47]. The former is located in a Paleoproterozoic rift-related supracrustal succession and was metamorphosed at medium-grade metamorphic conditions during the Svecofennian orogeny. The V mineralization in the Deadhorse Creek deposit, however, is situated in an Archean greenstone belt and has a Mesoproterozoic age. This is thought to have resulted from an interaction between earlier mineral assemblage that was developed at the invasion of "granitic" magmatic melts/fluids in a fault and later Nb-Ti-V-Cr-bearing alkaline fluids introduced into the same fault.

So, the vanadium mineralization in sulfide ores in which major minerals are vanadium oxides is linked with metamorphic recrystallization of primary sulfide ores at upper greenschist-facies/upper amphibolite-facies conditions. The vanadium mineralization was accompanied by chromium and scandium mineralizations. The simultaneous development of these three mineralizations is a unique feature of sulfide ores only in the Kola region. Scandium is thought to have been brought by metasomatizing fluids which introduced vanadium into shear zones. The scandium mineralization that is not related to sulfide ores is reported from the Biggejavri deposit in Norway [45] and the Deadhorse Creek deposit in Canada [47].

6.3. Source of Vanadium

Supracrustal rocks which host almost all the listed sulfide ore deposits and occurrences contain carbon-bearing black schists. Sedimentary protoliths of these rocks are accumulated in marine basins, and their formation is associated with underwater volcanic eruptions and biological activity during the Earth's evolution since the Archean [81,82]. Seawater extracts vanadium from basalt and then organic matter absorbs vanadium [83,84]. Thus, the source of vanadium and some other elements in the sediments can be sea water and an organic suspension in it, which precipitates together with sulfides and is captured by them [85]. A source of vanadium in sea water also is detritus of weathered continental rocks. Diagenesis, catagenesis and metamorphism result in the change of the organic form of vanadium into the mineral form, as well as in the removal of water from sediments and possible vanadium transportation by these fluids. Organic matter in schists is considered as a source of vanadium in the Rampura Agucha deposit [18] and the Sedex deposits [46].

A source of chromium which accompanies vanadium in all the listed deposits could be chromite often found in massive sulfide ores which are connected with mafic-ultramafic complexes similar to, for example, complexes in the Urals in Russia [86]. A primary enrichment of sediments in chromium associated with vanadium is described in [18,46]. As to scandium, the Sc_2O_3 concentration is 0.00004 ppm in seawater, whereas the concentration of scandium originated from weathered igneous rocks should have been much higher, c. 3 ppm [87]. This suggests that scandium seems to be almost completely extracted by sediments from sea water.

Metamorphism of primary pyrite ores formed in oceanic basalts led to the mobilization of ore elements located in sulfides in the dispersed state and to their concentration in ore minerals up to the formation of their own mineral phases [86,88,89]. Secondary hydrothermal stages and local metamorphism of sulfide ores resulted in redistribution of base and precious metals, refining of common sulfides, the appearance of submicroscopic and microscopic inclusions of Au–Ag alloys, and segregation of trace elements into new, discrete minerals [88]. It is quite possible that metamorphic recrystallization of sulfides can form vanadium oxides.

The findings of V-Sc-bearing minerals in relics of primary hydrothermal veins in the Bragino sulfide ores suggest that these veins were formed at a hydrothermal event that preceded the metamorphic recrystallization of the sulfide ores. A similar genesis of vanadium minerals is attributed to a metamorphic recrystallization of the Vihanti sulfide ores originally enriched in vanadium [16,46]. We think that the hydrothermal event in Bragino was more likely related to a hydrothermal alteration of basalts that occurred immediately after their eruption and resulted in extraction from them of vanadium and scandium. In basalts of the Mennel Formation, however, the contents of vanadium, scandium and chromium are 160–450 ppm, 19–36 ppm and 180–2300 ppm, respectively [48]. So, this vanadium abundance is just slightly higher than that in the Earth's crust, 138 ppm [90], whereas this scandium concentration practically coincides with its average content in the Earth's crust varying from 14 to 31 ppm [91]. These data suggest that the rocks hosting the Pyrrhotite Ravine and Bragino sulfide ores can hardly have been the only source of vanadium and scandium.

The Pyrrhotite Ravine and Bragino vanadium-bearing sulfide ores host a scandium mineralization which is not mentioned in other vanadium-bearing sulfide ore deposits described above (Table 1). The Pyrrhotite Ravine sulfide ores contain Nb-V-W-bearing rutile (up to 12, 13 and 4 mass percent of V_2O_3, WO_3 and Nb_2O_5, respectively) and vanadium-bearing muscovite, and the Bragino sulfide ores bear roscoelite (Table 1). Taking into account the metamorphic recrystallization of the sulfide ores during terminational events of the Lapland–Kola orogeny 1.86–1.90 Ga ago and a simultaneous metasomatic reworking of their host, we believe that hydrothermal metasomatizing fluids that circulated in shear zones could have been an additional source of vanadium and scandium.

7. Conclusions

1. Two types of vanadium mineralization which is rare in nature have been discovered in the Kola region. The mineralization includes both vanadium and vanadium-bearing minerals. Both mineralization types are located in supracrustal rocks of the Paleoproterozoic Pechenga-Varzuga riftogenic belt.

2. The first vanadium mineralization type is characteristic of sulfide ores and was studied in the Pyrrhotite Ravine deposit and the Bragino occurrence in the Imandra-Varzuga and Pechenga structures, respectively. These sulfide ores are hosted by island-arc-related basic metavolcanics of the Tominga Group (Imandra-Varzuga) and the South Pechenga Group (Pechenga). The rocks were metamorphosed under upper greenschist/lower amphibolite-facies conditions.

3. The first type of vanadium mineralization is localized in shear zones of the transpressional (late collisional) stage of the Paleoproterozoic Lapland–Kola collisional orogeny and was developed 1.90–1.86 Ga ago. This mineralization originated from a metamorphic recrystallization of sulfide ores that were originally enriched in vanadium and are thought to have formed during hydrothermal alteration of marine basalt. An additional input of vanadium is suggested to have been provided by metasomatizing fluids that circulated in the shear zones.

4. A unique feature of the first vanadium mineralization type is that it is accompanied by chromium and scandium mineralization. Scandium is believed to have been brought simultaneously with vanadium by metasomatizing fluids introduced in Paleoproterozoic shear zones. Scandium mineralization is not mentioned in sulfide ores that bear vanadium mineralization in other regions of the world.

5. The second type of vanadium mineralization is reported from the western part of the Imandra-Varzuga structure. It is localized in xenoliths of the Imandra-Varzuga basic metavolcanics in the 365 Ma Khibiny alkaline massif [47,48] and mineralized rocks are similar to fenites alkaline metasomatic rocks at a contact between island-arc-related basic metavolcanics and dolomites of the Tominga Group [49]. The formation of this vanadium mineralization type is related to a contact metamorphism and alkaline metasomatism produced by the giant Khibiny alkaline intrusion.

Author Contributions: Mineralogical investigation of the Bragino occurrence, mineralogical literature review and illustration preparation, A.A.K.; mineralogical investigation of the Pyrrhotite Ravine deposit and other objects in the Imandra-Varzuga structure, A.V.V.; regional geology review, structural position and age of the studied sulfide ores, V.V.B.

Funding: This research was funded by the Geological Institute of the Kola Science Centre of the RAS in the context of scientific themes 0231-2015-0001 and 0231-2015-0004; a study of the Paleoproterozoic metamorphic and deformational history in the Kola region was partly funded by RFBR grant 16-05-01031A.

Acknowledgments: The authors thank A.V. Bazai, Ye.E. Savchenko, E.A. Selivanova, and M.Yu. Glazunova from the Geological Institute of the Federal Research Center Kola Science Center of the Russian Academy of Sciences, and N.S. Vlasenko, V.V. Shilovskikh, and V.N. Bocharov from the Resousce Center "Geomodel" of the Saint Petersburg State University for helping in mineralogical investigations. R.V. Kislitsyn from the Dalhousie University in Halifax is thanked for correcting the manuscript. We are grateful to two anonymous reviewers and the Academic Editor whose notes helped us to make the manuscript stricter and more understandable.

Conflicts of Interest: The authors declare no conflict of interest.

References

1. Emsley, J. *The Elements*, 2nd ed.; Clarendon Press: Wotton-under-Edge, UK, 1991; p. 260, ISBN 0198555687.
2. Voytekhovsky, Y.L.; Neradovsky, Y.N.; Grishin, N.N.; Rakitina, E.Y.; Kasikov, A.G. The Kolvitsa ore deposit (geology, chemical and mineral composition of ores). *Proc. Murm. State Tech. Univ.* **2014**, *17*, 271–278. (In Russian)
3. Cawthorn, R.G. The Bushveld Complex, South Africa. In *Intrusions*; Charlier, B., Namur, O., Latypov, R., Tegner, C., Eds.; Springer Science+Business Media: Dordrecht, The Netherlands, 2015; pp. 517–587.

4. Fischer, L.A.; Yuan, Q. Fe-Ti-V-(P) resources in the upper zone of the Bushveld complex, South Africa. *Pap. Proc. R. Soc. Tasman.* **2016**, *150*, 15–22. [CrossRef]

5. Yao, Y.; Viljoen, M.J.; Viljoen, R.P.; Wilson, A.H.; Zhong, H.; Liu, B.G.; Ying, H.L.; Tu, G.Z.; Luo, N. Geological characteristics of PGE-bearing layered intrusions in Southwest Sichuan province, China. In *Economic Geology Research Unit. Information Circular*; University of the Witwatersrand: Johannesburg, South Africa, 2001; p. 17, ISBN 1-86838-302-4.

6. Zhou, M.F.; Wang, C.Y.; Pang, K.N.; Shellnutt, G.J.; Ma, Y. Origin of giant Fe-Ti-V oxide deposits in layered gabbroic intrusions, Pan-Xi district, Sichuan Province, SW China. In *Mineral Deposit Research: Meeting the Global Challenge, Proceedings of the Eighth Biennial SGA Meeting, Beijing, China, 18–21 August 2005*; Mao, J., Bierlein, F.P., Eds.; Springer: Berlin/Heidelberg, Germeny, 2005; pp. 511–513.

7. Radtke, A.S. Coulsonite, FeV_2O_4, a spinel-type mineral from Lovelock, Nevada. *Am. Mineral.* **1962**, *47*, 1284–1291.

8. Harris, C.; Hlongwane, W.; Gule, N.; Scheepers, R. Origin of tanzanite and associated gemstone mineralization at Merelani, Tanzania. *S. Afr. J. Geol.* **2014**, *117*, 15–30. [CrossRef]

9. Reznitsky, L.Z.; Skliarov, E.V.; Ushchapovskaia, Z.F. Natalyite $Na(V,Cr)Si_2O_6$—A new chromium-vanadium pyroxene from Slyudianka. *Proc. Russ. Mineral. Soc.* **1985**, *114*, 630–635. (In Russian)

10. Reznitsky, L.Z.; Sklyarov, E.V.; Ushchapovskaya, Z.F. Magnesiocoulsonite MgV_2O_4—A new mineral species in the spinel group. *Proc. Russ. Mineral. Soc.* **1995**, *124*, 91–98. (In Russian)

11. Reznitsky, L.Z.; Sklyarov, E.V.; Armbruster, T.; Ushchapovskaya, Z.F.; Galuskin, E.V.; Polekhovsky, Y.S.; Barash, I.G. The new mineral oxyvanite V_3O_5 and the oxyvanite-berdesinskiite V_2TiO_5 isomorphic join in metamorphic rocks of Sludyanka complex (South Baikal region). *Proc. Russ. Mineral. Soc.* **2009**, *138*, 70–81. (In Russian)

12. Kompanchenko, A.A.; Voloshin, A.V.; Bazai, A.V.; Polekhovsky, Yu.S. Evolution of a chromium-vanadium mineralization in massive sulfide ores at the Bragino occurrence in the South Pechenga structural zone (Kola region): An example of the spinel group minerals. *Proc. Russ. Mineral. Soc.* **2017**, *146*, 44–59. (In Russian)

13. Karpov, S.M.; Voloshin, A.V.; Savchenko, Ye.E.; Selivanova, E.A. Vanadium-bearing minerals in ores of the Pyrrhotite Ravine massive sulfide deposit (Khibiny region, Kola peninsula). *Proc. Russ. Mineral. Soc.* **2013**, *142*, 83–99. (In Russian)

14. Karpov, S.M.; Voloshin, A.V.; Kompanchenko, A.A.; Savchenko, Y.E.; Bazai, A.V. Crichtonite group minerals in massive sulfide ores and ore metasomatites of the Proterozoic structures of the Kola region. *Proc. Russ. Mineral. Soc.* **2016**, *145*, 39–56. (In Russian)

15. Long, J.V.P.; Vourelainen, Y.; Kuovo, O. Karelianite, a new vanadium mineral. *Am. Mineral.* **1963**, *48*, 33–41.

16. Sergeeva, N.E.; Eremin, N.I.; Dergachev, A.L. Vanadium mineralization in ore of the Vihanti massive sulfide base-metal deposit, Finland. *Dokl. Earth Sci.* **2011**, *436*, 210–212. [CrossRef]

17. Zakrzewski, M.A.; Burke, E.A.J.; Lustenhouwer, W.J. Vourelainenite, a new spinel, and associated minerals from the Sätra (Doverstorp) pyrite deposit, central Sweden. *Can. Mineral.* **1982**, *20*, 281–290.

18. Höller, W.; Stumpfl, E.F. Cr-V oxides from the Rampura Agucha Pb-Zn-(Ag) deposit, Rajasthan, India. *Can. Mineral.* **1995**, *33*, 745–752.

19. Balagansky, V.V.; Glaznev, V.N.; Osipenko, L.G. Early Proterozoic evolution of the northeastern Baltic Shield: Terrane analysis. *Geotectonics* **1998**, *32*, 81–93.

20. Daly, J.S.; Balagansky, V.V.; Timmerman, M.J.; Whitehouse, M.J. The Lapland-Kola Orogen: Palaeoproterozoic collision and accretion of the northern Fennoscandian lithosphere. In *European Lithosphere Dynamics*; Gee, D.G., Stephenson, R.A., Eds.; Memoir 32; Geological Society: London, UK, 2006; pp. 579–598.

21. Glebovitsky, V.A. (Ed.) *Early Precambrian of the Baltic Shield*; Nauka: Saint-Petersburg, Russia, 2005; p. 711, ISBN 5-02-024950-5. (In Russian)

22. Melezhik, V.A.; Sturt, B.A. General geology and evolutionary history of the early Proterozoic Polmak-Pasvik-Pechenga-Imandra/Varzuga-Ust'Ponoy Greenstone Belt in the north-eastern Baltic Shield. *Earth Sci. Rev.* **1994**, *36*, 205–241. [CrossRef]

23. Melezhik, V.A.; Prave, A.R.; Hanski, E.J.; Fallick, A.E.; Lepland, A.; Kump, L.R.; Srauss, H. (Eds.) The Palaeoproterozoic of Fennoscandia as Context for the Fennoscandian Arctic Russia—Drilling Early Earth Project. In *Reading the Archive of Earth's Oxygenation*; Springer: Heidelberg, Germany, 2013; p. 490.

24. Mints, M.V.; Dokukina, K.A.; Konilov, A.N.; Philippova, I.B.; Zlobin, V.L.; Babayants, P.S.; Belousova, E.A.; Blokh, V.I.; Bogina, M.M.; Bush, D.A.; et al. *East European Craton: Early Precambrian History and 3D Models of Deep Crustal Structure*; Geological Society of America Special Paper 510; Geological Society of America: Boulder, CO, USA, 2015; p. 433.

25. Rundqvist, D.V.; Mitrofanov, F.P. (Eds.) Precambrian Geology of the USSR. In *Developments in Precambrian Geology 9*; Elsevier: Amsterdam, The Netherlands, 1993; p. 528, ISBN 978-0-444-89380-2.

26. Zagorodny, V.G.; Predovsky, A.A.; Basalaev, A.A.; Batieva, I.D.; Borisov, A.E.; Vetrin, V.R.; Voloshina, Z.M.; Dokuchaeva, V.S.; Zhangurov, A.A.; Kozlova, N.E.; et al. *Imandra-Varzuga Zone of the Karelides (Geology, Geochemistry, History of Development)*; Nauka: Leningrad, Russia, 1982; p. 280. (In Russian)

27. Petrov, V.P.; Belyaev, O.A.; Voloshina, Z.M.; Balagansky, V.V.; Glazunkov, A.N.; Pozhilenko, V.I. *Endogenic Regimes of Early Precambrian Metamorphism*; Nauka: Leningrad, Russia, 1990; p. 184, ISBN 5-02-024410-4. (In Russian)

28. Skuf'in, P.K.; Bayanova, T.B.; Mitrofanov, F.P. Isotope age of subvolcanic graditoid rocks of the early Proterozoic Panarechka volcanotectonic structure, Kola Peninsula. *Dokl. Earth Sci.* **2006**, *409*, 774–778. [CrossRef]

29. Mitrofanov, F.P.; Balashov, Y.A.; Balagansky, V.V. New geochronological data on lower Precambrian complexes of the Kola Peninsula. In *Correlation of Lower Precambrian Formations of the Karelia-Kola Region, USSR, and Finland*; Mitrofanov, F.P., Balagansky, V.V., Eds.; Kola Science Centre of the RAS: Apatity, Russia, 1991; pp. 12–16.

30. Smol'kin, V.F.; Mitrofanov, F.P.; Avedisyan, A.A.; Balashov, Y.A.; Balagansky, V.V.; Borisov, A.E.; Borisova, V.V.; Voloshina, Z.M.; Kozlova, N.E.; Kravtsov, N.A.; et al. *Magmatism, Sedimentogenesis and Geodynamics of the Pechenga Paleorift*; Kola Science Centre of the Russian Academy of Sciences: Apatity, Russia, 1995; p. 256. (In Russian)

31. Kozlovsky, E.A. (Ed.) *The Superdeep Well in the Kola Peninsula*; Springer: Berlin, Germany, 1987; p. 490.

32. Zagorodny, V.G.; Radchenko, A.T. *Tectonics of the Karelides of the Northeastern Baltic Shield*; Nauka: Leningrad, Russia, 1988; p. 110, ISBN 5-02-024361-2. (In Russian)

33. Hanski, E.J.; Huhma, H.; Melezhik, V.A. New isotopic and geochemical data from the Palaeoproterozoic Pechenga Greenstone Belt, NW Russia: Implication for basin development and duration of the volcanism. *Precambrian Res.* **2014**, *245*, 51–65. [CrossRef]

34. Martin, A.P.; Condon, D.J.; Prave, A.R.; Melezhik, V.A.; Lepland, A.; Fallick, A.E. Dating the termination of the Palaeoproterozoic Lomagundi-Jatuli carbon isotopic event in the North Transfennoscandian Greenstone Belt. *Precambrian Res.* **2013**, *224*, 160–168. [CrossRef]

35. Melezhik, V.A.; Huhma, H.; Condon, D.J.; Fallick, A.E.; Whitehouse, M.J. Temporal constraints on the Paleoproterozoic Lomagundi-Jatuli carbon isotopic event. *Geology* **2007**, *35*, 655–658. [CrossRef]

36. Balashov, Y.A. Paleoproterozoic geochronology of the Pechenga-Varzuga structure, Kola Peninsula. *Petrology* **1996**, *4*, 3–25.

37. Akhmedov, A.V.; Voronyaeva, L.V.; Pavlov, V.A.; Krupenik, V.A.; Kuznezov, V.A.; Sveshnikova, K.Yu. Gold potential of the South-Pechenga structural zone (Kola Peninsula): Occurrence types and prospects of discovery of economic gold content. *Reg. Geol. Metallog.* **2004**, *20*, 139–151. (In Russian)

38. Skuf'in, P.K.; Elizarov, D.V.; Zhavkov, V.A. Geological and geochemical pecularities of volcanics of the the South Pechenga structural zone. *Proc. Murm. State Tech. Univ.* **2009**, *12*, 416–435. (In Russian)

39. Skuf'in, P.K.; Bayanova, T.B.; Mitrofanov, F.P.; Apanasevich, E.A.; Levkovich, N.V. The absolute age of granitoids from the Shuoniyarvi pluton in the southern framework of the Pechenga structure, the Kola Peninsula. *Dokl. Earth Sci.* **2000**, *370*, 114–117.

40. Vetrin, V.R.; Turkina, O.M.; Rodionov, N.V. U-Pb age and genesis of granitoids in the southern framing of the Pechenga structure, Baltic Shield. *Dokl. Earth Sci.* **2008**, *219*, 806–810. [CrossRef]

41. Gorstka, V.N. *The Contact Zone of the Khibiny Alkaline Massif (Geological and Petrographical Peculiarities, Chemistry and Petrology)*; Nauka: Leningrad, Russia, 1971; p. 99. (In Russian)

42. Fedotov, Z.A. *The Evolution of Proterozoic Volcanism of the Eastern Pechenga-Varzuga Belt (Petrogeochemical Aspects)*; Kola Science Centre of the RAS: Apatity, Russia, 1985; p. 120. (In Russian)

43. Konstantov, S.V.; Sobolev, I.I.; Surovtseva, O.E. *The South Contact of the Khibiny Massif. Aykuayvenchorr–Vudyavrchorr–Takhtarvumchorr Areas. Report on Geological Exploration of Iron Sulphides in 1931–1935*; Scientific Archive of Geological Institute of the Kola Science Centre of the RAS: Apatity, Russia, 1935; p. 408. (In Russian)

44. Whitney, D.L.; Evans, B.W. Abbreviations for names of rock-forming minerals. *Am. Mineral.* **2010**, *95*, 185–187. [CrossRef]

45. Olerud, S. Davidite-loveringite in early Proterozoic albite felsites in Finnmark, north Norway. *Mineral. Mag.* **1988**, *52*, 400–402.

46. Canet, C.; Alfonso, P.; Melgarejo, J.-C. V-rich minerals in contact-metamorphosed Silurian sedex deposit in the Poblet area, Southwestern Catalonia, Spain. *Can. Mineral.* **2003**, *41*, 561–579. [CrossRef]

47. Potter, E.G.; Mitchell, R.H. Mineralogy of the Deadhorse Creek volcaniclastic breccia complex, northwestern Ontario, Canada. *Contrib. Mineral. Petrol.* **2005**, *150*, 212–229. [CrossRef]

48. Skuf'in, P.K.; Theart, H.F.J. Geochemical and tectono-magmatic evolution of the volcano-sedimentary rocks of Pechenga and other greenstone fragments within the Kola Greenstone Belt, Russia. *Precambrian Res.* **2005**, *141*, 1–48. [CrossRef]

49. Raade, G.; Balić-Žunić, T.; Stanley, C.J. Byrudite, $(Be,\Box)(V^{3+},Ti)_3O_6$, a new mineral from the Byrud emerald mine, South Norway. *Mineral. Mag.* **2015**, *79*, 261–268. [CrossRef]

50. Mikhailova, Y.A.; Pakhomovsky, Y.A.; Menshikov, Y.P. Tausonite, baddeleyite and vuorelainenite from hornfels of mt. Kaskasnyunchorr (Khibiny massif). In *Proceedings of the 1st Fersman Scientific Session*; Geological Institute of the Kola Science Center: Apatity, Russia, 2004; pp. 28–29. (In Russian)

51. Mikhailova, Y.A.; Konopleva, N.G.; Yakovenchuk, V.N.; Ivanyuk, G.Y.; Men'shikov, Y.P.; Pakhomovsky, Y.A. Corundum-group minerals in rocks of the Khibiny alkaline pluton, Kola Peninsula. *Geol. Ore Depos.* **2007**, *49*, 590–598. [CrossRef]

52. Korchak, Y.A.; Pakhomovsky, Y.A.; Men'shikov, Y.P.; Ivanyuk, G.Y.; Yakovenchuk, V.N. Minerals of crichtonite group in hornfels of Khibiny massif. In *Proceedings of the 5th Fersman Scientific Session*; Geological Institute of the Kola Science Center: Apatity, Russia, 2008; pp. 264–265. (In Russian)

53. Karpov, S.M.; Voloshin, A.V.; Telezhkin, A.A. The natalyite-aegirine series in alkaline metasomatites of the Imandra-Varzuga belt, Kola region. *Proc. Russ. Mineral. Soc.* **2018**, *147*. in press (In Russian)

54. Rauhamäki, E.; Mäkelä, T.; Isomäki, O.-P. Geology of the Vihanti mine. In *Precambrian ores of Finland. Guide to Excursions 078 A + C, Part 2 (Finland), Proceedings of the 26th International Geological Congress, Paris, France 7 July 1980*; Häkli, T.A., Ed.; Geological Survey of Finland: Espoo, Finland, 1980; pp. 14–24.

55. Rouhunkoski, P. On the geology and geochemistry of the Vihanti zinc ore deposit, Finland. *Bull. Comm. Geol. Finl.* **1968**, *236*, 121.

56. Peltola, E. Origin of Precambrian Copper Sulfides of the Outokumpu Disctrict, Finland. *Econ. Geol.* **1978**, *73*, 461–477. [CrossRef]

57. Peltonen, P. Ophiolites. In *Precambrian Geology of Finland—Key to the Evolution of the Fennoscandian Shield*; Lehtinen, M.; Nurmi, P.A., Rämö, O.T., Eds.; Elsevier: Amsterdam, The Netherlands, 2005; pp. 237–278.

58. Peltola, E. Geology of the Vuonos ore deposit. In *Precambrian ores of Finland. Guide to excursions 078 A + C, Part 2 (Finland), Proceedings of the 26th International Geological Congress, Paris, France, 7 July 1980*; Häkli, T.A., Ed.; Geological Survey of Finland: Espoo, Finland, 1980; pp. 33–41.

59. Wikström, A. Beskrivning till berggrundskartan Katrineholm SV. *Sver. Geol. Unders. Ser. Af.* **1976**, *116*, 88.

60. Allen, R.L.; Lunström, I.; Ripa, M.; Simeonov, A.; Christofferson, H. Facies analysis of a 1.9 Ga, contintental margin, back-arc, felsic caldera province with diverse Zn-Pb-Ag-(Cu-Au) sulfide and Feoxide deposits, Bergslagen region, Sweden. *Econ. Geol.* **1996**, *91*, 979–1008. [CrossRef]

61. Höller, W.; Gandhi, S.M. Origin of tourmaline and oxide minerals from the metamorphosed Rampura Agucha Zn-Pb-(Ag) deposit, Rajasthan, India. *Mineral. Petrol.* **1997**, *60*, 99–110. [CrossRef]

62. Deb, M.; Thorpe, R.L.; Cumming, G.L.; Wagner, P.A. Age, source and stratigraphic implications of Pb isotope data for conformable, sediment-hosted, base metal deposits in the Proterozoic Aravalli-Delhi orogenic belt, northwestern India. *Precambrian Res.* **1989**, *43*, 1–22. [CrossRef]

63. Deb, M. Lithogeochemistry of rocks around Rampura Agucha massive zinc sulfide ore-body, NW India – implications for the evolution of a Proterozoic "Aulakogen". In *Metallogeny Related to Tectonics of the Proterozoic Mobile Belts*; Sarkar, S.C., Ed.; Balkhema: Rotterdam, The Netherlands, 1992; pp. 1–35.

64. Gandhi, S.M.; Paliwal, H.V.; Bhatnagar, S.N. Geology and ore reserve estimates of Rampura Agucha lead zinc deposit, Bhilwara District. *J. Geol. Soc. India* **1984**, *25*, 689–705.
65. Henderson, H.C.; Viola, G.; Nasuti, A. A new tectonic model for the Palaeoproterozoic Kautokeino Greenstone Belt, northern Norway, based on high-resolution airborne magnetic data and field structural analysis and implications for mineral potential. *Nor. J. Geol.* **2015**, *95*, 339–363. [CrossRef]
66. Platt, R.G.; Mitchell, R.H. Transition metal rutiles and titanates from the Deadhorse Creek diatreme complex, northwestern Ontario, Canada. *Mineral. Mag.* **1996**, *60*, 403–413. [CrossRef]
67. Bridgwater, D.; Marker, M.; Mengel, F. The eastern extension of the early Proterozoic Torngat Orogenic Zone across the Atlantic. In *Lithoprobe, Eastern Canadian Shield Onshore–Offshore Transect (ECSOOT)*; Wardle, R.J., Hall, J., Eds.; Report 27; Memorial University of Newfoundland: St. John's, NL, Canada, 1992; pp. 76–91.
68. Tuisku, P.; Huhma, H.; Whitehouse, M. Geochronology and geochemistry of the enderbite series in the Lapland Granulite Belt: Generation, tectonic setting, and correlation of the belt. *Can. J. Earth Sci.* **2012**, *49*, 1297–1315. [CrossRef]
69. Cox, S.F. Deformational controls on the dynamic of fluid flow in mesothermal gold system. In *Fractures, Fluid Flow and Mineralization*; McCaffrey, K.J.W., Lonergan, L., Wilkinson, J.J., Eds.; Geological Society London Special Publication: London, UK, 1999; pp. 123–140.
70. Ashworth, J.R.; Brown, M. An overview of diverse responses to diverse processes at high crustal temperatures. In *High-Temperatuire Metamorphism and Crustal Anatexis*; Ashworth, J.R., Brown, M., Eds.; Kluwer Academic Publishers: Dordrecht, The Netherlands, 1990; pp. 1–18.
71. Balagansky, V.V.; Mudruk, S.V. On the age of a Paleoproterozoic collision in the southeastern Kola region, Baltic shield. In *Geology and Geochronology of Rock-Forming and Ore-Forming Processes in Crystalline Shields*; Mitrofanov, F.P., Bayanova, T.B., Eds.; Kola Science Centre of the RAS: Apatity, Russia, 2013; pp. 13–16. (In Russian)
72. Goldfarb, R.J.; Groves, D.I.; Gardoll, S. Orogenic gold and geological time: A global synthesis. *Ore Geol. Rev.* **2001**, *18*, 1–75. [CrossRef]
73. Scott, K.M.; Radford, N.W. Rutile compositions at the Big Bell Au deposit as a guide for exploration. *Geochem. Explor. Environ. Anal.* **2007**, *7*, 353–361. [CrossRef]
74. Scott, K.M.; Radford, N.W.; Hough, R.M.; Reddy, S.M. Rutile compositions in the Kalgoorlie Goldfields and their implications for exploration. *Aust. J. Earth Sci.* **2011**, *58*, 803–812. [CrossRef]
75. Urban, A.J.; Hoskins, B.F.; Grey, I.E. Characterization of V-Sb-W-bearing rutile from the Hemlo gold deposit, Ontario. *Can. Mineral.* **1992**, *30*, 319–326.
76. Balagansky, V.V.; Belyaev, O.A. Gold-bearing shear zones in the Early Precambrian of the Kola Peninsula: A prognosis and the first results. In *Petrography of the XXI Century. Petrology and Ore-Potential of the CIS Regions and the Baltic Shield*; Mitrofanov, F.P., Fedotov, Z.A., Eds.; Kola Science Centre of the Russian Academy of Sciences: Apatity, Russia, 2005; Volume 3, pp. 37–38. (In Russian)
77. Daly, J.S.; Balagansky, V.V.; Timmerman, M.J.; Whitehouse, M.J.; de Jong, K.; Guise, P.; Bogdanova, S.; Gorbatschev, R.; Bridgwater, D. Ion microprobe U-Pb zircon geochronology and isotopic evidence supporting a trans-crustal suture in the Lapland Kola Orogen, northern Fennoscandian Shield. *Precambrian Res.* **2001**, *105*, 289–314. [CrossRef]
78. Molnár, F.; Middleton, A.; Stein, H.; O'Brien, H.; Lahaye, Y.; Huhma, H.; Pakkanen, L.; Johanson, B. Repeated syn- and post-orogenic gold mineralization events between 1.92 and 1.76 Ga along the Kiistala Shear Zone in the Central Lapland Greenstone Belt, northern Finland. *Ore Geol. Rev.* **2018**, *101*, 936–959. [CrossRef]
79. Gill, S.B.; Piercey, S.J.; Layton-Matthews, D. Mineralogy and metal zoning of the Cambrian Zn-Pb-Cu-Ag-Au Lemarchant volcanogenic massive sulfide (VMS) deposit, Newfoundland. *Can. Mineral.* **2016**, *54*, 1307–1344. [CrossRef]
80. Wagner, T.; Monecke, T. Germanium-bearing colusite from the Waterloo volcanic-rock-hosted massive sulfide deposit, Australia: Crystal chemistry and formation of colusite-group minerals. *Can. Mineral.* **2005**, *43*, 655–669. [CrossRef]
81. Altermann, W.; Kazmierczak, J. Archean microfossils: A reappraisal of early life on Earth. *Res. Microbiol.* **2003**, *154*, 611–617. [CrossRef] [PubMed]
82. Ohkouchi, N.; Kuroda, J.; Taira, A. The origin of the Cretaceous black shales in the surface ocean ecosystem and its triggers. *Proc. Jpn. Acad. Ser. B* **2015**, *91*, 273–291. [CrossRef] [PubMed]
83. Yudovich, Y.E.; Ketris, M.P. *Trace Elements in Black Shales*; Nauka: Ekaterinburg, Russia, 1994; p. 304.

84. Vine, J.D.; Tourtelot, E.B. Geochemistry of black shale deposits—A summary report. *Econ. Geol.* **1970**, *65*, 253–272. [CrossRef]

85. Butler, I.B.; Nesbitt, R.W. Trace element distribution in the chalcopyrite wall of a black smoker chimney: Insights from laser ablation inductively coupled plasma mass spectrometry (LA-ICP-MS). *Earth Planet. Sci. Lett.* **1999**, *167*, 335–345. [CrossRef]

86. Zaykov, V.V.; Maslennikov, V.V.; Zaykova, E.V.; Herrington, R. *Ore-Formation and Ore-Facies Analyses of Massive Sulphide Deposits of the Ural Paleoocean*; Institute of Mineralogy, the Ural Branch of Russian Academy of Sciences: Miass, Russia, 2001; p. 315. (In Russian)

87. Borisenko, L.F. *Scandium: Main Features of its Geochemistry, Mineralogy, and Genetic Types of Ore Deposits*; Academy of Sciences of the USSR: Moscow, Russia, 1961; p. 132. (in Russian)

88. Vikentiev, I.V. Precious metal and telluride mineralogy of large volcanic-hosted massive sulfide deposits in the Ural. *Mineral. Petrol.* **2006**, *87*, 305–326. [CrossRef]

89. Vikent'ev, I.V.; Yudovskaya, M.A.; Moloshag, V.P. Speciation of noble metals and conditions of their concentration in massive sulfide ores of the Urals. *Geol. Ore Depos.* **2006**, *48*, 77–107. [CrossRef]

90. Pouret, O.; Dia, A. Vanadium. In *Encyclopedia of Geochemistry*; White, W.M., Ed.; Springer: Cham, Switzerland, 2016.

91. Rudnick, R.L.; Gao, S. Composition of the continental crust. In *Treatise on Geochemistry*; Turekian, K.K., Holland, H.D., Eds.; Elsevier: Amsterdam, The Netherlands, 2003; pp. 1–64.

minerals

MDPI

Article

Age and Formation Conditions of U Mineralization in the Litsa Area and the Salla-Kuolajarvi Zone (Kola Region, Russia)

Tatiana V. Kaulina *, Arkady A. Kalinin, Vadim L. Il'chenko, Marja A. Gannibal, Anaid A. Avedisyan, Dmitry V. Elizarov, Lyudmila I. Nerovich and Elena A. Nitkina

Geological Institute, Kola Science Centre, Russian Academy of Sciences, Fersman Str. 14, 184209 Apatity, Russia; kalinin@geoksc.apatity (A.A.K.); vadim@geoksc.apatity.ru (V.L.I.); m.a.gannibal@gmail.com (M.A.G.); avedisyan@geoksc.apatity.ru (A.A.A.); elizarov@geoksc.apatity.ru (D.V.E.); nerovich@geoksc.apatity.ru (L.I.N.); nitkina@geoksc.apatity.ru (E.A.N.)
* Correspondence: kaulina@geoksc.apatity.ru; Tel.: +7-921-042-7983

Received: 29 September 2018; Accepted: 29 November 2018; Published: 1 December 2018

check for updates

Abstract: The Kola region (NE of Fennoscandian Shield) has high uranium potential. The most promising structures within the Kola region in respect to uranium enrichment are the Litsa area and the Salla-Kuolajarvi zone. The principal objective of the present study was to define sequence and timing of uranium deposition within these areas. Isotopic (U-Pb and Rb-Sr) exploration of the rocks from Skal'noe and Dikoe U occurrences of the Litsa area and Ozernoe occurrences of the Salla-Kuolajarvi zone was carried out. As it follows from isotopic dating, the principal stages of uranium mineralization had taken place 2.3–2.2, 1.75–1.65, and 0.40–0.38 Ga ago, simultaneously with the stages of alkaline magmatism in the Kola region, which provided the uranium input. Uranium mineralization was related to hydrothermal and metasomatic events under medium to low temperature of ~550 °C at 2.3 Ga to ~280 °C at 0.4 Ga.

Keywords: U mineralization; hydrothermal deposits; U-Pb; Rb-Sr; Kola region

1. Introduction

On the Fennoscandian Shield in Russia, the areas of high uranium potential are the Onezhskaya structure and Ladoga zone in Karelia, and several structures of the Kola region [1,2]. The most promising structures within the Kola region in respect to uranium enrichment are the Litsa area in the northwestern part of the Murmansk region, and the Salla-Kuolajarvi zone at the boundary with Northern Karelia ([3], Figure 1).

The Litsa uranium-ore zone was selected by Savitsky et al. [4] in the northeastern surroundings of the Pechenga rift and bounded by a chain of the Litsa-Araguba granitoid bodies in the east (Figure 1a,b). About 40 uranium ore occurrences have been discovered here, with a total estimate of 102,000 tons at 0.01% average grade in Speculative Resources (IAEA nomenclature) [5]. The Litsa area has long attracted the attention of geologists and is relatively well studied [1,4,5]. The uranium occurrences of the Litsa area are related to hydrothermal type in the long-lived faults in areas of re-activation of Precambrian shields [6], or to metasomatic type in granites [7].

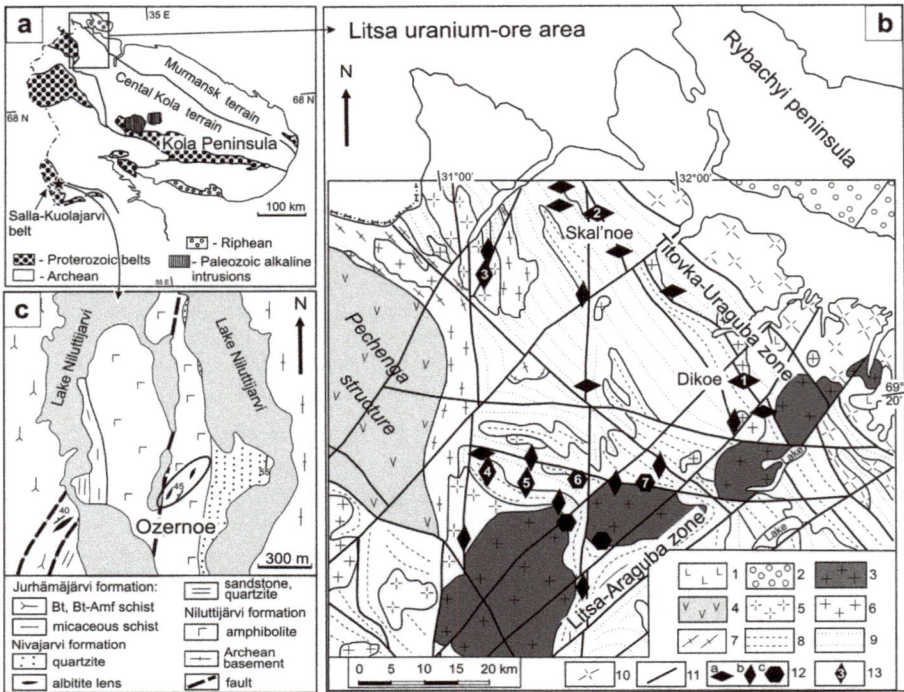

Figure 1. (**a**) Geological overview of the Kola region showing the locations of the Litsa area and the Salla belt; (**b**) Local geological map of the Litsa area (after [4]) with uranium occurrences. 1, 2—Riphean: 1—gabbro-dolerites and dolerites of the Murmansk complex, 2—sandstones, siltstones, and mudstones of the Kil'da Group; 3–5—Lower Proterozoic: 3—granite-granodiorites of the Litsa-Araguba complex, 4—volcanosedimentary rocks of the Pechenga Group, 5—granites of the Kaskel'javr complex; 6–10—Upper Archean: 6—granites of the Voron'ya complex, 7—diorite-plagiogranites of the Porojarvi complex; 8, 9—gneisses of the Kola Group: 8—amphibole-biotite, 9—high-alumina and garnet-biotite; 10—Lower Archean: tonalites and plagiogranites; 11—faults; 12—types of U mineralization (a—REE-Th-U, in pegmatoid granites and quartz-plagioclase metasomatites, b—U, in chlorite-albite metasomatites and albitites, c—U, in albite-hydromica-chlorite metasomatites); 13—the largest U occurrences (1—Dikoe, 2—Skal'noe, 3—Polyarnoe, 4—Namvara, 5—Cheptjavr, 6—Litsevskoe, 7—Beregovoe); (**c**) sketch geological setting of the Ozernoe uranium occurrence area.

Savitsky et al. [4] described four types of U mineralization in the Litsa area (Figure 1b):

(1) The early REE-Th-U mineralization (2750–2650 Ma) in pegmatoid granites and quartz-plagioclase metasomatites (Skal'noe and Dikoe occurrences);

(2) U mineralization (2200–2100 Ma) in chlorite-albite metasomatites and albitites (Polyarnoe, Namvara and Cheptjavr occurrences);

(3) Th-U (1850–1750 Ma) mineralization in quartz-albite-microcline and quartz-microcline metasomatites (Beregovoe area);

(4) U mineralization (400–300 Ma) in chlorite-hydromica-albite metasomatites (Litsevskoe, Beregovoe areas). The last two types of mineralization are manifested to a lower extent at all occurrences of the Litsa area, but developed predominantly at Litsevskoe and Beregovoe areas. The Paleozoic uranium mineralization is the most significant and well-studied one [4,5]).

The Archean age of uranium mineralization in Skal'noe and Dikoe areas of 2700 ± 50 Ma [4] is based on U-Pb age of uraninite from vein granitoids, obtained by Isotope Dilution Thermal Ionization mass-spectrometry (ID TIMS method). The oldest U-Pb age obtained for uraninite from the Dikoe ore occurrence by secondary ion mass-spectrometry (SIMS) SHRIMP II method is 1825 ± 20 Ma [5]. U–Pb (ID TIMS) uraninite age for the Polyarnoe area is 2165 ± 42 Ma [4]. This is supported by an age of 2185 ± 81 Ma, obtained by the SHRIMP II method for uraninite of Namvara occurrence of the same mineralization type [5]. Uraninite from the Cheptjavr occurrence was dated at 1771 ± 34 Ma (SIMS CAMECA IMS-3F [5]), yet uranium mineralization of the Cheptjavr occurrence was believed to be of the same age as the Polyarnoe one [4]. U-Pb (ID TIMS) age of pitchblende is 375 ± 18 Ma for the Polyarnoe area, 370 ± 20 Ma for the Litsevskoe area [4] and 455 ± 6 Ma (SHRIMP II) for the Dikoe area [5]. Uranium-bearing metasomatites of this age are concentrated in the intersections of tectonic zones with transverse faults, where Paleozoic alkaline, alkaline-magnesian metasomatites and lamprophyre dykes also occur [4].

Thus, we have four stages of uranium mineralization distinguished in [4] with three of them confirmed by SIMS dating [5]. Tables of analytical isotope data are published neither in [4] nor in [5], which makes it impossible to estimate the quality of analyses. The first stage of uranium mineralization remains the most doubtful one since the Archean gneisses and plagiogranites of the region with low U content can hardly be a source for the oldest U mineralization.

The principal objective of the present study was to define a sequence of events and timing of the uranium deposition within the Litsa area and Salla-Kuolajarvi zone. Mineralogical and isotopic (U-Pb and Rb-Sr) study of the rocks from Skal'noe and Dikoe ore occurrences of the Litsa area and Ozernoe occurrence in the Salla-Kuolajarvi zone was carried out.

2. Geological Setting

2.1. The Litsa Area

The Litsa area is located in the northwestern part of the Kola Region in the conjunction zone of the Central-Kola and Murmansk terranes. Lying at the intersection of Titovka-Uraguba and Litsa-Araguba tectonic zones bounded by N-S and E-W striking faults, the Litsa area has a complex mosaic-block structure (Figure 1a,b). Structural control is of crucial importance for the spatial distribution of the uranium mineralization in the area. Most of the Litsa ore occurrences are localized in zones of variable permeability, such as deep faults and shear zones with inherent ductile and brittle deformations, which provide favorable conditions for uranium concentration. The largest part of the Litsa area belongs to the Central Kola terrane composed of repeatedly metamorphosed 2.9–2.8 Ga gneisses of the Archean Kola Group, and orthometamorphic 2.8–2.7 Ga tonalite-trondhjemite-granodiorite (TTG) and granite-gneiss rocks [8]. The geodynamic evolution of the Litsa area as part of the Kola terrane began in the Late Archean and included several tectonic-magmatic cycles.

The Skal'noe and Dikoe areas are the largest ore occurrences of the REE-Th-U mineralization type of supposed Archean age (Figure 1a,b). The Skal'noe area lies within the Titovka–Uraguba tectonic zone, while Dikoe is located at the intersection of the Titovka-Uraguba and Litsa–Araguba tectonic zones (Figure 1b). The areas are composed of migmatized biotite, garnet–biotite, two-mica gneisses of the Kola Group strongly deformed in the Dikoe area, and plagiogranites with intercalates of amphibole gneisses, amphibolite bodies, and gabbro-diabase dykes.

Uranium anomalies were detected in fault-related fold zones (1 × 5 and 1.5 × 6 km) in strongly deformed gneisses where veins of pegmatoid granites and quartz-feldspathic metasomatites are developed. The uranium mineralization was traced by drilling to a depth of 400 m [4]. The average uranium content in the Late Archean gneisses and granitoids is mainly below 2.5 ppm. The maximum content of U in vein pegmatoid granites and metasomatites is up to 0.4 wt. % in the Skal'noe area and up to 0.2 wt. % in the Dikoe area [4,5].

2.2. The Salla-Kuolajarvi Zone

The Salla-Kuolajarvi zone constitutes the central part of the Paleoproterozoic Lapland greenstone belt (Figure 1a). Volcanic-sedimentary complexes of the belt are of Sumian age (2.5–2.4 Ga) [9,10]. During the Svecofennian metamorphic event (1.9–1.6 Ga) they were metamorphosed under greenschist to amphibolite facies. Two quartz gold deposits (Mayskoe and Kairaly) [11] and four occurrences of U-Mo ores with gold (Ozernoe, Alim-Kursujarvi, Alakurti and Lagernoe) were found at Russian territory of the belt [3].

Our study was aimed to the Ozernoe occurrence of U-Mo ores [3], located in the eastern flank of the belt (Figure 1c). Geology of the Ozernoe area, as well as mineral composition of the ore, was described in details earlier [3]. Albitite and associated chlorite-albite, carbonate-albite, dolomite, and quartz metasomatites form an echelon structure of lenses of the size up to 10 × 90 m in plagioamphibolites—metamorphosed gabbroids. Carbonate-albite and dolomite metasomatites form small veins or bodies of irregular shape up to 20 cm thick in the central part of albitite lenses. Quartz and dolomite-quartz metasomatites were formed at the final stage of the alteration processes.

Albitite and associated carbonate and chlorite metasomatites are highly radioactive, which is provided by high uranium content reaching up to 0.12 wt. %, registered in carbonate-albite and carbonate metasomatites. Metasomatites of the Ozernoe occurrence, as well as the host plagioamphibolites, displays some extent of sulfide and oxide mineralization. Altered plagioamphibolites contain pyrite, chalcopyrite, ilmenite, and magnetite. In albitite, hematite and rutile substitute magnetite and ilmenite, respectively. Both albitite and carbonate metasomatites contain molybdenite, marcasite, uraninite and brannerite.

3. Materials and Methods

3.1. Rock Samples

Rocks for the present study were sampled in two most representative early U ore occurrences of the Litsa area: Dikoe and Skal'noe, which belong to REE-Th-U type of mineralization (Figure 1a,b) and from albitites of the Ozernoe occurrence (Figure 1c).

3.1.1. Dikoe Occurrence

Veins of pegmatoid granitoids were sampled on two outcrops of the Dikoe area. Pegmatoid granitoids are massive coarse-grained rocks, which form 20–70 cm thick veins in biotite gneisses. Veins are highly radioactive—up to 800 µR/h. Samples TK-19 and TK-20 were taken within one outcrop (5 m from each other) and sample TK-22—in another outcrop—about 1 km to the east from the first one. Samples TK-19 and TK-20 show magmatic hypidiomorphic textures and are composed of plagioclase (50–55%), quartz (5–10%), microcline (20–25%), biotite (5–15%), muscovite (5–10%), apatite, monazite, zircon, and sulfides. Biotite is partly substituted by chlorite. Uranium minerals are mainly observed in biotite.

Sample TK-22 contains less plagioclase (15%), more quartz (40%), microcline (30%), biotite (5%), muscovite (20%), titanite, monazite, and sulfide minerals. The sample has allotriomorphic texture with areas of fine-grained rock. The irregular shape of magmatic feldspar grains is due to resorbed boundaries with quartz. Some plagioclase grains are altered by sericite.

3.1.2. Skal'noe Occurrence

Samples were collected from plagiogranites with varying degrees of schistocity. Sample KT-1—massive plagiogranite. The rock consists of plagioclase (60%), quartz (35%), and biotite (3–5%); and microcline, muscovite, and chlorite (<1%). Accessory phases are represented by apatite, zircon, and iron-oxides. Some biotite crystals are showing alteration to chlorite. This is a massive fine- to medium-grained rock with hypidiomorphic texture and with areas of granoblastic, which occupy at least 10% of the total rock. The secondary minerals are represented by muscovite and microcline. Microcline is observed in areas of granulation, rarely replacing plagioclase.

Sample KT-2—gneissic plagiogranite. Mineral composition: plagioclase (58%), quartz (30%), biotite (7%); and microcline (5%), muscovite, apatite, zircon and ore mineral. The rock is fine- to medium-grained with relicts of hypidiomorphic texture with areas of granoblastic, which occupy about 30% of the total rock. Magmatic plagioclase is altered, epidote is developed after plagioclase in areas of granulation.

3.1.3. Ozernoe Occurrence

KP-19—albitite (metasomatite). It is a fine-grained pink rock, massive in pure albitite, with schistose areas of carbonate and chlorite alteration. The rock has a granoblastic texture and consists of albite (80%), dolomite, and calcite, with minor chlorite and biotite. Accessory phases are hematite, rutile, molybdenite, and uranium minerals (uraninite and brannerite).

3.2. Analytical Methods

Mineralogical and isotope analyses were performed in the Geological Institute, Kola Science Center, Russian Academy of Sciences. Minerals for the analysis were prepared using magnetic separation in a Frantz separator and density separation in heavy liquids with subsequent handpicking of minerals under a binocular microscope. Monazite and uraninite grains were embedded in epoxy mounts and polished. The back-scattered electron (BSE) images were studied on a LEO-1450 Scanning Electron Microscope equipped with an XFlash-5010 Bruker energy dispersive spectrometer (GI KSC RAS, Apatity, Russia) with QUANTAX 200 software.

U-Pb (ID-TIMS) analysis was used to date monazite, uraninite, brannerite, and rutile. Digestion of minerals and U and Pb extraction followed method of Krogh [12]. Lead and uranium concentration were determined using the mixed ^{208}Pb + ^{235}U tracer. The measurements were performed on MI-1201T and Finnigan MAT 262 mass spectrometers. Pb and U were loaded together on outgassed single Re filaments with H_3PO_4 and silica gel. The temperatures of measurement were 1300 °C for Pb and 1500 °C for U. The laboratory blanks were 0.3 ng for Pb and 0.01 ng for U. The Pb isotope ratios were corrected for mass fractionation with a factor of 0.18% per amu (for MI-1201T) and 0.11% per amu (for Finnigan MAT-262) based on repeat analyses of the standard NBS SRM 982 [13]. The U analyses were corrected for mass fractionation with a factor of 0.08% and 0.003% per amu for MI-1201T and Finningan MAT-262, respectively, based on repeated analyses of the NBS U 500 standard. Reproducibility of the U-Pb ratios was estimated as 0.7% from the analysis of standard zircon IGFM-87 (Ukraine) at 95% confidence level.

Rb-Sr dating of whole rocks and rock-forming minerals (biotite, muscovite, K feldspar, plagioclase, apatite) was carried out via MI–1201T mass-spectrometer using Ta filaments. The Rb and Sr contents were determined by isotopic dilution method. The Sr isotope ratios were normalized to the NISTSRM–987 value of 0.71034 ± 0.00026. The uncertainties of the ^{87}Sr/^{86}Sr and ^{87}Rb/^{86}Sr ratios do not exceed ±0.04% and ±0.5% (2σ), respectively. The laboratory blanks were 2.5 ng for Rb and 1.2 ng for Sr. All calculations were made with the program Isoplot/Ex 3.70 [14].

4. Results

4.1. The Litsa Area

The Skal'noe and Dikoe areas contain two major types of mineralization: monazite-zircon-uraninite and molybdenite-thorite-uraninite [4]. Mineral composition of samples from pegmatoid veins show that the first association is represented by monazite (35%), apatite (30%), zircon (25%), and uraninite (10%). Uraninite occurs as intergrowths with apatite and zircon (Figure 2a). Uraninite seems to be the later phase since it overgrows intergrowths of monazite and apatite. Monazite and apatite definitely grew together from viscous melt because they do not have crystallographic faces and form clusters of monazite with apatite domains and apatite inclusions. Then these clusters were overgrown with uraninite grain. Zircon form independent well faceted crystals. Galgenbergite $Ca(Ce,La)_2(CO_3)_4 \cdot H_2O$ and anglesite rims develop between uraninite and plagioclase (Figure 2b).

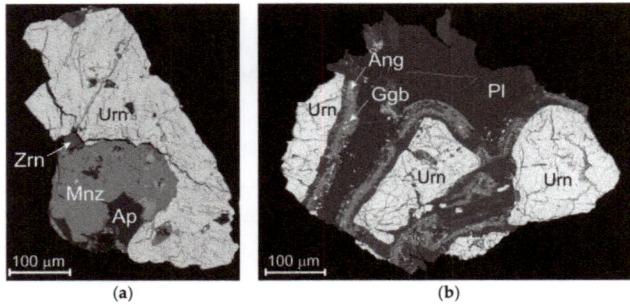

(a)   (b)

Figure 2. Uranium minerals in pegmatoid granitoid veins in the Dikoe U-occurrence of the Litsa area: (a) uraninite (Urn) intergrowth with monazite (Mnz), apatite (Ap), and zircon (Zrn); (b) galgenbergite (Ggb) and anglesite (Ang) rims between plagioclase (Pl) and uraninite.

Rare small inclusions of thorite (about 10 μm) were found within monazite grains. Molybdenite forms thin rims on uraninite grains and together with galenite fills fractures in uraninite.

U-Pb analysis of monazite from pegmatoid veins (samples TK-20 and TK-22) of the Dikoe uranium occurrence have shown that three monazite fractions yielded near concordant upper intercept age of 2549 ± 7 Ma (MSWD = 0.13) with a lower intercept at 204 ± 55 Ma (Table 1, Figure 3). U-Pb dating of uraninite was carried out for two pegmatoid veins of the same outcrop (samples TK-19 and TK-20). The uraninite grains were extracted from the epoxy mount after electron microscope analysis. A regression line through five points defines an upper intercept age of 2276 ± 21 Ma with a lower intercept at 402 ± 88 Ma (Table 1, Figure 3). High mean square weighted deviation (MSWD) of 12 shows chemical variation of uraninite grains.

Table 1. U-Pb data for uranium minerals from vein pegmatoid granitoids of the Dikoe ore-occurrence (Litsa area) and albitites of the Ozernoe occurrence (Salla-Kuolajarvi zone).

Fraction No	Fraction Weight, mg	Lead Isotope Composition				Isotope Ratios				Rho [1]	Age, Ma
		$\frac{^{206}Pb}{^{204}Pb}$	$\frac{^{206}Pb}{^{207}Pb}$	$\frac{^{206}Pb}{^{208}Pb}$	$\frac{^{207}Pb}{^{235}U}$	±2σ%	$\frac{^{206}Pb}{^{238}U}$	±2σ%			$\frac{^{207}Pb}{^{206}Pb}$
Uraninite (samples TK-19 and TK-20)											
TK-20-1	0.2	53,119	8.3205	131.7	6.782	0.7	0.3505	0.7	0.97	2231 + 1	
TK-20-2	0.7	81,150	7.5912	116.1	5.201	0.7	0.2779	0.7	0.79	2173 + 5	
TK-20-3	0.2	78,525	7.5842	131.4	4.058	1.0	0.2269	0.8	0.95	2094 + 3	
TK-19-1	0.1	30,018	7.2421	78.5	6.208	1.6	0.3271	1.6	0.99	2198 + 2	
TK-19-2	0.2	16,000	7.3829	19.4	5.402	0.7	0.2910	0.7	0.60	2159 + 8	
Monazite (samples TK-20 and TK-22)											
TK-20	2.0	11,860	5.9406	0.4441	9.943	3.6	0.4311	3.6	0.99	2531 + 3	
TK-22-1	1.2	19,200	5.8612	0.4491	11.592	5.0	0.4947	5.0	0.99	2556 + 2	
TK-22-2	0.9	10,450	5.8733	0.4460	11.197	4.3	0.4803	4.3	0.99	2548 + 2	
Rutile (sample KP-19)											
Rt 1	2.2	2049	11.1467	41.4056	1.3217	1.2	0.1153	0.5	0.49	1273 ± 20	
Rt 2	1.5	1042	8.4938	22.9735	3.8689	0.5	0.2673	0.5	0.65	1714 ± 7	
Rt 3	2.5	1220	8.9060	23.8900	4.3402	0.5	0.2951	0.5	0.77	1743 ± 8	
Brannerite (sample KP-19)											
Br 1	0.2	11,350	16.1249	149.95	0.5732	0.5	0.0683	0.5	0.95	634 ± 3	
Br 2	0.6	17,300	16.0905	159.12	0.6501	0.5	0.0729	0.5	0.99	764 ± 2	
Br 3	0.1	11,100	18.2989	120.33	0.4473	0.5	0.0607	0.5	0.85	347 ± 4	

[1] Rho, error correlation coefficient of Pb/U ratios.

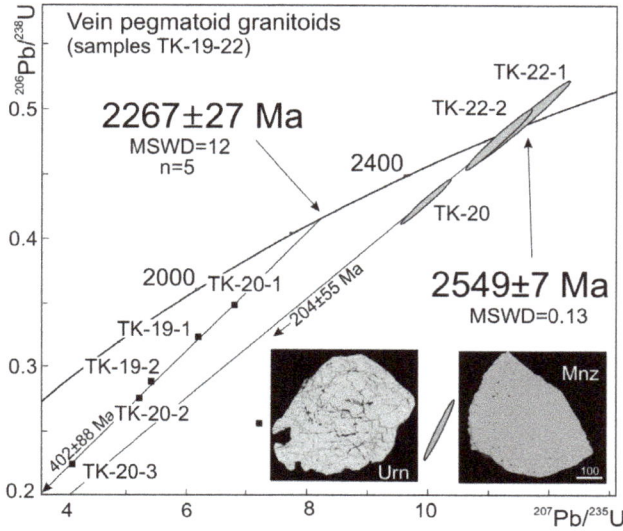

Figure 3. U-Pb isotope data for monazite and uraninite from vein pegmatoid granitoids of the Dikoe occurrence of the Litsa area.

Rb-Sr dating of minerals from the same vein (sample TK-20) defined an age of 1964 ± 21 Ma for primary biotite and whole rock and of 1701 ± 15 Ma for secondary muscovite, plagioclase and microcline (Table 2, Figure 4a). The first age is probably a cooling age at temperature of about 350 °C—closure temperature for an Rb-Sr biotite system [15]. An age of 1701 Ma corresponds to the development of secondary minerals at hydrothermal-metasomatic impact.

Mineral Rb-Sr isochrons (Ap-Pl-WR-Bt) for plagiogranites from Skal'noe area (sample KT-1 and KT-2) defined two ages: 1902 ± 21 Ma and 2370 ± 25 Ma (Table 2, Figure 4b). Older age of 2370 ± 25 Ma most likely has resulted from dating of coarser-grained biotite fraction where retention of Sr is better due to larger size of crystals.

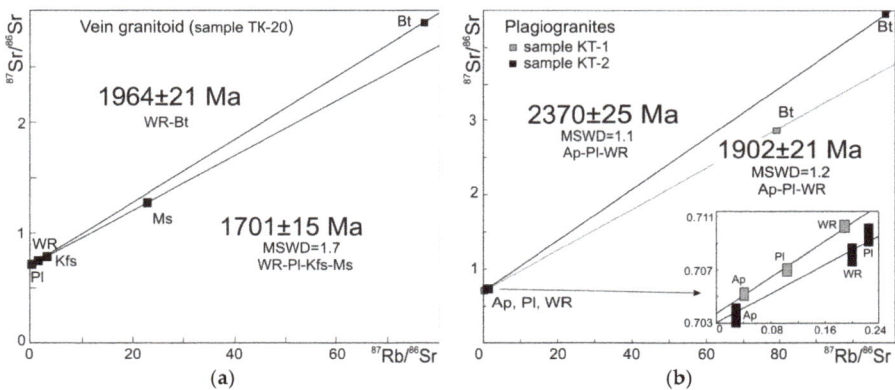

Figure 4. Rb-Sr isotope data for rocks and minerals of the Litsa area: (**a**) vein granitoid of the Dikoe ore occurrence; (**b**) plagiogranites of the Skal'noe occurrence. WR—whole rock, Pl—plagioclase, Ap—apatite, Bi—biotite, Kfs—K feldspar.

Table 2. Rb-Sr data for rocks and minerals of uranium occurrences in the Litsa area and the Salla-Kuolajarvi belt.

Mineral	Concentration, ppm		Rb⁸⁷/Sr⁸⁶	Sr⁸⁷/Sr⁸⁶
	Rb	Sr	Rb^{87}/Sr^{86}	Sr^{87}/Sr^{86}
Vein granitoid (sample TK-20, Dikoe area)				
WR	147.2	284.6	1.45918	0.75449
Bt (biotite)	489.2	17.9	77.2309	2.89697
Kfs (K-feldspar)	322.6	294.7	3.08889	0.76814
Ms (muscovite)	263.4	32.4	22.9352	1.27698
Pl (plagioclase)	32.3	516.5	0.17640	0.72165
Plagiogranite (sample KT-1, Skal'noe area)				
WR	38.9	549.4	0.19992	0.70817
Ap (apatite)	4.3	431.6	0.02837	0.70360
Pl (plagioclase)	58.4	736.1	0.22396	0.70969
Bt (biotite)	572.5	20.5	78.7900	2.86556
Plagiogranite (sample KT-2, Skal'noe area)				
WR	38.3	572.1	0.18898	0.71027
Pl (plagioclase)	30.6	832.6	0.10376	0.70696
Ap (apatite)	6.1	421.7	0.04081	0.70519
Bt (biotite)	519.9	13.4	109.222	4.44183
Albitite (sample KP-19, Ozernoe area)				
WR (whole rock)	9.49	66.5	0.402633	0.72601
Ab (albite)	2.26	64.5	0.098858	0.71819
Ap (apatite)	1.33	390	0.009617	0.71585
Bt (biotite)	234	4.36	151.4238	4.52693
Dol (dolomite)	7.55	40.5	0.525965	0.71778

4.2. The Salla-Kuolajarvi Zone

Uraninite is associated with marcasite, molybdenite, melonite, and rarely altaite, which fill fractures and inclusions or form micrograins at the boundary of uraninite with silicates and carbonates (Figure 5a). Xenomorphic grains of brannerite were found in a form of a chain at the boundary of carbonate metasomatite and albitite (Figure 5b). Brannerite grains are X-ray amorphous, the mineral was identified by X-ray diffraction method after annealing and thereby regaining crystalline structure.

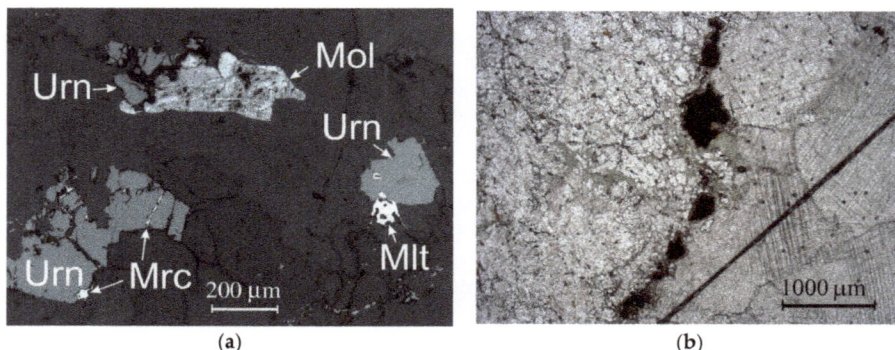

(a)   (b)

Figure 5. Uranium minerals in metasomatites of the Ozernoe U-occurrence: (**a**) uraninite (Urn) with molybdenite (Mol), marcasite (Mrc), melonite (Mlt); (**b**) a chain of brannerite grains at the boundary of albitite (on the left) and dolomite metasomatite.

The three rutile fractions from albitite (sample KP-19) yielded U-Pb upper intercept age of 1757 ± 7 Ma with the lower intercept at 416 ± 14 Ma (Figure 6a; Table 2). Paragenesis of hematite and rutile implies high-oxidation conditions [16], which are known to be unfavorable for uranium deposition, U-minerals therefore, were formed later, probably, at the same time as carbonate metasomatites.

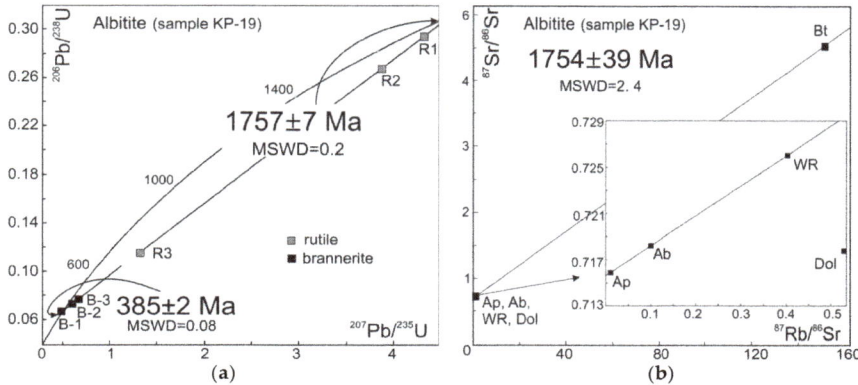

Figure 6. Isotope data for albitite of the Ozernoe area: (**a**) U-Pb Concordia diagrams for rutile and brannerite; (**b**) Rb-Sr isochron for whole rock (WR) and minerals: apatite (Ap), albite (Ab), dolomite (Dol).

The uraninite age obtained by CHIME method [17] is 1627 ± 42 Ma. This value corresponds to the formation time of carbonate-quartz and carbonate metasomatites, which controls distribution of uraninite mineralization. This age coincides within the error limits with Rb-Sr age of 1610 ± 30 Ma, obtained for the Mayskoe gold deposit—located 8 km to the east from the Ozernoe, where it was interpreted as the age of metasomatic alteration of rocks [11].

The brannerite (UTi_2O_6) age of 385 ± 2 Ma was obtained for three brannerite fractions as a lower intercept age (Figure 6a; Table 2). The upper intercept of the discordia line is in a good agreement with the rutile age.

Rb-Sr isochron data for biotite, apatite, albite, and whole rock defines albitite age at 1754 ± 39 (Figure 6b; Table 2) and coincide with rutile age. The dolomite point lies outside of the isochron and confirms the field evidence that carbonate mineralization occurred later. A similar age (1728 ± 39 Ma) was obtained using the Sm-Nd method for quartz-albite-carbonate-amphibole metasomatite from Alim-Kursujarvi occurrence, located 12 km north from the Ozernoe area.

5. Discussion

According to the obtained isotope data we have the following sequence of events. The monazite age of 2549 ± 7 Ma, obtained from two pegmatoid veins of the Dikoe uranium occurrence, most likely, corresponds to the time of pegmatoid veins crystallization. The rock-samples have a magmatic hypidiomorphic texture, where monazite belongs to magmatic mineral assemblage typical for granites. According to [4] the first mineralization in the Litsa area was described as REE-Th-U in pegmatoid granites and metasomatites. The obtained monazite age shows that REE-Th mineralization (since monazite is the main concentrator of REE and Th) occurred at 2.55 Ga as a result of magmatic crystallization of pegmatoid veins. U-Pb age of uraninite from the same pegmatoid veins determines the first stage of uranium mineralization at 2267 ± 27 Ma, associated with hydrothermal processes. Quite possibly, this uranium mineralization belongs to the same stage as U mineralization in chlorite-albite metasomatites and albitites of the Polyarnoe, Namvara and Cheptjavr occurrences, dated with high uncertainties at 2165 ± 42 Ma [4] and 2185 ± 81 Ma [5].

The Rb-Sr data obtained for pegmatoid vein of the Dikoe area and plagiogranites of the Skal'noe area have shown 1964 ± 21 and 1902 ± 21 Ma ages for primary magmatic minerals (biotite, apatite and plagioclase). These data probably represent cooling ages, since the crystallization age of pegmatoid veins and plagiogranites are 2.55 (monazite age) and 2.8 Ga [8], correspondingly. The closure temperature for an Rb-Sr biotite system is of about 350 °C [15] and any high-temperature event can disturb it. The thermal event of this time can be 1.94 Ga intrusion of the Kaskel'javr granitoid complex (Figure 1b) [18]. The older age of 2370 ± 25 Ma obtained for apatite, plagioclase, and coarser-grained biotite fraction probably reflects the first hydrothermal event of rocks at 2.3 Ga, which is masked by subsequent events in other samples. The temperature of formation of uranium metasomatites in vein pegmatoid granitoids is estimated at 500–550 °C using mineral geochemistry [19]. An Rb-Sr system of biotite, which essentially determines isochrons, shows that since 1964–1902 Ma, the temperature did not exceed 350 °C.

An age of 1701 ± 15 Ma (Rb-Sr) for secondary muscovite, plagioclase and microcline from pegmatoid vein corresponds to a hydrothermal-metasomatic event. This stage was also manifested in the Salla-Kuolajarvi zone where it resulted in albitite formation (1757 ± 7 Ma U-Pb rutile age). The closure temperature for U-Pb rutile system is 400–450 [20], Rb-Sr isochron for biotite, apatite, albite, and whole rock defines the same age of 1754 ± 39 Ma. Thus it is not a cooling age but a real age of albitite formation at temperature not higher than 350 °C—the closure temperature for an Rb-Sr biotite system.

The brannerite formation at 385 ± 2 Ma in the Ozernoe area was related to Paleozoic hydrothermal processes. Uranium mineralization of this age is widely developed in the Litsa area as well, where it is represented by a pitchblende formation [4,5]. The influence of Paleozoic processes is also reflected in the lower Discordia intercepts for U-Pb uraninite and monazite data.

Analysis of the chemical composition of rocks of the Litsa area [5] shows that uranium content is low in the Archean gneisses and plagiogranites (1–2.5 ppm), and a bit higher in the proterozoic granites (Kaskel'javr and Litsa-Araguba complexes)—3–5 ppm. At the same time, as it was mentioned in [4], the uranium content as well as amount of other incompatible elements increases dramatically in the host gneisses and plagiogranites in areas of concentration of uraniferous pegmatites and metasomatites. It means that the formation of uranium concentrations is not associated with uranium redistribution from the host rocks, and we need to look for a U source outside the host area.

Discussing the uranium mineralization of the Litsa area, it must be kept in mind that the Litsa area is located in close proximity to the Pechenga structure, and the processes of ore formation in the region should not be considered separately [21]. Thus, a formation of 2.3–2.2 Ga subalkaline volcanics of the Pirtijarvi suite [22,23] within the Pechenga structure might have triggered the oldest stage of Litsa U mineralization.

The next stage of mineralization described by [4] as "Th-U mineralization at 1850–1750 Ma in quartz-albite-microcline and quartz-microcline metasomatites" seemed to consist of two different types connected with different regional processes. The first stage is dated by 1825 ± 20 Ma uraninite from quartz-feldspar metasomatites of the Dikoe area [5]. Brannerite ores of 1.83 Ga within Litsevskoe occurrence are reported in [21] but without isotopic data. The source of the uranium-enriched fluid of this age could be the mantle melts, which gave rise to the foidolite-carbonatite series of the Proterozoic Gramyaha-Vyrmes massif of 1884 ± 6 Ma [24], whose initial magmas are similar in composition to the volcanics of the fourth suite of the Pechenga structure.

The next stage took place 1.75–1.65 Ga ago—uraninite formed in the quartz-albite-microcline metasomatites of the Litsa area (Beregovoe [4] and Cheptjavr [5] occurrences) and in the carbonate-chlorite metasomatites of the Salla-Kuolajarvi area. This was the period of last magmatic activity in the region, when emplacement of 1.77–1.75 Ga granitoids of the Litsa-Araguba complex [25]; granite veins of similar age and composition in the basement of the Pechenga structure and 1.71 Ga lamprophyre dykes had taken place [8]. This stage was accompanied with extensive hydrothermal activity, widely manifested in the Pechenga structure [26] and throughout the Kola region, which was inferred from Rb-Sr and K-Ar data [8]. According to our Rb-Sr data the temperature of U ore deposition during this period did not exceed 350–300 °C.

The last (Paleozoic) 0.40–0.38 Ga stage of uranium mineralization was associated with the formation of U-bearing albite-hydromica-chlorite metasomatites typical for the Litsa area [1,4,5]. Activation of hydrothermal and metasomatic processes at this time was connected with the Paleozoic 408–360 Ma alkaline magmatism (alkaline and nepheline syenite intrusions with carbonatites) [27]. The temperature of this stage according to mineral geochemistry [19] was 280–220 °C.

6. Conclusions

The earliest stage of uranium mineralization in the Litsa area occurred 2267 ± 27 Ma ago according to U-Pb age of uraninite from vein pegmatoid granitoids of the Dikoe area. The magmatic age of these veins is 2549 ± 7 Ma based on U-Pb monazite dating. Monazite as the main concentrator of REE and Th probably defines the development of the oldest REE-Th mineralization in the area.

Isotopic U-Pb and Rb-Sr data both obtained in this study and published earlier show that the principal stages of uranium mineralization in the Litsa area and Salla-Kuolajarvi zone had taken place 2.3–2.2, 1.75–1.65, and 0.40–0.38 Ga ago.

Comparison of the sequences of endogenous events in the Litsa area and Salla-Kuolajarvi zone with those in adjacent structures in the time span from 2.5 to 0.4 Ga allows us to link the stages of uranium mineralization to the regional cycles of endogenous activity, accompanied by the mantle alkaline magmatism in the Kola region, which provided the uranium input.

Uranium mineralization was related to hydrothermal events under medium to low temperatures from 550–500 °C at 2.27 Ga to 350–300 °C at 1.75–1.65 Ga and 280–220 °C at 0.40–0.38 Ga. A question of why uranium was concentrated in the Litsa area should be the subject of a future study.

Author Contributions: Conceptualization, T.V.K., A.A.K. and V.L.I.; Methods, T.V.K., A.A.K. and L.I.N.; Investigation, T.V.K., A.A.K., V.L.I., A.A.A., D.V.E., L.I.N., and E.A.N.; Writing—Original Draft Preparation, T.V.K., V.L.I., M.A.G., E.A.N.

Funding: This research was performed in the framework of the State contract No 0231-2015-0006 of GI KSC RAS and was partly funded by the program No. 19 "Fundamental Problems of Geological and Geophysical Studies of Lithospheric Processes" of Presidium of Russian Academy of Sciences.

Acknowledgments: The U-Pb and Rb-Sr isotope analyses of rocks and minerals were performed at the Geological Institute, Kola Science Center, Russian Academy of Sciences.

Conflicts of Interest: The authors declare no conflict of interest.

References

1. Afanas'eva, E.N.; Mikhailov, V.A.; Bylinskaya, L.V.; Lipner, A.A.; Serov, L.V. The Uranium Potential of the Kola Peninsula. *Mater. Geol. Uranium Rare- Rare-Earth-Metal Depos. Inf. Dig.* **2009**, *153*, 18–26. (In Russian)

2. Afanas'eva, E.N.; Mironov, Y.B. Metallogeny of uranium of the Baltic shield. *Explor. Prot. Bowels Earth* **2015**, No. 10. 82–88. (In Russian)

3. Kalinin, A.A.; Savchenko, E.E.; Selivanova, E.A. Renium- and selenium-containing molybdenite of the Ozernoe ore occurrence in the Salla-Kuolajarvi zone, North Karelia. *Proc. RMS* **2013**, *142*, 105–115. (In Russian)

4. Savitsky, A.V.; Gromov, Y.A.; Mel'nikov, E.V.; Sharikov, P.I. Uranium mineralization in the Litsa district of the Kola Peninsula (Russia). *Geol. Ore Depos.* **1995**, *37*, 403–416. (In Russian)

5. Serov, L. Métallogenèse de l'uranium dans la région de Litsa (Péninsule de Kola, Russie). Ph.D. Thesis, Nancy Université Henry Poincaré (en Géosciences), Nancy, France, 24 June 2011.

6. Konstantinov, A.K.; Maskovtsev, G.A.; Miguta, A.K.; Shumilin, M.V.; Schetochkin, V.N. *Uranium Ores of Russia*; VIMS: Moscow, Russia, 2010; 850p, ISBN 978-5-901837-63-4. (In Russian)

7. Dahlkamp, F.J. *Uranium Deposits of the World: USA and Latin America*; Springer: Berlin/Heidelberg, Germany, 2010; 520p.

8.   Glebovitsky, V.A. *Early Precambrian of the Baltic Shield*; Nauka: S.-Petersburg, Russia, 2005; 711p, ISBN 5-02-024950-5. (In Russian)

9.   Manninen, T.; Huhma, H. Radiometric age determinations from Finnish Lapland and their bearing on the timing of Precambrian volcano-sedimentary sequences. In *Geological Survey of Finland Special Paper*; Vaasjoki, M., Ed.; Geological Survey of Finland: Rovaniemy, Finland, 2001; Volume 33, pp. 201–209.

10.   Lehtinen, M.; Nurmi, P.A.; Rämö, O.T. (Eds.) *Precambrian Geology of Finland—Key to the Evolution of the Fennoscandian Shield*; Elsevier: Amsterdam, The Netherlands, 2005; p. 736, ISBN 13:9780444514219.

11.   Safonov, Y.G.; Volkov, A.B.; Wolfson, A.A.; Genkin, A.D.; Krylova, T.L.; Chugaev, A.V. The Maisk quartz gold deposit (northern Karelia): Geological, mineralogical, and geochemical studies and some genetic problems. *Geol. Ore Depos.* **2003**, *45*, 429–451.

12.   Krogh, T.E. A low-contamination method for hydrothermal decomposition of zircons and extraction of U and Pb for isotopic age determinations. *Geochim. Cosmochim. Acta* **1973**, *37*, 485–494. [CrossRef]

13.   *NBS SRM 982: Equal-Atom Lead Isotopic Standard*; National Bureau of Standards (Now as National Institute of Standards and Technology): Gaithersburg, MD, USA, 1979–1980; p. 53.

14.   Ludwig, K.R. *User's Manual for Isoplot, v. 3.0, A Geochronological Toolkit for Microsoft Excel*; Berkeley Geochronolgy Center: Berkeley, CA, USA, 2008.

15.   Faure, G. *Principles of Isotope Geology*; John Wiley and Sons, Inc.: Hoboken, NJ, USA, 1986; p. 589, ISBN 0-471-86412-9.

16.   Robb, L. *Introduction to Ore-Forming Processes*; Blackwell Publishing Company: Malden, MA, USA, 2005; p. 343, ISBN 0-632-06378-5.

17.   Kato, T.; Suzuki, K.; Adachi, M. Computer program for the CHIME age calculation. *J. Earth Planet. Sci. Nagoya Univ.* **1999**, *46*, 49–56.

18.   Skuf'in, P.K.; Bayanova, T.B.; Smolkin, V.F.; Apanasevich, E.A.; Levkovich, N.V. The Problem of granite genesis in Early Proterozoic riftogenic belts with reference to the Southern Pechenga zone, Kola Peninsula. *Geochem. Int.* **2003**, *41*, 236–244.

19.   Vinogradov, A.I.; Vinogradova, G.V. Evolution of ultrametagenetic and diaphthoric processes and associated U–Th and REE mineral genesis in the polymetamorphic complex of the Kola gneisses. In *Early Precambrian Metamorphism and Metamorphogenetic Ore Formation*; KFAN: Apatity, Russia, 1984; pp. 37–46. (In Russian)

20.   Mezger, K.; Hanson, G.N.; Bohlen, S.R. High-precision U-Pb ages of metamorphic rutile: Application to the cooling history of high-grade terranes. *Earth Planet. Sci. Lett.* **1989**, *96*, 106–118. [CrossRef]

21.   Kazansky, V.I.; Lobanov, K.V. On boundaries and metallogeny of Pechenga ore area. *Geol. Ore Depos.* **1996**, *38*, 103–109. (In Russian)

22.   Vetrin, V.F. Proterozoic processes of magmatism and metasomatism in the Archaean rocks of the Pechenga paleorift basement. *Vestnik MSTU* **2007**, *10*, 116–129. (In Russian)

23.   Smolkin, V.F. Kola (Pechenga-Varzuga) riftogenic system. In *Magmatism and Metallogeny of Riftogenic Systems of the East Baltic Shield*; Scheglov, A.D., Ed.; Nedra: S.-Petersburg, Russia, 1993; pp. 24–63, ISBN 5-247-03093-1. (In Russian)

24.   Bea, F.; Arzamastsev, A.; Montero, P.; Arzamastseva, L. Anomalous alkaline rocks of Soustov, Kola: evidence of mantle-derived metasomatic fluids affecting crustal materials. *Contrib. Mineral. Petrol.* **2001**, *140*, 554–566. [CrossRef]

25.   Vetrin, V.F. Duration of the formation and sources of the granitoids of the Litsk-Araguba Complex, Kola Peninsula. *Geochem. Int.* **2014**, *52*, 33–45. [CrossRef]

26.   Smolkin, V.F.; Lokhov, K.I.; Skublov, S.G.; Sergeeva, L.Y.; Lokhov, D.K.; Sergeev, S.A. Paleoproterozoic Keulik–Kenirim Ore-Bearing Gabbro–Peridotite Complex, Kola Region: A New Occurrence of Ferropicritic Magmatism. *Geol. Ore Depos.* **2018**, *60*, 142–171. [CrossRef]

27.   Kramm, U.; Kogarko, L.N.; Kononova, V.A.; Vartiainen, H. The Kola Alkaline Province of the CIS and Finland: Precise Rb-Sr ages define 380–360 age range for all magmatism. *Lithos* **1993**, *30*, 33–44. [CrossRef]

minerals

MDPI

Article

Long-Lived Mantle Plume and Polyphase Evolution of Palaeoproterozoic PGE Intrusions in the Fennoscandian Shield

Tamara Bayanova [1,*], Aleksey Korchagin [1], Alexander Mitrofanov [2], Pavel Serov [1], Nadezhda Ekimova [1], Elena Nitkina [1], Igor Kamensky [1], Dmitry Elizarov [1] and Milosh Huber [3]

[1]   Geological Institute, Kola Science Centre, Russian Academy of Sciences, 184209 Apatity, Russia;
      pana@geoksc.apatity.ru (A.K.); serov@geoksc.apatity.ru (P.S.); ekimova@geoksc.apatity.ru (N.E.);
      nitkina@geoksc.apatity.ru (E.N.); iglkam@mail.ru (I.K.); elizarov@geoksc.apatity.ru (D.E.)
[2]   SRK Consulting, Toronto, ON M4C 1T2, Canada; mitalex1987@gmail.com
[3]   Department of Earth Sciences and Spatial Management, Maria Curie-Skłodowska University, 520-031 Lublin,
      Poland; mhuber@poczta.umcs.lublin.pl
*   Correspondence: tamara@geoksc.apatity.ru; Tel.: +7-81555-7-92-18

Received: 18 October 2018; Accepted: 15 January 2019; Published: 18 January 2019

check for updates

Abstract: The NE Fennoscandian Shield comprises the Northern Belt in Finland and the Southern Belt in Karelia. They host mafic-ultramafic layered Cu-Ni-Cr and Pt-Pd-bearing intrusions. Precise U-Pb and Sm-Nd analyses indicate the 130-Ma evolution of these intrusions, with major events at 2.53, 2.50, 2.45, and 2.40 Ga. Barren phases were dated at 2.53 Ga for orthopyroxenites and olivine gabbro in the Fedorovo-Pansky massif. PGE-bearing phases of gabbronorites (Pechenga, Fedorovo-Pansky, Monchetundra massifs) and norites (Monchepluton) are 2.50 Ga old. Anorthosites of Mt. Generalskaya (Pechenga), the Fedorovo-Pansky, and Monchetundra massifs occurred at 2.45 Ga. This event produced layered PGE-bearing intrusions in Finland (Penikat, Kemi, Koitelainen) and mafic intrusions in Karelia. The Imandra lopolith dikes occurred at the final phase (2.40 Ga). Slightly negative εNd and I_{Sr} values (0.703–0.704) suggest that intrusions originated from an enriched mantle reservoir. Low ^{3}He/^{4}He ratios in accessory minerals (ilmenite and magnetite) indicate an upper mantle source. Large-scale correlations link the Fennoscandian Shield with the Superior and Wyoming cratons.

Keywords: Plume; LIP; PGE; Palaeoproterozoic; mafic intrusion; U-Pb; isotopes

1. Introduction

The NE Fennoscandian (Baltic) Shield covers the Archaean crust formed at 3.5–2.7 Ga [1,2]. The post-Archaean evolution is mainly reflected in rifting and emplacement of mafic-ultramafic complexes along NE-trending structures. The NE Baltic Shield hosts the Southern (Fenno-Karelian) Belt (FKB) and the Northern (Kola) Belt (KB) that we studied in detail in our profound research [3].

FKB extends 350 km along the northern edge of the Karelian craton. It comprises Palaeoproterozoic mafic-ultramafic layered bodies in Finland (e.g., Penikat, Kemi [4–6], Sweden (Tornio intrusion), and Russia (Olanga group) [7] that occurred in similar geodynamic settings [8].

KB strikes northwestwards for over 500 km. It borders the SW edge of the Archaean Kola-Norwegian Block and the Northern and Southern edges of the Palaeoproterozoic Pechenga-Imandra-Varzuga rift. There are several layered mafic-ultramafic bodies [7]. Layered PGE-bearing pyroxenite-norite-gabbro-anorthosite intrusions occur at the boundaries between early Proterozoic rifts lying over the Archaean basement [9,10].

The purpose of this work is a complex study of Palaeoproterozoic Pt-Pd and Cu-Ni intrusions that are widespread in the NE Baltic Shield. We use a set of isotope techniques to define precision ages of formation of reefs, lower and upper parts of layered intrusions, host rocks, and cutting dyke series. We also aimed at studying sources of primary magmas and primary reservoirs. The main objective of this research is to compare new isotope data with those on other Palaeoproterozoic intrusions of the Finnish group, Northern America, and Canada. This would corroborate the model of the plume mantle source for all Palaeoproterozoic intrusions.

The current study continues our investigations [3], but substantiates it with new isotope geochronological data. It allowed us to define that PGE-bearing intrusions of the Fennoscandian Shield referred to the enriched mantle reservoir EM-1. This paper provides new data showing that magmatism in the Kola-Karelian Province correlates with the peak of mafic-ultramafic magmatism in the Superior and Wyoming provinces (2.45 Ga).

2. Geological Setting

The Palaeoproterozoic East Scandinavian Large Igneous Province (LIP) occurs in the NE Fennoscandian Shield (Figure 1). Its granulite and gneiss-migmatite basement originated at >2550 Ma. This province had several stages of magmatism and sedimentation separated by breaks (conglomerates). The Sumian stage (2550–2400 Ma) was crucial for the production of Pt-Pd ores [9,10]. In FKB, the magmatism was most active at 2450 to 2400 Ma [4,10–16]. In KB, PGE-bearing intrusions formed at 2530–2450 Ma.

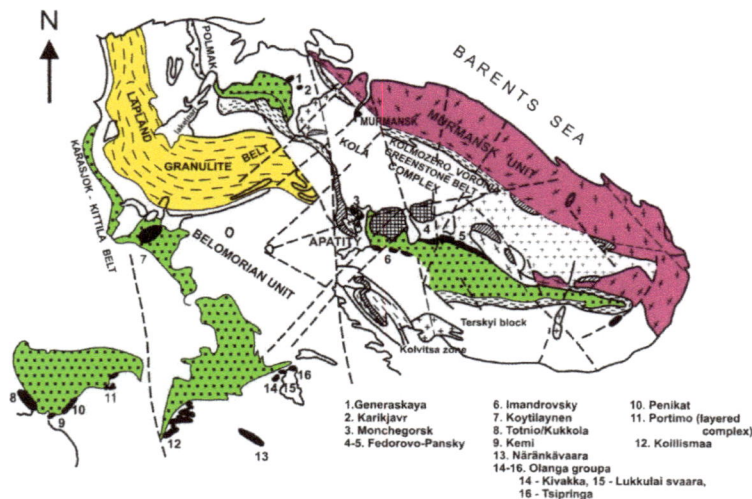

Figure 1. Sketch map of the Palaeoproterozoic East Scandinavian LIP in the Northern Fennoscandian Shield [15].

The Fedorovo-Pansky Layered Complex strikes north-westwards for >60 km and dips Southwestwards at an angle of 30–35° (Figure 2). The Fedorov, Lastjavr, Western and Eastern Pansky are its major blocks [9]. This complex is bordered by the Archaean Keivy terrain and the Palaeoproterozoic Imandra-Varzuga rift. In the North, the complex borders alkaline granites of the White Tundra intrusion. Their U-Pb age on zircon is 2654 + 15 Ma [17]. The contact of the Western Pansky Block with the Imandra-Varzuga volcano-sedimentary sequence mostly has Quaternary deposits. However, drill and excavation works to the South of Mt. Kamennik reveal a strongly sheared and metamorphosed contact between the intrusion and overlying Palaeoproterozoic volcano-sedimentary rocks.

The authors had earlier studied the Fedorovo-Pansky massif using a full set of geological, mineralogical, petrographical, and geochemical methods [3]. This allowed us to identify the block structure of the massif and define zones with different composition, age, and mineralization. Provided below is a compilation of previously obtained data [3] and results of this study.

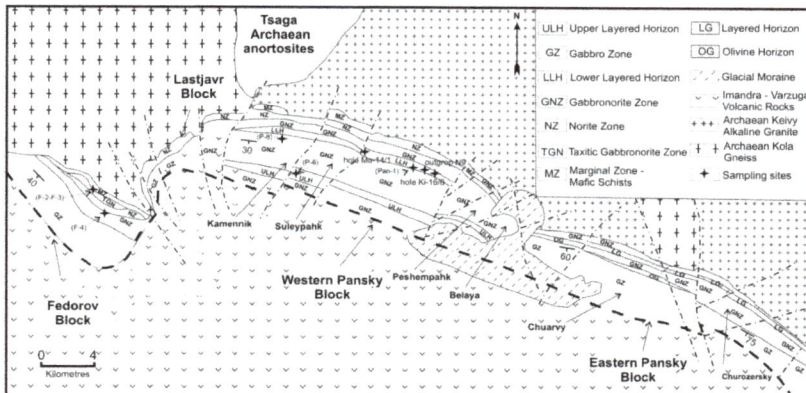

Figure 2. Sketch geological map of the Fedorovo-Pansky massif [9].

The Fedorovo-Pansky Complex consists of Marginal, Taxitic, and Norite zones (Figure 2). The Marginal Zone contains plagioclase-amphibole schists, massive fine-grained norites, and gabbronorites. The Taxitic Zone contains ore-bearing gabbronorites (2485 Ma), xenoliths of plagioclase-bearing pyroxenite and norite (2526–2516 Ma), syngenetic and magmatic ores (Pt and Pd sulfides, Cu and Ni sulfides with Pt, Pd and Au, bismuth-tellurides and arsenides). The Norite Zone hosts harzburgite and plagioclase-bearing pyroxenite with Cu-Ni-PGE mineralization in the lower part. These rocks are chromium-rich (up to 1000 ppm) and contain chromite. The same mineralization occur in FKB intrusions [5].

The Fedorov Block is a part of the Fedorovo-Pansky Complex. The Layered Lower Horizon (LLH) of its Main Gabbronorite Zone is composed of gabbronorite, norite, pyroxenite and interlayers of leucocratic gabbro and anorthosite, and reef-type PGE deposit. The Upper Layered Horizon (ULH) consists of olivine-bearing troctolite, norite, gabbronorite, and anorthosite. It comprises several layers of PGE-rich ore (Pd > Pt) [9]. The U-Pb age of the ULH rocks on single-grain zircon and baddeleyite is 2447 + 12 Ma. This age shows that ULH rocks are the youngest in the Fedorovo-Pansky Complex.

The Fedorov Tundra Block is in the West of the Fedorovo-Pansky massif. It contains mafic and rare ultramafic rocks, mostly gabbronorites with highly heterogeneous mineral grains. The contact-type Cu, Ni, and PGE mineralization is economically valuable.

There are three mineral types of the sulfide mineralization: pyrrhotite, pyrrhotite-chalcopyrite, and pentlandite-chalcopyrite-pyrrhotite. The latter type is best-valued commercially, since it contains the complex PGE mineralization.

Major ore minerals are chalcopyrite, pyrrhotite and pentlandite. Their average proportions recalculated for 100% sulfide are 41%, 35%, and 24%, respectively. In the total amount of sulfides, their portion is 95–100%. Secondary and accessory minerals are pyrite, ilmenite, magnetite, mackinawite, marcasite, cubanite, millerite, bornite, sphalerite, and violarite.

Besides pentlandite, chalcopyrite and pyrrhotite may also host minor PGE, though lower than pentlandite [18], i.e., kotulskite, merenskyite, braggite, stillwaterite, vysotskite, and sobolevskite for palladium; braggite, moncheite, merenskyite, and sperrylite for platinum; and gold-silver alloys for gold [19,20]. In total, 29 PGE and gold minerals, as well as eight PGE phases with no status of individual minerals, have been recorded in the ore. Microprobe analyses of their chemical compositions have been conducted in the Geological Institute KSC RAS [20,21].

3. Materials and Methods

U-Pb (TIMS) Method with $^{208}Pb/^{235}U$ Tracer [12,22–25], Zircon ID-TIMS U-Pb Analyses (with a ^{205}Pb Spike) [12,24,25], ID-TIMS Sm-Nd Analyses of Silicates [12,22,23,26,27], ID-TIMS Rb-Sr

Analyses [12,28], ID-TIMS Sm-Nd Analyses of Sulfides [12,13,23,24,26,29–31], Electron Microprobe and LA-ICP-MS Major and Trace Element Analyses [32] have been used to study the samples.

3.1. Isotope Analyses

The U-Pb concordia plot and Sm-Nd isochron methods are applied to define the age of crystallization of the rocks. U-Pb zircon and baddeleyite ages for the same sample are commonly coherent. They indicate a similar age of the magma crystallization and subsequent transformations. Coordinates of baddeleyites are near the concordia line. However, it is difficult to apply this method due to minor amounts of zircon and baddeleyite grains in these rocks.

Compared to the precise U-Pb systematics with an error of ca. 0.1%, the Sm-Nd method is not an accurate geochronometer (error of ca. 2–3%). However, it allows defining the crystallization age of mafic rocks using major rock-forming minerals. The method is especially important for dating rocks with syngenetic ore minerals. It has been used to date (2482 ± 36 Ma) the early gabbronorite and second anorthosite (2467 ± 39 Ma) ore bodies in the Fedorovo-Pansky deposit. This deposit is considered to be economically important [33].

Sm-Nd ages of the KB mafic-ultramafic intrusions overlap because of large errors. The estimations are compatible with U-Pb ages obtained on zircon and baddeleyite. They are especially important for marginal fast-crystallizing rocks of the Taxitic Zone within the Fedorovo-Pansky Complex. The ages of early barren orthopyroxenite and gabbro are 2521 ± 42 Ma and 2516 ± 35 Ma (Sm-Nd method) and 2526 ± 6 Ma and 2516 ± 7 Ma (U-Pb method). The ore-bearing norite of the Fedorov Block yields ages of 2482 ± 36 Ma (Sm-Nd method) and 2485 ± 9 Ma (U-Pb method) [34,35].

Importantly, the Sm-Nd method provides valuable petrological and geochemical markers, i.e., εNd(T) and TDM. The εNd value shows a degree of the mantle source depletion, while TDM indicates an approximate age of melt extraction from the mantle protolith [35].

The Rb-Sr whole-rock and mineral-isochron method, specific trace elements (Cu, Ni, Ti, V, and LREE), ε_{Nd} (2.5 Ga) and $^4He/^3He$ data, values of initial $^{87}Sr/^{86}Sr$ (I_{Sr}[2.5 Ga]) indicate an enriched 2.5 Ga-old mantle reservoir [36]. It is similar with the modern EM-I reservoir [37,38] and corroborated by the Re-Os systematics [39].

3.2. LA-ICP-MS

The distribution of rare and precious metals in sulfide parageneses has been first studied in detail using the laser ablation inductively coupled magma mass spectrometry (LA-ICP-MS). It enables detecting regular patterns of the element distribution with a high degree of accuracy. The results indicate that pentlandite in sulfide parageneses of the Fedorov Tundra deposit has commercially valuable PGE mineralization.

The method of LA-ICP-MS (UP-213 laser, high-resolution Element-XR mass spectrometer with ionization) has been applied to analyze concentrations of Cr, Co, As, Se, Ru, Rh, Pd, Ag, Cd, Sb, Re, Os, Ir, Pt, Au, Tl, Pb, and Bi in sulfides. The following parameters have been accepted for the analysis: crater diameter is 40 μm, impulse frequency of laser radiation is 4 Hz. The samples have been analyzed by blocks, which were prepared using the Element XR software. Standard samples have been measured at the beginning and end of each block. Internal laboratory sulfide standards have been used for analysis. The deviation by calibration standards is 10–20%. Fe has been applied, having quite high concentrations in relation to background values. It occurs in all of the studied samples and is the most homogeneously distributed in phases.

Regional PGE-bearing deposits are represented by the basal and reef-like types. According to modern economic evaluation, the basal type is preferable for mining, even if the PGE content (1–3 ppm) is lower than in reef-type deposits (>5 ppm). Basal deposits are thicker and contain more Pt, Cu, and especially, Ni.

In the Eastern Fennoscandian Shield, the Palaeoproterozoic magmatic activity is associated with the formation of Cu-Ni (±PGE), Pt-Pd (Rh, ±Cu, Ni, Au), Cr, and Ti-V deposits [40,41]. The Fedorov deposit is best-valued for PGE (Pt, Pd, Rh), but Ni, Cu, and Au are also economically important [10]. Ore-forming magmatic and post-magmatic processes are closely related to the Taxitic Zone gabbronorite of the 2485 ± 9 Ma magmatic pulse. Reef-type deposits (Pt-Pd (±Cu, Ni, Rh, Au) and ore occurrences of the Western Pansky Block (Fedorovo-Pansky Complex) are genetically associated with pegmatoid leucogabbro and anorthosite rich in late-stage fluids. Portions of this magma produce additional injections of ca. 2500 Ma, ca. 2470 Ma (the Lower, Northern PGE reef) and ca. 2450 Ma (the Upper, Southern PGE reef of the Western Pansky Block and PGE-bearing mineralization of Mt. Generalskaya intrusion). These magma injections are quite similar in their prevalence of Pd over Pt, ore mineral composition [9] and isotope geochemistry of the Sm-Nd and Rb-Sr systems. εNd values for these rocks vary from −2.1 to −2.3. This may indicate a single long-lived enriched magmatic source.

High contents of Cr (>1000 ppm) are typical of lower mafic-ultramafic rocks from layered intrusions in the Baltic Shield [4,5]. The chromite mineralization occurs in basal series of the Monchepluton, Fedorovo-Pansky Complex, Imandra lopolith (Russia), Penikat [42] and Narkaus intrusions (Finland), chromite deposits of the Kemi intrusion (Finland) [43] and Dunite Block (Monchepluton, Russia). In contrast, the Fe-Ti-V mineralization of the Mustavaara intrusion (Finland) tends to occur in the leucocratic-most parts of layered series, as well as in leucogabbro-anorthosite and gabbro-diorite of the Imandra lopolith (Russia) and Koillismaa Complex (Finland).

4. Results

4.1. Fedorovo-Pansky Massif: U-Pb and Sm-Nd Isotope Data

Several large samples have been selected in the Fedorovo-Pansky Complex for further isotope analyses. Medium- and coarse-grained gabbronorite have been sampled in LLH of the Eastern Kievey area. Three types of zircons (transparent, with a vitreous luster) have been separated [3]: Pan-1, regular bipyramidal-prismatic crystals of up to 120 μm; Pan-2, fragments of prismatic crystals; and Pan-3, pyramidal apices of crystals of 80–100 μm. In immersion view, all the zircons display a simple structure with fine zoning and cross jointing.

The previously obtained data show that the crystallization time of the main gabbronorite phase of LLH is 2491 ± 1.5 Ma, since the upper intersection age is 2491 ± 1.5 Ma (MSWD = 0.05) and the lower intersection is at zero and indicates the loss of Pb (Figure 3a, Table 1) [3,7,44]. The same zircon sample has been analyzed at the Royal Ontario Museum laboratory (Canada). It proved slightly older (2501.5 ± 1.7 Ma) [45].

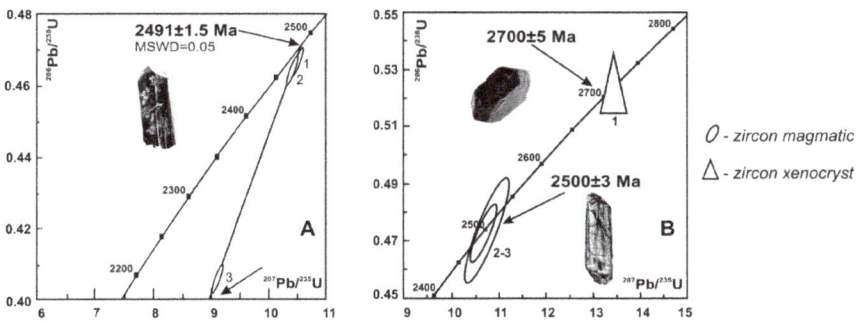

Figure 3. (**a**) Isotope U-Pb concordia diagrams for magmatic zircon (2-3) and (**b**) zircon xenocryst (1) from gabbronorite of the Western-Pansky Block of the Fedorovo-Pansky massif.

Table 1. Isotope U-Pb data on baddeleyite (Bd) and zircon (Zr) on the Western-Pansky and Fedorov Blocks of the Fedorovo-Pansky massif.

Sample No	Weight (mg)	Content (ppm)		Pb Isotope Composition [1]			Isotope Ratios [2]		Age [2] (Ma)
		Pb	U	$\frac{^{206}Pb}{^{204}Pb}$	$\frac{^{206}Pb}{^{207}Pb}$	$\frac{^{206}Pb}{^{208}Pb}$	$\frac{^{207}Pb}{^{235}U}$	$\frac{^{206}Pb}{^{238}U}$	$\frac{^{207}Pb}{^{206}Pb}$
Western-Pansky Block, gabbronorites (Pan-1)									
1	3.30	95.0	144	11740	6.091	3.551	10.510	0.4666	2491
2	1.60	84.0	144	6720	6.062	3.552	10.473	0.4650	2491
3	1.90	70.0	142	10300	6.100	4.220	9.135	0.4061	2489
Western-Pansky Block, gabbropegmatite (P-8)									
1	5.90	95.0	158	3240	5.991	3.081	10.435	0.4681	2471
2	7.30	181.0	287	8870	6.161	2.260	10.092	0.4554	2465
3	1.25	125.0	200	3400	6.012	2.312	10.082	0.4532	2468
Western-Pansky Block, anorthosite (P-6)									
1	0.75	218.0	322	5740	6.230	3.263	11.682	0.5352	2438
2	0.10	743.0	1331	3960	6.191	3.151	9.588	0.4393	2438
3	0.20	286.0	577	2980	6.021	3.192	8.643	0.3874	2474
4 (bd)	1.00	176.0	396	14780	6.290	63.610	9.548	0.4380	2435
5(bd)	0.26	259.0	560	3360	6.132	54.950	9.956	0.4533	2443
Fedorov Block, orthopyroxenite (F-3)									
1	0.75	48.0	60.9	825	4.9191	1.3039	10.0461	0.44249	2504
2	0.80	374.0	598.6	4588	6.0459	1.9650	9.6782	0.43153	2484
3	0.85	410.2	630.2	4521	6.0281	1.6592	9.5667	0.42539	2488
4	1.00	271.0	373.1	2552	5.9916	1.2393	9.4700	0.42406	2476
Fedorov Block, olivine gabbro (F-4)									
1	1.80	725.3	1322.8	14649	6.1121	3.8177	10.0132	0.44622	2484
2	2.00	731.3	1382.8	8781	6.1522	3.5517	9.4306	0.42454	2467
3	1.95	680.9	1374.0	7155	6.2645	3.6939	8.7401	0.40155	2433
Fedorov Block, PGE-bearing gabbronorite (F-2)									
1	0.30	498.0	833.4	2081	5.9502	2.2111	9.49201	0.42493	2477
2	0.65	513.8	932.2	5274	6.1519	2.6371	9.1373	0.41378	2458
3	0.55	583.2	999.3	3194	6.1132	2.0528	8.9869	0.40832	2452
4	0.80	622.5	1134.5	4114	6.1161	2.1914	8.6638	0.39165	2460

[1] All ratios are corrected for blanks of 0.1 ng for Pb and 0.04 ng for U, mass discrimination 0.17 ± 0.05%. [2] Correction for common Pb has been determined for the age according to [25].

In the framework of this research, the authors have conducted new measurements using single zircon grains from the main PGE gabbronorite phase from the old collection (Figure 3b, Table 2) and obtained the similar age of 2500 ± 3 Ma. The isotope Sm-Nd data show the age of the same gabbronorite on the main rock-forming minerals and whole-rock is 2487 ± 51 Ma (MSWD = 1.5) [16].

Table 2. U-Pb data on single zircon xenocryst (1) and magmatic zircon (2,3) from gabbronorite of the Fedorovo-Pansky massif.

No.	Weight mg	Content (ppm)		$^{206}Pb/^{204}Pb$	Isotope Composition [1]			Isotope Ratios and Age in Ma [2]			% Dis
		Pb	U		$^{206}Pb/^{238}U$ $\pm 2\sigma$	$^{207}Pb/^{235}U$ $\pm 2\sigma$	$^{207}Pb/^{206}Pb$ $\pm 2\sigma$	$^{206}Pb/^{238}U$ $\pm 2\sigma$	$^{207}Pb/^{235}U$ $\pm 2\sigma$	$^{207}Pb/^{206}Pb$ $\pm 2\sigma$	
1	0.0457	21.20	6.33	833.59	0.521 ± 0.018	13.327 ± 0.634	0.186 ± 0.006	2702 ± 95	2703 ± 129	2704 ± 82	0.1
2	0.0536	12.87	5.92	461.17	0.474 ± 0.015	10.768 ± 0.393	0.165 ± 0.003	2499 ± 80	2503 ± 91	2507 ± 42	0.3
3	0.0567	16.10	10.96	111.20	0.472 ± 0.008	10.688 ± 0.215	0.164 ± 0.002	2495 ± 42	2496 ± 50	2498 ± 26	0.1

[1] Ratios are corrected for blanks of 1 pg for Pb and 10 pg for U, mass discrimination 0.12 ± 0.04%. [2] Correction for common Pb has been determined for the age according to Reference [25].

In contrast with the zircon age, the rock-forming minerals originate at the post-magmatic stage. However, considering measurement errors, Sm-Nd ages are suggested to be close to the U-Pb ones.

Three types of zircon have been extracted from PGE-bearing gabbro-pegmatite (LLH). Zircons from samples P-8, D-15 and D-18 have been used for isotope analyses [3]. The concordant U-Pb age on zircons from PGE-bearing gabbro-pegmatite (LLH) is 2470 ± 9 Ma (MSWD = 0.37) (Figure 3a, Table 1). The lower intersection of the discordia line (c. 300 Ma) indicates a loss of Pb. This indicates the Palaeozoic tectonic activity in the Eastern Baltic Shield [3,46].

Analyses of whole-rock and rock-forming silicate and sulfide minerals from gabbro pegmatite give a Sm-Nd age of 2467 ± 39 Ma, MSWD = 1.8. εNd(T) is negative with −1.4 ± 0.5 (Figure 4a, Table 3). Noteworthy, these ages are quite similar within the measurement error.

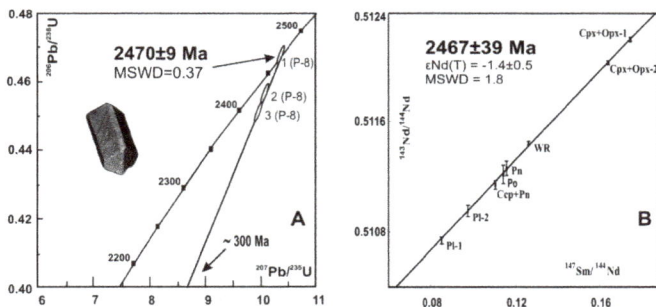

Figure 4. Isotope U-Pb diagram on zircon (**a**) and Sm-Nd mineral isochron for rock-forming and sulfides minerals (**b**) from gabbronorite (LLH) of the Fedorovo-Pansky massif.

Table 3. Sm-Nd data on rock-forming and sulfide minerals from gabbronorite (LLH) of the Fedorovo-Pansky massif.

Sample No.	Content, ppm		Isotope Ratios		T_{DM}, Ma	$\varepsilon_{Nd}(T)$
	Sm	Nd	$^{147}Sm/^{144}Nd$	$^{143}Nd/^{144}Nd$		
			Gabbro-pegmatite LLH			
WR	1.038	4.99	0.1263	0.511441 ± 10	2967	−1.3
Po	0.033	0.147	0.1144	0.511217 ± 69		
Pn	0.011	0.041	0.1160	0.511259 ± 53		
Pl-1	0.332	2.30	0.0853	0.510738 ± 24		
Pl-2	0.398	2.25	0.0977	0.510957 ± 39		
Cpx + Opx-1	4.75	16.44	0.1747	0.512209 ± 7		
Cpx + Opx-2	2.54	9.34	0.1641	0.512033 ± 9		
Ccp + Pn	0.022	0.124	0.1106	0.511143 ± 27		

Average standard values during measurements: N = 11 (La Jolla: = 0.511833 ± 10); N = 100 (JNdi1: = 0.512098 ± 15).

The U-Pb age is identical to the Sm-Nd one. Hence, we can suggest a very narrow closure temperature in the U-Pb (zircon) and Sm-Nd (rock-forming and sulfide minerals) systematics.

Three types of zircon and two varieties of baddeleyite (Sample P6-1) have been separated in ULH of the Southern Suleypahk area. The authors had previously studied and described zircons from samples Pb-1, Pb-2 and Pb-3 [3]. The U-Pb results for these samples define a discordia line intersecting the concordia curve at 2447 ± 12 Ma, MSWD = 2.7 (Figure 5a, Table 1). This age indicates the origin of late phase anorthosite, since baddeleyite forms mainly in residual igneous melts [3,47].

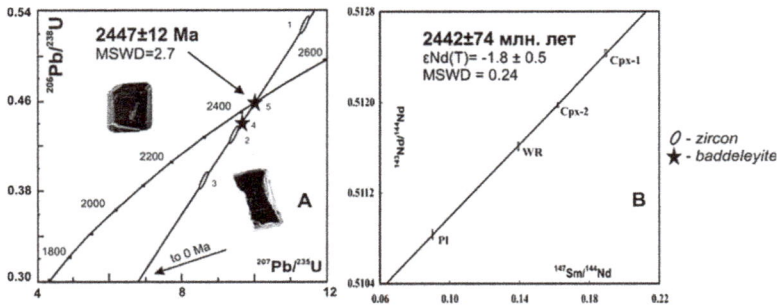

Figure 5. Isotope U-Pb age (**a**) on baddeleyite (4-5) and zircon (1-2-3); Sm-Nd mineral isochron (**b**) for rock-forming minerals and WR from anorthosites in ULH of the Fedorovo-Pansky massif.

A similar age is obtained from Sm-Nd data on rock-forming minerals of ULH anorthosite, i.e., 2442 ± 74 Ma with slightly negative εNd(T) = −1.8 ± 0.5, MSWD = 0.24 (Figure 5b, Table 4).

Table 4. Sm-Nd data on whole rock and rock-forming minerals from ULH anorthosite of the Fedorovo-Pansky massif.

Sample No.	Content, ppm		Isotope Ratios		T_{DM}, Ma	ε_{Nd}(T)
	Sm	Nd	$^{147}Sm/^{144}Nd$	$^{143}Nd/^{144}Nd$		
ULH anorthosite						
WR	0.271	1.176	0.1393	0.511613 ± 34	3131	−0.8
Pl	0.107	0.719	0.0901	0.510833 ± 39		
Cpx-1	0.921	2.94	0.1896	0.512436 ± 32		
Cpx-2	0.801	2.99	0.1618	0.511978 ± 20		

Average standard values during measurements: N = 7 (La Jolla: = 0.511837 ± 8); and N = 32 (JNdi1: = 0.512097 ± 19).

The U-Pb zircon age of early barren orthopyroxenite (F-3) from the Fedorov Block (2526 ± 6 Ma) marks the emplacement time (Figure 6a, Table 1). The similar U-Pb age (Figure 6b, Table 1) has also been obtained using zircon from barren olivine gabbro (F-4). Cu-Ni-PGE-bearing taxitic gabbronorites (F-2) from the Fedorov Block (Figure 6c, Table 1) yield a U-Pb age on zircon of 2485 ± 9 Ma [34].

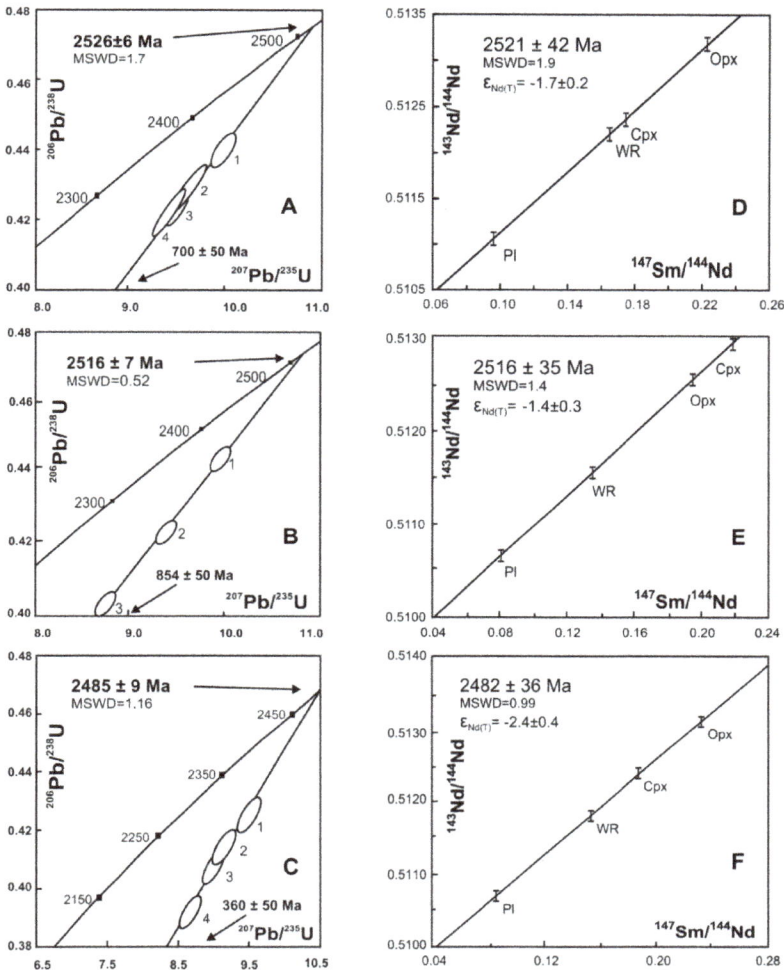

Figure 6. Isotope U-Pb ages on zircon (**A–C**) and Sm-Nd isochrones (**D–F**) for rock-forming minerals from the Fedorov Block of the Fedorovo-Pansky massif.

Isotope analyses of these samples using the whole rock, pyroxene and plagioclase yield ages that correspond to the respective U-Pb ages on zircon (Figure 6d–f). Table 5 provides a compilation of the previously obtained results [3] and new results of isotope geochemical research.

Table 5. Sm-Nd data on whole rock and rock-forming minerals from the Fedorov Block of the Fedorovo-Pansky massif.

Sample No.	Content (ppm)		Isotope Ratios		T_{DM} (Ga)	Sm-Nd (Ma)	$\varepsilon_{Nd\,(2.5Ga)}$
	Sm	Nd	$^{147}Sm/^{144}Nd$	$^{143}Nd/^{144}Nd$			
D-orthopyroxenite (F-3)							
WR	0.32	1.17	0.1648	0.512196 ± 12	3.05	2521 ± 42	−1.73
Opx	0.12	0.38	0.2228	0.513182 ± 16			
Cpx	2.21	7.67	0.1745	0.512349 ± 17			
Pl	0.26	1.62	0.0960	0.511071 ± 29			
E-olivine gabbro (F-4)							
WR	0.63	2.80	0.1357	0.511548 ± 8	2.94	2516 ± 35	−1.53
Opx	0.23	0.72	0.1951	0.512555 ± 15			
Cpx	0.83	2.28	0.2187	0.512947 ± 16			
Pl	0.24	1.77	0.0815	0.510677 ± 14			
F-PGE-bearing gabbronorite (F-2)							
WR	0.42	1.66	0.1537	0.511807 ± 20	3.18	2482 ± 36	−2.50
Pl	0.41	2.88	0.0865	0.510709 ± 14			
Cpx	1.78	5.73	0.1876	0.512387 ± 8			
Opx	0.13	0.33	0.2323	0.513088 ± 40			

Average standard values during measurements: N = 10 (La Jolla: = 0.511835 ± 11); and N = 22 (JNdi1: = 0.512097 ± 16).

4.2. Fedorovo Tundra Deposit: Trace Element Compositions

Three samples from different areas of the Fedorovo Tundra deposit have been investigated: sample BGF-237-132 from olivine-bearing gabbronorite of the Bolshoy Ikhtegipakhk, samples BGF-487-50.5 and BGF-495-76.5 from gabbronorite of the Pakhkvaraka area.

Mantle normalized diagrams (Figures 7 and 8, Table 6) show ratios of the obtained contents for each measured component normalized to the primitive mantle. Dashed areas indicate dispersion of the obtained results.

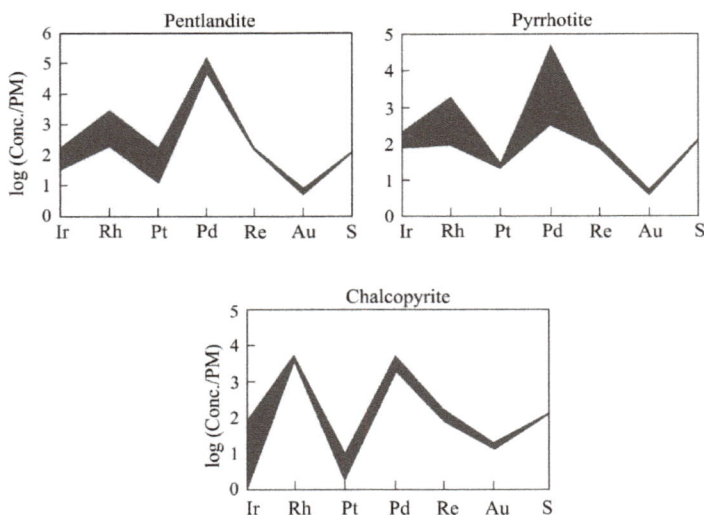

Figure 7. Mantle normalized diagram showing the distribution of PGEs, Au, and Re in sulfides (composition of the primitive mantle is according to Reference [48]).

Table 6. Contents (ppm) of different elements in sulfide parageneses of the Fedorovo Tundra deposit.

Sample	Mineral	Number of Analyses	Cr	Co	As	Se	Ru	Rh	Pd	Ag	Cd	Sb	Re	Os	Ir	Pt	Au	Tl	Pb	Bi
237/132.0	Pentlandite	8	2.8-16.5	6811-10681	0.3-0.7	92-183	0.5-1.0	0.1-6.9	258-1221	0.4-5.8	-	0.1-0.1	-	0.1	0.01-0.2	0.01-1.2	0.01	0.1-0.8	0.2-8.1	0.02-0.2
	Chalcopyrite	6	2.8	0.3-142	0.3	178	0.2	0.1-9.1	19	0.2-5.3	14	0.1-0.2	0.02	0.1	0.3	0.01	0.03	-	4.5-8.0	-
487/50.5	Pentlandite	10	0.9-5.9	8395	0.3-0.4	99	0.8	1.5-9.1	442	0.7	-	0.1	0.05	0.1-0.3	0.6	0.03-0.2	-	0.1-0.4	0.1-2.3	0.02
	Pyrrhotite	8	0.6-453	59-15612	0.2-2.5	102-222	0.3-1.5	1.2-2.9	0.1-636	0.1-1.6	-	0.1	0.03-0.09	0.2-0.6	0.6-1.1	0.2	0.01	0.1-1.4	0.1-5.8	0.01-0.04
	Chalcopyrite	5	-	-	0.5-0.7	98	0.2	8.8	18	0.4	5	0.1	-	-	-	-	-	-	1.5	-
495/76.5	Pentlandite	10	2.0	5499-12079	0.4	118	0.6-1.2	0.1-0.6	247-1487	0.3-1	-	0.1	-	0.2	0.3	0.2-2	0.01	0.1-0.9	0.1-1.5	0.03-0.1
	Pyrrhotite	9	0.7-15	66-119	-	132	0.2	0.1-0.2	0.1-4	0.4-2	-	0.1	0.03	0.1	0.1-0.3	0.04-0.3	0.01	0.02-0.05	0.1-6.3	0.02-0.6
	Chalcopyrite	8	2.6	0.5-487	0.7	142	0.1-0.2	11	23-125	0.6-1133	2-9	-	0.04	-	0.02	0.02-0.1	0.02	0.04-0.7	0.6-4.1	0.04-0.2

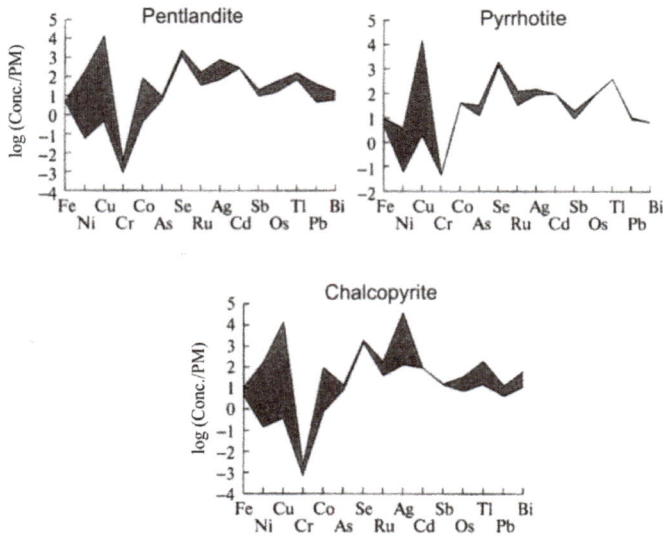

Figure 8. Mantle normalized diagram showing the distribution of some trace elements in sulfides (composition of the primitive mantle is according to References [49,50]).

Pentlandite is the main concentrator of Co and PGEs (Pd, Ru, and Pt). Since Pt and Pd are the best-valued commercial minerals of the Fedorovo Tundra deposit, pentlandite can be considered the most important among economic sulfide minerals of this deposit, along with PGMs. However, the abundance of pentlandite is much higher.

Pyrrhotite has a significant content of Cr and PGEs (Ir and Os). Pyrrhotite is the major concentrator of As and Bi.

Chalcopyrite accumulates almost all Ag and a significant proportion of Au. This mineral has high contents of Pb and Rh either.

Se, Re, Sn, and Ta are quite homogeneously distributed in all of the three mineral phases.

4.3. Penikat (Sompuyarvi reef) and Kemi Intrusions: Sm-Nd and U-Pb Geochronology

Gabbronorites (about 30 kg) were sampled in a geological field trip headed by M. Iljina (Geological Survey of Finland) to the Penikat and Kemi intrusions in 2008. The concordant age on 2 types of zircon from gabbronorite is 2430 ± 2 Ma. The concordant age on three types of zircon from the Kemi sample is 2447 ± 4 Ma (Figure 9b,c, Table 7).

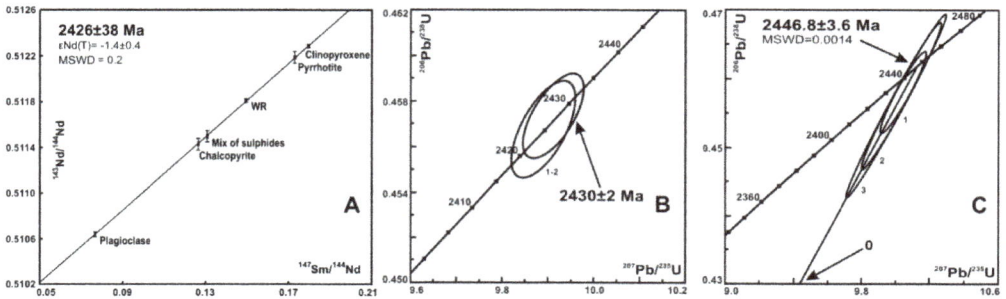

Figure 9. Mineral Sm-Nd isochron for PGE-bearing gabbronorites (**a**), the U-Pb age on single zircon (**b**) from the Penikat intrusion and the U-Pb age on single zircon grains (**c**) from the Kemi intrusion (Finland), FKB.

Table 7. U-Pb data on single zircon from layered PGE intrusions in Finland.

No.	Weight mg	Content (ppm)		Isotope Composition [1]				Isotope Ratios and Age in Ma [2]			% Dis
		Pb	U	$^{206}Pb/^{204}Pb$	$^{206}Pb/^{238}U \pm 2\sigma$	$^{207}Pb/^{235}U \pm 2\sigma$	$^{207}Pb/^{206}Pb \pm 2\sigma$	$^{206}Pb/^{238}U \pm 2\sigma$	$^{207}Pb/^{235}U \pm 2\sigma$	$^{207}Pb/^{206}Pb \pm 2\sigma$	
					Gabbronorite of Penikat						
1	0.1360	24.87	33.26	260.12	0.458 ± 0.004	9.929 ± 0.138	0.157 ± 0.002	2429 ± 21	2428 ± 34	2428 ± 24	−0.04
2	0.1840	28.14	42.74	254.48	0.457 ± 0.004	9.845 ± 0.138	0.156 ± 0.002	2426 ± 21	2420 ± 34	2416 ± 24	−0.41
					Gabbronorite of Kemi						
1	0.027	13.84	18.52	358.23	0.463 ± 0.009	10.155 ± 0.205	0.1592 ± 0.0005	2452 ± 49	2449 ± 49	2447 ± 7	−0.2
2	0.057	13.61	19.18	758.23	0.459 ± 0.009	10.069 ± 0.202	0.1592 ± 0.0005	2434 ± 48	2441 ± 49	2447 ± 7	0.5
3	0.032	11.51	16.10	404.27	0.453 ± 0.009	9.948 ± 0.196	0.1592 ± 0.0005	2410 ± 47	2430 ± 48	2447 ± 7	1.3

[1] Ratios are corrected for blanks of 1 pg for Pb and 10 pg for U, mass discrimination of 0.12 ± 0.04%. [2] Correction for common Pb has been determined for the age according to Reference [25].

The Sm-Nd isotope ages of rock-forming and sulfide minerals are much coeval to the U-Pb zircon data that suggested a very narrow closure temperature at the time of origination.

The whole rock, silicates and sulfides of the Penikat intrusion have a Sm-Nd age of 2426 ± 38 Ma (MSWD = 0.2) (Figure 9a, Table 8). It agrees with the Sm-Nd age of 2410 ± 64 Ma determined for the Penikat intrusion earlier [51]. Importantly, sulfide data lie on the isochron as well. It testifies to the coincidence of sulfide formation and the rock crystallization. Table 9 provides compiled isotope data obtained in previous [3] and the current research.

Table 8. Sm-Nd data on whole rock, rock-forming and sulfide minerals from ore-bearing gabbronorite of the Penikat layered intrusion.

	Content, ppm		Isotope Ratios			Model Age, Ma		ε_{Nd}
	Sm	Nd	$^{147}Sm/^{144}Nd$	$^{143}Nd/^{144}Nd$	Err.	CHUR	DM	
WR	2.004	10.066	0.14938	0.511811	28	2655	3155	−1.4
Plagioclase	0.654	3.655	0.07654	0.510639	22			
Clinopyroxene	1.901	6.398	0.17956	0.512285	9			
Chalcopyrite	0.109	0.647	0.13085	0.511499	53			
Pyrrhotite	0.301	2.017	0.17299	0.512185	47			
Sulfide mix	0.114	0.709	0.12648	0.511431	46			

Average standard values during measurements: N =7 (La Jolla: = 0.511836 ± 12); N = 21 (JNdi1: = 0.512097 ± 15).

Table 9. Sm-Nd and Rb-Sr data on whole rock from layered Palaeoproterozoic complexes of the Fennoscandian Shield.

| Sample No. | Content (ppm) | | Isotopic Ratios | | ε_{Nd} (2.5 Ga) | T_{DM} (Ga) | $^{87}Rb/^{86}Sr$ | $^{87}Sr/^{86}Sr$ ($\pm 2\sigma$), 2.5 Ga |
	Sm	Nd	$^{147}Sm/^{144}Nd$	$^{143}Nd/^{144}Nd$ ($\pm 2\sigma$)				
Fedorovo-Pansky intrusion								
Pan-1, gabbronorite	0.762	3.293	0.139980	0.511669 ± 7	−2.00	2.98	0.00135	0.7032 ± 1
Pan-2, gabbronorite	0.423	1.662	0.153714	0.511807 ± 20	−2.50	3.18	0.00174	0.7029 ± 2
F-4, olivine gabbro	0.629	2.801	0.135695	0.511548 ± 8	−1.53	2.94	0.00144	0.7029 ± 2
F-3, orthopyroxenite	0.318	1.166	0.164803	0.512196 ± 12	−1.73	3.05	0.00205	0.7033 ± 2
Monchetundra								
MT-10, medium-grained pyroxenite	0.483	1.913	0.152689	0.511925 ± 33	−0.36	2.81	0.00495	0.7039 ± 2
Mt. Generalskaya								
S-3464, gabbronorite	1.147	5.362	0.129320	0.511449 ± 14	−2.30	2.91	0.00534	0.7042 ± 2
Imandra lopolith								
6-57, gabbronorite	2.156	10.910	0.119130	0.511380 ± 3	−2.00	2.88	0.00339	0.7046 ± 3
Penikat								
gabbronorite	2.004	10.066	0.149380	0.511810 ± 15	−1.40	3.16	0.00527	0.7039 ± 5
Kemi								
gabbronorite	0.532	3.162	0.134950	0.511821 ± 9	−1.90	2.98	0.00431	0.7041 ± 8
Monchepluton								
M-1, quartz norite	1.750	8.040	0.131957	0.511493 ± 3	−1.51	2.91	0.01053	0.7034 ± 9
H-7, gabbronorite	0.920	4.150	0.134055	0.511537 ± 4	−1.37	2.90	0.00227	0.7037 ± 2

5. Discussion

A full set of advanced isotope geochronological methods has been applied to study the largest and ore-richest deposits of the Fedorovo-Pansky Complex. Table 10 is a data base of U-Pb and Sm-Nd ages of PGE layered intrusions in the Fennoscandian Shield. It includes both available and new data obtained by the authors of this study.

Previously, the authors had suggested that these PGE intrusions occurred at the same interval [3]. Based on U-Pb and Sm-Nd data, the authors defined zones with different ages in the Fedorov Block, Western Pansky Block of the Fedorovo-Pansky Complex, Monchetundra massif, Mt. Generalskaya and Imandra lopolith [3]. This study corroborates and substantiates results of the previous research. New data prove that, in the suggested hypothesis, there were four pulses of the magmatic activity associated with the formation of PGE intrusions of the Fennoscandian Shield in the interval 2.53–2.40 Ga.

The Fedorov Block of the Fedorovo-Pansky Complex represents an independent magma chamber. Its rocks and ores differ significantly from those of the Western Pansky Block [3,10]. A 2 km-thick rock sequence stretches from as a layered or differentiated syngenetic series. The age of melanocratic pyroxenite-norite-gabbronorite-gabbro between the Marginal Zone to the Lower Gabbro Zone is 2526 ± 6 and 2516 ± 7 Ma. The Taxitic Zone is intruded by concordant Cu-Ni-PGE-bearing gabbronorite (Fedorov deposit) of the second-pulse magmatic injection (2485 ± 9 Ma).

Table 10. U-Pb and Sm-Nd data on PGE layered intrusions of the Fennoscandian Shield.

Layered Intrusions	Age (Ma)		$\varepsilon_{Nd(T)}$ [2] U/Pb age
	U-Pb	Sm-Nd	
KB			
Mt. Generalskaya			
Gabbronorite	2496 ± 10 [1] (2505 ± 1.6) [2]	2453 ± 42 [1]	−2.3
Anorthosite	2446 ± 10 [1]		
Monchepluton			
Dunite block, dike	2505 ± 1.7 [19]		
Mt. Travyanaya, norite	2507 ± 9 [15]		
Dunite block, gabbronorite dike	2506 ± 10 [15]; 2496 ± 14 [15]		
Nyud Terrace, gabbronorite	2500±5 [14]		
Nyud Terrace, gabbronorite	2493±7 [1] (2504 ± 1.5) [2]	2492 ± 31 [3]	−1.4
Nyud Terrace, gabbronorite	2503.5 ± 4.6 [19]		
Monchepluton, gabbronorite	2498.2 ± 6.7 [19]		
Vurechuaivench foothills, metagabbronorite	2497 ± 21 [15]; 2498.2 ± 6.7 [17]		
Olenegorsk deposit, quartz diorite, dike	2495 ± 13 [15]		
Main Ridge			
Monchetundra, plagiopyroxenite	2502.3 ± 5.9 [16]		
Monchetundra, gabbronorite	2504 ± 7.4 [16]		
Monchetundra, gabbro	2463 ± 25 [4]; 2453 ± 4 [5]		
Monchetundra, gabbronorite	2501 ± 8 [14]; 2505 ± 6 [14]		
Monchetundra, gabbropegmatite	2445.1 ± 1.7 [16]		
Chunatundra, anorthosite	2467 ± 7 [15]		
Ostrovsky intrusion, gabbronorite-pegmatite	2445 ± 11 [15]		
Fedorovo-Pansky massif			
Orthopyroxenite	2526 ± 6 [12]	2521 ± 42 [13]	−1.7
olivine gabbro	2516 ± 7 [12]	2516 ± 35 [13]	−1.4
magnetite gabbro	2498 ± 5 [6]; 2500 ± 10 [16]		
Gabbronorite	2491 ± 1.5 [7] (2501 ± 1.7) [2]	2487 ± 51 [7]	−2.1
	2500 ± 3 [19]		
Cu-Ni PGE-bearing gabbronorite	2485 ± 9 [12]; 2500 ± 3 [21]	2482 ± 36 [13]	−2.4
PGE-gabbropegmatite	2470 ± 9 [7]	2467 ± 39 [21]	−1.4
PGE-anorthosite	2447 ± 12 [7]	2442 ± 74 [21]	−1.8
Imandra lopolith			
Gabbronorite	2446 ± 39 [7] (2441 ± 1.6) [2]	2444 ± 77 [7]	−2.0
gabbrodiorite-pegmatite	2440 ± 4 [6]		
Norite	2437 ± 7 [6]		
leucogabbro-anorthosite	2437±11 [6]		
Granophyre	2434 ± 15 [6]		
olivine gabbronorite (dike)	2395 ± 5 [6]		
monzodiorite dike	2398 ± 21 [6]		
FKB			
Kivakka, olivine gabbronorite	2445 ± 2 [7]	2439 ± 29 [8]	−1.2
Lukkulaisvaara, pyroxenite	2439 ± 11 [7] (2442 ± 1.9) [2]	2388 ± 59 [8]	−2.4
Tsipringa, gabbro	2441 ± 1.2 [2]	2430 ± 26 [8]	−1.1
Burakovskaya intrusion, gabbronorite	2449 ± 1.1 [2]	2365 ± 90 [8]	−2.0
Aganozero body		2372 ± 22 [20]	−3.2
Shalozero-Burakovo Inrusive Body		2433 ± 28 [20]	−3.1
Kovdozero intrusion, pegmatoid gabbronorite	2436 ± 9 [6]		
FINNISN GROUP			
Koitelainen	2433 ± 8 [9]	2437 ± 49 [11]	−2.0
Koilismaa	2436 ± 5 [10]		
Nyaryankavaara	2440 ± 16 [10]		
Penikat	2430 ± 2 [21]	2410 ± 64 [9]; 2426 ± 38 [18]	−1.6; −1.4
Akanvaara	2437 ± 7 [11]	2423 ± 49 [11]	−2.1
Kemi	2446.8 ± 3.6 [21]		

[1] [52]   [5] [55]   [9] [57]   [13] [35]   [17] [62]
[2] [45]   [6] [56]   [10] [58]   [14] [60]   [18] [33]
[3] [53]   [7] [16]   [11] [59]   [15] [12]   [19] [63]   [21] Present study
[4] [54]   [8] [56]   [12] [34]   [16] [61]   [20] [64]

The Western Pansky Block in the Main Gabbronorite Zone, excluding LLH and probably the upper part (above 3000 m), is a single syngenetic series of relatively leucocratic, mainly olivine-free gabbronorite-gabbro. They crystallized at 2526–2485 Ma. The Norite and Marginal Zones occur in the lower part of the Block. The Marginal Zone contains the poor disseminated Cu-Ni-PGE mineralization.

This rock series can be correlated with certain parts of the Fedorov Block. LLH is 40–80 m-thick. It has a unique structure with dominant leucocratic anorthositic rocks. The exposed part of the horizon strikes for almost 15 km and can be traced in boreholes down to a depth of 500 m [8]. Morphologically, the horizon seems to be a part of a single layered series. Nevertheless, there are anorthositic bodies with cross-cutting contacts and apophyses in outcrops [19]. Cumulus plagioclase compositions of the rocks in the horizon are different from those in surrounding rocks. The age of PGE-bearing leucogabbro-pegmatite (2470 ± 9 Ma) is precisely defined by concordant and near-concordant U-Pb data on zircon. It is slightly younger than ages of surrounding rocks (e.g., 2491 ± 1.5 Ma and 2500 ± 3 Ma). The LLH rocks, especially the anorthosite and PGE mineralization, are likely to represent an independent magmatic pulse.

The upper part and olivine-bearing rocks of the Western Pansky Block differ from the main layered units of the Block in rock, mineral and PGE mineralization composition [9]. Up to date, only one reliable U-Pb age (2447 ± 12 Ma) has been obtained for PGE-bearing anorthosite in the block. New Sm-Nd estimates yield an age of 2467 ± 39 Ma, which is much coeval to the U-Pb data.

The early magmatic activity at ca. 2.5 Ga reflects in gabbronorites of the Monchetundra (2505 ± 6 and 2501 ± 8 Ma) and Mt. Generalskaya (2496 ± 10 Ma). The magmatic activity that produced anorthosite took place at ca. 2470 and 2450 Ma. It also contributed to a layered series of the Chunatundra (2467 ± 7 Ma) and Mt. Generalskaya (2446 ± 10 Ma), Monchetundra gabbro (2453 ± 4 Ma) and pegmatoid gabbronorite of the Ostrovsky intrusion (2445 ± 11 Ma) (Table 10) [12].

The Imandra lopolith is the youngest large layered intrusion in KB. It differs from other KB intrusions both in its emplacement age and metallogeny. There are five U-Pb zircon and baddeleyite ages for the rocks of the main magmatic pulse. They were formed at 2445–2434 Ma.

Thus, several intrusions have been established in KB, including at least four intrusions (or phases) in the Fedorovo-Pansky Complex: a 2526–2516 Ma barren intrusion and three ore-bearing ones of 2505–2485, 2470, and 2450 Ma. For similar intrusions of FKB, e.g., the Penikat intrusion in Finland, five intrusions varying in geochemistry only have been distinguished from the same deep chamber [5].

A total duration for magmatic processes of over 130 Ma in the KB intrusions is surprising. The multi-phase magmatic duration of the FKB intrusions was short-term and took place about 2.44 Ga years ago. However, there are only a few U-Pb precise age estimations for the FKB intrusions [5]. Results for the Kola prove that layering of the intrusions with thinly-differentiated horizons and PGE reefs was not contemporaneous (or syngenetic). It was defined that each intrusion has its own metallogentic trend in time and space.

The magmatic activity revealed from 2.53 to 2.40 Ga with intrusions at 2.53, 2.50, 2.45, and 2.40 Ga. These four intrusions are based on precise U-Pb ages on single zircon and baddeleyite grains. The first three intrusions are corroborated by the Sm-Nd mineral isochron ages within the measurement error.

Since the Palaeoproterozoic (2.53 Ga), magmatic processes have affected most of the Kola-Lapland-Karelian province, after the continental crust became mature by ca. 2.55 Ga in the Neoarchaean [65]. Here, up to 3 km-thick basaltic volcanites of the Sumian age (2.53–2.40 Ga) cover an area of more than 200 000 km^2. In the north, magmatic analogues of these volcanic rocks are represented by two belts of layered intrusions and numerous dike swarms [14,66,67]. These units have similar geological, compositional and metallogenic features [68]. Jointly, they compose the East-Scandinavian LIP. Its geological settings indicate anorogenic rift-like intraplate arrangements. They link volcano-plutonic belts connecting different domains of the Palaeoarchaean Kola-Lapland-Karelia protocontinent. This resembles the early advection extensional geodynamics of passive rifting that is typical of intraplate plume processes [69,70]. Geochemical and isotope-geochemical data (εNd values −1.1 to −2.4, I_{Sr} values 0.703–0.704) indicate the single homogenous mantle source of the LIP rocks (Figure 10). This mantle source was enriched with both siderophile and lithophile elements, including LREE. This reservoir resembles the modern EM-1 source [37,38].

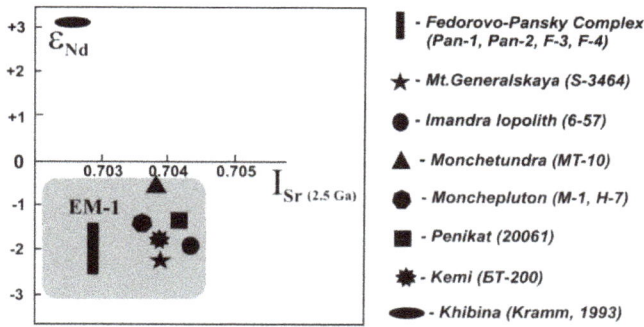

Figure 10. Plot of εNd-ISr for the KB and FKB layered intrusions. Grey color indicates an EM-1 reservoir at the time of the layered intrusions formation. Sm-Nd and Rb-Sr isotope data are provided in Table 9.

Isotope $^4He/^3He$ ratios are another reliable isotope tracer of mantle plume processes [56,70,71]. Their use in the Precambrian study requires further consideration. He isotope data on rocks and minerals of the KB intrusions show that the $^4He/^3He$ isotope ratios of $n \times 10^6$ and $n \times 10^5$ correspond to ratios of the upper mantle and differ from those of the crust ($n \times 10^8$) and lower mantle ($n \times 10^4$) [3,12,71]. According to these data, crustal contamination was local.

The LIP layered intrusions are directly related to the Fennoscandian Shield metallogeny [39]. The evolution of the two belts of layered mafic complex (2.53 to 2.40 Ga) encompasses a group of younger (2.44 Ga) intrusions within FKB [5,72].

The peak of the mafic-ultramafic magmatic activity in the Kola-Karelian, Superior and Wyoming provinces was at ca. 2.45 Ga (Figure 11) [3,12,67,69,72–74]. This study proves that layered intrusions in KB and FKB (2.53–2.40 Ga) had an intraplate nature. The Kola is considered as a part of the "greater" Karelia craton (Figure 11).

We consider a LIP as a product of a vast long-lived plume, according to geochemical data (Table 10). The studied East-Scandinavian LIP, or mafic LIP, has the following features:

- gravity anomalies caused by a crust-mantle layer at the bottom of the crust;
- riftogenic (anorogenic) structural ensembles with multipath extensional fault tectonics identified by the distribution of grabens and volcanic belts, dike swarms, and intrusive belts;
- protracted polyphase tectonics and magmatism, continental discontinuities and erosion with early stages of tholeiitic-basalt (trappean), boninite-like, and subalkaline magmatism in the continental crust, possible closing stages of the Red Sea-type spreading magmatism;
- intrusive sills, lopoliths, sheet-like bodies, large dikes and dike swarms. The intrusions are often layered and differ from rocks formed in subduction and spreading zones, with trends of thin differentiation layering, limited development of intermediate and felsic rocks, often with leucogabbro and anorthosite ends and abundant pegmatoid mafic varieties;
- typical mantle geochemistry of rocks and ores, as registered by isotope mantle tracers: $^{143}Nd/^{144}Nd$, $^{87}Sr/^{86}Sr$, $^{187}Os/^{188}Os$, and $^3He/^4He$;
- large orthomagmatic Cr, Ni, Cu, Co, PGE (±Au), Ti, and V deposits [18,69,72,75,76].

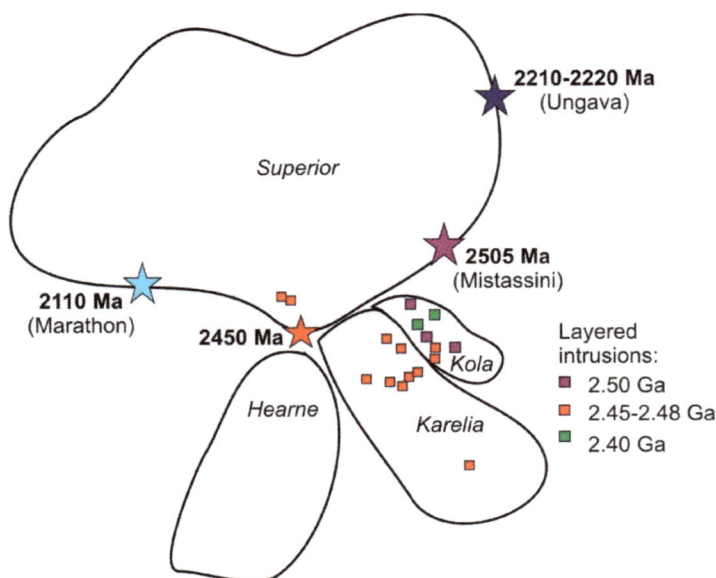

Figure 11. Correlation for the Superior, Karelia and Hearne cratons (based on Reference [72] with amendments).

Author Contributions: Conceptualization, T.B., A.K. and P.S.; Methodology, T.B., A.M., P.S., I.K. and D.E.; Validation, T.B.; Formal Analysis, N.E., E.N. and P.S.; Investigation, T.B., A.M., P.S., N.E., E.N., I.K. and D.E.; Sampling A.K. and M.H.; Writing—Original Draft Preparation, T.B.; Writing—Review & Editing, T.B. and M.H.; Supervision, T.B.

Funding: This research was funded by the Russian Foundation of Basic Researches (RFBR), grant numbers 18-35-00246, 16-05-00305 and 18-05-70082, Program of the Presidium of RAS No. 4.

Acknowledgments: The authors thank their colleagues J. Ludden, F. Corfu, W. Todt, U. Poller, L. Koval, L. Lyalina and N. Levkovich for their kind assistance in conducting the current research. The authors express their special gratitude to T.A. Miroshnichenko and E.N Steshenko for editing the manuscript and unknown reviewers for their contribution to the improvement of this paper.

Conflicts of Interest: The authors declare no conflict of interest.

References

1. Bogdanova, S.V.; Bingen, B.; Gorbatschev, R.; Kheraskova, T.N.; Kozlov, V.I.; Puchkov, V.N.; Volozh, Y.U. The East European Craton (Baltica) before and during the assembly of Rodinia. *Precambrian Res.* **2008**, *160*, 23–45. [CrossRef]

2. Slabunov, A.I.; Lobach-Zhuchenko, S.B.; Slabunov, A.I.; Lobach-Zhuchenko, S.B.; Bibikova, E.V.; Sorjonen-Ward, P.; Balagansky, V.V.; Volodichev, O.I.; Shchipansky, A.A.; Svetov, S.A.; et al. The Archaean nucleus of the Fennoscandian (Baltic) Shield. In *European Lithosphere Dynamics*; Gee, D.G., Stephenson, R.A., Eds.; Geological Society: London, UK, 2006; Volume 32, pp. 627–644.

3. Bayanova, T.; Mitrofanov, F.; Serov, P.; Nerovich, L.; Yekimova, N.; Nitkina, E.; Kamensky, I. Layered PGE Paleoproterozoic (LIP) Intrusions in the N-E Part of the Fennoscandian Shield—Isotope Nd-Sr and 3He/4He Data, Summarizing U-Pb Ages (on Baddeleyite and Zircon), Sm-Nd Data (on Rock-Forming and Sulphide Minerals), Duration and Mineralization. In *Geochronology—Methods and Case Studies/Edited by Nils-Axel Mörner*; INTECH: Houston, TX, USA, 2014; pp. 143–193. [CrossRef]

4. Alapieti, T.T. The Koillismaa layered igneous complex, Finland: Its structure, mineralogy and geochemistry, with emphasis on the distribution of chromium. *Geol. Surv. Finl.* **1982**, *319*, 116.

5. Iljina, M.; Hanski, E. Layered mafic intrusions of the Tornio-Näränkävaara belt. In *Precambrian Geology of Finland—Key to the Evolution of the Fennoscandian Shield*; Lehtinen, M., Nurmi, P.A., Rämo, O.T., Eds.; Elsevier: Amsterdam, The Netherland, 2005; pp. 101–138.

6. Vogel, D.C.; Vuollo, J.I.; Alapieti, T.T.; James, R.S. Tectonic, stratigraphic and geochemical comparison between ca. 2500-2440 Ma mafic igneous events in the Canadian and Fennoscandian Shields. *Precambrian Res.* **1998**, *92*, 89–116. [CrossRef]

7. Mitrofanov, F.P.; Balabonin, N.L.; Bayanova, T.B. Main results from the study of the Kola PGE-bearing province, Russia. In *Mineral Deposits*; Papunen, H., Gorbunov, G.I., Eds.; Balkema: Rotterdam, The Netherlands, 1997; pp. 483–486.

8. Ernst, R.E. *Large Igneous Provinces*; Cambridge University Press: Cambridge, UK, 2014; Volume 653.

9. Mitrofanov, F.P.; Korchagin, A.U.; Dudkin, K.O.; Rundkvist, T.V. Fedorovo-Pana layered mafic intrusion (Kola peninsula, Russia): Approaches, methods and criteria for prospecting PGEs. In *Exploration for Platinum-Group Elements Deposits*; Short Course Delivered on Behalf of the Mineralogical Association of Canada in Oulu, Finland; University of Toronto: Toronto, ON, Canada, 2005; Volume 35, pp. 343–358.

10. Schissel, D.; Tsvetkov, A.A.; Mitrofanov, F.P.; Korchagin, A.U. Basal platinum-group element mineralization in the Fedorov Pansky layered mafic intrusion, Kola Peninsula, Russia. *Econ. Geol.* **2002**, *97*, 1657–1677. [CrossRef]

11. Sharkov, Y.V. *Formation of Layered Intrusions and Related Mineralization*; Scientific World: Moscow, Russia, 2006; p. 364.

12. Bayanova, T.B.; Ludden, J.; Mitrofanov, F.P. Timing and duration of Palaeoproterozoic events producing ore-bearing layered intrusions of the Baltic Shield: Metallogenic, petrological and geodynamic implications. In *Palaeoproterozoic Supercontinents and Global Evolution*; Reddy, S.M., Mazumder, R., Evans, D.A.D., Collins, A.S., Eds.; Geological Society: London, UK, 2009; Volume 323, pp. 165–198.

13. Ekimova, N.A.; Serov, P.A.; Bayanova, T.B.; Elizarova, I.R.; Mitrofanov, F.P. New data on distribution of REEs in sulfide minerals and Sm-Nd dating of ore genesis of layered mafic intrusions. *Dokl. Earth Sci.* **2011**, *436*, 28–31. [CrossRef]

14. Kullerud, K.; Skjerlie, K.P.; Corfu, F.; De La Rosa, J. The 2.40 Ga Ringvassøy mafic dikes, West Troms Basement Complex, Norway: The concluding act of early Palaeoproterozoic continental breakup. *Precambrian Res.* **2006**, *150*, 183–200. [CrossRef]

15. Mitrofanov, F.P.; Bayanova, T.B.; Korchagin, A.U.; Groshev, N.Y.; Malitch, K.N.; Zhirov, D.V.; Mitrofanov, A.F. East Scandinavian and Noril'sk plume mafic Large Igneous Provinces of Pd-Pt ores: Geological and metallogenic comparison. *Geol. Ore Depos.* **2013**, *55*, 305–319. [CrossRef]

16. Balashov, Y.A.; Bayanova, T.B.; Mitrofanov, F.P. Isotope data on the age and genesis of layered mafic-ultramafic intrusions in the Kola Peninsula and northern Karelia, northeastern Baltic Shield. *Precambrian Res.* **1993**, *64*, 197–205. [CrossRef]

17. Zozulya, D.R.; Bayanova, T.B.; Nelson, E.G. Geology and age of the late Archaean Keivy alkaline province, NE Baltic Shield. *Geology* **2005**, *113*, 601–608. [CrossRef]

18. Godel, B.; Barnes, S.-J.; Maier, W.-D. Platinum-group elements in sulphide minerals, platinum-group minerals, and whole-rocks of the Merensky Reef (Bushveld Complex, South Africa): Implications for the formation of the reef. *J. Petrol.* **2007**, *48*, 1569–1604. [CrossRef]

19. Latypov, R.M.; Chistyakova, S.Y. *Mechanism for Differentiation of the Western-Pana Layered Intrusion*; Publ. of KSC RAS: Apatity, Russia, 2000; Volume 315.

20. Balabonin, N.L.; Subbotin, V.V.; Skiba, V.I.; Voitekhovsky, Y.L.; Savchenko, E.E.; Pakhomovsky, Y.A. Forms of occurrence and balance of precious metals in the Fedorovo-Pansky intrusion ores (the Kola Peninsula). *Obog. Rud.* **1998**, *6*, 24–30.

21. Subbotin, V.V.; Korchagin, A.U.; Savchenko, E.E. Platinometal mineralization of Fedorovo-Pansky ore region: Type of ore-bearing, mineral composition, particulary of genesis. *Vestn. KSC RAS* **2012**, *1*, 55–65.

22. Krogh, T.E. A low-contamination method for hydro-thermal dissolution of zircon and extraction of U and Pb for isotopic age determinations. *Geochim. Cosmochim. Acta* **1973**, *37*, 485–494. [CrossRef]

23. Ludwig, K.R. *PBDAT—A Computer Program for Processing Pb-U-Th Isotope Data*; Version 1.22. Open-File Report 88-542; U.S. Geological Survey: Reston, VA, USA, 1991; p. 38.

24. Ludwig, K.R. *ISOPLOT/Ex—A Geochronological Toolkit for Microsoft Excel*; Version 2.05; Berkeley Geochronology Center Special Publication: Berkeley, CA, USA, 1999; Volume 1a, p. 49.

25. Stacey, J.S.; Kramers, J.D. Approximation of terrestrial lead isotope evolution by a two-stage model. *Earth Planet. Sci. Lett.* **1975**, *26*, 207–221. [CrossRef]

26. Steiger, R.H.; Jäger, E. Subcommission on Geochronology: Convention on the use of decay constants in geo- and cosmochronology. *Earth Planet. Sci. Lett.* **1977**, *36*, 359–362. [CrossRef]

27. Jacobsen, S.B.; Wasserburg, G.J. Sm-Nd isotopic evolution of chondrites and achondrites. II. *Earth Planet. Sci. Lett.* **1984**, *67*, 137–150. [CrossRef]

28. Goldstein, S.J.; Jacobsen, S.B. Nd and Sr isotopic systematics of river water suspended material implications for crystal evolution. *Earth Plan. Sci. Lett.* **1988**, *87*, 249–265. [CrossRef]

29. Zhuravlyov, A.Z.; Zhuravlyov, D.Z.; Kostitsyn, Y.A.; Chernyshov, I.V. Determination of the Sm-Nd ratio for geochronological purposes. *Geochemistry* **1987**, *8*, 1115–1129.

30. DePaolo, D.J. Neodymium isotopes in the Colorado Front Range and crust-mantle evolution in the Proterozoic. *Nature* **1981**, *291*, 193–196. [CrossRef]

31. Liew, I.C.; Hofmann, A.W. Precambrian crustal components, plutonic associations, plate environment of the Hercinian Fold Belt of central Europe: Indications from a Nd and Sr isotopic study. *Contrib. Mineral. Petrol.* **1988**, *98*, 129–138. [CrossRef]

32. Van Achterbergh, E.; Ryanm, C.G.; Griffin, W.L. GLITTER: On-line interactive data reduction for the laser ablation ICP-MS microprobe. In Proceedings of the IX V.M. Goldschmidt Conference, Cambridge, MA, USA, 22–27 August 1999; Lunar and Planetary Institute Contribution No. 791: Houston, TX, USA, 1999; Volume 305.

33. Ekimova, N.A.; Serov, P.A.; Bayanova, T.B.; Yelizarova, I.R.; Mitrofanov, F.P. The REE distribution in sulphide minerals and Sm-Nd age determination for the ore-forming processes in mafic layered intrusions. *Dokl. RAS* **2011**, *436*, 75–78.

34. Nitkina, E.A. U-Pb zircon dating of rocks of the platiniferous Fedorova-Pana Layered Massif, Kola Peninsula. *Dokl. Earth Sci.* **2006**, *408*, 551–554. [CrossRef]

35. Serov, P.A.; Nitkina, E.A.; Bayanova, T.B.; Mitrofanov, F.P. Comparison of the new data on dating using U-Pb and Sm-Nd isotope methods of early barren phase rocks and basal ore-hosting rocks of the Pt-bearing Fedorovo-Pansky layered intrusion (Kola peninsula). *Dokl. Earth Sci.* **2007**, *415*, 1–3.

36. Faure, G. *Principles of Isotope Geology*; Wiley: New York, NY, USA, 1986; Volume 460.

37. Eisele, J.; Sharma, M.; Galer, S.J.G. The role of sediment recycling in EM-1 inferred from Os, Pb, Hf, Nd, Sr isotope and trace element systematic of the Pitcairn Hotspot, Earth Planet. *Sci. Lett.* **2002**, *196*, 197–212. [CrossRef]

38. Hofmann, A.W. Mantle geochemistry: The message from oceanic volcanism. *Nature* **1997**, *385*, 219–229. [CrossRef]

39. Yang, S.-H.; Hanski, E.; Li, C.; Maier, W.-D.; Huhma, H.; Mokrushin, A.V.; Latypov, R.; Lahaye, Y.; O'Brien, H.; Qu, W.-J. Mantle source of the 2.44–2.50 Ga mantle plume-related magmatism in the Fennoscandian Shield: Evidence from Os, Nd and Sr isotope compositions of the Monchepluton and Kemi intrusions. *Miner. Depos.* **2016**, *51*, 20. [CrossRef]

40. Mitrofanov, F.; Golubev, A. Russian Fennoscandian metallogeny. In Proceedings of the Abstracts of the 33 IGC, Oslo, Norway, 6–14 August 2008.

41. Richardson, S.H.; Shirey, S.B. Continental mantle signature of Bushveld magmas and coeval diamonds. *Nature* **2008**, *453*, 910–913. [CrossRef] [PubMed]

42. Maier, W.-D.; Halkoaho, T.; Huhma, H.; Hanski, E.; Barnes, S.-J. The Penikat Intrusion, Finland: Geochemistry, geochronology, and origin of platinum-palladium reefs. *J. Petrol.* **2018**, *59*, 967–1006. [CrossRef]

43. Perttunen, V.; Vaasjoki, M. U-Pb geochronology of the Peräpohja Schist Belt, northwestern Finland. In *Radiometric Age Determinations from Finnish Lapland and Their Bearing on the Timing of Precambrian Volcano-Sedimentary Sequences*; Vaasjoki, M., Ed.; Special Paper 33; Geological Survey of Finland: Espoo, Finland, 2001; pp. 45–84.

44. Mitrofanov, F.P.; Bayanova, T.B. Duration and timing of ore-bearing Palaeoproterozoic intrusions of Kola province. In *Mineral Deposits: Processes to Processing*; Stanley, C.J., Ed.; Balkema: Rotterdam, The Netherlands, 2002; pp. 1275–1278.

45. Amelin, Y.V.; Heaman, L.M.; Semenov, V.S. U-Pb geochronology of layered mafic intrusions in the eastern Baltic Shield: Implications for the timing and duration of Palaeoproterozoic continental rifting. *Precambrian Res.* **1995**, *75*, 31–46. [CrossRef]

46.	Kramm, U. Mantle components of carbonatites from the Kola Alkaline Province, Russia and Finland: A Nd-Sr study. *Eur. J. Mineral.* **1993**, *5*, 985–989. [CrossRef]

47.	Heaman, L.M.; Le Cheminant, A.N. Paragenesis and U-Pb systematics of baddeleyite (ZrO). *Chem. Geol.* **1993**, *110*, 95–126. [CrossRef]

48.	Fischer-Godde, M.; Becker, H.; Wombacher, F. Rhodium, gold and other highly siderophile element abundances in chondritic meteorites. *Geochim. Cosmochim. Acta* **2010**, *74*, 356–379. [CrossRef]

49.	Lyubetskaya, T.; Korenaga, J. Chemical composition of Earth's primitive mantle and its variance: 2. Implications for global geodynamics. *J. Geophys. Res.* **2007**, *112*. [CrossRef]

50.	Lyubetskaya, T.; Korenaga, J. Chemical composition of Earth's primitive mantle and its variance: 1. Method and results. *J. Geophys. Res.* **2007**, *112*. [CrossRef]

51.	Amelin, Y.V.; Semenov, V.S. U-Nd and Sr isotopic deochemistry of mafic layered intrusions in the eastern Baltic Shield: Implications for the evolution of Paleoproterozoic continental mafic magmas. *Contrib. Mineral. Petrol.* **1996**, *124*, 255–272. [CrossRef]

52.	Bayanova, T.B.; Smolkin, V.F.; Levkovich, N.V. U-Pb geochronological study of Mount Generalskaya layered intrusion, northwestern Kola Peninsula, Russia. In *Transactions of the Institution of Mining and Metallurgy*; Institution of Mining and Metallurgy: Vladikavkaz, Russia, 1999; Volume 108, pp. B83–B90.

53.	Tolstikhin, I.N.; Dokuchaeva, V.S.; Kamensky, I.L.; Amelin, Y.V. Juvenile helium in ancient rocks: II. U-He, K-Ar, Sm-Nd, and Rb-Sr systematics in the Monchepluton. ^{3}He/^{4}He ratios frozen in uranium-free ultramafic rocks. *Geochim. Cosmochim. Acta* **1992**, *56*, 987–999. [CrossRef]

54.	Vrevsky, A.B.; Levchenkov, O.A. Geological-geochronological scale of the endogenous processes operated within the Precambrian complexes of the central part of the Kola Peninsula. In *Geodynamics and Deep Structure of the Soviet Baltic Schield*; Mitrofanov, F.P., Bolotov, V.I., Eds.; KSC RAS: Apatity, Russia, 1992; Volume 150.

55.	Mitrofanov, F.P.; Balagansky, V.V.; Balashov, Y.A.; Gannibal, L.F.; Dokuchaeva, V.S.; Nerovich, L.I.; Radchenko, M.K.; Ryungenen, G.I. U-Pb age for gabbro-anorthosite of the Kola Peninsula. *Dokl. RAS* **1993**, *331*, 95–98.

56.	Bayanova, T.B. Baddeleyite: A promising geochronometer for alkaline and basic magmatism. *Petrology* **2006**, *14*, 187–200. [CrossRef]

57.	Huhma, H.; Cliff, R.; Perttunen, V.; Sakko, M. Sm-Nd and Pb isotopic study of mafic rocks associated with early Proterozoic continental rifting: The Perapohja schist belt in northern Finland. *Contrib. Mineral. Petrol.* **1990**, *104*, 369–379. [CrossRef]

58.	Alapieti, T.T.; Filen, B.A.; Lahtinen, J.J.; Lavrov, M.M.; Smolkin, V.F.; Voitekhovsky, Y.L. Early Proterozoic layered intrusions in the Northeastern part of the Fennoscandian Shield. *Contrib. Mineral. Petrol.* **1990**, *42*, 1–22. [CrossRef]

59.	Hanski, E.; Walker, R.J.; Huhma, H.; Suominen, I. The Os and Nd isotopic systematics of c. 2.44 Ga Akanvaara and Koitelainen mafic layered intrusions in northern Finland. *Precambrian Res.* **2001**, *109*, 73–102. [CrossRef]

60.	Bayanova, T.B.; Mitrofanov, F.P. Layered Proterozoic PGE intrusions in Kola region: New isotope data. In Proceedings of the Extended abstracts of the X International Symp. Platinum "Platinum-Group Elements—From Genesis to Beneficiation and Environmental Impact", Oulu, Finland, 8–11 August 2005; pp. 289–291.

61.	Nerovich, L.I.; Bayanova, T.B.; Serov, P.A.; Elizarov, D.V. Magmatic sources of dikes and veins in the Moncha Tundra Massif, Baltic Shield: Isotopic-geochronologic and geochemical evidence. *Geochem. Intern.* **2014**, *52*, 548–566. [CrossRef]

62.	Rundkist, T.V.; Bayanova, T.B.; Sergeev, S.A.; Pripachkin, P.V.; Grebnev, R.A. The Paleoproterozoic Vurechuaivench layered Pt-bearing pluton, Kola Peninsula: New results of the U-Pb (ID-TIMS, SHRIMP) dating of baddeleyite and zircon. *Dokl. Earth Sci.* **2014**, *454*, 1–6. [CrossRef]

63.	Mitrofanov, F.P.; Bayanova, T.B.; Zhirov, D.V.; Serov, P.A.; Golubev, A. Geological and isotope-geochemical characteristics of prediction and search method for the PGE-bearing mafic-ultramafic layered intrusions of the East-Scandinavian LIP. In Proceedings of the Abstracts of the XII Intern. Platinum Symp., Yekaterinburg, Russia, 11–14 August 2014; pp. 113–114.

64.	Chistyakov, A.V.; Bogatikov, O.A.; Grochovskaya, T.L.; Sharkov, E.V.; Belyatskiy, B.V.; Ovchinnikova, G.V. Burakovskiy layered pluton (S. Karelia) as a result of space combination double intrusive bodies: Petrological and isotope-geochemical data. *Dokl. RAS* **2000**, *372*, 228–235.

65.	Gorbatschev, R.; Bogdanova, S. Frontiers in the Baltic Shield. *Precambrian Res.* **1993**, *64*, 3–21. [CrossRef]

66. Vuollo, J.I.; Huhma, H. Palaeoproterozoic mafic dikes in NE Finland. In *Precambrian Geology of Finland*; Lehtinen, M., Nurmi, P.A., Rämö, O.T., Eds.; Elsevier: Amsterdam, The Netherlands, 2005; pp. 195–236.

67. Vuollo, J.I.; Huhma, H.; Stepana, V.; Fedotov, G.A. Geochemistry and Sm-Nd isotope studies of a 2.45 Ga dike swarm: Hints at parental magma compositions and PGE potential to Fennoscandian layered intrusions. In Proceedings of the IX Intern. Platinum Symp., Billings, MT, USA, 21–25 July 2002; pp. 469–470.

68. Coffin, M.F.; Eldholm, O. Large igneous provinces: Crustal structure, dimensions and external consequences. *Rev. Geophys.* **1994**, *32*, 1–36. [CrossRef]

69. Ernst, R.E.; Buchan, K.L. Recognizing mantle plumes in the Geological Record. *Earth Planet. Sci.* **2003**, *31*, 469–523. [CrossRef]

70. Pirajno, F. Mantle plumes, associated intraplate tectono-magmatic processes and ore systems. *Episodes* **2007**, *30*, 6–19.

71. Tolstikhin, I.N.; Marty, B. The evolution of terrestrial volatiles: A view from helium, neon, argon and nitrogen isotope modeling. *Chem. Geol.* **1998**, *147*, 27–52. [CrossRef]

72. Bleeker, W.; Ernst, R. Short-lived mantle generated magmatic events and their dike swarms: The key unlocking Earth's Paleogeographic record back to 2.6 Ga. In *Dike Swarms-Time Marker of Crustal Evolution*; Balkema Publishers: Zürich, Switzerland, 2006; pp. 1–20.

73. Campbell, I.H. Identification of ancient mantle plumes. In *Mantle Plumes: Their Identification Through Time*; Ernst, R.E., Buchan, K.L., Eds.; Geological Society of America: Boulder, CO, USA, 2001; Volume 352, pp. 5–22.

74. Heaman, L.M. Global mafic magmatism at 2.45 Ga: Remnants of an ancient large igneous province? *Geology* **1997**, *25*, 299–302. [CrossRef]

75. Bogatikov, O.A.; Kovalenko, V.I.; Sharkov, Y.V. *Magmatism, Tectonics, and Geodynamics of the Earth*; Nauka: Moscow, Russia, 2010; 606p.

76. Grachyov, A.F. Identification of mantle plumes on the basis of studying composition and isotope geochemistry of volcanic rocks. *Petrology* **2003**, *11*, 618–654.

![minerals logo] *minerals*

MDPI

Article

The Main Anorthosite Layer of the West-Pana Intrusion, Kola Region: Geology and U-Pb Age Dating

Nikolay Y. Groshev [1],* and Bartosz T. Karykowski [2]

[1] Geological Institute of the Kola Science Center of the Russian Academy of Sciences, 184209 Apatity, Russia
[2] Fugro Germany Land GmbH, 12555 Berlin, Germany; bkarykowski@yahoo.com
*   Correspondence: nikolaygroshev@gmail.com; Tel.: +8-951-296-2355

Received: 19 December 2018; Accepted: 22 January 2019; Published: 26 January 2019

check for updates

Abstract: The West-Pana intrusion belongs to the Paleoproterozoic Fedorova-Pana Complex of the Kola Region in NW Russia, which represents one of Europe's most significant layered complexes in terms of total platinum group element (PGE) endowment. Numerous studies on the age of the West-Pana intrusion have been carried out in the past; however, all published U-Pb isotope ages were determined using multi-grain ID-TIMS. In this study, the mineralized Main Anorthosite Layer from the upper portion of the intrusion was dated using SHRIMP-II for the first time. High Th/U (0.9–3.7) zircons gave an upper intercept age of 2509.4 ± 6.2 Ma (2σ), whereas the lower portion of the intrusion was previously dated at 2501.5 ± 1.7 Ma, which suggests an out-of-sequence emplacement of the West-Pana intrusion. Furthermore, high-grade PGE mineralization hosted by the anorthosite layer, known as "South Reef", can be attributed to (1) downward percolation of PGE-enriched sulfide liquid from the overlying gabbronoritic magma or (2) secondary redistribution of PGEs, which may coincide with a post-magmatic alteration event recorded by low Th/U (0.1–0.9) zircon and baddeleyite at 2476 ± 13 Ma (upper intercept).

Keywords: PGE; South Reef; West-Pana intrusion; Fedorova-Pana Complex; zircon dating; U-Pb

1. Introduction

The West-Pana intrusion represents the central block of the Paleoproterozoic Fedorova-Pana Complex located in the central part of the Kola Peninsula. It hosts several platinum group element (PGE) deposits that are interpreted to represent contact- and reef-style mineralization (Figure 1). The Fedorova-Tundra deposit occurs at the basal contact of the complex [1], whereas the North Kamennik, Kievey, and East Chuarvy deposits [2–4] are located at different stratigraphic levels of the complex and comprise one of Europe's largest PGE resource, exceeding 400 t of precious metals [5,6]. The West-Pana intrusion is the first layered intrusion known in Russia that hosts low-sulfide PGE mineralization at several stratigraphic levels, which share many similarities with the well-known Merensky Reef of Bushveld Complex and the J-M Reef of the Stillwater Complex, respectively [7].

In general, PGE mineralization at West-Pana occurs in two distinct layered horizons, both consisting of interlayered pyroxenite, norite, gabbronorite, and anorthosite. Anorthosites of the Lower Layered Horizon form relatively thin and discontinuous layers and host continuous PGE mineralization known as the "North Reef" [8]. Among the anorthosites of the Upper Layered Horizon, the thickest layer is the "Main Anorthosite Layer" [9], hosting highly discontinuous PGE-rich sulfide mineralization in the its upper two meters, which is referred to as the "South Reef" [10,11].

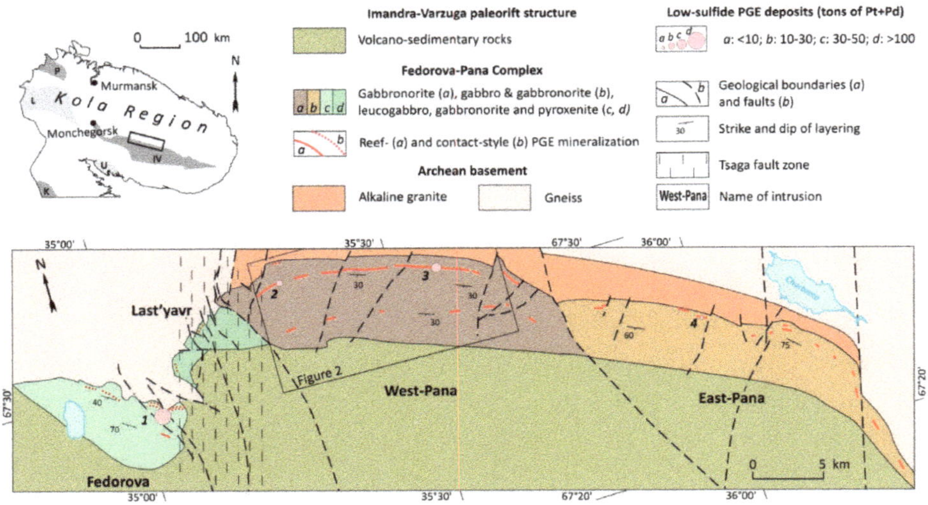

Figure 1. Simplified geological map of the Fedorova-Pana Complex, showing the location of low-sulfide PGE deposits (**1**, Fedorova Tundra; **2**, North Kamennik; **3**, Kievey; **4**, East Chuarvy). Modified after [12]. Proterozoic structures of the predominantly Archean Kola Region (inset): Imandra-Varzuga (IV), Kuolajarvi (K), and Pechenga (P) paleorift structures; Lapland (L) and Umba (U) granulite belts.

It is generally believed that the formation of PGE mineralization in most Russian Paleoproterozoic layered intrusions is related to the long duration of magmatism associated with prolonged igneous activity in response to a long-lived mantle plume affecting the Kola Craton for more than 50 Ma, which is mainly based on ID-TIMS U-Pb age dating [13,14]. This interpretation is generally at odds with the current paradigm of relatively short-lived mantle plume magmatism and the duration of cooling of basaltic magma chambers, such as the Bushveld Complex, which crystallized in less than 1 Ma [15].

The oldest published age for the West-Pana intrusion is 2501.5 ± 1.7 Ma for a gabbronorite from the Lower Layered Horizon [16], whereas the youngest age is 2447 ± 12 Ma for the Main Anorthosite Layer [13]. In this study, we provide new insights into the geology and the petrogenesis of the Main Anorthosite Layer based on recent drilling and U-Pb dating of zircon using SHRIMP-II. The results are discussed in the context of previous age dates from the complex, thus constraining the emplacement and crystallization history of the Fedorova-Pana Complex and anorthosite-hosted PGE mineralization. The study emphasizes the need for more high-precision zircon dating to elucidate the entire emplacement history of the complex.

2. Geological Setting

The Fedorova-Pana Complex includes an almost continuous strip of NW–SE-trending layered intrusions, located at the northern edge of the Imandra-Varzuga paleorift structure (Figure 1). The total extent of the strip is about 90 km in length and up to 6–7 km wide. The northern contact zone of all intrusions is composed of fine-grained gabbroic rocks that intruded the basement lithologies. The rocks in these zones are generally foliated and modified to epidote-amphibolite facies. The southern contact of the complex is defined by a northwest-trending fault with a dip angle of 40–50°, along which younger volcano-sedimentary rocks were thrust onto the complex. The Fedorova-Pana Complex consists of four intrusions, which are from west to east: Fedorova, Last'yavr, West-Pana, and East-Pana (Figure 1). Although all these intrusions are generally presented and explained as a single entity [1,8,16], most researchers believe that each intrusion represents a separate magma chamber with a distinct stratigraphy and formation history [4,17–21].

The West-Pana intrusion is a sheet-like 4 km-thick body, extending for more than 25 km along strike (Figure 2). The magmatic layering dips southwest at an angle of approximately 30–35° [22]. The stratigraphy of West-Pana is rather simple: the lowermost portion is represented by a thin Norite Zone (50 m) that is underlain by a marginal zone comprised of fine-grained gabbronorite, which is often strongly altered due to tectonic activity along the lower intrusion contact. The remainder of the intrusion is essentially unaltered and consists of massive gabbroic rocks of the Gabbronorite Zone except for two distinct horizons: the lower and upper layered horizons. The Lower Layered Horizon (LLH) is located some 600–800 m above the lower intrusion contact and is composed of several cyclic units, consisting of pyroxenite, gabbronorite, leucogabbro, and anorthosite with an average total thickness of 40 m [23]. Significant low-sulfide Pt-Pd mineralization is predominantly concentrated in the second cycle of the LLH, which is referred to as the "North Reef" [24]. Moreover, the LLH and the overlying massive gabbronorites are intruded by late magnetite gabbro [25]. The Upper Layered Horizon (ULH) is situated about 3000 m above the base of the intrusion and consists of two distinct parts with a total thickness of 300 m [26]. The lower part is characterized by a 100 m-thick zone of interlayered norite, gabbronorite, and anorthosite, whereas the upper part consists of cyclically interlayered olivine gabbronorite, troctolite, and anorthosite, which is often referred to as the "Olivine Horizon". The low-sulfide PGE mineralization is associated with both parts of the ULH, but it does not form a continuous ore body. The most significant PGE mineralization is hosted by the "South Reef", which occurs within the Main Anorthosite Layer, representing the thickest anorthosite layer in the lower part of the ULH.

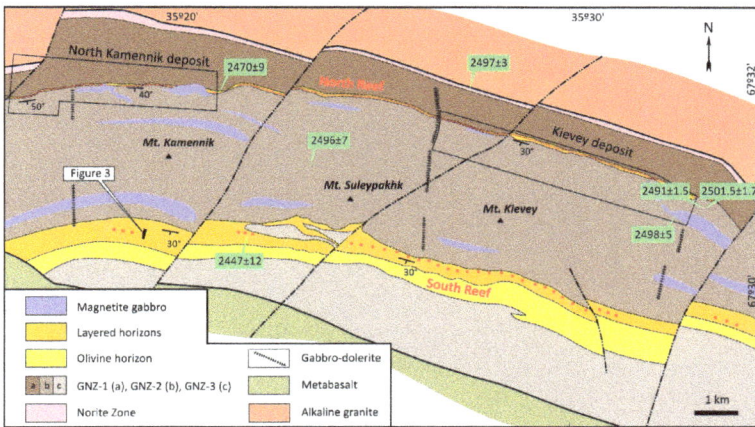

Figure 2. Simplified geological map of the West-Pana intrusion. Published U-Pb ages are shown in green rectangles. Note that the intrusion was explored for PGE mainly along strike of the North and South Reefs, but sub-economic to economic deposits were only discovered in the former. Abbreviations: GNZ, Gabbronorite Zone. Modified after [3].

The 10–17-m-thick Main Anorthosite Layer on the southern slope of Mt. Kamennik can be traced for up to 2 km based on drilling and outcrop mapping. In contrast, the eastern portion of the Main Anorthosite Layer at Mts. Suleypakhk and Kievey has a confirmed strike length of at least 10 km (Figures 2 and 3). The underlying lithology is a medium-grained gabbronorite (Figure 4A), and the contact between the gabbronorite and the anorthosite is gradational (Figure 4B). The overlying unit is composed of medium-grained, sometimes inequigranular gabbronorite that has a sharp contact with the underlying anorthosite (Figure 4C–F). Locally, a discontinuous, 1–2-m-thick norite layer occurs at the base of the overlying gabbronorite. These norites contain traces of PGE mineralization (<0.8 ppm Pd), as well as inequigranular gabbronorites associated with hornfels xenoliths (Figure 3) [27]. High-grade PGE mineralization with up to 33 ppm Pd is concentrated in the uppermost two meters of the anorthosite layer and is known as the "South Reef" [8,11].

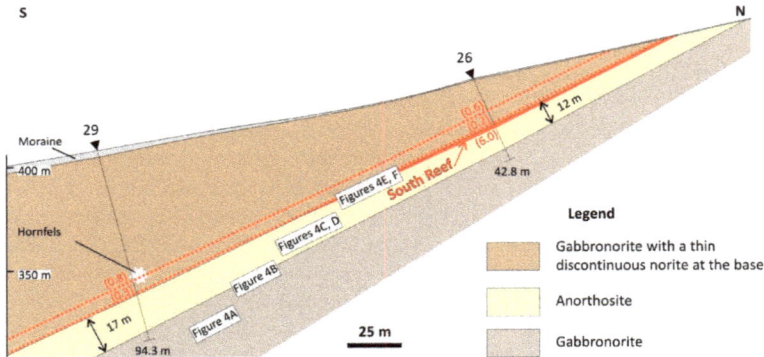

Figure 3. Simplified cross-section of the Main Anorthosite Layer from the West-Pana intrusion based on internal data from JSC Pana. The South Reef is shown as a solid red line, whereas PGE mineralization is indicated by red dotted lines. Note that the hornfels-hosted PGE mineralization (Borehole 29) is traced up-dip in the inequigranular gabbronorites (Borehole 26). Maximum Pd concentrations in drill core samples are shown in parentheses.

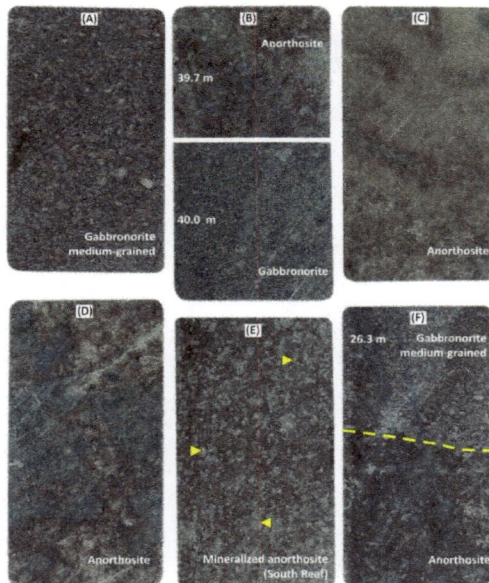

Figure 4. Different rock types from the Main Anorthosite Layer and its host rocks (Borehole 30). (**A**) Medium-grained gabbronorite from the footwall of the Main Anorthosite Layer (depth: 42.15–42.30 m). (**B**) Thirty centimeter-thick gradational lower contact of the Main Anorthosite Layer (depth: 39.7–40.0 m). (**C**) Monomineralic anorthosite (depth: 33.20–33.35 m). (**D**) Mottled anorthosite (depth: 29.45–29.6 m). (**E**) Mineralized anorthosite from the "South Reef". The sample contains 2 ppm Au, 3 ppm Pt, and 33 ppm Pd, respectively. Note the replacement of interstitial minerals by secondary epidote and sulfides (depth: 27.75–27.90 m). (**F**) Sharp upper contact (dashed line) between anorthosite and the overlying medium-grained gabbronorite (depth: 26.30–26.45 m).

These "South Reef" anorthosites are coarse-grained cumulate rocks with a mottled texture, containing some 75–98 vol. % plagioclase, intercumulus quartz, ortho-, and clino-pyroxene, as well as secondary amphibole, biotite, epidote with minor amounts of chalcopyrite, bornite, millerite, pentlandite, pyrrhotite, magnetite, and ilmenite. Accessory minerals include zircon, baddeleyite, apatite, titanite, and rutile. More than three tens PGE and Au minerals occur in the mineralized anorthosite [11,27]. The mottled texture of the anorthosite is defined by the local concentration of plagioclase crystals in distinct areas (Figure 4C), whereas other parts are strongly affected by autometamorphic processes, leading to the complete replacement of the initial intercumulus mineral assemblages by secondary amphibole and epidote (Figure 4D). The mineralized anorthosite contains 2–5 vol. % disseminated sulfide (Figure 4E), mostly hosted by secondary epidote interstitial to cumulus plagioclase (Figure 5A).

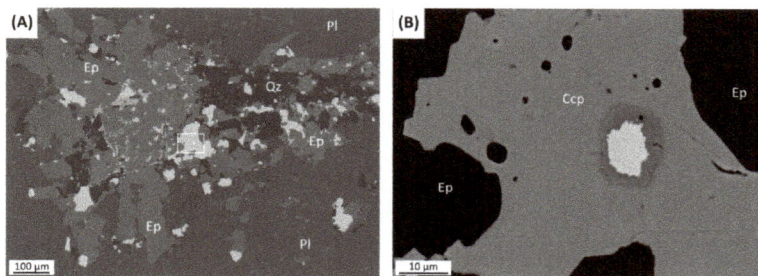

Figure 5. Back-scatter electron images of (**A**) finely-disseminated sulfides (light) intergrown with epidote, replacing the intercumulus space in the mineralized anorthosite and (**B**) baddeleyite rimmed by zircon hosted by chalcopyrite (Ccp). Abbreviations: Ep, epidote; Qz, quartz; Pl, plagioclase.

Moreover, the stratigraphic position of the "South Reef" PGE mineralization appears to be unrelated to changes in the mineral composition of cumulus rock-forming minerals, whereas the position of the "North Reef" coincides with a distinct increase in the anorthite content of plagioclase [9,28,29]. Unlike the barren anorthosite with unzoned plagioclase, the mineralized anorthosites of the "South Reef" are characterized by pronounced zonation of cumulus plagioclase, showing distinct brown rims that have a similar compositions to plagioclase rims from the overlying cumulate [9]. Furthermore, the composition of braggite and vysotskite from the "South Reef" indicates that the crystallization temperature of these minerals is about 750 °C, which is well below the crystallization temperature of these minerals from other deposits across the Fedorova-Pana Complex (830–920 °C) [30]. Thus, several lines of evidence indicate that the low-sulfide PGE mineralization of the "South Reef" is secondary in nature, either post-magmatic or locally remobilized, but the source and processes leading to sulfide concentration in the uppermost portion of the anorthosite layer remain unknown.

3. Materials and Methods

Three drill cores (26, 29, 30) intersecting the Main Anorthosite Layer were used for this study (JSC Pana, 2012–2013). A detailed overview of the petrography and mineral chemistry of the Main Anorthosite Layer is given in [9,11,12,26].

About 30 zircon grains with a size of 50–250 μm were separated from sample BG29 (~10 kg; Borehole 29 in Figure 3) using the methodology described in [31]. The zircon textures were investigated using optical microscopy, cathodoluminescence (CL), and back-scatter electron (BSE) images (Figure 6). The CL and BSE imaging were performed on a CamScan MX2500 scanning electron microscope equipped with a CLI/QUA2 system at the Centre of Isotopic Research of the Russian Geological Research Institute (CIR VSEGEI) in St. Petersburg, Russia.

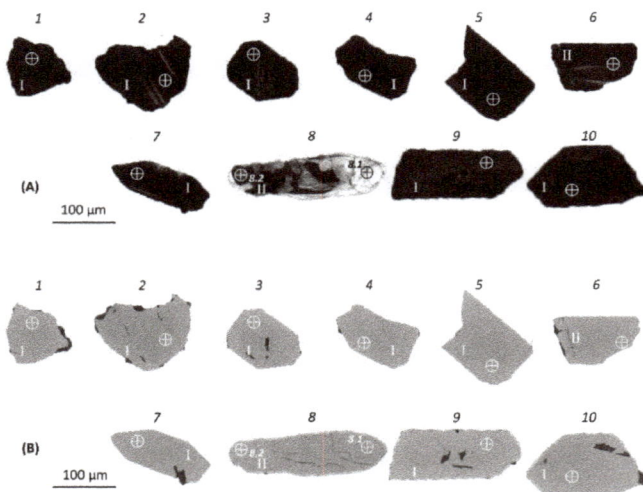

Figure 6. Images of different zircons from Sample BG29. (**A**) Cathodoluminescence (CL) images. (**B**) BSE images. Crosshairs on grains and Arabic numerals correspond to the points of analyses in Table 1; Roman numerals show zircon groups.

The analyses of U-Pb isotope ratios in zircon were carried out on a SHRIMP-II secondary-ion mass spectrometer at CIR VSEGEI using the method outlined in [32,33]. The intensity of the primary molecular oxygen beam was 4 nA; the size of the sampling crater was 20×25 μm with a depth of 2 μm. Correction for non-radiogenic Pb was carried out using the measured ^{204}Pb and the modern isotopic composition of Pb from [34]. The data processing was conducted using the software SQUID 1 [35], including concordia age calculation. The analytical results are shown in Table 1.

Table 1. SHRIMP and ID-TIMS U-Pb data for zircon and baddeleyite (bd) from the Main Anorthosite Layer of the West-Pana intrusion. Errors in isotopic ratios and ages are at the 1σ level. Pb_C and Pb^* indicate the common and radiogenic lead portions, respectively. The error in standard calibration was 0.51%. Common Pb corrected using measured ^{204}Pb. Abbreviations: D, discordance; Rho, correlation coefficient; nd, no data.

Spot/Fraction Name	$^{206}Pb_c$ (%)	Concentrations/Ratios				$^{206}Pb/^{238}U$-age (Ma)	$^{207}Pb/^{206}Pb$-age (Ma)	D (%)	Isotope Ratios						
		U (ppm)	Th (ppm)	Th/U	$^{206}Pb^*$ (ppm)				$\frac{^{207}Pb^*}{^{206}Pb^*}$	1σ (%)	$\frac{^{207}Pb^*}{^{235}U}$	1σ (%)	$\frac{^{206}Pb^*}{^{238}U}$	1σ (%)	Rho
SHRIMP data, this study															
1.1	0.01	787	1003	1.32	302	2383 ± 34	2497.0 ± 3.8	5	0.1640	0.22	10.110	1.7	0.4472	1.7	0.992
2.1	0.01	304	310	1.06	119	2416 ± 29	2507.6 ± 6.2	4	0.1650	0.37	10.350	1.5	0.4547	1.5	0.969
3.1	0.05	186	141	0.78	74	2448 ± 30	2506.0 ± 7.9	2	0.1649	0.47	10.500	1.5	0.4618	1.5	0.952
4.1	0.03	444	373	0.87	179	2475 ± 29	2501.3 ± 4.9	1	0.1644	0.29	10.610	1.5	0.4681	1.4	0.980
5.1	0.05	399	1436	3.72	146	2292 ± 46	2497.7 ± 5.5	9	0.1640	0.33	9.660	2.4	0.4270	2.4	0.991
6.1	0.04	210	28	0.14	89	2586 ± 31	2499.6 ± 7.3	−3	0.1642	0.43	11.170	1.5	0.4935	1.5	0.959
7.1	0.10	228	268	1.22	92	2480 ± 30	2511.6 ± 7.2	1	0.1654	0.43	10.700	1.5	0.4691	1.5	0.960
8.1	0.24	29	25	0.87	10	2160 ± 36	2340.0 ± 25.0	8	0.1495	1.50	8.200	2.4	0.3980	1.9	0.794
8.2	0.05	416	120	0.30	116	1818 ± 23	2159.0 ± 14.0	19	0.1346	0.83	6.046	1.7	0.3258	1.4	0.865
9.1	0.06	200	405	2.09	82	2516 ± 30	2513.2 ± 7.7	0	0.1656	0.46	10.900	1.5	0.4775	1.5	0.954
10.1	0.04	371	333	0.93	151	2498 ± 30	2506.4 ± 8.6	0	0.1649	0.51	10.760	1.5	0.4733	1.4	0.942
ID-TIMS data from [13]															
P6-2	nd	1331	nd	nd	743	nd	2438	nd	nd	nd	9.588	0.5	0.4393	0.5	0.865
P6-3	nd	577	nd	nd	286	nd	2474	nd	nd	nd	8.643	0.5	0.3874	0.5	0.794
P5-bd	nd	396	nd	nd	176	nd	2435	nd	nd	nd	9.548	0.5	0.4380	0.5	0.865
P6-bd	nd	560	nd	nd	259	nd	2443	nd	nd	nd	9.956	0.5	0.4533	0.5	0.794

4. Results of Zircon Imaging and U-Pb SHRIMP Dating

The studied set of zircons is relatively heterogeneous and can be divided into two distinct groups based on age, morphology, texture, and composition (Figure 6). The analytical results of the U-Pb isotope dating are given in Table 1 and plotted in the concordia diagram in Figure 7.

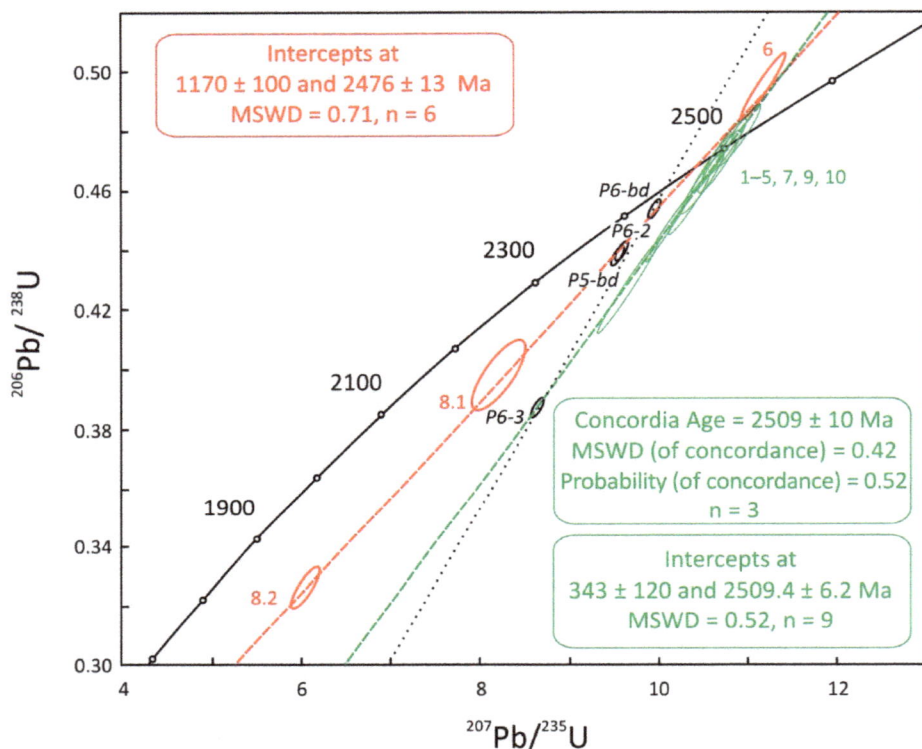

Figure 7. Concordia diagram for zircons separated from sample BG29 (coarse-grained anorthosite) with zircon and baddeleyite data from samples P5 and P6 (italic). Green ellipses show zircon with high Th/U ratios (mainly 0.9–3.7); red ellipses represent zircon with low Th/U ratios (0.1–0.9); black ellipses show zircon and baddeleyite (bd) from [13]. The dotted line shows the discordia exclusively based on ID-TIMS data.

The first group of zircons (eight grains: 1–5, 7, 9, 10) comprises fragments of large columnar crystals, showing a weak zonation in CL images. Most of these zircons have a low discordance (Table 1).Three of them (7, 9, 10) have a discordance close to zeroand plot on the concordia with a calculated age of 2509 ± 10 Ma (including decay constant errors; MSWD = 0.42; concordance probability is 0.52). Three relatively discordant zircon grains of this group (1, 2, 5) together with concordant zircons form a discordia, which includes the ID-TIMS multi-grain zircon analysis P6-3 (Figure 7). This data point was taken from a previous study and represents a zircon from the same lithology [13]. The upper intercept age of the resulting composite discordia (n = 9) is 2509.4 ± 6.2 Ma (MSWD = 0.52), whereas the lower intercept corresponds to an age of 343 ± 120 Ma.

The second group of zircon is represented by two grains (6, 8): the first grain strongly resembles zircon from the first group in terms of morphology and internal texture, but due to its composition, it was included in this group (cf. Th/U ratios in Table 1); the second zircon is elongated with a round shape and shows distinct internal domaining (Figure 6). Figure 7 shows that these Group 2 zircons form a discordia together with zircon (P6-2) and baddeleyite (P5-bd, P6-bd) from the same rocks analyzed by ID-TIMS in a previous study [13]. The upper intercept age of this composite discordia is 2476 ± 13 Ma (MSWD = 0.71).

5. Discussion

5.1. U-Pb Age of the Main Anorthosite Layer and Crystallization History of the West-Pana Intrusion

The existing U-Pb age of 2447 ± 12 Ma for the Main Anorthosite Layer (Table 2) was determined by multi-grain ID-TIMS on zircon and baddeleyite [13]. Since baddeleyite generally crystallizes as a late-stage mineral in layered intrusions [36], this age was considered to indicate the crystallization age of the anorthosite. Therefore, the Main Anorthosite Layer was interpreted to represent an additional sill-like intrusion that was emplaced some 50 Ma after the crystallization of the West-Pana intrusion [13,37].

Table 2. Published isotope U-Pb rock ages of the Fedorova-Pana Complex.

Intrusions	Rock Type	Age (Ma)	Mineral	References
Fedorova	gabbronorite min.	2485 ± 9	4 Zrn, SD	[38]
	gabbronorite min.	2493 ± 8	4 Zrn, SD	[39]
	orthopyroxenite	2526 ± 6	4 Zrn, SD	[38]
	leucogabbro min.	2518 ± 9	3 Zrn, SD	[39]
	leucogabbro	2515 ± 12	4 Zrn, SD	[39]
	leucogabbro	2516 ± 7	3 Zrn, SD	[38]
	leucogabbronorite	2507 ± 11	6 Zrn, D	[39]
West-Pana	norite	2497 ± 3	4 Zrn, SD	[38]
	gabbronorite	2496 ± 7	3 Zrn, D	[38]
	gabbronorite	2491 ± 1.5	3 Zrn, D	[13]
	gabbronorite	2501.5 ± 1.7	3 Zrn, C	[16]
	gabbro-pegmatite	2470 ± 9	3 Zrn, DC	[40]
	magnetite gabbro	2498 ± 5	3 Zrn, DC	[13]
	anorthosite	2447 ± 12	3 Zrn + 2Bdy, DC	[13]
East-Pana	gabbro	2487 ± 10	4 Zrn, SD	[12]
	gabbro-pegmatite	2464 ± 12	2 Zrn + 2Bdy, SD	[41]

C, concordant zircons; DC, discordant zircons with concordant zircon(s); D, discordant zircons; SD, strongly-discordant zircons; min., mineralized. All errors are reported as 2σ.

Based on the U-Pb SHRIMP-II dating of zircon from the Main Anorthosite Layer, two stages in the formation of the layer can be distinguished: (1) a magmatic stage and (2) a post-magmatic metasomatic stage. The magmatic stage is mainly represented by zircons from the first group and characterized by relatively high Th/U ratio, ranging from 0.9–3.7 (Table 1). Therefore, the calculated concordia age of 2509 ± 10 Ma and the slightly more precise upper intercept age of 2509.4 ± 6.2 Ma most likely represent the actual crystallization age of the anorthosite. In contrast, the second group of zircons has lower Th/U ratios, ranging from 0.1–0.9 (Table 1) and belongs to the post-magmatic metasomatic stage. Consequently, the upper intercept age of 2476 ± 13 Ma records the autometasomatic overprint of the anorthosites. It should be noted that late-stage baddeleyite in these anorthosites is unlikely to be magmatic as it mostly occurs together with secondary amphibole, epidote, and with presumably remobilized sulfide mineralization (Figure 5B), rather than with a typical magmatic interstitial mineral assemblage. This may explain the large difference between the 2509.4 ± 6.2 Ma age determined by SHRIMP-II and the 2447 ± 12 Ma ID-TIMS age for the same lithological unit. A potential mechanism for the autometasomatic overprint of the anorthosites may have been the downward infiltration of residual melts from the overlying gabbronoritic unit (Figure 3).

Geological relationships coupled with U-Pb dating of the Main Anorthosite Layer suggest that the layer crystallized coevally with the adjacent rocks of the ULH, but earlier than the LLH and other rocks from the lower portions of the West-Pana intrusion (Figures 2 and 8). This conclusion is consistent with available age dates from the layered series of the Fedorova intrusion that range from 2526–2507 Ma, whereas the basal marginal series is younger with 2493–2485 Ma (Table 2), although secondary overprinting may have obscured the actual crystallization ages. Notably, the Main Anorthosite Layer shares many similarities with the Anorthosite zones in the Middle Banded Series of the Stillwater Complex as these anorthosites are older than the underlying rock sequences, which also host the J-M Reef. This was interpreted to suggest an out-of-sequence emplacement of the Stillwater Complex, which could also apply to the Fedorova-Pana Complex [42].

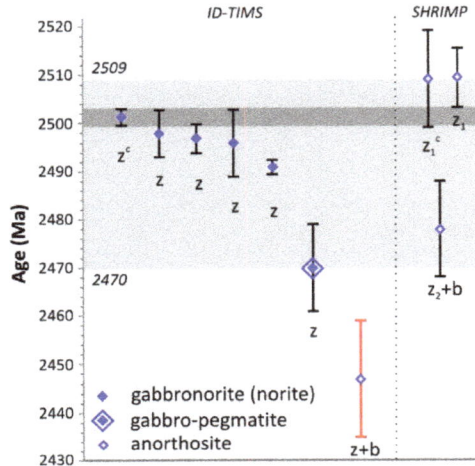

Figure 8. Overview of published U-Pb zircon (z) and baddeleyite (b) ages from different West-Pana lithologies. See Table 2 for references; C, mean ^{207}Pb/^{206}Pb age or concordia age. Note that the upper intercept age for magmatic zircon (z_1) from the Main Anorthosite Layer (2509.4 ± 6.2 Ma) is older than the mean ^{207}Pb/^{206}Pb age of the lower part of the intrusion (2501.5 ± 1.7 Ma, dark grey field), potentially indicating that (1) the upper portion of the intrusion is older than the lower portion and (2) the previous ID-TIMS date for this lithological unit (red error bar) did not record the actual crystallization age.

In terms of the crystallization history of the West-Pana intrusion and its long duration based on available age dating (Figure 8), this study shows that the geochronological questions are currently far from being conclusively answered. More modern and reliable high-precision age dating is needed to be able to resolve the entire crystallization history of not only the West-Pana intrusion, but the Fedorova-Pana Complex as a whole (Table 2, Figure 8). Considering the results of this study, the 2470 ± 9 Ma age for the gabbro-pegmatite from the LLH should be regarded as an upper temporal boundary for the crystallization of the intrusion (Figure 8). It appears, however, that this age most likely records the timing of late- to post-magmatic overprinting rather than the actual timing of emplacement, taking into account that high-precision dating of other large layered intrusions, such as the Bushveld or Stillwater Complexes, suggest a much shorter duration of magmatism, lasting for a few million years at most [15,42].

The main challenge associated with establishing a sound geochronological emplacement history for the Fedorova-Pana Complex is the precise dating of different mineralized and unmineralized lithologies from the Fedorova Tundra, the Northern Kamennik, and the Kievey deposits using the same methodology. This may potentially show that all these PGE deposits belong to the same mineral system and that they formed at the same time. These types of studies are necessary and feasible as was demonstrated for the PGE deposits of the Bushveld [15,43] and Stillwater Complexes [44].

5.2. Implications for the Formation of the South Reef

The formation of the South Reef PGE mineralization hosted in the uppermost portions of the Main Anorthosite Layer is one of the most important unresolved issues associated with the Fedorova-Pana Complex. Immediately after the discovery of mineralized rocks, containing tens of ppm Pd, the South Reef was considered to be highly prospective for further exploration [8]. Additional work on the South Reef, however, showed that the continuity of the high-grade PGE mineralization was generally limited to a few meters along strike of the Main Anorthosite Layer (cf. Boreholes 26 and 29 in Figure 3).

Based on the notion that the anorthosites in the West-Pana intrusion represented late sill-like bodies [13,36] and the presence of PGE-enriched rocks in the overlying and underlying gabbronoritic units [12], it was assumed that the PGEs were derived from the older gabbronorites that initially contained low-grade PGE mineralization. Upon intrusion of the anorthosites, this low-grade PGE mineralization was assimilated and enriched in the uppermost portions of the Main Anorthosite Layer [37].

The results of this study indicate that the anorthosites likely represent a regular part of the stratigraphy of West-Pana rather than late sill-like intrusions. This is further supported by the gradational lower contact of the anorthosite layer, which is characterized by progressively-increasing modal abundances of plagioclase (Figure 4B). Moreover, the anorthite content of cumulus plagioclase from the host gabbronorites and the anorthosite ranges from 73–75 mol. %, showing little variation across the contact [9]. It appears that the underlying gabbronorites together with the anorthosite represent the same cyclic unit of the ULH, while the overlying gabbronorites defines the base of another cyclic unit. The most likely source of the high-grade PGE mineralization hosted by the South Reef is the overlying gabbronoritic unit, which contains low-grade PGE mineralization associated with hornfels xenoliths, inequigranular gabbronorites, and norites. Two processes can be envisaged as potential mechanisms for the concentration of PGE in the anorthosite layer: (1) downward percolation of sulfide liquid from the overlying gabbronoritic magma into the uppermost portion of the anorthosite layer or (2) secondary redistribution of PGEs, which may coincide with the younger ages recorded by post-magmatic zircon and baddeleyite. The typical magmatic sulfide assemblage and the relatively high IPGE/PPGE ratios, however, argue strongly against a secondary origin of the mineralization.

Assuming that the most significant PGE enrichments in the Kola Region are associated with additional intrusions of sulfide-saturated and somewhat PGE-enriched magmas, as suggested for the Monchegorsk Complex [45,46] and the Fedorova intrusion [1], it is likely that the gabbronorites, overlying the Main Anorthosite Layer, also formed as a result of a late-stage intrusion of sulfide-saturated, PGE-enriched magma into the pre-existing cumulate pile. Further evidence for this mechanism is provided by the presence of a thin, discontinuous norite layer, as well as abundant hornfels xenoliths directly above the anorthosite layer (Figure 3). This late-stage intrusion may have led to the infiltration of PGE-enriched sulfide melt into the interstitial space of the underlying anorthosite cumulates [47,48], somewhat similar to contact-style sulfide mineralization, infiltrating basement lithologies that are in direct contact with the intrusion [46,49].

6. Concluding Remarks

The mineralized Main Anorthosite Layer is a plagioclase-rich cumulate that belongs to the cyclical Upper Layered Horizon of the West-Pana intrusion. The layer representing a leucocratic part of the cycle is overlain by slightly PGE-enriched gabbronorites of the next cyclic unit, which is characterized by abundant hornfels xenoliths and a discontinuous basal norite layer.

U-Pb SHRIMP-II dating of magmatic zircon with relatively high Th/U (0.9 to 3.7) from the anorthosite layer gives an upper intercept age of 2509.4 ± 6.2 Ma (2σ) and a concordia age of 2509 ± 10 Ma. The anorthosite and the related rocks of the Upper Layered Horizon are generally older than the lower portions of the West-Pana intrusion, suggesting an out-of-sequence emplacement of the intrusion. Secondary baddeleyite and zircon with relatively low Th/U (0.1–0.9) from the same anorthosite layer that were previously analyzed by multi-grain ID-TIMS yielded a significantly younger age of 2476 ± 13 Ma. Our study indicates that this age does not record the actual timing of emplacement, but a secondary, post-magmatic alteration event.

Author Contributions: Conceptualization, N.Y.G. and B.K.; investigation, N.Y.G.; writing, original draft preparation, N.Y.G.; writing, review and editing, B.K.

Funding: This research was carried out under the scientific theme No. 0226-2019-0053 and was partly funded by the Russian Foundation for Basic Research (RFBR projects 15-35-20501, 16-05-00367).

Acknowledgments: The authors thank A.U. Korchagin and JSC Pana for the drilling and assay data; L.I. Koval for help with zircon separation; N.V. Rodionov and CIR VSEGEI for conducting SHRIMP analyses; A.V. Antonov and E.E. Savchenko for BSE imaging; T.V. Rundquist for discussions of the results; and the anonymous reviewers for constructive criticism that improved the quality of the manuscript.

Conflicts of Interest: The authors declare no conflict of interest.

References

1. Schissel, D.; Tsvetkov, A.A.; Mitrofanov, F.P.; Korchagin, A.U. Basal Platinum-Group Element Mineralization in the Federov Pansky Layered Mafic Intrusion, Kola Peninsula, Russia. *Econ. Geol.* **2002**, *97*, 1657–1677. [CrossRef]
2. Korchagin, A.U.; Subbotin, V.V.; Mitrofanov, F.P.; Mineev, S.D. Kievey PGE-Bearing Deposit in the West-Pana Layered Intrusion. *Strat. Min. Resour. Lapland Apatity* **2009**, *2*, 12–32.
3. Korchagin, A.U.; Goncharov, Y.V.; Subbotin, V.V.; Groshev, N.Y.; Gabov, D.A.; Ivanov, A.N.; Savchenko, Y.E. Geology and Composition of the Ores of the Low-Sulfide North Kamennik PGE Deposit in the West-Pana Intrusion. *Ores Met.* **2016**, *1*, 42–51. (In Russian)
4. Kazanov, O.V.; Kalinin, A. The Structure and PGE Mineralization of the East Pansky Layered Massif. *Strat. Min. Resour. Lapland-Base Sustain. Dev. North. Interreg-Tacis Proj. Apatity Russ.* **2008**, *1*, 57–68.
5. Rasilainen, K.; Eilu, P.; Halkoaho, T.; Iljina, M.; Karinen, T. Quantitative Mineral Resource Assessment of Undiscovered PGE Resources in Finland. *Ore Geol. Rev.* **2010**, *38*, 270–287. [CrossRef]
6. Gurskaya, L.I.; Dodin, D.A. Mineral Resources of Platinum Group Metals in Russia: Expansion Prospects. *Reg. Geol. Met.* **2015**, *64*, 84–93. (In Russian)
7. Balabonin, N.L.; Korchagin, A.U.; Latypov, R.M.; Subbotin, V.V. Fedorovo-Pansky Intrusion. In *Kola Belt of Layered Intrusions. Geological Institute Kola Science Centre, Russian Academy of Sciences, Apatity, 7th International Platinum Symposium. Guide to the Pre-Symposium Field Trip*; KSC RAS: Apatity, Russia, 1994; pp. 9–41.
8. Mitrofanov, F.P.; Korchagin, A.U.; Dudkin, K.O.; Rundkvist, T.V. Fedorova-Pana Layered Mafic Intrusion (Kola Peninsula, Russia): Approaches, Methods, and Criteria for Prospecting PGEs. In *Exploration for Platinum-Group Elements Deposits*; Short Course Delivered on Behalf of the Mineralogical Association of Canada in Oulu, Finland; University of Toronto: Toronto, ON, Canada, 2005; Volume 35, pp. 343–358.
9. Groshev, N.Y.; Borisenko, Y.S.; Savchenko, Y.E. Plagioclase Composition through the Section of the Main Anorthosite Layer of the West-Pana Massif (Kola Peninsula, Russia): New Data. *Vestn. KSC* **2017**, *1*, 1–15. (In Russian)
10. Subbotin, V.V.; Korchagin, A.U.; Savchenko, E.E. Platinum Mineralization of the Fedorova-Pana Ore Node: Types of Ores, Mineral Compositions and Genetic Features. *Vestn. KSC* **2012**, *1*, 54–65. (In Russian)
11. Chernyavsky, A.V.; Groshev, N.Y.; Korchagin, A.U.; Shilovskikh, V.V. Minerals of Platinum Group Elements through the South Reef, West-Pana Intrusion. *Tr. FNS* **2018**, *15*, 392–395. (In Russian) [CrossRef]
12. Karpov, S.M. Geological Structure of the Pana Intrusion and Features of Localization of Complex PGE Mineralization. Ph.D. Thesis, KSC RAS, Apatity, Russia, 2004. (In Russian)
13. Bayanova, T.; Ludden, J.; Mitrofanov, F. Timing and Duration of Palaeoproterozoic Events Producing Ore-Bearing Layered Intrusions of the Baltic Shield: Metallogenic, Petrological and Geodynamic Implications. *Geol. Soc. Lond. Spec. Publ.* **2010**, *323*, 165–198. [CrossRef]
14. Mitrofanov, F.P.; Bayanova, T.B.; Korchagin, A.U.; Groshev, N.Y.; Malitch, K.N.; Zhirov, D.V.; Mitrofanov, A.F. East Scandinavian and Noril'sk Plume Mafic Large Igneous Provinces of Pd-Pt Ores: Geological and Metallogenic Comparison. *Geol. Ore Depos.* **2013**, *55*, 305–319. [CrossRef]
15. Zeh, A.; Ovtcharova, M.; Wilson, A.H.; Schaltegger, U. The Bushveld Complex Was Emplaced and Cooled in Less than One Million Years—Results of Zirconology, and Geotectonic Implications. *Earth Planet. Sci. Lett.* **2015**, *418*, 103–114. [CrossRef]

16. Amelin, Y.V.; Heaman, L.M.; Semenov, V.S. U-Pb Geochronology of Layered Mafic Intrusions in the Eastern Baltic Shield: Implications for the Timing and Duration of Paleoproterozoic Continental Rifting. *Precambrian Res.* **1995**, *75*, 31–46. [CrossRef]

17. Dokuchaeva, V.S. Petrology and Ore Genesis in the Federov Pansky Intrusion. *Geol. Genes. Platin.-Gr. Met. Depos. Mosc. Nauk. Publ. House* **1994**, *1*, 87–100. (In Russian)

18. Odinets, A.Y. The Petrology of the Pansky Massif of Mafic Rocks, Kola Peninsula. Ph.D. Thesis, Kola Branch AS USSR, Apatity, USSR, 1971. (In Russian)

19. Kozlov, E.K. *Natural Series of Nickel-Bearing Intrusion Rocks and Their Metallogeny*; Nauka: Leningrad, USSR, 1973. (In Russian)

20. Staritsina, G.N. Fedorova Tundra Massif of Basic-Ultrabasic Rocks. In *Issues of Geology and Mineralogy, Kola Peninsula*; Kola Branch AS USSR: Apatity, Russia, 1978; pp. 50–91. (In Russian)

21. Latypov, R.M.; Chistyakova, S.Y. *Mechanism of Differentiation of the West Pansky Tundra Layered Intrusion*; Kola Science Center: Apatity, Russia, 2000. (In Russian)

22. Latypov, R.M.; Mitrofanov, F.P.; Skiba, V.I.; Alapieti, T.T. The Western Pansky Tundra Layered Intrusion, Kola Peninsula: Differentiation Mechanism and Solidification Sequence. *Petrology* **2001**, *9*, 214–251.

23. Latypov, R.M.; Mitrofanov, F.P.; Alapieti, T.T.; Halkoaho, T.A.A. Petrology of the Lower Layered Horizon of the Western Pansky Tundra Intrusion, Kola Peninsula. *Petrology* **1999**, *7*, 482–508.

24. Korchagin, A.U.; Subbotin, V.V.; Mitrofanov, F.P.; Mineev, S.D. Kievey PGE-Bearing Deposit of the West-Pana Layered Intrusion: Geological Structure and Ore Composition. In *Strategic Mineral Resources of Lapland—Base for the Sustainable Development of the North*; KSC RAS: Apatity, Russia, 2009; pp. 12–33.

25. Latypov, R.M.; Chistyakova, S.Y. Physicochemical Aspects of Magnetite Gabbro Formation in the Layered Intrusion of the Western Pansky Tundra, Kola Peninsula, Russia. *Petrology* **2001**, *9*, 25–45.

26. Latypov, R.M.; Mitrofanov, F.P.; Alapieti, T.T.; Kaukonen, R.J. Petrology of the Upper Layered Horizon of the West-Pansky Tundra Intrusion (Kola Peninsula, Russia). *Geol. I Geofiz.* **1999**, *40*, 1434–1456.

27. Groshev, N.Y.; Rundkvist, T.V.; Bazay, A.V. Find of Cordierite Hornfels in the Upper Layered Horizon of the West-Pana Intrusion, Kola Peninsula. *Zap. RMO* **2015**, *144*, 82–98.

28. Subbotin, V.V.; Korchagin, A.U.; Savchenko, E.E. PGE Mineralization of the Fedorova-Pana Ore Complex: Types of Mineralization, Mineral Composition, Features of Genesis. *Her. Kola Sci. Cent. RAS* **2012**, *1*, 55–66.

29. Groshev, N.Y. Plagioclase Composition as an Indicator of the Economic Potential of a PGE Reef in a Layered Basic Intrusion. *Tr. FNS* **2017**, *14*, 86–88.

30. Subbotin, V.V.; Korchagin, A.U.; Gabov, D.A.; Savchenko, E.E. Localization and Composition of Low-Sulfide PGE Mineralization in the West-Pana Intrusion. *Tr. FNS* **2012**, *9*, 302–307.

31. Bayanova, T.B. *Age of Reference Geological Complexes of the Kola Region and Duration of Magmatic Processes*; Nauka: St. Petersburg, Russia, 2004.

32. Williams, I.S. U-Th-Pb Geochronology by Ion Microprobe. *Rev. Econ. Geol.* **1998**, *7*, 1–35.

33. Schuth, S.; Gornyy, V.; Berndt, J.; Shevchenko, S.; Sergeev, S.; Karpuzov, A.; Mansfeld, T. Early Proterozoic U-Pb Zircon Ages from Basement Gneiss at the Solovetsky Archipelago, White Sea, Russia. *Int. J. Geosci.* **2012**, *03*, 289–296. [CrossRef]

34. Stacey, J.S.; Kramers, J.D. Approximation of Terrestrial Lead Isotope Evolution by a Two-Stage Model. *Earth Planet. Sci. Lett.* **1975**, *26*, 207–221. [CrossRef]

35. Ludwig, K.R. *User's Manual for Isoplot 3.75: A Geochronological Toolkit for Microsoft Excel*; Berkeley Geochronology Center: Berkeley, CA, USA, 2012.

36. Heaman, L.M.; LeCheminant, A.N. Paragenesis and U-Pb Systematics of Baddeleyite (ZrO_2). *Chem. Geol.* **1993**, *110*, 95–126. [CrossRef]

37. Chashchin, V.V.; Mitrofanov, F.P. The Paleoproterozoic Imandra-Varzuga Rifting Structure (Kola Peninsula): Intrusive Magmatism and Minerageny. *Geodin. I Tektonofiz.* **2015**, *5*, 231–256. [CrossRef]

38. Nitkina, E.A. U-Pb Zircon Dating of Rocks of the Platiniferous Fedorova-Pana Layered Massif, Kola Peninsula. *Dokl. Earth Sci.* **2006**, *408*, 551–554. [CrossRef]

39. Groshev, N.Y.; Nitkina, E.A.; Mitrofanov, F.P. Two-Phase Mechanism of the Formation of Platinum-Metal Basites of the Fedorova Tundra Intrusion on the Kola Peninsula: New Data on Geology and Isotope Geochronology. *Dokl. Earth Sci.* **2009**, *427*, 1012–1016. [CrossRef]

40. Balashov, Y.A.; Bayanova, T.B.; Mitrofanov, F.P. Isotope Data on the Age and Genesis of Layered Basic-Ultrabasic Intrusions in the Kola Peninsula and Northern Karelia, Northeastern Baltic Shield. *Precambrian Res.* **1993**, *64*, 197–205. [CrossRef]

41. Bayanova, T.B.; Rundquist, T.V.; Serov, P.A.; Korchagin, A.U.; Karpov, S.M. The Paleoproterozoic Fedorov-Pana Layered PGE Complex of the Northeastern Baltic Shield, Arctic Region: New U-Pb (Baddeleyite) and Sm-Nd (Sulfide) Data. *Dokl. Earth Sci.* **2017**, *472*, 1–5. [CrossRef]

42. Wall, C.J.; Scoates, J.S.; Weis, D.; Friedman, R.M.; Amini, M.; Meurer, W.P. The Stillwater Complex: Integrating Zircon Geochronological and Geochemical Constraints on the Age, Emplacement History and Crystallization of a Large, Open-System Layered Intrusion. *J. Pet.* **2018**, *59*, 153–190. [CrossRef]

43. Scoates, J.S.; Friedman, R.M. Precise Age of the Platiniferous Merensky Reef, Bushveld Complex, South Africa, by the U-Pb Zircon Chemical Abrasion ID-TIMS Technique. *Econ. Geol.* **2008**, *103*, 465–471. [CrossRef]

44. Wall, C.J.; Scoates, J.S. High-Precision U-Pb Zircon-Baddeleyite Dating of the JM Reef Platinum Group Element Deposit in the Stillwater Complex, Montana (USA). *Econ. Geol.* **2016**, *111*, 771–782. [CrossRef]

45. Karykowski, B.T.; Maier, W.D.; Groshev, N.Y.; Barnes, S.J.; Pripachkin, P.V.; McDonald, I. Origin of Reef-Style Pge Mineralization in the Paleoproterozoic Monchegorsk Complex, Kola Region, Russia. *Econ. Geol.* **2018**, *113*, 1333–1358. [CrossRef]

46. Karykowski, B.T.; Maier, W.D.; Groshev, N.Y.; Barnes, S.J.; Pripachkin, P.V.; McDonald, I.; Savard, D. Critical Controls on the Formation of Contact-Style PGE-Ni-Cu Mineralization: Evidence from the Paleoproterozoic Monchegorsk Complex, Kola Region, Russia. *Econ. Geol.* **2018**, *113*, 911–935. [CrossRef]

47. Karykowski, B.T.; Maier, W.D. Microtextural Characterisation of the Lower Zone in the Western Limb of the Bushveld Complex, South Africa: Evidence for Extensive Melt Migration within a Sill Complex. *Contrib. Miner. Pet.* **2017**, *172*, 1–18. [CrossRef]

48. Ivanov, A.N.; Groshev, N.Y.; Korchagin, A.U. Transgressive Structures of the Lower Layered Horizon in the Area of North Kamennik Low-Sulfide PGE Deposit. *Tr. FNS* **2018**, *15*, 124–127. [CrossRef]

49. Groshev, N.Y.; Pripachkin, P.V. Geological setting and platinum potential of the Gabbro-10 massif, Monchegorsk Complex, Kola Region. *Ores Met.* **2018**, *4*, 4–13. [CrossRef]

Article

Study of the Cu-Ni Productive Suite of the Pechenga Structure on the Russian-Norway Border Zone with the Use of MHD Installation "Khibiny"

Abdulkhai A. Zhamaletdinov [1,2]

[1]   Geological Institute, Kola Science Centre, Russian Academy of Sciences, 184209 Apatity, Russia;
       abd.zham@mail.ru; Tel.: +7-921-169-2104

[2]   St. Petersburg Branch of the Institute of Terrestrial Magnetism, Ionosphere and Radio Waves
       Propagation (SPbB IZMIRAN), 190034 St Petersburg, Russia

Received: 28 December 2018; Accepted: 3 February 2019; Published: 8 February 2019

Abstract: The tracing of current-conducting channels of the Pechenga structure from Russian to Norwegian territory was the main task of this research. The study was carried out in the framework of the Soviet-Norwegian cooperation "Northern Region" to estimate the prospects for discovery of Cu-Ni deposits in northern Norway. In addition to previous publications of technical character, the emphasis here is on geological description. Experimental measurements have been performed in the field of the "Khibiny" dipole and with the use of DC electrical profiling. The "Khibiny" dipole consists of 160-ton aluminum cable flooded in the Barents Sea bays on opposite sides of the Sredny and Rybachy peninsulas. Measurements were implemented as in the mode of single pulses generated by 80 MW magneto-hydrodynamic (MHD) generator "Khibiny" ("hot" launches) and in the accumulation mode of rectangular current pulses of 0.125 Hz frequency generated by a 29 kW car generator ("cold" launches). From results of measurements, it was concluded that the most promising potential for Cu-Ni deposits Pil'gujarvi formation of the Northern wing of the Pechenga structure is rather quickly wedged out in Norway, while the conductive horizons of the Southern part of Pechenga, which have a weak prospect for Cu-Ni ores, follow into Norway nearly without a loss of power and integral electrical conductivity.

Keywords: Khibiny promising structures; conductive layers; MHD-source "Khibiny"; Pechenga structure; Kola Peninsula

1. Introduction

In the early 1970s, a new scientific direction appeared in Russia (former Soviet Union)—the deep electromagnetic soundings on the base of a unique technology—impulsive magnetohydrodynamic (MHD) generators with a power of tens and up to one hundred MegaWatts. The initiator and supervisor of the new geophysical direction was Academician E.P. Velikhov, vice president of the Russian Atomic Energy Institute [1,2]. The need to develop a new direction was determined by the increased requirements for the depth and reliability of geoelectrical research. For a relatively short time, several MHD experiments were performed for the purposes of earthquake prediction (the Garm test site in the Pamirs and the Frunze test site in the Tien Shan), to study the structure of the Earth's crust and to predict ore deposits (the Chelyabinsk test site in the Southern Urals and the Khibiny test site on the Kola Peninsula), to search for oil and gas deposits (Vanavaar test site in Siberia and the Astrakhan test site in the Southern Volga region) and to solve some problems of extra low-frequency (ELF) radio communications [3,4].

Among the noted research projects, the MHD experiment "Khibiny" on the Kola Peninsula, which lasted from 1974 to 1991, was one of the most significant. The task of the Khibiny MHD experiment included studying the structure of the Earth's crust and upper mantle up to the depth of 100–150 km, the study of the block structure of the Earth's crust conductivity of the Baltic Shield over aterritory of about 200 thousand square kilometers, the sounding of the seafloor in the adjacent water area of the Barents sea [5]. The study of ore prospective, high conductive structures (current conductive channels) in the Kola Peninsula and in the neighboring countries (Northern Norway and Finland) has been accepted as one promising task of the Khibiny experiment. In this article, the general concepts of theory and methodology of this study are presented. Specific application is presented on the example of volcanogenic-sedimentary high-conductive horizons tracing on the western flank of the Pechenga structure in Norway. The basis for the formulation of the problem was the hypothesis that in the West, in the territory of Northern Norway, the thickness of the Pil'guarvi (so-called "productive") high conductive sulfide and carbon-bearing stratum should increase and with it, the prospects for discovery of copper-nickel deposits.

The Pechenga structure of the Low Proterozoic age consists of Northern and Southern volcanogenic-sedimentary series. The most extensive sedimentary formations are presented by the Pil'gujarvi series in the Northern Pechenga and by the Kallojavr-Porjitash, Bragino and Langvannet-Porojarvi series in the Southern Pechenga [6]. They are similar to each other by composition but different in their Cu-Ni potential [7]. Pil'gujarvi formation (the fourth sedimentary layer of the Northern Pechenga) contains many Cu-Ni deposits (Kaula, Kammikivi, Semiletka, Zhdanovka, etc.), which are linked to differentiated massifs of basic-ultrabasic rocks. Thus, this suite has been called the Productive stratum [8]. In the Southern series of the Pechenga structure with its major Kallojavr-Porjitash, Bragino and Langvannet-Porojarvi formations, on the contrary, differentiated massifs of basic-ultrabasic rocks are nearly absent. Thus, their potentials for Cu-Ni ores are fairly low. All sedimentary suites of the Pechenga structure (northern and southern) possess high electric conductivity due to the overabundance of sulfide-carbonaceous schists and phyllites that contain electronically conductive minerals, such as pyrite, pyrrhotite, graphite and sometimes titan magnetite. In the field of the MHD-installation "Khibiny", the Northern and Southern series of the Pechenga structure reveal themselves as two individuals (insulated) current-conducting channels [9].

The "Northern Region" project headed by Brian Sturt (Norway) and Felix Mitrofanov (Russia) has aimed at thetracing of sedimentary volcanic formations of the Pechenga structure with the purpose of investigating perspectives for Cu-Ni deposits further to the West, in the Norwegian territory. The hypothesis on the possible increase of the Productive layer in the Northern Norway area has been accepted on the base of project "Northern Region". As a solution, the study of high electrically conductive structures was included in this project. Experimental measurements have been performed in the field of "Khibiny" dipole and with the use of DC electrical profiling. The "Khibiny" dipole consists of 160-ton aluminum cable flooded in the Barents Sea bays on opposite sides of the Sredny and Rybachy peninsulas. Measurements were implemented as in the mode of single pulses generated by the 80 MW magneto-hydrodynamic (MHD) generator "Khibiny" in the accumulation mode of rectangular current pulses of 0.125 Hz frequency generated by 29 kW car generator. In addition to previous publications of technical character [10,11], the emphasis here is on geological description.

The first section of this paper briefly highlights the parameters and techniques of "hot" and "cold" launches. The second section describes the physical model and technique of data processing that has been taken as the basis to study current-conducting channels. The third section provides analyses of the results obtained. The final part is devoted to a discussion and summary.

2. Technique of Sounding with MHD-Installation "Khibiny"

Technique and main results of the deep electromagnetic soundings with the use of the "Khibiny" MHD-source are described in numerous scientific and educational publications. Below, only a brief description of the technique necessary to understand the results is presented. Detailed data can be gathered from [2–5].

2.1. "Hot" Launches

The summary diagram illustrating the MHD source "Khibiny" is shown in Figure 1. The main peculiarity of the "Khibiny" source is a dual MHD generator of 80 MW power (Figure 1e) and transmitting antenna made of aluminum cable of 7.5 km length and 160 tons weight that has been immersed in the sea bays on the opposite sides of the isthmus between the Kola and Srednypeninsulas (Figure 1a). A schematic diagram of the operation of one of two generators is given in Figure 1b. The MHD-generator works based on the Faraday's law [9]. In a plasma generator (Figure 1b-1), which is a capsule with powder fuel of 700 kg weight, the conductive "cold" plasma is produced by fuel combustion with some cesium added. The conductive plasma flows into an MHD-channel with a velocity of 3 km/s (Figure 1b-2).

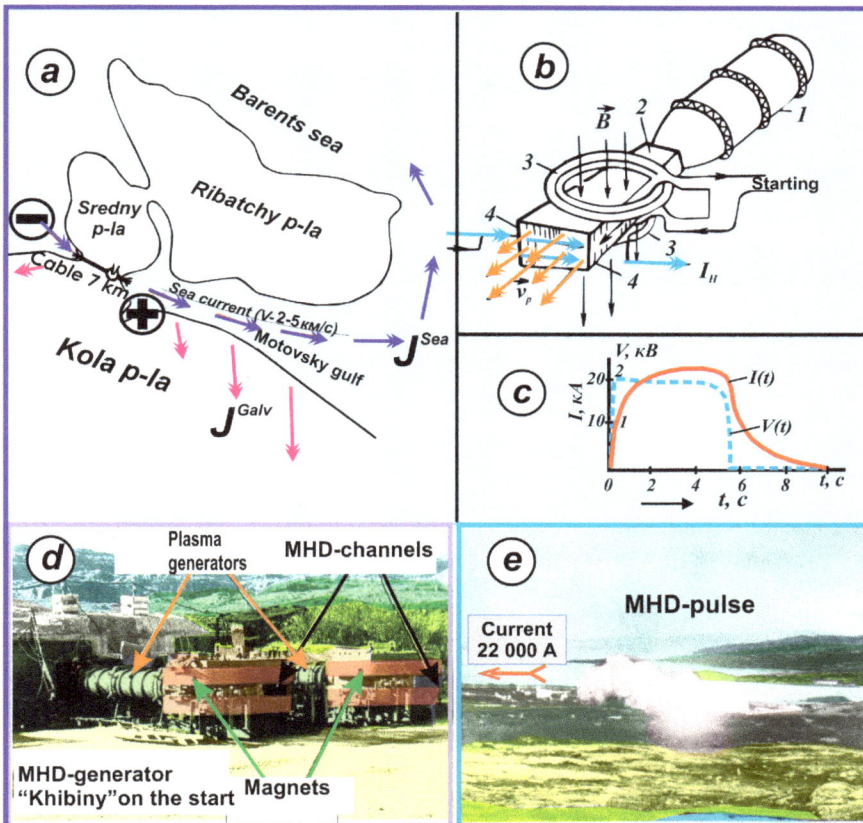

Figure 1. The MHD-installation "Khibiny". (**a**) Location of 160 ton cable of 7 km length and currents in the sea, J^{sea} and currents in the ground, J^{galv}; (**b**) principal scheme of pulse MHD-generator "Khibiny"; (**c**) voltage and current intensity in the output of MHD generator "Khibiny"; (**d**) the view of the dual MHD generator "Khibiny"in the working position; (**e**) the gas outbreak that appeared from the MHD pulse producing the current of 22 kA in the Motovsky sea gulf.

The MHD-channel is placed between two coreless magnets (Figure 1b-3). Force $\vec{F}$, induced by plasma flow, is exerted perpendicularly both towards the direction of the conductor's movement and towards the direction of the magnetic induction vector $\vec{B}$. Force $\vec{F}$ pushes charges from the conductor ls 2019, 9, 96*

(the Hall effect). The force $\vec{F}$ of interaction between charges and the magnetic field can be found by the Lorentz equation as: $\vec{F} = \vec{v} \cdot Q \cdot \vec{B}$, where $\vec{v}$ [m/s]—is velocity of conductor's movement, $Q[A \cdot m]$—charge in Coulomb, $\vec{F}[\frac{V \cdot s \cdot A}{m}]$—the force in Newton, $\vec{B}[\frac{V \cdot s}{m^2}]$—magnetic induction vector in Tesla.

The current flows through conductive sidewalls of MHD-channels and passes to an external load, by the cable to the sea (Figure 1b-4). One from two MHD-generators of the "Khibiny" installation works in the auto-excitation mode. It induces the current for the feeding of its own electric magnets. Half of the current is channeled to electric magnets of another generator that sends all of the current to a power cable connecting opposite gulfs of the Sredny and Rybachypeninsulas (Figure 1a). The maximum voltage is 2 kV, and the maximum current is 22 kA. Therefore, the maximum power exerted from the MHD generator is about 44 MW. The launch of the MHD-generator is followed by a huge gas outbreak, approximately 500 m, marked on Figure 1e. Figure 1c shows that the current in the external load increases gradually despite a rectangular voltage waveform. This is due to a high inductance of the circuit (up to 70 mHn). The current reaches its maximum value ("ceiling") in 1–2 s after switching on the rectangular pulse of voltage.

The current passes from West to the East since the positive charge is always supplied to the Motovsky Gulf, where the eastern end of the cable is flooded and grounded (Figure 1a). The current diffuses from the coastline towards the sea. The diffusion parameter of the electromagnetic field follows from the first two Maxwell equations and has the form $M = \frac{1}{\sigma \cdot \mu_0}$, $[\frac{m^2}{s}]$, where σ $[\frac{1}{\Omega \cdot m}]$—specific conductivity and $\mu_0 = 4\pi \cdot 10^{-7}$ $[\frac{V \cdot s}{A \cdot m}]$—vacuum permeability. The sea can be represented as a thin S-film with thefixed one coordinate—the depth of the sea, h. In this case, the expression for the diffusion rate of the electromagnetic field will take the form $v = \frac{1}{\sigma \cdot \mu_0 \cdot h}$, $[\frac{m}{s}] = \frac{800}{S}$, $[\frac{km}{s}]$, where $S = \sigma \cdot h$, $[Sm]$ is longitudinal conductivity in siemences. In the coastal zone, the sea depth is in the range of 50–100 m. The longitudinal conductivity of the sea is in the range of 150–300 Sm. From here, the average speed of current diffusion in the sea is in the range of 3–5 km/s. The drifting currents produce current loops (J^{sea}) around the Sredny and Rybachy peninsulas, creating induction-type (H^{ind}) magnetic fields. Some current (about 20%) leaks through the sea bottom into the Earth crust by the galvanic way (J^{galv}) and produces galvanic-type magnetic fields (H^{galv}). The ratio between the currents, penetrating into the earth by galvanic way, and the currents diffusing into the sea, was established with the use of experimental observations. For this purpose, electromagnetic frequency soundings were carried out with the use of a "dry" dipole, grounded to crystalline rocks in the inner part of the isthmus between the Sredny and Kola Peninsulas, and a "wet" dipole, representing the "MHD-Khibiny" cable that has been immersed into opposite sea gulfs. The similar parameters of two measured geoelectrical sections were obtained after taking into account that only 20% of the total current of the MHD generator penetrates the earth by galvanical way and forms electric type fields [9]. This ratio was confirmed later by theoretical calculations and physical modeling [2].

2.2. "Cold" Launches

In Northern Norway, because of short distances between transmitter and receiver (60–130 km), EM measurements were performed in the "cold" mode with the use of station ERS-67 of 29-kW power mounted on the chassis of a ZIL-131 car and connected to the "Khibiny" circuit. A ballast of 10 Ohm resistivity has been connected in series to the feeding cable to supply to the concordance between the high internal resistivity of ERS-67 generator (10 Ohm) and low total resistivity of cable and groundings of the circuit "Khibiny" (0.095 Ohm). As a result, only one-hundredth of the generator's power (about 290 W) dissipated into the ground, but it was enough to make successful measurements in Northern Norway at a distance of up to 130 km from the source. There was almost no impulsive noise in the study area; the natural noise was low and homogenous ("white noise"). Thus, signal sampling of the current for 20 min duration enlarged the ratio "signal-noise" about 20 times, which proved sufficient to register the electric and magnetic components of the field [11].

154

The current was transmitted as bipolar square pulses with a frequency of 0.125 Hz and a strength close to 100 A from peak to peak. The sampling has been made by the first fronts of each half period of 4 s duration. The current variation *I(t)* in the cable when switching on can be described by the formula:

$$J(t) = J_0 \cdot [1 - \exp(-\frac{t}{\tau})], \tag{1}$$

where $J_0 = \frac{V}{R}$—the totalDC current in the output of generator at $t \to \infty$; *V*—voltage of the car generator; *R*—the total active resistance of the cable and water groundings; $\tau = \frac{L}{R}$—the time constant; *L*—self-inductance of the "Khibiny" circuit, in Henri. τ corresponds to the time when the current *I(t)* reaches $0.7\ I_0$ after switching on the voltage. The self-induction *L* of the circuit Khibiny depends on frequency (time in terms of time domain soundings). With decreasing frequency from 10 Hz to 0.1 Hz, the induction *L* of the circuit Khibiny increases from 18 to 80 mHn, based on experimental data [9]. This increase in *L* can be explained by the increase in the length of the circuit "Khibiny" due to a spreading of the current in the sea (J^{sea}) along the coastline and by offshore drifting (diffusion). The ballast resistance *R*, connected in series with the "Khibiny" feeding line, reduces the time constant from 1 s up to 0.01 s. As a result, it makes the shape of the current fronts of "cold" appearsharper and more suitable for time domain measurements. Measurements were made in the time domain regime based on5components *Ex*, *Ey*, *Hx*, *Hy*, *Hz*.

3. The Model of Current-Conducting Channel Detection

The electromagnetic field of the "Khibiny" source is considered as a superposition of magnetic and electric sources. The magnetic source exists due to currents channeled through the cable to the sea (J^{sea}). This current produces magnetic loops around the Sredny and Rybachy peninsulas. The electric source is exerted by galvanic currents dissipating into the ground through the sea bottom (J^{galv}). By its nature, component Hz should reflect the form of MHD pulses, i.e., it should be unipolar (Figure 1c). All MHD pulses have been executed at the same positive polarity of the current, penetrating to the sea from the Motovsky gulf (Figure 1a). The positive polarity of the vertical component of the magnetic field complies with its penetration from the top to the down. In this case, by the low of the right screw, the magnetic field at the central parts of the Rybachy and Sredny peninsulas and at the most sea area should extend from bottom to top, i.e., in the negative direction. In the receiving points, situated outside the sea coastline (on the Kola peninsula), the magnetic field propagates in the direction from top to down, i.e., in the positive direction. The nature of negative signals Hz appearing in some receiving sites is connected with galvanic currents (J^{galv}) flowing in current-conducting channels represented by sulfide- and carbon-bearing rocks, fracture zones, grounded power lines, railway lines, etc. In receiving points, situated near current-conducting channels, the superposition happens in magnetic fields produced by induction owing to currents flowing in the sea (J^{sea}) and those produced by galvanic currents (J^{galv}) leaking into the Earth's crust through the sea bottom (Figure 1a).

The synthetic sketch of superposition of magnetic fields produced by induction and galvanic sources in the "Khibiny" experiment is shown in Figure 2.

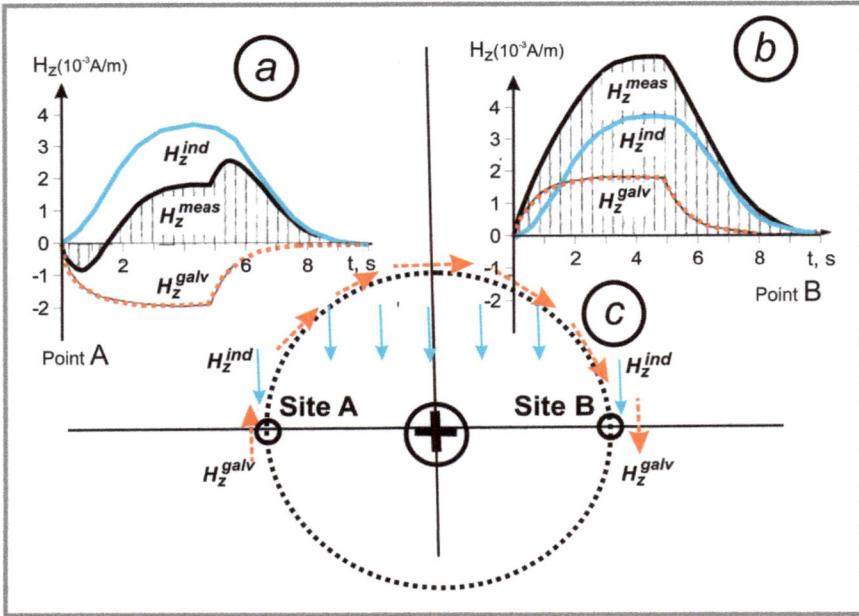

Figure 2. Scheme of magnetic fields, measured from MHD-installation "Khibiny" in opposite sides of the current-conducting channel. (**a**) Curves of the vertical magnetic field for the left site "A": H_z^{ind} (blue)—induced by sea currents, Hz^{galv} (red, dashed)—based on galvanic currents in the ground and $Hz^{meas} = H_z^{ind} + Hz^{galv}$ (black, solid) measured magnetic field; (**b**) same for right site "B"; (**c**) vectors of magnetic field: Hz^{galv} from current "+", flowing away from observer (dashed arrows) and H_z^{ind} from distant currents in the sea (solid arrows).

In Figure 2, the registration of the vertical magnetic component *Hz* is presented on the example at sites A and B situated symmetrically on opposite sides regarding the current conductor of endless length. Sign "+" indicates that the current flows away from the observer. Curves and arrows Hz^{ind} (Figure 2a–c) show the behavior of magnetic field produced by remote current streams in the sea. Since the velocity of current diffusion in the sea is low, the respective electromagnetic signals of the magnetic source have flat frames with a low-frequency range. They penetrate to the Earth's crust without responding to the heterogeneous electric conductivity of the upper Earth's crust. Therefore, graphs Hz^{ind} on opposite sides of the conductor (Figure 2a,b) are identical.

The amplitude of the vertical magnetic component produced by the current in the sea can be approximated by the formula for the stationary vertical magnetic dipole.

$$Hz^{sea}(t) = \frac{M}{4\pi r^3}, \quad \left[\frac{A}{m}\right], \tag{2}$$

where the magnetic moment of the source $M = J_{sea} \cdot \pi R^2$, J_{sea} is thecurrent strength in the sea, *R* is the radius of the magnetic loop in the sea (about 50 km), and *r* is the distance from the receiver to the center of the magnetic loop.

Curves Hz^{galv} (Figure 2a,b) reflect the behavior of the magnetic field produced by an endlessly long conductor with the current in the ground. The magnetic field of the galvanic nature has frames that are as steep as the current in the source. It also has different polarities on opposite sides of the current. The polarity (direction) of the magnetic field Hz^{galv} is defined by the right-hand rule. The amplitude of the field in the stationary state is found by Amper's law to be

$$Hz^{galv}(t) = \frac{J_{galv}(t)}{2\pi \cdot r}, \left[\frac{A}{m}\right]. \tag{3}$$

The measured vertical magnetic component (Hz^{meas}) is defined as a sum of fields of the induction and galvanic origin:

$$Hz^{meas}(t) = Hz^{sea}(t) + Hz^{galv}(t). \tag{4}$$

Curves Hz^{meas} indicating the behavior of the measured (summarized) vertical magnetic field on opposite sides of the conductor have absolutely different-looking shapeson sites "A" and "B" (Figure 2). Their behavior, character and amplitude allow tracing the location of the current-conducting channel and its parameters. If we estimate the distance to the center of the current-conducting channel and the distance from it to the reference point, we can easily define the current strength of the galvanic origin in the conductor, using the equation

$$Jtot^{galv} = 2\pi \cdot r^{galv} \cdot Htot^{galv}, \, [A]. \tag{5}$$

4. Results of Field Experiments

The common scheme of the current-conducting channels'distribution on the Kola Peninsula obtained with the use of MHD-source "Khibiny" in the mode of "hot" pulsesis shown in Figure 3a.

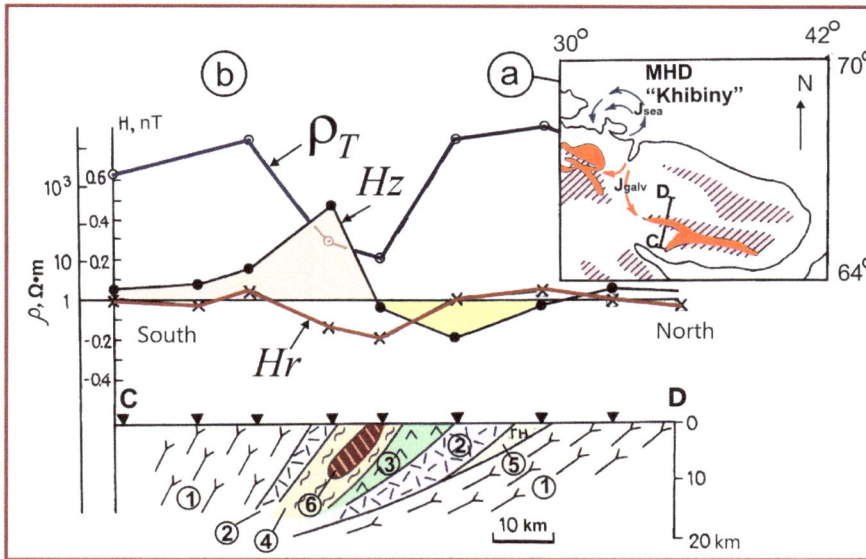

Figure 3. Example of current-conducting channels study on the Kola peninsula with the use of MHD-source "Khibiny" in the mode of "hot" pulses. (**a**) The view of conductive structures and MHD currents in the sea (blue arrows) and the Earth crust (red arrows); (**b**) current-conducting channel's study over the Imandra-Varzuga structure by profile CD. *Legend* (numerals in circles): 1—granite-gneiss; 2—volcanites; 3—effusives; 4—schists and volcanites of the Tominga Series; 5—gabbro-norites; 6—location of the conductive channel

The direction of galvanic currents is shown by hatched (red) arrows that flow from the MHD-source "Khibiny" and then concentrate on conductive structures of the Pechenga-Alarechensky region (on the western part of the Kola peninsula) and on the Imandra-Varzuga structure (on the eastern part of the Kola peninsula). It is apparent that the in rocks of the Imandra-Varzuga and Pechenga-Alarechensky zones galvanic currents run in opposite directions. It is important evidence of their galvanic nature caused by conductive currents in the earth's crust. The galvanic current's intensity in the Pechenga conductive suites, which are located closer to MHD-source, is of 40 A, and that of the Imandra-Varzuga zone is 12.5 A. These values make up about one percent of current intensity of the MHD-generator, penetrating into the earth galvanically and amounting about of 4×10^3 A. Nevertheless, that fact that the galvanic currents are found in the conducting geological structures is important to gain a better insight into the model of electric conductivity in the basement of the Baltic Shield. It is thus clear that, in the upper part of the earth's crust, in spite of the absence of the sedimentary deposits and high resistivity of the exposed crystalline rocks, there are channels for ultralow-frequency, virtually direct current running in the horizontal direction over hundreds of kilometers. A detailed scheme of a current-conducting channel study is presented in Figure 3b in the example of measurements carried out along the profile CD, crossing the Imandra-Varzuga structure.

The Imandra-Varzuga is a riftogene structure composed of the Lower Proterozoic volcano-sedimentary rocks, which are similar by composition and structure to the Pechenga formation. It extends for 350 km in the latitudinal direction (Figure 3a). It contains 10–12 volcano-sedimentary layers, some of which include high carbon- and sulphide-bearing phyllite-like rocks (black schists), enriched in pyrite-pyrrhotite mineralization and widespread in the section of the Tominga Series, which is located in the central part of the profile CD (Figure 3b).

Results of current-conducting channel study in Figure 3b are given by graphs of anomalous vertical magnetic component $Hz^{galv}(t)$ (signed as Hz), a module of the total horizontal magnetic field Hr and curve of apparent resistivity ρ_T, calculated by the total input impedance. The component $Hz^{galv}(t)$ has been estimated using the Equation (6) by subtracting the regional magnetic field $Hz^{sea}(t)$ produced by current loops in the sea from the measured field $Hz^{meas}(t)$:

$$Hz^{galv}(t) = Hz^{meas}(t) - Hz^{sea}(t) \qquad (6)$$

Values $Hz^{sea}(t)$ are given by the Equation (2) for the field of the vertical magnetic dipole. Values of the total horizontal magnetic field module Hr have been compiled from measured data after extraction of the regional low-frequency trend. An apparent resistivity curve ρ_T has been calculated for the period $T = 1$ s with the use of magnetotelluric Equation (7).

$$\rho_T = 0.127T \cdot (Z_{tot})^2, \qquad (7)$$

where the total (effective) impedance $Z_{tot} = \frac{Er}{Hr}$. Er, Hr are modules of the total horizontal electric and magnetic fields accordingly.

In the central part of profile CD, electric and magnetic components manifest anomalous behavior. The vertical magnetic component Hz, crossing the zero, has a negative sign at the north side from the current-conducting channel (6 on Figure 3b) anda positive sign at the south from conductive body. Horizontal magnetic field Hr has a negative sign over the anomalous body. All these features, if we compare them with Figure 2, denote that the anomalous current runs from the west to the eastwards in the graphitic schist of the Tominga series.

The quantitative 2D interpretation of the magnetic field (Figure 3a) points out the southern dipping of the conductive body and its continuation to the depth of about 10 km [9,12,13]. This estimation agrees well with results of numerical modeling of MHD-signals using the technique of electromagnetic migration [14] and with results of magnetovariation profiling implemented at the western flank of Imandra-Varzuga structure, near to profile CD [15].

Results of the tracing of current-conducting channels on the western flank of the Pechenga structure are presentedin Figure 4c. The data processing procedure was the same as has been applied above for the Imandra-Varzuga structure (Figure 3b). Current-conducting channels on Figure 4c are shown by dashed lines. Channels, which are confidently identified in the field of the MHD installation "Khibiny", are shown by bold dashed lines. Weaker conductors, which are mainly fixed from results of electrical profiling with the method of internal sliding contact (MISC) [16], are shown by thinner dashed lines. From a comparison of Figures 3b and 4c, we can notice sufficient difference in the behavior of magnetic fields. The vertical magnetic component *Hz* has a positive sign onthe northern side from the current-conducting channels anda negative sign at the south. Horizontal magnetic field *Hr* has a positive sign over the conductive channels. All these features point out that the anomalous current in the West Pechenga runs from the east side to the west, i.e., in opposite directions compared to the Imandra-Varzuga structure that has been shown above in Figure 3b.

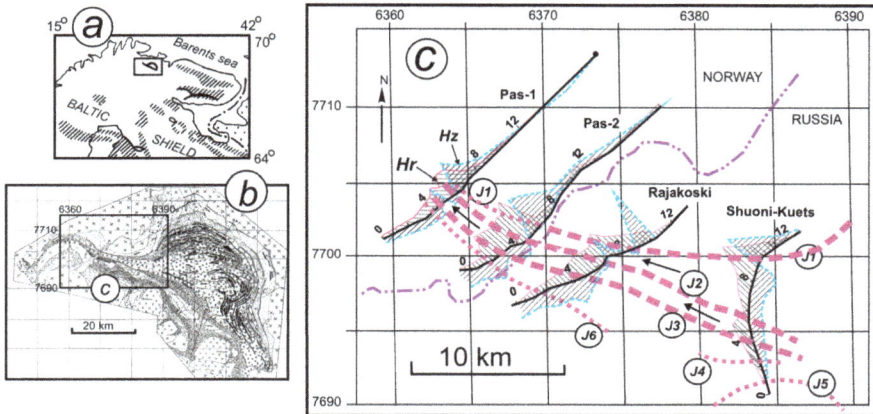

Figure 4. Results of the west Pechenga study in the field of MHD-installation "Khibiny". (**a**) Thecommon view of the study area; (**b**) geology of Pechenga structure by [17]; (**c**) current-conducting channels (J1–J6) of the west Pechenga and graphs of the anomalous magnetic field of the source "Khibiny"—vertical, Hz and horizontal, Hr. Black arrows show the direction of current flow in conductive channels.

Figure 4a,b point out in an uninterrupted continuation of current-conducting channels and geological formations of the western flank of the Pechenga structure. Channel J1 coinciding with Cu-Ni productive Pil'gujarvisuite creates the most intensive anomalous effect on the regional profile Shuoni-Kuets. Going further west, the intensity of the anomalous field above the productive suite decreases from one profile to another. The anomalous effect fades on profile Pas-1. Compared to results of MISC electrical profiling, this study suggests that near the profile Pas-1 the thickness of the conductive body equated to the productive suite decreases to 20–30 m. Its longitudinal conductivity accordingly decreases from 1000 S on the east profile Shuoni-Kuets to 3–5 S on the western profile Pas-1.

Geological interpretation of results is shown in Figure 5 on the base of the map from [17]. As noted above the Cu-Ni productive stratum, the fourth sedimentary horizon of Pechenga is most effectively detected on Shuoni-Kuets profile (channel J1). Further to the west, the anomalous effect from channel J1 gradually fades. At the interval between profiles PAS-1 and PAS-2, it can be traced only by electrical profiling results.

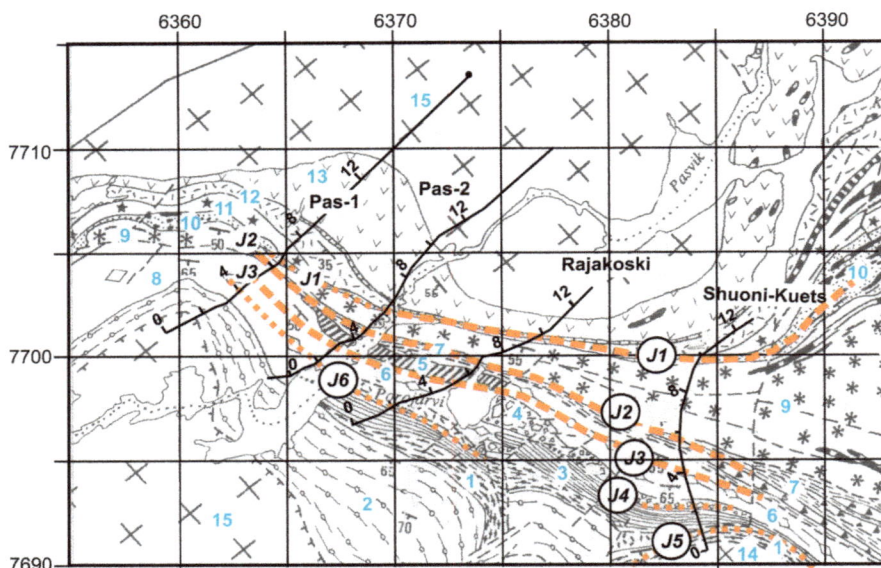

Figure 5. Location of conductive layers J1–J6 on the geological map by V.A. Melezhik [17]. *Legend shown by blue figures:* South Pechenga (1–8): 1—Pestchanoe Lake (toleit-basalt); 2—Talja (siltstone); 3—Kasesjoki (andesitic sandstone); 4—Kaplya (tuff, andesite); 5—Faregmo (andesit lava and tuff); 6—Bragino (tuffit, andesite); 7—Kallojavr, Porjitash (C$_{org}$ and sulfide-rich sandstone, basalt, andesite); 8—Langvannet, Porojarvi (C$_{org}$ and sulfide-rich sandstone, basalt, andesite).North Pechenga (9–13): 9—Pil'gujarvi volcanic (4th diabas, basalt layer); 10—Pil'gujarvi sedimantari (4th sedimentary layer, C$_{org}$ and sulfide-rich sandstone); 11—Kolasjoki (3rd diabas, basalt layer); 12—Kuetsjarvi (2nd diabas, basalt layer); 13—Ahmalahti (1st diabas, basalt layer); 14—Shuonijarvisinorogenic Granite Archaean Basement (2.9–2.5 Ga): 15—tonalit, granite, amfibolite.

Channels J2 and J3 create much stronger anomalous effects. Channel J2 is connected with the Kallojavr-Porjitash formation, composed of sandstones (rich of Corg and sulphides) with interlayers of basalts and andesites. Channel J3 has a rather complex structure. In the east, at the intersection with the Shooni-Kuets profile, it coincides with Bragino formation, composed of tuffs and andesites, but registered high electrical conductivity does not cause doubt in the presence of electronically-conductiveminerals like sulphides (pyrite, pyrrhotite), carbon rocks, and graphite in its composition.

Further in the west, channel J3 in Figure 5 becomes as a part of the Langvannet-Porojarvi formation, rich in sulfides and carbonaceous schists with interbeds of andesites and basalts. The total longitudinal conductivity of channels J2 and J3 in the region of the profiles Pas-1 and Pas-2 is about 2000 S, whereas in the channel J1 (productive stratum), this value is a thousand times less and is in the range of 2–3 S.

In Figure 6, results of numerical 2D modeling are presented altogether with graphs of horizontal and vertical magnetic field induced by "Khibiny" dipole and with resistivity curves of DC electrical profiling.

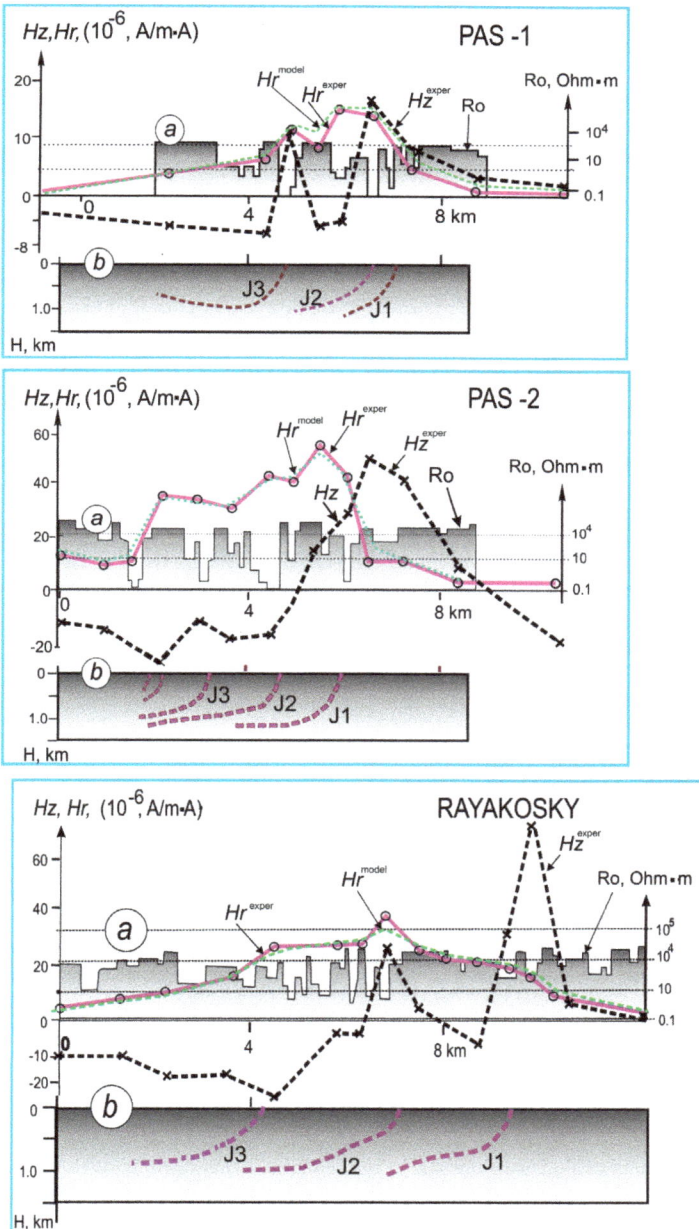

Figure 6. Results of the study of conductive layers of the Pechenga structure's Western flank by profiles PAS-1, PAS-2 andRayakosky. Location of profilesis shown in Figures 4 and 5. Letters in circles: *a*—graphs of resistivity Ro from DC electrical profiling and graphs of measured horizontal Hr^{meas} andvertical Hz^{meas} magnetic field from MHD "Khibiny" dipole and modeled horizontal magnetic field Hr^{model}; *b*—the deep sections of conductive layers from the results of 2D digital modeling. J1, J2 and J3—current conductive channels shown on Figures 5 and 6.

The 2D numerical modeling has been performed to estimate the bed position and the conductor length to the depth. The special "Magnet-2" program has been elaborated for this purpose by V.E. Asming [10]. This program allows detecting the magnetic field by summing a certain set of endlessly long horizontal current lines (DC) in a free half-space, ignoring any impact of extrinsic currents. An interpreter device detects the amount and location of currentlines in the 2D model on-line. An optimal solution of the inverse problem is solved automatically when model and experimental data are adjusted with the least-square method.

Results of DC profiling are obtained with the use of Method of Internal Sliding Contact (MISC), which combines sounding and profiling techniques [16]. Therefore, resulting resistivity curves present 1D inversion data and reflect the resistivity of crystalline rocks under overburden at a depth of about 20 m. Results of 2D modeling of the data from the study of current-conducting channels in the west flank of the Pechenga structure with the use of MHD-installation "Khibiny" make it possible to get an idea of the behavior of conductive horizons to the depth. Against the background of a general narrowing of the thickness of the Pechenga volcanic-sedimentary strata as they move westward (Figures 4 and 5), the conductor length is also reduced to a depth of 1–1.5 km. At the same time, the southern fall of rocks remains, subvertical near the surface of the day and flattening with depth. It is important to note from Figure 6 that the intensity of the anomalous magnetic field (as horizontal *Hr* so vertical *Hz)* sharply (nearly twice) decreased in the profile PAS-1 compared to PAS-2. It can be explained by the sharp bending of conducting layers on the westward (Figure 5) and by the changing of the geological structure from 2D to the 3D model.

5. Discussion

Results presented in the article of conductive horizons study of the Pechenga structure in the field of MHD-installation "Khibiny" are based on an informal, phenomenological approach. This is due, first of all, to the fact that parameters of the MHD source "Khibiny" do not fall under a well-defined physical-mathematical model. This unique transmitter consists of two types of sources—the magnetic (induction) one created by currents flowing in the Barents sea and the electric (galvanic) one that happened due to currents leaking to the ground from the sea gulfs' bottom. The intensity and correlation of these currents are established empirically on experimental measurements of electric and magnetic components of the field generated by MHD-source [9]. The phenomenological approach means the application of the strict well-known physical laws to study the heterogeneous media with the uncertain distribution of the current parameters. The current-conducting channels' existence and methods for their quantitative interpretation were developed with the use of experimental measurements and the data treatment experience in similar situations, for example, Viese vector study in the magnetovariation method [18]. Along with this, we note that the phenomenological approach is widespread in the practice of many other geophysical methods, and it brings its tangible fruits. In particular, the physical-mathematical models of the induced polarization method and the method of induction time domain sounding in the near-field zone are based on the phenomenological approach [19,20]. An important result of the phenomenological interpretation of the "Khibiny" experiment is connected with the finding that conductive channels are of limited penetration to the depth—from a few to 5–10 km and with conclusion about the concentration of galvanic currents in the upper film of the crystalline Earth's crust of about 10 km thickness [21]. One of the most powerful current-conducting channels (with a thickness of up to 10–12 km and a length of up to 80 km) is connected with the Cu-Ni productive formation Pil'gujarvi (the fourth sedimentary horizon) of the Pechenga structure. Within the framework of the "Northern Region" project, special observations were made to trace the continuation of the Cu-Ni productive strata Pil'gujarvi of the Pechenga structure to the territory of Northern Norway. Results of these works showed that the thickness of Pil'gujarvi Cu-Ni productive strata on the territory of Northern Norway is reduced to 20–30 m and accordingly decreases by approximately three orders of magnitude (from 2000 S to 2–3 S) in its longitudinal conductivity. At the same time, the conducting horizons of the southern part of the Pechenga structure

(Kallojavr-Porjitash, Bragino, Langvannet-Porojarvi), distinguished by low prospects for the discovery of copper-nickel deposits, are traced in Norway almost without a reduction in thickness (reaching 1–2 km) and in conductivity (reaching 2000 S). Thus, the results of the work with the MHD installation "Khibiny" testifies to the actual wedging of the Cu-Ni productive strata of the Pechenga structure in Northern Norway.

The novelty of the work, in comparison with previous publications [10,11], is based on a more complete description of technique of the work using not only "cold", but also "hot" launches of the MHD installation "Khibiny". The description is made for the entire territory of the Kola Peninsula. This allows us to more fully and visually assess the physical meaning of the original method of conductive channels tracing by means of dividing the field of the Khibiny MHD source into galvanic and induction modes. An important element of novelty is associated with a more rigorous description of geological results based on a rectangular Gauss-Kruger grid. This made it possible to make a more concrete and substantiated conclusion about the pinching out of the Pil'guyarvi (Cu-Ni productive) strata of the Pechenga on the territory of Northern Norway and corresponding decrease of perspectives on the search of copper-nickel deposits. Finally, the paper emphasizes the connection of the obtained results of current conductive channels study with modern concepts of the two-layered structure of the continental Earth's crust with its division into the upper, brittle heterogeneous part, about 10 km thick, and a lower ductile homogeneous part that extends to the Moho border and deeper [21].

Funding: This work was supported by the project of the Russian-Norwegian cooperation "Northern Region", the Russian Fund for Basic Research (RFBR), grant No 18-05-00528 (methods) and partly by the support of the State Mission, GI KSC RAS, the subject of research 0226-2019-0052 (interpretation).

Acknowledgments: The author is deeply grateful to colleagues whose participation in field research and data processing has been reflected in a number of joint articles on this topic: A.D. Tokarev, Yu.A. Vinogradov, T.G. Korotkova, V.E. Asming, A.N. Shevtsov, N.A. Ochkur, A.G. Yampolsky, O.B. Lile, Yan Ronning, H. Moxness. The author dedicates this work to the bright memory of Academician Felix Mitrofanov with gratitude for the support and high appreciation ofresults. The author is deeply grateful to the anonymous reviewers for the careful review of the manuscript and for the many comments that contributed to a more accurate and complete presentation of the material.

Conflicts of Interest: The authors declare no conflict of interest.

References

1. Velikhov, E.P.; Volkov, Y.M. Prospects for the development of pulsed MHD—Energy and its application in geology and geophysics. In *Deep Electromagnetic Soundings Using Pulse MHD Generators*; The Kol'sky Philial Akademii Nauk of USSR (KFAN SSSR): Apatity, Russia, 1982; pp. 5–25. (In Russian)
2. Velikhov, E.P. (Ed.) *Geoelectrical Investigations with the Powerful Source of Current on the Baltic Shield*; Nauka: Moscow, Russia, 1989; p. 272. (In Russian)
3. Zhamaletdinov, A.A. (Ed.) *Theory and Method of Deep Electromagnetic Sounding on Crystalline Shields*; Devoted to MHD Soundings; Kola Science Center RAS: Apatity, Russia, 2006; Volume 1, p. 240. (In Russian)
4. Velikhov, E.P.; Panchenko, V.P. Large-scale geophysical surveys of the earth's crust using high-power electromagnetic pulses. In *Active Geophysical Monitoring*; Kasahara, J., Korneev, V., Zhdanov, M., Eds.; Elsevier: Amsterdam, The Netherlands, 2010; pp. 29–52.
5. Zhamaletdinov, A.A. «Khibiny» MHD Experiment: The 30th Anniversary. *Izv. Phys. Solid Earth* **2005**, *41*, 737–742.
6. Smol'kin, V.F.; Skuf'in, P.K.; Mokrousov, V.A. Stratigraphic Position, Geochemistry and Genesis of Volcanic Association of the Early Proterozoic Pechenga Area. In *Geology of the Eastern Finnmark-Western Kola Peninsul*; Region, O., Nordgulen, D., Eds.; Special Publication No.7; Geolodical Survey of Norvay: Trondheim, Norway, 1995; pp. 93–110.
7. Zagorodny, V.G.; Mirskaya, D.; Suslova, S.N. *Geological Structure of the Pechenga Sedimentary-volcanic Series*; Nauka: Moscow-Leningrad, Russia, 1964; p. 207. (In Russian)
8. Gorbunov, G.I. *Geology and Genesis of Sulphide Copper-Nickel Deposits of Pechenga*; Nedra: Moscow, Russia, 1968; p. 352. (In Russian)

9.   Zhamaletdinov, A.A. *The Model of the Lithosphere Electro-Conductivity from the Results of Researches with Controlled Sources*; Nauka: Leningrad, Russia, 1990; p. 156. (In Russian)

10. Zhamaletdinov, A.A.; Tokarev, A.D.; Vinogradov, Y.N.; Asming, V.E.; Otchkur, N.A.; Ronning, I.S.; Lile, O.B. Deep geoelectrical studies in the Finnmark and the Pechenga area by means of the "Khibiny" source. *Phys. Earth Planet. Int.* **1993**, *81*, 277–287. [CrossRef]

11. Zhamaletdinov, A.A.; Ronning, J.S.; Lile, O.B.; Tokarev, A.D.; Smolkin, V.F.; Vinogradov, Y.A. Geoelectrical Investigation with the 'Khibiny' Source in the Pechenga-Pasvik Area. In *Geology of the Eastern Finnmark-Western Kola Peninsula*; Region, D., Nordgulen, O., Eds.; Special Publication No.7; Geolodical Survey of Norvay: Trondheim, Norway, 1995; pp. 339–347. ISBN 82-7385-158-3.

12. Velikhov, E.P.; Zhamaletdinov, A.A.; Zhdanov, M.S. The Khibiny experiment. *Earth Planets* **1989**, *15*, 12–18. (In Russian)

13. Kirillov, S.K.; Osipenko, L.G. Study of the Imandra-Varzuga Conductive Zone (Kola Peninsula) with the Use of MHD Generator. In *Crustal Anomalies of Electrical Conductivity*; Nauka: Leningrad, Russia, 1984; pp. 79–86. (In Russian)

14. Zhdanov, M.S.; Frenkel, M.A. *Migration of Electromagnetic Fields in Solving Inverse Problems of Geoelectrics*; Moscow, Doklady Akademii Nauk (DAN USSR): Moscow, Russia, 1983; Volume 271, pp. 589–594.

15. Zhamaletdinov, A.A.; Kulik, S.N.; Pavlovsky, V.I.; Rokityansky, I.I.; Tanatchev, G.S. Anomaly of short-period geomagnetic variations over the Imandra-Varzuga structure (Kola Peninsula). *Geofiz. J.* **1980**, *2*, 91–96. (In Russian)

16. Zhamaletdinov, A.A.; Ronning, J.S.; Vinogradov, Y.A.B. Electrical profiling by the MISC and Slingram methods in the Pechenga-Pasvik area. In *Geology of the Eastern Finnmark-Western Kola Peninsula Region*; Roberts, D., Nordgulen, O., Eds.; Special Publication No.7; Geolodical Survey of Norvay: Trondheim, Norway, 1995; pp. 333–338.

17. Melezhik, V. Geological Map of Pechenga. In *Geology of the Eastern Finnmark-Western Kola Peninsula Region*; Roberts, D., Nordgulen, O., Eds.; Special Publication No.7; Geolodical Survey of Norvay: Trondheim, Norway, 1995; (map is enclosed as attachment).

18. Rokitjansky, I.I. *Inductive soundings of the Earth*; Naukova Dumka: Kyiv, Ukraine, 1981; p. 296. (In Russian)

19. Komarov, V.A. *Electrical Exploration by Induced Polarization*; Nedra: Moscow, Russia, 1980; p. 391. (In Russian)

20. Kamenetsky, F.M. *Electromagnetic Geophysical Studies by the Method of Transient Processes*; Moscow, Geos: Moscow, Russia, 1997; p. 162. (In Russian)

21. Zhamaletdinov, A.A. The Nature of the Conrad Discontinuity with Respect to the Results of Kola Superdeep Wel Drilling and the Data of a Deep Geoelectrical Survey. *Dokl. Earth Sci.* **2014**, *455 Pt 1*, 350–354. [CrossRef]

Article

Chemical Composition and Petrogenetic Implications of Apatite in the Khibiny Apatite-Nepheline Deposits (Kola Peninsula)

Lia Kogarko

Vernadsky Institute of Geochemistry and Analytical Chemistry, Russian Academy of Sciences, Moscow 119991, Russia; kogarko@geokhi.ru

Received: 4 October 2018; Accepted: 8 November 2018; Published: 16 November 2018

Abstract: Khibiny, one of the largest of the world's peralkaline intrusions, hosts gigantic apatite deposits. Apatite is represented by F-apatite and it contains exceptionally high concentration of SrO. (4.5 wt % on average) and increased amounts of rare earth elements (REEs; up to 8891 ppm). Such enrichment of apatite ores in REEs defined Khibiny deposit as world-class deposit with resources reaching several millions tons REE_2O_3. Apatite from the Khibina alkaline complex is characterized by the significant enrichment in light REEs relative to the heavy REEs (with average Ce/Yb ratio of 682) and the absence of a negative Eu anomaly. The obtained geochemical signature of apatite suggests a residual character of the Khibiny alkaline magma and it indicates that the differentiation of the primary olivine-melanephelinitic magma developed without fractionation of plagioclase which is the main mineral-concentrator of Sr and Eu in basaltic magmatic systems. The compositional evolution of the Khibiny apatite in the vertical section of the intrusion reflects primary fractionation processes in the alkaline magma that differentiated in situ. The main mechanism for the formation of the apatite-nepheline deposits was the gravitational settling of large nepheline crystals in the lower part of the magma chamber, while very small apatite crystals were suspended in a convective magma, and, together with the melt, were concentrated in its upper part of the magmatic chamber.

Keywords: apatite; Khibiny; apatite-nepheline deposit; phase diagram apatite-nepheline-diopside

1. Introduction

Apatite is a common accessory mineral and it has been extensively used to obtain significant genetic information [1–4]. This mineral is found in virtually all igneous rocks due to expanded crystallization fields. Experimental studies of the apatite solubility in the wide variety of silicate melt compositions have shown strong dependence of apatite saturation level upon silica activity and to a lesser extent concentration of Al, Fe, alkalis and oxygen fugacity [5]. Apatite is soluble in basic melts as compared to leucocratic magmas and its solubility in magmas decreases markedly with the increasing silica content and falling temperature [5,6]. Our experimental studies demonstrated the extremely high solubility of apatite in olivine melilite nephelinitic melts [6]. At a temperature of 1250 °C, basaltic melt, containing 50% SiO_2 dissolves 3–4 wt % P_2O_5, [5] and olivine melilite nephelinite, containing 41% SiO_2 under the same conditions of 8–9% P_2O_5. Previous detailed studies of Kola Alkaline Carbonatite Province (KACP) [7,8] have established that the composition of primary magma corresponded to a sodic melilitite or olivine melanephelinite (24.5–26.7% SiO_2) [7,9,10]. The mantle is generally P-depleted (86 ppm [11] and it only ultra-alkaline silica-undersaturated magmas produced at very low degrees of partial melting of metasomatised P-enriched mantle could then be saturated in apatite at near liquidus conditions. Highly undersaturated character of Kola primary magmas suggests the significant potential of P_2O_5 in alkaline rocks and carbonatites. Giant apatite deposits are associated with Khibiny peralkaline nepheline syenites. Khibiny ore deposits had been mined since 1930s.

The formation of apatite deposits within KACP has been addressed in a large number of studies [4,12–17]. The, two leading models include separation of primary iiolite-urtite magma into two immiscible melts (phosphate and aluminosilicate) and the second hypothesis relates to the formation of apatite ores with processes of crystallization differentiation.

This paper is focused on the trace element composition of apatite from several Khibiny apatite-nepheline deposits. The trace elements, particularly Sr and REE, are used as effective monitors of the magmatic evolution leading to ore formation in Khibiny complex.

2. Geological Setting

The geology of the Khibiny alkaline complex has been described in a large number of publications [4,14,16,18], and only a brief summary of the south-west apatite deposits is given here (Figures 1 and 2).

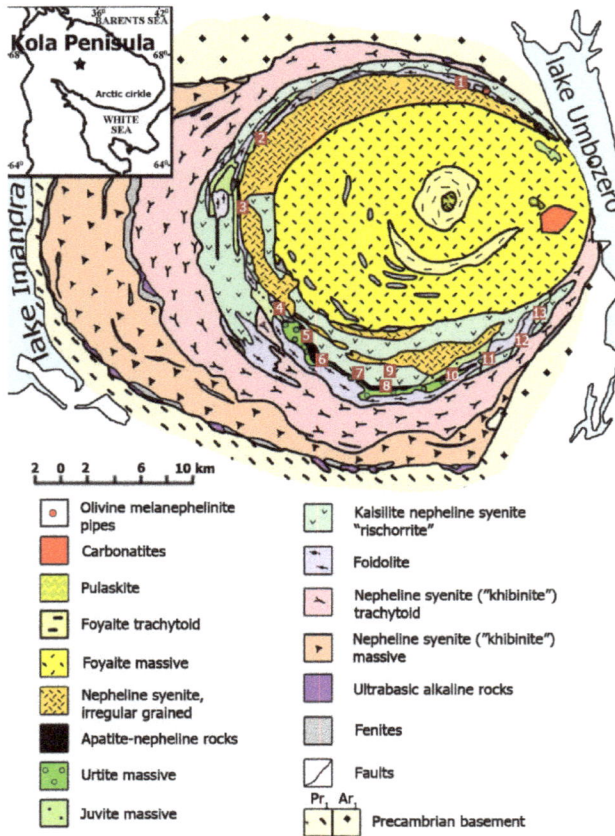

Figure 1. Geological map of the Khibiny massif generalized from the map of MGRE PGO "Sevzapgeologiya" (V.P.Pavlov) [2]. The geology of the Khibiny alkaline massif modified after using data from, and include references here [2]. ■—Titanite-Apatite deposits: 1—Valepakhk; 2—Partomchorr; 3—Kuelporr; 4—Snezhny Tsyrk; 5—Kukisvumchorr; 6—Yuksporr; 7—Apatitovy Tsyrk; 8—Rasvumchorr; 9—Eveslogchorr; 10—Koashva; 11—Vuonemyok; 12—Nyorkpakhk; 13—Oleny Ruchey [19].

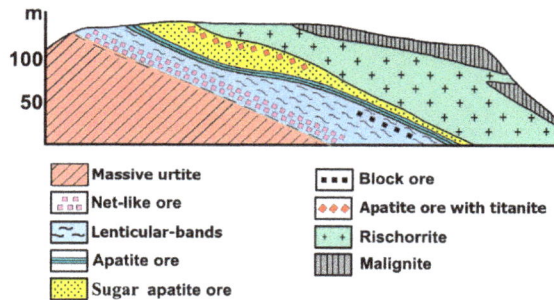

Figure 2. Cross-section of the Kukisvumchorr apatite deposit [14].

This complex is located in the central part of the Kola Peninsula, Russia, where it occurs as a ring intrusion of about 1325 km^2.

The Khibiny alkaline massif is a complex multiphased body built up from a number of ring-like and conical intrusions (Figure 1). The oldest rocks tend to occur towards the margins with successively younger intrusions being encountered towards the center.

From the oldest to youngest, the igneous units are: I—nepheline and alkali syenites, nepheline syenite-porphyries, II—massive khibinites (coarse-grained nepheline syenites), III—trachytoid khibinites, IV—rischorrites (potassium-rich nepheline syenites), V—ijolites, melteigites, and urtites. (This intrusion comprises the well-known stratified complex of rocks which contains the world-class apatite ore deposits.) VI—medium-grained nepheline syenites, VII—foyaites, VIII—carbonatites.

The intrusion of ijolite-urtite is of special interest, because it hosts the apatite ores. The intrusion is cone-shaped and outcrops as a discontinuous ring with a diameter of 26–29 km and length of the order of 75 km. The apatite-bearing intrusion has been separated into three subphases [14]. These are preore (I), ore (II), and post-ore (III) units. The rocks of subphase (I) consist of a series of ijolites interlayered with subordinate amounts of melteigite, urtite, juvite and malignite. Total thickness is less than 800 m. Subphase (II) consists of massive feldspathic urtite, ijolite-urtite and apatite ore with a total thickness of 200 to 800 m. The rocks of subphase (III) are from 10 to 1400 m thick and include lujavrite.

The principal phosphate ore deposits are found in subphase (II), where the apatite-rich rocks are found in the hanging wall of an ijolite-urtite intrusion (Figure 2).

The deposit is characterized by layering, as evidenced by distinct petrographic and geochemical features. The most upper parts of the apatite-rich bodies are composed of titanite–apatite ores (∼18% titanite, 80% apatite, 2% titanomagnetite). The upper zone (rich apatite ore) is represented by patchy and patchy-banded, so called sugar apatite ore [14]. The rock consists of 60–90% euhedral apatite crystals of several tenths of mm in size (Figure 3). Interstitial minerals are clinopyroxene, titanite, K-feldspar, titananagnetite, and nepheline. In some cases, the monomineralic layers of nepheline alternate with monomineralic apatite. The lower zone (poor apatite ore) is lenticular-banded, net-like and block ore. Lenticular-banded ore consists of fine-grained ijolite separated by layers of apatite and fine-grained urtite. Ijolite mainly consists of nepheline (up to 0.6 mm across) and pyroxene (up to 0.5 mm across). Net-like ore is texturally and structurally similar to lenticular-banded ore and differs from it only by the smaller proportion of urtite and apatite bands. Block ore appears pegmatitic. Occasional large crystals of nepheline (up to 15 mm across) occur in nepheline-apatite rock and in monomineralic apatite aggregates. The lowest zone grades into underlying massive urtite (Figure 2). The urtite consists of 75–90% large euhedral nepheline (up to 1–3 mm across) with intergranular acmitic clinopyroxene, titanite, feldspar, titanomagnetite, and aenigmatite (Figure 3). Extremely small grains of euhedral apatite are also found in the mesostasis. Massive urtites occupy about 89% of the thickness of the whole cross-section, whereas poor-ores and rich apatite ores occupy 8% and 3%, respectively. The average composition of the apatite-bearing intrusion that was obtained as result of detail mapping is given in Table 1. Our isotopic study has established the age of the rocks and apatite ores Khibiny massif-370 Ma [20,21]. The initial Hf, Sr, Nd isotope ratios are similar to the isotopic signatures of OIB indicating depleted mantle as a source [21]. This leads to the suggestion that the origin of this gigantic alkaline intrusion and super large apatite deposits is connected to a deep seated mantle source and metasomatic interaction between mantle material and fluid-melts transporting phosphorus and rare elements into magma-generation zones [20].

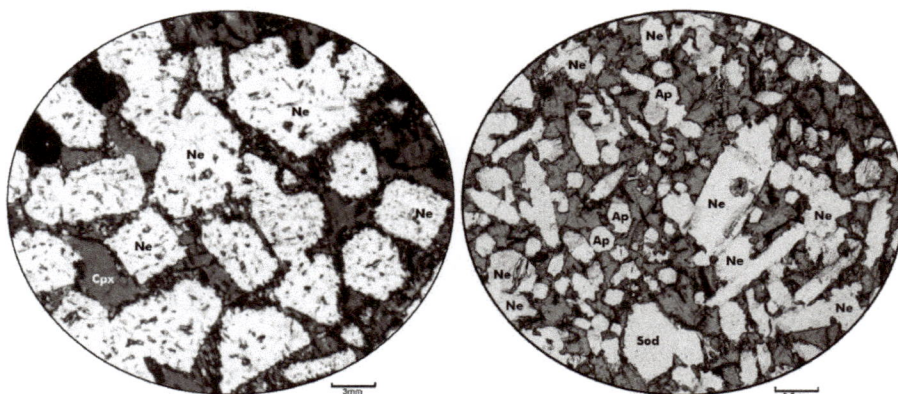

Figure 3. Thin sections of Khibiny apatite ore and massive urtite. Ne—nepheline Ap—apatite, Cpx—pyroxene, Sod—sodalite.

Table 1. Average composition of the apatite- bearing intrusion, wt %: [22].

Element	SiO_2	TiO_2	Al_2O_3	Fe_2O_3	FeO	MnO	MgO	CaO	Na_2O	KO	SrO	BaO	P_2O_5	F	H_2O	Total
Average	49.93	1.76	22.71	3.12	2.2	0.12	0.89	5.22	9.79	6.63	0.12	0.1	2.3	0.19	0.16	99.24

3. Composition of Khibiny Apatite

Over 900 apatite grains from several Khibiny apatite deposits have been analyzed. The specimens were collected from drill holes of four Khibiny apaite deposits (Table 2). Rusvumchorr drill hole sampled a complete stratigraphic section through deposit. Samples of apatite of Kukisvumchorr and Nuorkpakhk deposits were obtained from surface outcrops.

Table 2. Examples of the distribution of elements in the apatites of various fields of Khibiny massif. (1200 * analyses).

Element	Rasvumchorr (38 Samples) Average	Oleniy Ruchey (43 Samples) Average	Koashva (62 Samples) Average	Yuksporr (12 Samples) Average	Khibiny Massif Average
Ce/Yb	441	705	940	416	682
ΣREE	7451	7379	12,979	7218	8891
Ce	3334	3283	6086	3202	4040
La	2294	2247	4293	2152	2799
Pr	314	308	521	309	367
Nd	1085	1101	1637	1111	1243
Sm	150	157	182	156	162
Eu	41	44	45	45	44
Gd	113	120	117	122	118
Tb	13	14	12	14	13
Dy	62	64	52	64	60
Ho	10	10	8.3	10	9.8
Er	21	21	16	21	20
Tm	2.2	2.0	1.5	2.0	1.9
Yb	10	10	7.0	9.3	8.9

Table 2. *Cont.*

Element	Rasvumchorr (38 Samples) Average	Oleniy Ruchey (43 Samples) Average	Koashva (62 Samples) Average	Yuksporr (12 Samples) Average	Khibiny Massif Average
Lu	1.0	1.0	0.7	1.0	0.9
Si	1123	-	-	-	1123
Na	1100	-	1200	-	1150
Sr	30,149	30,543	64,771	18,947	38,520
Hf	0.02	-	0.1	0.1	0.1
Ta	0.003	-	0.1	0.1	0.0
Pb	1.4	-	1.0	1.4	1.2
Th	19	-	23	25	22
U	2.0	-	1.3	2.5	1.8
Mg	-	4181	22	58	2225
Sc	0.04	3	0.6	1.2	1.6
Mn	150	157	114	136	142
Fe	88	313	89	99	182
Y	283	265	171	277	245
Zr	2.3	8	7.2	7.1	6.6
Nb	0.1	1	0.6	0.1	0.4
Ba	331	418	318	326	363
Zn	0.4	-	-	-	0.4

* The average is obtained from all apatite analyses, ppm.

Samples were analyzed at the Vernadsky Institute of Geochemistry and Analytical Chemistry RAS, Moscow. Mineral compositions were determined using CAMECA SX 100 electron probe microanalyser (CAMECA, Gennevilliers, France) with four diffraction spectrometers at an accelerating voltage of 15 kV and probe current of 60 nA. We used next natural and synthetic reference samples: for P and Ca—apatite, for Sr—celestine, for F—LiF, for Na—jadeite, for La—$La_2(PO_4)_2$, for Ce—$Ce_2(PO_4)_2$. All standarts are from Smithsonian Institution. Concentrations were calculated from relative peak intensities using the PAP-correction.

Trace element analyses were performed in Frakfurt University and Max Plank Institute in Minze using a Thermo Scientific Element 2 ICPMS (Thermo Fisher, Waltham, MA, USA) coupled with a Resonetics Resolution M-50 excimer laser. The laser spot size varied from 20 to 60 μm. The following isotopes were used for determining the abundances of the elements: ^{29}Si, ^{45}Sc, ^{88}Sr, ^{89}Y, ^{90}Zr, ^{93}Nb, ^{137}Ba, ^{139}La, ^{140}Ce, ^{141}Pr, ^{146}Nd, ^{147}Sm, ^{151}Eu, ^{157}Gd, ^{159}Tb, ^{161}Dy, ^{165}Ho, ^{167}Er, ^{169}Tm, ^{172}Yb, ^{175}Lu, ^{178}Hf, ^{181}Ta, ^{208}Pb, ^{232}Th, ^{55}Mn, ^{57}Fe, and ^{24}Mg.

As the standarts we used NIST glasses. All samples were analysed on polished sections and done by the author with the help of Institutes staff.

Apatites have been extracted from different rocks of deposits and all types of ores. Apatite typically occurs as small (0.1–0.3 mm) well-defined idiomorphic rounded or acicular grains (Figure 3). The apatite is light yellow in colour or transparent. Apatite often contains numerous primary magmatic micro inclusions, mainly containing nepheline, K-feldspar, clinopyroxene, apatite, and titanite [23]. In the apatite ore, there are zones of large late apatite of dark color. Previous studies [24–27] have established that Khibiny apatite corresponds to fluorapatite according to new nomenclature [28]. In this paper we investigated in detail the trace element composition of Khibiny apatite from several deposits and various sections through ore bodies. We estimated average concentrations of Sr, REE (La, Ce, Yb), Na, Si, Th, U in apatites of some Khibiny deposits. (Table 2). Previous studies [24] have shown that the principal substitution mechanisms involve replacement of Ca^{2+} by Sr^{2+}, and to a much smaller extent, Mn^{2+} and Fe^{2+} [24]. The main REE substitution is $REE^{3+} + Si^{4+} = Ca^{2+} + P^{5+}$, and $REE^{3+} + Na^+ = 2Ca^{2+}$. The substitution $2Ca^{2+} < Na^+ + LREE^{3+}$ is quite rare in apatites and it occurs in highly alkaline igneous systems [29].

Petrographic study demonstrated that apatite was an early liquidus phase in ores and rocks of ijolite-urtite intrusion.

Our study showed that the majority of analyzed apatite grains are homogeneous with respect to the distribution of trace elements. The pronounced zonation of apatite crystals was not detected. In some cases, a subtle zoning is observed in apatites from the Oleniy Ruchey and Yukspor deposits where Sr and Ce levels increase from central to the marginal zones of apatite crystals. (SrO—2.1–3.5 wt %, 1.7–2.1 wt %, Ce_2O_3—0.25–0.37 wt %, and 0.3–0.36 wt %).

In several apatite grains from the Oleniy Ruchey ores a reverse zonation with respect to Ce_2O_3 was recorded (0.32–0.25 wt %, 0.13–0.096 wt %).

Khibiny apatite contains exeptionaly high levels of SrO (4.5 wt % on average), which varies markedly within 0.16–9.8 wt % in different ore deposits. The concentration of Sr in apatite is much higher compared to the host magma (0.2 wt % [9]). We investigated the composition of apatite in the vertical section (up to 600 m depth) of the Rasvumchorr deposit. Despite significant variations in the concentrations of Sr in apatites at the same level, an increase in the content of Sr upwards is observed along the borehole crossection (Figure 4). Cryptic layering of Sr in Khibiny apatite is not as pronounced as for example in loparite in a nearby alkaline Lovozero massif.

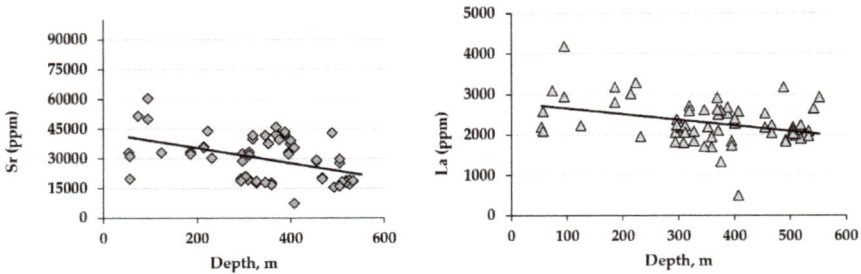

Figure 4. Distribution of Sr, La in apatite in the vertical section of Rasvumchorr deposit.

Apatite and titanite are the main minerals that host REE in the Khibiny deposits. The total REE content in apatite ranges from hundredths of percent to over 1.5% with an average of 8891 ppm. Such concentration of rare earth in millions of tons of apatite ores defined Khibiny deposit as world-class with resources 5.5 Mt REE_2O_3 [30]. Chondrite-normalized REE plots (Figure 5) of Khibiny apatites demonstrated significant enrichment of the light rare earths over heavy REE, Ce/Yb is very high with an average of 682. It is worth noting that all chondrite-normalized REE plots of apatite lack negative Eu anomaly (Figure 5). This confirms our conclusion [31] that the oxygen fugacities of apatite-bearing intrusion of the Khibiny complex were close to the QFM buffer system and apatite contained Eu^{3+}.

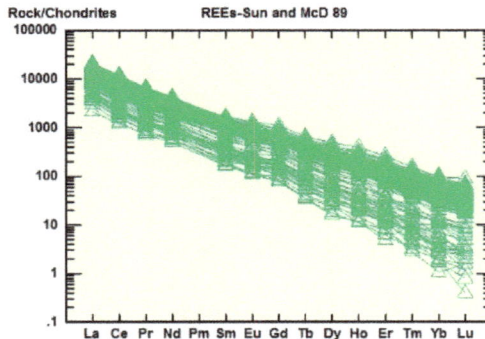

Figure 5. Distribution of REE in Khibiny apatite.

The Eu anomaly and levels of Sr are also controlled by the plagioclase which concentrate Eu^{2+} from the melt (Kd^{Eu} in plagioclase—3.8–7.9 [32], Kd^{Sr}—2.7–10 [33]. In the process of crystallization differentiation of KACP primary magma of olivine nephelinites composition plagioclase did not crystallize plagioclase. Main minerals of this process, such as olivine, melilite and clinopyroxene do not typically concentrate Eu and Sr, because of the very low distribution coefficients of Eu and Sr in these minerals [32,34]. We studied the distribution of Ce in the vertical section of the Rasvumchorr deposit drillhole. Ce contents in apatite increase systematically upwards through 600 m of the apatite-nepheline deposit (Figure 4). The compositional evolution of apatite reflects primary fractionation processes in the alkaline magma that differentiated in situ from the bottom to the top of the magma chamber. The contents of Mn, Fe, and Mg in apatites are low and vary within a considerable range (Mn—30–250 ppm, Fe—89–1153 ppm, Mg—22–4420 ppm). The majority of coarse-grained dark color apatite grains contains slightly elevated levels of Fe, Mn, and Mg, while the concentration of rare earths is similar to that of the light-yellow apatite, which is most common.

The concentration of radiactive elements in Khibiny apatite is low (average Th—22 ppm, U—1.8 ppm) relative to other alkaline complexes (e.g., Lovozero, Pilansberg) Table 2.

In some cases, Sr apatites may contain some Ba, for example, the Pilansberg apatite from alkaline complex (South Africa) contains up to 2500 ppm Ba [29]. Khibiny apatite contains significantly lower concentrations of barium on average 363 ppm. Relative to apatites from several other localities, including Pilansberg [29] and Lovozero [25], apatite from Khibiny is depleted in rare earths and radioactive elements (Table 2). The concentrations of Sr in the Khibiny apatite are comparable to those from Lovozero, Pilansberg and Ilimaussaq. Notably, the highly alkaline complexes typically host belovite, this mineral was not found in the apatite-nepheline ores of Khibiny.

Compositionally, most of the studied apatites correspond to fluorapatites [28]. Fluorine content varies from 4.73 to 3.33 wt %. The chlorine concentration is very low-hundredths of a percent.

The large amount of data accumulated during recent decades has shown that composition of apatite from different rocks of the world have distinctive absolute abundances of many trace elements and chondrite-normalized trace-element patterns [4,35,36]. Apatite global data-base [37] can be used for the identification of apatites from different rock types and processes of mineralization.

To establish the geochemical signature of Khibiny apatites, we used previously published data [37–39] and several discrimination plots, including Sr versus Y, Sr versus Mn, and Ce/Yb versus REE (Figure 6). Apatites from different rock types plot within well-defined compositional fields on the majority of these diagrams [37]. According to our data, apatite from that the Khibiny complex is characterized by the highest levels of Sr among all rock varieties (Figure 6). It should be noted that apatite from the Khibiny complex shows characteristically low levels of Y and heavy REEs, especially in comparison with the Si rich granitoids (Figure 6). Mn value is close to that of apatite from carbonatites and much lower relative to granitoids, granitic pegmatites, jacupirangites, and dolirites [37].

Khibiny apatite is considerably enriched in light relative to heavy REEs and has the highest Ce/Yb ratio (Figure 6).

The obtained geochemical signature of Khibiny apatite suggests a residual character of the Khibiny alkaline magma and it indicates that the differentiation of the primary magma developed without fractionation of mineral-concentrators of Sr and Eu (such as plagioclase). Our data confirm previous findings regarding the olivine-nephelinitic nature of the primary magma of Khibiny apatite intrusion [7–10].

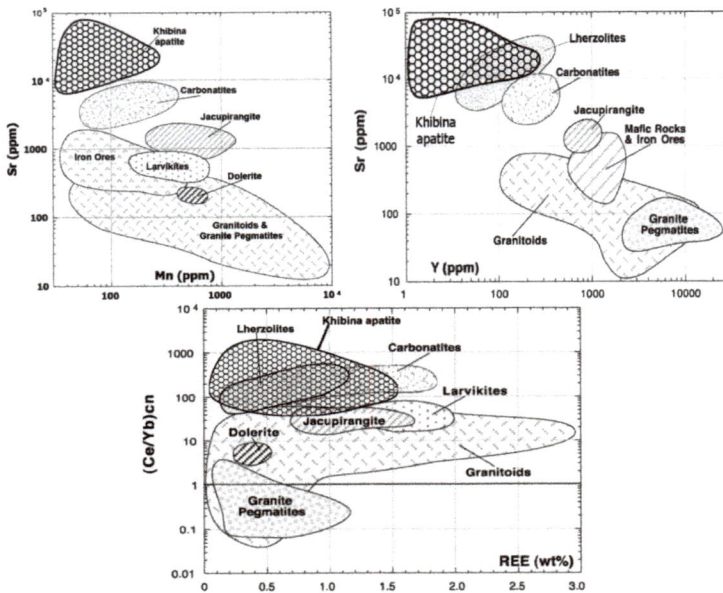

Figure 6. Compositional Field of Khibiny apatite on the discriminant diagrams proposed by [37].

4. Genetic Problems of the Khibiny Apatite Deposits

Apatite saturation in the magma mainly depends on its P_2O_5 content and to a lesser extent on composition of magma. At upper mantle pressure and a temperature of 1250 °C basaltic melt (50% SiO_2) will dissolve up to 4% P_2O_5 before apatite saturation is reached [6]. According to [7,8,10,40], the P_2O_5 content of Devonian Kola primitive magmas (28–32% SiO_2) was in the range from 0.3 to 1.2 wt % P_2O_5 and was not saturated in apatite. In the process of long evolution of primary alkaline magmas, the phosphorus content increased significantly and in Khibiny apatite intrusion it reaching 2.4% [22].

Extensive studies of fluid microinclusions in rock-forming minerals and phase equilibria of apatite-bearing systems were performed previously in order to establish the physico-chemical conditions of apatite ore formation [17,23]. According to these studies, primary inclusions were formed as a result of melt entrapment and they contain polyphase aggregates composed of sodalite, microcline, apatite, lepidomelan, pyroxene, iron sulfide, and villiaumite [23]. Thermometric experiments demonstrated that the melt appeared at a temperature of about 700 °C; at gradual heating up to 900 °C, villiaumite, microcline, mica, sulfides, sodalite, pyroxene, and sphene were completely resorbed and the microinclusions contained the equilibrium association melt + apatite + nepheline + gas. All phases were completely dissolved at 970 °C, while the inclusions were homogenized and they contained only aluminasilicate melt. The microprobe analysis of the homogenized inclusions showed that aluminasilicate melt contained about 2 wt % P_2O_5 [23], this value is close to the average for apatite-nepheline intrusion.

The phase equilibria of the apatite-bearing ijolite-urtite rocks of Khibiny can be approximated in the system $NaAlSiO_4$-$CaMgSi_2O_6$-$Ca_5(PO_4)_3F$ (Figure 7) [9]. Composition (Table 1) represents a weighted average of the bulk composition of the apatite-bearing complex under consideration (Table 1 and Figure 2). From melt of such a composition nepheline would crystallize first. Apatite and nepheline would be the next to crystallize as the temperature falls. Apatite, nepheline, and pyroxene begin to precipitate as the temperature is lowered further. This crystallization sequence is in agreement with petrographic observations of the rocks [16]. Thus, from the viewpoint of experimental phase equilibria, an alumina-silicate melt with 2.4 wt % P_2O_5 could have been the parental magma of the ore apatite complex. Such a magma would crystallize about 10–15% of nepheline before reaching the nepheline apatite-pyroxene cotectic and all these minerals would crystallize simultaneously. The close to eutectic character of apatite-bearing intrusion and the coincidence of the order of crystallization of apatite ores with the regularities of crystallization of experimentally studied phosphate-silicate systems indicate that the main process in the formation of apatite deposits was crystallization differentiation. It should be pointed out that field of existence of two liquids-phosphate and aluminosilicate (Figure 7) is separated from the average composition of apatite intrusion by the temperature barrier, which means that the "immiscibility" model [12,13] is not realistic. In addition, the temperature of melting of monomineral apatite ore should be higher than 1500 °C.

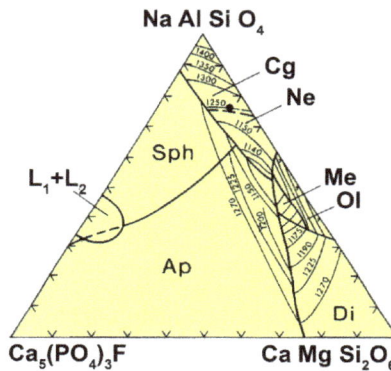

Figure 7. Phase relations in the system $NaAlSiO_4$—$CaMgSi_2O_6$-$Ca_5(PO_4)_3F$ at 1 atm. pressure [9]. Cg—carnegieite, Ne—nepheline, Ol—olivine, Me—melilite, Di—diopside, Ap—apatite, Sph—silicophosphate, L_1 + L_2—two immiscible liquids, ●—average composition of the apatite-bearing intrusion.

As it has been demonstrated by a number of authors [41–44], a convective regime must exist in magmatic reservoirs with thickness exceeding 10 m. However, the style of the convection with crystallization is likely steady-state conditions [45].

One outstanding problem that is presented by the apatite-ore bodies concerns the manner in which accumulation of nepheline, apatite and pyroxene occurs in terms of steady-state convection.

The consideration of the values of the velocity of sedimentation calculated according to Stokes law and viscosity, temperature difference, densities, and heat-conductivity gives [44]:

$$ln\frac{N_{p_1}}{N_{p_2}} = -0.66\Delta\rho\rho^{-\frac{2}{3}}\eta^{-2}k^{-\frac{2}{3}}\alpha^{-\frac{1}{3}}\Delta T^{-\frac{1}{3}}C_p^{-\frac{1}{3}}g^{\frac{1}{3}}a^2 \qquad (1)$$

where:

$\Delta\rho = \rho_s - \rho_l$ the difference between densities of melt and crystals (g cm^{-3});
η viscosity (poise);
k coefficient of heat-conductivity (cal cm^{-1} s^{-1} k^{-1});
α thermal expansion coefficient (k^{-1});
ΔT temperature difference between roof and foot-wall;
C_p heat-capacity (cal.g^{-1} k^{-1});
g gravity acceleration (980 cm s^{-2}); and,
a dimensions of crystals (cm).
N_{P1}/N_{P2}—mineral particle number ratio near the roof and the floor of the magma chamber.

This equation shows that the strongest influence on the distribution of particles is that of their size; with particles above a certain size-nepheline (up to 3–5 mm across), the stirring effect of the convection ceases to act and the particles are settled to the bottom of the magma chamber, forming a lower cumulus layer (massive urtite), while smaller particles-apatite (up to several tenths of mm) are stirred more efficiently and enrich the later (upper) cumulative layers (rich apatite ore).

Very important is the presence of sorting of minerals [2] which suggests that the mechanism of accumulation of minerals and formation of apatite ores was the gravitational differentiation of the close to eutectic apatite- intrusion in conditions of convective motion. Sorting coefficients of apatite, nepheline, and pyroxene (Figure 8) are close to unit. Such values of sorting coefficients are characteristic of well-sorted sedimentary deposits, for instance, sands. According to the calculations, pyroxenes should accumulate in the middle part of apatite deposits, which corresponds to the distribution of this mineral in the vertical section of deposits. In the formation of massive urtites, very small apatite crystals were suspended and captured by the interstitial melt, demonstrating a significant difference in the sizes of these minerals in the early stages of crystallization (Figure 9). Thus crystals of apatite remained in suspension until the settling velocity is small as compared with the velocity of convective currents. At cooling of intrusion convection falls and even the small crystals of apatite forms accumulations and ores. This model is supported by the composition of Khibiny apatite. Very homogeneous character of Khibiny apatite and weakly expressed zoning suggest crystallization of apatite in a large volume of magmatic chamber, in which inevitably there is strong convection leading to active mixing.

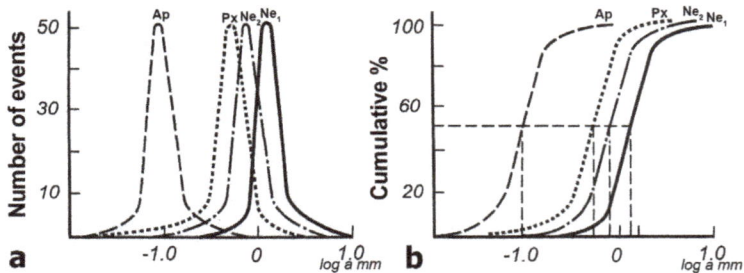

Figure 8. (**a**) Distribution diagrams for various minerals; (**b**) Calculated cumulative diagrams for minerals. Ap—apatite from rich apatite ore; Px—clinopyroxene from lenticular-banded and net- like ore (poor apatite ore); Ne$_2$—nepheline from ores; Ne$_1$—nepheline from massive urtite; a—diameter of mineral.

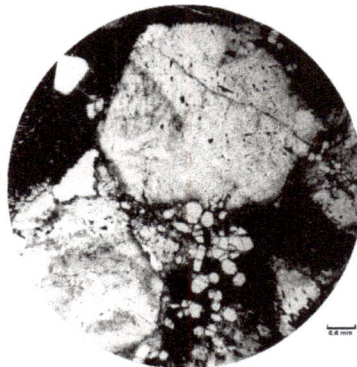

Figure 9. Thin section of massif urtite. Small crystals of apatite in interstitial between large crystals of nepheline.

Our study demonstrated that in the cross-section of Rasvumchorr deposit the concentrations of Sr, Ce, and La in apatite systematically increase with the increasing stratigraphic height (Figure 4). In the huge alkaline magmatic chamber of Etna (Italy) volcanic system, the content of REE in apatite also increases with the degree of differentiation [3]. The compositional evolution of the Khibiny apatite reflects the primary fractionation processes in the alkaline magma that differentiated from the bottom to the top of the magma chamber as a result of magmatic convection, coupled with the settling of minerals with different settling velocities. Compositional variations in apatite record subtle changes in the composition of alkaline magma. The cryptic variation observed in the Khibiny apatites appears to be very similar to that in minerals from basic layered intrusions [46,47]. The character of this cryptic layering suggests that fractional crystallization in situ of a single batch of alkaline magma was the main process governing the formation of the layered Khibiny apatite-nepheline intrusion. It should be noted that the ijolite dyke containing angular xenoliths of apatite ores and massive urtite cross-cuts the apatite intrusion. Also, active processes of post-ore tectonics took place and they caused the formation of numerous folds in the apatite deposits and displacements in the plastic state.

5. Conclusions

1. Our data show that apatite from the Khibiny complex is enriched in SrO (4.5 wt %) and REEs—(up to 8891 ppm). Such concentration of rare earth in millions of tons of apatite ores defined Khibiny deposit as world-class with resources several millions tons REE_2O_3.
2. Statistical analysis showed that Khibiny apatites demonstrated a significant enrichment of the light rare earths over heavy REE (average Ce/Yb ratio 682) and absence of negative Eu anomaly.
3. The obtained geochemical signature of Khibiny apatite suggests a residual character of the Khibiny alkaline magma and indicate that the differentiation of the primary olivine-melanephelinitic magma developed without fractionation of plagioclase which is the main mineral-concentrator of Sr and Eu of basaltic magmatic systems.
4. The compositional evolution of the Khibiny apatite in the vertical section of the intrusion reflects primary fractionation processes in the alkaline magma that differentiated in situ from the bottom to the top of the magma chamber as a result of magmatic convection, coupled with the precipitation of minerals with different settling velocities.
5. Our data suggest that the main mechanism for the formation of the apatite-nepheline deposits was the gravitational settling of large nepheline crystals the low part, while very small apatite crystals were suspended in a convective magma and, together with the melt, were concentrated in the upper part of the magmatic chamber.

Acknowledgments: I thank Kononkova N.N., Lahaye Y, Kuzmin D. for the help for sample analyses. The comments by 3 anonymous referees are much appreciated. This work is supported by the Program of RAS I.48 no. 0137-2018-0041.

Conflicts of Interest: The authors declare no conflict of interest.

References

1. Chakhmouradian, A.R.; Reguir, E.P.; Mitchell, R.H. Strontium-apatite: New occurrences, and the extent of Sr-for-Ca substitution in apatite-group minerals. *Can. Miner.* **2002**, *40*, 121–136. [CrossRef]
2. Kalashnikov, A.O.; Konopleva, N.G.; Pakhomovsky, Y.A.; Ivanyuk, G.Y. Rare Earth Deposits of the Murmansk Region, Russia—A Review. *Soc. Econ. Geol. Inc. Econ. Geol.* **2016**, *111*, 1529–1559. [CrossRef]
3. Busa, T.; Clochiatti, R.; Cristofolini, R. The role of apatite fractionation and REE distribution in alkaline rocks from Mt. Etna, Sicily. *Miner. Pet.* **2002**, *74*, 95–114. [CrossRef]

4. Borutzky, B.E. Modern Understanding of the Nature and Geological History of the Formation of Rocks of the Khibiny Alkaline Massif. (Critical Comparison of the Proposed Hypotheses and Comments to Them). In Proceedings of the Materials of the All-Russian (with International Participation) Field Conference Dedicated to the 80th Anniversary of the Kola Science Centre RAS, Apatity, Russia, 20–23 June 2010; Geological Institute KSC RAS, Kola Branch of Russian Mineralogical Society: Apatity, Russia, 2010.

5. Watson, E.B. Apatite and phosphorus in mantle source regions: An experimental study of Apatite/melt equilibria at pressures to 25 kbar. *Earth Planet. Sci. Lett.* **1980**, *51*, 322–335. [CrossRef]

6. Kogarko, L.N.; Krigman, L.D.; Krot, T.V.; Ignatenko, K.I. Influence of chemical composition of magmatic melt on the solubility of P_2O_5. *Geochem. Int.* **1986**, *10*, 138.

7. Arzamastsev, A.A.; Bea, F.; Glaznev, V.N.; Arzamastseva, L.V.; Montero, P. Kola alkaline province in the Paleozoic: Evaluation of primary mantle magma composition and magma generation conditions. *Russ. J. Earth Sci.* **2001**. [CrossRef]

8. Downesa, H.; Balaganskaya, E.; Beard, A.; Liferovich, R.; Demaiffe, D. Petrogenetic processes in the ultramafic, alkaline and carbonatitic magmatism in the Kola Alkaline Province: A review. *Lithos* **2005**, *85*, 48–75. [CrossRef]

9. Kogarko, L.N. *Genetic Problems of Agpaitic Magmas*; Nauka: Moscow, Russia, 1977; 294p. (In Russian)

10. Dunworth, E.; Bell, K. The Turiy Massif, Kola Peninsula, Russia: Isotopic and Geochemical Evidence for Multi-source Evolution. *J. Pet.* **2001**, *42*, 377–405. [CrossRef]

11. Palme, H.; O'Neill, H.S.C. Cosmochemical Estimates of Mantle Composition. In *Treatise on Geochemistry*; Holland, H.D., Turekian, K.K., Eds.; Elsevier: Amsterdam, The Netherlands, 2003.

12. Marakushev, A.A.; Suk, N.I. Experimental modeling of the Khibiny layered nepheline-syenite massif to clarify the genesis of apatite deposits. *Russ. J. Earth Sci.* **1993**, *330*, 241–244.

13. Suk, N.I. Liquid immiscibility in P-bearing melts as for the genesis of apatite deposits. *Petrologiya* **1993**, *1*, 282–291.

14. Zak, S.L.; Kamenev, E.A.; Minakov, P.V.; Armand, A.L.; Mikheichev, A.S.; Petersilie, J.A. *The Khibiny Alkaline Massif*; Nedra Publishers: Leningrad, Russia, 1972. (In Russian)

15. Kogarko, L.N.; Khapaev, V.V. The modelling of formation of apatite deposits of the Khibiny massif (Kola peninsula). In *Origins of Igneous Layering*; D. Reidel Publishing Company: Dordrecht, The Netherlands, 1987; pp. 589–611.

16. Ivanova, T.N.; Dudkin, O.B.; Kosyreva, L.V.; Polyakov, K.I. *Ijolite–Urtites of the Khibiny Massif*; Nauka Publish. House: Leningrad, Russia, 1970. (In Russian)

17. Kogarko, L.N. Ore-forming potential of alkaline magmas. *Lithos* **1990**, *26*, 167–175. [CrossRef]

18. Eliseev, N.A. The Khibine apatite deposits. *Zap. RMO* **1937**, *66*, 491–516. (In Russian)

19. Arzamastsev, A.A.; Yakovenchuk, V.N.; Pakhomovsky, Y.A.; Ivanyuk, G.Y. *The Khibiny and Lovozero Alkaline Massifs: Geology and Unique Mineralization*; 33 IGS Excursion, 47; Geological Institute of the Russian Academy of Science: Apatity, Russia, 2008.

20. Kogarko, L.N.; Lahaye, Y.; Brey, G.P. Plume-related mantle source of super-large rare metal deposits from the Lovozero and Khibiny massifs on the Kola Peninsula, Eastern part of Baltic Shield: Sr, Nd and Hf isotope systematics. *Miner. Pet.* **2010**, *98*, 197–208. [CrossRef]

21. Kramm, U.; Kogarko, L. Nd and Sr isotope signatures of the Khibiny and Lovozero agpaitic centres, Kola Alkaline Province, Russia. *Lithos* **1994**, *32*, 225–242. [CrossRef]

22. Minakov, F.V.; Kamenev, E.A.; Kalinkin, M.M. On original composition and evolution of ijolite-urtite magma from the Khibiny alkaline massif. *Int. Geochim.* **1967**, *8*, 901–915.

23. Kogarko, L.N.; Romanchev, B.P. Geochemical criteria of ore potential of alkaline magmas. *Geochem. Int.* **1986**, *10*, 1423–1430.

24. Pushcharovsky, D.Y.; Nadezhina, T.N.; Khomyakov, A.P. Crystal structure of strontium-apatite from Khibiny. *Crystallogr. Rep.* **1987**, *32*, 891–895. (In Russian)

25. Chakhmouradian, A.R.; Reguir, E.P.; Zaitsev, A.N.; Couëslan, C.; Xu, C.; Kynický, J.; Mumin, A.H.; Yang, P. Apatite in carbonatitic rocks: Compositional variation, zoning, element partitioning and petrogenetic significance. *Lithos* **2017**, *274*, 188–213. [CrossRef]

26. Pekov, I.V.; Chukanov, N.V.; Eletskaya, O.V.; Khomyakov, A.P.; Menshikov, Y.P. Belovite-(Ce): New data, refined formula, and relationships with other minerals of apatite group. *Zap. Vserossiiskogo Miner. Obshchestva* **1995**, *124*, 98–110. (In Russian)

27. Khomyakov, A.P. *Mineralogy of Hyperagpaitic Alkaline Rocks*; Clarendon Press: Oxford, UK, 1995.
28. Pasero, M.; Kampf, A.R.; Ferraris, C.; Pekov, I.V.; Rakovan, J.; White, T.J. Nomenclature of the apatite supergroup minerals. *Eur. J. Miner.* **2010**, *22*, 163–179. [CrossRef]
29. Liferovich, R.P.; Mitchell, R.H. Apatite-group minerals from nepheline syenite, Pilansberg alkaline complex, South Africa. *Miner. Mag.* **2006**, *70*, 463–484. [CrossRef]
30. Zaitsev, A.; Williams, T.; Jeffries, T.; Strekopytov, S.; Moutte, J.; Ivashchenkova, O.; Spratt, J.; Petrovd, S.; Wall, F.; Seltmann, R.; et al. Rare earth elements in phoscorites and carbonatites of the Devonian Kola Alkaline Province, Russia: Examples from Kovdor, Khibiny, Vuoriyarvi and Turiy Mys complexes. *Ore Geol. Rev.* **2014**, *61*, 204–225. [CrossRef]
31. Ryabchikov, I.D.; Kogarko, L.N. Magnetite compositions and oxygen fugacities of the Khibiny magmatic system. *Lithos* **2006**, *91*, 35–45. [CrossRef]
32. Villemant, B.; Jaffrezic, H.; Joron, J.L.; Treuil, M. Distribution Coefficients of Major and Trace-Elements—Fractional Crystallization in the Alkali Basalt Series of Chaine-Des-Puys (Massif Central, France). *Geochim. Cosmochim. Acta* **1981**, *45*, 1997–2016. [CrossRef]
33. Kuehner, S.M.; Laughlin, J.R.; Grossman, L.; Johnson, M.L.; Burnett, D.S. Determination of trace element mineralliquid partition coefficients in melilite and diopside by ion and electron microprobe techniques. *Geochim. Cosmochim. Acta* **1989**, *53*, 3115–3130. [CrossRef]
34. Nash, W.P.; Crecraft, H.R. Partition coefficients for trace elements in silicic magmas. *Geochim. Cosmochim. Acta.* **1985**, *49*, 2309–2322. [CrossRef]
35. Chu, M.F.; Wang, K.L.; Griffin, W.L.; Chung, S.L.; O'Reilly, S.Y.; Pearson, N.J.; Iizuka, Y. Apatite Composition: Tracing Petrogenetic Processes inTranshimalayan Granitoids. *J. Pet.* **2009**, *50*, 1829–1855. [CrossRef]
36. Hoshino, M.; Kimata, M.; Shimizu, M.; Nishida, N. Minor-element systematics of fluorapatite and zircon inclusions in allanite-(ce) of felsic volcanic rocks from three orogenic belts: Implications for the origin of their host magmas. *Can. Miner.* **2007**, *45*, 1337–1353. [CrossRef]
37. Belousova, E.A.; Griffin, W.L.; O'Reilly, S.Y.; Fisher, N.I. Apatites as an indicator mineral for mineral exploration: Trace-element compositions and their relationship to host rock type. *J. Geochem. Explor.* **2002**, *76*, 45–69. [CrossRef]
38. Roeder, P.L.; MacArthur, D.; Ma, X.D.; Palmer, G.R.; Mariano, A.N. Cathodoluminescence and microprobe study of rare earth elements in apatite. *Am. Miner.* **1987**, *72*, 801–811.
39. O'Reilly, S.Y.; Griffin, W.L. Apatite in the mantle: Implications for metasomatic processes and high heat production in Phanerozoic mantle. *Lithos* **2000**, *53*, 217–232. [CrossRef]
40. Kogarko, L.N. Alkaline Magmatism and Enriched Mantle Reservoirs: Mechanisms, Time, and Depth of Formation. *Geochem. Int.* **2006**, *44*, 3–10. [CrossRef]
41. Turner, J.S.; Campbell, I.H. The influence of viscosity on fountains in magma chambers. *J. Pet.* **1986**, *27*, 1–30.
42. Parsons, I. (Ed.) *Origins of Igneous Layering*; NATO ASI Series C196; D. Reidel Publishing Company: Dordrecht, The Netherlands, 1987; ISBN 978-90-481-8435-4.
43. Sparks, R.S.J.; Huppert, H.E.; Koyaguchi, T.; Hallworth, M.A. Origin of modal and rhythmic igneous layering by sedimentation in a convecting magma chamber. *Nature* **1993**, *361*, 246–249. [CrossRef]
44. Bartlett, R.W. Magma convection, temperature distribution and differentiation. *Am. J. Sci.* **1969**, *267*, 1067–1082. [CrossRef]
45. Spera, F.J.; Oldenburg, C.M.; Christensen, C.; Todesco, M. Simulations of convection with crystallization in the system $KAlSi_2O_6$-$CaMgSi_2O_6$: Implications for compositionally zoned magma bodies. *Am. Miner.* **1995**, *80*, 1188–1207. [CrossRef]
46. Cawthorn, R.G. *Layering Intrusions. Developments in Petrology*; Elsevier Science B.V.: Amsterdam, The Netherlands; ALausanne, Switzerland; New York, NY, USA; Oxford, UK; Tokyo, Japan, 1996.
47. Tegner, C.; Cawthorn, G.; Kruger, J. Cyclicity in the Main and Upper Zones of the Bushveld Complex, South Africa: Crystallization from a Zoned Magma Sheet. *J. Pet.* **2006**, *47*, 2257–2279. [CrossRef]

![minerals logo]

MDPI

Article

Mineralogical and Geochemical Constraints on Magma Evolution and Late-Stage Crystallization History of the Breivikbotn Silicocarbonatite, Seiland Igneous Province in Northern Norway: Prerequisites for Zeolite Deposits in Carbonatite Complexes

Dmitry R. Zozulya [1,*], Kåre Kullerud [2,3], Erling K. Ravna [2], Yevgeny E. Savchenko [1], Ekaterina A. Selivanova [1] and Marina G. Timofeeva [1]

[1] Geological Institute, Kola Science Centre, 14 Fersman Str, 184209 Apatity, Russia;
evsav@geoksc.apatity.ru (Y.E.S.); selivanova@geoksc.apatity.ru (E.A.S.); marchim@mail.ru (M.G.T.)
[2] Department of Geology, University of Tromsø, N-9037 Tromsø, Norway; Kare.Kullerud@bvm.no (K.K.);
ekr001@post.uit.no (E.K.R.)
[3] Norwegian Mining Museum, N-3616 Kongsberg, Norway
* Correspondence: zozulya@geoksc.apatity.ru; Tel.: +7-81555-79742

Received: 22 September 2018; Accepted: 16 November 2018; Published: 20 November 2018

check for updates

Abstract: The present work reports on new mineralogical and whole-rock geochemical data from the Breivikbotn silicocarbonatite (Seiland igneous province, North Norway), allowing conclusions to be drawn concerning its origin and the role of late fluid alteration. The rock shows a rare mineral association: calcite + pyroxene + amphibole + zeolite group minerals + garnet + titanite, with apatite, allanite, magnetite and zircon as minor and accessory minerals, and it is classified as silicocarbonatite. Calcite, titanite and pyroxene (Di_{36-46} Acm_{22-37} Hd_{14-21}) are primarily magmatic minerals. Amphibole of mainly hastingsitic composition has formed after pyroxene at a late-magmatic stage. Zeolite group minerals (natrolite, gonnardite, Sr-rich thomsonite-(Ca)) were formed during hydrothermal alteration of primary nepheline by fluids/solutions with high Si-Al-Ca activities. Poikilitic garnet (Ti-bearing andradite) has inclusions of all primary minerals, amphibole and zeolites, and presumably crystallized metasomatically during a late metamorphic event (Caledonian orogeny). Whole-rock chemical compositions of the silicocarbonatite differs from the global average of calciocarbonatites by elevated silica, aluminium, sodium and iron, but show comparable contents of trace elements (REE, Sr, Ba). Trace element distributions and abundances indicate within-plate tectonic setting of the carbonatite. The spatial proximity of carbonatite and alkaline ultramafic rock (melteigite), the presence of "primary nepheline" in carbonatite together with the trace element distributions indicate that the carbonatite was derived by crystal fractionation of a parental carbonated foidite magma. The main prerequisites for the extensive formation of zeolite group minerals in silicocarbonatite are revealed.

Keywords: silicocarbonatite; melteigite; calcite; nepheline; zeolite group minerals; garnet; crystal fractionation; Breivikbotn; Northern Norway

1. Introduction

Alkaline rocks and carbonatites represent less than 1% of all igneous rocks of the Earth's crust. However, the petrogenesis of these rocks is particularly interesting, in part due to their great variability and in part because they are economically important, containing most of the global reserves of, for example, rare earth elements (REE), zirconium, niobium and phosphorus (apatite). Moreover, alkaline igneous rocks and carbonatites have great petrological and geodynamical significances. Most common point of view for their generation is low-volume partial melting of mantle domains enriched in trace elements and volatiles. Owing to their deep source and rapid ascent to the crust, the rocks of this clan bear information on the composition of the deep Earth. Combining observations on the geochemical characteristics of natural rocks with experimental results may provide insight into processes of mantle metasomatism and melt generation in high-pressure conditions. Metasomatism by carbonatite melts has been recognized as an important mechanism for enrichment of mantle domains.

Although it has long been recognized that alkaline rocks occur in various tectonic settings worldwide, study of these rocks, particularly carbonatites, has focused mainly on within-plate continental rift environments. Due to the typically low volume and high reactivity of these magmas, it may nonetheless be assumed that an extensional tectonic environment is prerequisite to their emplacement into the upper crust. In post-collisional extensional settings, alkaline rocks and carbonatites have the potential to provide information about the effects of convergent tectonic processes on the geochemical evolution of the upper mantle. Although the majority of known carbonatites are found in rift or near-rift settings, they may rarely occur in off-craton settings where extension may be localized in back-arc regimes or as a consequence of widespread orogenic collapse [1]. Nevertheless, recently, the carbonatites in pure collisional and subduction tectonic settings were reported from several localities worldwide [2–6]. Enrichment in high field strength elements (HFSE) such as niobium and zirconium, once considered an essential characteristic of within-plate carbonatite, is conspicuously absent from carbonatites in collisional and subduction tectonic settings. Furthermore, carbonatite melts in within-plate setting are highly reactive and cause alkaline metasomatic alteration (fenitization) of the surrounding rocks [7].

In this paper, new geochemical and mineralogical data on the Breivikbotn carbonatite are presented and its possible origin and late-, post-crystallization processes are proposed. The occurrence is remarkable for the high zeolite content, and the main prerequisites for zeolite formation in carbonatite complexes are substantiated.

2. Geological Setting

2.1. Seiland Igneous Province

The Seiland Igneous Province (SIP) consists of contemporaneous mafic, ultramafic, intermediate, granitic and alkaline intrusions emplaced into a 50 × 100-km area 570–560 Ma ago. These intrusions are constrained to a single nappe within the Kalak Nappe complex of Northern Norway. This nappe complex has been generally assumed to be a parautochthonous terrane within the 420 Ma Norwegian Caledonides [8], but more recent work has indicated that the terrane may be exotic and allochthonous [9]. The largest portion of the SIP consists of numerous mafic plutons, commonly layered, which comprise at least 50–60% of the province. Large ultramafic complexes comprise a further 25–35% of the complex, while intermediate rock types such as monzonite and diorite make up 10%. Alkaline intrusions occur throughout the province, covering about 5% of the area. Granitic rocks are restricted to a few small, insignificant bodies on Øksfjord and Sørøy.

Sturt et al. [10] and Ramsay and Sturt [11] suggested that the magmatic activity in the SIP was synorogenic and related to the "Finnmarkian Orogeny", an early phase of the Caledonian Orogeny. This was based on field observations suggesting that the foliated igneous rocks were cut by younger intrusions that in turn were overprinted by a later metamorphic fabric. According to Sturt and Ramsay [12] and Sturt et al. [10], the late stage alkaline rocks in the province were both intruded and deformed during the youngest phase of the Finnmarkian deformation in the Sørøy Nappe. Krill and Zwaan [13] re-evaluated the field relationships in Sørøy and suggested that the igneous rocks of SIP were preorogenic, rather than synorogenic. They proposed that the magmatic activity was related to crustal attenuation and rift formation along the margin of Fennoscandia. Based on new U-Pb zircon ages from SIP, Roberts et al. [14] proposed an extensional setting for SIP and interpreted the younger ages (460–420 Ma) as evidence for superimposed metamorphism during the main stages of the Caledonian orogeny.

2.2. Breivikbotn Carbonatite

This complex consists of carbonatite, malignite (named shonkinite [12]), nepheline syenite, aplitic syenite and pyroxenite. It occurs as a deformed, 2 km long and 500 m wide sill hosted within the Klubben Psammite, a metasedimentary unit of the Kalak Nappe Complex [15]. At the northern side of the bay at Haraldseng, the carbonatite has intruded one limb of a N-S-trending fold, and stands out clearly in contrast to the surrounding layered sediments. Single layers of the intrusion extend north–south, with the eastern edge marked by a thin (<10 m) aegirine-augite pyroxenite body, which appears to lie conformably on the steeply dipping psammites. Some shearing has taken place along this contact, but there is nothing to contradict the conclusion that this is the bottom contact of the intrusion [12]. The pyroxenite, like the rest of the complex, is intruded by numerous centimetre-thick nepheline syenite and dolerite dykes, and is extremely variable in appearance.

Overlying the pyroxenite, across a thin band of carbonatitic breccia, there is a coarse-grained malignite, dominated by feldspar, but also containing pyroxene (aegirine-augite) and amphibole. Some localities are rich in melanitic garnet [15]. Malignite from the Breivikbotn complex can be divided in two varieties based on the dominant mafic mineral. The most common type is rich in melanite and the other type is rich in clinopyroxene. Both types occur within the central parts of the Breivikbotn complex. Present are also zeolites (pseudomorphosed after nepheline; [15]), calcite, titanite and zircon. Melanite (up 30–70 modal %), pyroxene (40–60 modal %), feldspar (10–40 modal %, in parts absent) and "nepheline" are obviously magmatic minerals. In less deformed malignite, melanite shows oscillatory zoned crystals. The zoning is characterized by an alternation of light- and dark-brown zones, probably reflecting variations in the Ti-content (overall, the TiO_2 content in garnet varies from 2.1 to 4.1 wt % [15]). Commonly, melanite shows dark central parts and lighter marginal parts. The crystals are euhedral, and may be overgrown by later garnet. Characteristic for melanite is the presence of inclusions of zeolite aggregates. Furthermore, melanite is overgrown by clinopyroxene, which in turn is overgrown by amphibole. Clinopyroxene occurs as prismatic, subhedral to euhedral crystals. The size of the crystal varies; they can be up to 3 cm long. Inclusions of titanite, calcite and an opaque mineral are observed. In addition, grains are observed. Pyroxene is often overgrown by amphibole and light yellow garnet. In places, clinopyroxene has both inclusions, and a rim of amphibole. Pyroxene is occasionally observed as inclusions in melanite. Based on its optical properties, in addition to mineral chemical analyses [15], it is assumed that the mineral is aegirine-augite. To our opinion, the petrography (particularly, the presence of garnet and low modal content of nepheline) of this rock does not fit well to *s.s.* malignite and below the term "malignite" is used.

The "malignite" has a banded appearance, caused by changes in grain-size and mineralogy. The underlying carbonatitic breccia comprises a network of thin carbonatite veins enclosing large angular fragments of both "malignite" and pyroxenite, and generally sheared along the edges.

"Malignite" intrudes alkali pyroxenite, which can also be observed as inclusions in the «malignite». The "malignite" has been carbonatized, and it is cross-cut by 4–5 cm thick carbonate veins. "Malignite" also occurs as xenoliths in the carbonatite. The «malignite» grades into carbonatite, with no clear boundary between the two rock types. The carbonatite is very variable in both texture and composition, and generally occurs as sheets. In some layers, particularly at the top of the intrusion, the carbonatite contains fragments of country rock [12]. At the top of the intrusion the carbonatite truncates the sedimentary banding in the psammites, but there is a ubiquitous metasomatic alteration [12,15].

Nepheline syenite occurs mainly as thin carbonated dykes but seems to develop into a more extensive unit further north [15]. It commonly contains biotite, or locally clinopyroxene. The nepheline syenite is often foliated and has extremely variable nepheline and carbonate contents that allow interpreting it as a product of alkaline metasomatism (fenitization) of aplitic syenite.

The dolerite dykes are apparently the last magmatic event. They intrude the carbonatite in numerous sites, and somewhere folded and boudinaged.

3. Analytical Methods

3.1. Mineral Analyses

The chemical compositions of minerals from the carbonatite were carried out at the Geological Institute, Kola Science Centre, by means of an electron microprobe Cameca MS-46, Cameca, Paris, France (WDS mode, 22 kV, 30–40 nA, with 50 s counting time). The following calibrating materials (and analytical lines) were used: wollastonite (Si$K\alpha$, Ca$K\alpha$), hematite (Fe$K\alpha$), apatite (P$K\alpha$), lorenzenite (Na$K\alpha$), thorite (Th$M\alpha$), $MnCO_3$ (Mn$K\alpha$), $Y_3Al_5O_{12}$ (Y$L\alpha$), (La,Ce)S (La$L\alpha$), CeS (Ce$L\alpha$), $Pr_3Al_5O_{12}$ (Pr$L\beta_1$), $LiNd(MoO_4)_2$ (Nd$L\alpha$), $SmFeO_3$ (Sm$L\alpha$), $EuFeO_3$ (Eu$L\alpha$), GdS (Gd$L\alpha$), $TbPO_4$ (Tb$L\alpha$), $Dy_3Al_5O_{12}$ (Dy$L\alpha$), $Ho_3Ga_5O_{12}$ (Ho$L\beta_1$), $ErPO_4$ (Er$L\alpha$), $Tm_3Al_5O_{12}$ (Tm$L\alpha$), $Yb_3Al_5O_{12}$ (Yb$L\alpha$), and $Y_{2.8}Lu_{0.2}Al_5O_{12}$ (Lu$L\alpha$). Detection limits for Fe, Mn are 0.01%; Si, Al, Cl, Ca, K, Cl—0.02%; P, Na, Y, Sr, La, Ce, Nd—0.03%; Ba—0.05%; Nb, Zr—0.1%. Representative electron microprobe data for minerals are given in Tables 1–4.

The accessory mineral identification was performed using a LEO-1450 SEM (scanning electron microscope, Carl Zeiss AG, Oberkochen, Germany) equipped with XFlash-5010 Bruker Nano GmbH EDS (energy-dispersive X-ray spectroscopy, Bruker Nano GmbH, Berlin, Germany). The system was operated at 20 kV acceleration voltage, 0.5 nA beam current, with 200 s accumulation time.

Materials from small areas of zeolite group minerals close to points analysed by microprobe were examined by the X-ray powder diffraction (XRPD) method (Debye-Scherer) by means of an URS-1 operated at 40 kV and 16 mA with RKU-114.7 mm camera and Fe$K\alpha$-radiation.

3.2. Whole-Rock Analyses

Whole-rock composition data for carbonatite-like rock were obtained at the Kola Science Centre in Apatity, Russia. Most of the major elements were determined by atomic absorption spectrophotometry; TiO_2 by colorimetry; K_2O and Na_2O by flame photometry; FeO and CO_2 by titration (volumetric analysis); and F and Cl by potentiometry using an ion-selective electrode (for a description of the methods, see [16]). Trace elements were determined by Inductively Coupled Plasma-Mass Spectrometry on a PerkinElmer Elan 9000 DRC-e (PerkinElmer Inc., Waltham, MA, USA).

The additional whole-rock composition data for the carbonatite-like and alkaline rocks from the occurrence were obtained at the Department of Biology and Geology, University of Tromsø. Two parallels of each sample were analyzed for major, minor and trace elements by X-ray fluorescence (XRF) on a Philips PW 1400 instrument (Philips, Amsterdam, The Netherlands). For major and minor elements analyses, fused pellets containing a mixture of rock powder and lithium tetraborate flux were used (mixed in ratio of 1:6). Trace element analyses were carried out on pressed powder pellets. The calibration of the analytical instrument was checked against the international standards GH, GM and NIM-S [17].

Major and minor element compositions of selected samples are presented in Table 5, and trace elements in Table 6. The whole dataset is presented in the Supplementary Data Table S1.

4. Petrography and Mineral Compositions: Results and Interpretation

The Breivikbotn carbonatite is a massive rock of porphyritic texture with hypidiomorphic, lesser idiomorphic phenocrysts of garnet and pyroxene. The groundmass is medium and coarse grained and composed of carbonate, pyroxene, amphibole and zeolite. The textural relationships of the studied samples are shown on Figure 1. The mineral content is variable and consists of carbonate (20–50 vol. %), amphibole (5–20 vol. %), pyroxene (5–20 vol. %), zeolite group minerals (0–20 vol. %), garnet (0–30 vol. %). Minor and accessory minerals are apatite (1–3 vol. %), titanite (1–5 vol. %), allanite, magnetite, zircon, pyrite, pyrrhotite, chalcopyrite, scheelite, celestine, barite, and baddeleyite.

In several thin sections, single grains of quartz were observed. Its interstitial character points on deuteric nature.

The relationships between different minerals and their internal structure are shown on Figures 2–4. Chemical analyses for major and accessory minerals are given in Tables 1–4 and in text.

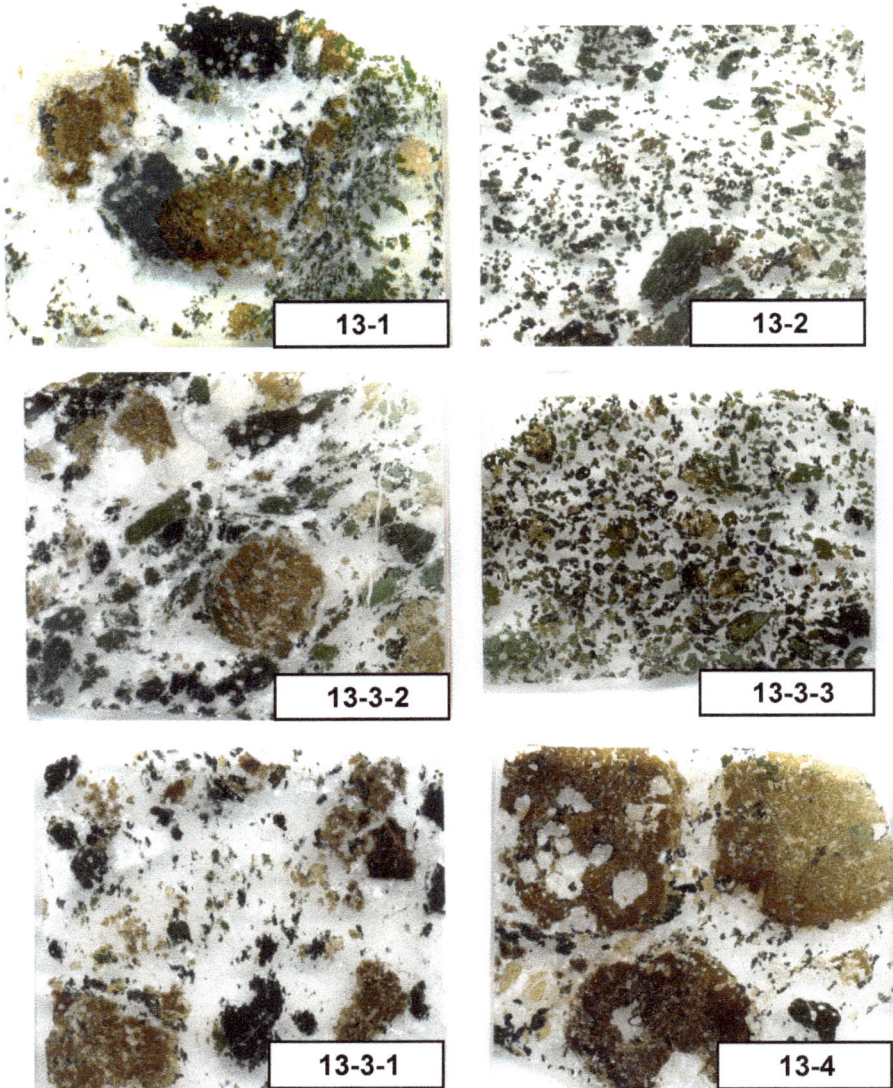

Figure 1. Scans of the selected samples of the Breivikbotn carbonatite, showing extremely heterogeneous textures. Carbonate and zeolite group minerals (white and transparent) form the groundmass; garnet (dark brown) forms rounded poikilitic grains; titanite (yellow) forms small angular and elongated euhedral crystals, pyroxene (greenish) forms subhedral and euhedral crystals, amphibole (bluish, dark-green, and indigo) forms anhedral grains and large laths. Width of all of the scans is about 2.5 cm.

Carbonate is clearly a primary mineral as it forms euhedral and subhedral crystals of 0.15–5 mm size with triple junctions between grains (Figure 2c). The carbonate is calcite (average formulae (($Ca_{0.953}$ $Sr_{0.014}$ $Mn_{0.003}$ $Fe_{0.002}$ $Mg_{0.001}$)$_{0.973}$ CO_3) with negligible contents of Mg, Fe, and Mn (Table 1). Calcite contains elevated SrO (up to 2.15 wt %) that is characteristic of magmatic calcite from carbonatites.

Pyroxene occurs as subhedral grains 0.3–4 mm in size, rarely as phenocrysts up to 1 cm in size. Pyroxene rims are often resorbed, and amphibole growth along the rims is observed. Pyroxene contains calcite inclusions. Representative compositions of pyroxene are given in Table 2. Compositionally, pyroxene shows a high content of the diopside component, with increased quantities of acmite and hedenbergite (Di_{36-46} Acm_{22-37} Hd_{14-21}). The average formula is ($Ca_{0.75}Na_{0.27}$)$_{1.02}$($Mg_{0.42}Fe^{3+}_{0.27}Fe^{2+}_{0.20}Al_{0.07}Ti_{0.02}$)$_{0.98}$[$Si_{1.9}Al_{0.1}O_6$]. Pyroxene always shows minor content of TiO_2 (up to 1 wt %).

Amphibole occurs as anhedral and subhedral grains of 0.3–5 mm size. The mineral occurs as individual grains and as overgrowths on pyroxene (Figure 2b). Rarely, amphibole forms up to 1 cm poikilitic grains with inclusions of calcite and pyroxene. Amphibole is mainly hastingsitic with an essential proportion of magnesiohastingsite, rarely mineral is closed to taramite and sadanagaite groups (sample 13–4, see Table 2). The average formula is ($Na_{0.88}$ $K_{0.43}$)$_{1.31}$ ($Ca_{1.68}$ $Na_{0.14}$ $Fe^{2+}_{0.18}$)$_2$ ($Ti_{0.15}$ $Fe^{2+}_{2.73}$ $Mg_{1.56}$ $Al_{0.43}$ $Mn_{0.12}$)$_{4.99}$ [$Si_{6.1}$ $Al_{1.9}$ O_{23}]. The mineral shows elevated contents of K_2O (2–2.3 wt %) and TiO_2 (1.2–1.4 wt %). Fluorine and Cl are below detection limit.

Garnet usually occurs as porphyritic subhedral rounded grains of 0.5–1 cm size. The mineral has poikilitic texture and contains inclusions of calcite, pyroxene, amphibole, titanite, zeolite group minerals (Figure 2a,d,f and Figure 4). Rims of garnet overgrowing amphibole can also be observed (Figure 2c). Garnet is patchy-zoned; in BSE images, garnet is generally bright along the rims, with darker central parts, however, patches of bright garnet are also observed within the darker central parts (Figure 2a). The bright patches appear to reflect elevated Fe contents. Garnet texture and morphology suggest porphyroblastic growth. Representative chemical compositions are given in Table 3. Garnet can be classified as Ti-bearing andradite with the average formula ($Ca_{2.90}Na_{0.01}Y_{0.01}$)$_{2.92}$($Fe^{3+}_{1.50}Al_{0.3}Ti_{0.15}Mn_{0.09}Mg_{0.03}Zr_{0.02}$)$_{2.1}$[$Si_{2.91}Al_{0.09}O_{12}$]. The mineral contains V, Zr and Y as minor constituents. The content of TiO_2 varies in the range of 1.3–3.2 wt %, which is low compared to titaniferous garnets from carbonatites and alkaline rocks (>5 wt %, according to [18–22], and even lower than for melanite from the «malignite» from the Breivikbotn occurrence.

Zeolite group minerals and "altered nepheline". Clusters of zeolite group minerals (ZGM) have a stubby rectangular or equant rounded (roughly hexagonal) form, up to 2–3 mm in diameter (Figures 2–4). Most clusters are composed of natrolite and gonnardite; natrolite often occurs in the central parts of gonnardite aggregates, and it is inferred that natrolite is the earliest phase (Figure 3e,f).

Natrolite forms colourless, white, smooth anhedral grains of 1–2 mm size. The average chemical composition of natrolite, $Na_{1.98}Ca_{0.03}Si_{3.01}Al_{1.98}O_{10} \cdot 2H_2O$ (Table 4), is very close to the stoichiometric formula ($Na_2(Si_3Al_2)O_{10} \cdot 2H_2O$, IMA-list 09-2017). The Si/(Si+Al) ratio varies from 0.57 to 0.65, while sodium is in the range 1.81–2.22 apfu and calcium does not exceed 0.1 apfu.

Gonnardite occurs as colourless, uneven, cracky aggregates up to 1–3 mm in size, the individual grains are anhedral and 100–500 μm in diameter. The average composition of gonnardite is calculated as ($Na_{1.57}Ca_{0.38}$)$_{2.05}$($Si_{2.67}Al_{2.38}$)$_{5.05}O_{10} \cdot 3H_2O$, which is close the stoichiometric formula ((Na,Ca)$_2(Si,Al)_5O_{10} \cdot 3H_2O$, IMA-list 09-2017). The Si/(Si+Al) ratio varies from 0.52 to 0.54, while Na/(Na+Ca) is from 0.76 to 0.85.

Thomsonite-(Ca) forms colourless and white rectangular grains (Figure 3d). It is irregularly zoned, and in BSE images characterized by brighter and darker zones. The mineral appears as partly fibrous. The average composition is $Na_{1.13}Ca_{1.7}(Al_{4.98}Si_{5.1})O_{20} \cdot 6H_2O$, which is close to ideal formula ($NaCa_2(Al_5Si_5)O_{20} \cdot 6H_2O$, IMA-list 09-2017). The mineral is characterized by elevated Sr content (0.02–0.34 apfu, with average 0.09 apfu). The Sr content may vary within a single crystal as indicated by the brighter and darker zones in BSE images. The Si/(Si+Al) ratio varies from 0.49 to 0.52, while Na/(Na+Ca+Sr) varies from 0.34 to 0.42.

Overall, the ZGM of Breivikbotn carbonatite show successively increasing Ca and Al contents from natrolite, through gonnardite to thomsonite-(Ca).

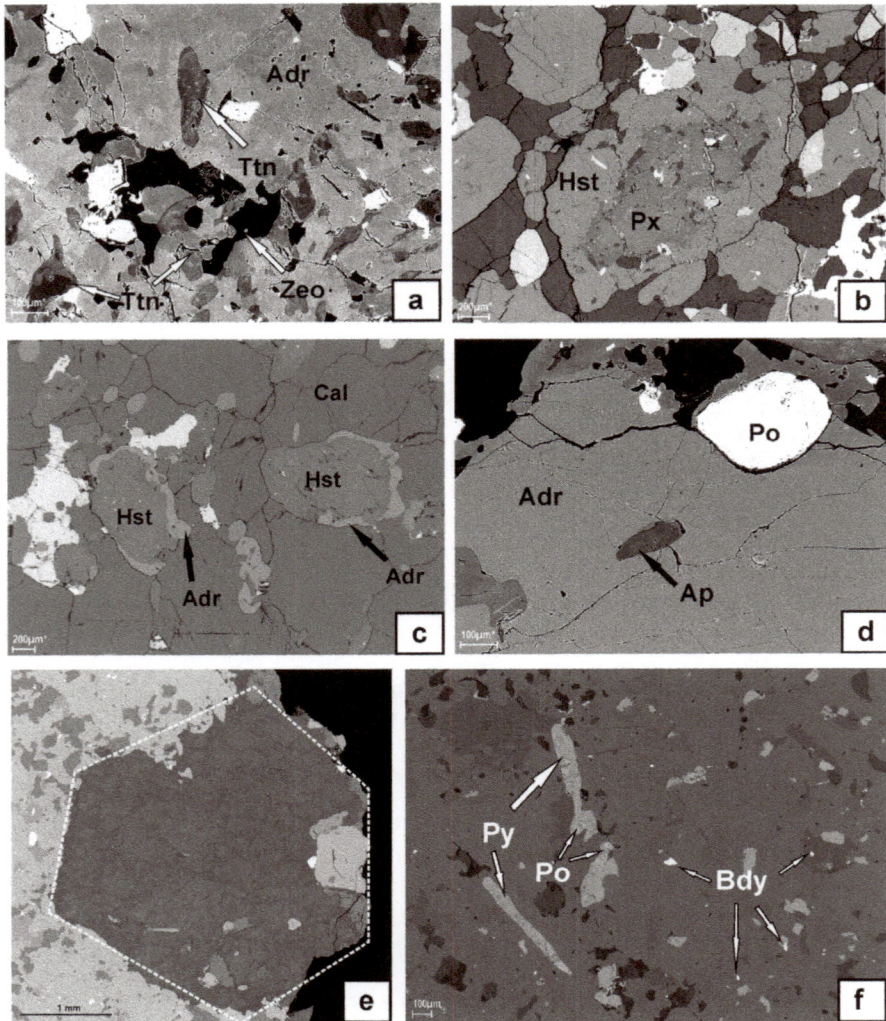

Figure 2. Back-scattered-electron (BSE) images showing the relationships between different minerals in the Breivikbotn carbonatite: (**a**) titanite, zeolite and magnetite (brightest) included in andradite; (**b**) hastingsite rimming pyroxene; (**c**) andradite rim around hastingsite, calcite grains show typical triple junctions; (**d**) apatite and pyrrhotite included in andradite; (**e**) roughly hexagonal habit of zeolite aggregate; (**f**) inclusions of accessory pyrite, pyrrhotite and baddeleyite in andradite. Mineral abbreviations are from [23].

In some natrolite-gonnardite clusters, water-absent Na-Al silicates with chemical compositions close to nepheline were found (Table 4). These compositions in combination with the textural appearance of the natrolite-gonnardite aggregates suggest that the aggregates are pseudomorphs after nepheline. Thomsonite-(Ca) can also be inferred as an alteration product of nepheline.

The water-absent nepheline-like mineral is characterized by compositions corresponding to $Na_{0.53-0.7}Ca_{0.01-0.16}Al_{1.07-1.24}Si_{1.06-1.22}O_4$, which is somewhat different from the stoichiometric formula of nepheline, with lower Na and higher Ca. We suppose that the mineral initially crystallized as nepheline from a carbonatite magma, and subsequently underwent alteration in a high-Ca environment. Ca-bearing and Ca-rich nephelines have been found in alkaline rocks from the Messum complex, Namibia [24], from the Marangudzi Complex, Zimbabwe [25], and from the Allende meteorite [26,27].

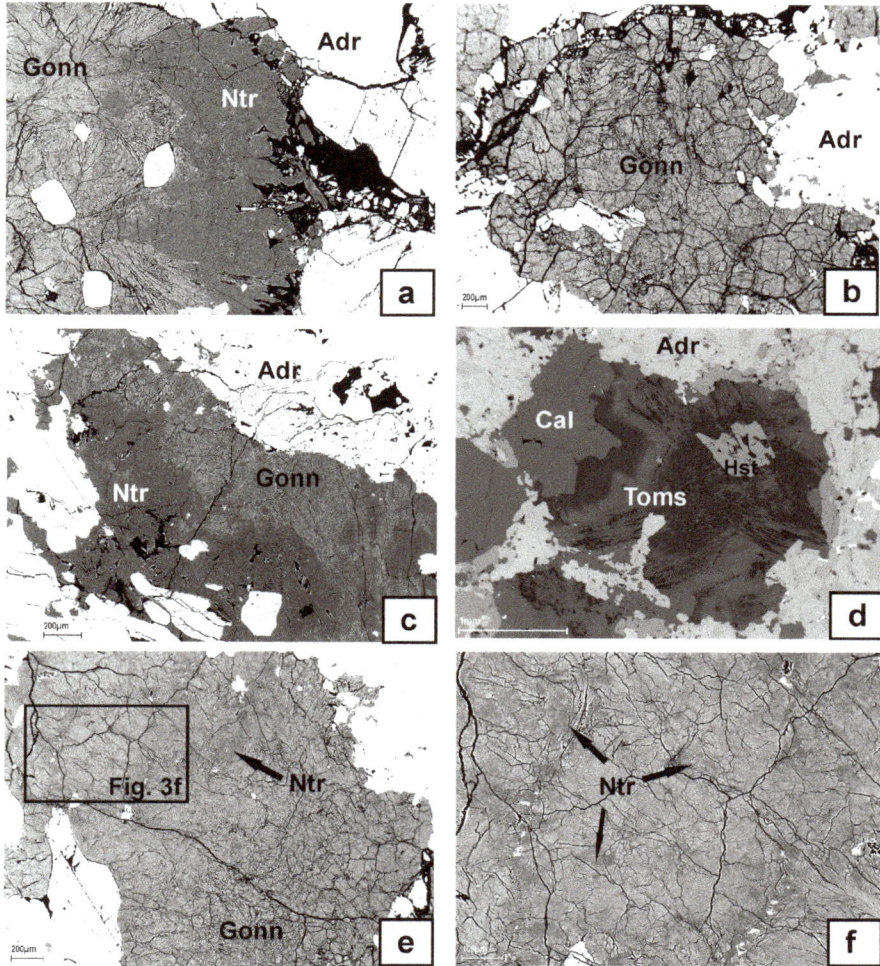

Figure 3. BSE images showing the morphology and internal textures of zeolite group minerals: (**a**–**c**) intergrowths and aggregates of natrolite and gonnardite; (**d**) zonal structure of thomsonite-(Ca) with low (dark gray) and high (gray) Sr content; (**e**,**f**) "shadow"-type domains of natrolite (dark-gray) in gonnardite (light-gray), illustrating the early crystallization of natrolite relative to gonnardite (XRPD of the sample indicated a mix of both minerals). Mineral abbreviations are from [23]. Toms–thomsonite-(Ca), Gonn–gonnardite.

Titanite occurs as euhedral and subhedral elongated grains. It is often associated with garnet, i.e., included in garnet and occurring adjacent to garnet. The average chemical composition of titanite is

$(Ca_{0.97}Y_{0.01}Ce_{0.01})$ $(Ti_{0.93}Fe_{0.05}Zr_{0.01}Nb_{0.01})$ $(Si_{0.97}Al_{0.04}O_5)$. Titanite shows elevated contents of REE, Zr, Nb, Fe and Al. Incorporation of iron and aluminium in the titanite structure requires the coupled substitutions: $Ti^{4+} + O^{2-} = (Al,Fe^{3+}) + (F,OH)^-$ and takes place at high-P metamorphic conditions [28].

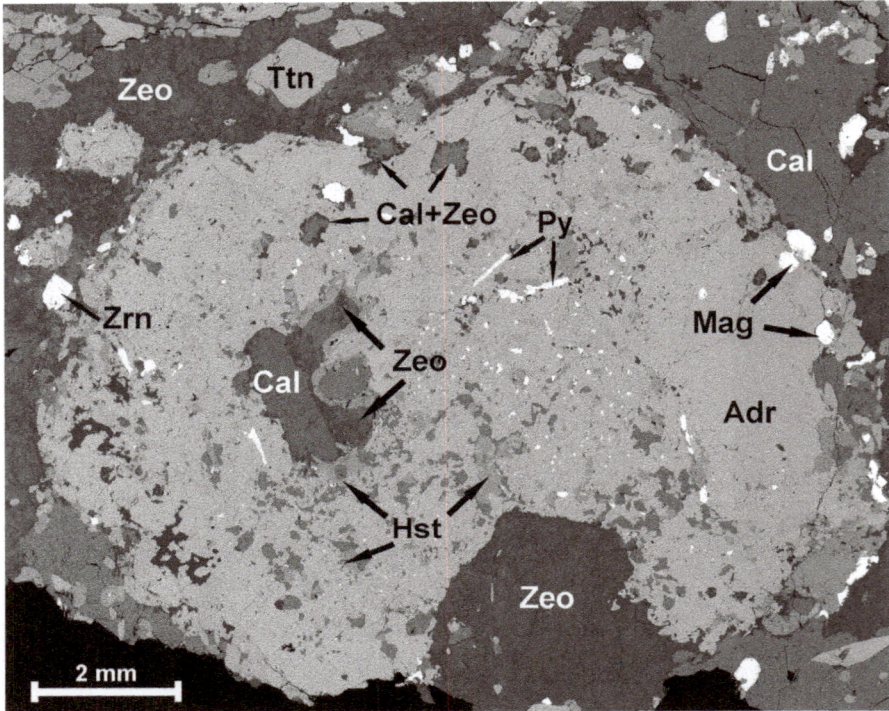

Figure 4. BSE image of typical poikilitic andradite with inclusions of calcite, hastingsite, zeolites, magnetite and pyrite. Zeolites often form intergrowths and aggregates of rectangular shape (bottom). A large number of tiny grains of baddeleyite (bright grains, $\leq$50 μm) occur in the central part of the andradite grain. Mineral abbreviations are from [23]. The image is a mosaic of 160 small 1.2 × 0.8 mm BSE images.

Table 1. Representative chemical compositions and mineral formulae of carbonate from the Breivikbotn carbonatite.

Sample No.	13-4-1-5-1	13-4-1-5-2	13-2-1a
	wt %		
FeO	0.16	0.00	0.16
MnO	0.25	0.22	0.27
MgO	0.07	0.00	-
CaO	53.65	55.29	51.32
SrO	2.07	0.11	2.15
	Formulae on the basis of 1 cation		
Fe	0.002	-	0.002
Mn	0.004	0.003	0.004
Mg	0.002	-	-
Ca	0.957	0.986	0.972
Sr	0.020	0.001	0.022

Celestine is the only accessory mineral that was analysed (SO_3 = 43.42 wt %; BaO = 0.35 wt %; CaO = 0.17 wt %; SrO = 56.45 wt %), and its occurrence together with barite and Fe-Cu-sulphides indicate high S fugacity of the system during crystallization of the carbonatite.

Baddeleyite occurs as tiny angular grains of 10–60 μm size as numerous inclusions in garnet (Figure 2f). Baddeleyite is a characteristic mineral of carbonatite-ultramafic intrusions worldwide, indicative of the Si-undersaturated environment during formation of such rocks [29]. Zircon formed apparently during late Si-saturated stages.

Table 2. Chemical analyses of pyroxene and amphibole from Breivikbotn carbonatite.

Mineral	Pyroxene					Amphibole				
Sample	13-4	13-2	13-3	13-3a	13-3b	13-4a	13-4b	13-2	13-3a	13-3b
	wt %									
SiO_2	49.74	50.99	48.47	51.16	51.91	37.90	37.46	39.33	39.14	39.28
Al_2O_3	3.18	3.44	4.97	3.70	3.26	12.32	12.95	11.72	12.98	12.76
TiO_2	0.49	0.54	1.02	0.45	0.34	1.37	1.31	1.16	1.12	1.25
FeO	17.75	14.16	14.16	13.45	12.62	25.61	24.30	21.18	19.89	19.52
MnO	0.85	0.69	0.92	0.60	0.60	1.02	0.99	0.89	1.01	0.92
MgO	6.47	7.59	7.06	7.65	8.15	5.26	4.98	7.62	7.64	7.72
CaO	17.33	18.59	20.59	17.86	18.91	9.04	9.86	9.85	10.47	10.36
Na_2O	4.17	3.59	2.66	4.33	3.73	3.75	3.50	2.85	3.17	3.32
K_2O	-	-	-	-	-	2.21	2.24	1.96	2.28	2.06
ZnO	-	-	-	-	-	0.09	0.10	0.07	0.09	0.06
Total	99.98	99.59	99.85	99.20	99.51	98.57	97.66	96.62	97.80	97.25
	apfu (4 cations)					apfu (23 oxygen atoms)				
Si	1.882	1.923	1.838	1.921	1.946	6.031	5.993	6.215	6.100	6.135
Al(iv)	0.118	0.077	0.162	0.079	0.054	1.969	2.007	1.785	1.900	1.865
Al(vi)	0.024	0.076	0.060	0.084	0.090	0.342	0.434	0.398	0.485	0.484
Al(tot)	0.142	0.153	0.222	0.164	0.144	2.311	2.442	2.183	2.385	2.349
Ti	0.014	0.015	0.029	0.013	0.010	0.164	0.158	0.138	0.132	0.146
Fe^{3+}	0.373	0.233	0.238	0.285	0.216	-	-	-	-	-
Fe^{2+}	0.189	0.213	0.211	0.137	0.180	3.408	3.251	2.800	2.592	2.550
Mn	0.027	0.022	0.030	0.019	0.019	0.138	0.134	0.119	0.133	0.122
Mg	0.365	0.427	0.399	0.428	0.455	1.247	1.187	1.794	1.774	1.797
Ca	0.703	0.751	0.837	0.718	0.759	1.541	1.689	1.667	1.749	1.733
Na	0.306	0.262	0.196	0.315	0.271	1.158	1.084	0.874	0.959	1.005
K	-	-	-	-	-	0.448	0.456	0.394	0.453	0.411

Note. Fe^{3+} in pyroxene was calculated by charge balancing to 6 oxygen atoms.

Table 3. Chemical analyses of garnet and titanite from Breivikbotn carbonatite.

Mineral	Garnet									Titanite		
Sample	13-4 C	13-4 R	13-4	13-4	13-4 C	13-4 R	13-4 C	13-4 R	13-1	13-4	13-4	13-1
wt %												
SiO_2	34.43	34.68	33.98	35.06	34.87	34.63	35.68	34.83	36.44	28.86	29.02	30.17
Al_2O_3	4.65	3.39	3.65	3.56	4.43	3.52	5.62	3.51	3.19	1.11	1.19	1.11
TiO_2	3.19	2.88	2.58	1.53	2.95	2.06	1.25	2.63	2.03	37.23	38.28	37.19
FeO	19.95	21.79	22.25	22.52	20.25	22.25	20.65	21.60	21.36	1.88	1.58	1.68
MnO	1.28	1.41	1.48	1.20	1.10	1.25	1.70	1.35	1.25	0.08	0.06	0.05
MgO	0.34	0.21	0.17	0.15	0.27	0.18	0.14	0.29	0.15	0.00	0.00	0.00
CaO	32.67	32.87	31.64	32.76	32.70	32.19	32.14	31.63	32.81	26.79	27.82	27.92
Na_2O	0.16	-	0.22	-	-	0.11	-	-	-	0.00	0.00	0.00
ZnO	-	-	-	-	-	-	-	-	-	0.07	0.00	-
Y_2O_3	-	0.17	0.17	-	-	-	0.18	0.15	-	0.91	0.00	-
ZrO_2	0.64	0.29	0.31	0.35	0.78	0.42	0.14	0.76	0.21	0.69	0.65	0.53
Yb_2O_3	-	0.04	-	-	-	-	-	-	-	-	-	-
V_2O_5	0.06	0.06	0.06	0.06	-	0.06	0.04	0.04	0.10	0.00	0.00	0.17
Nb_2O_5	-	-	-	-	-	-	-	-	-	0.37	0.10	0.63
La_2O_3	-	-	-	-	-	-	-	-	-	0.12	0.06	0.26
Ce_2O_3	-	-	-	-	-	-	-	-	-	0.30	0.26	0.71
Nd_2O_3	-	-	-	-	-	-	-	-	-	0.45	0.33	-
Sm_2O_3	-	-	-	-	-	-	-	-	-	0.24	0.12	-
Gd_2O_3	-	-	-	-	-	-	-	-	-	0.36	0.00	-
Total	99.59	100.23	98.97	99.70	99.61	99.16	99.83	99.19	99.92	99.45	99.45	100.40
apfu (8 cations)										apfu (3 cations)		
Si	2.862	2.888	2.863	2.928	2.906	2.909	2.949	2.935	3.031	0.961	0.956	0.986
Al	0.456	0.333	0.362	0.350	0.435	0.349	0.547	0.349	0.312	0.043	0.046	0.043
Ti	0.199	0.181	0.163	0.096	0.185	0.130	0.078	0.167	0.127	0.933	0.949	0.914
Fe^{2+}	1.387	1.517	1.568	1.573	1.412	1.564	1.428	1.522	1.486	0.052	0.043	0.046
Mn	0.090	0.100	0.106	0.085	0.077	0.089	0.119	0.096	0.088	0.002	0.002	0.001
Mg	0.042	0.026	0.021	0.019	0.034	0.023	0.018	0.036	0.019	-	-	-
Ca	2.909	2.933	2.857	2.931	2.920	2.898	2.846	2.855	2.923	0.956	0.982	0.978
Na	0.026	-	0.036	-	-	0.018	-	-	-	-	-	-
Zn	-	-	-	-	-	-	-	-	-	0.002	-	-
Y	-	0.008	0.007	-	-	-	0.008	0.007	-	0.016	-	-
Zr	0.026	0.012	0.013	0.014	0.032	0.017	0.006	0.031	0.008	0.011	0.010	0.008
Nb	-	-	-	-	-	-	-	-	-	0.006	0.002	0.009
V	0.003	0.003	0.003	0.003		0.003	0.002	0.002	0.006	-	-	0.004
La	-	-	-	-	-	-	-	-	-	0.001	0.001	0.003
Ce	-	-	-	-	-	-	-	-	-	0.004	0.003	0.008
Nd	-	-	-	-	-	-	-	-	-	0.005	0.004	-
Sm	-	-	-	-	-	-	-	-	-	0.003	0.001	-
Gd	-	-	-	-	-	-	-	-	-	0.004	-	-
Yb	-	0.001	-	-	-	-	-	-	-	-	-	-

Note. C—core; R—rim.

Table 4. Chemical analyses of zeolite group minerals and "altered nepheline" from Breivikbotn carbonatite.

Mineral	Natrolite				Gonnardite					Thomsonite-(Ca)								Altered Nepheline		
Sample	13-1	13-3 *	13-3 *	13-3 *	13-3 *	13-3 *	13-3 *	13-3 *	13-3 *	13-4a	13-4	13-4	14-3 *	14-3 *	14-3 *	14-3 *	14-3 *	13-1	13-1	13-1
										wt %										
SiO_2	45.22	46.05	46.35	45.04	41.28	39.47	38.85	40.44	38.40	36.34	35.58	34.93	37.28	37.07	36.84	37.59	37.54	49.35	42.87	41.36
Al_2O_3	20.38	26.46	27.33	28.36	30.20	30.50	30.56	28.89	30.16	32.06	31.56	30.36	29.57	30.23	30.04	29.54	29.47	36.67	42.32	41.09
CaO	0.08	0.12	0.31	1.48	4.17	5.12	6.43	4.34	6.25	13.16	12.19	7.08	11.36	12.61	12.08	12.06	11.14	0.24	6.05	5.97
Na_2O	16.61	15.67	15.13	14.41	12.67	12.96	11.41	11.80	11.62	3.82	3.75	3.68	4.09	4.38	4.35	5.07	4.57	14.57	10.86	10.80
K_2O	-	0.02	0.03	0.02	0.03	0.04	0.02	0.00	0.02	-	-	-	-	-	-	-	-	-	-	-
SrO	-	-	-	-	-	-	-	-	-	0.73	2.69	11.76	3.21	1.75	2.14	1.80	1.71	-	-	-
Total	82.29	88.32	89.15	89.31	88.36	88.09	87.28	85.47	86.44	86.12	85.78	87.81	85.51	86.05	85.45	86.07	84.43	100.83	102.09	99.22
	apfu (7 cations)				apfu (7 cations)					apfu (13 cations)								apfu (3 cations)		
Si	3.118	2.991	2.997	2.919	2.727	2.605	2.617	2.775	2.605	4.932	4.940	5.176	5.220	5.070	5.092	5.116	5.225	1.223	1.069	1.059
Al	1.656	2.026	2.083	2.166	2.352	2.372	2.426	2.336	2.411	5.129	5.164	5.302	4.878	4.874	4.894	4.739	4.835	1.071	1.244	1.240
Ca	0.006	0.008	0.021	0.103	0.295	0.362	0.464	0.319	0.454	1.914	1.813	1.123	1.705	1.848	1.790	1.759	1.662	0.006	0.162	0.164
Na	2.220	1.974	1.897	1.810	1.622	1.658	1.490	1.570	1.528	1.005	1.009	1.058	1.109	1.161	1.166	1.338	1.232	0.700	0.525	0.537
K	-	0.002	0.002	0.002	0.003	0.003	0.002	0.000	0.002	-	-	-	-	-	-	-	-	-	-	-
Sr	-	-	-	-	-	-	-	-	-	0.019	0.073	0.341	0.088	0.047	0.058	0.048	0.047	-	-	-

Note. *—mineral species were confirmed by XRPD.

5. Whole-Rock Compositions

5.1. Major Elements

Major element concentrations of 28 rock samples (17 carbonatitic and 11 "malignite" samples) have been analysed. Representative analyses are given in Table 5, while all analyses and CIPW norms are available in the Supplementary Data Table S1.

Carbonatite has SiO_2 contents in the range of 20–36 wt % (average 31 wt %), Al_2O_3 = 2.5–15 wt %, MgO = 1.1–4.2 wt %, CaO = 17–34 wt %, TiO_2 = 1.0–2.3 wt %, Na_2O = 1.8–6.5 wt %, CO_2 = 6–15.7 wt % and P_2O_5 = 0.27–1.44 wt %. Relatively large variations in aluminium, sodium, phosphorous and carbon oxide reflect variable modal contents of ZGM, apatite and calcite. The high contents of Fe_2O_3 (4–9.7 wt %) and FeO (3–7 wt %) can be explained by elevated contents of andradite, magnetite and possibly pyroxene. The content of K_2O is low and varies from 0.27 to 1.1 wt % (average 0.54 wt %). High LOI in several samples (2.06–5.35 wt %) indicates on crystallization water in ZGM. The Mg# ranges from 20–56. The CIPW composition of carbonatite is characterized by the prevalence of *calcite* (15–36 wt %), *nepheline* (6–20 wt %), *diopside* (6–29 wt %), *hedenbergite* (5–16 wt %), *magnetite* (6–14 wt %) and appearance of *acmite* (up to 3.3 wt %).

"Malignite" is characterized by SiO_2 contents in the range of 35–39 wt %, which is much lower than in true malignite worldwide, and reflects its melanocratic features. Al_2O_3 varies in the range 10–17 wt %, MgO = 0.2–1.1 wt %, CaO = 17–26 wt %, TiO_2 = 0.8–2.1 wt %, Na_2O = 2–6 wt %, K_2O = 0.26–1.38 wt %, CO_2 = 0.8–3.35 wt % and P_2O_5 = 0.17–0.26 wt %. "Malignite" also shows elevated contents of Fe_2O_3 (6.7–13.6 wt %) and FeO (1.2–3.5 wt %). The Mg# ranges from 10–20. Compared to carbonatite, "malignite" has higher silica, iron, potassium and lower phosphorus. The CIPW norms of rock are characterized by the appearance of *nepheline* (9–21 wt %), *wollastonite* (20–41 wt %), *diopside* (1.3–8.7 wt %), *magnetite* (3–10 wt %), *hematite* (2–9 wt %) and *orthoclase* (up to 8.2 wt %).

5.2. Trace Elements

Trace element analyses of representative samples are given in Table 6 (the complete data set is available in the Supplementary Data Table S1). Five samples of carbonatite were analyzed by ICP-MS for a broad range of elements, while the Rb, Sr, Y, Zr, Nb analyses are available for the rest of the carbonatite samples and the "malignite" samples.

The *carbonatite* is strongly enriched in large-ion lithophile elements (LILE), particularly LREE (880–1900 ppm), Sr (2700–8900 ppm) and Ba (200–1000 ppm) (Figure 5), as compared to the primitive mantle [30]. Mantle-normalized patterns show strong to moderate negative anomalies of K, Pb, P and Ti (Figure 5). Compared to average calico-carbonatite, the Breivikbotn carbonatite has the lower contents of most incompatible elements, except of K, Zr and Hf. Chondrite-normalized REE patterns (Figure 6) show negative slopes ((La/Yb)n = 6–70), but not as steep as in "average" carbonatite. The REE patterns and the large variations in the REE content of the rocks reflect variations in the modal content of garnet, which is responsible for the accumulation of HREE. The carbonatite does not show any Eu anomalies (Eu/Eu* = 0.9–1.1).

The *"malignite"* shows elevated concentrations of Sr (650–3900 ppm, average 1580), Zr (1050–1350 ppm, average 1230) and Nb (35–125 ppm, average 80). Compared to the carbonatite, the "malignite" is characterized by higher Zr, but lower Nb, Y and Sr.

Table 5. Representative major and minor element analyses of the Breivikbotn carbonatite and alkaline rocks (wt %).

Sample	13_1	13_2	13_3	13_4	13_5	B9.5	B19.4	B21.4	B11.1	H6.2	H40.3	H10.7	H41.3	H11.6	H15.5	H31.3	H48.2
Rock					Carbonatite									"Malignite"			
SiO_2	31.74	26.4	32.1	28.97	26.51	31.54	28.67	28.96	29.76	20.47	34.95	37.1	37.46	37.53	37.71	39.18	39.43
TiO_2	1.73	1.1	1.86	1.89	1.51	0.87	0.99	0.98	1.62	1.21	2.01	1.94	1.59	1.61	1.39	1.49	0.83
Al_2O_3	11.7	3.35	11.27	14.9	2.46	13.11	13.82	13.63	11.71	6.87	9.72	11.89	12.91	12.66	13.74	13.05	17.35
Fe_2O_3	6.12	8.92	6.93	9.69	5.25	5.46	4.41	4.37	5.19	9.62	13.59	12.39	10.85	10.39	10.23	10.27	6.73
FeO	7	6.19	5.74	5.48	3.63	4.62	3.46	3.43	3.64	6.39	2.79	2.55	2.96	3.5	2.86	2.89	1.16
MnO	0.52	0.57	0.56	0.45	0.36	0.5	0.43	0.44	0.47	0.69	1.19	0.99	0.91	0.91	0.84	0.87	0.55
MgO	2.22	3.98	2.17	1.1	4.22	2.33	1.99	2.13	2.08	1.49	0.77	0.49	1.02	1.08	0.82	0.92	0.24
CaO	20.46	30.22	21.49	19.98	33.6	21.32	22.54	22.5	24.16	30.41	26.42	23.08	21.28	21.41	20.41	20.16	17.12
Na_2O	4.08	1.8	4.52	3.1	1.78	4.77	5.82	5.75	5.13	2.44	2.04	2.94	3.82	3.53	3.97	3.74	5.97
K_2O	1.1	0.41	0.67	0.41	0.27	0.56	0.33	0.33	0.29	0.27	0.45	0.77	0.66	0.77	0.77	1.16	1.15
H_2O	0.89	0.46	0.91	0.64	0.4	-	-	-	-	-	-	-	-	-	-	-	-
LOI	2.94	0.26	3.83	5.35	2.06	-	-	-	-	-	-	-	-	-	-	-	-
P_2O_5	0.45	0.95	0.52	0.27	1.44	0.76	0.78	0.79	0.63	0.68	0.23	0.20	0.26	0.26	0.33	0.25	0.17
F	0.068	0.086	0.067	0.042	0.1	-	-	-	-	-	-	-	-	-	-	-	-
Cl	0.011	0.011	0.029	0.014	0.02	-	-	-	-	-	-	-	-	-	-	-	-
CO_2	7.91	13.17	6.57	6.07	15.35	9.65	10.72	10.57	10.41	15.74	1.39	0.84	1.49	1.59	1.6	1.42	2.96
Total	98.94	97.88	99.24	98.36	98.96	95.49	93.96	93.88	95.09	96.28	95.55	95.18	95.21	95.24	94.67	95.40	93.66

Table 6. Representative trace elements and REE analyses of the Breivikbotn carbonatite and alkaline rocks (ppm).

Sample	13_1	13_2	13_3	13_4	13_5	B19.4	B21.4	B11.1	H6.2	H40.3	H10.7	H41.3	H11.6	H31.3	H48.2
Rock	Carbonatite									"Malignite"					
La	214.4	265.7	190.5	150.4	470.2	-	-	-	-	-	-	-	-	-	-
Ce	449.7	516.8	421.8	341.1	930.9	-	-	-	-	-	-	-	-	-	-
Pr	49.2	51.8	47.6	39	96.4	-	-	-	-	-	-	-	-	-	-
Nd	189.6	174.4	209.4	169.7	324.2	-	-	-	-	-	-	-	-	-	-
Sm	40	25.3	49.4	42.4	56.2	-	-	-	-	-	-	-	-	-	-
Eu	12.1	6.66	16	13.9	15.1	-	-	-	-	-	-	-	-	-	-
Gd	29.7	19.8	35.7	31.1	42.5	-	-	-	-	-	-	-	-	-	-
Tb	4.86	2.2	6.87	6.06	5.77	-	-	-	-	-	-	-	-	-	-
Dy	24.5	7.83	35.4	31.9	23.7	-	-	-	-	-	-	-	-	-	-
Ho	4.57	1.26	6.79	6.32	4.12	-	-	-	-	-	-	-	-	-	-
Er	12.5	2.95	18.9	17.2	9.17	-	-	-	-	-	-	-	-	-	-
Tm	1.84	0.37	2.8	2.58	1.14	-	-	-	-	-	-	-	-	-	-
Yb	12.7	2.59	19.2	18.3	7.14	-	-	-	-	-	-	-	-	-	-
Lu	1.87	0.43	2.88	2.72	0.97	-	-	-	-	-	-	-	-	-	-
Y	111.7	30.8	149.8	148.3	84	30	32	124	28	43	60	46	44	42	14
Ta	3.23	0.8	3.27	4.19	18	-	-	-	-	-	-	-	-	-	-
Nb	111.2	20.7	106.3	124	231.4	188	187	136	42	88	103	78	85	92	34
Hf	13.7	8.05	19.9	17.9	9.84	-	-	-	-	-	-	-	-	-	-
Zr	492.5	363.2	696.7	716	473.1	493	505	840	616	1359	1350	1133	1137	1145	1043
Sr	4735	5803	3290	8856	4656	6169	6173	3780	8985	645	958	1211	1235	1505	3993
Rb	16.4	5.82	10	5.23	6.41	bd	bd	bd	bd	8	10	6	8	12	9
Ba	618.5	638.4	204.7	178.2	1003	-	-	-	-	-	-	-	-	-	-
U	3.46	0.8	3.99	3.13	1.86	-	-	-	-	-	-	-	-	-	-
Th	12.7	5.63	13.4	9.43	25.8	-	-	-	-	-	-	-	-	-	-
Pb	4.17	1.86	3.53	3.2	2.81	-	-	-	-	-	-	-	-	-	-
Mo	5.02	1.1	8.9	13.8	0.62	-	-	-	-	-	-	-	-	-	-
Cs	0.16	0.016	0.11	0.02	0.091	-	-	-	-	-	-	-	-	-	-

Note. bd—below detection limit.

6. Discussion

6.1. Petrographical Classification of the Breivkbotn Carbonatite and Alkaline Rocks

The recommendations for the classification of carbonatites by the IUGS Sub-commission on the Systematics of Igneous Rocks [31–33] defined carbonatites as "igneous rocks, intrusive as well as extrusive, which contain more than 50% by volume of carbonate minerals". The samples of the Breivikbotn carbonate rocks studied here contain 20–50 vol. % of carbonate mineral. Strictly speaking the rocks should be referred to as calcitic ijolites, i.e., igneous rocks containing between 10 and 50% of igneous calcite, with additional other primary minerals, such as pyroxene and nepheline. Nevertheless, the rock is extremely heterogeneous with proportions of carbonate in parts close to the classification boundary at 50 vol. % (e.g., sample 13-3-1, see Figure 1). In terms of the IUGS chemical classification, the rock can be referred to as silicocarbonatite if the SiO_2 content of the rock exceeds 20%. For this reason, we prefer to name the Breivikbotn carbonatite-like rocks as nepheline-diopside-calcite-carbonatite, or for simplicity, silicocarbonatite.

Nepheline is an unusual and rare mineral in classic carbonatite complexes world-wide. Nevertheless nepheline-bearing carbonatites have been reported from a number of carbonatite localities; e.g., Laacher See, Germany and Alnø, Sweden (both mentioned in [34]), Fen [35] and Lillebukt [36] in Norway, Chilwa Island and Kangankunde, Malawi [37], Budeda Hill and Homa Bay, Uganda [38,39], Walloway, Australia [40], Dicker Willem, Namibia [41], and in Ilmeny–Vishnevogorsky, Urals, Russia [42] and from other Russian occurrences (summarized by [43]. In addition, other carbonatites contain natrolite, analcime or cancrinite, formed by breakdown of nepheline; e.g., Oka, Canada [44]; Legetet Hills, Kenya, Nachendezwaya, Tanzania, Tororo and Bukusu, Uganda (all summarized by [34]). On the basis of this distribution, and the petrographic and mineralogical evidence for the above described localities, it is possible to infer that nepheline-bearing carbonatites are often associated with the "nephelenitic-clan" rather than with the "melilititic-clan" carbonatites (using the terminology of [45]).

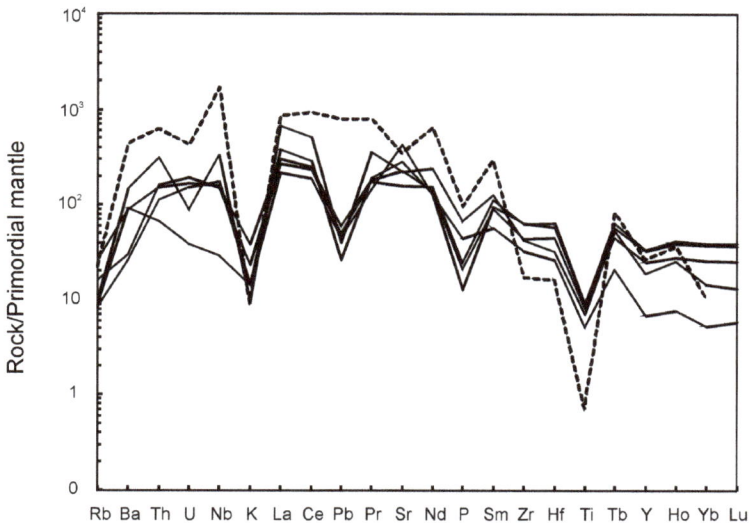

Figure 5. Mantle normalized incompatible elements patterns of silicocarbonatite from Breivikbotn occurrence. Curve for average carbonatite [32] is dotted and shown for reference. Primordial mantle composition is taken from [30].

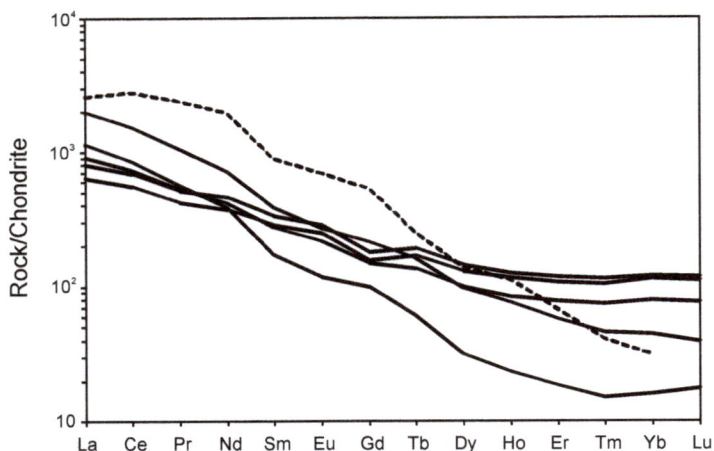

Figure 6. Chondrite normalized REE patterns of the silicocarbonatite from Breivikbotn. Curve for average carbonatite [32] is dotted and shown for reference. Chondrite composition is taken from [30].

The "malignite" from the Breivikbotn occurrence has an unusual mineralogical composition among magmatic rocks. Therefore, the classification based on the chemical composition was applied. Based on the SiO_2 and $Na_2O + K_2O$ contents, the rock belongs to the clans of melilitolite or ultramafic foidolite. The last clan include melteigite, jacupirangite, ijolite, urtite and missurite, and the Na-rich (Na > K) and melanocratic varieties of the clan are comparable to the rock discussed here, namely melteigite and jacupirangite. Furthermore, the low MgO and relatively high Al_2O_3 of the Breivikbotn rock suggest that it could be classified as melteigite, but the elevated CO_2 content (0.8–3.4 wt %) justify that the rock is referred to as a carbonated melteigite. Broadly speaking, it can be stated that Breivikbotn "malignite" crystallized from a carbonated foidite melt.

Thus, the Breivikbotn occurrence provides good field material for studying the genetic link between carbonatite and foidite magma (i.e., the derivation of carbonatite magma, either by liquid immiscibility or crystal fractionation from carbonated foidite).

6.2. Geochemical Constraints on Primary Melts, Magma Evolution and Geodynamic Setting of Breivikbotn Complex

The proposed origins of carbonatites associated with silicate alkaline rocks include three main processes: (1) direct melting from carbonate-rich mantle peridotite or wehrlite [46–49]; (2) low-pressure liquid immiscibility from parental nephelinite (e.g., [50]); (3) extensive crystal fractionation from carbonated alkaline silicate magmas, i.e., carbonated nephelinite or melilitite [48,51–53].

Robins [36] suggested that the primary Breivikbotn carbonatite melt was derived from carbonated nephelinite magma through liquid immiscibility. According to the model of Robins [36], nephelinite magma fractionated through mela-phonolite magma to «malignite» and phonolite (yielding nepheline syenite), and through trachyte magma to alkali syenite. Parts of the model are questionable because of its complexity, and the absence at the outcrops of a number of hypothetical magmatic phases. Moreover, if the model were correct, it would be expected that the Breivikbotn carbonatitic and alkaline rocks contain significantly more highly incompatible HFS elements than nephelinite [50]. It is therefore clear that neither liquid immiscibility nor fractional crystallization can explain the formation of the studied rocks from primary nephelinite melt. We consider it more likely that the primary melt of the studied rocks corresponds to "malignite" or carbonated melteigite.

For the Breivikbotn case, direct melting of silicocarbonatite from carbonate-rich mantle rock seems unlikely because of: (1) low #Mg (17–56, average 37), (2) the association of carbonatite and alkaline silicate rocks in outcrops, (3) through this process, Mg-carbonate would be expected to dominate.

The derivation of melt from carbonated foidite by liquid immiscibility or crystal fractionation can be identified by geological, petrographic and mineralogical criteria, but also on the basis of trace-element geochemistry, particularly in relation to the associated silicate lithologies. The HFSE are highly informative in this respect. Data available for both the silicocarbonatitic and the alkaline silicate rocks from Breivikbotn include Zr, Nb and Y. Nb is a highly incompatible element, whereas Zr and Y are more compatible. Therefore, it can be expected that Nb partitioned into the carbonatitic melt relative to Zr and Y both during liquid immiscibility and crystal fractionation. The prevalence of Nb in the Breivikbotn silicocarbonatite compared to the alkaline silicate rock supports the proposed petrogenetic processes. Noteworthy, the absence of any gap in the Nb-Zr/Nb diagram (Figure 7) indicates crystal fractionation from a carbonated foidite melt (of "malignite" or melteigite composition, as proposed above). Zr is strongly partitioned into garnet, and would be incorporated in the early magmatic melanite in the Breivikbotn melteigite, what is reflected in Y-Zr/Y diagram (Figure 7). The role of crystal fractionation during formation of the silicocarbonatite is also supported by its relatively high Th/U ratios (3–14; average 6.2), low Zr/Hf (35–48; 41) and Nb/Ta (13–34; 27) comparing to earlier magmatic phases [54–56]. Fractionation of zircon, titanite and clinopyroxene could explain such element distributions.

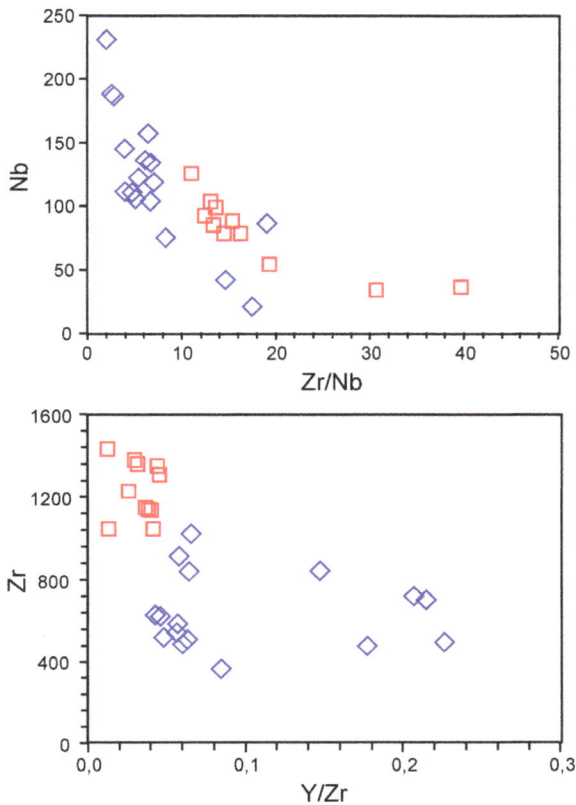

Figure 7. Selected paired trace-element ratios for the Breivikbotn silicocarbonatite and associated melteigite (blue diamonds and red squares, respectively).

Field observations, such as gradual transitions between silicocarbonatite and melteigite, dykes of pure carbonatite cutting the melteigite, and xenoliths of melteigite in silicocarbonatite is in agreement with a process of crystal fractionation of the carbonatite from a foidite magma.

The concept of synorogenic versus extensional origin of the Seiland igneous province (see above) can also be addressed using geochemical data. Trace element concentrations provide important information about the tectonic setting. Most carbonatites in within-plate settings, e.g., continental rifts, are characterized by high concentrations of LILE (especially Sr and Ba) and HFSE; the average values for Zr, Nb, Hf and Ta are 256.4, 308.9, 4.3 and 8.9 ppm, respectively [54]. Subduction- and collision-related carbonatites in off-craton settings have higher LILE contents, but much lower Zr, Nb, Hf and Ta. For example, Chakhmouradian et al. [55] reported for the Eden Lake carbonatite the following values (ppm): 47–98 for Zr, 4.0 for Nb, 1.5–2.4 for Hf, and 0.2 for Ta. Similar values are reported from the subduction-related carbonatites in Italy, Antarctica, north-eastern China, Northern Norway [3,6,57,58], and even much lower from collision related carbonatites in southwestern and northern China [2,4]. The Breivikbotn carbonatite with average values for Sr = 5100 ppm, Ba = 530 ppm, Zr = 700 ppm, Nb = 122 ppm, Hf = 14 ppm, and Ta = 6 ppm has the signature of within-plate carbonatites.

6.3. Late- to Post-Magmatic Alteration and Possible Metamorphic Overprints

Because of the alkalic and volatile-rich character, the intrusion of silicocarbonatite resulted in the extensive fenitization that can be observed at the Breivikbotn outcrops. Moreover, the presence of late-magmatic amphibole and biotite suggests that the silicocarbonatite magma contained sufficient water to produce hydrothermal fluids during late stages of crystallization. Such deuteric fluids would be able to alter the primary mineral assemblage producing the abundant water-bearing phases. The metasomatic alteration can be considered to be the result of subsolidus autometasomatism, during which early magmatic phases reacted with residual fluids to form a suite of lower-temperature minerals such as natrolite, gonnardite, and thomsonite. Experimental studies, quantitative estimates and field observations indicate that natrolite-group and thomsonite-group zeolites formed in the temperature range of 100–200 °C [59–62].

Earlier petrographic studies have shown that ZGM could be the products of nepheline alteration. Several reactions for nepheline alteration at successive enrichment of H_2O, CaO Al_2O_3 and SiO_2 in hydrothermal fluid may be possible:

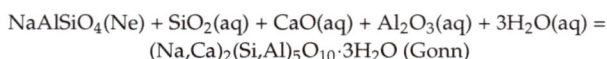

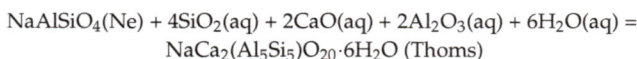

$$2NaAlSiO_4(Ne) + SiO_2(aq) + 2H_2O(aq) = Na_2(Si_3Al_2)O_{10} \cdot 2H_2O \ (Ntr)$$

$$NaAlSiO_4(Ne) + SiO_2(aq) + CaO(aq) + Al_2O_3(aq) + 3H_2O(aq) = (Na,Ca)_2(Si,Al)_5O_{10} \cdot 3H_2O \ (Gonn)$$

$$NaAlSiO_4(Ne) + 4SiO_2(aq) + 2CaO(aq) + 2Al_2O_3(aq) + 6H_2O(aq) = NaCa_2(Al_5Si_5)O_{20} \cdot 6H_2O \ (Thoms)$$

Textural evidence for replacements of natrolite by gonnardite is supported by the successive enrichment of the fluid phase by H_2O, CaO, Al_2O_3 and SiO_2 according to the following reaction:

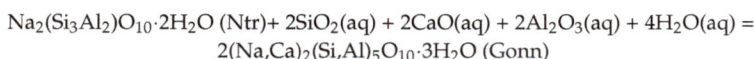

$$Na_2(Si_3Al_2)O_{10} \cdot 2H_2O \ (Ntr) + 2SiO_2(aq) + 2CaO(aq) + 2Al_2O_3(aq) + 4H_2O(aq) = 2(Na,Ca)_2(Si,Al)_5O_{10} \cdot 3H_2O \ (Gonn)$$

Ti-rich garnets can form over a wide range of temperatures and pressures. Particularly in alkaline rocks, they appear to reflect complex metasomatic reactions between earlier mafic minerals and late-stage fluids [63,64]. This idea is consistent with the petrographic observations from the garnet-bearing rocks of the Breivikbotn outcrops, which indicate that Ti-rich garnets formation from clinopyroxene and amphibole was possibly driven by metasomatic fluids. We suggest that these reactions either occurred as a result of alkali metasomatism in a fluid-enriched environment, or in response to a late metamorphic event coeval with the Caledonian orogeny. Metamorphic

Minerals **2018**, *8*, 537

overprint of the Breivikbotn rocks can also explain the elevated contents of Al and Fe in titanite from the silicocarbonatite.

7. Conclusions

The Breivikbotn carbonatite-alkaline complex consists of silicocarbonatite and alkaline ultramafic rock of an unusual mineralogic composition. The silicocarbonatite is composed mainly of primary magmatic calcite, pyroxene, altered nepheline, late-magmatic amphibole, post-magmatic garnet (andradite) and zeolite group minerals. The alkaline ultramafic rock, which has a composition similar to melteigite, is composed of garnet (melanite), pyroxene, alkali feldspar and altered nepheline. Geological observations and geochemical data suggest that the melteigite is an early magmatic phase parental for the silicocarbonatite, which formed by fractional crystallization from the carbonated foidite melt.

Zeolites are the products of late/post-magmatic alteration of magmatic nepheline at low temperature (<200 °C) by Ca-, Al- and Si-bearing hydrous fluids. The alteration of nepheline to zeolite group minerals progressed through several steps: (1) "altered nepheline", (2) natrolite, (3) gonnardite, and (4) Sr-rich thomsonite-(Ca). The main prerequisites for zeolite formation in the carbonatite complex were (1) silicocarbonatite composition of the parent magma and crystallization of nepheline, (2) fractional crystallization of primary magmas leading to fluid enrichment of residual melts/hydrothermal solutions, and (3) extensive fluid alteration of the rock at late- and post-magmatic stages.

Textural and compositional data for andradite distinctly point to its formation either due to alkali metasomatism at late-, post-magmatic stage of the Breivikbotn complex, or during the Caledonian metamorphic event.

Supplementary Materials: The following is available online at http://www.mdpi.com/2075-163X/8/11/537/s1, Table S1: Chemical composition and CIPW norms of the Breivikbotn carbonatites and alkaline rocks.

Author Contributions: Conceptualization, D.R.Z. and K.K.; Methodology, E.A.S., Y.E.S. and M.G.T.; Investigation, D.R.Z., K.K. and E.K.R.; Writing-Original Draft Preparation, D.R.Z.; Visualization, D.R.Z. and Y.E.S.

Funding: This research was funded by Russian Government grant 0231-2015-0009.

Acknowledgments: We are grateful to two anonymous referees whose comments improve the manuscript greatly.

Conflicts of Interest: The authors declare no conflict of interest.

References

1. Woolley, A.R.; Kjarsgaard, B.A. Paragenetic types of carbonatite as indicated by the diversity and relative abundances of associated silicate rocks: Evidence from a global database. *Can. Mineral.* **2008**, *46*, 741–752. [CrossRef]

2. Hou, Z.Q.; Tian, S.H.; Yuan, Z.X.; Xie, Y.L.; Yin, S.P.; Yi, L.S.; Fei, H.C.; Yang, Z.M. The Himalayan collision zone carbonatites in western Sichuan, SW china: Petrogenesis, mantle source and tectonic implication. *Earth Planet. Sci. Lett.* **2006**, *244*, 234–250. [CrossRef]

3. D'Orazio, M.; Innocenti, F.; Tonarini, S.; Doglioni, C. Carbonatites in a subduction system: The Pleistocene alvikites from Mt. Vulture (southern Italy). *Lithos* **2007**, *98*, 313–334. [CrossRef]

4. Xu, C.; Taylor, R.N.; Kynicky, J.; Chakhmouradian, A.R.; Song, W.L.; Wang, L.J. The origin of enriched mantle beneath north China block: Evidence from young carbonatites. *Lithos* **2011**, *127*, 1–9. [CrossRef]

5. Hagen-Peter, G.; Cottle, J.M. Synchronous alkaline and subalkaline magmatism during the late Neoproterozoic-Early Paleozoic Ross orogeny, Antarctica: Insights into magmatic sources and processes within a continental arc. *Lithos* **2016**, *262*, 677–698. [CrossRef]

6. Ravna, E.K.; Zozulya, D.; Kullerud, K.; Corfu, F.; Nabelek, P.I.; Janak, M.; Slagstad, T.; Davidsen, B.; Selbekk, R.S.; Schertl, H.P. Deep-seated carbonatite intrusion and metasomatism in the UHP Tromso Nappe, northern Scandinavian Caledonides—A natural example of generation of carbonatite from carbonated eclogite. *J. Petrol.* **2017**, *58*, 2403–2428. [CrossRef]

7. Elliott, H.A.L.; Wall, F.; Chakhmouradian, A.R.; Siegfried, P.R.; Dahlgren, S.; Weatherly, S.; Finch, A.A.; Marks, M.A.W.; Dowman, E.; Deady, E. Fenites associated with carbonatite complexes: A review. *Ore Geol. Rev.* **2018**, *93*, 38–59. [CrossRef]

8. Ramsay, D.M.; Sturt, B.A.; Zwaan, K.B.; Roberts, D. Caledonides of Northern Norway. In *The Caledonian Orogen: Scandinavia and Related Areas*; Gee, D.G., Sturt, B.A., Eds.; Wiley: New York, NY, USA, 1985; pp. 163–184.

9. Corfu, F.; Torsvik, T.H.; Andersen, T.B.; Ashwal, L.D.; Ramsay, D.M.; Roberts, R.J. Early Silurian mafic-ultramafic and granitic plutonism in contemporaneous flysch, Mageroy, northern Norway: U-Pb ages and regional significance. *J. Geol. Soc.* **2006**, *163*, 291–301. [CrossRef]

10. Sturt, B.A.; Pringle, I.R.; Ramsay, D.M. The Finnmarkian phase of the Caledonian Orogeny. *J. Geol. Soc. Lond.* **1978**, *135*, 597–610. [CrossRef]

11. Ramsay, D.M.; Sturt, B.A. The contribution of Finnmarkian Orogeny to the framework of the Scandinavian Caledonides. In *Synthesis of the Caledonian Rocks of Britain*; Fettes, D.J., Harris, A.L., Eds.; D. Reidel Publishing Company: Dordrecht, The Netherlands, 1986.

12. Sturt, B.A.; Ramsay, D.M. The alkaline complex of the Breivikbotn area, Sørøy, Northern Norway. *Nor. Geol. Unders. Bull.* **1965**, *231*, 1–142.

13. Krill, A.G.; Zwaan, B. Reinterpretation of Finnmarkian deformation on western Soroy, northern Norway. *Nor. Geol. Tidsskr.* **1987**, *67*, 15–24.

14. Roberts, R.J.; Corfu, F.; Torsvik, T.H.; Hetherington, C.J.; Ashwal, L.D. Age of alkaline rocks in the Seiland igneous province, northern Norway. *J. Geol. Soc.* **2010**, *167*, 71–81. [CrossRef]

15. Jonassen, A. Geologiske og petrologiske undersøkelser av alkaline bergarter og metasedimenter tilknyttet Breivikbotnkomplekset pa Sørøy, Vest Finnmark. Cand. Scient. Master's Thesis, University of Tromsø, Tromsø, Norway, 1996. (In Norwegian)

16. Cazes, J. *Ewing's Analytical Instrumentation Handbook*, 3rd ed.; Marcel Dekker: New York, NY, USA, 2005; p. 1037.

17. Govindaraju, K. 1994 compilation of working values and sample description for 383 geostandards (vol 18, pg 53, 1994). *Geostand. Newsl.* **1994**, *18*, 331. [CrossRef]

18. Chakhmouradian, A.R.; McCammon, C.A. Schorlomite: A discussion of the crystal chemistry, formula, and inter-species boundaries. *Phys. Chem. Miner.* **2005**, *32*, 277–289. [CrossRef]

19. Huggins, F.E.; Virgo, D.; Huckenholz, H.G. Titanium-containing silicate garnets. 1. Distribution of Al, Fe^{3+}, and Ti^{4+} between octahedral and tetrahedral sites. *Am. Mineral.* **1977**, *62*, 475–490.

20. Lupini, L.; Williams, C.T.; Woolley, A.R. Zr-rich garnet and Zr-rich and Th-rich perovskite from the Polino carbonatite, Italy. *Mineral. Mag.* **1992**, *56*, 581–586. [CrossRef]

21. Russell, J.K.; Dipple, G.M.; Lang, J.R.; Lueck, B. Major-element discrimination of titanian andradite from magmatic and hydrothermal environments: An example from the Canadian Cordillera. *Eur. J. Mineral.* **1999**, *11*, 919–935. [CrossRef]

22. Saha, A.; Ray, J.; Ganguly, S.; Chatterjee, N. Occurrence of melanite garnet in syenite and ijolite-melteigite rocks of samchampi-samteran alkaline complex, mikir hills, northeastern india. *Curr. Sci.* **2011**, *101*, 95–100.

23. Whitney, D.L.; Evans, B.W. Abbreviations for names of rock-forming minerals. *Am. Mineral.* **2010**, *95*, 185–187. [CrossRef]

24. Blancher, S.B.; D'Arco, P.; Fonteilles, M.; Pascal, M.L. Evolution of nepheline from mafic to highly differentiated members of the alkaline series: The Messum complex, Namibia. *Mineral. Mag.* **2010**, *74*, 415–432. [CrossRef]

25. Henderson, C.M.B.; Gibb, F.G.F. Plagioclase-Ca-rich-nepheline intergrowths in a syenite from the Marangudzi complex, Rhodesia. *Mineral. Mag.* **1972**, *38*, 670–677. [CrossRef]

26. Lumpkin, G.R. Nepheline and sodalite in a barred olivine chondrule from the Allende meteorite. *Meteoritics* **1980**, *15*, 139–147. [CrossRef]

27. Ross, D.K.; Simon, J.I.; Simon, S.B.; Grossman, L. Two generations of sodic metasomatism in an Allende Type B CAI. In Proceedings of the 46th Lunar and Planetary Science Conference, The Woodlands, TX, USA, 16–20 March 2015.

28. Carswell, D.A.; Wilson, R.N.; Zhai, M. Ultra-high pressure aluminous titanites in carbonate-bearing eclogites at Shuanghe in Dabieshan, central China. *Mineral. Mag.* **1996**, *60*, 461–471. [CrossRef]

29. Mitchell, R.H. *Undersaturated Alkaline Rocks: Mineralogy, Petrogenesis, and Economic Potential*; Mineralogical Association of Canada, Short Courses: Quebec City, QC, Canada, 1996; Volume 24, p. 312.

30. Sun, S.S.; McDonough, W.F. Chemical and isotopic systematics of oceanic basalts: Implications for mantle composition and processes. *Geol. Soc. Lond. Spec. Publ.* **1989**, *42*, 313–345. [CrossRef]
31. Streckeisen, A. Classification and nomenclature of volcanic-rocks, lamprophyres, carbonatites, and melilitic rocks—Recommendations and suggestions of the IUGS sub-commission on the systematics of igneous rocks. *Geology* **1979**, *7*, 331–335.
32. Woolley, A.R.; Kempe, D.R.C. Carbonatites: Nomenclature, average compositions, and element distribution. In *Carbonatites: Genesis and Evolution*; Bell, K., Ed.; Unwin Hyman: London, UK, 1989; pp. 1–14.
33. LeMaitre, R.W. *Igneous Rocks: A classification and Glossary of Terms*; Cambridge University Press: Cambridge, UK, 2002; p. 236.
34. Heinrich, E.W. *The Geology of Carbonatites*; Rand McNally: Chicago, IL, USA, 1966; p. 555.
35. Andersen, T. Evolution of peralkaline calcite carbonatite magma in the Fen complex, southeast Norway. *Lithos* **1988**, *22*, 99–112. [CrossRef]
36. Robins, B. The Seiland Igneous Province, N. Norway: General Geology and Magmatic Evolution. Field Trip Guidebook, Part II, IGCP Project 336. Norges Geologiske Undersøkelse Report. 1996, p. 34. Available online: http://www.ngu.no/upload/Publikasjoner/Rapporter/1996/96_127.pdf (accessed on 22 September 2018).
37. Garson, M.S. Carbonatites in Malawi. In *Carbonatites*; Tuttle, O.F., Gittins, J., Eds.; Interscience: New York, NY, USA, 1966; pp. 33–71.
38. King, B.C.; Sutherland, D.S. The carbonatite complexes of Eastern Uganda. In *Carbonatites*; Tuttle, O.F., Gittins, J., Eds.; Interscience: New York, NY, USA, 1966; pp. 73–126.
39. Le Bas, M.J. *Carbonatite–Nephelinite Volcanism: An African Case History*; John Wiley: London, UK, 1977.
40. Nelson, D.R.; Chivas, A.R.; Chappell, B.W.; McCulloch, M.T. Geochemical and isotopic systematics in carbonatites and implications for the evolution of ocean-island sources. *Geochim. Cosmochim. Acta* **1988**, *52*, 1–17. [CrossRef]
41. Cooper, A.F.; Reid, D.L. Nepheline sovites as parental magmas in carbonatite complexes: Evidence from Dicker Willem, southwest Namibia. *J. Petrol.* **1998**, *39*, 2123–2136. [CrossRef]
42. Nedosekova, I.L.; Belousova, E.A.; Sharygin, V.V.; Belyatsky, B.V.; Bayanova, T.B. Origin and evolution of the Ilmeny-Vishnevogorsky carbonatites (Urals, Russia): Insights from trace-element compositions, and Rb-Sr, Sm-Nd, U-Pb, Lu-Hf isotope data. *Mineral. Petrol.* **2013**, *107*, 101–123. [CrossRef]
43. Kapustin, Y.L. The origin of early calcitic carbonatites. *Int. Geol. Rev.* **1986**, *28*, 1031–1044. [CrossRef]
44. Treiman, A.H.; Essene, E.J. The Oka carbonatite complex, Quebec—Geology and evidence for silicate-carbonate liquid immiscibility. *Am. Mineral.* **1985**, *70*, 1101–1113.
45. Mitchell, R.H. Carbonatites and carbonatites and carbonatites. *Can. Mineral.* **2005**, *43*, 2049–2068. [CrossRef]
46. Wallace, M.E.; Green, D.H. An experimental-determination of primary carbonatite magma composition. *Nature* **1988**, *335*, 343–346. [CrossRef]
47. Eggler, D.H. Carbonatites, primary melts, and mantle dynamics. In *Carbonatites: Genesis and Evolution*; Bell, K., Ed.; Unwin Hyman: London, UK, 1989; pp. 561–579.
48. Wyllie, P.J.; Lee, W.J. Model system controls on conditions for formation of magnesiocarbonatite and calciocarbonatite magmas from the mantle. *J. Petrol.* **1998**, *39*, 1885–1893. [CrossRef]
49. Dalton, J.A.; Wood, B.J. The compositions of primary carbonate melts and their evolution through wallrock reaction in the mantle. *Earth Planet. Sci. Lett.* **1993**, *119*, 511–525. [CrossRef]
50. Le Bas, M.J. Nephelinites and carbonatites. In *Alkaline Igneous Rocks*; Fitton, J.G., Upton, B.G.J., Eds.; Geological Society Special Publications; Blackwell Scientific Publications: Oxford, UK, 1987; pp. 85–94.
51. Gittins, J. The origin and evolution of carbonatite magmas. In *Carbonatites: Genesis and Evolution*; Bell, K., Ed.; Unwin Hyman: London, UK, 1989; pp. 580–600.
52. Gittins, J.; Jago, B.C. Differentiation of natrocarbonatite magma at Oldoinyo Lengai volcano, Tanzania. *Mineral. Mag.* **1998**, *62*, 759–768. [CrossRef]
53. Veksler, I.V.; Petibon, C.; Jenner, G.A.; Dorfman, A.M.; Dingwell, D.B. Trace element partitioning in immiscible silicate-carbonate liquid systems: An initial experimental study using a centrifuge autoclave. *J. Petrol.* **1998**, *39*, 2095–2104. [CrossRef]
54. Chakhmouradian, A.R. High-field-strength elements in carbonatitic rocks: Geochemistry, crystal chemistry and significance for constraining the sources of carbonatites. *Chem. Geol.* **2006**, *235*, 138–160. [CrossRef]

55. Chakhmouradian, A.R.; Mumin, A.H.; Demeny, A.; Elliott, B. Postorogenic carbonatites at Eden Lake, Trans-Hudson orogen (northern Manitoba, Canada): Geological setting, mineralogy and geochemistry. *Lithos* **2008**, *103*, 503–526. [CrossRef]

56. Kogarko, L.N. Geochemistry of radioactive elements in the rocks of the Guli massif, Polar Siberia. *Geochem. Int.* **2012**, *50*, 719–725. [CrossRef]

57. Ying, J.; Zhou, X.; Zhang, H. Geochemical and isotopic investigation of the Laiwu-Zibo carbonatites from western Shandong Province, China, and implications for their petrogenesis and enriched mantle source. *Lithos* **2004**, *75*, 413–426. [CrossRef]

58. Stoppa, F.; Woolley, A.R. The Italian carbonatites: Field occurrence, petrology and regional significance. *Mineral. Petrol.* **1997**, *59*, 43–67. [CrossRef]

59. Gottardi, G. The genesis of zeolites. *Eur. J. Mineral.* **1989**, *1*, 479–487. [CrossRef]

60. Senderov, E.E.; Khitarov, N.I. Synthesis of thermodynamically stable zeolites in the Na_2O-Al_2O_3-SiO_2-H_2O system. *Am. Chem. Soc. Adv. Chem. Ser.* **1971**, *101*, 149–154.

61. Kristmannsdóttir, H.; Tómasson, J. Zeolite zones in geothermal areas in Iceland. In *Natural Zeolites, Occurrence, Properties, and Use*; Sand, L.B., Mumpton, F.A., Eds.; Pergamon: New York, NY, USA, 1978; pp. 277–284.

62. Carpenter, A.B. Graphical analysis of zeolite mineral assemblages from the Bay of Fundy area, Nova Scotia. *Am. Chem. Soc. Adv. Chem. Ser.* **1971**, *101*, 328–333.

63. Dingwell, D.B.; Brearley, M. Mineral chemistry of igneous melanite garnets from analcite-bearing volcanic-rocks, Alberta, Canada. *Contrib. Mineral. Petrol.* **1985**, *90*, 29–35. [CrossRef]

64. Gwalani, L.G.; Rock, N.M.S.; Ramasamy, R.; Griffin, B.J.; Mulai, B.P. Complexly zoned Ti-rich melanite-schorlomite garnets from Ambadungar carbonatite-alkalic complex, Deccan igneous province, Gujarat state, western India. *J. Asian Earth Sci.* **2000**, *18*, 163–176. [CrossRef]

Article

Three-D Mineralogical Mapping of the Kovdor Phoscorite–Carbonatite Complex, NW Russia: I. Forsterite

Julia A. Mikhailova [1,2], Gregory Yu. Ivanyuk [1,2],*, Andrey O. Kalashnikov [2],
Yakov A. Pakhomovsky [1,2], Ayya V. Bazai [1,2], Taras L. Panikorovskii [1], Victor N. Yakovenchuk [1,2],
Nataly G. Konopleva [1] and Pavel M. Goryainov [2]

[1] Nanomaterials Research Centre of Kola Science Centre, Russian Academy of Sciences, 14 Fersman Street, Apatity 184209, Russia; ylya_korchak@mail.ru (J.A.M.); pakhom@geoksc.apatity.ru (Y.A.P.); bazai@geoksc.apatity.ru (A.V.B.); taras.panikorovsky@spbu.ru (T.L.P.); yakovenchuk@geoksc.apatity.ru (V.N.Y.); konoplyova55@mail.ru (N.G.K.)
[2] Geological Institute of Kola Science Centre, Russian Academy of Sciences, 14 Fersman Street, Apatity 184209, Russia; kalashnikov@geoksc.apatity.ru (A.O.K.); pgor@geoksc.apatity.ru (P.M.G.)
* Correspondence: g.ivanyuk@gmail.com; Tel.: +7-81555-79531

Received: 30 May 2018; Accepted: 16 June 2018; Published: 20 June 2018

Abstract: The Kovdor alkaline-ultrabasic massif (NW Russia) is formed by three consequent intrusions: peridotite, foidolite–melilitolite and phoscorite–carbonatite. Forsterite is the earliest mineral of both peridotite and phoscorite–carbonatite, and its crystallization governed evolution of magmatic systems. Crystallization of forsterite from Ca-Fe-rich peridotite melt produced Si-Al-Na-K-rich residual melt-I corresponding to foidolite–melilitolite. In turn, consolidation of foidolite and melilitolite resulted in Fe-Ca-C-P-F-rich residual melt-II that emplaced in silicate rocks as a phoscorite–carbonatite pipe. Crystallization of phoscorite began from forsterite, which launched destruction of silicate-carbonate-ferri-phosphate subnetworks of melt-II, and further precipitation of apatite and magnetite from the pipe wall to its axis with formation of carbonatite melt-III in the pipe axial zone. This petrogenetic model is based on petrography, mineral chemistry, crystal size distribution and crystallochemistry of forsterite. Marginal forsterite-rich phoscorite consists of Fe^{2+}-Mn-Ni-Ti-rich forsterite similar to olivine from peridotite, intermediate low-carbonate magnetite-rich phoscorite includes Mg-Fe^{3+}-rich forsterite, and axial carbonate-rich phoscorite and carbonatites contain Fe^{2+}-Mn-rich forsterite. Incorporation of trivalent iron in the octahedral $M1$ and $M2$ sites reduced volume of these polyhedra; while volume of tetrahedral set has not changed. Thus, trivalent iron incorporates into forsterite by schema $(3Fe^{2+})_{oct} \rightarrow (2Fe^{3+} + \square)_{oct}$ that reflects redox conditions of the rock formation resulting in good agreement between compositions of apatite, magnetite, calcite and forsterite.

Keywords: forsterite; typochemistry; crystal structure; Kovdor phoscorite–carbonatite complex

1. Introduction

Phoscorite and carbonatite are igneous rocks genetically affined with alkaline massifs [1]. Many (phoscorite)-carbonatite complexes contain economic concentrations of REE (Bayan Obo, Cummins Range, Kovdor, Maoniuping, Mt. Pass, Mt. Weld, Mushgai Khudag, Tomtor, etc.), P (Catalão, Jacupiranga, Palabora, Kovdor, Seligdar, Sokli, Tapira, etc.), Nb (Araxá, Catalão, Fen, Lueshe, Mt. Weld, Oka, Panda Hill, St. Honoré, Tomtor, etc.), Cu (Palabora), Fe (Kovdor, Palabora, etc.), Zr (Kovdor, Palabora, etc.), U (Araxá, Palabora, etc.), Au, PGE (Catalão, Ipanema, Palabora, etc.), F (Amba Dongar, Maoniuping, etc.) with considerable amounts of phlogopite, vermiculite, calcite and dolomite [2–8]. The Kovdor phoscorite–carbonatite complex in the Murmansk Region (Russia) has large resources of Fe (as magnetite), P (as hydroxylapatite), Zr and Sc (as baddeleyite), and also contains forsterite, calcite, dolomite, pyrochlore and copper sulfides with potential economic significance. Early, we have described in detail the geology and petrography of the Kovdor phoscorite–carbonatite complex [9,10] and the main economic minerals: magnetite, apatite and baddeleyite [11–13]. In this series of articles, we would like to show results of our study of potential economic minerals, namely forsterite, sulfides and pyrochlore.

Phoscorite is a rock composed of magnetite, olivine and apatite and is usually associated with carbonatites [1]. Between the phoscorite and carbonatite, there are both gradual transitions (when carbonate content in phoscorite exceeds 50 modal % the rock formally obtains name carbonatite [1]) and sharp contact (carbonatite veins in phoscorite). However, temporal relations between rocks of marginal and internal parts of phoscorite–carbonatite complexes as well as the processes that caused formation of such zonation are still unclear. The mechanism of formation of phoscorite–carbonatite rock series is widely discussed (see reviews e.g., in [8,14–19]). Most of researchers believe that phoscorite as a typical rock occurring « ... around a core of carbonatite» is a result of a separate magmatic event preceding carbonatite magmatism (e.g., [20–22]). Some researchers suggest that carbonate-free phoscorite enriched by apatite and silicates (mainly forsterite) is the earlier rock in this sequence, while later carbonate-rich phoscorite and phoscorite-related carbonatite (i.e., the same phoscorite with carbonate content above 50 vol %) are formed due to the reaction between phosphate-silicate-rich phoscorite and carbonate-rich fluid or melt [21,23–28]. Some researchers divide phoscorite–carbonatite process into numerous separate intrusive events. They mainly substantiate their approach with the presence of sharp contacts between the rock varieties [29,30].

We believe that 3D mineralogical mapping is the best way to reconstruct genesis of any geological complex including phoscorite–carbonatite. This approach enabled us to establish a clear concentric zonation of the Kovdor phoscorite–carbonatite complex in terms of content, composition and properties of all economic minerals [11,13,24]. In general, the pipe marginal zone consists of (apatite)-forsterite phoscorite carrying fine grains of Ti-rich magnetite (with exsolution lamellae of ilmenite), FeMg-bearing hydroxylapatite and FeSi-bearing baddeleyite; the intermediate zone contains carbonate-free magnetite-rich phoscorite with medium to coarse grains of MgAl-bearing magnetite (with exsolution inclusions of spinel), pure hydroxylapatite and baddeleyite; and the axial zone of carbonate-rich phoscorite and phoscorite-related carbonatite includes medium- to fine-grained Ti-rich magnetite (with exsolution inclusions of geikielite–ilmenite), Sr-Ba-REE-bearing hydroxylapatite and Sc-Nb-bearing baddeleyite [11].

Consequently, phoscorite and phoscorite-related carbonatite of the Kovdor alkaline-ultrabasic massif consist of four main minerals belonging to separate classes of compounds: silicate–forsterite, phosphate–apatite, oxide–magnetite and carbonate–calcite, which compositions do not intercross (besides Ca in calcite and apatite and Fe in olivine and magnetite). Therefore, we can use content, composition and grain-size distribution, etc. of apatite for phosphorus behavior analysis, magnetite characteristics for iron and oxygen activity estimation, and forsterite and calcite characteristics for silicon and carbon evolution studies.

Forsterite can be the main key to understanding genesis and geology of the whole Kovdor alkaline-ultrabasic massif as its formation started from peridotite intrusion and finished with late carbonatites. In addition, forsterite is another economic mineral concentrated within two separate deposits [9]: the Baddeleyite-Apatite-Magnetite deposit within the phoscorite–carbonatite pipe and the Olivinite deposit within the peridotite core of the massif. For this reason, studied in details forsterite from the Kovdor phoscorite–carbonatite complex will be also compared with forsterite of the peridotite stock.

2. Geological Setting

The Kovdor massif of alkaline and ultrabasic rocks, phoscorite and carbonatites is situated in the SW of Murmansk Region, Russia (Figure 1a). It is a central-type intrusive complex with an area of 40.5 km^2 at the day surface emplaced in Archean granite-gneiss [31–33]. The geological setting of the Kovdor massif has been described by [9,21,28,30,34]. The massif consists of a central stock of earlier peridotite rimmed by later foidolite (predominantly) and melilitolite (Figure 1). In cross-section, the massif is an almost vertical stock, slightly narrowing with depth at the expense of foidolite and melilitolite [35]. There is a complex of metasomatic rocks between peridotite core and foidolite–melilitolite rim: diopsidite; phlogopitite; melilite-, monticellite-, vesuvianite-, and andradite-rich skarn-like rocks. Host gneiss transforms into fenite at the distance of 0.2–2 km from the alkaline ring intrusion. Numerous dikes and veins (up to 5 m thick) of nepheline and cancrinite syenite, (micro)ijolite, phonolite, alnoite, shonkinite, calcite, calcite and dolomite carbonatites cut into all the above mentioned intrusive and metasomatic rocks [9].

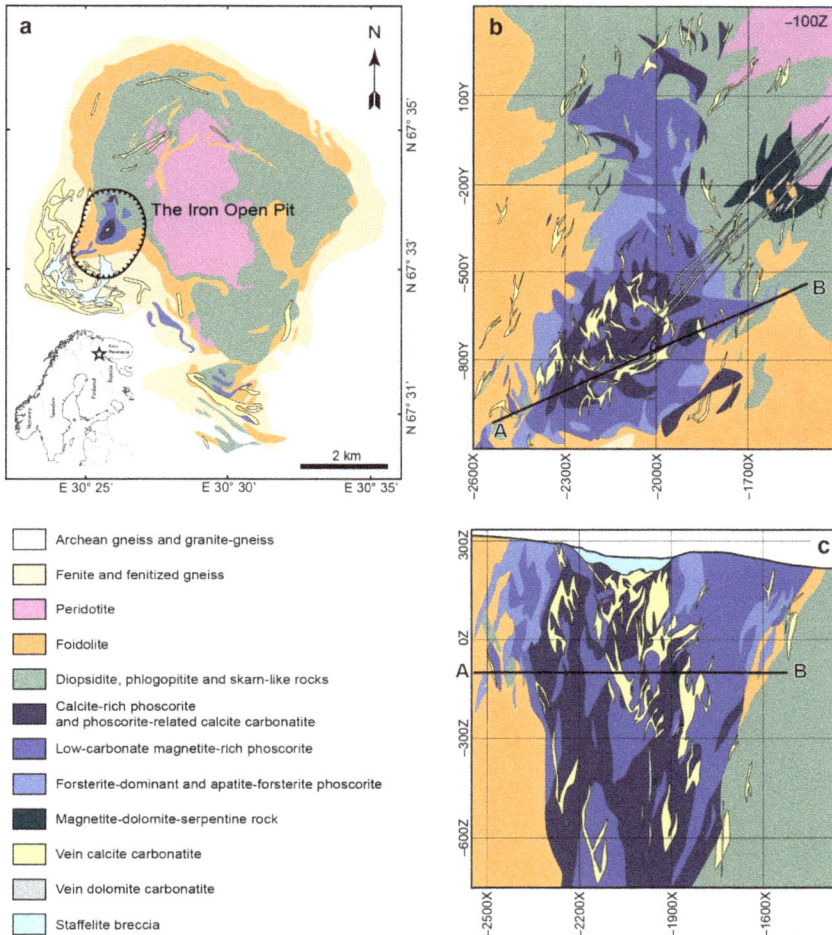

Figure 1. (**a**) Geological map of the Kovdor alkaline-ultrabasic massif, after Afanasyev and Pan'shin, modified after [9]; cross-sections of the Kovdor phoscorite–carbonatite complex: (**b**) horizontal (−100 m, Y axis shows the North) and (**c**) vertical along A-B line, after [11].

At the western contact of peridotite and foidolite, there is a concentrically zoned pipe of phoscorite and carbonatites (Figure 1b,c) highly enriched in magnetite, hydroxylapatite and baddeleyite.

The marginal zone of this pipe is composed of (apatite)-forsterite phoscorite (Figure 2a,b), the intermediate zone consists of low-carbonate magnetite-rich phoscorite (Figure 2c) and the axial zone contains calcite-rich phoscorite (Figure 2d) and phoscorite-related calcite carbonatite (non-vein bodies characterized by transient contact with phoscorite). Numerous carbonatite veins cut the phoscorite body, with the highest concentration of veins encountered in its axial, calcite-rich zone (Figure 1b,c and Figure 2e,g). Main varieties of phoscorite and phoscorite-related carbonatite are shown in Table 1. However, there are no distinct boundaries between these rocks, and artificial boundaries between them are quite conventional [10]. Zone of linear veins of dolomite carbonatite (Figure 1b,c) extends from the central part of the phoscorite–carbonatite pipe to the north-east and associates with metasomatic magnetite-dolomite-serpentine rock, which replaced peridotite or forsterite-rich phoscorite [9,11,24].

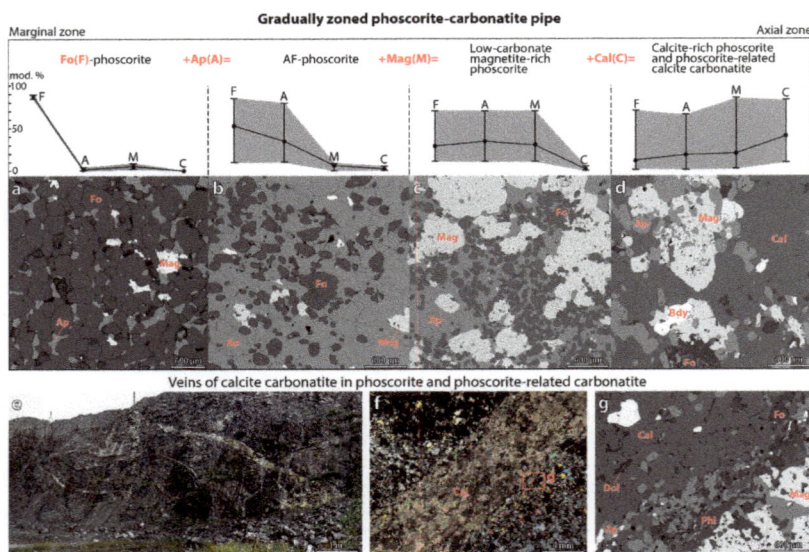

Figure 2. Relations of major rocks within the Kovdor phoscorite–carbonatite pipe. BSE-images (**a–d,g**) of main rock types, photo of outcrop (**e**) and image of thin section in transmitted light (**f**). (**a**) 914/185.2; (**b**) 993/132.3; (**c**) 981/217.1; (**d**) 1006/436.1; (**f,g**) 927/21.7.

Table 1. Main varieties of phoscorite and phoscorite-related carbonatite [10].

Group of Rock	Rock	Mineral Content, Modal %			
		Fo	Ap	Mag	Cal
Forsterite-rich phoscorite (Cal < 10 modal %, Mag < 10 modal %)	Forsteritite (F)	85–90	0–5	1–8	–
	Apatite-forsterite phoscorite (AF)	10–85	10–80	0–8	0–5
Low-carbonate magnetite-rich phoscorite (Cal < 10 modal %, Mag > 10 modal %)	Magnetite-forsterite phoscorite (MF)	10–70	0–5	15–85	0–5
	Magnetite-apatite-forsterite phoscorite (MAF)	10–70	10–70	10–70	0–8
	Magnetite-apatite phoscorite (MA)	0–5	5–50	40–85	0–5
	Magnetitite (M)	0–8	0–5	80–95	0–5
Calcite-rich phoscorite (10 modal % < Cal < 50 modal %)	Calcite-magnetite-apatite-forsterite phoscorite (CMAF)	10–60	10–60	10–55	10–40
	Calcite-magnetite-forsterite phoscorite (CMF)	10–70	0–5	15–60	10–45
	Calcite-apatite-forsterite phoscorite (CAF)	10–45	20–55	2–8	20–40
	Calcite-magnetite-apatite phoscorite (CMA)	0–6	10–63	11–79	10–45
	Calcite-apatite phoscorite (CA)	2–6	50–65	1–6	27–41
	Calcite-magnetite phoscorite (CM)	0–5	0–5	70–84	16–20
Phoscorite-related carbonatite (Cal > 50 modal %)	Calcite carbonatite (C)	0–35	2–40	1–35	50–82

3. Materials and Methods

For this study, we used 540 thin polished sections of phoscorite (mainly), carbonatites and host rocks from 108 exploration holes drilled within the Kovdor phoscorite–carbonatite complex. The thin polished sections were analyzed using the LEO-1450 scanning electron microscope (Carl Zeiss Microscopy, Oberkochen, Germany) with energy-dispersive analyzer Röntek to obtain BSE-images of representative regions and pre-analyze all minerals found in the samples. The ImageJ open source image processing program [36] was used to create digital images from the BSE-images, and determine forsterite grain size (equivalent circular diameter) and orientation of the grain long axis.

Chemical composition of forsterite was analyzed in the Geological Institute of the Kola Science Center, Russian Academy of Sciences, with the Cameca MS-46 electron microprobe (Cameca, Gennevilliers, France) operating in WDS-mode at 22 kV with beam diameter 10 mm, beam current 30 nA and counting times 20 s (for a peak) and 2×10 s (for background before and after the peak), with 5–10 counts for every element in each point. The analytical precision (reproducibility) of forsterite analyses is 0.2–0.05 wt % (2 standard deviations) for the major element and about 0.01 wt % for impurities. Used standards and detection limits are given in Table 2. The systematic errors are within the random errors.

Table 2. Parameters of EPMA.

Element	Type of Crystal	Standards	DL, wt %
Mg	KAP	Forsterite	0.1
Al	KAP	Pyrope	0.05
Si	KAP	Forsterite	0.05
Ca	PET	Diopside	0.03
Sc	PET	Thortveitite	0.02
Ti	PET	Lorenzenite	0.02
Cr	Quartz	Chromite	0.02
Mn	Quartz	Synthetic $MnCO_3$	0.01
Fe	Quartz	Hematite	0.01
Ni	LiF	Metal nickel	0.01

n = 5–10 counts for each point (depending on dispersion).

At the X-ray Diffraction Centre of Saint-Petersburg State University, single-crystal X-ray diffraction experiments were performed on forsterite crystals 919/18.5 (1), 924/26.7 (2), 924/169.1 (3), 966/62.9 (4), 987/67.2 (5) with the Agilent Technologies Xcalibur Eos diffractometer operated at 50 kV and 40 mA. A hemisphere of three-dimensional data was collected at room temperature, using monochromatic Mo$K\alpha$ X-radiation with frame widths of 1° and 10 s count for each frame. Crystal structures were refined in the standard setting (space group *Pnma*) by means of the *SHELX* program [37] incorporated in the *OLEX2* program package [38]. Empirical absorption correction was applied in the CrysAlis PRO [39] program using spherical harmonics, implemented in the SCALE3 ABSPACK scaling algorithm. Volumes of coordination polyhedra are calculated with the VESTA 3 program [40]. Crystal structures were visualized with the Diamond 3.2f program [41].

Cation contents were calculated in the MINAL program by D. V. Dolivo-Dobrovolsky [42]. Statistical analyses were implemented with the STATISTICA 8.0 [43] and TableCurve 2D [44] programs. Geostatisical studies and 3D modeling were conducted with the MICROMINE 16.1 [45] program. Interpolation was performed by ordinary kriging.

Abbreviations used include Ap (hydroxylapatite), Bdy (baddeleyite), Cal (calcite), Cb (carbonate), Chu (clinohumite), Clc (clinochlore), Di (diopside), Dol (dolomite), Fo (forsterite), Mag (magnetite), Nph (nepheline), Phl (phlogopite), Po (pyrrhotite), Spl (spinel), Srp (serpentine) and Val (valleriite).

4. Results

4.1. Content, Morphology and Grain Size of Forsterite

Peridotite contains 40–90 modal % of forsterite that has rounded isometric grains (Figure 3a) up to 12 cm in diameter. Interstices within forsterite aggregate are filled with short prismatic grains of diopside (up to 50 modal %), phlogopite plates (up to 15 modal %), anhedral grains of (titano)magnetite (up to 10 modal %), and fine-granular nests of hydroxylapatite (up to 5 modal %). Within the forsterite grains, there are lens-like inclusions of diopside with skeletal or tabular crystals of relatively pure magnetite inside (Figure 3b), as well as rounded inclusions of calcite. Typical products of forsterite alteration include serpentine (lizardite and clinochrysotile), clinochlore and, rarely, clinohumite. Near the contact with foidiolite intrusion, forsterite is intensively replaced with newly formed diopside, phlogopite (Figure 3a) and richterite, up to transformation of peridotite into diopsidite and/or phlogopite glimmerite.

Phoscorite contains 0–90 modal % of forsterite (Figure 4a). The highest content usually occurs in marginal forsteritite and apatite-forsterite phoscorite (89 and 53 modal % correspondingly, Figure 2a,b and Figure 3c,d). In intermediate low-carbonate magnetite-rich phoscorite, average content of forsterite decreases from 42 modal % in magnetite-forsterite (MF) phoscorite (Figure 3f) to 28 modal % in predominant magnetite-apatite-forsterite (MAF) phoscorite (Figures 2c and 3e), and then to 4 modal % in apatite-magnetite (AM) phoscorite and 2 modal % in magnetitite (M). Average content of forsterite in axial calcite-rich phoscorite varies from 28–21 modal % in calcite-magnetite-forsterite (CMF) and calcite-magnetite-apatite-forsterite (CMAF) phoscorite (Figure 2d) to 3 modal % in calcite-magnetite-apatite (CMA) and calcite-magnetite (CM) phoscorite. Lastly, phoscorite-related carbonatite contains 5 modal % of forsterite [24].

Gradual decrease of forsterite content from earlier forsteritite of the marginal zone to later carbonatites is accompanied by regular changes in morphology and grain size of forsterite as well as in its relations with other rock-forming minerals. In the marginal forsterite-rich phoscorite, forsterite forms spherical (small) to ellipsoidal (large) grains (Figure 2a,b) or, more rarely, well-shaped short prismatic crystals with $a:c$ = 1:1.3. Average equivalent circular diameter of forsterite grains is 0.18 mm (Figure 4b), and grain size distribution is negative-exponential (Figure 5), when cumulative frequencies are concave down in log-log space (Figure 5d), and linear in semilog space (Figure 5c). There is insufficient anisotropy in grain orientation (Figure 5e). When forsterite content sufficiently exceeds apatite content, then hydroxylapatite fills interstices between forsterite grains. In this case, forsterite grains contain numerous inclusions of hydroxylapatite. Its content increases in the vicinity of large segregation of hydroxylapatite (Figure 3d). If the amount of hydroxylapatite increases, then spatial separation of forsterite and hydroxylapatite is observed (Figure 3e). Such monomineral segregations randomly alternate with areas evenly filled with apatite and forsterite (Figure 2b). Moreover, there are indications of co-crystallization of forsterite and hydroxylapatite. In this case, forsterite grains contain numerous inclusions of apatite and have sinuous boundaries (Figure 3c). In addition, forsterite grains sometimes contain prismatic inclusions of baddeleyite (up to 20 μm long, Figure 2f) as well as spherical inclusions ("drops") of calcite (up to 60 μm in diameter, Figure 3c) and, rarely, dolomite (up to 20 μm in diameter). Magnetite occurs mainly within apatite segregations or fills interstices between forsterite grains together with hydroxylapatite.

In the intermediate low-carbonate magnetite-rich phoscorite, average size of forsterite grains is 0.19 mm (Figure 4b), grain size distribution is the same as in the marginal forsterite-rich phoscorite (Figure 5h,i), but without anisotropy in grain orientation (Figure 5j). Magnetite content growth leads to concentration of magnetite and forsterite in separate monomineralic nests (Figure 3f); however, hydroxylapatite still closely associates with magnetite. Forsterite grains obtain mirror-like faces at the boundary with calcite nests and veinlets. Similar to marginal (apatite)-forsterite phoscorite, forsterite grains usually contain ellipsoidal inclusions of hydroxylapatite, prismatic inclusions of baddeleyite and spherical "drops" of calcite and dolomite (Figure 3g) in the pipe intermediate zone.

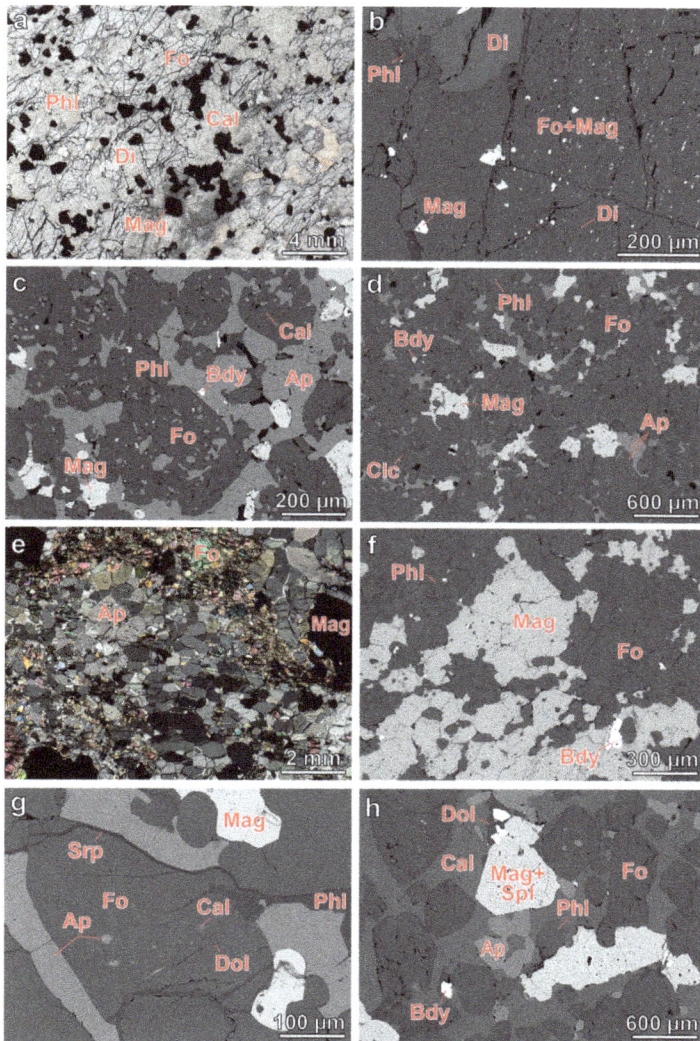

Figure 3. Morphology of forsterite and its relations with other minerals in rocks of the Kovdor alkaline-ultrabasic massif: (**a**) replacement of forsterite with diopside and phlogopite in peridotite 10p/76.01; (**b**) grain of forsterite with inclusions of diopside and magnetite in peridotite 912/231.6; (**c**) co-crystallization of forsterite (with inclusions of calcite) and hydroxylapatite in AF phoscorite 934/112.1; (**d**) interstitial segregations of magnetite and hydroxylapatite in AF phoscorite 976/33.1; (**e**) network of forsterite grains in MAF phoscorite 932/205.9; (**f**) typical MF phoscorite 983/64.6; (**g**) inclusions of hydroxylapatite, calcite and dolomite in forsterite grains of MAF phoscorite 938/30.6; (**h**) well shaped crystals of forsterite with inclusions of hydroxylapatite and calcite in CMF phoscorite 953/6.0. (**a**,**e**) images of polished thin section in transmitted light; (**b–d**,**f–h**) BSE images. Mineral abbreviations see in the Section 3.

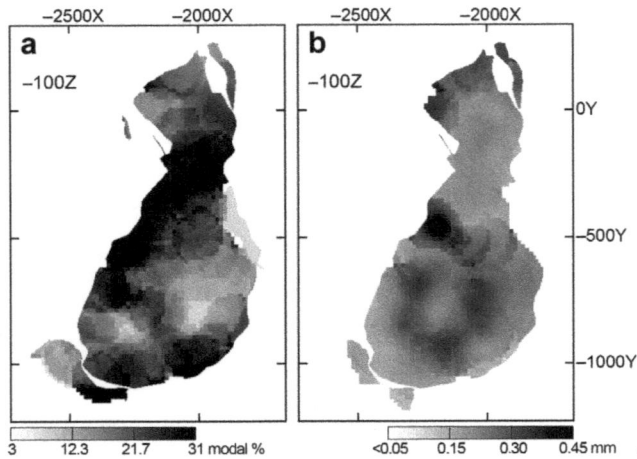

Figure 4. Distribution of forsterite content (**a**) and average grain size (**b**) within the baddeleyite-apatite-magnetite deposit.

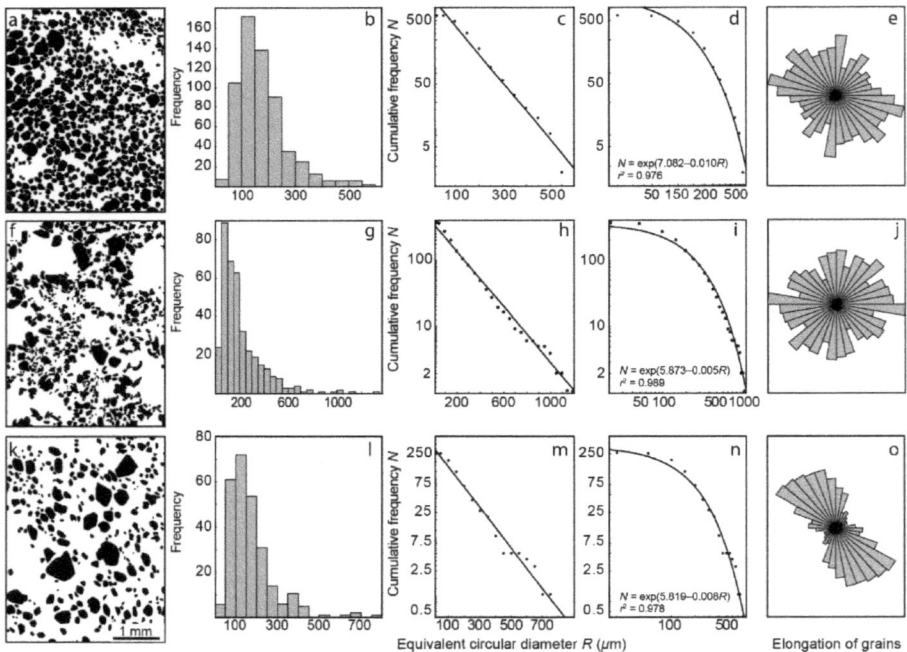

Figure 5. Grains of forsterite in thin sections of AF phoscorite 937/177.0 (**a**), MAF phoscorite 899/82.9 (**f**) and CMAF phoscorite 996/241.2 (**k**), corresponding histograms of equivalent circular diameter (**b,g,l**), cumulative frequency diagrams in semilog (**c,h,m**) and double logarithmic (**d,i,n**) coordinates, and orientation of elongated forsterite grains in the sections (**e,j,o**).

In the axial calcite-rich phoscorite and phoscorite-related carbonatite, minor forsterite occurs as xenomorph rounded grains within its monomineralic nests, and well-shaped short prismatic crystals (up to 15 cm long) at the contact with calcite segregations (Figure 3h). Average size of forsterite grains is 0.2 mm (Figure 4b), grain size distribution is exponential (Figure 5m,n), and anisotropy of grain orientation is strong (Figure 5o). Inclusions in forsterite grains become rarer and smaller, and mainly consist of rounded hydroxylapatite and prismatic baddeleyite.

Vein calcite and dolomite carbonatites include only 1 modal % of small idiomorphic grains of forsterite (Figure 2g). The mineral concentrates in marginal parts of the veins intersecting forsterite-rich host phoscorite. Usually, such crystals are free of inclusions.

In all rocks of the Kovdor phoscorite–carbonatite complex, forsterite is usually partially replaced by secondary phlogopite, clinochlore, clinohumite, valleriite, serpentine, and dolomite, with clear correspondence to the pipe zonation. Apo-forsterite phlogopite occurs throughout the pipe volume; but in phoscorite-related carbonatite, it is more sparsely spread (Figure 6a). At first, mica forms polycrystalline rims around forsterite grains, and later, it forms large flexural plates with forsterite relics inside. Secondary clinochlore closely associates with phlogopite, forming rims around resorbed forsterite grains (Figure 3d), and its distribution within the phoscorite–carbonatite pipe is similar to mica (Figure 6a). Dolomite and serpentine replace forsterite predominantly within linear zone of dolomite carbonatite (Figures 2a and 6b), finally forming magnetite-dolomite-serpentine rocks after peridotite and forsterite-rich phoscorite [11,24]. Besides, serpentine associates closely with valleriite (Figure 6c), which content predictably increases in sulphide-bearing phoscorite. Apo-forsterite clinohumite occurs mainly in axial carbonate-rich phoscorite and carbonatites (Figure 6d): at first, as thin rims around forsterite grains, then, as comparatively large grains (up to 2 cm in diameter) with rare relics of forsterite.

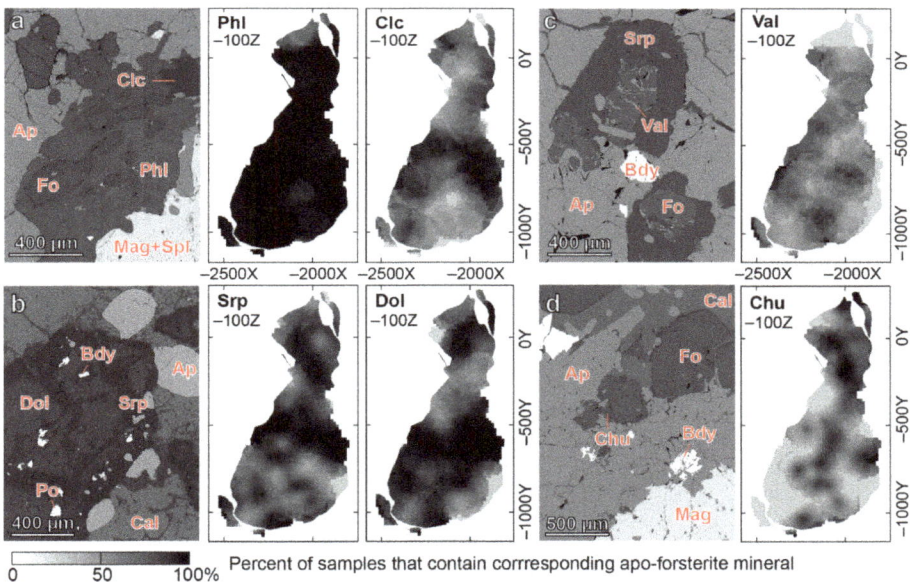

Figure 6. Relation of apo-forsterite phlogopite, clinochlore, serpentine, dolomite, valleriite and clinohumite with forsterite and distribution of these minerals within the Kovdor baddeleyite-apatite-magnetite deposit at horizon −100 m. BSE-images: (**a**) MAF phoscorite 992/0.9; (**b**) AF phoscorite 970/93.1; (**c**) AF phoscorite 966/165.1; (**d**) MAF phoscorite 986/82.2. Mineral abbreviations see in the Section 3.

4.2. Chemical Composition

Despite the long history of the Kovdor study, only 16 chemical analyzes of forsterite from this massif can be found in the literature [9,21,30,46,47]. Average data on chemical composition of forsterite are listed in Table 3. As compared with forsterite from phoscorite and carbonatites, forsterite in peridotite is relatively enriched in CaO and NiO. In the phoscorite–carbonatite complex, forsterite contains minor amounts of the substitutions; however, this is enough to define zonation of the phoscorite–carbonatite pipe. In particular, the highest content of FeO (the average content 10 wt %) characterizes forsterite from marginal (apatite)-forsterite phoscorite, the highest content of CaO (about 0.2 wt %) is predictably found in forsterite from carbonatites. The highest content of minor substitutions of TiO_2 and NiO (up to 0.07 and 0.06 wt %, correspondingly) occur in forsterite from marginal forsterite-rich phoscorite, and comparatively high content of Cr_2O_3 and Sc_2O_3 (up to 0.04 and 0.11 wt %, correspondingly) is typical for forsterite from axial calcite-rich phoscorite and phoscorite-related carbonatite. Usually, forsterite grains do not have any chemical zonation; however, some crystals of this mineral from forsterite-rich phoscorite contain iron-rich core and iron-poor marginal zone that differ by approx. 3 wt % in terms of MgO content.

Table 3. Chemical composition of forsterite (average ± SD/min–max).

Rock	Peridotite	Phoscorite			Carbonatites	
		(Ap)-Fo	Low-Cb Mag-Rich	Cal-Rich	Phoscorite-Related	Vein
n	7	39	176	117	20	7
SiO_2, wt %	41.1 ± 0.9 40.37–42.63	40.8 ± 0.6 39.73–42.55	40.9 ± 0.6 38.48–42.21	40.8 ± 0.6 39.33–43.98	40.8 ± 0.6 39.60–42.01	40.8 ± 0.7 39.65–41.68
MgO	47 ± 2 44.12–50.26	52 ± 1 47.73–53.87	53 ± 1 47.10–55.25	52 ± 1 46.65–55.93	53 ± 1 48.51–54.43	52 ± 2 48.53–53.69
FeO	10 ± 1 8.68–12.11	7 ± 1 4.43–11.10	6 ± 1 3.48–8.82	6 ± 1 1.53–10.89	6 ± 1 3.73–10.22	6 ± 1 4.22–8.04
MnO	0.4 ± 0.2 0.10–0.55	0.34 ± 0.07 0.23–0.56	0.3 ± 0.3 0.14–0.49	0.33 ± 0.06 0.19–0.53	0.34 ± 0.09 0.25–0.66	0.33 ± 0.04 0.28–0.39
CaO	0.3 ± 0.1 0.13–0.36	0.13 ± 0.08 <0.03–0.40	0.12 ± 0.08 <0.03–0.55	0.13 ± 0.08 <0.03–0.60	0.17 ± 0.06 0.09–0.31	0.19 ± 0.16 0.05–0.49
TiO_2	<0.02 <0.02–0.02	<0.02 <0.02–0.07	<0.02 <0.02–0.05	<0.02 <0.02–0.04	<0.02 <0.02–0.03	<0.02 <0.02–0.03
Al_2O_3	<0.05	<0.05	<0.05	<0.05	<0.05	<0.05
Cr_2O_3	<0.02 <0.02–0.03	<0.02	<0.02	≤0.02	<0.02 <0.02–0.04	<0.02
NiO	0.10 ± 0.05 <0.02–0.17	<0.02 <0.01–0.06	<0.01 <0.01–0.03	<0.01 <0.01–0.03	<0.01	<0.01
Sc_2O_3	<0.02	<0.02	<0.02 <0.02–0.03	<0.02 <0.02–0.11	<0.02	<0.02
Mg, *apfu*	1.75 ± 0.06 1.66–1.82	1.87 ± 0.04 1.77–1.92	1.89 ± 0.03 1.74–1.95	1.89 ± 0.04 1.75–1.99	1.89 ± 0.03 1.79–1.93	1.88 ± 0.04 1.8–1.93
Fe^{2+}	0.21 ± 0.03 0.15–0.24	0.11 ± 0.04 0.05–0.21	0.09 ± 0.03 0.00–0.18	0.08 ± 0.04 0.00–0.23	0.09 ± 0.03 0.03–0.17	0.09 ± 0.04 0.04–0.17
Fe^{3+}	0.01 ± 0.01 0.00–0.03	0.02 ± 0.02 0.00–0.07	0.03 ± 0.02 0.00–0.11	0.03 ± 0.02 0.00–0.09	0.03 ± 0.02 0.00–0.05	0.03 ± 0.02 0.00–0.06
Mn	0.01 0.00–0.01	0.01 0.00–0.01	0.01+0.01 0.00–0.01	0.01 0.00–0.01	0.01 0.00–0.01	0.01 0.00–0.01
Ca	0.01 0.00–0.01	0.00 0.00–0.01	0.00 0.00–0.01	0.00 0.00–0.02	0.00 0.00–0.01	0.00 0.00–0.01
Si	1.02 ± 0.03 0.98–1.08	0.99 ± 0.01 0.97–1.04	0.99 ± 0.01 0.93–1.03	0.99 ± 0.01 0.95–1.06	0.98 ± 0.01 0.97–1.00	0.99 ± 0.01 0.97–1.01

Forsterite composition was recalculated at three cations per formula unit and O = 4 *apfu*, which permitted to obtain more realistic results than calculations based on 3 cations per formula unit or

O = 4 *apfu* (no Fe^{3+}-Fe^{2+} re-distribution), Si = 1 and O = 4 *apfu* (excess of cations in the octahedral *M* position), *M* = 2 and O = 4 *apfu* (deficit of Si). The result showed that in the average 30% of iron is in the three-valent form. There are significant correlations between Mg, Fe^{2+}, Fe^{3+} and Mn ($r = \pm 0.49$–0.96, $p = 0.0000$) and weak correlations of these elements with Ca (Figure 7). Factor analysis of the cation contents (in *apfu*) was performed according to the principal components analysis with normalization and varimax rotation (Table 4). The analysis enabled us to reveal the following isomorphic substitutions:

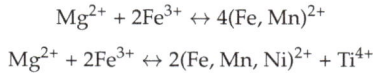

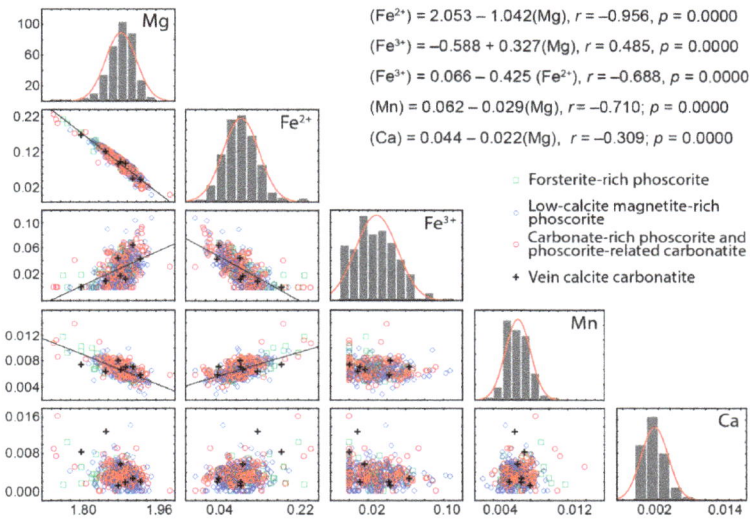

$$Mg^{2+} + 2Fe^{3+} \leftrightarrow 4(Fe, Mn)^{2+}$$

$$Mg^{2+} + 2Fe^{3+} \leftrightarrow 2(Fe, Mn, Ni)^{2+} + Ti^{4+}$$

$(Fe^{2+}) = 2.053 - 1.042(Mg)$, $r = -0.956$, $p = 0.0000$

$(Fe^{3+}) = -0.588 + 0.327(Mg)$, $r = 0.485$, $p = 0.0000$

$(Fe^{3+}) = 0.066 - 0.425 (Fe^{2+})$, $r = -0.688$, $p = 0.0000$

$(Mn) = 0.062 - 0.029(Mg)$, $r = -0.710$; $p = 0.0000$

$(Ca) = 0.044 - 0.022(Mg)$, $r = -0.309$; $p = 0.0000$

□ Forsterite-rich phoscorite

○ Low-calcite magnetite-rich phoscorite

○ Carbonate-rich phoscorite and phoscorite-related carbonatite

+ Vein calcite carbonatite

Figure 7. Matrix diagram for major octahedral cations of forsterite (*apfu*).

Table 4. Results of factor analysis of forsterite composition.

Variables	Factor Loadings	
	Factor 1	Factor 2
Mg	−0.927	−0.248
Fe^{2+}	0.961	0.179
Fe^{3+}	−0.680	−0.021
Mn	0.676	0.219
Ca	0.357	0.014
Ti	0.164	0.844
Ni	0.090	0.861
Explained variance	2.863	1.596
% of total variance	40.9	22.8

Factor loadings above 0.6 are shown in bold.

These substitutions result in clear concentric zonation of the Kovdor phoscorite–carbonatite complex in terms of forsterite composition (Figure 8). The features of forsterite composition (see above) and these figures show that the pipe marginal zone consists of Fe^{2+}-Mn-Ni-Ti-rich forsterite similar to olivine from peridotite, the intermediate zone includes Mg-Fe^{3+}-rich forsterite, and the axial zone contains Fe^{2+}-Mn-rich forsterite. In addition, the content of Fe^{3+} in forsterite increases with depth. The tendency is accompanied by growth of Mg and Mn cumulative concentration with depth; however, these elements themselves vary in inverse proportion to each other. Ca content in forsterite increases sufficiently in carbonate-rich rocks located deeper than −500 m.

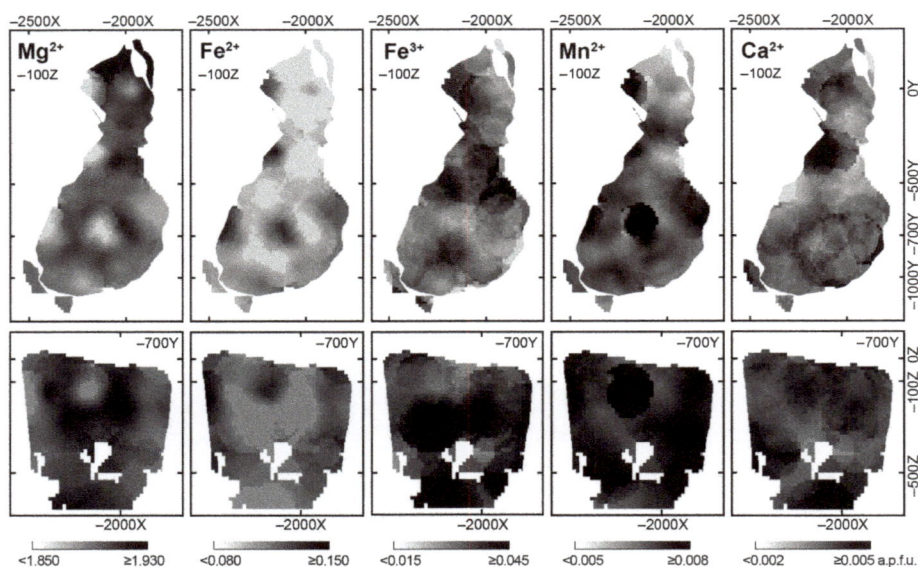

Figure 8. Variations in forsterite composition (*apfu*) within the Kovdor baddeleyite-apatite-magnetite deposit.

As noted earlier [9,11,13,24], chemical compositions of other minerals also change in accordance with a concentric zonation of the Kovdor phoscorite–carbonatite complex. Therefore, compositions of forsterite and other rock-forming and accessory minerals must be interdependent. Figure 9 shows correlation coefficients between main components of forsterite and co-existing rock-forming minerals. Forsterite composition closely correlates with composition of apatite, magnetite and calcite, with fundamental role of the main scheme of isomorphism, $Mg^{2+} + 2Fe^3 + \square \leftrightarrow 4Fe^{2+}$ that reflects redox conditions of the rock formation. In fact, oxidized condition results in presence of Fe^{3+} instead of Fe^{2+} in melt/fluid/solution, and thus in crystallization of $Mg-Fe^{3+}$-rich members of the corresponding mineral series, while reduced conditions cause domination of Fe^{2+} and formation of ferrous members of these series.

As a result, in marginal forsterite-rich phoscorite, predominant «ferrous forsterite» associates with Fe^{2+}-Si-rich hydroxylapatite, Mn-Si-Ti-Zn-Cr-rich magnetite, and Fe^{2+}-rich calcite. In the intermediate low-carbonate magnetite-rich phoscorite, predominant Fe^{3+}-bearing forsterite occurs together with pure hydroxylapatite, Mg-rich magnetite and pure calcite. In the axial calcite-rich phoscorite and phoscorite-related carbonatite, «ferrous forsterite» again predominates in the associations with Fe^{2+}-Mn-rich hydroxylapatite, Ca-V-rich magnetite and Fe^{2+}-rich calcite. Comparison of the maps shown in Figure 8 with the corresponding schemas for associated rock-forming minerals [11] also confirms the above conclusion.

Electron spin resonance spectroscopy performed by Zeira et al. [48] demonstrated incorporation of Fe^{3+} in forsterite structure into *M*1 and *M*2 octahedral sites. According to Janney and Banfield [49], during oxidation under acidic conditions, incorporation of Fe^{3+} into octahedral sets of olivine is compensated by vacancies in octahedral sets: $(3Fe^{2+})_{oct} \rightarrow (\square + 2Fe^{3+})_{oct}$ (laihunite schema). Under alkaline conditions, olivine oxidation is accompanied by leaching of SiO_4-tetrahedra: $(4Fe^{2+})_{oct} + (4Si^{4+})_{tet} \rightarrow (4Fe^{3+})_{oct} + (\square + 3Si^{4+})_{tet}$. Due to permanent deficit of tetrahedral cations in forsterite (see Table 3) and alkaline nature of the Kovdor phoscorite–carbonatite complex, we assume that forsterite oxidation follows the latter schema. However, this assumption should be confirmed with X-ray crystal study.

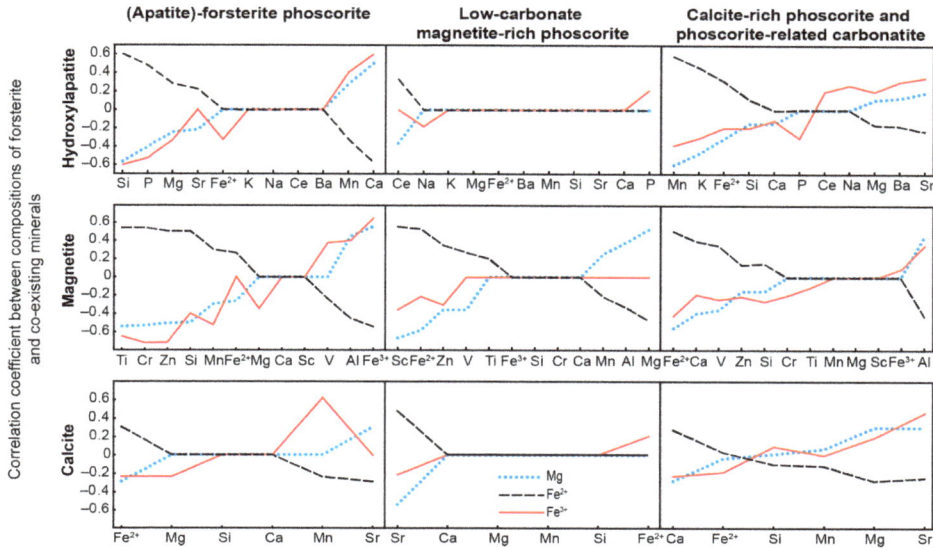

Figure 9. Correlation coefficients of forsterite composition with hydroxylapatite, magnetite and calcite compositions.

4.3. Single Crystal X-ray Diffraction

For the X-ray crystal study, we selected 5 forsterite crystals with various content of Mg, Fe^{2+} and Fe^{3+} from different zones of the phoscorite–carbonatite pipe (Table 5). The study details and crystallographic parameters obtained are shown in Table 6. Final atomic coordinates and isotropic displacement parameters selected interatomic distances and anisotropic displacement parameters are specified in the supplementary electronic materials (Tables S1–S20 in Supplementary Materials, CIF data available).

Table 5. Chemical composition of forsterite analyzed with single crystal X-ray diffraction.

Sample	1	2	3	4	5
Drill hole	919	924	924	966	987
Depth, m	18.5	26.7	169.1	62.9	67.2
Phoscorite	Mag-Ap-Fo	Cal-Mag-Ap-Fo	Ap-Fo	Mag-Ap-Fo	Cal-Mag-Ap-Fo
SiO_2, wt %	39.37	40.69	40.71	40.16	40.74
TiO_2	bd	0.01	0.07	bd	bd
FeO	6.20	8.96	8.51	8.42	6.89
MnO	0.30	0.46	0.49	4.27	0.39
MgO	53.37	47.79	48.96	47.10	52.18
CaO	0.13	0.34	0.06	0.14	0.09
NiO	bd	bd	0.06	bd	bd
Total	99.37	98.25	98.86	100.09	100.29
Mg, *apfu*	1.917	1.778	1.804	1.738	1.871
Fe^{2+}	0.022	0.187	0.176	0.163	0.098
Fe^{3+}	0.103	–	–	0.012	0.040
Mn^{2+}	0.006	0.010	0.010	0.090	0.008
Ca^{2+}	0.003	0.009	0.002	0.004	0.002
Ni^{2+}	–	–	0.001	–	–
Ti^{4+}	–	–	0.001	–	–
Si^{4+}	0.949	1.016	1.006	0.994	0.980

bd—below detection limit.

Minerals **2018**, *8*, 260

Table 6. Crystal data, data collection and structure refinement parameters of forsterite.

Sample	1	2	3	4	5
Refined formula	$Mg_{1.94}Fe_{0.06}SiO_4$	$Mg_{1.87}Fe_{0.13}SiO_4$	$Mg_{1.84}Fe_{0.16}SiO_4$	$Mg_{1.89}Fe_{0.11}SiO_4$	$Fe_{0.10}Mg_{1.90}SiO_4$
Temperature/K			293(2)		
Crystal system			orthorhombic		
Space group			$Pnma$		
a, (Å)	10.1899(6)	10.2165(4)	10.2097(4)	10.2027(4)	10.1980(4)
b, (Å)	5.9730(4)	5.9911(2)	5.9876(3)	5.9775(3)	5.9810(2)
c, (Å)	4.7403(3)	4.76168(14)	4.7600(2)	4.7541(2)	4.75403(16)
$\alpha = \beta = \gamma$, (°)	90	90	90	90	90
Volume, (Å^3)	288.51(3)	291.453(18)	290.99(2)	289.94(2)	289.970(18)
Z	4	4	4	4	4
ρ_{calc}, (g/cm^3)	3.279	3.298	3.331	3.299	3.299
μ/mm^{-1}	1.321	1.639	1.813	1.544	1.544
Crystal size, (mm^3)	0.23 × 0.18 × 0.16	0.27 × 0.21 × 0.18	0.29 × 0.25 × 0.16	0.18 × 0.15 × 0.14	0.19 × 0.17 × 0.16
Radiation			MoKα (λ = 0.71073)		
2Θ range for data collection, (°)	7.99–54.916	7.978–54.914	7.984–54.942	7.99–54.844	7.992–54.86
Index ranges	$-13 \leq h \leq 11$, $-7 \leq k \leq 4$, $-6 \leq l \leq 5$	$-13 \leq h \leq 5$, $-4 \leq k \leq 7$, $-6 \leq l \leq 5$	$-13 \leq h \leq 3$, $-6 \leq k \leq 7$, $-6 \leq l \leq 3$	$-13 \leq h \leq 9$, $-6 \leq k \leq 7$, $-5 \leq l \leq 6$	$-6 \leq h \leq 13$, $-5 \leq k \leq 7$, $-6 \leq l \leq 3$
Reflections collected	725	755	794	753	738
Independent reflections	[R_{int} = 0.0222, R_{sigma} = 0.0333]	[R_{int} = 0.0161, R_{sigma} = 0.0228]	[R_{int} = 0.0291, R_{sigm} = 0.0414]	[R_{int} = 0.0212, R_{sigma} = 0.0290]	[R_{int} = 0.0149, R_{sigma} = 0.0214]
Data/restraints/parameters	361/0/42	364/0/36	364/0/42	363/0/42	363/0/42
Goodness-of-fit on F^2	1.040	1.146	1.077	1.149	1.253
Final R indexes [I >= 2σ (I)]	R_1 = 0.0334, wR_2 = 0.0852	R_1 = 0.0274, wR_2 = 0.0713	R_1 = 0.0260, wR_2 = 0.0593	R_1 = 0.0239, wR_2 = 0.0602	R_1 = 0.0258, wR_2 = 0.0683
Final R indexes [all data]	R_1 = 0.0359, wR_2 = 0.0873	R_1 = 0.0283, wR_2 = 0.0722	R_1 = 0.0336, wR_2 = 0.0634	R_1 = 0.0261, wR_2 = 0.0619	R_1 = 0.0273, wR_2 = 0.0690
Largest diff. peak/hole, (e/Å^{-3})	0.57/−0.55	0.53/−0.69	0.50/−0.48	0.47/−0.60	0.46/−0.65

Forsterite crystal structure (Figure 10a) was firstly described by Bragg and Brown [50]. Ideally, it consists of a hexagonal close packing of oxygen atoms, where one-half of octahedral interstices is occupied by $M1$ and $M2$ sites and one-eighth of tetrahedral interstices is occupied by $Z1$ sites. This structure can be described as heteropolyhedral framework consisting of stacking of identical sheets parallel to the (001) plane. The sheets, in turn, are based upon chains of edge-sharing $M1$ octahedra with adjacent $M2$ octahedra connected by vertex-shared $Z1$ tetrahedra (Figure 10b).

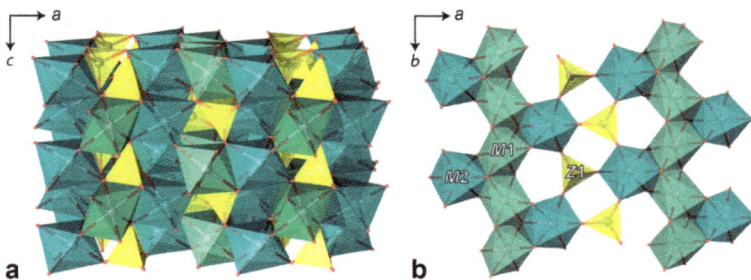

Figure 10. Crystal structure of forsterite (1): a general view (**a**) and heteropolyhedral sheet based on $M1$, $M2$ octahedra (green) and $Z1$ tetrahedra (yellow) projected along c axis (**b**).

There are few reports on non-equivalent distribution of magnesium and iron at octahedral $M1$ and $M2$ positions of the olivine-type structure [51–54]. This type of cation ordering does not reveal a correlation between cation distribution and genesis of the crystals, which is typical for amphiboles and pyroxenes [53,55,56]. In all forsterite analyzed, refined occupations of $M1$ and $M2$ sites provide domination of iron at the "large" $M2$ site (Table 7, Figure 11a). From the crystal-chemical point of view, the substitution $^{M2}Mg^{2+} \to {}^{M2}Fe^{2+}$ is more reasonable than $^{M1}Mg^{2+} \to {}^{M1}Fe^{2+}$ because the observed <*M2*-O> bond lengths of 2.128–2.135 Å are closer to ideal <Fe^{2+}-O> distance of 2.180 Å [57] than to <*M1*-O> distances (2.091–2.099 Å). For the same reason, *M1* site is theoretically more suitable for incorporation of Fe^{3+} (ideal <Fe^{3+}-O> bond length is 2.045 Å).

Table 7. Refined iron content of octahedral *M1* and *M2* sites for **1–5** samples (*apfu*).

Sample	M1	M2
1	0.020	0.035
2	0.057	0.070
3	0.080	0.084
4	0.047	0.057
5	0.04	0.065

The average <*M*-O> distances increase statistically irregularly with increasing content of Fe^{2+} (Figure 11b), which results in alignment of "small" *M1* and "large" *M2* octahedra in the Fo–Fa series (the average <*M1*-O> and <*M2*-O> distances are 2.094 and 2.130 Å correspondingly in forsterite, and 2.161 and 2.179 Å correspondingly in fayalite [58]). In case of sufficient difference between ionic radii of Mg^{2+} and incorporated elements [e.g., Fe^{3+} (−10.4%) or Mn^{2+}(+15.3%)], trivalent iron occupies firstly "large" *M2* site, and Mn^{2+} incorporates into "small" *M1* site [59,60]. Since the most significant difference in sizes of *M1* and *M2* polyhedra is observed in forsterite $Fo_{1.00}$–$Fo_{0.8}Fa_{0.2}$, incorporation of Fe^{3+} at the octahedral sites will have maximum impact on the *M1* and *M2* polyhedra volume in such forsterites.

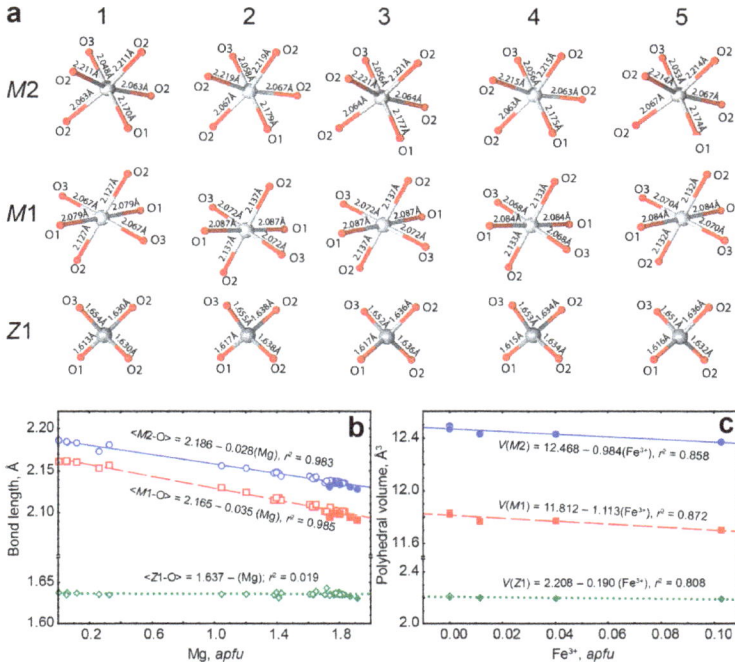

Figure 11. Geometry of coordination polyhedra in crystal structure of forsterites 1–5 (**a**), scatterplot of average <*M1*-O>, <*M2*-O> and <*Z1*-O> bond lengths against magnesium content in olivine according to [61–64], and present data (filled squares and circles) (**b**), scatterplot of polyhedral volumes (*M1*, *M2* and *Z1*) against ferric iron content (**c**).

In crystal structure of Kovdor's forsterite, the average <Z1-O> distances range between 1.631–1.637 Å, and scattering factors of Z1 sites vary in the range of 13.30–14.00 electrons per formula unit, which is in good agreement with full occupation of Z1 site by Si atoms only (Figure 11a). Polyhedral volumes of Z1 tetrahedra are actually unchanged, and this fact does not confirm incorporation of Fe^{3+} into tetrahedral sites (Figure 11c). The polyhedral volume decreasing with growth of Fe^{3+} content in our samples confirms incorporation of trivalent iron into octahedral M1 and M2 sites via laihulite-like substitution $(3Fe^{2+})_{oct} \leftrightarrow (2Fe^{3+} + \square)_{oct}$. Unconstrained refinement of forsterites with significant amounts of Fe^{3+} (samples 1 and 5) with full occupancies of octahedral (M11.00, M21.00) sites results in significant underestimation of Fe content. This fact also proves presence of vacancies at octahedral sites only. Consequently, data of crystal structure refinement is in good agreement with factor analysis and chemical data. Presence of vacations at octahedral sites of partially "oxidized" olivine questions applicability of distribution coefficients KD and associated variables [53,65,66].

5. Discussion

We would like to express that forsterite from peridotite has an important feature—bimineral exsolution lamellae of magnetite and diopside (Figures 3b and 12b). Such lamellae are not found in forsterite from phoscorite and carbonatites of the Kovdor alkaline-ultrabasic massif; but they are common in other (ultra)basic complexes where forsterite is enriched in Fe^{2+} and Ca [67,68]. In turn, Ca-rich forsterite crystallizes from melt with relatively low mg# value MgO/(MgO + FeO) and high contents of Ca and Na [69]. The fact that ultrabasic melt of the Kovdor massif was enriched in Ca is confirmed by co-crystallization of forsterite and diopside as well as by presence of numerous calcite inclusions ("drops") within forsterite grains (Figure 12b). In addition, Ca-rich foidolite (Figure 2d) and melilitholite formed later than peridotites contain calcite "drops" inside grains of all the main minerals including nepheline [24]. Low value of mg# in this melt causes crystallization of titanomagnetite in interstices of olivine grains (Figure 3a) up to formation of magnetite-rich peridotite (see Figure 1) that have economic importance [9].

Alkaline melts have relatively higher $Fe^{3+}/(Fe^{3+} + Fe^{2+})$ ratio, then non-alkaline melts [70,71], and Fe^{3+} is partly included in forsterite. The rock cooling causes exsolution of Fe^{3+}-rich forsterite into diopside and magnetite [72]:

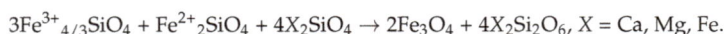

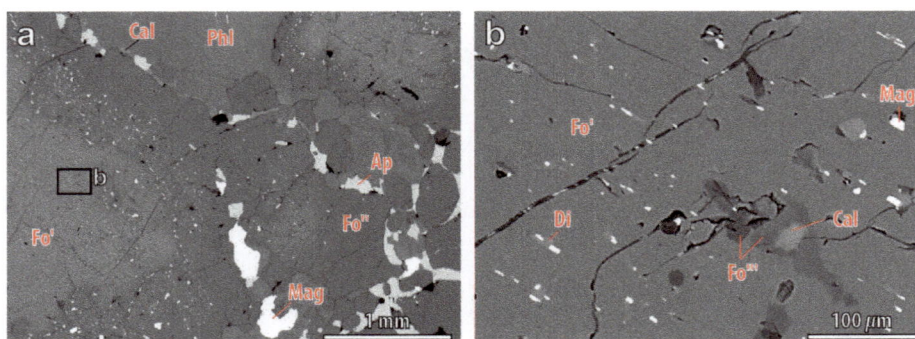

$$3Fe^{3+}_{4/3}SiO_4 + Fe^{2+}_2SiO_4 + 4X_2SiO_4 \rightarrow 2Fe_3O_4 + 4X_2Si_2O_6, X = Ca, Mg, Fe.$$

Figure 12. BSE-images of a rim of newly formed "phoscoritic" forsterite-II (Fo" = $Fo_{92}Fa_8$) around relict grain of "peridotitic" forsterite-I (Fo' = $Fo_{93}Fa_7$) in AF-phoscorite 937/114.6 (**a**) and the enlarged region demonstrating diopside lamellae and calcite "drops" in forsterite (**b**). Fo'" = $Fo_{95}Fa_5$. Other mineral abbreviations see in the Section 3.

The pyroxene phase acts as a sink for elements not compatible with the olivine structure, such as Ca. Pyroxene is formed as long as there is sufficient Ca present and the temperature is high enough for it to diffuse to the reaction front.

After crystallization of forsterite and diopside in peridotite, residual melt (melt-I) became comparatively rich in Si, Al, Na and K [21,73]. The melt-I emplaced into contact zone between peridotite stock and host gneisses and formed ring of foidolite and melilitolite. The next residual melt-II contained insignificant amount of Si (and Mg), but it was strongly enriched in Fe, Ca, C, P, F and also Nb, Zr, REE, Th and U. So, it was a real residual melt that was not caused by liquid immiscibility, because «ore-bearing rare metal carbonatites that are found in association with ultramafic and alkaline silicate rocks are likely to have formed from a residual liquid after extensive fractional crystallization of carbonated silicate magma rather than by silicate–carbonate liquid immiscibility» [74]. The possible existence of carbonatite magmas was experimentally confirmed in the system $CaO-CO_2-H_2O$ by Wyllie and Tuttle [75]. There were discovered the liquid immiscibility between albite-rich silicate and sodium carbonate-rich liquids [76,77], between ijolitic and alkali carbonatitic liquids in experiments on whole-rock compositions [78], and between alkali-poor silicate and carbonate liquids in the system albite/anorthite-calcite [79,80]. Studies in $NaAlSi_3O_8-NaAlSiO_4-CaCO_3-H_2O$ system by [81] and in $Mg_2SiO_4-CaCO_3-Ca(OH)_2$ system by [82,83] indicated that carbonatite magmas could be produced by crystal fractionation of silicate magmas of appropriate compositions (for example, SiO_2-undersaturated alkalic liquids with H_2O and CO_2). For other compositions, this is precluded by the presence of thermal barriers between high-temperature liquids precipitating silicates, and low-temperature liquids precipitating carbonates and hydrous minerals [82,83]. Reasons of phosphorus concentration in the residual melt-II include enrichment of the melt in Fe^{3+} with further formation of stable complex $Fe^{3+}(PO_4)^{3-}$ [84,85] as well as high content of Ca and Mg forming stable complexes $(Ca, Mg)-PO_4$ [86]. We believe that residual melt-II intruded into the foidolite–peridotite contact, and rapidly crystallized from the pipe walls towards its axial zone due to both cooling and blast-like degassing [24]. On this reason, hydroxylapatite co-crystallized with forsterite contains numerous liquid-vapor inclusions [87].

The crystallization of phoscorite–carbonatite rock series can be considered in systems that are extensions of well-studied $CaO-CO_2-H_2O$ [75]. The crystallization of apatite from low-temperature melts in $CaO-CaF_2-P_2O_5-H_2O$ and $CaO-P_2O_5-CO_2-H_2O$ systems was investigated by Biggar [88]. There is a large field for the primary crystallization of apatite in the ternary systems $Ca_3(PO_4)_2-CaF_2-Ca(OH)_2$ and $Ca_3(PO_4)-CaCO_3-Ca(OH)_2$, and the liquid precipitating the apatite persists down to 675 °C and 654 °C at the respective ternary eutectics [89]. Addition of other components to the system $CaO-CO_2-H_2O$ produces suitable conditions for the crystallization of other calcium-bearing minerals from low-temperature liquids in the presence of an aqueous vapor phase. Fields for the crystallization of hydrated and carbonated calcium silicates are found in the system $CaO-SiO_2-CO_2-H_2O$ [90].

According to Moussallam et al. [91], silica and carbonate form two separate subnetworks. Phosphorus in silicate melts also forms separate clusters that confine iron within stable complexes $Fe^{3+}(PO_4)$ [86]. It seems likely that structure of the phoscorite melt was constituted by interconnected subnetworks of SiO_4-tetrahedra and CO_3-triangles with local domains of PO_4-tetrahedra and $Fe^{3+}(PO_4)$-clusters, without a liquid immiscibility [92]. Interaction of this melt with silicate rock launched forsterite crystallization, sometimes with grains of primary "peridotitic" forsterite as seed crystals (Figure 12a). Exponential distribution of forsterite grain size (see Figure 5) shows slower diffusion rates of magnesium and silica, which seems to be the main factor of size-independent (constant) crystal growth [93]. At the contact with the carbonates (predominantly calcite), forsterite grains of phoscorite and phoscorite-related carbonatite became larger (see Figure 4b) and well-shaped (see Figure 3h) due to collective recrystallization. Similar processes are typical for magnetite [12].

Crystallization of forsterite from residual melt-II near the pipe wall and top resulted in depletion of the melt in Mg, which launched apatite precipitation with the melt cooling. In turn, formation of apatite destroyed $Fe^{3+}(PO_4)$-complexes and launched magnetite crystallization. Consequently, crystallization front moved rapidly from the pipe wall towards its axis accompanied by separation of volatiles. This process resulted in concentration of residual carbonate melt-III in the pipe axial zone.

Carbonate melts are low-viscous and remain interconnected up to 0.05 wt % melt [94]. In addition, silicate melt selectively wets the grain-edge channels between solid phases, excluding the carbonate melt to the center of melt pockets, away from grain edges [95]. These features of carbonate melts enable us to understand why forsterite grains can crystallize from carbonate-rich melt according to low-rate diffusion mechanism, and why they obtain predominant orientation in carbonate melt flow (see Figure 5).

Water solubility in carbonate melts is significantly higher than in alkaline silicate melts, reaching values of nearly 15 wt % at 100 MPa and 900 °C [96]. The depth of -200 to -400 m is probably the interval of separation of water, CO_2, F and other volatiles from phoscorite–carbonatite melt. These volatiles reacted with early crystallized phoscorites and phoscorite-related carbonatites in the pipe axial zone, with formation of later water/fluor-containing apo-forsterite minerals (phlogopite, clinochlore, clinohumite, etc.—see Figure 6). The final products of this process were staffelite breccias (fragments of altered phoscorite and carbonatites cemented by colloform carbonate-rich fluorapatite) filling several funnels in apical part of the phoscorite–carbonatite pipe axial zone [24,97].

6. Conclusions

Three-D mineralogical mapping was used to establish spatial distribution of forsterite content, morphology, grain size, composition and alteration products within the Kovdor phoscorite–carbonatite pipe. This work pursues our study of "through minerals" of the Kovdor complex and enables us to make some interesting conclusions:

(1) Forsterite is the earliest mineral of both peridotite and phoscorite–carbonatite complexes, and its crystallization governed the further evolution of corresponding magmatic systems. Thus, crystallization of forsterite from the Ca-Fe-rich peridotite melt produces Si-Al-Na-K-rich residual melt-I corresponding to foidolite–melilitolite. In turn, consolidation of foidolite and melilitolite produced Fe-Ca-C-P-F-rich residual melt-II that emplaced in silicate rocks as the phoscorite–carbonatite pipe. Phoscorite crystallization started from forsterite, which launched destruction of silicate-carbonate-ferriphosphate subnetworks of the melt followed by precipitation of apatite and magnetite from the pipe wall to its axis with formation of carbonatite melt-III in the pipe axial zone;

(2) Growth of forsterite grains from phoscorite–carbonatite melt was diffusion-limited, which causes constant growth rate of each grain and exponential distribution of grain size;

(3) Chemical composition of forsterite in phoscorite–carbonatite pipe is determined by two schemas of isomorphism: $Mg^{2+} + 2Fe^{3+} \leftrightarrow 4(Fe, Mn)^{2+}$ and $Mg^{2+} + 2Fe^{3+} \leftrightarrow 2(Fe, Mn, Ni)^{2+} + Ti^{4+}$. Marginal forsterite-rich phoscorite consists of Fe^{2+}-Mn-Ni-Ti-rich forsterite similar to olivine from peridotite, intermediate low-carbonate magnetie-rich phoscorite includes $Mg-Fe^{3+}$-rich forsterite, and axial carbonate-rich phoscorite and carbonatites contain Fe^{2+}-Mn-rich forsterite;

(4) Trivalent iron incorporates into forsterite by scheme $(3Fe^{2+})_{oct} \rightarrow 2Fe^{3+} + (\square)_{oct}$ that reflects redox conditions of the rock formation causing significant agreement between compositions of apatite, magnetite, calcite and forsterite;

(5) Incorporation of trivalent iron at the octahedral $M1$ and $M2$ sites decreases the volume of these polyhedra, while volume of tetrahedral set does not change. Thus, the assumed substitution $(4Fe^{2+})_{oct} + (4Si^{4+})_{tet} \rightarrow (4Fe^{3+})_{oct} + (3Si^{4+} + \square)_{tet}$ proposed by D. E. Janney and J. F. Banfield [49] was not confirmed. Our data show that laihunite-like isomorphism is more common in forsterite than it was considered to be.

Supplementary Materials: The following are available online at http://www.mdpi.com/2075-163X/8/6/260/s1, Table S1: Fractional atomic coordinates ($\times 10^4$) and equivalent isotropic displacement parameters ($\text{Å}^2 \times 10^3$) for Forsterite_1. Table S2: Anisotropic displacement parameters ($\text{Å}^2 \times 10^3$) for Forsterite_1. Table S3: Bond lengths for Forsterite_1, Table S4: Atomic occupancy for Forsterite_1. Table S5: Fractional atomic coordinates ($\times 10^4$) and equivalent isotropic displacement parameters ($\text{Å}^2 \times 10^3$) for Forsterite_2. Table S6: Anisotropic displacement parameters ($\text{Å}^2 \times 10^3$) for Forsterite_2. Table S7: Bond lengths for Forsterite_2, Table S8: Atomic occupancy for Forsterite_2, Table S9: Fractional atomic coordinates ($\times 10^4$) and equivalent isotropic displacement parameters ($\text{Å}^2 \times 10^3$) for Forsterite_3. Table S10: Anisotropic displacement parameters ($\text{Å}^2 \times 10^3$) for Forsterite_3. Table S11: Bond lengths for Forsterite_3, Table S12: Atomic occupancy for Forsterite_3, Table S13: Fractional atomic coordinates ($\times 10^4$) and equivalent isotropic displacement parameters ($\text{Å}^2 \times 10^3$) for Forsterite_4. Table S14: Anisotropic displacement parameters ($\text{Å}^2 \times 10^3$) for Forsterite_4. Table S15: Bond lengths for Forsterite_4, Table S16: Atomic occupancy for Forsterite_4, Table S17: Fractional atomic coordinates ($\times 10^4$) and equivalent isotropic displacement parameters ($\text{Å}^2 \times 10^3$) for Forsterite_5. Table S18: Anisotropic displacement parameters ($\text{Å}^2 \times 10^3$) for Forsterite_5. Table S19: Bond lengths for Forsterite_5, Table S20: Atomic occupancy for Forsterite_5.

Author Contributions: J.A.M. designed the experiments, carried out petrographical investigations and crystal size distribution analyses, and wrote the manuscript. G.Y.I. designed the experiments, performed statistical investigations, and reviewed the manuscript. A.O.K. performed geostatistical investigation, drew maps and took samples. Y.A.P. and A.V.B. took BSE images and performed electron microscope investigations. T.L.P. performed crystallographic investigations and formulated crystal-chemical conclusions. V.N.Y. and N.G.K. conceived of the work, took and prepared samples. P.M.G. drew maps. All authors discussed the manuscript.

Funding: The research is supported by the Russian Science Foundation, grant 16-17-10173.

Acknowledgments: Samples were taken during exploration of deep levels of the Kovdor deposit implemented by JSC Kovdorskiy GOK in 2007–11. X-ray crystal studies were carried out with the equipment provided by the X-ray Diffraction Centre of Saint-Petersburg State University. The comments by anonymous reviewers helped us to significantly improve this paper.

Conflicts of Interest: The authors declare no conflict of interest. The founding sponsors had no role in the design of the study; in the collection, analyses, or interpretation of data; in the writing of the manuscript, and in the decision to publish the results.

References

1. Igneous Rocks. *A Classification and Glossary of Terms. Recommendations of the International Union of Geological Sciences Subcommission on the Systematics of Igneous Rocks*; Le Maitre, R.W., Ed.; Cambridge University Press: New York, NY, USA, 2002; ISBN 9780521662154.
2. Jaireth, S.; Hoatson, D.M.; Miezitis, Y. Geological setting and resources of the major rare-earth-element deposits in Australia. *Ore Geol. Rev.* **2014**, *62*, 72–128. [CrossRef]
3. Lazareva, E.V.; Zhmodik, S.M.; Dobretsov, N.L.; Tolstov, A.V.; Shcherbov, B.L.; Karmanov, N.S.; Gerasimov, E.Y.; Bryanskaya, A.V. Main minerals of abnormally high-grade ores of the Tomtor deposit (Arctic Siberia). *Russ. Geol. Geophys.* **2015**, *56*, 844–873. [CrossRef]
4. Liu, Y.-L.; Ling, M.-X.; Williams, I.S.; Yang, X.-Y.; Wang, C.Y.; Sun, W. The formation of the giant Bayan Obo REE-Nb-Fe deposit, North China, Mesoproterozoic carbonatite and overprinted Paleozoic dolomitization. *Ore Geol. Rev.* **2018**, *92*, 73–83. [CrossRef]
5. Mackay, D.A.R.; Simandl, G.J. Geology, market and supply chain of niobium and tantalum—A review. *Miner. Depos.* **2014**, *49*, 1025–1047. [CrossRef]
6. Mitchell, R.H. Primary and secondary niobium mineral deposits associated with carbonatites. *Ore Geol. Rev.* **2015**, *64*, 626–641. [CrossRef]
7. Smith, M.P.; Moore, K.; Kavecsánszki, D.; Finch, A.A.; Kynicky, J.; Wall, F. From mantle to critical zone: A review of large and giant sized deposits of the rare earth elements. *Geosci. Front.* **2016**. [CrossRef]
8. Wall, F.; Zaitsev, A.N. (Eds.) *Phoscorites and Carbonatites from Mantle to Mine: The Key Example of the Kola Alkaline Province*; Mineralogical Society: London, UK, 2004.
9. Ivanyuk, G.Y.; Yakovenchuk, V.N.; Pakhomovsky, Y.A. *Kovdor*; Laplandia Minerals: Apatity, Russia, 2002; ISBN 5900395413.
10. Mikhailova, J.A.; Kalashnikov, A.O.; Sokharev, V.A.; Pakhomovsky, Y.A.; Konopleva, N.G.; Yakovenchuk, V.N.; Bazai, A.V.; Goryainov, P.M.; Ivanyuk, G.Y. 3D mineralogical mapping of the Kovdor phoscorite–carbonatite complex (Russia). *Miner. Depos.* **2016**, *51*, 131–149. [CrossRef]

11. Ivanyuk, G.Y.; Kalashnikov, A.O.; Pakhomovsky, Y.A.; Mikhailova, J.A.; Yakovenchuk, V.N.; Konopleva, N.G.; Sokharev, V.A.; Bazai, A.V.; Goryainov, P.M. Economic minerals of the Kovdor baddeleyite-apatite-magnetite deposit, Russia: Mineralogy, spatial distribution and ore processing optimization. *Ore Geol. Rev.* **2016**, *77*, 279–311. [CrossRef]

12. Ivanyuk, G.Y.; Kalashnikov, A.O.; Pakhomovsky, Y.A.; Bazai, A.V.; Goryainov, P.M.; Mikhailova, J.A.; Yakovenchuk, V.N.; Konopleva, N.G. Subsolidus Evolution of the Magnetite-Spinel-Ulvöspinel Solid Solutions in the Kovdor phoscorite–carbonatite Complex, NW Russia. *Minerals* **2017**, *7*, 215. [CrossRef]

13. Kalashnikov, A.O.; Yakovenchuk, V.N.; Pakhomovsky, Y.A.; Bazai, A.V.; Sokharev, V.A.; Konopleva, N.G.; Mikhailova, J.A.; Goryainov, P.M.; Ivanyuk, G.Y. Scandium of the Kovdor baddeleyite–apatite–magnetite deposit (Murmansk Region, Russia): Mineralogy, spatial distribution, and potential resource. *Ore Geol. Rev.* **2016**, *72*, 532–537. [CrossRef]

14. Bell, K.; Kjarsgaard, B.A.; Simonetti, A. Carbonatites—Into the Twenty-First Century. *J. Petrol.* **1998**, *39*, 1839–1845. [CrossRef]

15. Gittins, J.; Harmer, R.E.; Barker, D.S. The bimodal composition of carbonatites: Reality or misconception? *Lithos* **2005**, *85*, 129–139. [CrossRef]

16. Mitchell, R.H. Carbonatites and carbonatites and carbonatites. *Can. Mineral.* **2005**, *43*, 2049–2068. [CrossRef]

17. Woolley, A.R.; Kjarsgaard, B.A. Paragenetic types of carbonatite as indicated by the diversity and relative abundances of associated silicate rocks: Evidence from a global database. *Can. Mineral.* **2008**, *46*, 741–752. [CrossRef]

18. Woolley, A.R.; Bailey, D.K. The crucial role of lithospheric structure in the generation and release of carbonatites: Geological evidence. *Mineral. Mag.* **2012**, *76*, 259–270. [CrossRef]

19. Jones, A.P.; Genge, M.; Carmody, L. Carbonate Melts and Carbonatites. *Rev. Mineral. Geochem.* **2013**, *75*, 289–322. [CrossRef]

20. Russell, H.D.; Hiemstra, S.A.; Groeneveld, D. The mineralogy and petrology of the Carbonatite at Loolekop, Eastern Transvaal. *S. Afr. J. Geol.* **1954**, *57*, 197–208.

21. Kukharenko, A.A.; Orlova, M.P.; Bulakh, A.G.; Bagdasarov, E.A.; Rimskaya-Korsakova, O.M.; Nefedov, E.I.; Ilyinsky, G.A.; Sergeev, A.S.; Abakumova, N.B. *Caledonian Complex of Ultrabasic, Alkaline Rocks and Carbonatites of Kola Peninsula and Northern Karelia (Geology, Petrology, Mineralogy and Geochemistry) (in Russian)*; Nedra: Moscow, Russia, 1965.

22. Yegorov, L.S. Phoscorites of the Maymecha-Kotuy ijolite-carbonatite association. *Int. Geol. Rev.* **1993**, *35*, 346–358. [CrossRef]

23. Rimskaya-Korsakova, O.M. On Question about Genesis of the Kovdor Iron-Ore Deposit. In *Problems of Magmatism and Metamorphism (in Russian)*; Leningrad State University Publishing: Leningrad, Russia, 1963; pp. 125–142.

24. Kalashnikov, A.O.; Ivanyuk, G.Y.; Mikhailova, J.A.; Sokharev, V.A. Approach of automatic 3D geological mapping: The case of the Kovdor phoscorite–carbonatite complex, NW Russia. *Sci. Rep.* **2017**, *7*, 1–13. [CrossRef] [PubMed]

25. Ternovoy, V.I. *Carbonatite Massifs and Their Mineral Resources (in Russian)*; Leningrad State University: Leningrad, Russia, 1977.

26. Afanasyev, B. *Mineral Resources of Alkaline-Ultrabasic Massifs of the Kola Peninsula (in Russian)*; Roza Vetrov Publishing: Saint-Petersburg, Russia, 2011.

27. Dunaev, V.A. Structure of the Kovdor deposit (in Russian). *Geol. Ore Depos.* **1982**, *3*, 28–36.

28. Kapustin, Y.L. *Mineralogy of Carbonatites*; Amerind Publishing: New Delhi, India, 1980.

29. Krasnova, N.I.; Kopylova, L.N. The Geologic Basis for Mineral-Technological Mapping at the Kovdor Ore Deposit. *Int. Geol. Rev.* **1988**, *30*, 307–319. [CrossRef]

30. Krasnova, N.I.; Petrov, T.G.; Balaganskaya, E.G.; García, D.; Moutte, J.; Zaitsev, A.N.; Wall, F. Introduction to phoscorites: Occurrence, composition, nomenclature and petrogenesis. In *Phoscorites and Carbonatites from Mantle to Mine: The Key Example of the Kola Alkaline Province*; Zaitsev, A.N., Wall, F., Eds.; Mineralogical Society: London, UK, 2004; pp. 43–72. ISBN 0-903056-22-4.

31. Bayanova, T.B.; Kirnarsky, Y.M.; Levkovich, N.V. U-Pb dating of baddeleyite from Kovdor massif (in Russian). *Dokl. Earth Sci.* **1997**, *356*, 509–511.

32. Amelin, Y.; Zaitsev, A.N. Precise geochronology of phoscorites and carbonatites: The critical role of U-series disequilibrium in age interpretations. *Geochim. Cosmochim. Acta* **2002**, *66*, 2399–2419. [CrossRef]

33. Rodionov, N.V.; Belyatsky, B.V.; Antonov, A.V.; Kapitonov, I.N.; Sergeev, S.A. Comparative in-situ U–Th–Pb geochronology and trace element composition of baddeleyite and low-U zircon from carbonatites of the Palaeozoic Kovdor alkaline–ultramafic complex, Kola Peninsula, Russia. *Gondwana Res.* **2012**, *21*, 728–744. [CrossRef]

34. Rimskaya-Korsakova, O.M.; Krasnova, N.I. *Geology of Deposits of the Kovdor Massif (in Russian)*; St. Petersburg University Press: Saint Petersburg, Russia, 2002.

35. Shats, L.; Sorokina, I.; Kalinkin, M.; Kornyushin, A. *Report on Geophysical Works Made by the Kovdor Geological Party in the Area of Kovdor in 1966 (in Russian)*; Archives of the Natural Reserves Department of the Murmansk region: Apatity, Russia, 1967.

36. ImageJ, Open Source Image Processing Software. Available online: http://imagej.net/ (accessed on 19 June 2018).

37. Sheldrick, G.M. A short history of SHELX. *Acta Crystallogr. Sect. A Found. Crystallogr.* **2008**, *64*, 112–122. [CrossRef] [PubMed]

38. Dolomanov, O.V.; Bourhis, L.J.; Gildea, R.J.; Howard, J.A.K.; Puschmann, H. OLEX2: A complete structure solution, refinement and analysis program. *J. Appl. Crystallogr.* **2009**, *42*, 339–341. [CrossRef]

39. Agilent CrysAlis PRO. 2014. Available online: https://www.rigaku.com/en/products/smc/crysalis (accessed on 19 June 2018).

40. Momma, K.; Izumi, F. VESTA 3 for three-dimensional visualization of crystal, volumetric and morphology data. *J. Appl. Crystallogr.* **2011**, *44*, 1272–1276. [CrossRef]

41. Putz, H.; Brandenburg, K. *Diamond–Crystal and Molecular Structure Visualization*; Crystal Impact GbR: Bonn, Germany, 2012.

42. Dolivo-Dobrovolsky, D.D. MINAL, Free Software. Available online: http://www.dimadd.ru (accessed on 8 July 2013).

43. StatSoft Inc, Statistica 8. Available online: www.statsoft.ru (accessed on 19 June 2018).

44. TableCurve 2D. Available online: www.sigmaplot.co.uk/products/tablecurve2d/tablecurve2d.php (accessed on 19 June 2018).

45. Micromine Pty Ltd. Micromine 16.1. Available online: https://www.micromine.com/ (accessed on 19 June 2018).

46. Veksler, I.V.; Nielsen, T.F.D.; Sokolov, S.V. Mineralogy of Crystallized Melt Inclusions from Gardiner and Kovdor Ultramafic Alkaline Complexes: Implications for Carbonatite Genesis. *J. Petrol.* **1998**, *39*, 2015–2031. [CrossRef]

47. Tarasenko, Y.; Litsarev, M.A.; Tretyakova, L.I.; Vokhmentsev, A.Y. Chrysolite of the Kovdor phlogopite deposit (in Russian). *Izv. AN SSSR Seriya Geol.* **1986**, *9*, 67–80.

48. Zeira, S.; Hafner, S.S. The location of Fe^{3+} ions in forsterite (Mg_2SiO_4). *Earth Planet. Sci. Lett.* **1974**, *21*, 201–208. [CrossRef]

49. Janney, D.E.; Banfield, J.F. Distribution of cations and vacancies and the structure of defects in oxidized intermediate olivine by atomic-resolution TEM and image simulation. *Am. Mineral.* **1998**, *83*, 799–810. [CrossRef]

50. Bragg, W.L.; Brown, G.B. XXX. Die Struktur des Olivins. *Z. Krist. Cryst. Mater.* **1926**, *63*, 538–556. [CrossRef]

51. Brown, G.E. *Crystal Chemistry of the Olivines*; Virginia Polytechnic Institute and State University: Blacksburg, VA, USA, 1970.

52. Huggins, F.E. Cation order in olivines: Evidence from vibrational spectra. *Chem. Geol.* **1973**, *11*, 99–108. [CrossRef]

53. Nover, G.; Will, G. Structure refinements of seven natural olivine crystals and the influence of the oxygen partial pressure on the cation distribution. *Z. Krist. Cryst. Mater.* **1981**, *155*. [CrossRef]

54. Francis, C.A. New data on the forsterite-tephroite series. *Am. Mineral.* **1985**, *70*, 568–575.

55. Seifert, F.A.; Virgo, D. Kinetics of the Fe^{2+}-Mg, Order-Disorder Reaction in Anthophyllites: Quantitative Cooling Rates. *Science* **1975**, *188*, 1107–1109. [CrossRef] [PubMed]

56. Seifert, F.; Virgo, D. Temperature dependence of intracrystalline Fe^{2+}-Mg distribution in a natural anthophyllite. In *Carnegie Institute of Washington Year Book*; Carnegie Institute of Washington: Washington, DC, USA, 1974; pp. 405–411.

57. Shannon, R.D. Revised effective ionic radii and systematic studies of interatomic distances in halides and chalcogenides. *Acta Crystallogr. Sect. A* **1976**, *32*, 751–767. [CrossRef]

58. Riekel, C.; Weiss, A. Cation-Ordering in Synthetic $Mg_{2-x}Fe_xSiO_4$-Olivines. *Z. Naturforsch. B* **1978**, *33*. [CrossRef]

59. Redfern, S.A.T.; Henderson, C.M.B.; Knight, K.S.; Wood, B.J. High-temperature order-disorder in $(Fe_{0.5}Mn_{0.5})_2SiO_4$ and $(Mg_{0.5}Mn_{0.5})_2SiO_4$ olivines: An in situ neutron diffraction study. *Eur. J. Mineral.* **1997**, *9*, 287–300. [CrossRef]

60. Shen, B.; Tamada, O.; Kitamura, M.; Morimoto, N. Superstructure of laihunite-3M ($\square_{0.40}Fe^{2+}_{0.80}Fe^{3+}_{0.80}SiO_4$). *Am. Mineral.* **1986**, *71*, 1455–1460.

61. Brown, G.E.; Prewitt, C.T. High-temperature crystal chemistry of hortonolite. *Am. Mineral.* **1973**, *58*, 577–587.

62. Princivalle, F.; Secco, L. Crystal structure refinement of 13 olivines in the forsterite-fayalite series from volcanic rocks and ultramafic nodules. *TMPM* **1985**, *34*, 105–115. [CrossRef]

63. Motoyama, T.; Matsumoto, T. The crystal structures and the cation distributions of Mg and Fe of natural olivines. *Miner. J.* **1989**, *14*, 338–350. [CrossRef]

64. Princivalle, F. Influence of temperature and composition on Mg-Fe^{2+} intracrystalline distribution in olivines. *Mineral. Petrol.* **1990**, *43*, 121–129. [CrossRef]

65. Heinemann, R.; Kroll, H.; Kirfel, A.; Barbier, B. Order and anti-order in olivine III: Variation of the cation distribution in the Fe,Mg olivine solid solution series with temperature and composition. *Eur. J. Mineral.* **2007**, *19*, 15–27. [CrossRef]

66. Kroll, H.; Kirfel, A.; Heinemann, R.; Barbier, B. Volume thermal expansion and related thermophysical parameters in the Mg, Fe olivine solid-solution series. *Eur. J. Mineral.* **2012**, *24*, 935–956. [CrossRef]

67. Ren, Y.; Chen, F.; Yang, J.; Gao, Y. Exsolutions of Diopside and Magnetite in Olivine from Mantle Dunite, Luobusa Ophiolite, Tibet, China. *Acta Geol. Sin. Engl. Ed.* **2010**, *82*, 377–384. [CrossRef]

68. Markl, G.; Marks, M.A.W.; Wirth, R. The influence of T, $aSiO_2$, and fO_2 on exsolution textures in Fe-Mg olivine: An example from augite syenites of the Ilimaussaq Intrusion, South Greenland. *Am. Mineral.* **2001**, *86*, 36–46. [CrossRef]

69. Libourel, G. Systematics of calcium partitioning between olivine and silicate melt: Implications for melt structure and calcium content of magmatic olivines. *Contrib. Mineral. Petrol.* **1999**, *136*, 63–80. [CrossRef]

70. Carmichael, I.S.E.; Nicholls, J. Iron-titanium oxides and oxygen fugacities in volcanic rocks. *J. Geophys. Res.* **1967**, *72*, 4665–4687. [CrossRef]

71. Mysen, B.O.; Richet, P. *Silicate Glasses and Melts. Properties and Structure*; Elsevier: New York, NY, USA, 2005; ISBN 0-444-52011-2.

72. Moseley, D. Symplectic exsolution in olivine. *Am. Mineral.* **1984**, *69*, 139–153.

73. Wyllie, P.J.; Baker, M.B.; White, B.S. Experimental boundaries for the origin and evolution of carbonatites. *Lithos* **1990**, *26*, 3–19. [CrossRef]

74. Veksler, I.V.; Petibon, C.; Jenner, G.A.; Dorfman, A.M.; Dingwell, D.B. Trace Element Partitioning in Immiscible Silicate-Carbonate Liquid Systems: An Initial Experimental Study Using a Centrifuge Autoclave. *J. Petrol.* **1998**, *39*, 2095–2104. [CrossRef]

75. Wyllie, P.J.; Tuttle, O.F. The System CaO-CO_2-H_2O and the Origin of Carbonatites. *J. Petrol.* **1960**, *1*, 1–46. [CrossRef]

76. Koster van Groos, A.F.; Wyllie, P.J. Liquid immiscibility in the system Na_2O-Al_2O_3-SiO_2-CO_2 at pressures to 1 kilobar. *Am. J. Sci.* **1966**, *264*, 234–255. [CrossRef]

77. Koster van Groos, A.F.; Wyllie, P.J. Liquid immiscibility in the join $NaAlSi_3O_8$-Na_2CO_3-H_2O and its bearing on the genesis of carbonatites. *Am. J. Sci.* **1968**, *266*, 932–967. [CrossRef]

78. Freestone, I.C.; Hamilton, D.L. The role of liquid immiscibility in the genesis of carbonatites? An experimental study. *Contrib. Mineral. Petrol.* **1980**, *73*, 105–117. [CrossRef]

79. Kjarsgaard, B.A.; Hamilton, D.L. Liquid immiscibility and the origin of alkali-poor carbonatites. *Mineral. Mag.* **1988**, *52*, 43–55. [CrossRef]

80. Kjarsgaard, B.A.; Hamilton, D.L. The genesis of carbonatites by liquid immiscibility. In *Carbonatites: Genesis and Evolution*; Bell, K.E., Ed.; Unwin Hyman: London, UK, 1989; pp. 388–404.

81. Watkinson, D.H.; Wyllie, P.J. Experimental Study of the Composition Join $NaAlSiO_4$-$CaCO_3$-H_2O and the Genesis of Alkalic Rock—Carbonatite Complexes. *J. Petrol.* **1971**, *12*, 357–378. [CrossRef]

82. Franz, G.W. *Melting Relationships in the System CaO-MgO-SiO_2-CO_2-H_2O: A Study of Synthetic Kimberlites*; The Pennsylvania State University: State College, PA, USA, 1965.

83. Franz, G.W.; Wyllie, P.J. Experimental Studies in the system CaO-MgO-SiO$_2$-CO$_2$-H$_2$O. In *Ultramafic and Related Rocks*; Wyllie, P.J., Ed.; John Wiley and Sons: New York, NY, USA, 1967; pp. 323–326.

84. Toplis, M.J.; Libourel, G.; Carroll, M.R. The role of phosphorus in crystallisation processes of basalt: An experimental study. *Geochim. Cosmochim. Acta* **1994**, *58*, 797–810. [CrossRef]

85. Mysen, B.O. Iron and phosphorus in calcium silicate quenched melts. *Chem. Geol.* **1992**, *98*, 175–202. [CrossRef]

86. Mysen, B.O.; Ryerson, F.J.; Virgo, D. The structural role of phosphorus in silicate melts. *Am. Mineral.* **1981**, *66*, 106–117.

87. Mikhailova, J.A.; Krasnova, N.I.; Krezer, Y.L.; Wall, F. Inclusions in minerals of the Kovdor intrusion of ultrabasic, alkaline rocks and carbonatites as indicators of the endogenic evolution processes. In *Deep-Seated Magmatism, Magmatic Sources and the Problem of Plumes (in Russian)*; Vladykin, N.V., Ed.; Siberian Branch of the Russian Academy of Sciences: Irkutsk/Valdivostok, Russia, 2002; pp. 296–320.

88. Biggar, G.M. *High Pressure High Temperature Phase Equilibrium Studies in the System CaO-CaF$_2$-P$_2$O$_5$-H$_2$O-CO$_2$ with Special Reference to the Apatites*; University of Leeds: Leeds, UK, 1962.

89. Wyllie, P.J.; Biggar, G.M. Fractional Crystallization in the "Carbonatite Systems" CaO-MgO-CO$_2$-H$_2$O and CaO-CaF$_2$-P$_2$O$_5$-CO$_2$-H$_2$O. In *Papers and Proceedings of the Fourth General Meeting. International Mineralogical*; International Mineralogical Association: Gauteng, South Africa, 1966; pp. 92–105.

90. Wyllie, P.; Haas, J. The system CaO-SiO$_2$-CO$_2$-H$_2$O: 1. Melting relationships with excess vapor at 1 kilobar pressure. *Geochim. Cosmochim. Acta* **1965**, *29*, 871–892. [CrossRef]

91. Moussallam, Y.; Florian, P.; Corradini, D.; Morizet, Y.; Sator, N.; Vuilleumier, R.; Guillot, B.; Iacono-Marziano, G.; Schmidt, B.C.; Gaillard, F. The molecular structure of melts along the carbonatite-kimberlite-basalt compositional joint: CO$_2$ and polymerisation. *Earth Planet. Sci. Lett.* **2016**, *434*, 129–140. [CrossRef]

92. Klemme, S. Experimental constraints on the evolution of iron and phosphorus-rich melts: Experiments in the system CaO-MgO-Fe$_2$O$_3$-P$_2$O$_5$-SiO$_2$-H$_2$O-CO$_2$. *J. Mineral. Petrol. Sci.* **2010**, *105*, 1–8. [CrossRef]

93. Eberl, D.D.; Kile, D.E.; Drits, V.A. On geological interpretations of crystal size distributions: Constant vs. proportionate growth. *Am. Mineral.* **2002**, *87*, 1235–1241. [CrossRef]

94. Minarik, W.G.; Watson, E.B. Interconnectivity of carbonate melt at low melt fraction. *Earth Planet. Sci. Lett.* **1995**, *133*, 423–437. [CrossRef]

95. Minarik, W.G. Complications to Carbonate Melt Mobility due to the Presence of an Immiscible Silicate Melt. *J. Petrol.* **1998**, *39*, 1965–1973. [CrossRef]

96. Keppler, H. Water solubility in carbonatite melts. *Am. Mineral.* **2003**, *88*, 1822–1824. [CrossRef]

97. Krasnova, N.I. Geology, mineralogy and problems of genesis of apatite-francolite rocks of the Kovdor massif (in Russian). In *Composition of Phosphorites*; Nauka: Novosibirsk, Russia, 1979; pp. 164–172.

minerals

Article

Three-D Mineralogical Mapping of the Kovdor Phoscorite-Carbonatite Complex, NW Russia: II. Sulfides

Gregory Yu. Ivanyuk [1,2,*], Yakov A. Pakhomovsky [1,2], Taras L. Panikorovskii [1], Julia A. Mikhailova [1,2], Andrei O. Kalashnikov [2], Ayya V. Bazai [1,2], Victor N. Yakovenchuk [1,2], Nataly G. Konopleva [1] and Pavel M. Goryainov [2]

[1] Nanomaterials Research Centre of Kola Science Centre, Russian Academy of Sciences, 14 Fersman Street, Apatity 184209, Russia; pakhom@geoksc.apatity.ru (Y.A.P.); taras.panikorovsky@spbu.ru (T.L.P.); ylya_korchak@mail.ru (J.A.M.); bazai@geoksc.apatity.ru (A.V.B.); yakovenchuk@geoksc.apatity.ru (V.N.Y.); konoplyova55@mail.ru (N.G.K.)

[2] Geological Institute of Kola Science Centre, Russian Academy of Sciences, 14 Fersman Street, Apatity 184209, Russia; kalashnikov@geoksc.apatity.ru (A.O.K.); pgor@geoksc.apatity.ru (P.M.G.)

* Correspondence: g.ivanyuk@gmail.com; Tel.: +7-81555-79531

Received: 30 May 2018; Accepted: 5 July 2018; Published: 9 July 2018

Abstract: The world largest phoscorite-carbonatite complexes of the Kovdor (Russia) and Palabora (South Africa) alkaline-ultrabasic massifs have comparable composition, structure and metallogenic specialization, and can be considered close relatives. Distribution of rock-forming sulfides within the Kovdor phoscorite-carbonatite complex reflects gradual concentric zonation of the pipe: pyrrhotite with exsolution inclusions of pentlandite in marginal (apatite)-forsterite phoscorite, pyrrhotite with exsolution inclusions of cobaltpentlandite in intermediate low-carbonate magnetite-rich phoscorite and chalcopyrite (±pyrrhotite with exsolution inclusions of cobaltpentlandite) in axial carbonate-rich phoscorite and phoscorite-related carbonatite. Chalcopyrite (with relicts of earlier bornite and exsolution inclusions of cubanite and mackinawite) predominates in the axial carbonate-bearing phoscorite and carbonatite, where it crystallizes around grains of pyrrhotite (with inclusions of pentlandite-cobaltpentlandite and pyrite), and both of these minerals contain exsolution inclusions of sphalerite. In natural sequence of the Kovdor rocks, iron content in pyrrhotite gradually increases from Fe_7S_8 (pyrrhotite-4C, *Imm*2) to Fe_9S_{10} (pyrrhotite-5C, *C*2 and *P*2$_1$) and $Fe_{11}S_{12}$ (pyrrhotite-6C) due to gradual decrease of crystallization temperature and oxygen fugacity. Low-temperature pyrrhotite 2C (troilite) occurs as lens-like exsolution inclusions in grains of pyrrhotite-4C (in marginal phoscorite) and pyrrhotite-5C (in axial phoscorite-related carbonatite). Within the phoscorite-carbonatite complex, Co content in pyrrhotite gradually increases from host silicate rocks and marginal forsterite-dominant phoscorite to axial carbonate-rich phoscorite and carbonatite at the expense of Ni and Fe. Probably, this dependence reflects a gradually decreasing temperature of the primary monosulfide solid solutions crystallization from the pipe margin toward its axis. The Kovdor and Loolekop phoscorite-carbonatite pipes in the Palabora massif have similar sequences of sulfide formation, and the copper specialization of the Palabora massif can be caused by higher water content in its initial melt allowing it to dissolve much larger amounts of sulfur and, correspondingly, chalcophile metals.

Keywords: pyrrhotite; chalcopyrite; pentlandite; cobaltpentlandite; typochemistry; crystal structure; Kovdor phoscorite-carbonatite complex

1. Introduction

Sulfur, alongside H_2O, CO_2 and halogens, is an important volatile constituent in magmas [1–3]. Unlike all other major volatile elements, sulfur changes its oxidation state depending on the oxygen fugacity (fO_2) regime. The two most abundant forms of S in silicate melts are sulfide (S^{2-}) and sulfate (S^{6+}) [4–6]. The transition between these two oxidation states is very sharp and occurs over the range of oxygen fugacity between the FMQ equilibrium and 2logfO_2 units above (FMQ + 2). However, sulfur shows radically different behavior in these two oxidation states, and, consequently, the existing sulfur species may provide important insights into the evolution of magmas and mineralizing fluids.

Reduced sulfur actively reacts with iron forming FeS(melt) species; therefore, iron content determines "sulfide capacity" of silicate melts [4,7,8]. Besides, iron influences the sulfur oxidation state in silicate melts at given redox conditions [4,5]. Various sulfur-bearing species (H_2S, SO_2, HS^-, S_n^-, S_nS^{2-}, HSO_3^-, HSO_4^-, $S_2O_3^{2-}$, SO_3^{2-}, SO_4^{2-}) serve as ligands for transportation of chalcophile and highly siderophile elements (Cu, Ni, Co, Zn, Mo, Ag, Au and PGE) in silicate melts and magmatic volatiles [9,10], and also control precipitation of the ore-forming elements in magmatic-hydrothermal environment.

Sulfide mineralization is a typical component of the (phoscorite)-carbonatite complexes in alkaline-ultrabasic massifs. The Loolekop phoscorite-carbonatite pipe in the Palabora (Phalaborwa) massif, South Africa (RSA) is a typical example of such complexes, where sulfide mineralization reached the economic level. Sulfides are abundant in alkaline-ultrabasic massifs of the Kola Alkaline Province (NW Russia), and the Kovdor massif is one of the richest. Like the Palabora massif, the Kovdor massif predominantly concentrates sulfides within the Kovdor phoscorite-carbonatite complex, while host silicate rocks (peridotite and foidolite) contain much less sulfides [11,12]. Different aspects of sulfide mineralization in the Kovdor phoscorite-carbonatite complex were discussed in [11–17]. However, almost all of these works presented results of the studies implemented on few random samples with comparatively rare sulfides, while rock-forming sulfides remained almost unstudied.

The general geology of the Kovdor alkaline-ultrabasic massif and its phoscorite-carbonatitre complex has been described by [11,13,18–21], and there is a short geological digest of these works in the first article of this series [22]. In recent years, we have obtained new 3D data on petrography [23–25], composition and properties of economic [22,26–29] and rare minerals [30–36] of the Kovdor phoscorite-carbonatite complex and host ultrabasic and alkaline rocks. This data enabled us to establish formation sequences of the Kovdor massif and its phoscorite-carbonatite complex [21]: peridotite → foidolite and melilitolite → metasomatic rocks (fenite, diopsidite, phlogopitite, scarn-like rocks) → phoscorite and phoscorite-related carbonatite → vein calcite carbonatite → vein dolomite carbonatite and dolomite-magnetite-serpentine rock. The 3D data on distribution of content, grain size and composition of sulfides presented here will help us to understand the behavior of both sulfur and sulfur-related metals during crystallization and subsolidus evolution of the Kovdor massif.

2. Materials and Methods

About 550 core samples of phoscorite, carbonatites and host rocks were taken from 108 boreholes drilled within the Kovdor phoscorite-carbonatite complex. We analyzed thin polished sections of these core samples with a petrographic microscope to estimate textural characteristics, mineral relations and grain size of sulfides. Quantitative relations between magnetic and non-magnetic pyrrhotites were determined in polished sections with a nematic liquid crystal MBBA [37]. Sulfide grain sizes were estimated with the Image Tool 3.0 program [38] as a mean equivalent circular diameter.

Electron-microscope analyses were carried out using a LEO-1450 scanning electron microscope with a Röntek energy-dispersive spectrometer. Chemical compositions of sulfides were determined with the Cameca MS-46 electron microprobe (Geological Institute of the Kola Science Center, Russian Academy of Sciences) operating in a wavelength-dispersive mode at 20 kV and 20–30 nA. The electron beam diameter used was 1–10 μm. The applied standards and detection limits are listed in Table 1. Abbreviations used include Acn (acanthite), Ap (hydroxylapatite), Apn (argentopentlandite),

Bdy (baddeleyite), Bn (bornite), Brt (barite), Cal (calcite), Cbn (cubanite), Ccp (chalcopyrite), Clc (clinochlore), Cls (clausthalite), Djf (djerfisherite), Dol (dolomite), Fo (forsterite), Ght (goethite), Gn (galena), Hss (hessite), Hwl (hawleyite), Mag (magnetite), Mch (moncheite), Mck (mackinawite), Mrc (marcasite), Nph (nepheline), Pcl (pyrochlore), Phl (phlogopite), Pn (pentlandite), Po (pyrrhotite), Py (pyrite), Sgn (siegenite), Sp (sphalerite), Spl (spinel), Srp (serpentine), Tro (troilite), Ttn (titanite), Val (valleriite) and Vlt (violarite).

Bulk-rock samples were analyzed by the Tananaev Institute of Chemistry of KSC RAS (Apatity) by means of inductively coupled plasma-mass spectrometry (ICP-MS) performed with an ELAN 9000 DRC-e mass spectrometer (Perkin Elmer, Waltham, MA, USA). For the analyses, the samples were dissolved in a mixture of concentrated hydrofluoric and nitric acids with distillation of silicon and further addition of hydrogen peroxide to a cooled solution to suppress hydrolysis of polyvalent metals [39].

Cation contents were calculated with the MINAL program of D. Dolivo-Dobrovolsky [40]. Statistical analyses were carried out with the STATISTICA 8.0 [41] and TableCurve 2.0 [42] programs. For the statistics, resulting values of the analyses below the detection limit (see Table 1) were considered to be 10 times lower than the limit. Geostatistical studies and 3D modelling were conducted with the MICROMINE 16 program [43]. Interpolation was performed with ordinary kriging. Automatic 3D geological mapping (Figure 1a) was performed by means of conversion of the rocks chemical composition to mineral composition by logical computation [24].

Table 1. Parameters of EPMA analyses.

Element	Detection Limit, wt %	Standards for EPMA Analyses	Element	Detection Limit, wt %	Standards for EPMA Analyses
Mg	0.1	Pyrope	Se	0.08	Synthetic PbSe
Al	0.05	Pyrope	Mo	0.1	Metallic molybdenum
Si	0.05	Diopside	Pd	0.05	Metallic palladium
S	0.05	Synthetic $Fe_{10}S_{11}$	Ag	0.05	Metallic silver
K	0.03	Wadeite	Cd	0.05	Synthetic CdS
Ca	0.03	Diopside	Sn	0.05	Metallic tin
Mn	0.01	Synthetic $MnCO_3$	Sb	0.05	Antimony
Fe	0.01	Synthetic $Fe_{10}S_{11}$	Te	0.05	Synthetic PbTe
Co	0.01–0.03	Metallic cobalt	Pt	0.05	Metallic platinum
Ni	0.01	Synthetic NiAs	Au	0.05	Metallic gold
Cu	0.01	Metallic copper	Pb	0.05	Synthetic PbSe
Zn	0.01	Synthetic ZnO	Bi	0.06	Bismuth
As	0.05	Synthetic NiAs			

Crystal structures of pyrrhotite samples (00-01, 36/33, 00-10-41, 01-11-91, 00-51) were studied by means of Agilent Technologies Xcalibur Eos diffractometer operated at 50 kV and 40 mA. More than a hemisphere of three-dimensional data was collected at 100 K temperature by monochromatic Mo$K\alpha$ X-radiation with frame widths of 1° and 5–15 s count for each frame. The crystal structures were refined in standard and non-standard settings of different space groups: P-6m2, P312, P32, P-3m, P-3, P3m, Cmce, Cca, Cc, C222$_1$, C2, P2$_1$ for the 5C-polytype and P-6m2, Aem2, Amm2, C2/m, C/m, C2, F2/d, Imm2 for the 4C-polytype by means of the *SHELX* program [44] incorporated in the *OLEX2* program package [45]. The final models were chosen according to the following criteria: absence of violating reflections, lower means of R-factors and GOOF, absence or low number of atoms with physically unrealistic anisotropic displacement parameters (without restraints). Crystal structures were refined with a pseudomerohedral twin model with [−1 0 0 0 −1 0 0 0 1 2] twining matrix for the C2 and P2$_1$ models. Empirical absorption correction was applied in the CrysAlisPro program [46] using spherical harmonics implemented in the SCALE3 ABSPACK scaling algorithm. The crystal structures were visualized with the Diamond 3.2f program [47].

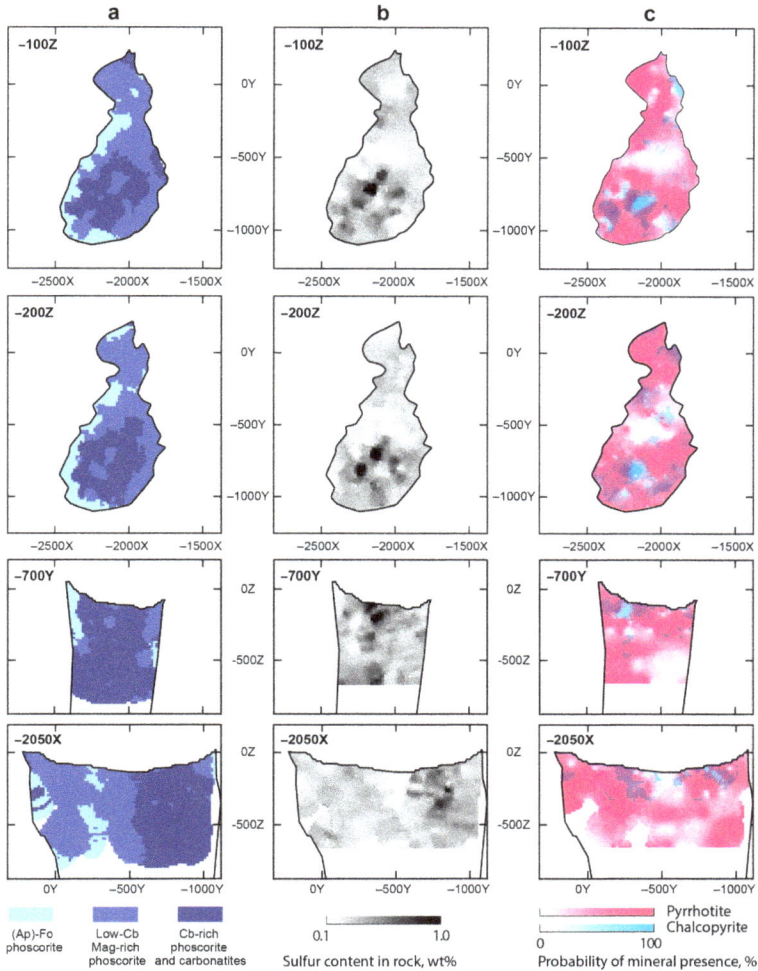

Figure 1. Distribution of rock types (**a**), sulfur content (**b**) and probability of pyrrhotite and chalcopyrite presence (**c**) within the Kovdor phoscorite-carbonatite complex.

3. Results

3.1. Sulfur Mineralization

In primary silicate rocks of the Kovdor massif, average sulfur content is comparatively low (Table 2), but it becomes higher in apo-peridotite metasomatic diopsidite and phlogopitite. Within the phoscorite-carbonatite pipe, sulfur content increases from marginal low-carbonate phoscorite to axial calcite-rich phoscorite and calcite carbonatite (Figure 1b). Besides sulfur, the Kovdor sulfuric compounds include Fe, Co, Ni, Cu, Zn, Ag, Pb and Ba (see Table 1). Many of these metals initially concentrated in rock-forming forsterite (Ni), apatite (Ba), magnetite (Fe, Co, Zn) and pyrochlore (Pb), and demonstrate unclear or even negative correlation with sulfur. Nevertheless, sulfuric compounds inherited initial distribution of these metals in earlier minerals resulting in concentric distribution corresponding to the secondary sulfides and sulfates within the phoscorite-carbonatite complex.

Table 2. Median contents of sulfide-forming elements in rocks of the Kovdor massif.

Rock	Host Silicate Rock	(Ap)-Fo Phoscorite	Low-Cb Mag-Rich Phoscorite	Cal-Rich Phoscorite and Related Carbonatite	Vein Calcite Carbonatite
n	44	17	90	50	16
S, wt %	0.22	0.20	0.15	0.29	1.53
Fe	6.49	7.88	29.79	17.48	3.85
Cu, ppm	56	44	37	78	55
Zn	89	100	198	117	28
Co	40	65	92	70	29
Ni	111	49	39	17	17
Ag	2	1	8	6	1
Pb	1	1	1	2	3

In the Kovdor rocks, major concentrators of sulfur are sulfides, mainly, pyrrhotite and chalcopyrite (Figure 1c), while sulfates, mainly, barite, are the products of low-temperature alteration of sulfides. Pyrrhotite is a common accessory to a rock-forming mineral of the most phoscorite and carbonatites varieties, apart from vein dolomite carbonatite and the related dolomite-magnetite-serpentine metasomatic rock. Chalcopyrite is closely associated with pyrrhotite, being predominantly concentrated in carbonate-rich phoscorite, phoscorite-related carbonatite and vein calcite carbonatite of the ore-pipe apical part (up to 20 modal %). In the apical part of the pipe axial zone, there is calcite-rich phoscorite with rock-forming chalcopyrite (up to 30 modal %).

All other sulfides and sulfates resulted from subsolidus exsolution and low-temperature hydrothermal alteration of pyrrhotite and chalcopyrite. Pyrrhotite contains exsolution inclusions of pentlandite–cobaltpentlandite (almost always) and troilite (rarely), and chalcopyrite carries exsolution inclusions of sphalerite (common) and cubanite (rare). Pyrite and valleriite are spread products of pyrrhotite and chalcopyrite alteration, while primary bornite sometimes occurs as relicts within secondary chalcopyrite. Rare secondary sulfides include djerfisherite, marcasite (after both pyrrhotite and chalcopyrite), chalcocite and covellite (after chalcopyrite only). In addition, galena, violarite, mackinawite, moncheite, petzite, hessite and clausthalite form separate inclusions in pyrrhotite, while acanthite, argentopentlandite, altaite, galena, hawleyite, siegenite, tsumoite, volynskite, wittichenite occur as inclusions in bornite and chalcopyrite.

3.2. Pyrrhotite and Products of Its Alteration

In the Kovdor massif, pyrrhotite-4C, Fe_7S_8, pyrrhotite-5C, Fe_9S_{10}, and pyrrhotite-2C (troilite), FeS were found and structurally verified, while presence of pyrrhotite-6C, $Fe_{10}S_{11}$ was established only according to its composition. Few pyrrhotite modifications often occur within one sample, and even within one grain of pyrrhotite (Figure 2a); however, non-magnetic monoclinic modification 5C always predominates (in average, 88 vol. %). Content of ferrimagnetic orthorhombic pyrrhotite-4C reaches 50 vol. % in host silicate rocks and marginal (apatite)-forsterite phoscorite, 30 vol. % in intermediate low-carbonate magnetite-rich phoscorite, 20 vol. % in axial calcite-rich phoscorite and related carbonatite, and 15 vol. % in vein calcite and dolomite carbonatite [14]. Pyrrhotite-6C is the latest modification that substitutes magnetite and chalcopyrite (mainly, in calcite carbonatite), and forms pyrrhotite-pyrite intergrowth resulting from exsolution of earlier pyrrhotite-5C or -4C. Pyrrhotite-4C forms separate grains, marginal zones of non-magnetic pyrrhotite crystals and thin (up to 50 μm) lens-like inclusions in pyrrhotite-5C (see Figure 2a). Exsolution hexagonal pyrrhotite-2C (troilite) is found in about 5% of the samples (mainly, in marginal apatite-forsterite phoscorite and axial phoscorite-related carbonatite) as lens-like inclusions in grains of pyrrhotite-4C and pyrrhotite-5C (Figure 2b).

In peridotite, rounded or irregularly shaped pyrrhotite grains (up to 120 μm in diameter) with exsolution inclusions of pentlandite fill interstices between rock-forming olivine and diopside. In foidolite, there are sporadic pyrrhotite grains in cancrinitized nepheline grains (Figure 2c) and late calcite veinlets. Apo-peridotite metasomatites, especially carbonatized and apatitized, contain up to 3 modal % of pyrrhotite that fills interstices between grains of rock-forming minerals in close association with chalcopyrite, magnetite and ilmenite (Figure 2d).

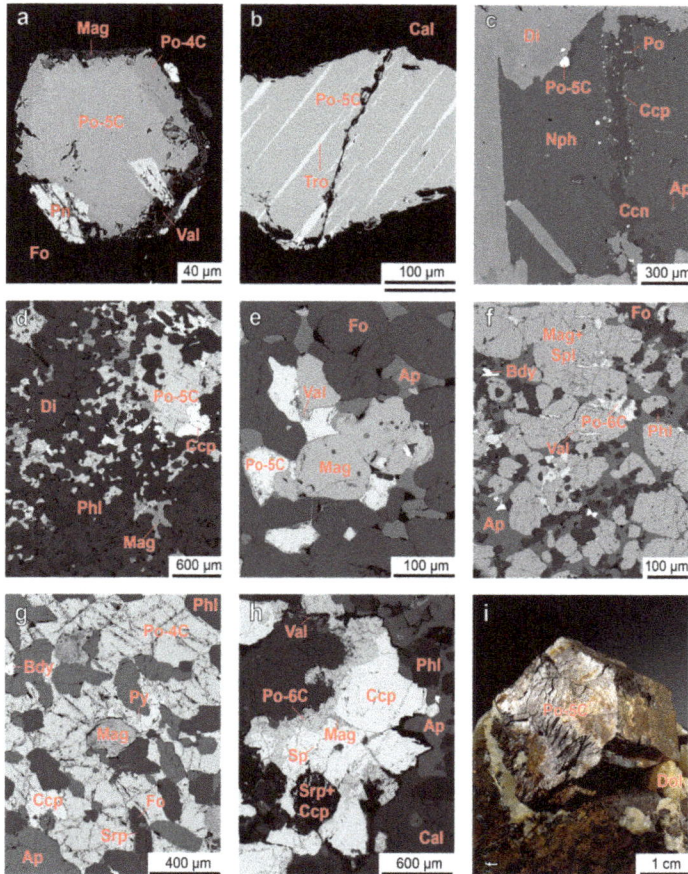

Figure 2. Typical pyrrhotite morphology in rocks of the Kovdor massif: (**a**) pseudohexagonal crystal of pyrrhotite-5C with inclusions of pentlandite replaced with pyrrhotite-4C (F-phoscorite 927/87.4); (**b**) exsolution inclusions of troilite in pyrrhotite-5C (a, phoscorite-related carbonatite 924/26.7); (**c**) plate crystals of pyrrhotite-5C in cancrinitized nepheline (foidolite 949/209.2); (**d**) irregularly shaped intergrowths of pyrrhotite-5C, magnetite and chalcopyrite (diopside-phlogopite rock 1010/619.4); (**e**) intergrowth of pyrrhotite-5C with magnetite (AF-phoscorite 956/70.7); (**f**) interstitial aggregates and veinlets of pyrrhotite-6C in magnetite (MAF-phoscorite 905/160.0); (**g**) aggregates of pyrrhotite-4C in a sulfide segregation (CMAF-phoscorite 1004/656.5); (**h**) granular rims of pyrrhotite-6C around chalcopyrite grains (phoscorite-related carbonatite 975/270.0); (**i**) well-shaped crystal of pyrrhotite-5C (vein dolomite carbonatite K-2011-5). BSE-images (**a–h**) and macrophoto (**i**). Mineral abbreviations see in the Section 2.

As mentioned above, pyrrhotite is a predominant accessory to rock-forming sulfide of phoscorite and carbonatite. In marginal forsteritite and apatite-forsterite phoscorite, pyrrhotite (up to 0.2 modal %) occurs within and between forsterite grains as pseudohexagonal plate crystals (up to 500 µm in diameter) with exsolution lamellae of pentlandite inside (see Figure 2a), as well as irregularly shaped grains in close intergrowth with magnetite (Figure 2e). Content of pyrrhotite in phoscorite increases with growth of magnetite amount and reaches 8 modal % in intermediate low-carbonate magnetite-rich phoscorite. Herein, pyrrhotite usually forms close intergrowths with magnetite, and, together with valleriite, replaces earlier magnetite and fills fractures in its grains (Figure 2f). Such secondary pyrrhotite often inherits cubic inclusions of spinel from exsolved Mg-Al-rich magnetite.

Gradual transition of low-carbonate magnetite-rich phoscorite into axial calcite-rich phoscorite and phoscorite-related carbonatite is accompanied by growth of pyrrhotite content and disappearance of its connection with magnetite. In this type of rock, there are sulfide-rich areas (up to 8 m in diameter), where pyrrhotite (up to 50 modal %) and chalcopyrite cement separate grains of all other minerals (Figure 2g). Besides, pyrrhotite forms here porous rims around chalcopyrite grains (Figure 2h) and tabular to short prismatic crystals (up to 2 cm in diameter) with hexagonal prismatic {10-10} and pinacoidal {0001} faces. In selvages of vein calcite and dolomite carbonatites, pyrrhotite occurs as bronze-yellow platy crystals (up to 15 cm in diameter and 1.5 cm thick), partially replaced with pyrite and goethite (Figure 2i).

Mean equivalent circular diameter of pyrrhotite grains (Table 3) increases insignificantly from 80 μm in marginal (apatite)-forsterite phoscorite to 160 μm in intermediate low-carbonate magnetite-rich phoscorite, and then to 200 μm in axial calcite-rich phoscorite and carbonatite. This trend is similar to that of co-existing rock-forming and accessory minerals [21,26,27].

Table 3. Grain size and chemical composition of the pyrrhotite group minerals (mean ± SD/min–max).

Rock	Host Rock		Phoscorite and Related Carbonatite			Vein Carbonatite	
	Foidolite	Diopsidite	(Ap)-Fo	Low-Cb Mag-rich	Cal-rich	Cal	Dol
n	9	27	13	60	94	26	25
D, μm	100 ± 80 15–300	140 ± 90 10–300	120 ± 90 50–400	200 ± 80 50–450	220 ± 90 10–600	230 ± 90 100–600	200 ± 100 50–600
S, wt %	39.2 ± 0.5 38.54–40.38	39.1 ± 0.5 37.67–39.98	39 ± 1 36.72–39.82	39.1 ± 0.6 36.31–40.52	38.9 ± 0.7 35.81–40.36	39.0 ± 0.4 38.27–39.77	39.0 ± 0.5 37.74–39.77
Fe	60.5 ± 0.5 59.83–61.11	60.6 ± 0.9 59.31–63.01	61 ± 2 59.77–64.13	60.8 ± 0.8 58.18–63.51	60.9 ± 0.8 58.93–63.37	60.8 ± 0.5 59.77–61.87	60.7 ± 0.6 59.37–61.92
Co	0.1 ± 0.1 <0.01–0.43	0.06 ± 0.06 <0.01–0.19	0.1 ± 0.1 <0.01–0.39	0.2 ± 0.1 <0.01–0.39	0.2 ± 0.1 <0.01–0.54	0.14 ± 0.09 0.02–0.36	0.13 ± 0.08 <0.01–0.33
Ni	0.3 ± 0.1 0.03–0.44	0.3 ± 0.2 <0.01–0.82	0.2 ± 0.1 <0.01–0.38	0.06 ± 0.09 <0.01–0.36	0.04 ± 0.09 <0.01–0.77	0.01 ± 0.01 <0.01–0.08	0.1 ± 0.1 <0.01–0.38
Fe, at. %	46.8 ± 0.4 45.85–47.35	47.0 ± 0.7 46.02–48.98	48 ± 1 46.34–50.02	47.1 ± 0.7 45.26–50.10	47.2 ± 0.8 45.80–50.22	47.2 ± 0.4 46.20–47.87	47.2 ± 0.5 46.21–48.08
Ni	0.2 ± 0.1 0.00–0.32	0.2 ± 0.2 0.00–0.59	0.1 ± 0.1 0.00–0.26	0.04 ± 0.07 0.00–0.26	0.03 ± 0.07 0.00–0.58	0.00 ± 0.01 0.00–0.05	0.06 ± 0.08 0.00–0.31
Co	0.1 ± 0.1 0.00–0.32	0.04 ± 0.05 0.00–0.16	0.05 ± 0.07 0.00–0.27	0.11 ± 0.08 0.00–0.27	0.14 ± 0.09 0.00–0.38	0.10 ± 0.06 0.00–0.27	0.10 ± 0.07 0.00–0.27
S	52.9 ± 0.4 52.55–53.88	52.8 ± 0.6 51.02–53.56	52 ± 1 49.98–53.39	52.7 ± 0.7 49.90–54.47	52.6 ± 0.7 49.73–53.82	52.7 ± 0.4 52.03–53.53	52.7 ± 0.5 51.76–53.36

Chemical composition of pyrrhotite (Table 3, Figure 3a) varies from Me_7S_8 (pyrrhotite-4C) to MeS (pyrrhotite-2C), with intensive maximum at Me_9S_{10} (pyrrhotite-5C). Approximation of total metal content histogram by the corresponding Gaussians with fixed positions at Me = 46.66 for Fe_7S_8, 47.36 for Fe_9S_{10}, 47.82 for $Fe_{11}S_{12}$ and 50.00 for FeS has confirmed a sharp predominance of pyrrhotite-5C with subordinate roles of 6C, 4C and, especially, 2C modifications. Contents of iron and sulfur are distributed in accordance with normal law (Figure 3b), while cobalt and nickel are characterized by lognormal and exponential distributions correspondingly (Figure 3c,d).

Iron content in pyrrhotite (exclusive of exsolution troilite) gradually increases from earlier host silicate rocks to the latest vein carbonatites (Figure 4a), which is fully in accordance with gradual increase of pyrrhotite-5C fraction due to presence of pyrrhotite-4C in this sequence. Taking into account pyrrhotite structural data according to [48], where $Fe_7S_8 = Fe^{2+}_5Fe^{3+}_2\square S^{2-}_8$, $Fe_9S_{10} = Fe^{2+}_7Fe^{3+}_2\square S^{2-}_{10}$ and $Fe_{11}S_{12} = Fe^{2+}_9Fe^{3+}_2\square S^{2-}_{12}$, one can see a gradual decrease of Fe^{3+}/Me^{2+} ratio in natural rock sequence of the Kovdor massif (Figure 4b), and from marginal (apatite)-forsterite phoscorite with pyrrhotite-4C to intermediate low-carbonate magnetite-rich phoscorite with pyrrhotite-5C and, finally, calcite-rich phoscorite and calcite carbonaite with pyrrhotite-5C and -6C (Figure 5a).

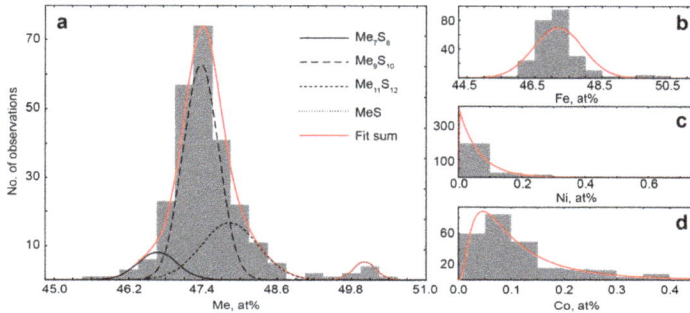

Figure 3. Frequency histograms of the total metals (**a**), Fe (**b**), Ni (**c**) and Co (**d**) contents in pyrrhotite and troilite of the Kovdor phoscorite-carbonatite complex.

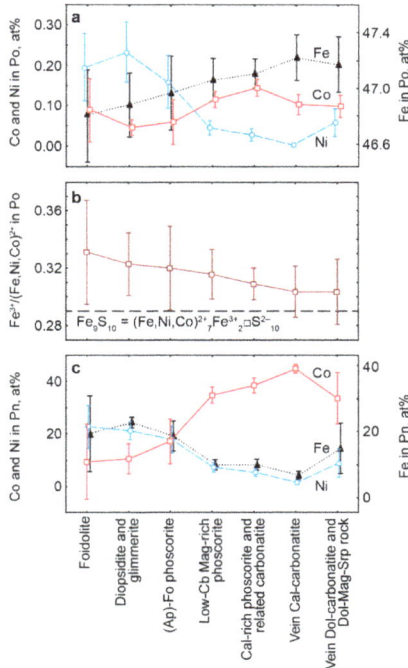

Figure 4. Mean contents of Fe, Ni and Co (at. %, mean ±95% confidence interval) in pyrrhotite ((**a**) without the troilite account) and exsolution pentlandite–cobaltpentlandite (**c**), as well as ratio Fe^{3+}/Me^{2+} in pyrrhotite (**b**), in the order of the rock formation sequence.

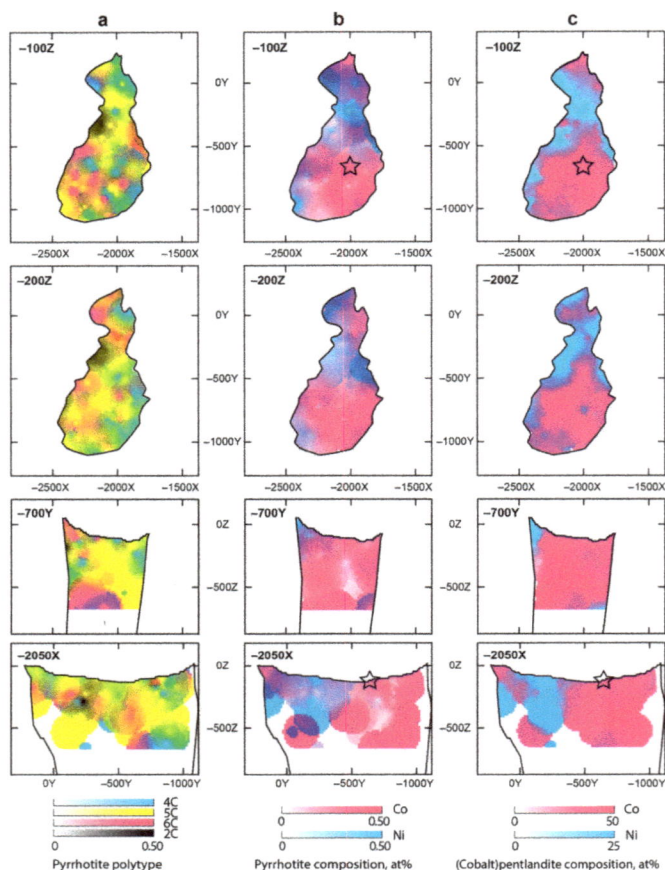

Figure 5. Distribution of pyrrhotite polytypes (**a**), content of Ni and Co in pyrrhotite (**b**) and pentlandite–cobaltpentlandite (**c**) within the phoscorite-carbonatite complex (black contour). Stars denote TL of pakhomovskyite, $Co_3[PO_4]_2 \cdot 8H_2O$.

Experimental details and crystallographic parameters of structurally different pyrrhotites are given in Table 4. The site occupancy factors (s.o.f.) were refined using the scattering curves for neutral atoms given in the International Tables for Crystallography [49]. Due to the large scope of structural information (atom coordinates, bond-lengths, displacement parameters for samples 1–5), these data can be obtained from the CIFs, which are available as Supplementary Materials.

The pyrrhotite crystal structure was firstly described by Nils Alsen [50] in the space group $P6_3/mmc$ with unit cell parameters $a = 3.43$, $c = 5.68$ Å. The description was based upon hexagonal-close packing stacked perpendicular to the c axis, where sulfur atoms approximately occupy the nodes, and iron atoms are regularly arranged in octahedral interstices of sulfur atoms [51]. The vacancies between iron sites are in ordered arrangement, and confined to alternate layers of iron atoms and normal to the c axis [52,53]. Formation of different pyrrhotite polytypes is caused by ordered arrangement of vacancies and/or Fe^{2+}/Fe^{3+} atoms that results in formation of superstructures [53]. Random arrangement of vacancies leads to the formation of hexagonal 1C polytype, their partial ordering causes orthorhombic 4C modification, and fully ordered vacancy distribution produces monoclinic 5C, 6C and modulated NC structures. Up today, 4C, 5C, 6C pyrrhotite polytypes and modulated *NC* structures with *N* = 4.88, 5.5 etc. were reported [52,54,55]. The *N* is defined as a number of supercell reflections along c^* direction (Figure 6) situating between bright reflections of a hexagonal cell.

Table 4. Structural data for studied pyrrhotites.

Sample	1 (00-01)	2 (36-33)	3 (00-10-41)	4 (01-11-91)	5 (00-51)
Modification	4C	5C	5C	5C	5C
Refined formula	$Fe_{6.78}S_8$	$Fe_{8.91}Ni_{0.25}S_{10}$	$Fe_{8.99}S_{10}$	$Fe_{8.84}S_{10}$	$Fe_{8.88}S_{10}$
Formula weight	634.86	833.04	822.41	814.45	816.41
Temperature/K	100 (2)	100 (2)	100 (2)	100 (2)	100 (2)
Crystal system	orthorhombic	monoclinic	monoclinic	monoclinic	monoclinic
Space group	*Imm*2	*C*2	*C*2	*C*2	*P*2$_1$
a (Å)	22.678 (4)	11.8624 (9)	11.8875 (11)	11.8706 (7)	6.8477 (5)
b (Å)	3.4131 (5)	6.8613 (5)	6.8667 (6)	6.8589 (5)	28.584 (4)
c (Å)	5.9083 (13)	28.593 (2)	28.661 (2)	28.5953 (16)	6.8518 (5)
α (°)	90	90	90	90	90
β (°)	90	89.897 (8)	90.023 (8)	89.982 (5)	119.972 (11)
γ (°)	90	90	90	90	90
Volume/Å^3	457.31 (15)	2327.3 (3)	2339.6 (4)	2328.2 (3)	1161.8 (2)
Z	2	8	8	8	4
ρ_{calc} (g/cm^3)	4.611	4.755	4.670	4.647	4.668
μ (mm^{-1})	12.202	12.914	12.542	12.432	12.499
F (000)	608.0	3190.0	3149.0	3119.0	1563.0
Crystal size (mm^3)	$0.15 \times 0.15 \times 0.15$	$0.18 \times 0.18 \times 0.10$	$0.13 \times 0.13 \times 0.13$	$0.13 \times 0.13 \times 0.05$	$0.12 \times 0.12 \times 0.12$
Radiation			MoKα (λ = 0.71073)		
2Θ range for data collection (°)	7.126–54.92	5.7–61.33	5.69–62.04	5.70–50.00	5.7–61.932
Index ranges	$-29 \leq h \leq 27$, $-3 \leq k \leq 4$, $-4 \leq l \leq 7$	$-15 \leq h \leq 16$, $-8 \leq k \leq 9$, $-40 \leq l \leq 23$	$-16 \leq h \leq 16$, $-6 \leq k \leq 9$, $-33 \leq l \leq 40$	$-14 \leq h \leq 14$, $-8 \leq k \leq 7$, $-34 \leq l \leq 33$	$-9 \leq h \leq 7$, $-38 \leq k \leq 32$, $-9 \leq l \leq 9$
Reflections collected	1018	5920	10,932	7368	6471
Independent reflections	475	4494	4797	3597	4708
R_{int}, R_{sigma}	0.0171, 0.0189	0.0354, 0.0397	0.0396, 0.0545	0.0607, 0.0571	0.0278, 0.0271
Data/restraints/parameters	475/31/53	4494/127/255	4797/1/136	3597/1/135	4708/211/216
Goodness-of-fit on F^2	1.202	1.074	1.145	1.608	1.125
Final R indexes [I>= 2σ(I)]	R_1 = 0.0780 wR$_2$ = 0.1173	R_1 = 0.0896, wR$_2$ = 0.2586	R_1 = 0.0884, wR$_2$ = 0.2361	R_1 = 0.1170, wR$_2$ = 0.3522	R_1 = 0.0881, wR$_2$ = 0.1802
Final R indexes [all data]	R_1 = 0.0817 wR$_2$ = 0.1197	R_1 = 0.1009, wR$_2$ = 0.2728	R_1 = 0.1311, wR$_2$ = 0.2941	R_1 = 0.1286, wR$_2$ = 0.3608	R_1 = 0.0990, wR$_2$ = 0.1874
Largest diff. peak/hole/e Å^{-3}	1.61/−2.06	4.02/−3.57	5.13/−5.73	4.01/−4.36	4.93/−4.66
Flack parameter	0.49 (12)	0.53 (19)	0.5 (2)	0.26 (15)	0.09 (6)

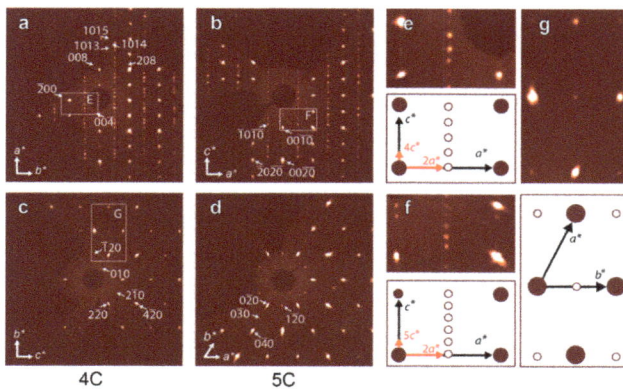

Figure 6. Reconstructed sections of reciprocal space obtained for (*h*0*l*) and (*hk*0) sections for 1 (**a**,**c**) and 4 pyrrhotite samples (**b**,**d**) and enlarged fragments of these sections (**e–g**). For sample 1, there was used the transformation matrix [010 001 100]. White arrows and numbers indicate reflections and its indices. On corresponding schemas, large dark red circles and small unfilled circles belong to the hexagonal cell (*a* = 3.43, *c* = 5.68 Å) and supercell respectively; black and red arrows indicate subcell and supercell vectors respectively.

The crystal structure of ferrimagnetic pyrrhotite-4C, $Fe_{6.78}S_8$, was refined in the *Imm2* space group. It differs from the previous refinements of crystal structure of stoichiometric Fe_7S_8 in *C2/c* and *C2* space groups [51,56] by distribution of vacancies in Fe sites (Figure 7) based on the observed differences in their chemical compositions. The Kovdor pyrrhotite-4C contains four independent iron sites with refined occupancies of 0.89, 0.75, 1 and 0.75 for Fe1, Fe2, Fe3 and Fe4 respectively.

In the crystal structure of the pyrrhotite-5C with the $P2_1$ symmetry, anomalously short distance 2.470 Å has been observed between Fe2 and Fe4 sites (Figure 8). There are two possible explanations for this fact: (1) one of these sites is vacant, while the second one is populated (their refined occupancies are 0.24 for the Fe2 site and 0.75 for the Fe4 site); (2) the Fe2-Fe4 interaction has at least partially bonding character. The same structural effect has been observed in [57].

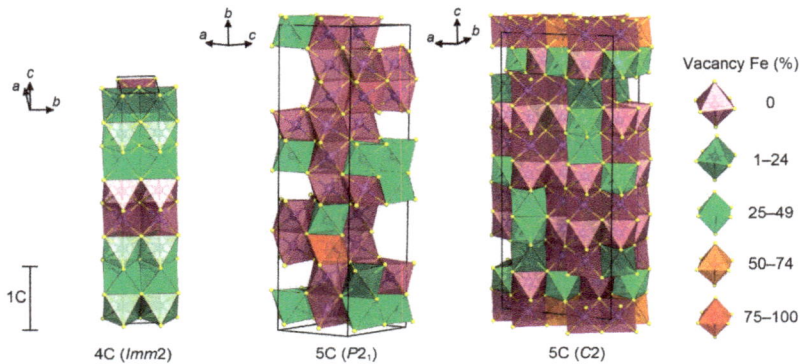

Figure 7. Three structural modifications of pyrrhotite from the Kovdor massif (for comparison, the transformation matrix [010 001 100] was used for 4C polytype).

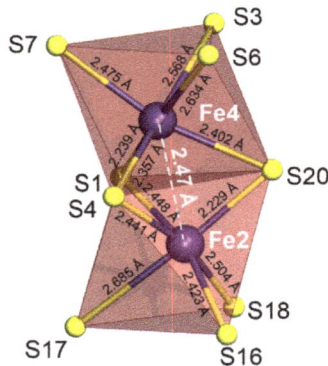

Figure 8. Local configuration around Fe2- and Fe4-centered octahedra in sample **5** crystal structure (pyrrhotite-5C with $P2_1$ space group).

Most samples of the Kovdor pyrrhotite-5C are crystallized in *C2* space group. Their chemical composition widely varies, $Fe_{8.84-8.99}S_{10}$, which causes differences in occupancy of Fe-sites (Figure 9). From 6 to 9 iron sites (from 22 ones in *C2* model) are observed. They have a different vacancy proportion ranging from 1% to 100%. Such variability of 5C polytype can explain its domination in the Kovdor massif (see Figure 3). In samples 1–5, the mean <Fe-S> distance ranges from 2.426 to 2.454 Å, and the expected elongation of bonds can be compensated by partial incorporation of Fe^{3+} in low-occupied sites [48,57].

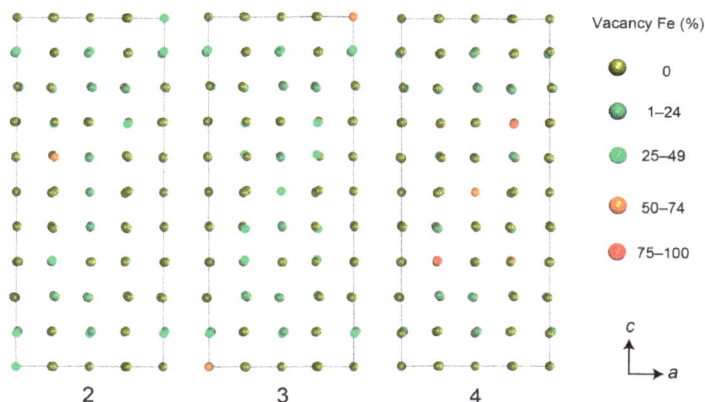

Figure 9. Arrangement of iron vacancies in pyrrhotite-5C (samples 2, 3 and 4). Sulfur atoms are omitted for clarity, unit cell is outlined.

Similar to nickel content in the rocks (see Table 2), the amount of Ni in pyrrhotite gradually decreases from host foidolite and diopsidite–glimmerite to marginal (apatite)-forsterite phoscorite, intermediate low-carbonate magnetite-rich phoscorite, calcite-rich phoscorite, phoscorite-related and vein calcite carbonatite, and then insignificantly increases in vein dolomite carbonatite and the related dolomite-magnetite-serpentine rock. In contrast, content of Co both in rock and pyrrhotite increases from host silicate rocks and marginal calcite-poor phoscorite to calcite-bearing phoscorite and phoscorite-related carbonatite and then slightly decreases in the vein carbonatites (Figure 4c). Correspondingly, Ni-dominant pyrrhotite is concentrated in the marginal zone of the phoscorite-carbonatite complex, Ni-Co-bearing pyrrhotite occurs in the intermediate zone, and Co-dominant pyrrhotite occupies the axial zone of the pipe (Figure 5b).

Almost all pyrrhotite grains contain exsolution inclusions of pentlandite, $(Ni_{4.5}Fe_{4.5})S_8$, or cobaltpentlandite, Co_9S_8, and, much rarely, troilite, FeS:

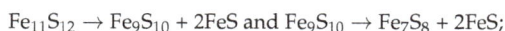

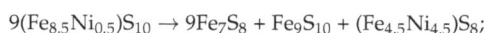

$$Fe_{11}S_{12} \rightarrow Fe_9S_{10} + 2FeS \text{ and } Fe_9S_{10} \rightarrow Fe_7S_8 + 2FeS;$$

$$9(Fe_{8.5}Ni_{0.5})S_{10} \rightarrow 9Fe_7S_8 + Fe_9S_{10} + (Fe_{4.5}Ni_{4.5})S_8;$$

$$9(Fe_8Co)S_{10} \rightarrow 9Fe_7S_8 + Fe_9S_{10} + Co_9S_8.$$

Troilite occurs only as lens-like lamellae along (001) planes of host pyrrhotite from marginal (apatite)-forsterite phoscorite (see Figure 5a) and, rarely, phoscorite-related carbonatite (Figure 10a). The host pyrrhotite is predominantly represented by its 4C modification in marginal (apatite)-forsterite phoscorite, and by 5C modification in axial phoscorite-related carbonatite. In host silicate rocks and vein calcite and dolomite carbonatites, troilite has not been found.

Exsolution inclusions of pentlandite in pyrrhotite occur mainly in host silicate rocks and (apatite)-forsterite phoscorite of the phoscorite-carbonatite complex marginal zone, and independent pentlandite grains (up to 120 μm in diameter) occur rarely in (apatite)-forsterite phoscorite and dolomite-magnetite-serpentine rock. Pyrrhotite with inclusions of cobaltpentlandite dominates in the pipe axial carbonate-rich zone, while intermediate low-carbonate magnetite-rich phoscorite usually contains pyrrhotite with inclusions of both pentlandite and cobaltpentlandite with approximately equal proportions of Co and Ni-Fe. Exsolution pentlandite–cobaltpentlandite forms distinctive flame-like (up to 200 μm long, Figure 2b) round or hexagonal lamellar inclusions (up to 100 μm in diameter and 10 μm thick, Figures 2a and 10b) that predominantly grow along {0001} planes of host pyrrhotite from boundaries of its grains and fissures. Sometimes, the oriented inclusions of pentlandite–cobaltpentlandite are found in pseudomorphs of pyrite, valleriite, carbonates (see Figure 10b) and secondary magnetite after pyrrhotite.

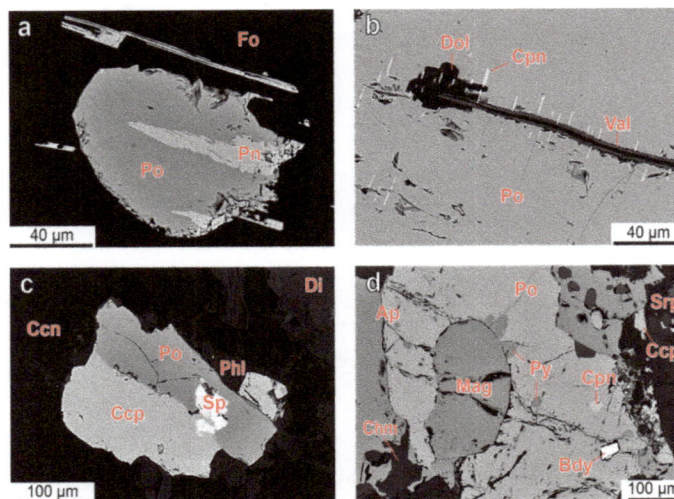

Figure 10. BSE-images of troilite exsolution inclusions ((**a**) phoscorite-related carbonatite 924/26.7), cobaltpentlandite ((**b**) CMAF-phoscorite 996/304.7), sphalerite ((**c**) diopsidite 1017/115.1) and pyrite ((**d**) CMAF-phoscorite 966/29) in pyrrhotite. Mineral abbreviations can be seen in Section 2.

Chemical composition of pentlandite–cobaltpentlandite varies in all possible ranges (Table 5). Increase of Co/Ni-ratio in pyrrhotite causes logarithmical growth of this ratio in exsolution pentlandite–cobaltpentlandite (Figure 11). In the sequence of rock formation, there is a gradual increase of Co content in pentlandite–cobaltpentlandite due to Ni and Fe (see Figure 4c), which causes corresponding spatial zonation of the phoscorite-carbonatite complex (see Figure 5c).

Table 5. Chemical composition of pentlandite–cobaltpentlandite (mean ± SD/min–max).

Rock	Host Rock		Phoscorite and Related Carbonatite			Vein Carbonatite	
	Foidolite	Diopsidite	(Ap)-Fo	Low-Cb Mag-rich	Cal-rich	Cal	Dol
n	6	24	13	51	67	17	8
S, wt %	33 ± 1	33 ± 1	33.0 ± 0.5	32.8 ± 0.7	32.7 ± 0.8	32.6 ± 0.4	32.4 ± 0.9
	32.34–34.80	27.51–34.73	32.06–33.74	31.84–35.24	31.61–36.12	31.80–33.57	30.64–33.66
Fe	26 ± 9	28 ± 3	23 ± 7	14 ± 7	13 ± 8	9 ± 3	17 ± 9
	11.64–30.64	15.52–32.87	9.25–30.94	4.81–28.00	3.52–35.02	4.24–15.42	4.58–28.42
Co	10 ± 20	10 ± 10	20 ± 20	40 ± 10	50 ± 20	57 ± 4	30 ± 20
	1.72–47.16	0.77–54.33	4.50–52.27	10.11–61.91	1.21–61.92	47.63–62.36	8.16–63.16
Ni	30 ± 10	30 ± 8	20 ± 10	10 ± 9	10 ± 10	2 ± 3	20 ± 10
	8.86–34.53	2.65–37.12	5.11–34.47	0.7–30.14	0.62–32.96	0.25–10.26	3.94–32.21
Ni, *apfu*	3 ± 2	4 ± 1	3 ± 1	1 ± 1	1 ± 1	0.3 ± 0.3	2 ± 2
	1.20–4.64	0.42–5.01	0.68–4.64	0.09–3.93	0.08–4.31	0.03–1.36	0.53–4.24
Fe	4 ± 1	3.9 ± 0.4	3.2 ± 0.9	1.9 ± 0.9	2 ± 1	1.2 ± 0.4	0.65 ± 4.07
	1.65–4.23	2.59–4.53	1.28–4.38	0.68–3.87	0.50–4.51	0.60–2.17	3.87–23.32
Co	2 ± 3	1 ± 2	3 ± 2	6 ± 2	6 ± 2	7.6 ± 0.6	4 ± 3
	0.22–6.35	0.10–8.60	0.60–6.88	1.31–8.38	0.15–8.26	6.30–8.45	1.07–8.52
S	8	8	8	8	8	8	8

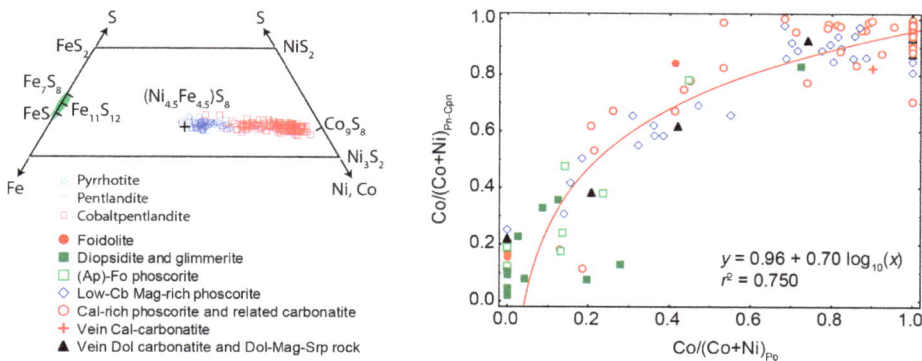

Figure 11. Compositions of host pyrrhotite and exsolution pentlandite–cobaltpentlandite.

On the M_9S_8 plane of the Fe-Ni-Co-S tetrahedron (Figure 12), chemical compositions of pentlandite are concentrated within the field of the solid-solution stability at ≥ 200 °C [58]. Chemical compositions of cobaltpentlandite correspond to the $Fe_{4.5}Ni_{4.5}S$–Co_9S_8 trend, whereas most samples from host silicate rocks, marginal forsterite-dominant and intermediate low-carbonate magnetite-rich phoscorite are within the stability field at 300–400 °C, and most samples from axial carbonate rich phoscorite and carbonatites are within the stability field at 200–400 °C.

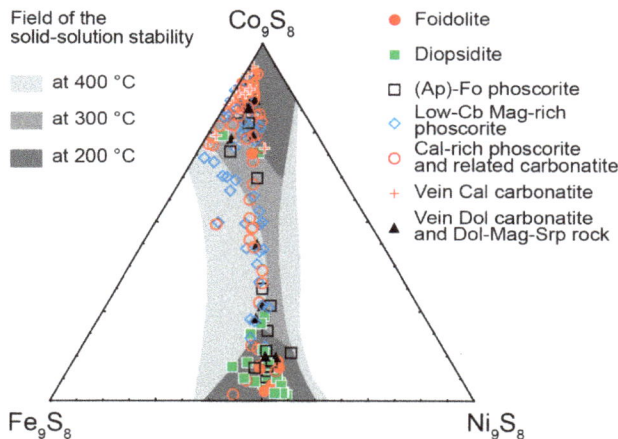

Figure 12. Relationship between composition of the Kovdor's pentlandite–cobaltpentlandite and fields of pentlandite–cobaltpentlandite stability after [58].

Comparatively rare exsolution sphalerite occurs as rounded, irregularly shaped, cross- and butterfly-like inclusions (up to 100 μm, Figure 10c) in pyrrhotite from any types of rocks. Sphalerite together with chalcopyrite also form rims (up to 120 μm thick) around pyrrhotite grains in dolomite-magnetite-serpentine rock. Since chemical composition of exsolution sphalerite (Table 6) at temperature below 500 °C depends on pressure only [59–61], we calculate the pressure using the following equation [62]:

$$FeS_{Sp} = 20.53 - 1.313P + 0.0271P^2,$$

where FeS_{Sp} is in mol %, and P is in kbar. The results (see Table 6) shows gradual pressure decrease from 4 kbar in marginal (apatite)-forsterite phoscorite to 2 kbar in axial carbonate-rich phoscorite and phoscorite-related carbonatite, and then again growth to 4 kbar in vein calcite carbonatite, and decrease to 2 kbar in vein dolomite carbonatite (i.e., decrease by 2 kbar from earlier to the latest rock within both pipe and vein series). It is necessary to note similar behavior of oxygen fugacity [26] estimated with the $Fe^{2+}Ti–Fe^{3+}_2$ magnetite-ilmenite exchange geothermometer/oxometer of Ghiorso and Evans [63].

Table 6. Chemical composition of sphalerite inclusions in pyrrhotite (mean ± SD/min–max) and pressure of their formation estimated with a sphalerite-pyrite-pyrrhotite geobarometer [62].

Rock	Host Rock		Phoscorite and Related Carbonatite			Vein Carbonatite	
	Foidolite	Diopsidite	(Ap)-Fo	Low-Cb Mag-rich	Cal-rich	Cal	Dol
n	3	13	3	14	24	9	17
S, wt %	31.6 ± 0.6 31.02–32.22	32.0 ± 0.6 30.96–32.83	32 ± 1 30.43–32.88	31.7 ± 0.9 29.60–33.22	31.7 ± 0.9 29.78–33.86	31.5 ± 0.7 30.70–32.59	32 ± 1 27.60–33.44
Mn	0.5 ± 0.5 <0.01–0.95	0.3 ± 0.5 <0.01–1.44	0.8 ± 0.2 0.57–0.99	0.6 ± 0.9 <0.01–3.51	1 ± 1 <0.01–6.17	1 ± 1 <0.01–2.89	0.5 ± 0.5 <0.01–1.52
Fe	10 ± 1 9.51–11.51	9 ± 1 6.18–11.33	8.8 ± 0.9 8.13–9.79	10 ± 2 6.67–15.06	10 ± 2 5.51–15.82	9 ± 2 5.69–13.23	10 ± 2 5.86–12.77
Co	<0.01	<0.01	0.1 ± 0.2 <0.01–0.40	0.2 ± 0.3 <0.01–0.99	0.1 ± 0.2 <0.01–0.63	0.2 ± 0.4 <0.01–1.15	0.2 ± 0.3 <0.01–1.04
Ni	<0.01	<0.01	<0.01	<0.01	0.01 ± 0.03 <0.01–0.13	<0.01	<0.01
Zn	56 ± 1 54.52–57.01	58 ± 2 54.04–61.50	58 ± 2 56.78–60.27	57 ± 3 51.71–62.99	54 ± 5 42.41–63.11	57 ± 5 49.17–62.79	57 ± 3 52.99–62.12
Cd	2 ± 1 0.73–3.37	1 ± 1 <0.01–3.57	<0.01	0.6 ± 0.8 <0.01–2.98	3 ± 3 <0.01–10.01	2 ± 2 <0.01–5.96	0.1 ± 0.3 <0.01–1.03
Zn, *apfu*	0.86 ± 0.01 0.85–0.87	0.89 ± 0.04 0.81–0.95	0.90 ± 0.06 0.86–0.97	0.89 ± 0.07 0.77–1.04	0.83 ± 0.09 0.63–0.99	0.88 ± 0.09 0.75–1.00	0.88 ± 0.07 0–0
Fe	0.19 ± 0.02 0.18–0.21	0.16 ± 0.03 0.11–0.20	0.16 ± 0.02 0.14–0.17	0.17 ± 0.04 0.12–0.28	0.19 ± 0.04 0.10–0.27	0.17 ± 0.05 0.11–0.25	0.19 ± 0.04 0.11–0.27
Mn	0.01 ± 0.01 0.00–0.02	0.01 ± 0.01 0.00–0.03	0.01 0.01–0.02	0.01 ± 0.02 0.00–0.06	0.03 ± 0.02 0.00–0.11	0.01 ± 0.02 0.00–0.05	0.01 ± 0.01 0.00–0.03
Cd	0.02 ± 0.01 0.01–0.03	0.01 ± 0.01 0.00–0.03	–	0.01 ± 0.01 0.00–0.03	0.03 ± 0.03 0.00–0.09	0.02 ± 0.02 0.00–0.05	0.00 0.00–0.01
Co	–	–	0.00 0.00–0.01	0.00 ± 0.01 0.00–0.02	0.00 0.00–0.01	0.00 ± 0.01 0.00–0.02	0.00 ± 0.01 0.00–0.02
S	1	1	1	1	1	1	1
P, kbar	2.2 ± 0.1 2.1–2.3	4 ± 2 0.5–8.9	4 ± 1 2.6–5.4	4 ± 2 1.1–7.6	3 ± 2 0.4–9.9	5 ± 3 1.8–9.4	3 ± 2 0.2–9.5

In intermediate low-carbonate magnetite-rich and, especially, axial calcite-rich phoscorite and phoscorite-related carbonatite, there are grains of pyrrhotite-6C with rounded inclusions of pyrite (up to 100 µm in diameter, Figure 10d) that can be regarded as products of exsolution of sulfur-rich pyrrhotite-4C and -5C [62,64,65]:

$$4Fe_7S_8 \leftrightarrow 3Fe_9S_{10} + FeS_2,$$

$$5Fe_7S_8 \leftrightarrow 4Fe_{11}S_{12} + FeS_2, \text{ or}$$

$$5Fe_9S_{10} \leftrightarrow Fe_{11}S_{12} + FeS_2.$$

In this case, distribution of impurities between the coexisting pyrrhotite and pyrite depends on temperature: with temperature decrease, the impurities undergo redistribution from pyrite into pyrrhotite, and vice versa [66]. Constant presence of cobalt impurity in the Kovdor pyrite-pyrrhotite pairs (Table 7) allows us to estimate temperature of their formation using the Co-Fe exchange pyrite-pyrrhotite geothermometer:

$$T = \frac{1000}{1.907 - 0.538 \log_{10} K_D^{Co}}$$

where T is temperature in K, K_D^{Co} = $(Co_{at.\%}/Fe_{at.\%})_{Po}$: $(Co_{at.\%}/Fe_{at.\%})_{Py}$ [66]. Our calculations showed (see Table 7) that temperature of Co equilibration between pyrite and pyrrhotite increases from 170 °C in intermediate low-carbonate magnetite-rich phoscorite to 300 °C in axial carbonate-rich phoscorite, and then decrease to 230 °C in vein dolomite carbonatite.

It is necessarily to note that pyrrhotite-4C may become stable at about 140 °C in nature [67], while 6C and 5C superstructures can be formed at temperatures below approx. 60 °C [62]. Therefore, in natural rock sequence, the ratio of pyrrhotite-5C + 6C to pyrrhotite-4C (see Figure 4a,b) gradually increases due to the gradual decrease of the pyrrhotite superstructure formation temperature. These hypotheses are in good agreement with the results of calcite-dolomite and ilmenite-magnetite geothermometry for this rock sequence [26], where the average temperature of the mineral equilibration gradually decreases from 500 °C to 400 °C for carbonates and from 500 °C to 300 °C for oxides.

Table 7. Chemical composition of pyrite inclusions in pyrrhotite (mean ± SD/min–max), and temperature of their formation estimated with a Co-exchange pyrite-pyrrhotite geothermometer [66].

Rock	Low-Cb Mag-Rich Phoscorite	Cal-Rich Phoscorite/Carbonatite	Dol-Carbonatite
n	2	8	2
S, wt %	52.4 ± 0.6/51.94–52.80	53.0 ± 0.2/52.76–53.25	53.1 ± 0.2/52.97–53.30
Fe	48 ± 1/47.86–48.65	46.8 ± 0.9/45.52–48.51	46.1 ± 0.2/45.93–46.27
Co	0.4 ± 0.2/0.26–0.49	0.1 ± 0.2/<0.01–0.01	0.8 ± 0.6/0.35–1.21
Ni	<0.01	0.03 ± 0.05/<0.01–0.13	0.02 ± 0.01/0.02–0.03
Fe, apfu	1.05 ± 0.04/1.02–1.08	1.02 ± 0.02/0.98–1.05	1.00 ± 0.01/0.99–1.00
Co	0.01	0.00/0.00–0.01	0.02 ± 0.01/0.01–0.03
S	2	2	2
T, °C	170 ± 20/159–188	300 ± 100/235–444	170 ± 90/106–228

Besides exsolution inclusions of troilite, pentlandite–cobaltpentlandite, sphalerite and pyrite, there are unit inclusions of galena, mackinawite, clausthalite, violarite, hessite, moncheite and petzite (Table 8) in pyrrhotite grains. Mackinawite is a typical mineral for axial calcite-rich phoscorite and phoscorite-related carbonatite, where it forms equant or wedge-shaped inclusions (up to 100 μm long) in pyrrhotite and, especially, chalcopyrite. Violarite is found together with pentlandite in dolomite-magnetite-serpentine rock as lens-like inclusions (up to 30 μm long and 2 μm thick) oriented along the {0001} planes of host pyrrhotite. Both these minerals are not independent from pyrrhotite. Galena forms rounded inclusions (up to 20 μm in diameter) in peripheral parts of pyrrhotite grains and pyrrhotite-chalcopyrite boundaries.

Table 8. Chemical composition of characteristic inclusions in pyrrhotite grains (wt %/*apfu*).

Mineral	Mck	Vlt	Cls
Sample	K-02-124	74/67.2	73/205.8
Rock	CM-phoscorite	Mag-Dol-Srp rock	Vein Cal-carbonatite
S	35.61/8.00	43.31/4.00	bd
Mn	bd	bd	bd
Fe	58.27/7.52	16.99/0.90	bd
Co	0.94/0.12	13.33/0.67	bd
Ni	5.56/0.68	26.40/1.33	bd
Zn	bd	bd	bd
Se	bd	bd	26.82/1.00
Cd	bd	bd	bd
Pb	bd	bd	73.98/1.05
Total	100.38/16.32	100.03/6.90	100.80/2.05

Table 8. *Cont.*

Mineral	Hss	Mch	Ptz
Sample		931/341.2	
Rock		Diopsidite	
Ag	56.57/1.97	bd	37.07/2.84
Te	34.04/1.00	44.54/1.54	30.84/2.00
Pd	bd	8.38/0.35	bd
Pt	bd	25.40/0.58	bd
Au	bd	bd	17.93/0.75
Pb	8.91/0.16	bd	14.40/0.58
Bi	bd	21.68/0.46	bd
Total	99.52/3.13	100.00/2.93	100.24/6.17

In nepheline-bearing diopsidite, pyrrhotite contains tabular inclusions of moncheite (up to 15 μm in diameter and 4 μm thick), and irregularly shaped polyphase inclusions of hessite, petzite, galena and clausthalite (up to 6 μm in diameter). All these inclusions may result from pyrrhotite self-cleaning from the corresponding impurities during rock cooling.

Under the influence of low-temperature hydrothermal solutions, pyrrhotite-4C (often) and pyrrhotite-5C (rarely) can be replaced by pyrite, marcasite, valleriite, djerfisherite, pyrite and goethite (Table 9):

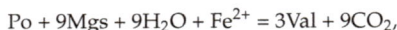

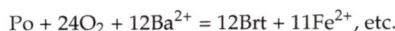

$$4Po + 10H_2O + 15O_2 = 24Py + 20Ght,$$

$$Po + 9Mgs + 9H_2O + Fe^{2+} = 3Val + 9CO_2,$$

$$13Po + 6NaCl + 36KCl + 7Fe^{2+} = 6Djf,$$

$$Po + 24O_2 + 12Ba^{2+} = 12Brt + 11Fe^{2+}, etc.$$

where Brt is barite, $BaSO_4$; Djf—djerfisherite, $K_6NaFe_{25}S_{26}Cl$; Gth—goethite, $FeO(OH)$; Mgs—magnesite, $MgCO_3$; Po—pyrrhotite, $Fe_{11}S_{12}$; Py—pyrite, FeS_2 and Val—valleriite, $Mg_3Fe_4S_4(OH)_6$.

Pyrite and other secondary minerals form partial to complete pseudomorphs after pyrrhotite (Figure 13), which are mostly spread within ring-like stockwork of vein calcite carbonatite and linear zone of vein dolomite carbonatite (see Figure 1), and also in well-shaped crystals in surrounding voids and fissures. In particular, pyrite forms cubic to pentagonal-dodecahedral crystals (up to 20 cm in diameter); marcasite composes sheaf-like aggregates of flattened prismatic crystals (up to 3 mm long); valleriite and tochilinite, $6Fe0.9S·5Mg(OH)_2$, occur as twisted platy crystals (up to 4 mm in diameter); and barite forms well shaped prismatic crystals (up to 4 cm long).

Table 9. Chemical composition of characteristic products of pyrrhotite alteration (wt %/*apfu*).

Mineral	Py	Mrc	Djf	Val	Brt
Sample	1011/79/6	K-96-19-1	1004/656.5	K-051	941/41.5
Rock	AF-phoscorite	Dol-carbonatite	Cal-carbonatite	Dol-carbonatite	MAF-foscorite
Mg	bd	bd	bd	11.08/2.53	bd
Al	bd	bd	bd	0.67/0.14	bd
S	53.05/2.00	53.34/2.00	32.86/26.00	23.16/4.00	13.65/1.00
Cl	bd	bd	1.43/1.02	bd	bd
K	bd	bd	9.33/6.05	bd	bd
Ca	bd	bd	bd	bd	0.37/0.02
Fe	46.19/1.00	46.79/1.01	35.21/15.99	31.81/3.15	bd
Co	0.39/0.01	bd	0.03/0.01	bd	bd
Ni	0.05/0.00	bd	0.05/0.02	2.93/0.28	bd
Cu	bd	bd	21.52/8.59	9.23/0.80	bd
Sr	bd	bd	bd	bd	0.87/0.02
Ba	bd	bd	bd	bd	57.02/0.96
Total	99.68/3.01	100.13/3.01	100.43/57.70	78.88/10.90	71.91/2.00

In addition, Fe^{2+} and Co^{2+} cations resulted from low-temperature hydrothermal alteration of pyrrhotite react with $(PO_4)^{3-}$ and $(CO_3)^{2-}$ anions, and form late hydrothermal phosphates and carbonates: mitridatite, $(Ca_{1.72}Fe^{2+}_{0.02})_{\Sigma1.74}Fe^{3+}_{3.10}[P_{3.04}O_{12}]O_2 \cdot 3H_2O$, strengite, $Fe^{3+}_{0.99}[P_{1.01}O_4]\cdot 2H_2O$, gladiusite, $(Fe^{2+}_{2.02}\ Mg_{1.61}Mn_{0.06})_{\Sigma3.69}Fe^{3+}_{2.17}[P_{1.02}O_4](OH)_{11}\cdot H_2O$, baricite, $(Mg_{1.75}Fe_{1.18})_{\Sigma2.93}[P_{2.03}O_8]\cdot 7H_2O$, vivianite, $(Fe_{2.88}Mg_{0.02})_{\Sigma2.90}[P_{2.04}O_8]\cdot 8H_2O$, pakhomovskyite, $(Co_{2.38}Mg_{0.38}Mn_{0.17}Ni_{0.04}Fe^{2+}_{0.03})_{\Sigma2.99}[P_{2.01}O_8]\cdot 8H_2O$, siderite, $(Fe_{0.97}Ca_{0.02}Mg_{0.01})_{\Sigma1.00}[CO_3]$, and pyroaurite $(Mg_{5.59}\ Fe^{2+}_{0.27})_{\Sigma5.86}(Fe^{3+}_{1.92}Al_{0.08})_{2.00}(OH)_{16}(CO_3)_{0.86}\cdot 4H_2O$ [11,33,35,68–70]. In particular, pakhomovskyite and gladiusite together with pyrite precipitated on walls of leached fissures in dolomite carbonatite vein (see Figure 5b,c) as a result of the following approximate reaction [35]:

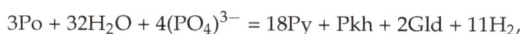

$$3Po + 32H_2O + 4(PO_4)^{3-} = 18Py + Pkh + 2Gld + 11H_2,$$

where Po is Co-bearing pyrrhotite, $(Fe_{10}Co)S_{12}$, Py—pyrite, Pkh—pakhomovskyite, Gld—gladiusite. Both phosphates form small (up to 0.5 mm in diameter) radiated aggregates of elongated plate crystals coloured dark brown (gladiusite) and bright pink (pakhomovskyite).

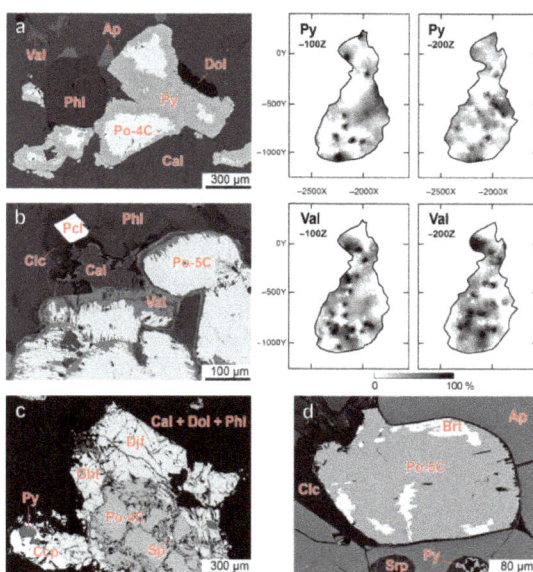

Figure 13. Replacement of pyrrhotite with pyrite ((**a**) phoscorite-related Cal-carbonatite 999/172.9), valleriite ((**b**) phoscorite-related Cal-carbonatite 972/86.9), djerfisherite ((**c**) phoscorite-related Cal-carbonatite 1004/656.5) and barite ((**d**) MAF-foscorite 941/41.5), BSE-images, and probability of apopyrrhotite pyrite and valleriite presence in the phoscorite-carbonatite complex. Mineral abbreviations are shown in Section 2.

3.3. Chalcopyrite and Products of Its Alteration

In natural sequence of the Kovdor rocks, content of copper gradually increases from host silicate rocks to earlier (apatite)-forsterite phoscorite, intermediate low-carbonate magnetite-rich phoscorite, and then to the latest calcite-rich phoscorite and carbonatites (see Table 2). On this reason, chalcopyrite content and grain size increase with growth of carbonate amount in the rock (up to 15 modal % and 1 cm in diameter, Figure 14a). In fact, this mineral is absent in peridotite and, rarer, in non-altered foidolite. In hydrothermally altered ijolite-urtite, it occurs as inclusions (up to 40 μm in diameter) in newly formed cancrinite (see Figure 2c) and calcite. Apo-peridotite diopsidite and phlogopitite contain irregularly shaped chalcopyrite grains (up to 1 mm in diameter) associated with pyrrhotite, magnetite and ilmenite in interstices of rock-forming silicates and apatite (see Figure 2d).

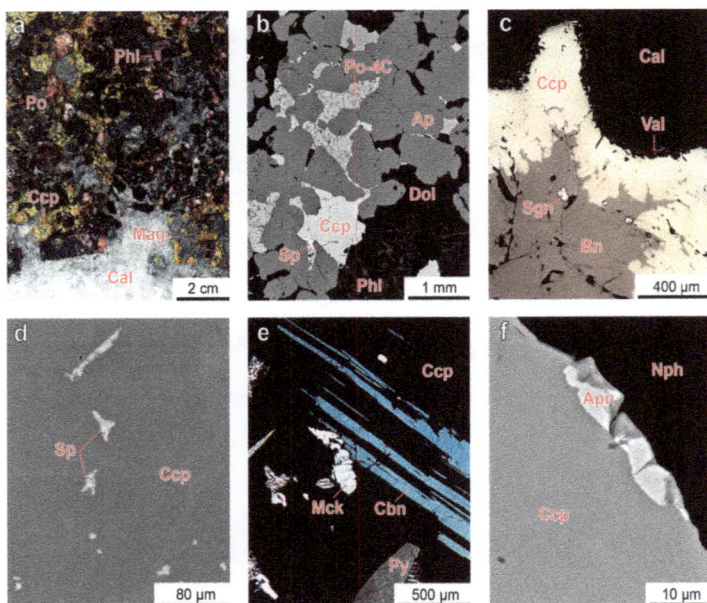

Figure 14. Chalcopyrite morphology and typical inclusions: (**a**) rich pyrrhotite-chalcopyrite mineralization in CMAF-phoscorite (K-2002, photo of hand specimen); (**b**) interstitial grains of sphalerite-containing chalcopyrite and pyrrhotite (MA-phoscorite 968/11.0, BSE-image); (**c**) bornite relict in chalcopyrite grain rimmed by valleriite (CMAF-phoscorite 2002/Bn2, photo in polarized reflected light); (**d**) exsolution inclusions of sphalerite in chalcopyrite (MA-phoscorite 968/11.0, BSE-image); (**e**) exsolution inclusions of cubanite in mackinawite-containing chalcopyrite (CMAF-phoscorite K-97-3, photo in polarized reflected light with crossed polarizers); (**f**) argentopentlandite inclusion in chalcopyrite (ijolite 949/209.2, BSE-image). Mineral abbreviations are shown in Section 2.

Marginal (apatite)-forsterite phoscorite is usually free from chalcopyrite. In intermediate low-carbonate phoscorite, chalcopyrite, together with pyrrhotite, fills interstices between grains of earlier forsterite, phlogopite, hydroxylapatite and magnetite (Figure 14b). Chalcopyrite usually occurs in marginal parts of pyrrhotite-chalcopyrite segregations or forms irregularly shaped gulf-like inclusions growing inside pyrrhotite grains from their margins. It also fills cleavage fractures in phlogopite, impregnates serpentine pseudomorphs after forsterite, accompanies late dolomite veinlets and segregations. In voids of vein dolomite carbonatite, late chalcopyrite forms druses of well-shaped tetrahedral crystals (up to 1 mm in diameter) in association with pyrite, anatase and titanite [11].

In calcite-rich phoscorite and carbonatite, chalcopyrite forms irregularly shaped chalcopyrite grains (up to 1 cm in diameter, see Figure 14a) partially replaced and rimmed by fine-grained pyrrhotite (see Figure 2h), as well as inclusions in magnetite, calcite and dolomite. Chalcopyrite grains sometimes contain irregularly shaped relicts (up to 15 μm in diameter, Figure 14c) of high bornite (cubic, $a = 5.47$ Å) accompanied by newly formed covellite [11].

Chemical composition of chalcopyrite varies insignificantly (Table 10), and its averaged formula corresponds to a theoretical one. Nevertheless, like pyrrhotite, chalcopyrite contains different exsolution inclusions (Table 11); therefore, we can assume that its initial composition is more complex. In particular, chalcopyrite grains usually carry sharply bounded cross- or star-like exsolution inclusions of sphalerite (up to 100 μm in diameter), as well as its gradually bounded rims around chalcopyrite grains and veinlets within them (up to 20 μm thick, Figure 14d). In axial carbonate-rich phoscorite and phoscorite-related carbonatite, there are exsolution inclusions of cubanite in mackinawite-bearing chalcopyrite (Figure 14e).

Cubanite forms brass-yellow, uniformly oriented, long prismatic inclusions (up to 3 mm × 1.5 mm) with split ends, as well as micro-granular aggregates, up to 1 mm in diameter [11,14]. Mackinawite occurs as wedge-shaped grains, skeletal crystals and stellate intergrowths (up to 100 μm in diameter, see Figure 14e), which are abundant in chalcopyrite, but relict or even absent in co-existing cubanite [11].

Table 10. Chemical composition of chalcopyrite (mean ± SD/min–max).

Rock	Phoscorite and Related Carbonatite			Vein Carbonatite	
	(Ap)-Fo	Low-Cb Mag-Rich	Cal-Rich	Cal	Dol
n	2	2	2	2	2
S, wt %	34.1 ± 0.8	34.7 ± 0.1	35.0 ± 0.9	34.9 ± 0.1	35.1 ± 0.3
	33.53–34.70	34.60–34.79	34.40–35.68	34.86–35.00	34.83–35.29
Fe	30.6 ± 0.2	29.9 ± 0.2	30.4 ± 0.3	30.1 ± 0.5	30.8 ± 0.1
	30.40–30.73	29.76–30.02	30.22–30.65	29.80–30.50	30.74–30.92
Co	<0.01	0.01 ± 0.01	0.03 ± 0.1	<0.01	<0.01
		<0.01–0.02	0.02–0.04		
Ni	<0.01	0.02 ± 0.02	0.01 ± 0.01	0.01 ± 0.02	0.02 ± 0.02
		<0.01–0.04	<0.01–0.02	<0.01–0.02	<0.01–0.03
Cu	35.0 ± 0.3	34.29 ± 0.05	34.3 ± 0.5	34.0 ± 0.3	34.5 ± 0.1
	34.80–35.29	34.26–34.32	33.93–34.58	33.80–34.24	34.39–34.53
Zn	<0.01	<0.01	<0.01	0.3 ± 0.3	<0.01
				0.00–0.50	
Cu, *apfu*	1.04 ± 0.04	1.00	0.99 ± 0.04	0.99 ± 0.01	0.99 ± 0.01
	1.01–1.06		0.96–1.01	0.98–0.99	0.99–1.00
Fe	1.03 ± 0.03	0.99	1.00 ± 0.02	0.99 ± 0.02	1.01 ± 0.01
	1.01–1.05		0.99–1.01	0.98–1.00	1.00–1.01
S	2	2	2	2	2

Table 11. Chemical composition of characteristic inclusions in chalcopyrite (wt %/*apfu*).

Mineral	Bn	Cbn	Mck	Sgn	Sp
Sample	KZh-25b	K-97-3	K-97-3	2002/Bn2	K-97-3
Rock	CM-phoscorite	CMF-phoscorite	CMF-phoscorite	CMAF-phoscorite	CMF-phoscorite
S	25.34/4.00	35.57/3.00	35.87/8.00	41.90/4.00	33.94/1.00
Mn	bd	bd	bd	bd	1.59/0.03
Fe	11.82/1.07	41.09/1.99	51.59/6.61	0.57/0.03	13.38/0.23
Co	bd	bd	12.50/1.52	19.72/1.02	bd
Ni	bd	bd	0.40/0.05	35.98/1.88	bd
Cu	62.67/4.99	23.70/1.01	bd	2.12/0.10	0.71/0.01
Zn	bd	bd	bd	bd	46.17/0.67
Cd	bd	bd	bd	bd	4.80/0.04
Total	99.83/10.06	100.36/6.00	100.36/16.18	100.29/7.03	100.59/1.98
Mineral	**Hwl**	**Apn**	**Acn**	**Vol**	**Tsu**
Sample	1009/121.4	910/348.2	999/76.6	913/57.1	913/57.1
Rock	MF-phoscorite	Diopsidite	CMAF-phoscorite	CM-phoscorite	CM-phoscorite
S	23.76/1.00	30.78/8.00	12.45/1.00	bd	bd
Fe	3.12/0.08	32.43/4.84	bd	bd	3.58/0.11
Co	bd	0.34/0.05	bd	bd	bd
Ni	bd	23.08/3.28	bd	bd	bd
Cu	6.50/0.14	bd	bd	bd	bd
Zn	8.29/0.17	bd	bd	bd	bd
Ag	bd	13.35/1.03	87.03/2.08	19.82/1.08	bd
Cd	58.11/0.70	bd	bd	bd	bd
Te	bd	bd	bd	43.52/2.00	38.07/1.00
Bi	bd	bd	bd	35.79/1.00	58.42/0.94
Total	99.78/2.09	99.98/17.20	99.48/3.08	99.13/4.08	100.07/2.05

In addition to the minerals described above, <50 μm sized inclusions of galena, hawleyite, argentopentlandite, siegenite and acanthite were found in chalcopyrite grains [11]. Galena and hawleyite form rounded grains at the contacts between chalcopyrite and pyrrhotite, bornite, covellite, carbonates. Argentopentlandite (Figure 14f) occurs in ijolite, diopsidite and phlogopitite as irregularly shaped inclusions (up to 30 μm) in marginal parts of chalcopyrite grains. Siegenite together with galena, hawleyite, wittichenite and covellite form small (10–70 μm) inclusions in bornite and chalcopyrite pseudomorphs after bornite (see Figure 14c). Irregularly shaped grains of acanthite (up to 20 μm) were found at the contact between pyritized chalcopyrite and calcite.

Besides, chalcopyrite may contain fine impregnation of volynskite, tsumoite and native silver. Volynskite and tsumoite were found in calcite-magnetite phoscorite as thin intergrowth (up to 5 μm in diameter) within chalcopyrite inclusions in magnetite. Native silver forms thin (<1 μm thick) veinlets and xenomorphic inclusions (up to 3 μm in diameter) in chalcopyrite grains, especially, at the contacts between chalcopyrite grains and pyrrhotite, sphalerite.

Typical products of chalcopyrite alteration include valleriite, djerfisherite, chalcocite, covellite, pyrite and goethite (Table 12), which replace chalcopyrite grains from margins and fractures (see Figure 14c), up to formation of complete pseudomorphs after chalcopyrite:

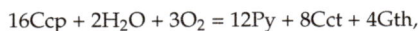

$$16Ccp + 2H_2O + 3O_2 = 12Py + 8Cct + 4Gth,$$

$$2Ccp + 3Mgs + 3H_2O = Val + CO_2,$$

$$25Ccp + 12KCl + 2NaCl + 2S + Cl_2 = 2Djf, \text{ etc.}$$

where Ccp is chalcopyrite, $FeCuS_2$; Cct—chalcocite, Cu_2S; Djf—djerfisherite, $K_6NaFe_{12.5}Cu_{12.5}S_{26}Cl$; Gth—goethite, $FeO(OH)$; Mgs—magnesite, $MgCO_3$; Py—pyrite, FeS_2 and Val—valleriite, $Mg_3Fe_2Cu_2S_4(OH)_6$.

Table 12. Chemical composition of characteristic products of chalcopyrite alteration (wt %/*apfu*).

Mineral	Val	Djf	Cct	Cv	Py
Sample	KZh-25b	K-97-3	K-01-1110	K-02-122	K-0042
Rock	CM-phoscorite	CMF-phoscorite	CMAF-phoscorite	CM-phoscorite	Dol-carbonatite
Mg	11.12/2.54	bd	bd	bd	bd
Al	3.57/0.73	bd	bd	bd	bd
S	23.14/4.00	32.91/26.00	19.64/1.00	31.68/1.00	53.20/2.00
Cl	bd	0.63/0.45	bd	bd	bd
K	bd	9.06/5.87	bd	bd	bd
Fe	25.66/2.55	35.55/16.13	bd	1.54/0.03	46.76/1.01
Co	bd	0.06/0.03	bd	bd	0.36/0.01
Ni	0.12/0.01	1.40/0.60	bd	bd	0.02/0.00
Cu	12.06/1.05	16.51/6.58	79.12/2.03	66.30/1.06	bd
Ag	bd	3.34/0.78	bd	bd	bd
Total	75.67/10.88	99.46/56.44	98.76/3.03	99.52/2.09	100.34/3.02

Different low-temperature copper minerals appear also in neighboring voids and fractures. In particular, several funnels of staffelite breccia in apical (now excavated) part of the phoscorite-carbonatite pipe axial zone are formed by fragments of altered chalcopyrite-rich phoscorite and carbonatite cemented by bluish green colloform fluorapatite with inclusions of brochantite, $Cu_4[SO_4](OH)_6$, malachite, $(Cu_{1.98}Fe_{0.02})_{2.00}[CO_3](OH)_2$, pseudomalachite, $(Cu_{4.85}Ca_{0.07})_{4.92}[P_{2.03}O_8](OH)_4$, chrisocolla, $(Cu_{1.85}Ca_{0.03})_{1.88}H_{2.00}Si_{2.06}O_5(OH)_4 \cdot nH_2O$, tenorite, $Cu_{1.00}O$, and goethite $(Fe_{0.86}Si_{0.11})_{0.97}O(OH)$ [11].

4. Discussion

The Kovdor phoscorite-carbonatite complex has gradual concentric zonation in terms of the rocks modal and chemical composition, as well as grain size, chemical composition, crystallochemical features and properties of rock-forming and accessory minerals [11,21,22,24,26–29]. This zonation mainly results from gradual change in chemical composition of the residual phoscorite melt from silicate-rich to carbonate-rich during crystallization from the pipe margins to its axial zone [22].

Geochemically, nickel behaves like Mg and Fe^{2+}, and readily substitutes them in most crystalline phases including olivine [71]. Besides, there is a strong negative dependence of Ni activity on FeO_{tot} content in silicate melt [72,73]. However, even in a high-magnesian system, like the Kovdor peridotite and forsteritite, nickel can remain in the residual silicate melt (SM) because its partition coefficient $D^{Ol/SM}$ depends significantly on sulfur content [74]. In fact, at fixed P–T conditions, the partition coefficients of chalcophile elements M between monosulfide solid solution (MSS), $(Fe,Ni,Co)_{1-x}S$, and silicate melt are controlled by the ratio fS_2/fO_2, which, in turn, depends on FeO_{tot} content in sulfide-saturated silicate melt [75–77]:

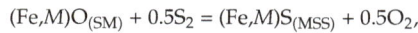

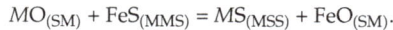

$$(Fe,M)O_{(SM)} + 0.5S_2 = (Fe,M)S_{(MSS)} + 0.5O_2,$$

$$MO_{(SM)} + FeS_{(MMS)} = MS_{(MSS)} + FeO_{(SM)}.$$

As a result, the MSS–SM partition coefficients for Ni, Cu and Co substantially increase with FeO_{tot} decrease in silicate melt (Figure 15a), and we can conclude that crystallization of Mg-rich (apatite)-forsterite phoscorite can produce Fe-Ni-Co-Cu-rich sulfide melt.

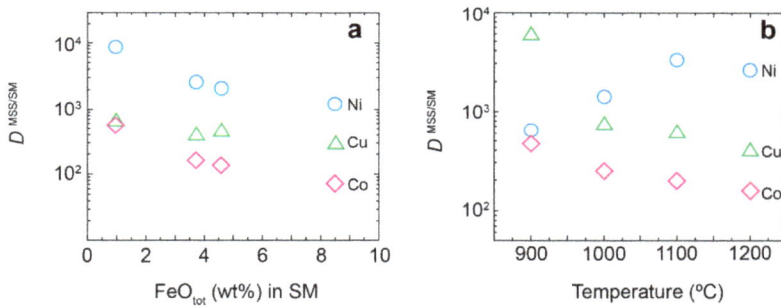

Figure 15. Effect of FeO_{tot} content in SM ((**a**) T = 1200 °C and P = 1.5 GPa) and melt temperature ((**b**) FeO_{tot} = 3.7 wt %) on partition coefficients of Ni, Co and Cu between MSS and SM [77].

In earlier low-carbonate phoscorite, pyrrhotite is formed at the final stage of magnetite crystallization, because comparatively low sulfur content (see Table 2) still requires melt saturation in FeS complexes. For this reason, ferrimagnetic pyrrhotite-4C (or the corresponding MSS) first closely associates with magnetite forming gulf-like ingrowths inside its grains. Then, with transition to carbonate-rich phoscorite and carbonatite, iron-rich pyrrhotite-5-6C (or the corresponding MSS) crystallizes independently of magnetite. Increase of mean equivalent circular diameter of pyrrhotite grains reflects a decrease of the rock crystallization rate from the pipe wall to its axis. This trend is similar to that of co-existing rock-forming and accessory minerals, in particular, forsterite, magnetite and baddeleyite [22,26,27].

Our data on crystal structure of the Kovdor pyrrhotites are consistent with numerous previous studies of the structure [48,51,54–57,78], which demonstrate different space groups for the most common 4C, 5C and 6C polytypes. Such structural diversity is caused by different P-T conditions of crystallization, amount of incorporated Fe^{3+} or different mechanisms of vacancies distribution in the structure. According to the structural complexity classification [79,80], 4C polytype is "simple", while 5C and 6C modifications are "intermediate" (Figure 16). Relatively low-temperature formation of pyrrhotite-5C in comparison with pyrrhotite-4C is in good agreement with a common tendency of structural complexity growth with crystallization temperature decrease. From this point of view, use of the term 'pyrrhotite-*n*C' for 5C and 6C polytypes is justified by the absence of significant differences between their complexity.

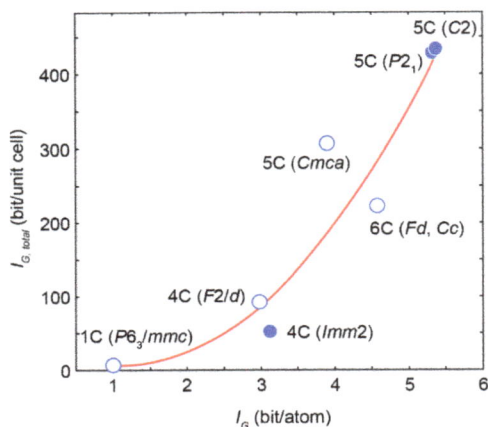

Figure 16. Total structural information content of pyrrhotite ($I_{G, total}$) plotted against the information content per atom (I_G) according to [48,50,51,54,57] (empty circles) and our data (full circles).

Decrease of monosulfide crystallization temperature causes growth of $D^{MSS/SM}$ for Cu and Co (Figure 15b), while Ni partition coefficient remains the same or even decreases [77,81], which enables fractionation of Ni and Cu-Co between low- and high-temperature sulfides respectively. Therefore, we believe that concentrically zoned distribution of pyrrhotite with different Fe, Ni and Co contents is a response to a gradual decrease of crystallization temperature from the ore-pipe marginal forsterite-dominant zone to its axial carbonate-rich zone. Exsolution of MSS or the corresponding higher-temperature pyrrhotite modifications produced lamella or flames of pentlandite in the ore-pipe marginal zone, and cobaltpentlandite in the pipe axial zone.

For the same reason, copper is concentrated in the pipe axial carbonate-rich zone. Carbonate melt appears to contain a significant amount of water [82] and copper as stable hydrosulfide complexes [83]. Their destruction by bornite and chalcopyrite crystallization after carbonates caused intensive hydrothermal alteration of associated minerals, and formation of secondary pyrite, valleriite and other phases.

It is interesting to compare sulfide mineralization of the phoscorite-carbonatite complexes of the Kovdor (Russia) and Palabora (South Africa) alkaline-ultrabasic massifs that have comparable composition, structure and metallogenic specialization [12]. However, the Loolekop phoscorite-carbonatite pipe in the Palabora massif is the largest carbonatite-hosted copper deposit in the world 850 Mt @ 0.5% Cu [84], and the Kovdor phoscorite-carbonatite complex has comparatively low copper content. What caused such a significant enrichment of the Loolekop phoscorite and carbonatites in copper?

The Loolekop phoscorite-carbonatite pipe is a close relative to the Kovdor one. The Loolekop pipe (1.4 km × 0.8 km) is situated in the eastern part of the Proterozoic central-type Palabora complex of apo-peridotite serpentinite (central stock), shonkinite (outer ring intrusion) and the related diopsidite and phlogopitite between them. It has a clear concentric zonation with marginal low-carbonate phoscorite, intermediate "banded" zone of interlayered carbonate-rich phoscorite and phoscorite-related calcite carbonatite and axial stockwork of "transgressive" calcite and dolomite carbonatite [85–87].

All these rocks are rich in pyrrhotite with exsolution inclusions of pentlandite–cobaltpentlandite [87], and contain also sufficient amount of copper sulfides: dominant bornite and chalcocite in phoscorite and banded carbonatite and dominant chalcopyrite with relicts of bornite and exsolution inclusions of cubanite in transgressive carbonatite [85,86,88,89]. Minor copper minerals include covellite, valleriite, tetrahedrite, and native copper [12,85,90]. As a result, copper content increases from <0.3 wt % in marginal phoscorite to 0.3–0.9 wt % in intermediate banded carbonatite and about 1 wt % in axial transgressive carbonatite [85,86].

In phoscorite and phoscorite-related "banded" carbonatite, sulfides fill interstices and thin fractures in magnetite-apatite-forsterite-calcite aggregate, and form irregularly shaped segregations with pyrrhotite in core and copper-bearing sulfides in marginal zone, while in the late "transgressive" carbonatite, they form subparallel vertical lenses and veinlets (usually, up to 3 cm thick) within a vertical ore-zone with cross-section about 200 m × 600 m [85,87]. In both rocks, pyrrhotite and pentlandite–cobaltpentlandite are the earliest sulfides, high bornite is the next one, and chalcopyrite and other minerals follow them.

Thus, the Kovdor and Loolekop phoscorite-carbonatite pipes have similar sequences of sulfide formation (Figure 17a), from earlier pyrrhotite to intermediate copper sulfides and late pyrite and valleriite. There are only differences in positions of argentopentlandite and mackinawite that are common products of chalcopyrite exsolution in the Kovdor complex, and substitute pentlandite in the Loolekop pipe [12,87,89]. It is necessary to note that secondary djerfisherite is widely spread in the Kovdor complex due to higher alkalinity of this massif.

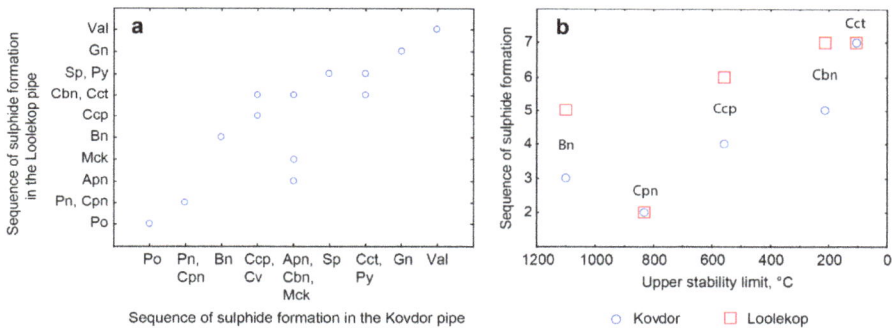

Figure 17. Order of sulfide formation in the Kovdor and Loolekop phoscorite-carbonatite pipes (**a**) and the corresponding stability temperature [62] for the Kovdor's sequence (**b**). Mineral abbreviations are shown in Section 2.

The sequences of sulfide formation in the phoscorite-carbonatite pipes are in a good accordance with the upper limits of their stability [62] that linearly decrease from 1100 °C for high bornite to 103 °C for chalcocite (Figure 17b). For the Kovdor complex, these relations can be supplemented with temperature of pyrite-pyrrhotite equilibration ranged from 170 °C in intermediate low-carbonate magnetite-rich phoscorite and dolomite carbonatite to 300 °C in axial carbonate-rich phoscorite and phoscorite-related carbonatite, and temperatures of the pyrrhotite superstructure formation from 140 °C for pyrrhotite-4C to 60 °C for pyrrhotite-5C [62].

One of the most likely reasons for sulfide specialization in the Loolekop phoscorite-carbonatite pipe in comparison with the Kovdor one is insignificant differences in oxygen fugacity of initial phoscorite melts (due to higher water content), because the larger oxygen fugacity causes more intensive copper extraction by melt from surrounding rocks [5,91–93]. In fact, even a slight increase in oxidation state (to ≥FMQ + 1) can shift the balance between sulfide and sulfate components dissolved in melt towards more soluble sulfates, allowing magmas to dissolve much larger amounts of sulfur than reduced melts [94]. Higher water content in the Palabora's magma is confirmed, in particular, by much deeper alteration of peridotite and phoscorite into serpentinite, much abundant pegmatites, much richer valleriite mineralization, etc. [85,90,95].

5. Conclusions

1. Primary silicate rocks of the Kovdor massif (peridotite, foidolite–melilitolite and forsteritite) are free of sulfides due to sulfur fractionation in fluid phase. For the same reason, hydrothermally altered parts of these rocks, including diopsidite, phlogopitite and skarn-like rocks, carry rich sulfide mineralization associated with secondary minerals (cancrinite, natrolite, phlogopite, clinochlore, serpentine, vesuvianite, etc.).

2. Distribution of rock-forming sulfides within the Kovdor phoscorite-carbonatite complex reflects gradual concentric zonation of the pipe: pyrrhotite with exsolution inclusions of pentlandite in marginal (apatite)-forsterite phoscorite, pyrrhotite with exsolution inclusions of cobaltpentlandite in intermediate low-carbonate magnetite-rich phoscorite, and chalcopyrite (±pentlandite with exsolution inclusions of cobaltpentlandite) in axial carbonate-rich phoscorite and phoscorite-related carbonatite;

3. Both pyrrhotite and chalcopyrite fill interstices between the main rock-forming minerals, including carbonates, and form late veinlets and irregularly shaped segregations with inclusions of surrounding minerals. Usually, chalcopyrite (with relicts of earlier bornite and exsolution inclusions of cubanite and mackinawite) crystallizes around grains of pyrrhotite (with inclusions of pentlandite-cobaltpentlandite and pyrite), and both these minerals contain common exsolution inclusions of sphalerite. Temperature of pyrite-pyrrhotite equilibration ranges from 100 °C to 400 °C, and pressure of pyrrhotite-sphalerite equilibration reaches 10 kbar;

4. For the most part, pyrrhotite corresponds to its non-magnetic 5C polytype. Ferrimagnetic pyrrhotite-4C forms individual grains, marginal zones of non-magnetic pyrrhotite crystals and thin lens-like inclusions in pyrrhotite-5C. Low-temperature pyrrhotite 2C (troilite) occurs as lens-like exsolution inclusions in grains of pyrrhotite-4C and -5C. In natural sequence of the Kovdor rocks, iron content in pyrrhotite gradually increases from Fe_7S_8 to Fe_9S_{10} and $Fe_{11}S_{12}$ in accordance with gradual decrease of crystallization temperature and oxygen fugacity.

5. High complexity of crystal structure of pyrrhotite-5C ($I_{G,total}$ = 429.75 bits per unit cell), which is the most common within the Kovdor massif, confirms the lowest temperature of its formation in comparison with other polytypes. Wide dissemination of this modification is associated with structural stability (variations in composition $Fe_{8.84-8.99}S_{10}$) due to different vacancies ordering mechanisms resulting in two structural modifications: $P2_1$ and $C2$.

6. Within the phoscorite-carbonatite complex, content of Co in pyrrhotite gradually increases from host silicate rocks and marginal forsterite-dominant phoscorite to axial carbonate-rich phoscorite and carbonatite due to Ni and Fe. This dependence probably reflects a gradual decrease of crystallyzation temperature in primary monosulfide solid solutions from the pipe margin toward its axis.

7. Low-temperature hydrothermal alteration of pyrrhotite and chalcopyrite first produces partial or complete pseudomorphs of djerfisherite, pyrite, chalcocite, valleriite and goethite after the rock-forming sulfides, and then numerous phosphates and carbonates of Fe, Co and Cu (baricite, gladiusite, malachite, mitridatite, pakhomovskyite, pseudomalachite, pyroaurite, siderite, strengite, vivianite) in surrounding voids and fissures.

8. The Kovdor and Loolekop phoscorite-carbonatite pipes have similar sequences of sulfide formation: pyrrhotite–pentlandite–cobaltpentlandite–bornite–chalcopyrite–covellite–cubanite–sphalerite–pyrite-chalcocite–galena–vallereite. This sequence corresponds to the sulfide stability limits ranged from 1100 °C (high bornite) to 103 °C (chalcocite) and even lower (for valleriite). Thus, copper specialization of the Loolekop pipe is not caused by any specific process but reflects stochastic specifics of the phoscorite-carbonatite genesis.

9. Sulfide specialization of the Palabora massif can be caused by higher water content in its initial melt allowing it to dissolve much larger amounts of sulfur and, consequently, concentrate more chalcophile metals.

Supplementary Materials: The following are available online at http://www.mdpi.com/2075-163X/8/7/292/s1, Table S1: Grain size and chemical composition of the pyrrhotite group minerals. The CIF files for structurally studied pyrrhotites 1–5.

Author Contributions: G.Y.I. designed the experiments, performed statistical investigations and crystal size distribution analyses, and wrote the manuscript. T.L.P. carried out X-ray investigations. A.O.K. and P.M.G. performed geostatistical investigation, drew maps, took samples and reviewed the manuscript. Y.A.P. and A.V.B. took BSE images and performed electron microscope investigations. J.A.M. carried out petrographical investigations and reviewed the manuscript. V.N.Y. and N.G.K. conceived of the work, took and prepared samples. All authors discussed the manuscript.

Funding: The research is supported by the Russian Science Foundation, grant 16-17-10173.

Acknowledgments: Samples were taken during exploration of deep levels of the Kovdor deposit implemented by JSC Kovdorskiy GOK in 2007–2011. X-ray crystal studies were carried out with the equipment provided by the X-ray Diffraction Centre of Saint-Petersburg State University. The authors would like to thank I.R. Elizarova who performed the ICP-MS analyses of bulk-rock samples. Comments and remarks of anonymous reviewers were very useful.

References

1. Bailey, D.K.; Hampton, C.M. Volatiles in alkaline magmatism. *Lithos* **1990**, *26*, 157–165. [CrossRef]
2. Samson, I.M.; Williams-Jones, A.E.; Liu, W. The chemistry of hydrothermal fluids in carbonatites: Evidence from leachate and SEM-decrepitate analysis of fluid inclusions from Oka, Quebec, Canada. *Geochim. Cosmochim. Acta* **1995**, *59*, 1979–1989. [CrossRef]
3. Brooker, R.A.; Sparks, R.S.J.; Kavanagh, J.L.; Field, M. The volatile content of hypabyssal kimberlite magmas: Some constraints from experiments on natural rock compositions. *Bull. Volcanol.* **2011**, *73*, 959–981. [CrossRef]
4. Klimm, K.; Kohn, S.C.; Botcharnikov, R.E. The dissolution mechanism of sulphur in hydrous silicate melts. II: Solubility and speciation of sulphur in hydrous silicate melts as a function of fO_2. *Chem. Geol.* **2012**, *322–323*, 250–267. [CrossRef]
5. Klimm, K.; Kohn, S.C.; Dell, L.A.O.; Botcharnikov, R.E.; Smith, M.E. The dissolution mechanism of sulphur in hydrous silicate melts. I: Assessment of analytical techniques in determining the sulphur speciation in iron-free to iron-poor glasses. *Chem. Geol.* **2012**, *322–323*, 237–249. [CrossRef]
6. Métrich, N.; Berry, A.J.; O'Neill, H.S.C.; Susini, J. The oxidation state of sulfur in synthetic and natural glasses determined by X-ray absorption spectroscopy. *Geochim. Cosmochim. Acta* **2009**, *73*, 2382–2399. [CrossRef]
7. O'Neill, H.S.C.; Mavrogenes, J.A. The Sulfide Capacity and the Sulfur Content at Sulfide Saturation of Silicate Melts at 1400 degrees C and 1 bar. *J. Petrol.* **2002**, *43*, 1049–1087. [CrossRef]
8. Klimm, K.; Botcharnikov, R.E. The determination of sulfate and sulfide species in hydrous silicate glasses using Raman spectroscopy. *Am. Mineral.* **2010**, *95*, 1574–1579. [CrossRef]
9. Seward, T.M.; Williams-Jones, A.E.; Migdisov, A.A. The Chemistry of Metal Transport and Deposition by Ore-Forming Hydrothermal Fluids. In *Treatise on Geochemistry*; Elsevier: Berlin, Germany, 2014; pp. 29–57. ISBN 9780080959757.
10. Reed, M.H.; Palandri, J. Sulfide Mineral Precipitation from Hydrothermal Fluids. *Rev. Mineral. Geochem.* **2006**, *61*, 609–631. [CrossRef]
11. Ivanyuk, G.Y.; Yakovenchuk, V.N.; Pakhomovsky, Y.A. *Kovdor*; Laplandia Minerals: Apatity, Russia, 2002; ISBN 5900395413.
12. Rudashevsky, N.S.; Kretser, Y.L.; Rudashevsky, V.N.; Sukharzhevskaya, E.S. A review and comparison of PGE, noble-metal and sulphide mineralization in phoscorites and carbonatites from Kovdor and Phalaborwa. In *Phoscorites and Carbonatites from Mantle to Mine: The Key Example of the Kola Alkaline Province*; Wall, F., Zaitsev, A.N., Eds.; Mineralogical Society: London, UK, 2004; pp. 375–405.

13. Kukharenko, A.A.; Orlova, M.P.; Bulakh, A.G.; Bagdasarov, E.A.; Rimskaya-Korsakova, O.M.; Nefedov, E.I.; Ilyinsky, G.A.; Sergeev, A.S.; Abakumova, N.B. *Caledonian Complex of Ultrabasic, Alkaline Rocks and Carbonatites of Kola Peninsula and Northern Karelia (Geology, Petrology, Mineralogy and Geochemistry)*; Nedra: Moscow, Russia, 1965. (In Russian)

14. Bykova, E. Sulphide mineralization in magnetite ores and carbonatites of the Kovdor massif. *Mineral. Geochem.* **1975**, *5*, 11–16. (In Russian)

15. Subbotina, G.F.; Subbotin, V.V.; Pakhomovsky, Y.A. Some features of a sulphide mineralization of apatite-magnetite ores and carbonatites of the Kovdor deposit. In *Substantial Composition of Alkaline Intrusive Complexes of the Kola Peninsula*; KFAN USSR: Apatity, Russia, 1981; pp. 88–95.

16. Balabonin, N.L.; Voloshin, A.V.; Pakhomovsky, Y.A. Rare sulphides in rocks of the Kovdor deposit. In *Mineral Complexes and Minerals of the Kola Peninsula*; KFAN USSR: Apatity, Russia, 1980; pp. 88–92. (In Russian)

17. Balabonin, N.L.; Voloshin, A.V.; Pakhomovsky, Y.A.; Polyakov, K.I. Composition of djerfisherite from alkaline complexes of the Kola Peninsula. *Mineral. Zhurnal* **1980**, *1*, 90–99. (In Russian)

18. Kapustin, Y.L. *Mineralogy of Carbonatites*; Amerind Publishing: New Delhi, India, 1980.

19. Rimskaya-Korsakova, O.M.; Krasnova, N.I. *Geology of Deposits of the Kovdor Massif*; St. Petersburg University Press: St. Petersburg, Russia, 2002. (In Russian)

20. Krasnova, N.I.; Petrov, T.G.; Balaganskaya, E.G.; García, D.; Moutte, J.; Zaitsev, A.N.; Wall, F. Introduction to phoscorites: Occurrence, composition, nomenclature and petrogenesis. In *Phoscorites and Carbonatites from Mantle to Mine: The Key Example of the Kola Alkaline Province*; Zaitsev, A.N., Wall, F., Eds.; Mineralogical Society: London, UK, 2004; pp. 43–72, ISBN 0903056224.

21. Afanasyev, B.V. *Mineral Resources of Alkaline-Ultrabasic Massifs of the Kola Peninsula*; Roza Vetrov Publishing: St. Petersburg, Russia, 2011. (In Russian)

22. Mikhailova, J.A.; Ivanyuk, G.Y.; Kalashnikov, A.O.; Pakhomovsky, Y.A.; Bazai, A.V.; Panikorovskii, T.L.; Yakovenchuk, V.N.; Konopleva, N.G.; Goryainov, P.M. Three-D Mineralogical Mapping of the Kovdor Phoscorite–Carbonatite Complex, NW Russia: I. Forsterite. *Minerals* **2018**, *8*, 260. [CrossRef]

23. Mikhailova, J.A.; Kalashnikov, A.O.; Sokharev, V.A.; Pakhomovsky, Y.A.; Konopleva, N.G.; Yakovenchuk, V.N.; Bazai, A.V.; Goryainov, P.M.; Ivanyuk, G.Y. 3D mineralogical mapping of the Kovdor phoscorite–carbonatite complex (Russia). *Miner. Depos.* **2016**, *51*, 131–149. [CrossRef]

24. Kalashnikov, A.O.; Ivanyuk, G.Y.; Mikhailova, J.A.; Sokharev, V.A. Approach of automatic 3D geological mapping: The case of the Kovdor phoscorite-carbonatite complex, NW Russia. *Sci. Rep.* **2017**, *7*, 6893. [CrossRef] [PubMed]

25. Kalashnikov, A.O.; Konopleva, N.G.; Pakhomovsky, Y.A.; Ivanyuk, G.Y. Rare earth deposits of the Murmansk Region, Russia—A review. *Econ. Geol.* **2016**, *111*, 1529–1559. [CrossRef]

26. Ivanyuk, G.; Kalashnikov, A.; Pakhomovsky, Y.; Bazai, A.; Goryainov, P.; Mikhailova, J.; Yakovenchuk, V.; Konopleva, N. Subsolidus Evolution of the Magnetite-Spinel-Ulvöspinel Solid Solutions in the Kovdor Phoscorite-Carbonatite Complex, NW Russia. *Minerals* **2017**, *7*, 215. [CrossRef]

27. Ivanyuk, G.Y.; Kalashnikov, A.O.; Pakhomovsky, Y.A.; Mikhailova, J.A.; Yakovenchuk, V.N.; Konopleva, N.G.; Sokharev, V.A.; Bazai, A.V.; Goryainov, P.M. Economic minerals of the Kovdor baddeleyite-apatite-magnetite deposit, Russia: Mineralogy, spatial distribution and ore processing optimization. *Ore Geol. Rev.* **2016**, *77*, 279–311. [CrossRef]

28. Kalashnikov, A.O.; Yakovenchuk, V.N.; Pakhomovsky, Y.A.; Bazai, A.V.; Sokharev, V.A.; Konopleva, N.G.; Mikhailova, J.A.; Goryainov, P.M.; Ivanyuk, G.Y. Scandium of the Kovdor baddeleyite-apatite-magnetite deposit (Murmansk Region, Russia): Mineralogy, spatial distribution, and potential resource. *Ore Geol. Rev.* **2016**, *72*, 532–537. [CrossRef]

29. Ivanyuk, G.Y.; Pakhomovsky, Y.A.; Panikorovsky, T.L.; Mikhailova, J.A.; Kalashnikov, A.O.; Bazai, A.V.; Yakovenchuk, V.N.; Konopleva, N.G.; Goryainov, P.M. Three-D Mineralogical Mapping of the Kovdor Phoscorite-Carbonatite Complex, NW Russia: III. Pyrochlore Supergroup Minerals. *Minerals* **2018**, *8*, 277. [CrossRef]

30. Krivovichev, S.V.; Yakovenchuk, V.N.; Zhitova, E.S.; Zolotarev, A.A.; Pakhomovsky, Y.A.; Ivanyuk, G.Y. Crystal chemistry of natural layered double hydroxides. 1. Quintinite-2H-3c from the Kovdor alkaline massif, Kola peninsula, Russia. *Mineral. Mag.* **2010**, *74*, 821–832. [CrossRef]

31. Krivovichev, S.V.; Yakovenchuk, V.N.; Zhitova, E.S.; Zolotarev, A.A.; Pakhomovsky, Y.A.; Ivanyuk, G.Y. Crystal chemistry of natural layered double hydroxides. 2. Quintinite-1M: First evidence of a monoclinic polytype in M2+-M3+ layered double hydroxides. *Mineral. Mag.* **2010**, *74*, 833–840. [CrossRef]

32. Zhitova, E.S.; Yakovenchuk, V.N.; Krivovichev, S.V.; Zolotarev, A.A.; Pakhomovsky, Y.A.; Ivanyuk, G.Y. Crystal chemistry of natural layered double hydroxides. 3. The crystal structure of Mg, Al-disordered quintinite-2H. *Mineral. Mag.* **2010**, *74*, 841–848. [CrossRef]

33. Zhitova, E.S.; Ivanyuk, G.Y.; Krivovichev, S.V.; Yakovenchuk, V.N.; Pakhomovsky, Y.A.; Mikhailova, Y.A. Crystal Chemistry of Pyroaurite from the Kovdor Pluton, Kola Peninsula, Russia, and the Långban Fe–Mn deposit, Värmland, Sweden. *Geol. Ore Depos.* **2017**, *59*, 652–661. [CrossRef]

34. Zhitova, E.S.; Krivovichev, S.V.; Yakovenchuk, V.N.; Ivanyuk, G.Y.; Pakhomovsky, Y.A.; Mikhailova, J.A. Crystal chemistry of natural layered double hydroxides. 4. Crystal structures and evolution of structural complexity of quintinite polytypes from the Kovdor alkaline massif, Kola peninsula, Russia. *Mineral. Mag.* **2018**, *82*, 329–346. [CrossRef]

35. Yakovenchuk, V.N.; Ivanyuk, G.Y.; Mikhailova, Y.A.; Selivanova, E.A.; Krivovichev, S.V. Pakhomovskyite, $Co_3(PO_4)_2 \cdot 8H_2O$, a new mineral species from Kovdor, Kola Peninsula, Russia. *Can. Mineral.* **2006**, *44*, 117–123. [CrossRef]

36. Yakovenchuk, V.N.; Ivanyuk, G.Y.; Pakhomovsky, Y.A.; Panikorovskii, T.L.; Britvin, S.N.; Krivovichev, S.V.; Shilovskikh, V.V.; Bocharov, V.N. Kampelite, $Ba_3Mg_{1.5}Sc_4(PO_4)_6(OH)_3 \cdot 4H_2O$, a new very complex Ba-Sc phosphate mineral from the Kovdor phoscorite-carbonatite complex (Kola Peninsula, Russia). *Mineral. Petrol.* **2017**, *4*, 111–121. [CrossRef]

37. Tomilin, M.G.; Ivanyuk, G.Y. The application of thin nematic liquid crystal layers to mineral analysis. *Liq. Cryst.* **1993**, *14*, 1599–1606. [CrossRef]

38. UTHSCSA. ImageTool 3.0. Available online: http://www.compdent.uthscsa.edu/dig/pub/IT (accessed on 8 July 2013).

39. Rosstandart FR.1.31.2016.25424. The Method for Determining the Content of Rare-Earth Elements (Y, La, Ce, Pr, Nd, Sm, Eu, Gd, Tb, Dy, Ho, Er, Tm, Yb, Lu), Sodium, Aluminum, Potassium, Calcium, Titanium, Iron, Thorium and Uranium in Apatite Mineral Raw Materials and Phosphogypsum by Inductively Coupled Plasma Mass Spectrometry. Available online: http://www.fundmetrology.ru/06_metod/2view_file.aspx?id=25424 (accessed on 27 June 2018).

40. Dolivo-Dobrovolsky, D.D. MINAL, Free Software. Available online: http://www.dimadd.ru (accessed on 8 July 2013).

41. StatSoft Inc. Statistica 8. Available online: www.statsoft.ru (accessed on 27 June 2018).

42. Systat Software Inc. TableCurve 2D. Available online: www.sigmaplot.co.uk/products/tablecurve2d/tablecurve2d.php (accessed on 19 June 2018).

43. Micromine Pty Ltd. Micromine 16.1. Available online: www.micromine.com (accessed on 27 June 2018).

44. Sheldrick, G.M. A short history of SHELX. *Acta Crystallogr. Sect. A Found. Crystallogr.* **2008**, *64*, 112–122. [CrossRef] [PubMed]

45. Dolomanov, O.V.; Bourhis, L.J.; Gildea, R.J.; Howard, J.A.K.; Puschmann, H. OLEX2: A complete structure solution, refinement and analysis program. *J. Appl. Crystallogr.* **2009**, *42*, 339–341. [CrossRef]

46. Agilent CrysAlis PRO 2014. Available online: https://www.rigaku.com/en/products/smc/crysalis (accessed on 19 June 2018).

47. Putz, H.; Brandenburg, K. *Diamond–Crystal and Molecular Structure Visualization*; Crystal Impact GbR: Bonn, Germany, 2012.

48. De Villiers, J.P.R.; Liles, D.C. The crystal-structure and vacancy distribution in 6C pyrrhotite. *Am. Mineral.* **2010**, *95*, 148–152. [CrossRef]

49. *International Tables for Crystallography. Volume C: Mathematical, Physical and Chemical Tables*; Wilson, A.J.C. (Ed.) Kluwer Academic: Dordrecht, The Netherlands, 1992.

50. Alsén, N. Röntgenographische Untersuchung der Kristallstrukturen von Magnetkies, Breithauptit, Pentlandit, Millerit und verwandten Verbindungen. *Geologiska Föreningen i Stockholm Förhandlingar* **1925**, *47*, 19–72. [CrossRef]

51. Tokonami, M.; Nishiguchi, K.; Morimoto, N. Crystal structure of a monoclinic pyrrhotite (Fe7S8). *Am. Mineral.* **1972**, *57*, 1066–1080.

52. Morimoto, N.; Gyobu, A.; Mukaiyama, H.; Izawa, E. Crystallography and stability of pyrrhotites. *Econ. Geol.* **1975**, *70*, 824–833. [CrossRef]

53. De Villiers, J.P.R.; Liles, D.C.; Becker, M. The crystal structure of a naturally occurring 5C pyrrhotite from Sudbury, its chemistry, and vacancy distribution. *Am. Mineral.* **2009**, *94*, 1405–1410. [CrossRef]

54. Koto, K.; Morimoto, N.; Gyobu, A. The superstructure of the intermediate pyrrhotite. I. Partially disordered distribution of metal vacancy in the 6 C type, Fe 11 S 12. *Acta Crystallogr. Sect. B Struct. Crystallogr. Cryst. Chem.* **1975**, *31*, 2759–2764. [CrossRef]

55. Yamamoto, A.; Nakazawa, H. Modulated structure of the NC-type (N = 5.5) pyrrhotite, $Fe_{1-x}S$. *Acta Crystallogr. Sect. A* **1982**, *38*, 79–86. [CrossRef]

56. De Villiers, J. The Composition and Crystal Structures of Pyrrhotite: A Common but Poorly Understood Mineral. Available online: http://www.mintek.co.za/Mintek75/Proceedings/L01-DeVilliers.pdf (accessed on 19 June 2018).

57. Liles, D.C.; de Villiers, J.P.R. Redetermination of the structure of 5C pyrrhotite at low temperature and at room temperature. *Am. Mineral.* **2012**, *97*, 257–261. [CrossRef]

58. Kaneda, H.; Takenouchi, S.; Shoji, T. Stability of pentlandite in the Fe-Ni-Co-S system. *Miner. Depos.* **1986**, *21*, 169–180. [CrossRef]

59. Barton, P.B.; Toulmin, P. Phase relations involving sphalerite in the Fe-Zn-S system. *Econ. Geol.* **1966**, *61*, 815–849. [CrossRef]

60. Lusk, J.; Ford, C.E. Experimental extension of the sphalerite geobarometer to 10 kbar. *Am. Mineral.* **1978**, *63*, 516–519.

61. Scott, S.D.; Barnes, H.L. Sphalerite geothermometry and geobarometry. *Econ. Geol.* **1971**, *66*, 653–669. [CrossRef]

62. Fleet, M.E.M.E. Phase Equilibria at High Temperatures. *Rev. Mineral. Geochem.* **2006**, *61*, 365–419. [CrossRef]

63. Ghiorso, M.S.; Evans, B.W. Thermodynamics of rhombohedral oxide solid solutions and a revision of the Fe-Ti two-oxide geothermometer and oxygen-barometer. *Am. J. Sci.* **2008**, *308*, 957–1039. [CrossRef]

64. Yund, R.A.; Hall, H.T. Kinetics and Mechanism of Pyrite Exsolution from Pyrrhotite. *J. Petrol.* **1970**, *11*, 381–404. [CrossRef]

65. Kissin, S.A.; Scott, S.D. Phase relations involving pyrrhotite below 350 degrees C. *Econ. Geol.* **1982**, *77*, 1739–1754. [CrossRef]

66. Nekrasov, I.J.; Besmen, N.I. Pyrite-pyrrhotite geothermometer. Distribution of cobalt, nickel and tin. *Phys. Chem. Earth* **1979**, *11*, 767–771. [CrossRef]

67. Lusk, J.; Scott, S.D.; Ford, C.E. Phase relations in the Fe-Zn-S system to 5 kbars and temperatures between 325 degrees and 150 degrees C. *Econ. Geol.* **1993**, *88*, 1880–1903. [CrossRef]

68. Kapustin, Y. *Mineralogy of the Weathering Crust of Carbonatites*; Nedra: Moscow, Russia, 1973. (In Russian)

69. Liferovich, R.P.; Pakhomovsky, Y.A.; Yakovenchuk, V.N.; Bogdanova, A.N.; Bakhchisaraitsev, A.Y. Vivianite minerals group and bobierrite from the Kovdor massif. *Zap. RMO* **1999**, *128*, 109–117.

70. Liferovich, R.P.; Sokolova, E.V.; Hawthorne, F.C.; Laajoki, K.V.O.; Gehor, S.; Pakhomovsky, Y.A.; Sorokhtina, N.V. Gladiusite, $Fe^{3+}_2(Fe^{2+},Mg)_4(PO_4)(OH)_{11}(H_2O)$, a new hydrothermal mineral species from the phoscorite carbonatite unit, Kovdor complex, Kola Peninsula, Russia. *Can. Mineral.* **2000**, *38*, 1477–1485. [CrossRef]

71. Doyle, C.D.; Naldrett, A.J. Ideal mixing of divalent cations in mafic magma. II. The solution of NiO and the partitioning of nickel between coexisting olivine and liquid. *Geochim. Cosmochim. Acta* **1987**, *51*, 213–219. [CrossRef]

72. Seifert, S.; O'Neill, H.S.C.; Brey, G. The partitioning of Fe, Ni and Co between olivine, metal, and basaltic liquid: An experimental and thermodynamic investigation, with application to the composition of the lunar core. *Geochim. Cosmochim. Acta* **1988**, *52*, 603–616. [CrossRef]

73. Ehlers, K.; Grove, T.L.; Sisson, T.W.; Recca, S.I.; Zervas, D.A. The effect of oxygen fugacity on the partitioning of nickel and cobalt between olivine, silicate melt, and metal. *Geochim. Cosmochim. Acta* **1992**, *56*, 3733–3743. [CrossRef]

74. Li, C.; Ripley, E.M.; Mathez, E.A. The effect of S on the partitioning of Ni between olivine and silicate melt in MORB. *Chem. Geol.* **2003**, *201*, 295–306. [CrossRef]

75. Gaetani, G.A.; Grove, T.L. Partitioning of moderately siderophile elements among olivine, silicate melt, and sulfide melt: Constraints on core formation in the Earth and Mars. *Geochim. Cosmochim. Acta* **1997**, *61*, 1829–1846. [CrossRef]

76. Kiseeva, E.S.; Wood, B.J. A simple model for chalcophile element partitioning between sulphide and silicate liquids with geochemical applications. *Earth Planet. Sci. Lett.* **2013**, *383*, 68–81. [CrossRef]

77. Li, Y.; Audétat, A. Effects of temperature, silicate melt composition, and oxygen fugacity on the partitioning of V, Mn, Co, Ni, Cu, Zn, As, Mo, Ag, Sn, Sb, W, Au, Pb, and Bi between sulfide phases and silicate melt. *Geochim. Cosmochim. Acta* **2015**, *162*, 25–45. [CrossRef]

78. Powell, A.V.; Vaqueiro, P.; Knight, K.S.; Chapon, L.C.; Sánchez, R.D. Structure and magnetism in synthetic pyrrhotite Fe7S8: A powder neutron-diffraction study. *Phys. Rev. B* **2004**, *70*, 014415. [CrossRef]

79. Krivovichev, S.V. Topological complexity of crystal structures: Quantitative approach. *Acta Crystallogr. Sect. A Found. Crystallogr.* **2012**, *68*, 393–398. [CrossRef] [PubMed]

80. Krivovichev, S.V. Structural complexity of minerals: Information storage and processing in the mineral world. *Mineral. Mag.* **2013**, *77*, 275–326. [CrossRef]

81. Holzheid, A.; Palme, H. The influence of FeO on the solubilities of cobalt and nickel in silicate melts. *Geochim. Cosmochim. Acta* **1996**, *60*, 1181–1193. [CrossRef]

82. Keppler, H. Water solubility in carbonatite melts. *Am. Mineral.* **2003**, *88*, 1822–1824. [CrossRef]

83. Zhong, R.; Brugger, J.; Chen, Y.; Li, W. Contrasting regimes of Cu, Zn and Pb transport in ore-forming hydrothermal fluids. *Chem. Geol.* **2015**, *395*, 154–164. [CrossRef]

84. Leroy, A. Palabora—Not just another copper mine. *Miner. Ind. Int.* **1992**, *1005*, 14–19.

85. Palabora_Mining_Company_Limited. The geology and the economic deposits of copper, iron, and vermiculite in the Palabora igneous complex: A brief review. *Econ. Geol.* **1976**, *71*, 177–192.

86. Vielreicher, N.M.; Groves, D.I.; Vielreicher, R.M. The Phalaborwa (Palabora) Deposit and its Potential Connection to Iron-Oxide Copper-Gold Deposits of Olympic Dam Type. In *Hydrothermal Iron Oxide Copper-Gold & Related Deposits: A Global Perspective, Volume 1*; Porter, T.M., Ed.; PGS Publishing: Adelaide, Australia, 2000; pp. 321–329.

87. Karchevsky, P.I. *Sulfide, Strontium, and REE Mineralozation in Phoscorites and Carbonatites of the Turiy Complex (Russia) and Loolekop Deposit (RSA)*; Kolo: St. Petersburg, Russia, 2005.

88. Verwoerd, W. Mineral deposits associated with carbonatites and alkaline rocks. In *Mineral Deposits of Southern Africa. Volume 2*; Anhaeusser, C.R., Maske, S., Eds.; Geological Society of South Africa: Johannesburg, South Africa, 1986; pp. 2173–2191.

89. Ericsson, S. Phalaborwa: A saga of magmatism, metasomatism and miscibility. In *Carbonatites: Genesis and Evolution*; Bell, K., Ed.; Unwin Hyman: London, UK, 1989; pp. 221–250.

90. Heinrich, E. The Palabora Carbonatitic Complex—A Unique Copper Deposit. *Can. Mineral.* **1970**, *10*, 585–598.

91. Gaillard, F.; Scaillet, B.; Pichavant, M.; Bény, J.-M. The effect of water and fO_2 on the ferric–ferrous ratio of silicic melts. *Chem. Geol.* **2001**, *174*, 255–273. [CrossRef]

92. Wilke, M.; Behrens, H.; Burkhard, D.J.; Rossano, S. The oxidation state of iron in silicic melt at 500 MPa water pressure. *Chem. Geol.* **2002**, *189*, 55–67. [CrossRef]

93. Schuessler, J.A.; Botcharnikov, R.E.; Behrens, H.; Misiti, V.; Freda, C. Amorphous Materials: Properties, structure, and Durability: Oxidation state of iron in hydrous phono-tephritic melts. *Am. Mineral.* **2008**, *93*, 1493–1504. [CrossRef]

94. Richards, J.P. The oxidation state, and sulfur and Cu contents of arc magmas: Implications for metallogeny. *Lithos* **2014**, *233*, 27–45. [CrossRef]

95. Groves, D.I.; Vielreicher, N.M. The Phalabowra (Palabora) carbonatite-hosted magnetite-copper sulfide deposit, South Africa: An end-member of the iron-oxide copper-gold-rare earth element deposit group? *Miner. Depos.* **2001**, *36*, 189–194. [CrossRef]

minerals

MDPI

Article

Three-D Mineralogical Mapping of the Kovdor Phoscorite-Carbonatite Complex, NW Russia: III. Pyrochlore Supergroup Minerals

Gregory Yu. Ivanyuk [1,2,*], Nataly G. Konopleva [1], Victor N. Yakovenchuk [1,2], Yakov A. Pakhomovsky [1,2], Taras L. Panikorovskii [1], Andrey O. Kalashnikov [2], Vladimir N. Bocharov [3], Ayya A. Bazai [1,2], Julia A. Mikhailova [1,2] and Pavel M. Goryainov [2]

[1] Nanomaterials Research Centre of Kola Science Centre, Russian Academy of Sciences, 14 Fersman Street, Apatity 184209, Russia; konoplyova55@mail.ru (N.G.K.); yakovenchuk@geoksc.apatity.ru (V.N.Y.); pakhom@geoksc.apatity.ru (Y.A.P.); taras.panikorovsky@spbu.ru (T.L.P.); bazai@geoksc.apatity.ru (A.A.B.); ylya_korchak@mail.ru (J.A.M.)

[2] Geological Institute of Kola Science Centre, Russian Academy of Sciences, 14 Fersman Street, Apatity 184209, Russia; kalashnikov@geoksc.apatity.ru (A.O.K.); pgor@geoksc.apatity.ru (P.M.G.)

[3] Geo Environmental Centre "Geomodel", St. Petersburg State University, 1 Ul'yanovskaya Street, St. Petersburg 198504, Russia; bocharov@molsp.phys.spbu.ru

*   Correspondence: g.ivanyuk@gmail.com; Tel.: +7-81555-79531

Received: 30 May 2018; Accepted: 26 June 2018; Published: 28 June 2018

check for updates

Abstract: The pyrochlore supergroup minerals (PSM) are typical secondary phases that replace (with zirconolite–laachite) earlier Sc-Nb-rich baddeleyite under the influence of F-bearing hydrothermal solutions, and form individual well-shaped crystals in surrounding carbonatites. Like primary Sc-Nb-rich baddeleyite, the PSM are concentrated in the axial carbonate-rich zone of the phoscorite-carbonatite complex, so their content, grain size and chemical diversity increase from the pipe margins to axis. There are 12 members of the PSM in the phoscorite-carbonatite complex. Fluorine- and oxygen-dominant phases are spread in host silicate rocks and marginal carbonate-poor phoscorite, while hydroxide-dominant PSM occur mainly in the axial carbonate-rich zone of the ore-pipe. Ti-rich PSM (up to oxycalciobetafite) occur in host silicate rocks and calcite carbonatite veins, and Ta-rich phases (up to microlites) are spread in intermediate and axial magnetite-rich phoscorite. In marginal (apatite)-forsterite phoscorite, there are only Ca-dominant PSM, and the rest of the rocks include Ca-, Na- and vacancy-dominant phases. The crystal structures of oxycalciopyrochlore and hydroxynatropyrochlore were refined in the $Fd\bar{3}m$ space group with R_1 values of 0.032 and 0.054 respectively. The total difference in scattering parameters of B sites are in agreement with substitution scheme $^B Ti^{4+} + {}^Y OH^- = {}^B Nb^{5+} + {}^Y O^{2-}$. The perspective process flow diagram for rare-metal "anomalous ore" processing includes sulfur-acidic cleaning of baddeleyite concentrate from PSM and zirconolite–laachite impurities followed by deep metal recovery from baddeleyite concentrate and Nb-Ta-Zr-U-Th-rich sulfatic product from its cleaning.

Keywords: pyrochlore supergroup minerals; typochemistry; crystal structure; Kovdor phoscorite-carbonatite complex

1. Introduction

The Kovdor phoscorite-carbonatite complex is the largest source of magnetite, hydroxylapatite and baddeleyite in the NW of Russia [1–4]. Besides, the deposit contains significant amounts of Sc, Ln, Ta, Nb and U concentrated mainly in baddeleyite and the products of its alteration, first of all, in zirconolite and minerals of the pyroclore supergroup [5–10]. Therefore, it is important to highlight the fact that all aforementioned rare metals can be produced during chemical cleaning of baddeleyite concentrate [6]. Also, the pyrochlore supergroup minerals are sensitive indicators of mineral formation conditions, and their study can give us important information on subsolidus and postmagmatic evolution of the phoscorite-carbonatite complex.

The general formula of the pyrochlore supergroup minerals (PSM) is $A_{2-m}B_2X_{6-w}Y_{1-n}$, where $m = 0$–2, $w = 0$–0.7 and $n = 0$–1.0; the species are named according to the dominant cation (or anion) of the dominant valence at each site [11,12]. The position A can be occupied by Na, Ca, Sr, Pb^{2+}, Sn^{2+}, Sb^{3+}, Y, U, Ba, Fe^{2+}, Ag, Mn, Bi^{3+}, REE, Sc, Th, H_2O, or be vacant. The position B can be occupied by Nb, Ta, Ti, Sb^{5+}, W, Al and Mg (as main components), V^{5+}, Sn^{4+}, Zr, Hf, Fe^{3+} and Si (as impurities). The position X is occupied mainly by oxygen, but OH and F can also occur here. The position Y is occupied usually by OH^-, F^-, and O^{2-} anions (hydroxy-, fluor-, and oxy-compounds correspondingly), but also it can be vacant (keno-compounds) or occupied by water and/or large univalent cations of K, Cs and Rb.

Correspondingly, there are groups of pyrochlore (M^{5+} cations are dominant at the B site, and Nb is dominant among them, and O^{2-} is dominant at the X site), microlite (M^{5+} cations are dominant at the B site, and Ta is dominant among them, and O^{2-} is dominant at the X site), roméite (M^{5+} cations are dominant at the B site, and Sb is dominant among them, and O^{2-} is dominant at the X site), betafite (M^{4+} cations are dominant at the B site, and Ti is dominant among them, and O^{2-} is dominant at the X site), elsmoreite (M^{6+} cations are dominant at the B site, and W is dominant among them, and O^{2-} is dominant at the X site), ralstonite (M^{3+} cations are dominant at the B site, and Al is dominant among them, and F^- is dominant at the X site) and coulsellite (M^{2+} cations are dominant at the B site, and Mg is dominant among them, and F^- is dominant at the X site). Hereinafter, we will use generic terms pyrochlores, microlites and betafites for Nb-, Ta- and Ti-dominant members of the pyrochlore supergroup, respectively.

Data on the geology and petrography of the Kovdor alkaline-ultrabasic massif and the eponymous phoscorite-carbonatitre complex described in [1,2,4,13–16] is summarized in the first article of this series [17] that has shown that spatial distribution of forsterite content, morphology, grain size, composition and alteration products accent concentric zonation of the Kovdor phoscorite-carbonatite complex. This zonation includes marginal (apatite)-forsterite phoscorite, intermediate low-carbonate magnetite-rich phoscorite and axial carbonate-rich phoscorite and carbonatites. Presented in the second article, 3D data on distribution of content, grain size and composition of sulfides helped us to understand the behavior of sulfur-related metals during crystallization and subsolidus evolution of the phoscorite-carbonatite complex. Wide diversity of the pyrochlore supergroup minerals in the Kovdor phoscorite-carbonatite complex [1,4,13,16,18] permits a reconstruction of the latest hydrothermal episodes of the complex formation and draws attention to the economic significance of pyrochlore mineralization.

2. Materials and Methods

For this study, we used 548 samples of phoscorite, carbonatites and host rocks taken from 108 exploration holes drilled at the interval from −80 to −650 m within the Kovdor phoscorite-carbonatite complex [16]. Bulk-rock samples were analyzed in the Tananaev Institute of Chemistry of KSC RAS (Apatity) by means of inductively coupled plasma-mass spectrometry (ICP-MS) performed with an ELAN 9000 DRC-e mass spectrometer (Perkin Elmer, Waltham, MA, USA). For the analyses, the samples were dissolved in a mixture of concentrated hydrofluoric and nitric acids with distillation of silicon and further addition of hydrogen peroxide to a cooled solution to suppress hydrolysis of polyvalent metals [19].

Composition of PSM grains smaller than 20 μm in diameter was determined using a LEO-1450 scanning electron microscope (Carl Zeiss Microscopy, Oberkochen, Germany) with a Quantax 200 energy-dispersive X-ray spectrometer (Bruker, Ettlingen, Germany). The same equipment with 500-pA beam current and 20 kV acceleration voltage was used to obtain back-scattered electron

(BSE) images of thin polished sections. The Image Tool 3.04 (The University of Texas Health Science Center, San Antonio, TX, USA) was used to determine the equivalent circular diameter of studied PSM grains.

Grains larger 20 μm in diameter were then studied using a Cameca MS-46 electron probe microanalyzer (EPMA) (Cameca, Gennevilliers, France) operating in wavelength-dispersive mode at 20 kV and 20–30 nA. The analyzes were performed with the beam size of 5–10 μm and the counting time of 10–20/10 s on peaks/background for every chemical element and every of 5–10 measurement points. Table 1 presents the used standards, detection limits and precisions based on repeated analyses of standards. Fluorine was determined with a Quantax 200 energy-dispersion instrument and standard-less ZAF method, based on a detection limit of 0.5 wt %. Coefficients in crystallochemical formulas were calculated using the MINAL program by D. V. Dolivo-Dobrovolsky [20].

Table 1. Parameters of electron probe microanalyzer (EPMA) analyses.

Element	Detection Limit, wt %	Standard	Element	Detection Limit, wt %	Standard
Na	0.1	Lorenzenite	Y	0.1	Synthetic $Y_3Al_5O_{12}$
Mg	0.1	Forsterite	Zr	0.1	Synthetic $ZrSiO_4$
Al	0.05	Pyrope	Nb	0.05	Synthetic $LiNbO_3$
Si	0.05	Wollastonite	Ba	0.05	Barite
P	0.05	Fluorapatite	La	0.05	Synthetic $LaCeS_2$
K	0.03	Wadeite	Ce	0.05	Synthetic $LaCeS_2$
Ca	0.03	Wollastonite	Pr	0.1	Synthetic $LiPr(WO_4)_2$
Sc	0.02	Thortveitite	Nd	0.1	Synthetic $LiNd(MoO_4)_2$
Ti	0.02	Lorenzenite	Ta	0.05	Metallic tantalum
Mn	0.01	Synthetic $MnCO_3$	Pb	0.05	Synthetic PbSe
Fe	0.01	Hematite	Th	0.2	Thorite
Sr	0.1	Celestine	U	0.2	Metallic uranium

Single-crystal X-ray diffraction studies of oxycalciopyrochlore and hydroxynatropyrochlore were performed at 293 K using a Bruker Kappa APEX DUO diffractometer equipped with the IμS microfocus source (beam size of 0.11 mm, MoKα radiation, λ = 0.71073 Å and operated at 45 kV and 0.6 mA) and a CCD area detector. The intensity data was reduced and corrected for Lorentz, polarization and background effects. The APEX2 software (Bruker-AXS, Billerica, MA, USA, 2014) applied a multi-scan absorption-correction. Crystal structures were refined with the SHELX program [21] and drawn using the VESTA 3 program [22].

Raman spectra of 6 typical PSM were produced using a Jobin-Yvon LabRam HR 800 spectrometer (Horiba, Kyoto, Japan) with a 514 nm laser under the same conditions for each sample. The band component analysis was performed using the OriginPro 8.1 SR2 program [23], with Lorenzian peak function.

Statistical analyses were implemented with the STATIATICA 8.0 (StatSoft) program [24]. Geostatisical studies and 3D modeling were conducted with the MICROMINE 16 program [25]. Interpolation was performed by ordinary kriging. Automatic 3D geological mapping was developed by conversion of rock chemical composition to mineral composition using logical computation [26].

The following abbreviations were used: Ap (hydroxylapatite), Bdy (baddeleyite), Cal (calcite), Cb (carbonate), Clc (clinochlore), Dol (dolomite), Fo (forsterite), Gn (galena), Ilm (ilmenite), Mag (magnetite), Pcl (pyrochlore unspecified), Phl (phlogopite), Po (pyrrhotite), Py (pyrite), Spl (spinel), Srp (serpentine), Str (strontianite), Val (valleriite), and Zrl (zirconolite). Pyrochlore group minerals (PSM): OCP (oxycalciopyrochlore), ONP ("oxynatropyrochlore"), HCP (hydroxycalciopyrochlore), HNP (hydroxynatropyrochlore), HKP (hydroxykenopyrochlore), FCP (fluorcalciopyrochlore), FNP (fluornatropyrochlore) FKP ("fluorkenopyrochlore"); OCB ("oxycalciobetafite"), FCM (fluorcalciomicrolite), HCM (hydroxycalciomicrolite), HKM (hydroxykenomicrolite).

3. Results

3.1. Occurrence and Morphology

The pyrochlore supergroup minerals are common accessories in all rocks of the Kovdor alkaline-ultrabasic massif. The occurrence and content of PSM decrease from host foidolite and diopsidite–phlogopitite to earlier (apatite)-forsterite phoscorite, and then increase to intermediate phoscorite and, finally, to the latest carbonatites (Table 2). In the phoscorite-carbonatite complex, the pyrochlore supergroup minerals result mainly from the alteration of Nb-rich baddeleyite formed in the pipe axial zone due to the substitution $2Zr^{4+} \leftrightarrow Sc^{3+}Nb^{5+}$. On this reason, the pyrochlore areal coincides with that of Sc-Nb-rich baddeleyite (Figure 1). Carbonate-rich phoscorite and carbonatites enriched in PSM form an intensive radioactive anomaly (about 200 m in diameter and >900 m in depth), known as the "Anomalous Zone" [1]. Content of pyrochlore in rocks of the "Anomalous Zone" gradually increases with depth at the account of baddeleyite [27].

Table 2. Pyrochlore supergroup minerals (PSM) occurrence and grain size in the Kovdor massif.

Natural Sequence of Rock Formation, after [4]	Proportion of Samples with Identified PSM, %	Median Equivalent Circle Diameter of Grains (Min–Max), μm	Mean Nb Content in Baddeleyite, wt %
Foidolite	14	16 (8–30)	0.22
Diopsidite and phlogopitite	22	39 (3–400)	0.28
(Apatite)-forsterite phoscorite	9	40 (5–200)	0.08
Low-carbonate magnetite-rich phoscorite	21	50 (1–1090)	0.17
Calcite-rich phoscorite and phoscorite-related carbonatite	38	60 (4–535)	0.38
Vein calcite carbonatite	60	55 (1–400)	0.66
Vein dolomite carbonatite and magnetite-dolomite-serpentine rocks	47	40 (6–340)	0.39
Total	31	50 (1–1090)	

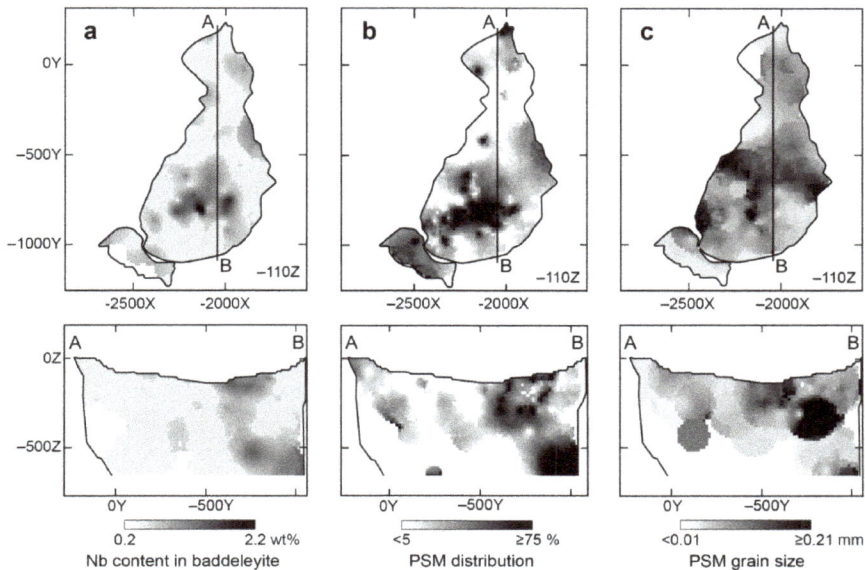

Figure 1. Distributions of Nb in baddeleyite (**a**), PSM-bearing rocks (**b**) and average size of PSM grains in a sample (**c**) within the Kovdor phoscorite-carbonatite complex.

The pyrochlore group minerals, namely oxycalciopyrochlore (OCP), oxynatropyrochlore (ONP), hydroxycalciopyrochlore (HCP), hydroxynatropyrochlore (HNP), hydroxykenopyrochlore (HKP), fluorcalciopyrochlore (FCP), fluornatropyrochlore (FNP) and fluorkenopyrochlore (FKP), are common in foidolite, diopsidite, phlogopitite, phoscorite and carbonatite (OCP, HCP and HNP are predominant). Minerals of microlite and betafite groups play a subordinate role, mainly, as separate parts of PSM inhomogeneous crystals. Oxycalciobetafite (OCB) and its cation-deficient analogue occur in phoscorite and carbonatite of the ore-pipe intermediate and axial zones; while fluorcalciomicrolite (FCM), hydroxycalciomicrolite (HCM) and hydroxykenomicrolite (HKM) are found only in calcite-rich phoscorite and calcite carbonatite.

Within the phoscorite-carbonatite complex, the equivalent circle diameter of the analyzed PSM grains widely varies from 1 to 1000 μm (see Table 2) increasing from host foidolite and diopsidite to marginal (apatite)-forsterite phoscorite, and then to intermediate low-carbonate magnetite-rich phoscorite and axial calcite-rich phoscorite and carbonatite (see Figure 1c). In fact, the PSM grain size is proportional to their content in the rock: in PSM-poor rocks, there are small (up to 100 μm in diameter) separate grains of these minerals; while PSM-rich rocks contain irregularly-shaped and lens-like segregations, bands and veinlets of much larger (up to 1 mm in diameter) crystals of the pyrochlore supergroup minerals.

The morphology and microstructure of pyrochlore particles show different aspects, such as irregularly shaped and rounded grains (Figure 2a,b), idiomorphic octahedral, cubic, cubooctahedral and truncated octahedral crystals (Figure 2c–e), poikilitic and skeletal (meta)crystals (Figure 2e,f), veinlets and filling of fractures (Figure 2g), rims around baddeleyite and lueshite grains followed by partial pseudomorphs (Figure 2h), epitaxial intergrowth with baddeleyite (Figure 2i), as well as the finest inclusions in exsolved titanomagnetite. In all rocks of the phoscorite-carbonatite complex, there are zoned PSM crystals with primary regular zonation corresponding to crystal shape, secondary irregular zoning caused by mineral alteration, and both of these zoning types, which is typical for PSM from carbonatites [28]. The primary zoning is caused usually by increases of Ca, Ta, Ti, Zr and F contents from the crystal core to rim at the expense of Na, Th, U, REE, Nb and (OH) amounts. Secondary zoning appears, first of all, due to leaching of Na, Ca and F from marginal parts of metamict grains of U-Th-rich PSM. Sometimes, separate zones of one crystal are formed by different minerals of the pyrochlore supergroup (see Figure 2f).

Irregularly shaped and drop-like yellowish-brown pyrochlore grains dominate in host silicate rocks and carbonate-poor phoscorite (see Figure 2a); while in carbonate-rich phoscorite and related carbonatite, they also form dark brown truncated octahedral crystals, sometimes with baddeleyite relics (see Figure 2e,i). In calcite carbonatite veins, there are reddish-brown to creamy and yellow octahedral, cubic, truncated octahedral and cubooctahedral crystals of pyrochlores (see Figure 2c), and their close intergrowth with baddeleyite and zirconolite [1,4,29,30]. In vein dolomite carbonatite, yellow to brown truncated octahedral to cubic pyrochlore crystals (see Figure 2d) occur in voids in typical association with zircon and endemic phosphates [1,27,31,32]. Betafite group minerals occur as separate irregularly shaped, drop-like and ellipsoidal grains (up to 100 μm) or form marginal zones of pyrochlore crystals (see Figure 2d) and rims around baddeleyite grains. Microlite group members form marginal zones of pyrochlore crystals and inclusions (up to 20 μm in diameter) in zirconolite, titanomagnetite and exsolution ilmenite.

Like PSM, minerals of the zirconolite–laachite series are typical products of baddeleyite alteration; however, they can also replace pyrochlores (Figure 3a). Besides, grains of U-Th-rich kenopyrochlores are often replaced/rimmed by pyrite (Figure 3b) and valleriite (Figure 3c) and sometimes by strontianite (Figure 3d), probably, due to radiolytic splitting of water into hydrogen peroxide and molecular hydrogen. In particular, with radiation dose growth, a water solution of H_2SO_4 (below 200 °C) becomes significantly rich in H_2 due to H_2O_2 [33,34], and this, probably, causes precipitation of sulfides around the radiation source.

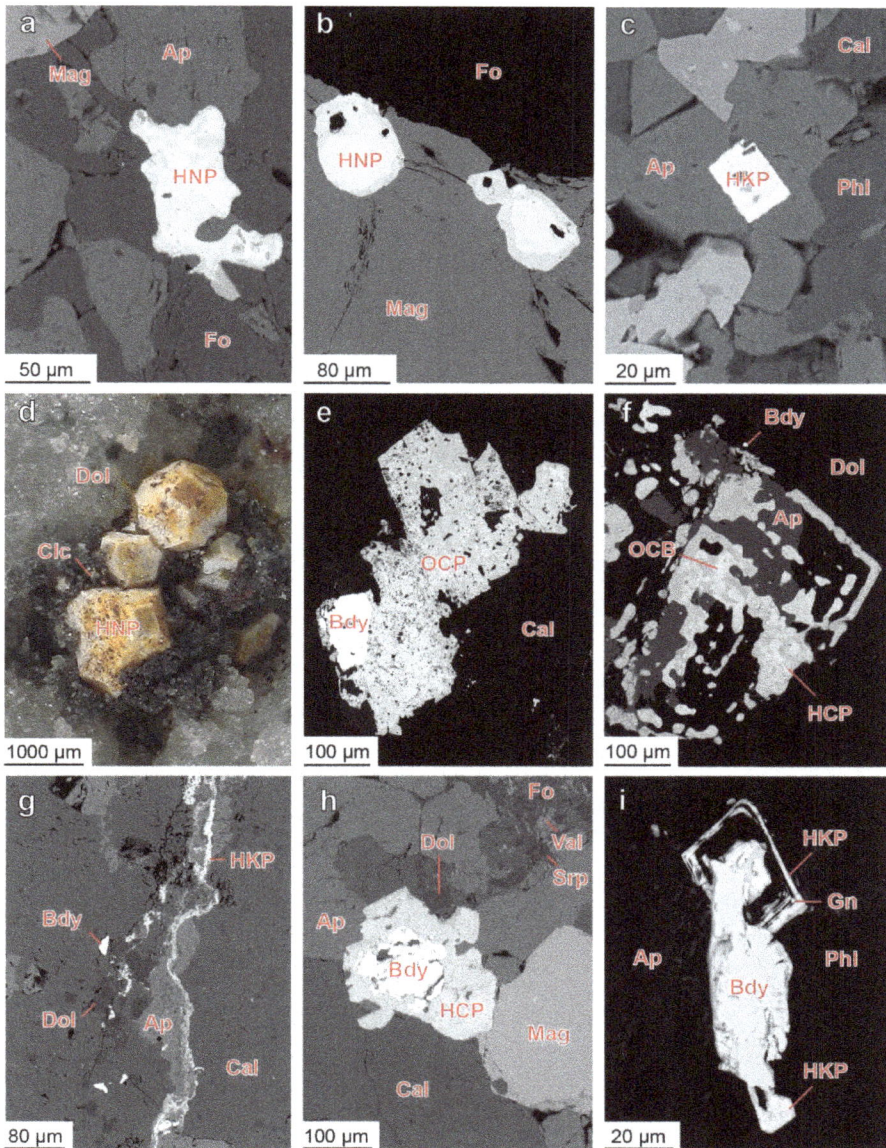

Figure 2. Typical PSM morphology: (**a**) irregularly shaped grain of hydroxycalciopyrochlore in forsterite-magnetite phoscorite 987/2.1; (**b**) drop-like grains of oxycalciopyrochlore in calcite-magnetite-apatite-forsterite phoscorite 987/99.6; (**c**) cubic crystal of hydroxykenopyrochlore in vein calcite carbonatite 963/61.3; (**d**) truncated octahedral crystals of hydroxynatropyrochlore in dolomite carbonatite K-017-4; (**e**) poikilitic metacrystal of oxycalciopyrochlore from calcite-magnetite-forsterite phoscorite 917/318.5; (**f**) skeletal crystal of oxycalciobetafite-hydroxycalciopyrochlore in magnetite-dolomite-serpentine rock 987/198.0; (**g**) veinlets of hydroxykenopyrochlore in vein calcite carbonatite 989/57.8; (**h**) hydroxycalciopyrochlore crystal with baddeleyite relics in calcite- magnetite-apatite-forsterite phoscorite 986/49.6; (**i**) epitaxial overgrowth of hydroxykenopyrochlore on baddeleyite in calcite-apatite-forsterite phoscorite 1009/186.6. Macrophoto (**d**) and back-scattered electron (BSE) images of thin polished sections (the rest). Mineral abbreviations are in Section 2.

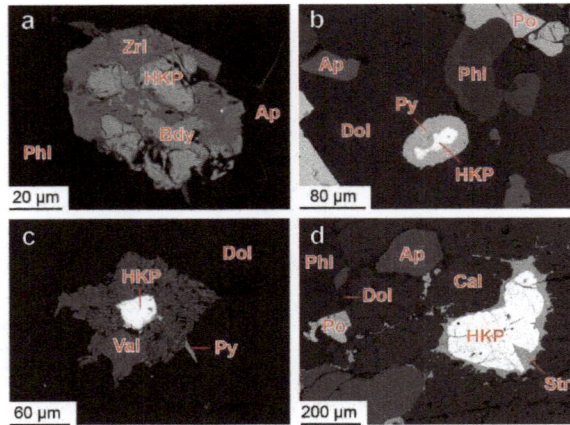

Figure 3. Typical products of hydroxykenopyrochlore low-temperature alteration: **a**—replacement of baddeleyite and hydroxykenopyrochlore by zirconolite in dolomite carbonatite 964/148.2; **b**—relic of hydroxykenopyrochlore within pyrite grain in phlogopitite 948/31.4; **c**—radiated aggregate vallereite around hydroxykenopyrochlore grain in vein calcite carbonatite 943/54.7; **d**—hydroxykenopyrochlore grain rimmed by strontianite in phoscorite-related calcite carbonatite 974/115.6. BSE-images of polished thin sections. Mineral abbreviations are in Section 2.

3.2. Chemical Composition

Table 3 shows the results of precision EPMA analyses conducted on 12 different members of the pyrochlore supergroup found in the Kovdor massif, and Table 4 presents statistical data on the PSM composition in different rocks of this massif. Most of the analyzed grains (92%) correspond to Ca-, Na- and vacancy-dominant members of the pyrochlore group and the rest 8% are represented by minerals of microlite and betafite groups (about 4% of each). Pyrochlores occur in all rocks of the Kovdor massif (Figure 4), and microlites and betafites co-exist with pyrochlore in the central part of the phoscorite-carbonatite complex.

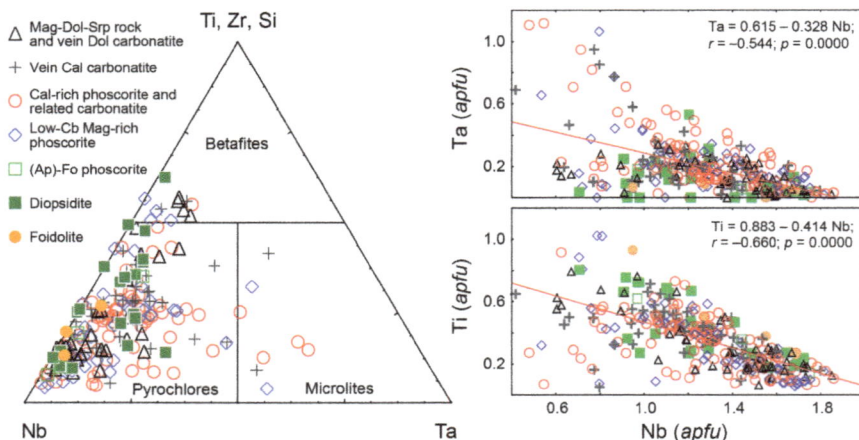

Figure 4. The Kovdor PSM classification diagram [11] and relations between Nb, Ta and Ti contents.

Table 3. Precision microprobe analyses of pyrochlore supergroup minerals (wt %) (see abbreviations above) and the corresponding crystallochemical formulas calculated on the basis of $B = 2$, $X = 6$ and $Y = 1$ atoms per formula unit (*apfu*).

PSM	OCP	ONP	HCP	HNP	HKP	FCP	FNP	FKP	HCM	HKM	FCM	OCB
Sample	1010/243.7	987/67.2	1010/243.7	972/86.9	987/2.1	901/109.9	987/198.0	974/50.6	901/109.9	981/520	973/533.3	957/141.0
F, wt %	1.51	bd	2.00	0.09	0.03	1.99	2.48	2.31	0.42	0.43	2.47	bd
Na_2O	3.59	5.12	7.31	8.03	0.89	3.45	9.90	bd	0.51	0.66	3.46	bd
MgO	bd	bd	bd	bd	bd	bd	bd	bd	bd	0.24	0.29	bd
Al_2O_3	0.21	bd	bd	bd	bd	0.11	bd	bd	0.19	bd	bd	bd
SiO_2	bd	0.19	bd	bd	0.35	bd	bd	bd	bd	1.62	bd	bd
CaO	13.70	8.83	15.18	10.05	8.47	15.78	8.60	8.58	9.80	4.41	11.33	12.30
TiO_2	3.50	6.71	1.92	3.83	8.32	0.90	5.66	4.36	1.03	5.00	1.00	16.07
MnO	bd	bd	bd	bd	bd	bd	bd	bd	bd	0.35	bd	bd
Fe_2O_3	2.09	1.16	0.83	bd	1.70	0.32	bd	1.62	0.50	2.23	7.01	4.60
ZrO_2	bd	1.32	0.73	bd	1.29	bd	bd	bd	bd	1.09	7.68	bd
Nb_2O_5	42.34	31.49	60.82	56.72	32.21	53.79	57.11	43.72	17.95	19.34	11.26	21.24
BaO	bd	bd	0.17	bd	bd	bd	bd	bd	bd	7.12	bd	bd
La_2O_3	bd	bd	bd	bd	bd	bd	bd	bd	bd	0.46	bd	3.59
Ce_2O_3	1.37	0.34	1.60	3.01	0.45	2.88	2.63	1.45	0.65	3.07	bd	6.70
Pr_2O_3	bd	bd	bd	bd	bd	bd	bd	bd	bd	0.51	bd	bd
Nd_2O_3	0.41	0.13	0.11	bd	0.10	bd	bd	bd	bd	1.03	bd	bd
Ta_2O_5	12.76	16.25	7.35	7.48	14.81	6.59	3.20	11.23	53.29	33.39	51.21	7.80
PbO	bd	bd	bd	bd	0.88	bd	bd	bd	bd	bd	bd	bd
ThO_2	14.35	1.48	bd	4.66	2.57	1.33	3.67	bd	8.09	6.11	bd	22.92
UO_2	0.45	24.64	bd	3.84	20.86	0.01	4.94	26.90	5.62	bd	bd	bd
$-O=F_2$	0.64	bd	0.84	0.04	0.01	0.84	1.04	0.97	0.18	0.18	1.04	bd
Total	95.64	97.67	97.18	97.67	92.91	86.31	97.16	99.20	97.86	86.86	94.66	95.22
Ca.*apfu*	1.08	0.74	1.02	0.71	0.67	1.24	0.60	0.67	0.88	0.36	0.83	0.97
Na	0.51	0.79	0.89	1.02	0.13	0.49	1.24	–	0.08	0.10	0.46	–
Ba	–	–	–	–	–	–	–	–	–	0.21	–	–
Mn	–	–	–	–	–	–	–	–	–	0.02	–	–
La	–	–	–	–	–	–	–	–	–	0.01	–	0.10
Ce	0.04	0.01	0.04	0.07	0.01	0.08	0.06	0.04	0.02	0.09	–	0.18
Pr	–	–	–	–	–	–	–	–	–	0.01	–	–
Nd	0.01	–	–	–	0.02	–	–	–	–	0.03	–	–
Pb	–	–	–	–	0.04	–	–	–	–	–	–	–
Th	0.24	0.03	–	0.07	0.04	0.02	0.05	–	0.15	0.11	–	0.38
U	0.01	0.43	–	0.06	0.34	–	0.07	0.85	0.10	–	–	–
vac	0.10	–	0.05	0.07	0.79	0.17	–	0.44	0.76	1.06	0.71	–
ΣA	2.00	2.00	2.00	2.00	2.00	2.00	2.02	2.00	2.00	2.00	2.00	2.00
Nb	1.41	1.13	1.72	1.68	1.07	1.79	1.67	1.45	0.68	0.68	0.71	0.70
Ta	0.26	0.35	0.13	0.13	0.30	0.13	0.06	0.22	1.21	0.71	0.95	0.16
Ti	0.19	0.40	0.09	0.19	0.46	0.05	0.27	0.24	0.06	0.29	0.05	0.89

Table 3. *Cont.*

PSM	OCP	ONP	HCP	HNP	HKP	FCP	FNP	FKP	HCM	HKM	FCM	OCB
Zr	–	0.05	0.02	–	0.05	–	–	–	–	0.04	0.26	–
Si	–	–	0.02	–	0.03	0.02	–	–	–	0.12	–	–
Fe^{3+}	0.12	0.07	0.04	–	0.09	0.01	–	0.09	0.03	0.13	0.36	0.25
Al	0.02	–	–	–	–	–	–	–	0.02	–	–	–
Mg	–	–	–	–	–	–	–	–	–	–	0.03	–
ΣB	**2.00**	**2.00**	**2.00**	**2.00**	**2.00**	**2.00**	**2.00**	**2.00**	**2.00**	**2.00**	**2.00**	**2.00**
*O^{2-}	6.00	6.00	5.86	5.94	5.36	6.00	5.83	5.80	5.74	4.24	4.00	5.91
*$(OH)^-$	–	–	0.14	0.06	0.64	–	0.17	0.20	0.26	1.76	2.00	0.09
ΣX	**6.00**	**6.00**	**6.00**	**6.00**	**6.00**	**6.00**	**6.00**	**6.00**	**6.00**	**6.00**	**6.00**	**6.00**
*O^{2-}	0.36	0.54	–	–	–	0.20	–	–	–	–	–	0.50
*$(OH)^-$	0.29	0.46	0.72	0.98	0.99	0.34	0.49	0.47	0.89	0.93	0.47	–
F^-	0.35	–	0.28	0.02	0.01	0.46	0.51	0.53	0.11	0.07	0.53	–
ΣY	**1.00**	**1.00**	**1.00**	**1.00**	**1.00**	**1.00**	**1.00**	**1.00**	**1.00**	**1.00**	**1.00**	**0.50**

* Calculated values. bd—below detection limit.

Table 4. The PSM mean composition in rocks of the Kovdor massif (mean ± standard deviation (SD)/min–max, *apfu*).

Rock	Foidolite	Diopsidite and Phlogopitite	(Ap)-Fo Phoscorite	Low-Cb Mag-Rich Phoscorite	Cal-Rich Phoscorite and Related Carbonatite	Vein Calcite Carbonatite	Vein Dol Carbonatite and Mag-Dol-Srp Rock
n	3	20	4	51	92	44	39
Ca	0.5 ± 0.2 0.29–0.67	0.6 ± 0.4 0.12–1.59	0.8 ± 0.4 0.38–1.14	0.7 ± 0.3 0.10–1.24	0.6 ± 0.3 0.08–1.61	0.5 ± 0.2 0.13–1.00	0.6 ± 0.3 0.09–1.11
Na	0.6 ± 0.5 0.00–0.90	0.4 ± 0.4 0.00–1.02	0.3 ± 0.3 0.00–0.70	0.3 ± 0.3 0.00–0.96	0.3 ± 0.4 0.00–1.32	0.2 ± 0.4 0.00–1.37	0.4 ± 0.4 0.00–1.31
Mn	–	0.00 ± 0.01 0.00–0.03	0.02 ± 0.05 0.00–0.09	0.01 ± 0.03 0.00–0.10	0.01 ± 0.02 0.00–0.10	0.01 ± 0.03 0.00–0.18	0.01 ± 0.02 0.00–0.08
Ni	–	–	0.01 ± 0.01 0.00–0.03	–	–	–	–
Ba	–	–	–	0.00 ± 0.01 0.00–0.04	0.03 ± 0.09 0.00–0.55	0.02 ± 0.07 0.00–0.36	0.02 ± 0.07 0.00–0.36
Sc	–	–	–	0.00 0.00–0.03	–	–	–
Cu	–	–	–	–	–	–	0.00 ± 0.01 0.00–0.08
Sr	–	–	0.1 ± 0.1 0.00–0.19	0.02 ± 0.05 0.00–0.26	0.04 ± 0.08 0.00–0.36	0.03 ± 0.07 0.00–0.29	0.1 ± 0.1 0.00–0.30
Y	–	0.01 ± 0.04 0.00–0.17	–	0.01 ± 0.03 0.00–0.16	0.01 ± 0.03 0.00–0.18	0.00 ± 0.03 0.00–0.18	–
La	0.05 ± 0.01 0.03–0.06	0.02 ± 0.02 0.00–0.06	0.01 ± 0.01 0.00–0.02	0.02 ± 0.05 0.00–0.25	0.01 ± 0.02 0.00–0.07	0.02 ± 0.02 0.00–0.06	0.01 ± 0.02 0.00–0.06

Table 4. Cont.

Rock	Foidolite	Diopsidite and Phlogopitite	(Ap)-Fo Phoscorite	Low-Cb Mag-Rich Phoscorite	Cal-Rich Phoscorite and Related Carbonatite	Vein Calcite Carbonatite	Vein Dol Carbonatite and Mag-Dol-Srp Rock
Ce	0.07 ± 0.03 0.05–0.10	0.10 ± 0.08 0.00–0.37	0.04 ± 0.05 0.00–0.08	0.06 ± 0.07 0.00–0.38	0.06 ± 0.05 0.00–0.19	0.08 ± 0.05 0.00–0.20	0.06 ± 0.04 0.00–0.16
Pr	–	–	–	–	0.00 0.00–0.01	–	–
Nd	–	0.01 ± 0.02 0.00–0.04	0.01 ± 0.02 0.00–0.03	0.01 ± 0.01 0.00–0.06	0.01 ± 0.01 0.00–0.07	0.02 ± 0.02 0.00–0.08	0.01 ± 0.02 0.00–0.07
Pb	–	–	–	0.01 ± 0.04 0.00–0.28	0.01 ± 0.01 0.00–0.04	–	–
Th	0.2 ± 0.1 0.06–0.33	0.07 ± 0.09 0.00–0.35	0.06 ± 0.05 0.00–0.12	0.1 ± 0.1 0.00–0.48	0.06 ± 0.05 0.00–0.25	0.04 ± 0.05 0.00–0.20	0.05 ± 0.05 0.00–0.25
U	0.1 ± 0.1 0.00–0.22	0.2 ± 0.1 0.00–0.41	0.1 ± 0.1 0.00–0.27	0.2 ± 0.2 0.00–0.57	0.2 ± 0.2 0.00–0.43	0.2 ± 0.1 0.00–0.44	0.1 ± 0.1 0.00–0.43
A	**1.4 ± 0.3** **1.11–1.57**	**1.3 ± 0.5** **0.57–2.10**	**1.3 ± 0.4** **0.93–1.85**	**1.4 ± 0.4** **0.51–2.14**	**1.4 ± 0.4** **0.63–2.14**	**1.1 ± 0.4** **0.41–2.03**	**1.4 ± 0.5** **0.65–2.14**
Nb	**1.5 ± 0.2** **1.27–1.69**	1.2 ± 0.3 0.71–1.72	1.4 ± 0.3 0.97–1.75	1.2 ± 0.3 0.53–1.79	1.2 ± 0.3 0.35–1.86	1.1 ± 0.3 0.42–1.80	1.3 ± 0.4 0.55–1.72
Ti	0.4 ± 0.1 0.27–0.50	0.5 ± 0.2 0.18–0.81	0.3 ± 0.2 0.14–0.62	0.4 ± 0.3 0.05–1.17	0.4 ± 0.2 0.05–0.92	0.4 ± 0.2 0.09–0.71	0.4 ± 0.2 0.16–0.80
Ta	0.05 ± 0.04 0.00–0.09	0.1 ± 0.1 0.00–0.53	0.08 ± 0.07 0.03–0.18	0.2 ± 0.2 0.00–1.21	0.3 ± 0.2 0.00–1.12	0.2 ± 0.2 0.02–0.95	0.13 ± 0.09 0.00–0.34
Fe	0.07 ± 0.07 0.00–0.14	0.08 ± 0.08 0.00–0.32	0.13 ± 0.06 0.08–0.20	0.12 ± 0.08 0.00–0.40	0.10 ± 0.07 0.00–0.45	0.12 ± 0.08 0.00–0.40	0.10 ± 0.06 0.00–0.28
Al	–	–	–	0.01 ± 0.03 0.00–0.21	0.00 ± 0.01 0.00–0.10	–	0.01 ± 0.02 0.00–0.08
Zr	–	0.01 ± 0.05 0.00–0.16	–	0.02 ± 0.05 0.00–0.24	0.04 ± 0.07 0.00–0.36	0.1 ± 0.1 0.00–0.36	0.1 ± 0.1 0.00–0.39
Mg	–	0.00 ± 0.02 0.00–0.08	0.04 ± 0.07 0.00–0.15	0.02 ± 0.06 0.00–0.32	0.01 ± 0.03 0.00–0.14	0.03 ± 0.05 0.00–0.21	0.03 ± 0.05 0.00–0.19
Si	–	0.1 ± 0.2 0.00–0.54	–	0.02 ± 0.06 0.00–0.31	0.02 ± 0.05 0.00–0.34	0.02 ± 0.05 0.00–0.23	0.01 ± 0.03 0.00–0.18
P	–	–	–	–	0.00 ± 0.01 0.00–0.10	–	0.01 ± 0.04 0.00–0.18
B	2	2	2	2	2	2	2
K	–	–	–	0.00 ± 0.02 0.00–0.13	0.01 ± 0.02 0.00–0.15	–	0.00 ± 0.01 0.00–0.03
F	–	–	–	0.2 ± 0.2 0.00–0.71	0.1 ± 0.1 0.00–0.53	0.1 ± 0.2 0.00–0.53	0.2 ± 0.3 0.00–0.51
Nb/(Nb + Ta)	0.96 ± 0.03 0.93–1.00	0.91 ± 0.09 0.69–1.00	0.94 ± 0.06 0.84–0.98	0.9 ± 0.1 0.43–1.00	0.9 ± 0.1 0.30–1.00	0.8 ± 0.1 0.38–0.98	0.90 ± 0.07 0.73–1.00
Ti/(Nb + Ta + Ti)	0.3 ± 0.1 0.13–0.48	0.3 ± 0.1 0.09–0.52	0.2 ± 0.1 0.07–0.35	0.2 ± 0.1 0.03–0.57	0.2 ± 0.1 0.03–0.52	0.2 ± 0.1 0.03–0.38	0.2 ± 0.1 0.04–0.49
Ca/(Ca + Na)	0.5 ± 0.3 0.26–1.00	0.7 ± 0.3 0.31–1.00	0.7 ± 0.2 0.58–1.00	0.6 ± 0.2 0.17–1.00	0.7 ± 0.2 0.08–1.00	0.8 ± 0.3 0.05–1.00	0.6 ± 0.2 0.26–1.00

Over 92% of the analyzed PSM are represented by O- and (OH)-dominant phases, and the rest are fluorine-dominant. Fluorine shows positive correlations with Ca and Nb and negative correlations with Ti and U (Figure 5). Correspondingly, fluorine-dominant phases occur mainly among calciopyroclores; while natropyrochlores, betafites and microlites are usually represented by hydroxyl-dominant and sometimes oxygen-dominant phases. As for pyrochlore, Ca-dominant phases prevail in ≈45% of the analyzed samples; about 20% of the samples include Na-dominant pyrochlore, and the rest are formed by kenopyrochlore (Figure 6a) as a result of heterovalent substitutions: $2Ca^{2+} \rightarrow Na^+REE^{3+}$, $2Ca^{2+} \rightarrow \square U^{4+}$, $Ca^{2+}O^{2-} \rightarrow Na^+(OH)^-$, etc. Simultaneously, Nb is replaced with Ta and Ti up to the formation of microlite and betafite.

In about 35% of the analyzed PSM, the sum of cations in the *A*-position does not exceed 1 *apfu* (see Figure 6a). Kenopyrochlores occur in 59% of the analyzed samples of vein calcite carbonatite, 45% of diopsidite and phlogopitite samples, 25% of phoscorite samples, and 23% of vein dolomite carbonatite samples. The deficit of cations in the *A*-position is caused by both the presence of high-charge cations of U^{4+}, Th^{4+}, REE^{3+} instead of Na^+ and Ca^{2+} ($4N^{a+} \leftrightarrow 3\,\square\,U^{4+}$, $3Ca^{2+} \leftrightarrow \square 2REE^{3+}$, etc.), and cation loss during pyrochlore metamictization and hydration. The last processes are typical for pyrochlore with >15 wt % of $(U,Th)O_2$, and accompanied by destruction of the mineral crystal structure and leaching of Na, then Ca, and finally REE, U and Th (Figure 6b). In kenopyrochlore, uranium is a predominant high-charge cation; while Th-dominant phases occur much rarely (Figure 6c). Rare-earth elements are represented mainly by light lanthanides La through Nd, with total average content of La and Ce of about 87%. The highest REE content in pyrochlore is typical for host diopsidite and phlogopitite, as well as vein calcite carbonatite.

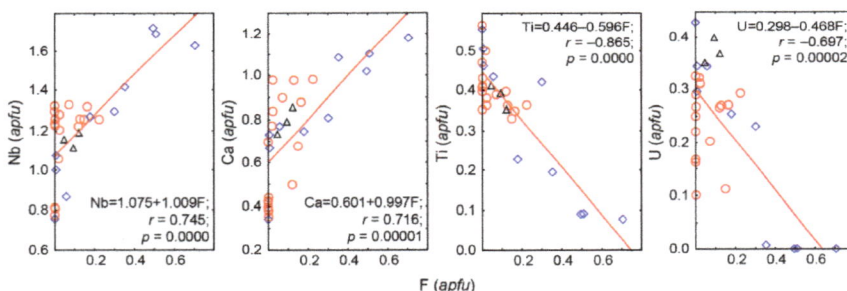

Figure 5. Nb, Ti, Ca and U content vs. fluorine amount in PSM.

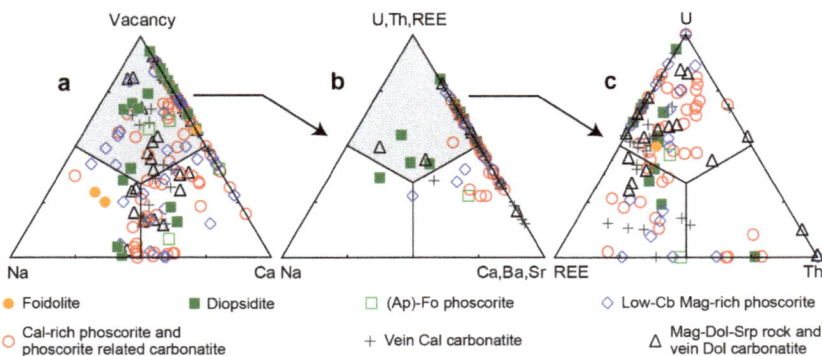

Figure 6. Relations between cations and vacancies in the *A*-site of the Kovdor PSM (**a**—all samples, **b**—kenopyrochlores, **c**—kenopyrochlores enriched in high-charge cations).

To determine the chemical evolution of the PSM in natural rock sequence, we implemented factor analysis of 314 PSM compositions using the method of principal components (with normalization and varimax rotation of factors) (Table 5). The resultant factors enable us to specify five schemas of isomorphic substitutions (elements with high factor loadings are bolded): (1) $\mathbf{Na^+Ca^{2+}Nb^{5+}} \leftrightarrow \mathbf{U^{4+}Ti^{4+}}$; (2) $\mathbf{(Th^{4+}, REE^{3+})} \leftrightarrow \mathbf{U^{4+}}$; (3) $\mathbf{Na^+Nb^{5+}} \leftrightarrow \mathbf{Sr^{2+}Zr^{4+}}$; (4) $\mathbf{Nb^{5+}} \leftrightarrow \mathbf{(Ta^{5+}, Fe^{3+})}$; (5) $\mathbf{Ca^{2+}} \leftrightarrow \mathbf{Ba^{2+}}$. Correspondingly, in the natural sequence of the Kovdor rocks (Figure 7), content of Ba, Sr, U, Ta, Fe and Zr in PSM gradually increases due to Na, Ca, REE, Th and Nb (F2–F5), while higher contents of Ti, U and vacancies are observed in host silicate and calcio-carbonatite rocks (F1).

Table 5. Result of factor analysis of PSM composition.

Variables	Factor Loadings				
	Factor 1	Factor 2	Factor 3	Factor 4	Factor 5
Na	**−0.828**	0.164	−0.135	0.059	−0.165
Ca	**−0.740**	0.129	−0.117	−0.182	−0.298
U	**0.594**	**0.624**	−0.214	0.034	−0.174
Th	0.213	**−0.628**	−0.279	0.075	0.033
REE	0.235	**−0.726**	0.073	−0.111	−0.265
Ba	0.021	0.032	0.128	0.035	**0.742**
Mn	0.288	0.132	−0.330	−0.026	0.452
Sr	0.103	−0.016	**0.620**	−0.243	0.315
Vacancy	**0.818**	−0.199	0.167	0.078	0.241
Nb	**−0.652**	−0.107	−0.186	**−0.657**	0.131
Ta	0.064	0.314	−0.130	**0.779**	−0.010
Ti	**0.832**	−0.015	−0.001	0.010	−0.256
Zr	0.256	0.108	**0.760**	0.152	−0.113
Fe	−0.017	−0.303	0.028	**0.626**	0.083
Expl. Var	3.638	1.627	1.328	1.575	1.233
Prp. Totl	0.260	0.116	0.095	0.113	0.088

Marked factor loadings exceed 0.5. Expl. Var—single factor variance explained, Prp. Totl—percentage of the total variance explained.

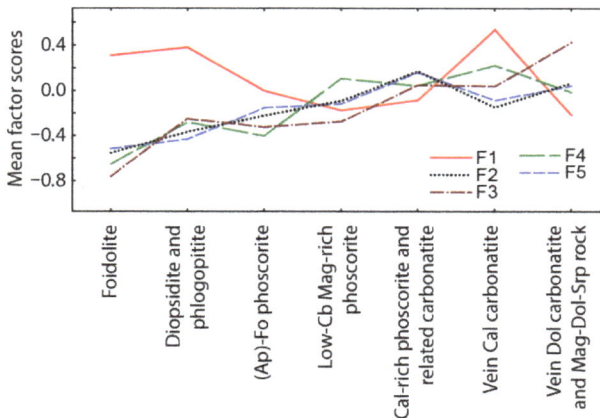

Figure 7. Changes of mean factor scores reflected PSM composition (see Table 5) in natural sequence of the Kovdor rocks.

All above substitutions cause complex zonation of the ore-pipe in terms of the PSM composition (Figure 8): marginal (apatite)-forsterite phoscorite contains Th-REE-rich (keno)pyrochlore and betafite, intermediate low-carbonate magnetite-rich phoscorite comprises pyrochlore with medium content of basic cations, and axial calcite-rich phoscorite and carbonatites accumulate (keno)pyrochlore, (keno)microlite and betafite comparatively enriched in U, Fe, Zr, Ba and Sr.

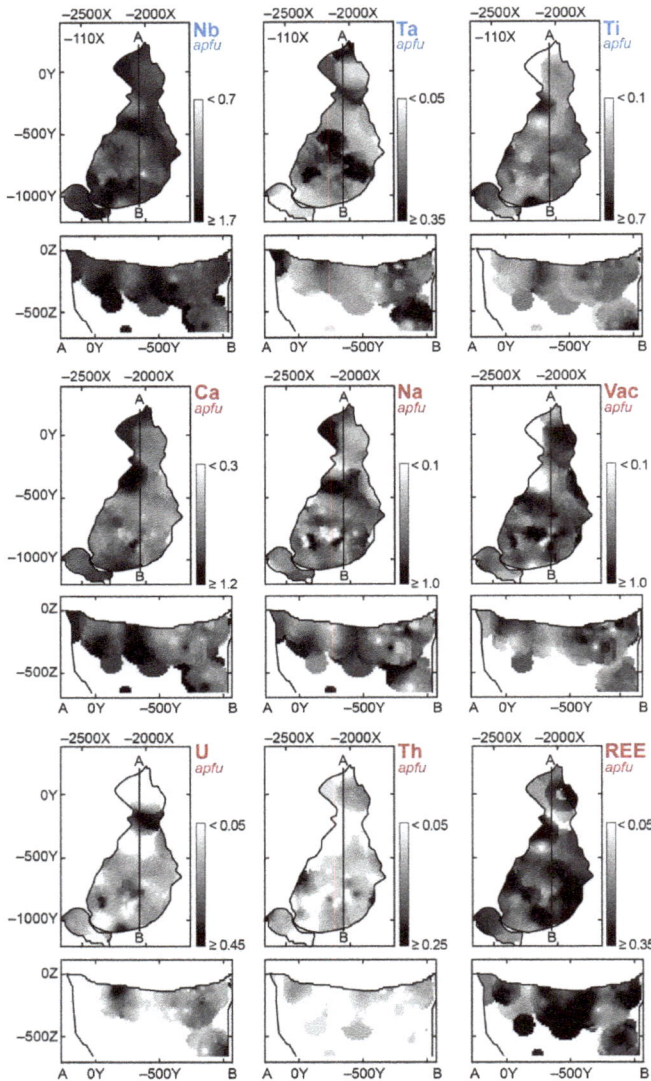

Figure 8. Distribution of the PSM basic constituents in the Kovdor phoscorite-carbonatite complex.

In the unaltered PSM crystals from diopsidite, phlogopitite, phoscorite and calcite carbonatite, there are irregular variations of chemical composition between separate zones, without any clear trends from cores to margins. However, in host foidolite, marginal zones of pyrochlore crystals are constantly enriched in Ca, Na, Th and Nb in comparison with REE-U-Ta-Ti-rich cores, and the latest dolomite carbonatite, pyrochlore grains have Na-Nb-dominant cores and Ca-U-Zr-rich margins. Besides, U-Th-rich (keno)pyrochlore grains often have secondary zonation due to leaching of *A*-cations and then *B*-cations from the grain marginal parts. Fluorine content increases from the core to rim in fresh pyrochlore crystals (Figure 9a). The PSM alteration under the influence of self-irradiation causes loss of fluorine (Figure 9b); therefore, the content of fluorine in the PSM is directly proportional to the amounts of Ca and Nb, and inversely proportional to the content of U and Th (see Figure 5).

X-ray powder diffraction of the Kovdor PSM showed good crystallinity of U/Th-poor calcio- and natropyrochlores; while all analyzed kenopyrochlores became amorphous. Therefore, a single-crystal X-ray study was performed only for low-vacant pyrochlore.

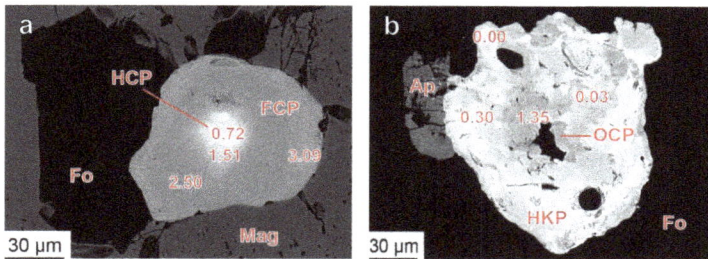

Figure 9. Zonal crystals of pyrochlore: (**a**) primary zonation of hydroxycalciopyrochlore–fluorcalciopyrochlore crystal from magnetite-forsterite phoscorite 1010/243.7; (**b**) secondary zonation in grain of oxycalciopyrochlore–U-rich hydroxykenopyrochlore from magnetite-forsterite phoscorite 987/2.1. BSE-images of thin polished sections with indicated fluorine contents. Mineral abbreviations are in Section 2.

3.3. Crystal Structure

Single-crystal X-ray diffraction data were obtained for well-crystalline oxycalciopyrochlore 917-318.5 (see Figure 2e) and hydroxynatropyrochlore K-017-4 (see Figure 2d). A quadrant of three-dimensional data was collected with frame widths of $1°$ in ω, and with 220 s in the range 2θ $6.8°$–$55°$. Scattering factors were calculated from initial model with all Ca and Na at *A*, all Nb and Ti at *B*, O at O1, and O at *Y*1 site. All cation site-occupancies are given in accordance with electron-microprobe-determined values (normalized on a basis of 2*B* cations per formula unit), the O-populated sites (O1 and *Y*1) was set at full occupancy.

Table 6 shows data collection and refinement parameters for the single-crystal X-ray experiments. Atom coordinates, displacement parameters and site occupancies are given in Table 7, and bond lengths in Table 8. Anisotropic displacement parameters are attached in Supplementary Materials (CIF data is available).

Crystal structures of oxycalciopyrochlore 917/318.5 and hydroxynatropyrochlore K-017-4 (Figure 10) were refined in the $Fd\bar{3}m$ space group with R_1 values of 0.032 and 0.054, respectively. Octahedral *B* site has scattering 39.7 and 33.2 *epfu*, respectively, which agrees well with occupancies $(Nb_{0.965}Ti_{0.02}Ta_{0.15})_{1.00}$ and $(Nb_{0.65}Ti_{0.32}Mg_{0.02}Ta_{0.01})_{1.00}$. The calculated scattering factor values of 22.1 and 18.7 *epfu* for the *A* site slightly exceed the observed values of 18.7 and 15.3 *epfu*, probably, due to variation of Th- and U-content. In the crystal structure of oxycalciopyrochlore 917/318.5 and hydroxynatropyrochlore K-017-4, eight-coordinated *A* sites are predominately occupied by calcium and sodium, respectively, and their final occupancies are $(Ca_{0.59}Na_{0.25}Y_{0.09}Fe_{0.02}Th_{0.02}Ce_{0.02}La_{0.01})_{1.00}$ and $(Na_{0.49}Ca_{0.20}\square_{0.13}Fe_{0.06}Sr_{0.06}Th_{0.04}U_{0.02})_{1.00}$ respectively. The mean *A*1-O bond ranges from 2.530 to 2.533 Å, which is more suitable for ideal Ca-O distance 2.54 Å than for ideal Na-O distance 2.60 Å [35]. For hydroxynatropyrochlore K-017-4, displacement parameters for the *Y* site are slightly higher than those for oxygen, and are consistent with its occupancy by (OH)-groups (Table 7); while for oxycalciopyrochlore 917-318-5, displacement parameters for *Y*1 and O1 are almost equal.

Based on the structure refinement, crystal-chemical formulas of oxycalciopyrochlore 917/318.5 and hydroxynatropyrochlore K-017-4 can be determined as $^A(Ca_{1.18}Na_{0.50}Y_{0.18}Fe_{0.04}Ce_{0.04}Th_{0.04}La_{0.02})_{2.00}$

^{B}(Nb$_{1.93}$Ti$_{0.04}$Ta$_{0.015}$)$_{2.00}$O$_{6.00}$^{Y}(O$_{0.78}$OH$_{0.22}$)$_{1.00}$ and ^{A}(Na$_{0.98}$Ca$_{0.40}$$\square_{0.26}Fe_{0.12}Sr_{0.12}Th_{0.08}U_{0.04}$)$_{2.00}$ ^{B}(Nb$_{1.30}$Ti$_{0.64}$ Mg$_{0.04}$ Ta$_{0.02}$)$_{2.00}$[O$_{4.98}$OH$_{1.02}$]$_{6.00}$^{Y}(OH$_{0.61}$F$_{0.39}$)$_{1.00}$ respectively. According to the structural data, the common difference in scattering parameters of the *B* site (39.7 and 33.2 *epfu*) lies in agreement with the substitution scheme ^{B}Ti^{4+} + ^{Y}OH$^-$ = ^{B}Nb^{5+} + ^{Y}O^{2-}. For oxycalciopyrochlore 917/318.5, lower means of *Y*1 site displacement parameters reflect lesser content of (OH)-groups as compared to hydroxynatropyrochlore K-017-4.

Table 6. Crystal data and structure refinement for oxycalciopyrochlore 917/318.5 and hydroxynatropyrochlore K-017-4.

Mineral Sample	Oxycalciopyrochlore 917/318.5	Hydroxynatropyrochlore K-017-4
Temperature (K)	293(2)	293(2)
Crystal system	cubic	cubic
Space group	$Fd\bar{3}m$	$Fd\bar{3}m$
a (Å)	10.4065(4)	10.3917(4)
Volume (Å^3)	1126.96(14)	1122.16(14)
Z	8	8
ρ_{calc} (g/cm^3)	4.310	3.829
μ (mm^{-1})	5.508	4.517
F(000)	1378.0	1225.0
Crystal size (mm^3)	0.15 × 0.15 × 0.15	0.13 × 0.13 × 0.13
Radiation	MoKα (λ = 0.71073)	MoKα (λ = 0.71073)
2Θ range for data collection (°)	6.782–54.54	6.792–54.62
Index ranges	$-8 \leq h \leq 13, -7 \leq k \leq 11, -12 \leq l \leq 13$	$-9 \leq h \leq 9, -12 \leq k \leq 13, -13 \leq l \leq 10$
Reflections collected	602	758
Independent reflections	83 [R_{int} = 0.0433, R_{sigma} = 0.0224]	84 [R_{int} = 0.0981, R_{sigma} = 0.0348]
Data/restraints/parameters	83/0/12	84/0/12
Goodness-of-fit on F^2	1.286	1.173
Final R indexes [I $\geq$ 2σ (I)]	R_1 = 0.0320, wR_2 = 0.0687	R_1 = 0.0536, wR_2 = 0.1346
Final R indexes [all data]	R_1 = 0.0363, wR_2 = 0.0716	R_1 = 0.0595, wR_2 = 0.1497
Largest diff. peak/hole (eÅ^{-3})	0.63/−0.50	0.93/−1.30

Table 7. Atom coordinates, displacement parameters (Å^2) and site occupancies for oxycalciopyrochlore 917/318.5 and hydroxynatropyrochlore K-017-4.

Sample	Site	Occupancy	x/a	y/b	z/c	Ueq	s.s. (*epfu*)	s.s.$_{calc}$ (*epfu*)
917/318.5	B1	Nb$_{0.965}$Ti$_{0.02}$Ta$_{0.15}$	0	1/2	0	0.014(1)	39.67	41.1
K-017-4	B1	Nb$_{0.65}$Ti$_{0.32}$Mg$_{0.02}$Ta$_{0.01}$	0	1/2	0	0.021(1)	33.2	34.7
917/318.5	A1	Ca$_{0.59}$Na$_{0.25}$Y$_{0.09}$Fe$_{0.02}$Th$_{0.02}$Ce$_{0.02}$La$_{0.01}$	−1/4	3/4	0	0.021(1)	18.65	22.11
K-017-4	A1	Na$_{0.49}$Ca$_{0.20}$$\square_{0.13}Fe_{0.06}Sr_{0.06}Th_{0.04}U_{0.02}$	−1/4	3/4	0	0.020(3)	15.3	18.7
917/318.5	Y1 *	O$_{0.78}$OH$_{0.22}$	−3/8	5/8	1/8	0.015(3)		
K-017-4	Y1 *	OH$_{0.61}$F$_{0.39}$	−3/8	5/8	1/8	0.044(4)		
917/318.5	O1 *	O$_{1.00}$	−0.0704(5)	5/8	1/8	0.017(2)		
K-017-4	O1 *	O$_{0.83}$OH$_{0.17}$	−0.0693(6)	5/8	1/8	0.031(2)		

* Relation of O/(OH, F) is calculated for the formula charge balance.

Table 8. Selected bond distances (Å) for oxycalciopyrochlore 917/318.5 and hydroxynatropyrochlore K-017-4 crystal structures.

Mineral Sample	Oxycalciopyrochlore 917/318.5		Hydroxynatropyrochlore K-017-4	
B1 - O1	1.980(2)	×6	B1 - O1	1.973(2)	×6
A1 - Y1	2.2530(1)	×2	A1 - Y1	2.2499(1)	×2
A1 - O1	2.622(4)	×6	A1 - O1	2.627(5)	×6
<A1 - O>	2.530		<A1 - O>	2.533	

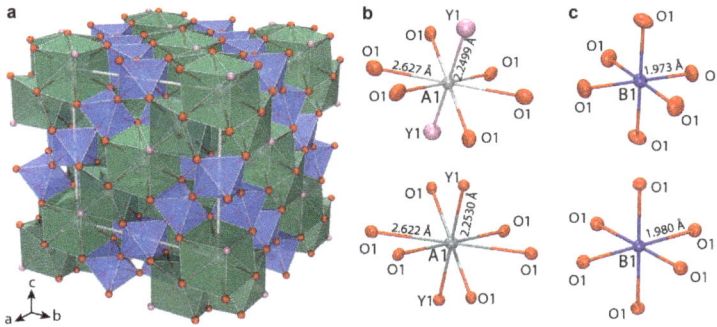

Figure 10. General view of hydroxynatropyrochlore K-017-4 crystal structure (**a**) and geometry of the coordination polyhedra in the crystal structures of hydroxynatropyrochlore K-017-4 (**b**) and oxycalciopyrochlore 917/318.5 (**c**). AO_8 polyhedra are green, BO_6 octahedra are blue, oxygen sites are represented by red circles, *Y*1 sites are shown as pink circles.

3.4. Raman Spectroscopy

The PSM Raman spectra obtained under the same conditions showed significantly different observed intensities I_{obs} of absorption bands (Figure 11a). The observed intensity of the spectrum depends on mineral crystallinity, which, in turn, gradually decreases with growth of U and Th total content from 0.13 *apfu* in HNP 972/86.9 to 0.47 *apfu* in ONP 966/62.9 (Table 9). For the same reason, the stability of the UTh-rich PSM decreases under the influence of laser beams. Corrected absorption band intensities I were calculated as $I = I_{obs}/(n(\omega) + 1)$, where $n(\omega)$ is the Bose (Einstein factor). Figure 11b shows the results for a zoned HKP-HNP crystal from calcite carbonatite 972/86.9.

The absorption bands (see Table 9) were assigned by analogy with other pyrochlore-like compounds [36–38] taking into account theoretical considerations by McCauley [39] and Arenas et al. [40]. According to McCauley [39], pyrochlore yields six Raman-active modes and one acoustic. These modes involve four vibrations of F_{1u}, F_{2g}, E_g, and A_{1g} symmetry. Theoretical calculations by [40] consider bands in the region of 70–180 cm^{-1} as related to acoustic modes (lattice vibrations or bending modes of O-A-O and stretching modes of A-BO_6). Bands in the region of 250–400 cm^{-1} can be assigned to different modes of A-O vibrations. The most intensive bands are related to bending vibrations of O-B-O bonds (400–680 cm^{-1}) and stretching vibrations of B-O bonds in BO_6 octahedra (680–900 cm^{-1}). Positions of typical absorption bands caused by different stretching vibrations depend on composition of the corresponding polyhedra (Figure 12), which enables us to estimate the content of major impurities using the PSM Raman spectra.

Table 9. The PSM chemical composition, Raman shifts and their assignment.

Sample		972/86.9	979/34.0	987/67.2	989/23.1	972/86.9	966/62.9
PSM		HNP	HKP	OCP	HCP	HKP	ONP
Bands assignment		Raman frequencies (cm^{-1})					
^1B-O	stretching	894				894	
^2B-O	stretching	828	842	786	819	830	836
^3B-O	stretching	734	778	707	777	756	755
^4B-O	stretching	681			715	688	
^{1}O-B-O	bending	628	637	622	627	631	644
^{2}O-B-O	bending	567	532	539	527	595	540
^{3}O-B-O	bending			449			
^1A-O	stretching	354	362	381	390	354	361
^2A-O	stretching		260	304	326		257

Table 9. *Cont.*

Sample		972/86.9	979/34.0	987/67.2	989/23.1	972/86.9	966/62.9
PSM		HNP	HKP	OCP	HCP	HKP	ONP
[1]O-*A*-O	bending	282			279	253	
A-BO$_6$	stretching	180	203	212	221	195	203
[2]O-*A*-O	bending		144	164	126	145	142
[3]O-*A*-O	bending	86	91			92	90
Chemical composition (*apfu*)							
Ca		0.70	0.15	1.21	0.62	0.18	0.78
Na		1.02	0.00	0.78	0.71	0.27	1.03
Ce		0.07	0.12	0.01	0.04	0.07	0.06
Th		0.07	0.03	0.03	0.00	0.05	0.06
U		0.06	0.31	0.43	0.43	0.35	0.41
Nb		1.68	1.03	1.12	1.16	1.21	1.31
Ti		0.19	0.72	0.40	0.46	0.48	0.35
Ta		0.13	0.14	0.35	0.26	0.26	0.34
Fe		0.00	0.08	0.07	0.12	0.05	0.00

OCP is oxycalciopyrochlore, ONP is oxynatropyrochlore, HCP is hydroxycalciopyrochlore, HNP is hydroxynatropyrochlore, HKP is hydroxykenopyrochlore.

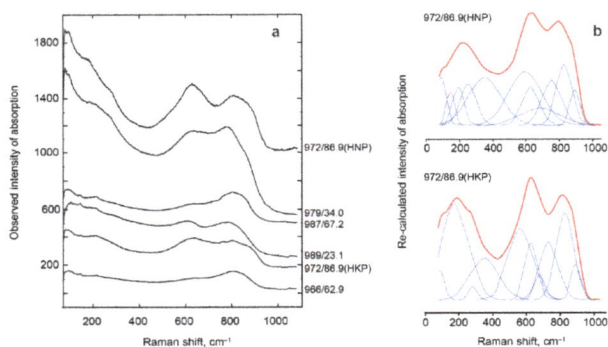

Figure 11. Initial (**a**) and re-calculated (**b**) Raman spectra of pyrochlore supergroup minerals. HNP is hydroxynatropyrochlore, HKP is hydroxykenopyrochlore.

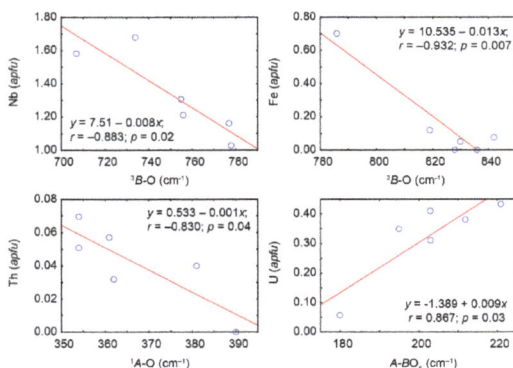

Figure 12. PSM composition vs. position of typical absorption bands in the corresponding Raman spectra.

4. Discussion

Within the Kovdor phoscorite-carbonatite complex, the PSM are concentrated in the axial carbonate-rich zone, so their content, grain size and diversity increase from the pipe margins towards its axis (see Figure 1). Besides, in this direction, we observe gradual growth of Ba, Sr, U, Ta, Fe and Zr content in PSM due to Na, Ca, REE, Th and Nb amounts (see Figures 7 and 8). This trend is the reverse of known ranges of carbonate/silicate melt partition coefficients [41–44]: Al < Si < Ti < Fe < Mg < K < Na < Ca < F < P < CO_2 (for major elements), and Hf < Zr < Th < U < Ta < Y < Nb < Nd < Sr < Ba (for trace elements). In other words, we can assume that most of the high-field-strength elements (HFSE: Ti, Zr, Hf, Nb, Ta and Y) are predominantly distributed in silicate melt; while alkaline-earth and most of rare-earth elements will be localized in carbonate melt. If we compare the earliest silicate rocks (including forsterite-dominant phoscorite) and the latest carbonatite veins (Table 10), we can see that this assumption is correct.

Table 10. Mean contents of the main rock constituents in the Kovdor phoscorite-carbonatite complex.

Rock	Host Silicate Rock	(Ap)-Fo Phoscorite	Low-Cb Mag-Rich Phoscorite	Cal-Rich Phoscorite and Related Carbonatite	Vein Cal-Carbonatite
n	23	11	59	34	9
Al_2O_3, wt %	5.19	1.25	1.44	0.54	0.20
SiO_2	41.91	25.60	7.00	5.06	0.73
TiO_2	0.94	0.19	0.45	0.36	0.04
Fe_2O_{3tot}	9.71	11.28	42.07	28.09	5.61
MgO	14.11	31.93	13.05	10.38	2.95
K_2O	1.59	0.40	0.12	0.12	0.14
Na_2O	1.97	0.20	0.13	0.16	0.13
CaO	17.32	10.20	13.60	24.79	45.50
F	0.04	0.11	0.14	0.13	0.08
P_2O_5	0.18	5.84	10.48	6.70	3.22
CO_2	0.42	0.40	0.40	3.27	9.62
Hf, ppm	7.21	5.17	31.63	21.43	1.37
Zr	237.19	192.78	1270.64	799.99	47.52
Th	4.37	3.12	3.75	3.70	4.01
U	0.85	0.35	0.43	0.65	0.41
Ta	8.33	5.62	15.08	17.65	5.65
Y	4.60	9.45	11.22	12.72	24.77
Nb	78.03	24.27	52.32	62.76	14.64
Nd	32.58	36.27	49.20	79.08	125.72
Sr	434.41	731.49	628.25	1680.81	5067.39
Ba	173.34	229.95	130.74	262.78	464.52

However, the highest concentrations of Zr, Hf, Ta and F as well as the local maximum of Nb content characterize magnetite-rich phoscorite, and can be caused by initial concentration of HFSE in Mg-Al-Ti-rich magnetite. Magnetite exsolution produces ScNb-rich baddeleyite (Figure 13) as a co-product of spinel and ilmenite–geikielite [45], and later alteration of such baddeleyite with fluorine-bearing hydrothermal solutions enriches them in Zr, Nb and Sc. When the concentration of HFSE in the solution reaches a critical level, the PSM start crystallizing. One part of the PSM is formed in situ as rims around baddeleyite grains, and the other part is crystallized as individual crystals in carbonatites that act as a geochemical barrier for $Nb(OH)_3F_2$, $Ta(OH)_3F_3^+$ and other HFSE complexes [46,47]. Besides zirconolite-laachite, the typical PSM associated minerals in carbonate-rich rocks include Sc-phosphates juonniite and kampelite [27,48]:

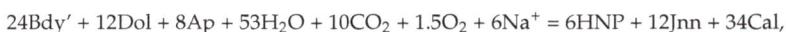

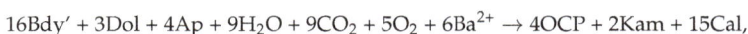

$$24Bdy' + 12Dol + 8Ap + 53H_2O + 10CO_2 + 1.5O_2 + 6Na^+ = 6HNP + 12Jnn + 34Cal,$$

$$16Bdy' + 3Dol + 4Ap + 9H_2O + 9CO_2 + 5O_2 + 6Ba^{2+} \rightarrow 4OCP + 2Kam + 15Cal,$$

where Ap—hydroxylapatite; Bdy'—ScNb end member of baddeleyite, $Sc_{0.5}Nb_{0.5}O_2$; Cal—calcite; Dol—dolomite; Jnn—juonniite, $CaMgScP_2O_8(OH)\cdot4H_2O$; Kam—kampelite, $Ba_3Mg_{1.5}Sc_4(PO_4)_6(OH)_3\cdot4H_2O$; HNP—hydroxynatropyrochlore, $NaCaNb_2O_6(OH)$; OCP—oxycalciopyrochlore, $Ca_2Nb_2O_7$.

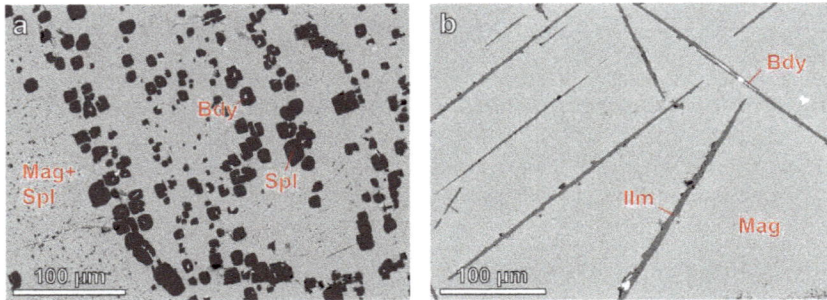

Figure 13. Particles of baddeleyite within exsolution inclusions of spinel and ilmenite in magnetite from calcite-magnetite-apatite phoscorite 956/138.9 (**a**) and magnetite-forsterite phoscorite 987/2.1 (**b**). Mineral abbreviations are in Section 2.

Since Sc-Nb-rich baddeleyite and U-Ta-rich PSM of the ore-pipe axial zone contain most of the Kovdor's Nb, Ta, U, as well as a major part of Sc, this "Anomalous" zone can be regarded as a complex rare-metal deposit [6–8]. Prevailing close intergrowths of Sc-rich baddeleyite and U-rich kenopyrochlore significantly complicate the production of a high-quality baddeleyite concentrate. A perspective approach for the deposit development includes selective mining of "anomalous ore", sulfur-acidic cleaning of baddeleyite concentrate from pyrochlore and zirconolite impurities (Figure 14a), followed by deep metal recovery from baddeleyite concentrate (Figure 14b) and Nb-Ta-Zr-U-Th-rich sulfatic product of its cleaning [49–51].

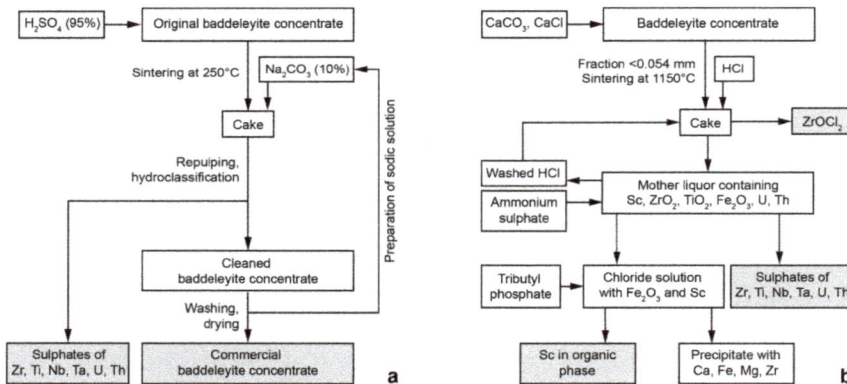

Figure 14. Process flow diagrams of the PSM-bearing baddeleyite concentrate acidic cleaning (**a**, after [49]) and deep processing (**b**, after [50,51]). Gray rectangles show final products.

5. Conclusions

According to mineralogical, geochemical, crystallochemical and spectroscopic data obtained for the PSM from the Kovdor phoscorite-carbonatite complex, we can make the following conclusions:

(1) High-temperature magmatic magnetite is a primary concentrator of Zr, Ti, Nb and Sc in the Kovdor phoscorite-carbonatite complex. The magnetite exsolution under cooling produces spinel and ilmenite-geikielite inclusions containing, in turn, the smallest baddeleyite particles. Besides, separate baddeleyite crystals are crystallized within and around magnetite grains due to their self-cleaning from listed impurities. The content of Nb and Sc in baddeleyite gradually increase

from marginal (apatite)-forsterite phoscorite to axial carbonate-magnetite-rich phoscorite and carbonatite, which is made possible following formation of apo-baddeleyite Nb-Sc-minerals including PSM;

(2)   The PSM are secondary minerals that replace (together with zirconolite–laachite) grains of Sc-Nb-rich baddeleyite and use them as seed crystals. Content, grain size and chemical diversity of PSM increase from the pipe margins to axis against the background of a gradual decreasing of temperature of subsolidus processes. In particular, Ca-(Nb,Ti)-F-rich PSM are spread in marginal (apatite)-forsterite phoscorite, Na-(Nb,Ta)-OH-rich phases occur mainly in intermediate low-carbonate magnetite-rich phoscorite, and U-(Nb,Ti)-OH-rich PSM are localized in axial carbonate-rich phoscorite and carbonatites. Subsolidus PSM crystals usually have primary zoning, with increases of Ca, Ta, Ti, Zr and F contents from core to rim at the expense of Na, Th, U, REE, Nb and (OH) amounts;

(3)   In addition to comparatively high-temperature subsolidus PSM, there are hydrothermal (Na,Ca)-OH-rich pyrochlores that form well-shaped crystals in voids and fractures (in characteristic association with low-temperature Sc-phosphates). Primary zoning of hydrothermal PSM crystals is characterized by growth of U and F contents at the expense of Na, Ca and (OH) amounts. Besides, high U and Th contents cause radiation destruction of the PSM crystal structure, with the following loss of Na, Ca and F under the influence of hydrothermal solutions and the formation of corresponding secondary zoning.

Supplementary Materials: The following are available online at http://www.mdpi.com/2075-163X/8/7/277/s1.

Author Contributions: G.Y.I. and N.G.K. designed the experiments, took samples, performed statistical investigations, and wrote the manuscript. V.N.Y. took and prepared samples and carried out mineralogical investigations. Y.A.P. and A.V.B. took BSE images and performed electron microscope investigations. A.O.K. performed geostatistical investigation, built 3D models, drew maps and took samples. T.L.P. performed crystallographic investigations and formulated crystal-chemical conclusions. V.N.B. performed Raman spectroscopy. J.A.M. carried out petrographical investigations and reviewed the manuscript. All authors discussed the manuscript.

Funding: The research is supported by the Russian Science Foundation, Grant 16-17-10173.

Acknowledgments: Samples were taken during exploration of deep levels of the Kovdor deposit implemented by JSC Kovdorskiy GOK in 2007-11. X-ray crystal studies were carried out with the equipment provided by the X-ray Diffraction Centre of Saint-Petersburg State University. The authors would like to thank I.R. Elizarova for ICP-MS analyses of bulk-rock samples. The anonymous reviewers helped us to considerably improve this paper.

Conflicts of Interest: The authors declare no conflict of interest.

References

1.   Ivanyuk, G.Y.; Yakovenchuk, V.N.; Pakhomovsky, Y.A. *Kovdor*; Laplandia Minerals: Apatity, Russia, 2002; ISBN 5900395413.

2.   Krasnova, N.I.; Balaganskaya, E.G.; Garcia, B. Kovdor—Classic phoscorites and carbonatites. In *Phoscorites and Carbonatites from Mantle to Mine: The Key Example of the Kola Alkaline Province. Mineralogical Society Series 10*; Wall, F., Zaitsev, A.N., Eds.; Mineralogical Society: London, UK, 2004; pp. 99–132. ISBN 0-903056-22-4.

3.   Afanasyev, B.V. *Mineral Resources of Alkaline-Ultrabasic Massifs of the Kola Peninsula*; Roza Vetrov Publishing: Saint-Petersburg, Russia, 2011. (In Russian)

4.   Ivanyuk, G.Y.; Kalashnikov, A.O.; Pakhomovsky, Y.A.; Mikhailova, J.A.; Yakovenchuk, V.N.; Konopleva, N.G.; Sokharev, V.A.; Bazai, A.V.; Goryainov, P.M. Economic minerals of the Kovdor baddeleyite-apatite-magnetite deposit, Russia: Mineralogy, spatial distribution and ore processing optimization. *Ore Geol. Rev.* **2016**, *77*, 279–311. [CrossRef]

5.   Kalashnikov, A.O.; Konopleva, N.G.; Pakhomovsky, Y.A.; Ivanyuk, G.Y. Rare earth deposits of the Murmansk Region, Russia—A review. *Econ. Geol.* **2016**, *111*, 1529–1559. [CrossRef]

6.   Kalashnikov, A.O.; Yakovenchuk, V.N.; Pakhomovsky, Y.A.; Bazai, A.V.; Sokharev, V.A.; Konopleva, N.G.; Mikhailova, J.A.; Goryainov, P.M.; Ivanyuk, G.Y. Scandium of the Kovdor baddeleyite-apatite-magnetite deposit (Murmansk Region, Russia): Mineralogy, spatial distribution, and potential resource. *Ore Geol. Rev.* **2016**, *72*, 532–537. [CrossRef]

7.   Epshteyn, E.M. Genesis of Kovdor apatite-magnetite ores. Reports on Mineralogy and Geochemistry for 1968. *Tr. VIMS* **1970**, 218–224. (In Russian)

8.   Zhuravleva, L.N.; Berezina, L.A.; Gulin, E.N. Peculiarities of the geochemistry of rare and radioactive elements in apatite-magnetite ores of ultrabasic-alkaline complexes. *Geochemistry* **1976**, *10*, 1512–1532. (In Russian)

9.   Zaitsev, A.N.; Williams, C.T.; Jeffries, T.E.; Strekopytov, S.; Moutte, J.; Ivashchenkova, O.V.; Spratt, J.; Petrov, S.V.; Wall, F.; Seltmann, R.; Borozdin, A.P. Rare earth elements in phoscorites and carbonatites of the Devonian Kola Alkaline Province, Russia: Examples from Kovdor, Khibina, Vuoriyarvi and Turiy Mys complexes. *Ore Geol. Rev.* **2014**, *61*, 204–225. [CrossRef]

10.   Bazai, A.V.; Goryainov, P.M.; Elizarova, I.R.; Ivanyuk, G.Y.; Kalashnikov, A.O.; Konopleva, N.G.; Mikhailova, J.A.; Pakhomovsky, Y.A.; Yakovenchuk, V.N. New data on REE resource potential of the Murmansk Region. *Vestn. Kola Sci. Cent.* **2014**, *4*, 50–65. (In Russian)

11.   Atencio, D.; Andrade, M.B.; Christy, A.G.; Giere, R.; Kartashov, P.M. The pyrochlore supergroup of minerals: Nomenclature. *Can. Mineral.* **2010**, *48*, 673–698. [CrossRef]

12.   Atencio, D.; Andrade, M.B.; Bastos Neto, A.C.; Pereira, V.P. Ralstonite Renamed Hydrokenoralstonite, Coulsellite Renamed Fluornatrocoulsellite, and Their Incorporation Into the Pyrochlore Supergroup. *Can. Mineral.* **2017**, *55*, 115–120. [CrossRef]

13.   Kukharenko, A.A.; Orlova, M.P.; Bulakh, A.G.; Bagdasarov, E.A.; Rimskaya-Korsakova, O.M.; Nefedov, E.I.; Ilyinsky, G.A.; Sergeev, A.S.; Abakumova, N.B. *Caledonian Complex of Ultrabasic, Alkaline Rocks and Carbonatites of Kola Peninsula and Northern Karelia (Geology, Petrology, Mineralogy and Geochemistry)*; Nedra: Moscow, Russia, 1965. (In Russian)

14.   Kapustin, Y.L. *Mineralogy of Carbonatites*; Amerind Publishing: New Delhi, India, 1980.

15.   Rimskaya-Korsakova, O.M.; Krasnova, N.I. *Geology of Deposits of the Kovdor Massif*; St. Petersburg University Press: Saint-Petersburg, Russia, 2002. (In Russian)

16.   Mikhailova, J.A.; Kalashnikov, A.O.; Sokharev, V.A.; Pakhomovsky, Y.A.; Konopleva, N.G.; Yakovenchuk, V.N.; Bazai, A.V.; Goryainov, P.M.; Ivanyuk, G.Y. 3D mineralogical mapping of the Kovdor phoscorite–carbonatite complex (Russia). *Miner. Depos.* **2016**, *51*, 131–149. [CrossRef]

17.   Mikhailova, J.A.; Ivanyuk, G.Y.; Kalashnikov, A.O.; Pakhomovsky, Y.A.; Bazai, A.V.; Panikorovskii, T.L.; Yakovenchuk, V.N.; Konopleva, N.G.; Goryainov, P.M. Three-D mineralogical mapping of the Kovdor phoscorite-carbonatite complex, NW Russia: I. Forsterite. *Minerals* **2018**, *8*, 260. [CrossRef]

18.   Williams, T.C. The occurrence of nbiobian zirconolite, pyrochlore and baddeleyite in the Kovdor carbonatite complex, Kola Peninsula, Russia. *Mineral. Mag.* **1996**, *60*, 639–646. [CrossRef]

19.   Rosstandart FR.1.31.2016.25424. The Method for Determining the Content of Rare-Earth Elements (Y, La, Ce, Pr, Nd, Sm, Eu, Gd, Tb, Dy, Ho, Er, Tm, Yb, Lu), Sodium, Aluminum, Potassium, Calcium, Titanium, Iron, Thorium and Uranium in Apatite Mineral Raw Materials and Phosphogypsum by Inductively Coupled Plasma Mass Spectrometry. Available online: http://www.fundmetrology.ru/06_metod/2view_file.aspx?id=25424 (accessed on 27 June 2018).

20.   Dolivo-Dobrovolsky, D.D. MINAL, Free Software. Available online: http://www.dimadd.ru (accessed on 8 July 2013).

21.   Sheldrick, G.M. A short history of SHELX. *Acta Crystallogr. Sect. A Found. Crystallogr.* **2008**, *64*, 112–122. [CrossRef] [PubMed]

22.   Momma, K.; Izumi, F. VESTA 3 for three-dimensional visualization of crystal, volumetric and morphology data. *J. Appl. Crystallogr.* **2011**, *44*, 1272–1276. [CrossRef]

23.   OriginLab Corporation. OriginPro 8.1 SR2 2010. Available online: www.OriginLab.com (accessed on 27 June 2018).

24.   StatSoft Inc. Statistica 8. Available online: www.statsoft.ru (accessed on 27 June 2018).

25.   Micromine Pty Ltd. Micromine 16.1. Available online: www.micromine.com (accessed on 27 June 2018).

26.   Kalashnikov, A.O.; Ivanyuk, G.Y.; Mikhailova, J.A.; Sokharev, V.A. Approach of automatic 3D geological mapping: The case of the Kovdor phoscorite-carbonatite complex, NW Russia. *Sci. Rep.* **2017**, *7*. [CrossRef] [PubMed]

27.   Ivanyuk, G.Y.; Yakovenchuk, V.N.; Panikorovskii, T.L.; Konoplyova, N.G.; Pakhomovsky, Y.A.; Bazai, A.V.; Bocharov, V.N.; Krivovichev, S.V. Hydroxynatropyrochlore, $(Na, Ca, Ce)_2Nb_2O_6(OH)$, a new member of the pyrochlore group from the Kovdor phoscorite-carbonatite pipe (Kola Peninsula, Russia). *Mineral. Mag.* **2018**. [CrossRef]

28. Hogarth, D.D.; Williams, C.T.; Jones, P. Primary zoning in pyrochlore group minerals from carbonatites. *Mineral. Mag.* **2000**, *64*, 683–697. [CrossRef]

29. Rimskaya-Korsakova, O.M.; Dinaburg, I.B. Baddeleyite in massifs of ultrabasic and alkaline rocks of the Kola Peninsula. *Mineral. Geochem.* **1964**, *1*, 13–30. (In Russian)

30. Kirnarsky, Y.M.; Afanas'ev, A.; Men'shikov, Y.P. Accessory Th-rich betafite from carbonatites. In *Materialy po Mineralogii Kol'skogo Poluostrova (Data on the Mineralogy of the Kola Peninsula)*; Nauka: Leningrad, Russia, 1968; Volume 6, pp. 155–163. (In Russian)

31. Osokin, A.S. Accessory rare-metal mineralization of carbonatites of a certain massif of alkaline-ultrabasic rocks (the Kola Peninsula). *Mineral. Geochem.* **1979**, *6*, 27–38. (In Russian)

32. Strelnikova, L.A.; Polezhaeva, L.I. Asseccory minerals of pyrochlore group from carbonatites of some alkali-ultrabasic massifs. In *Substantial Composition of Alkaline Intrusive Complexes of the Kola Peninsula*; KolFAN: Apatity, Russia, 1981; pp. 81–88. (In Russian)

33. Yamada, R.; Nagaishi, R.; Hatano, Y.; Yoshida, Z. Hydrogen production in the γ-radiolysis of aqueous sulfuric acid solutions containing Al_2O_3, SiO_2, TiO_2 or ZrO_2 fine particles. *Int. J. Hydrog. Energy* **2008**, *33*, 929–936. [CrossRef]

34. Dolin, P.I.; Ershler, B.V. Radiolysis of water in the presence of H_2 and O_2 due to reactor radiation, fission fragments and X-radiation. In *Investigations in Chemistry, Geology and Metallurgy. Reports of Soviet Delegation at the International Conference on the Peaceful Uses of Atomic Energy*; AN SSSR: Geneva, Switzerland, 1955; pp. 293–319. (In Russian)

35. Shannon, R.D. Revised effective ionic radii and systematic studies of interatomic distances in halides and chalcogenides. *Acta Crystallogr. Sect. A* **1976**, *32*, 751–767. [CrossRef]

36. Seikh, M.M.; Sood, A.K.; Narayana, C. Electronic and vibrational Raman spectroscopy of Nd0.5Sr0.5MnO3 through the phase transitions. *Pramana* **2005**, *64*, 119–128. [CrossRef]

37. Bahfenne, S.; Frost, R.L. Raman spectroscopic study of the antimonate mineral roméite. *Spectrochim. Acta Part A Mol. Biomol. Spectrosc.* **2010**, *75*, 637–639. [CrossRef] [PubMed]

38. Maczka, M.; Knyazev, A.V.; Kuznetsova, N.Y.; Ptak, M.; Macalik, L. Raman and IR studies of $TaWO_{5.5}$, $ASbWO_6$ (A = K, Rb, Cs, Tl), and $ASbWO_6 \cdot H_2O$ (A = H, NH_4, Li, Na) pyrochlore oxides. *J. Raman Spectrosc.* **2011**, *42*, 529–533. [CrossRef]

39. McCauley, R.A. Infrared-absorption characteristics of the pyrochlore structure. *J. Opt. Soc. Am.* **1973**, *63*, 721. [CrossRef]

40. Arenas, D.J.; Gasparov, L.V.; Qiu, W.; Nino, J.C.; Patterson, C.H.; Tanner, D.B. Raman study of phonon modes in bismuth pyrochlores. *Phys. Rev. B* **2010**, *82*, 214302. [CrossRef]

41. Jones, J.H.; Walker, D.; Pickett, D.A.; Murrell, M.T.; Beattie, P. Experimental investigations of the partitioning of Nb, Mo, Ba, Ce, Pb, Ra, Th, Pa, and U between immiscible carbonate and silicate liquids. *Geochim. Cosmochim. Acta* **1995**, *59*, 1307–1320. [CrossRef]

42. Brooker, R.A. The effect of CO_2 saturation on immiscibility between silicate and carbonate liquids: An experimental study. *J. Petrol.* **1998**, *39*, 1905–1915. [CrossRef]

43. Veksler, I.V.; Petibon, C.; Jenner, G.A.; Dorfman, A.M.; Dingwell, D.B. Trace Element Partitioning in Immiscible Silicate-Carbonate Liquid Systems: An Initial Experimental Study Using a Centrifuge Autoclave. *J. Petrol.* **1998**, *39*, 2095–2104. [CrossRef]

44. Lee, M.J.; Lee, J.I.; Garcia, D.; Moutte, J.; Williams, C.T.; Wall, F.; Kim, Y. Pyrochlore chemistry from the Sokli phoscorite-carbonatite complex, Finland: Implications for the genesis of phoscorite and carbonatite association. *Geochem. J.* **2006**, *40*, 1–13. [CrossRef]

45. Ivanyuk, G.; Kalashnikov, A.; Pakhomovsky, Y.; Bazai, A.; Goryainov, P.; Mikhailova, J.; Yakovenchuk, V.; Konopleva, N. Subsolidus Evolution of the Magnetite-Spinel-Ulvöspinel Solid Solutions in the Kovdor Phoscorite-Carbonatite Complex, NW Russia. *Minerals* **2017**, *7*, 215. [CrossRef]

46. Timofeev, A.; Migdisov, A.A.; Williams-Jones, A.E. An experimental study of the solubility and speciation of niobium in fluoride-bearing aqueous solutions at elevated temperature. *Geochim. Cosmochim. Acta* **2015**, *158*, 103–111. [CrossRef]

47. Timofeev, A.; Migdisov, A.A.; Williams-Jones, A.E. An experimental study of the solubility and speciation of tantalum in fluoride-bearing aqueous solutions at elevated temperature. *Geochim. Cosmochim. Acta* **2017**, *197*, 294–304. [CrossRef]

48. Yakovenchuk, V.N.N.; Ivanyuk, G.Y.Y.; Pakhomovsky, Y.A.A.; Panikorovskii, T.L.L.; Britvin, S.N.N.; Krivovichev, S.V.V.; Shilovskikh, V.V.V.; Bocharov, V.N.N. Kampelite, $Ba_3Mg_{1.5}Sc_4(PO_4)_6(OH)_3 \cdot 4H_2O$, a new very complex Ba-Sc phosphate mineral from the Kovdor phoscorite-carbonatite complex (Kola Peninsula, Russia). *Mineral. Petrol.* **2017**, *4*, 1–11. [CrossRef]
49. Lokshin, E.P.; Lebedev, V.N.; Bogdanovich, V.V.; Novozhilova, V.V.; Popovich, V.F. Method of Baddeleyite Concentrate Cleaning; Patent RU-2139250, 1998. Available online: https://patents.google.com/patent/RU2139250C1/ru (accessed on 27 June 2018).
50. Lebedev, V.N.; Lokshin, E.P.; Mel'nik, N.A.; Shchur, T.E.; Popova, L.A. Possibility of Integrated Processing of the Baddeleyite Concentrate. *Russ. J. Appl. Chem.* **2004**, *77*, 708–710. [CrossRef]
51. Lebedev, V.N. Extraction and refining of scandium upon the processing of baddeleyite concentrates. *Theor. Found. Chem. Eng.* **2007**, *41*, 718–722. [CrossRef]

minerals

MDPI

Article

Zircon Macrocrysts from the Drybones Bay Kimberlite Pipe (Northwest Territories, Canada): A High-Resolution Trace Element and Geochronological Study

Ekaterina P. Reguir [1,*], Anton R. Chakhmouradian [1], Barrett Elliott [2], Ankar R. Sheng [1] and Panseok Yang [1]

[1] Department of Geological Sciences, University of Manitoba, 125 Dysart Road, Winnipeg, MB R3T 2N2, Canada; chakhmou@cc.umanitoba.ca (A.R.C.); mr.ankar@gmail.com (A.R.S.); Panseok.Yang@umanitoba.ca (P.Y.)

[2] Northwest Territories Geological Survey, 4601-B 52 Avenue, Yellowknife, NT X1A 1K3, Canada; Barrett_Elliott@gov.nt.ca

* Correspondence: umreguir@cc.umanitoba.ca; Tel.: +1-204-474-8765

Received: 1 October 2018; Accepted: 22 October 2018; Published: 25 October 2018

check for updates

Abstract: Zircon macrocrysts in (sub)volcanic silica-undersaturated rocks are an important source of information about mantle processes and their relative timing with respect to magmatism. The present work describes variations in trace element (Sc, Ti, Y, Nb, lanthanides, Hf, Ta, Pb, Th, and U) and isotopic (U-Pb) composition of zircon from the Drybones Bay kimberlite, Northwest Territories, Canada. These data were acquired at a spatial resolution of $\leq$100 μm and correlated to the internal characteristics of macrocrysts (imaged using cathodoluminescence, CL). Six types of zircon were distinguished on the basis of its luminescence characteristics, with the majority of grains exhibiting more than one type of CL response. The oscillatory-zoned core and growth sectors of Drybones Bay zircon show consistent variations in rare-earth elements (REE), Hf, Th, and U. Their chondrite-normalized REE patterns are typical of macrocrystic zircon and exhibit extreme enrichment in heavy lanthanides and a positive Ce anomaly. Their Ti content decreases slightly from the core into growth sectors, but the Ti-in-zircon thermometry gives overlapping average crystallization temperatures (820 ± 26 °C to 781 ± 19 °C, respectively). There is no trace element or CL evidence for Pb loss or other forms of chemical re-equilibration. All distinct zircon types are concordant and give a U-Pb age of 445.6 ± 0.8 Ma. We interpret the examined macrocrysts as products of interaction between a shallow (<100 km) mantle source and transient kimberlitic melt.

Keywords: zircon; macrocrysts; kimberlite; trace elements; geochronology; cathodoluminescence; Ti-in-zircon geothermometry

1. Introduction

The Slave craton in northern Canada is a small (~200,000 km^2) but well-exposed fragment of Archean (4.05–2.55 Ga) continental crust that hosts a variety of gold, silver, base metal, critical metal, and gemstone deposits. Notably, the latter include >300 kimberlite bodies (vents, dikes, and diatremes) [1]. The discovery of first diamondiferous kimberlites in the Lac de Gras area in 1991 sparked the largest claim-staking rush in Canadian mining history [2], which ultimately paved the way to the development of five new mines to date (Ekati since 1998, Diavik since 2003, Jericho in 2006–2008, Snap Lake in 2008–2015, and Gahcho Kué since 2016). Nineteen individual bodies have been, or are presently, mined and about a dozen other kimberlite clusters are under active exploration (Figure 1).

An increase in diamond production in recent years has moved Canada up to the second place in the global market in volume terms (23.2 million carats valued at $2.06 billion in 2017).

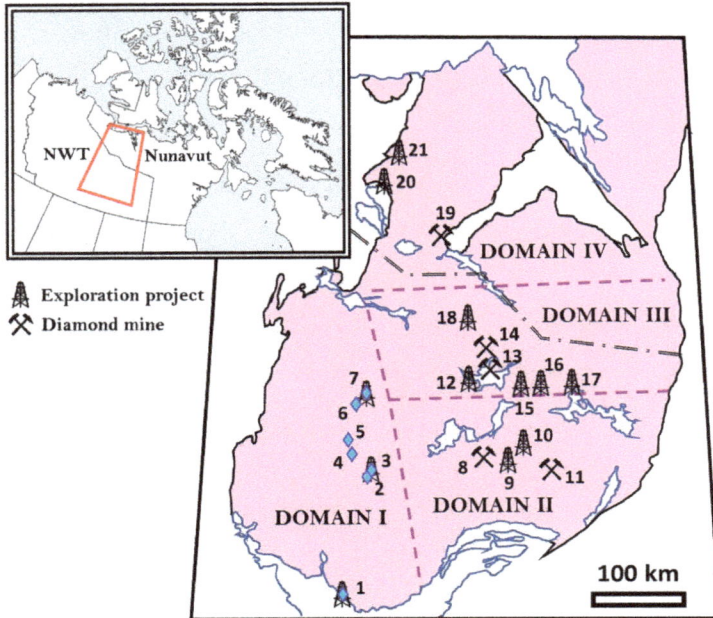

Figure 1. Schematic location map (outlined red in the inset) of the four major kimberlite domains in the Archean Slave craton (pink); based on Heaman et al. (2003), NWT Government (2016) and an unpublished kimberlite database (NWT Geological Survey). Siluro-Ordovician kimberlites are indicated by blue diamonds and include: (1) Drybones Bay and Mud Lake; (2) Ursa and Winny; (3) Orion and Cross; (4) Jean, Cirque, and Rich; (5) Willow; (6) Kent, Otter, PCE, Wolverine North, and South; (7) Aquila and Cygnus. Other localities: (8) Snap Lake mine; (9) Camsel Lake; (10) Munn Lake; (11) Gahcho Kué mine; (12) Monument Bay; (13) Diavik mine; (14) Ekati mine; (15) DO27; (16) Afridi Lake; (17) Nicholas Bay; (18) Yamba Lake; (19) Jericho mine; (20) Anuri; (21) Hammer. NWT–Nunavut territorial boundary is marked by a black dot-dashed line. NWT: Northwest Territories.

Only a small number of the 300-plus Slave kimberlites have been studied in significant detail. As can be expected, the bulk of published work is focused on the producing areas, especially the Ekati and Diavik mines at Lac de Gras. One of the least-studied areas with a poorly understood potential is the southwestern section of the craton hosting Early- to Mid-Paleozoic kimberlites, which are aligned in a roughly longitudinal fashion from Drybones Bay and Mud Lake along the northern shore Great Slave Lake to the Aquila and Cygnus pipes some 250 km to the north (Figure 1). Very little data is available on these rocks in the literature, which is practically limited to a study of mantle xenoliths from the Drybones Bay pipe [3] and U-Pb geochronological determinations for several of these bodies [4]. This lack of information is regrettable because (1) the majority of tested kimberlites from this part of the craton were found to be diamondiferous; (2) geologically, these localities are separated from the other Slave fields by a major crustal-scale fault (see Regional Geology), and (2) similar-aged economic pipes are known in the North China and Siberian cratons, where they have been studied extensively [5,6]. The present project was initiated to examine the petrology, mineralogy, and geochemistry of the SW Slave kimberlites using extensive exploration data and drill-core material available at the Geological Materials Storage facility operated by the Northwest Territories Geological Survey. This contribution is concerned with the trace element and isotopic characteristics of zircon from the Drybones Bay pipe.

This mineral was chosen because of its recognized value as a geochronometer, geothermometer and process indicator in the study of kimberlites and other mantle-derived rocks.

2. Geological Setting

2.1. Regional Geology

About one-third of the presently exposed Archean bedrock in the Slave craton is represented by metasedimentary and metavolcanic supracrustal rocks of the Yellowknife Supergroup (2.61–2.71 Ga), whereas some 65% is underlain by 2.58–2.64 Ga granitoid intrusions [7]. The Slave craton has been subdivided into eastern and western domains, based primarily on the presence of 2.85–4.05 Ga Mesoarchean rocks and Pb-Nd isotopic variations [8,9]. The Mesoarchean rocks comprise tonalitic to dioritic gneisses that so far have been reported only in the western domain [10]. Lower crustal xenoliths from kimberlites in the Lac de Gras area (Figure 1) indicate the presence of Mesoarchean rocks also east of the domain boundary [11]. The gneisses are overlain by a post-2.93 Ga autochthonous cover sequence, which forms the base of the Yellowknife Supergroup [10]. Supracrustal packages that can be correlated across the entire craton include 2.66–2.69 Ga calc-alkaline subvolcanic-to-extrusive units and younger (2.63–2.66 Ga) turbidites [12]. Syn- to late-kinematic granitoid plutons are voluminous and widespread [7], but intrusions older than 2.61 Ga are limited to the southwestern part of the craton, including the tonalite-granodioritic Defeat suite (ca. 2.62–2.63 Ga) at Drybones Bay. Widespread felsic magmatism spanning the entire craton occurred from ca. 2.61 to 2.58 Ga, and produced both I- and S-type granitoids [13]. Kimberlites are the most widespread form of post-Archean small-volume mantle-derived magmatism in the Slave craton. It spans a period of ~560 Ma from the emplacement of the 613-Ma Anuri pipe to the eruption of Eocene kimberlites in the Lac de Gras area [14,15]. There is a distinct correlation between the spatial distribution of kimberlites and the timing of magmatism, which allows the craton to be subdivided into four domains (Figure 1): (1) southwestern, hosting Siluro-Ordovician (435–459 Ma) pipes; (2) southeastern, characterized by Cambrian (523–542 Ma) kimberlites; (3) central, represented by Cretaceous to Eocene (48–74 Ma) localities; and (4) northern or "mixed", hosting the Jurassic (173 Ma) Jericho pipe, Permian-Triassic (245–286 Ma) Victoria Island and Ediacaran (613 Ma) Coronation Gulf fields [4,15]. The reason for this distribution of kimberlite ages is presently unknown. The sublongitudinal boundary of the southwestern domain roughly coincides with the exposed margin of the Mesoarchean rocks, and the kimberlites of this domain (including Drybones Bay) define a linear N-S trend, which parallels the crustal-scale Beniah Lake fault system [4].

2.2. Local Geology and Exploration History

The Drybones Bay kimberlite "pipe" is situated in a small bay on the northern shore of Great Slave Lake 45 km SE of Yellowknife. The body is submerged under 38 m of water and blanketed by 67 m of overburden, which consists of lacustrine clay, sand, and a basal layer of boulder conglomerate [16]. It intruded the 2.62–2.63 Ga Defeat pluton comprising predominantly biotite tonalite and granodiorite with minor metasedimentary xenoliths derived from the Yellowknife Supergroup [17]. The kimberlite body is interpreted to be a vent-like structure from drilling and geophysical surveys [18,19], but it is unknown how its morphology or size change with depth. The deepest drill hole (302 m) was stopped in volcaniclastic kimberlite. The structure is lobate in shape and measures ~900 × 400 m, covering an estimated area of 22 ha at the present erosion level [16]. The sidewall contacts dip at a lower angle in the western lobe relative to the eastern lobe. The large size and shallow dipping contacts are characteristic of Canadian Class-2 kimberlites, rather than Class-3 pipes characteristic of the central Slave domain [19,20]. The Drybones Bay structure comprises a core intrusion (termination of a feeder dike?), several crater-facies units and an epiclastic unit [16,17]. Unambiguous tuffisitic facies are notably absent.

During the staking rush of the 1990s, diamond exploration on the northern shore of Great Slave Lake led to the discovery of a geophysical anomaly submerged beneath Drybones Bay (Figure 1).

The anomaly was staked and drilled in the winter of 1994; 24 m of kimberlite was intersected under a blanket of lacustrine clay. Low-level airborne and ground-based magnetic surveys were conducted over the claim block over the following year to delineate the kimberlite body with more precision. This was followed by two extensive drill programs (1995 and 1996) that included 20 holes ranging from 131 to 302 m in length and produced over 5.5 km of drill-core. Most of the recovered kimberlite core was split and processed by caustic fusion. Three samples weighing 0.64, 1.87, and 7.58 tonnes yielded 16, 26, and 48 macrodiamonds, respectively, measuring >0.5 mm in at least one dimension [21]. Diamonds have been recovered from each of the lithostratigraphic units, but the highest exploration grades (24–39 carats per hundred tonnes) were measured for the core [16]. In the past 24 years, the Drybones Bay project has changed hands several times and been the subject of litigation between developers and Aboriginal groups [22].

3. Materials and Methods

Cathodoluminescence (CL) images were obtained using a Reliotron VII instrument (Relion Ind., Bedford, MA, USA) operated at 8.5 eV and 400–450 mA, and an electronically cooled Nikon DS-Ri1 camera (Nikon Canada Inc., Mississauga, ON, Canada). The concentrations of major elements were determined by wavelength-dispersive spectrometry (WDS) using a Cameca SX100 (Cameca SAS, Gennevilliers, France) automated electron microprobe operated at 15 kV and 20 nA. The CL and BSE images were used for the selection of areas suitable for further trace element and isotopic analyses.

The abundances of selected trace elements and isotopes were measured by laser-ablation inductively coupled-plasma mass-spectrometry (LA-ICP-MS) using a 213-nm Nd-YAG laser (Elemental Scientific Lasers, Bozeman, MT, USA) connected to an Element 2 sector-field mass-spectrometer (Thermo Fisher Scientific Inc., Bremen, Germany). The zircon grains were analyzed in situ using a spot ablation mode and the following laser parameters: 30 μm beam size, ~0.03 mJ incident pulse energy, 3.1–4.0 J/cm^2 energy density on a sample. Areas suitable for analysis were selected on the basis of reflected-light, BSE, and CL images. The ablation was performed in Ar and He atmospheres. The rate of oxide production was monitored during instrument tuning by measuring the ThO/Th ratio and kept below 0.2%. For LA-ICP-MS measurements, synthetic glass standard NIST SRM 610 [23] was employed for calibration and quality control. After considering potential spectral overlaps and molecular interferences, the following isotopes were chosen for analysis: ^{25}Mg, ^{29}Si, ^{45}Sc, ^{49}Ti, ^{51}V, ^{55}Mn, ^{88}Sr, ^{89}Y, ^{93}Nb, ^{139}La, ^{140}Ce, ^{141}Pr, ^{146}Nd, ^{147}Sm, ^{151}Eu, ^{157}Gd, ^{159}Tb, ^{163}Dy, ^{165}Ho, ^{167}Er, ^{169}Tm, ^{172}Yb, ^{175}Lu, ^{178}Hf, ^{181}Ta, ^{208}Pb, ^{232}Th, and ^{238}U. Silicon, measured by WDS, was employed as an internal standard. All analyses were performed in a low-resolution mode (~300) using Pt skimmer and sample cones.

The U-Pb geochronology of major zircon varieties was investigated using 100 μm spot analyses obtained at a repetition rate of 10 Hz with an incident laser beam pulse energy of 0.605–0.620 mJ and energy density on sample in the range 7.6–8.0 J/cm^2. The detector was operated in counting mode (^{235}U, ^{232}Th, ^{206}Pb, ^{207}Pb, ^{202}Pb, ^{202}Hg) and analogue mode (^{238}U). Natural zircon standard GJ-1 (609 Ma; [24]) was employed for calibration, and was run at the beginning, in the middle, and at the end of each analytical session. Natural zircon FC-1 (1099 ± 0.6 Ma; [25]) was used as a secondary standard, yielding accuracy and precision levels near 2%. The data reduction was done using the VisualAge DRS procedure [26] in the Iolite software [27]. A Concordia diagram showing 2σ error ellipses was produced using Isoplot/Ex 3.75 [28].

4. Results

4.1. CL Imaging

The examined zircon grains are macroscopically brown in color, homogeneous in transmitted light and BSE images, and vary from small angular fragments to large subhedral crystals reaching several mm in size. Cathodoluminescence revealed significant variations in the distribution of CL-active centers in the Drybones Bay zircons and their internal complexity (Figures 1 and 2). Based on their luminescence characteristics, several types of zircon were distinguished:

- Type 1: oscillatory-zoned yellow to brown in CL images (Figure 2a);
- Type 2: broad brown zones confined to blue-luminescing areas (Figure 2b);
- Type 3: light blue, homogeneous or subtly zoned (Figure 2c);
- Type 4: patchy yellow to greenish-yellowish (Figure 2d);
- Type 5: yellowish-grey (Figure 2e);
- Type 6: zoned dark-blue with brownish overtones (Figure 2f).

Figure 2. Cathodoluminescence images of zircon from the Drybones Bay kimberlite: (**a**) Oscillatory-zoned crystal of Type 1 (0.6 mm across); (**b**) zoned crystal comprising Type 2 (wide brown band 0.06 mm in thickness) and Type 3 (blue areas); (**c**) blue-luminescing zircon of Type 3 (width of the field of view 1.25 mm); (**d**) patchy (greenish) yellow grain of Type 4 (0.75 mm in length); (**e**) yellowish-grey Type 5 (0.65 mm across); (**f**) dark brownish-blue luminescence in Type 6 (0.8 mm in ength).

In many cases, more than one type of luminescence is present within a single zoned grain, commonly as concentric growth zones (e.g., Figure 2a,b,f and Figure 3a). In addition, several large zircon grains exhibit a sector pattern consisting of bright yellow and light blue sectors (types 3 and 4, respectively) separated by a well-defined boundary (Figure 3b). The CL features revealed in the present work, including oscillatory zoning (Type 1), broad bands (Type 2), and irregularly shaped areas (Type 5), have been previously observed in zircon from other kimberlite localities, often in the same grain [29–31].

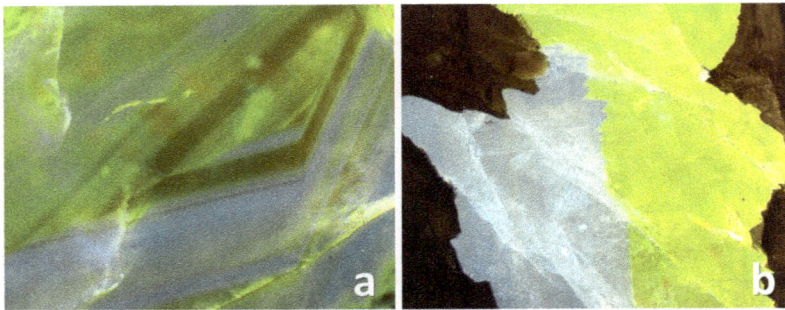

Figure 3. Cathodoluminescence images of zoned zircon from the Drybones Bay kimberlite (width of the field of view is 1.25 mm for both images): (**a**) oscillatory-zoned core (Type 1) rimmed by Type 3 zircon; (**b**) sector-zoned grain comprising Type 3 and Type 4 domains.

4.2. Trace Element Composition and Variations

In this work, we studied zircon grains with all types of CL response. In total, 90 trace-element analyses were done in samples imaged using CL. Our data show that the Drybones Bay zircon macrocrysts exhibit appreciable variations in trace element abundances and ratios, particularly with regard to rare-earth elements (REE = La-Lu). All samples are relatively enriched in Ta (1.8–12.3 ppm), but contain moderate levels of Hf (8200–10,900 ppm) at low levels of REE (<1.5 ppm Lu, <60 ppm Y). None of the seven types exhibit Eu anomalies. The average δEu ratio, which reflects deviation of the Eu concentration from a value interpolated from the chondrite-normalized Sm and Gd abundances, and is measured as $Eu_{cn}/[0.5 \times (Sm_{cn} + Gd_{cn})]$, is 1.0 ± 0.2. These compositional characteristics are consistent with kimberlitic zircon [32] and rule out crustal rocks or the Defeat granitoids as a potential source of these macrocrysts (cf. [33]). Another notable characteristic of the studied material is its enrichment in Sc (570–930 ppm), which is far in excess of what has been reported for felsic rocks [34].

The most significant compositional differences are observed between grains showing blue and yellow luminescence (Table 1), in accord with the observations made previously by Belousova et al. [30]. Areas luminescing blue show the lowest levels of REE and Y (5–10 ppm and 6–11 ppm, respectively, in Type 3; 12–18 and 14–22 ppm, respectively, in Type 6; Figure 4a). In contrast, those showing yellow or yellowish-brown response contain the highest levels of these elements among the examined samples (23–53 ppm REE and 26–59 ppm Y in Type 1; 19–33 and 21–36 ppm, respectively, in Type 4). These differences are particularly striking in sector-zoned grains (Figure 3b), where REE+Y levels increase two- to fourfold across the sector boundary (Figure 4a). These variations are not accompanied by any consistent changes in Sc content. Grains showing yellow luminescence are also enriched in Th and U (3–9 and 8–20 ppm, respectively, in Type 1) relative to blue areas ($\leq$2 ppm Th and $\leq$5 ppm U in Types 3 and 6; Figure 4b). The Th/U ratio (Figure 4c) tends to be somewhat higher in Type-1, -4, and -5 zircon (0.31–0.47) relative to the blue-luminescing material (0.22–0.40), but overall shows little variation (average Th/U = 0.34 ± 0.07). The abundances of Nb and Ta are comparable in sector-zoned crystals (Types 3 and 4 in Figure 4d) and reach their maximum values in brown growth zones, such as those shown in Figure 2a,b (up to 11 ppm Nb and 8 ppm Ta in Type 1, and up to 12 ppm of both in Type 2). The Nb/Ta ratio is strongly subchondritic and only slightly higher in the oscillatory-zoned core (1.1 ± 0.1) relative to the rest of the data (0.98 ± 0.08; Figure 4e). The highest

Ti contents (21–34 ppm) are observed in the core (Figure 4c). Titanium levels are lower (10–25 ppm) and comparable in sector-zoned (Type-3 and -4) and blue-luminescing zircon. The latter is richer in Hf (9500–10,900 ppm) than the oscillatory core and yellow sectors (8200–9400 ppm), which is reflected in Zr/Hf variations depicted in Figure 4e. As can be expected, the Pb content reaches maximum values in U- and Th-rich zones (0.09–0.26 ppm), and ranges from levels below the LA-ICP-MS detection limit to 0.11 ppm in U-Th-poor blue-luminescing zircon (Figure 4f). Type 5 is compositionally transitional between Types 3 and 4.

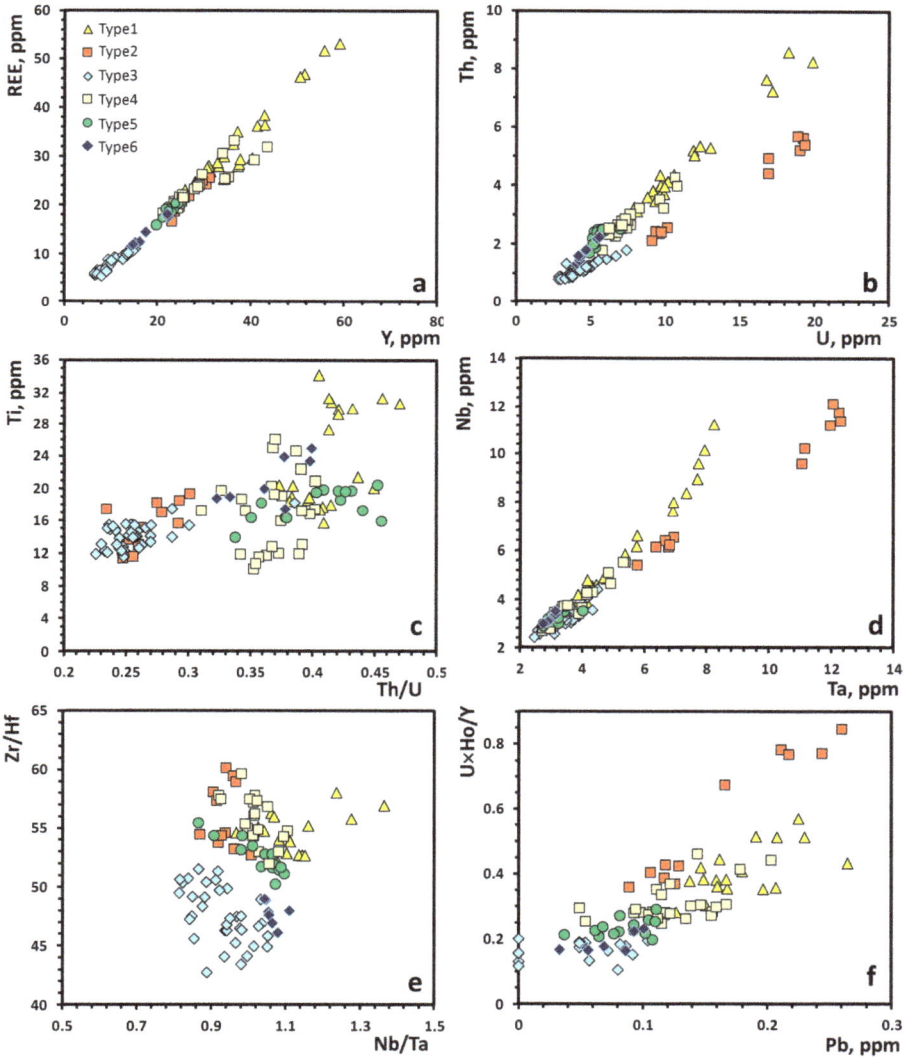

Figure 4. Variations in trace element composition among different types of zircon (distinguished on the basis of their cathodoluminescence (CL) characteristics).

Table 1. Trace element composition of zircon from Drybones Bay kimberlite.

Element	Type 1			Type 2			Type 3			Type 4			Type 5			Type 6		
	Min	Max	Av	Min	Max	Av	Min	Max	Av	Min	Max	Av	Min	Max	Av	Min	Max	Av
Ti	15.72	34.18	24.93	11.43	19.34	15.09	12.30	14.35	18.21	10.09	20.27	14.07	13.85	20.42	17.96	17.45	24.95	21.03
V	0.10	0.36	0.20	0.09	0.22	0.13	0.08	0.11	0.17	0.10	0.16	0.14	0.08	0.17	0.12	0.07	0.17	0.13
Mn	0.33	0.33	0.33	0.29	0.75	0.52	0.39	0.39	0.39	b.d.l.	b.d.l.	b.d.l.	0.27	0.27	0.27	b.d.l.	b.d.l.	b.d.l.
Sr	0.33	0.36	0.34	0.25	0.62	0.37	0.23	0.37	0.55	0.41	0.41	0.41	b.d.l.	b.d.l.	b.d.l.	b.d.l.	b.d.l.	b.d.l.
Y	25.99	59.04	39.12	23.54	31.15	26.89	6.51	8.73	11.01	21.20	36.43	26.34	19.79	28.87	23.16	14.47	22.32	17.57
Nb	3.67	11.24	6.69	6.14	12.14	8.89	2.40	2.98	3.54	2.68	5.51	3.35	2.86	3.53	3.20	2.95	3.50	3.19
La[1]	b.d.l.	0.09	0.04	b.d.l.	0.10	0.01	b.d.l.	b.d.l.	b.d.l.	b.d.l.	0.06	0.03	b.d.l.	0.02	b.d.l.	b.d.l.	b.d.l.	b.d.l.
Ce	1.20	3.13	1.84	1.10	1.77	1.44	0.67	0.75	0.94	0.99	1.49	1.17	0.80	1.07	1.00	0.85	1.10	0.94
Pr	0.03	0.16	0.09	b.d.l.	0.04	0.02	b.d.l.	0.01	0.03	b.d.l.	0.10	0.04	b.d.l.	0.06	0.02	b.d.l.	0.03	0.01
Nd	0.82	2.53	1.39	0.19	0.74	0.37	0.13	0.26	0.37	0.42	1.29	0.69	0.26	0.81	0.53	0.18	0.44	0.33
Sm	0.94	2.89	1.48	0.28	0.69	0.51	0.08	0.11	0.43	0.53	1.28	0.89	0.47	0.89	0.69	0.24	0.55	0.42
Eu	0.52	1.28	0.81	0.25	0.45	0.34	0.08	0.11	0.23	0.33	0.71	0.47	0.23	0.49	0.38	0.17	0.35	0.26
Gd	2.10	6.09	3.70	0.92	2.39	1.69	0.32	0.53	1.05	1.50	3.72	2.22	1.37	2.48	1.84	0.82	1.84	1.27
Tb	0.49	1.20	0.82	0.35	0.58	0.45	0.08	0.14	0.19	0.41	0.77	0.54	0.34	0.55	0.44	0.24	0.41	0.31
Dy	4.34	10.37	6.86	3.64	4.93	4.21	0.92	1.30	1.76	3.40	6.59	4.54	3.28	4.68	3.93	2.27	3.76	2.84
Ho	1.06	2.37	1.66	0.93	1.32	1.13	0.25	0.34	0.43	0.94	1.58	1.11	0.84	1.20	0.97	0.58	0.94	0.70
Er	3.05	7.30	4.96	2.90	4.17	3.52	0.78	1.13	1.48	2.43	4.88	3.31	2.44	3.62	2.93	1.78	2.91	2.22
Tm	0.62	1.38	0.98	0.59	0.87	0.74	0.16	0.23	0.34	0.58	0.98	0.67	0.47	0.73	0.57	0.35	0.49	0.42
Yb	6.61	14.36	9.49	5.96	8.01	6.87	1.55	2.22	2.93	5.52	9.74	6.43	3.95	7.33	5.46	3.21	5.05	3.98
Lu	0.51	1.25	0.84	0.57	0.72	0.65	0.12	0.22	0.30	0.46	0.73	0.57	0.42	0.69	0.51	0.32	0.48	0.39
Hf	8439	9411	8982	8154	9287	8753	9502	10,033	10,873	8211	9243	8689	8839	9738	9321	9985	10,595	10,237
Ta	3.79	8.23	5.86	6.36	12.30	9.48	2.46	3.33	4.34	2.73	5.38	3.32	2.76	4.04	3.13	2.77	3.15	2.99
Pb	0.13	0.38	0.23	0.11	0.31	0.18	0.03	0.05	0.08	0.05	0.18	0.11	0.04	0.11	0.09	0.03	0.10	0.07
Th	2.62	8.58	5.05	2.37	5.69	3.95	0.77	0.98	1.31	1.80	4.28	2.67	1.67	2.51	2.25	1.32	2.23	1.73
U	6.86	19.86	11.93	9.35	19.34	14.42	2.86	3.64	4.61	5.78	10.65	7.20	4.95	6.99	5.55	4.09	5.58	4.66
ΣREE	23.01	53.20	34.92	18.60	25.64	21.95	5.62	7.37	9.32	18.36	33.17	22.64	15.90	24.37	19.27	11.60	18.13	14.08
Nb/Ta	0.97	1.37	1.12	0.87	1.01	0.94	0.82	0.90	1.05	0.87	1.11	1.01	0.87	1.10	1.03	1.04	1.11	1.07
Zr/Hf	52	58	55	53	60	56	45	49	51	53	60	56	50	55	53	46	49	48
Y/Ho	22	26	24	22	25	24	22	26	28	22	26	24	21	26	24	24	27	25
Th/U	0.38	0.47	0.42	0.25	0.30	0.27	0.23	0.27	0.39	0.31	0.40	0.37	0.34	0.46	0.40	0.32	0.40	0.37
$(Nd/Yb)_{cn}$	0.04	0.07	0.05	0.01	0.04	0.02	0.01	0.04	0.08	0.03	0.05	0.04	0.02	0.05	0.04	0.02	0.04	0.03
δCe	3.9	9.0	5.3	6.2	15.7	11.4	5.3	7.0	7.8	3.5	9.1	6.3	4.1	10.3	7.5	6.8	13.2	10.8

[1] b.d.l.: below detection limit.

The REE budget of all six zircon varieties is very similar. Chondrite-normalized patterns show a steep slope rising towards heavy lanthanides ($Nd_{cn}/Yb_{cn} = 0.036 \pm 0.015$), with a conspicuous positive Ce anomaly (Figure 5). The average δCe value, measured as $Ce_{cn}/[0.5 \times (La_{cn} + Pr_{cn})]$, is 8.2 ± 3.9, and none of the individual varieties give δCe below 3.5. Neither Eu nor Y anomalies are observed, and the average Y/Ho ratio is essentially chondritic (25 ± 2). The REE characteristics of zircon (i.e., the shape of chondrite-normalized profiles, Ce, Eu, and Y anomalies) are sensitive to its interaction with fluids and melts [33,35,36]. Because the Y/Ho ratio shows the least variation across the dataset, we used it as a "common denominator" to gauge the extent of Pb loss (or gain). In Figure 4f, we compare variations in Pb to the measured U values normalized to the Y/Ho ratio in the same analysis. There is a clear positive correlation ($R^2 = 0.79$) between the two parameters, and no obvious outliers that would indicate variations in Pb content beyond those caused by heterogeneities in U distribution.

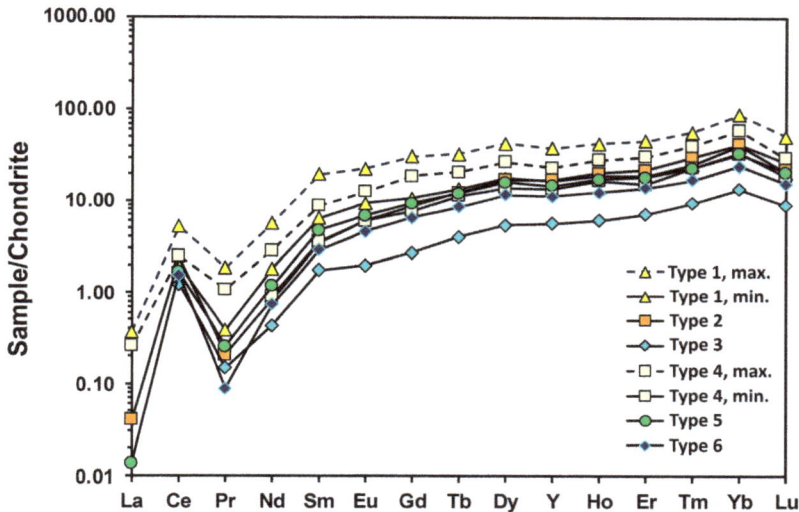

Figure 5. Chondrite-normalized patterns of the different types of zircon from the Drybones Bay kimberlite (lower limits of detection are plotted for La and Pr where these elements were not detectable by LA-ICP-MS).

4.3. Thermometry

The concentration of Ti in the analyzed macrocrysts ranges between 10 and 34 ppm (Figure 4c). The oscillatory-zoned core, which fluoresces from yellow to yellowish-brown, shows the highest levels of this element, suggesting its crystallization at relatively higher temperature. Crystals showing sector zoning (Types 3 and 4) and brown zones in these crystals (Types 2 and 6) have overlapping Ti levels. The Ti-in-zircon thermometer of Watson et al. [37] gives apparent crystallization temperatures spanning from 742 °C (yellow sector) to 862 °C (core; Table 2). The average estimated temperatures are 820 ± 26 °C for the core and 781 ± 19 °C for the rest of the data, i.e., essentially within the estimated standard deviation of each other.

Table 2. Variations in Ti content and estimated crystallization temperatures of zircon from the Drybones Bay kimberlite.

Parameter	Type 1	Type 2	Type 3	Type 4	Type 5	Type 6
Ti (ppm) ± SD	24.9 ± 6.2	15.1 ± 2.9	14.6 ± 1.9	14.1 ± 3.1	18.0 ± 2.0	21.0 ± 3.0
Average T (°C ± SD)	825 ± 27	777 ± 18	775 ± 12	770 ± 20	795 ± 11	810 ± 14
# analyses	18	11	18	15	14	7

4.4. U-Pb Geochronometry of Zircon from the Drybones Bay Kimberlite

The U-Pb isotopic study of zircon from the Drybones Bay kimberlite was performed on three large (several mm across) anhedral grains showing essentially the entire range of variations in CL. In total, 80 points were analyzed. Data reduction showed that, whereas the majority of the data points formed a fairly tight cluster on the Concordia diagram, some of the analyses were not concordant and corresponded either to notably younger or older ages than the rest of the data. Closer inspection of these points showed that most of these analyses were performed close to the visible cracks or grain boundaries. Eventually, 63 data points were used for the calculation of zircon crystallization age, yielding a well-constrained value of 445.6 ± 0.8 Ma (Figure 6). The isotopic ratios and calculated ages for all data points are presented in Table 3.

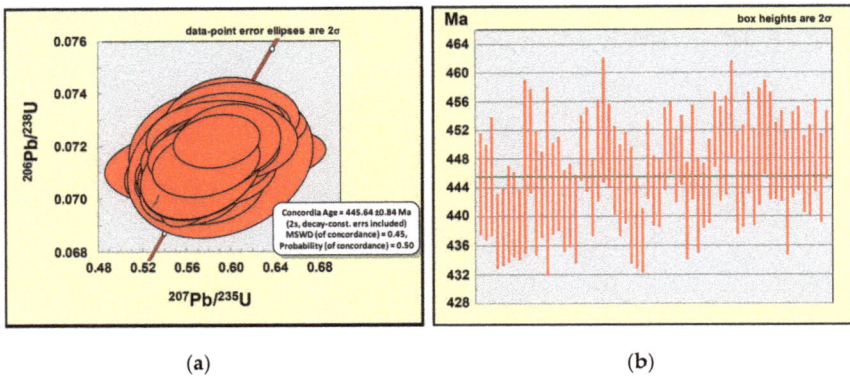

(a)   (b)

Figure 6. U-Pb data for zircon macrocrysts (all CL types) from the Drybones Bay kimberlite: (a) Concordia diagram; (b) weighted mean of $^{207}Pb/^{235}U$ ages.

Table 3. U-Pb isotopic data for zircon from the Drybones Bay kimberlite.

Analysis	Measured Ratios							Age, Ma	
	$^{207}Pb/^{235}U$	±2σ	$^{206}Pb/^{238}U$	±2σ	$^{207}Pb/^{206}Pb$	±2σ	$^{206}Pb/^{238}U$	±2σ
1	0.5500	0.0460	0.0714	0.0012	0.0558	0.0047	445	7
2	0.5560	0.0220	0.0712	0.0011	0.0564	0.0021	443	7
3	0.5500	0.0370	0.0716	0.0014	0.0557	0.0035	446	8
4	0.5380	0.0220	0.0703	0.0008	0.0554	0.0022	438	5
5	0.5410	0.0230	0.0704	0.0009	0.0555	0.0023	439	5
6	0.5400	0.0200	0.0707	0.0011	0.0553	0.0021	440	7
7	0.5410	0.0230	0.0707	0.0010	0.0556	0.0024	440	6
8	0.5390	0.0250	0.0705	0.0008	0.0554	0.0025	439	5
9	0.5550	0.0390	0.0718	0.0019	0.0565	0.0043	447	12
10	0.5600	0.0310	0.0724	0.0012	0.0560	0.0032	451	7
11	0.5530	0.0330	0.0712	0.0014	0.0565	0.0034	443	9
12	0.5470	0.0300	0.0712	0.0010	0.0557	0.0031	443	6
13	0.5590	0.0430	0.0715	0.0022	0.0566	0.0042	445	13
14	0.5460	0.0270	0.0713	0.0011	0.0555	0.0027	444	6
15	0.5510	0.0270	0.0714	0.0011	0.0559	0.0028	445	7
16	0.5450	0.0280	0.0708	0.0009	0.0557	0.0029	441	6
17	0.5410	0.0240	0.0709	0.0009	0.0553	0.0024	442	6
18	0.5410	0.0180	0.0706	0.0010	0.0553	0.0019	440	6
19	0.5550	0.0190	0.0722	0.0008	0.0556	0.0019	450	5
20	0.5570	0.0230	0.0722	0.0010	0.0558	0.0023	449	6
21	0.5450	0.0160	0.0711	0.0009	0.0555	0.0017	443	5
22	0.5640	0.0270	0.0722	0.0012	0.0563	0.0026	449	7
23	0.5610	0.0280	0.0729	0.0014	0.0560	0.0028	453	9
24	0.5570	0.0220	0.0723	0.0010	0.0559	0.0023	450	6
25	0.5530	0.0250	0.0717	0.0010	0.0557	0.0025	446	6

Table 3. Cont.

Analysis	Measured Ratios							Age, Ma	
	$^{207}Pb/^{235}U$	$\pm2\sigma$	$^{206}Pb/^{238}U$	$\pm2\sigma$	$^{207}Pb/^{206}Pb$	$\pm2\sigma$	$^{206}Pb/^{238}U$	$\pm2\sigma$
26	0.5490	0.0310	0.0713	0.0010	0.0558	0.0030	444	6
27	0.5500	0.0210	0.0715	0.0010	0.0557	0.0022	445	6
28	0.5490	0.0290	0.0709	0.0013	0.0561	0.0031	442	8
29	0.5430	0.0240	0.0705	0.0010	0.0559	0.0025	439	6
30	0.5410	0.0130	0.0701	0.0007	0.0557	0.0012	437	4
31	0.5550	0.0120	0.0720	0.0009	0.0558	0.0013	448	5
32	0.5480	0.0200	0.0713	0.0008	0.0556	0.0020	444	5
33	0.5480	0.0150	0.0712	0.0008	0.0557	0.0016	443	5
34	0.5560	0.0180	0.0722	0.0010	0.0558	0.0018	449	6
35	0.5610	0.0150	0.0725	0.0008	0.0561	0.0015	451	5
36	0.5550	0.0130	0.0718	0.0008	0.0558	0.0012	447	5
37	0.5570	0.0210	0.0722	0.0008	0.0558	0.0021	449	5
38	0.5440	0.0240	0.0708	0.0011	0.0557	0.0024	441	7
39	0.5560	0.0260	0.0721	0.0011	0.0557	0.0027	449	7
40	0.5430	0.0250	0.0709	0.0011	0.0555	0.0027	442	7
41	0.5500	0.0200	0.0711	0.0007	0.0558	0.0021	443	4
42	0.5510	0.0300	0.0715	0.0010	0.0559	0.0031	445	6
43	0.5640	0.0360	0.0727	0.0009	0.0562	0.0035	452	5
44	0.5590	0.0330	0.0721	0.0011	0.0560	0.0031	449	7
45	0.5680	0.0270	0.0723	0.0011	0.0568	0.0027	450	7
46	0.5660	0.0310	0.0731	0.0011	0.0562	0.0032	455	7
47	0.5540	0.0300	0.0714	0.0012	0.0563	0.0031	445	7
48	0.5510	0.0280	0.0716	0.0012	0.0559	0.0030	446	7
49	0.5610	0.0310	0.0723	0.0012	0.0558	0.0028	450	7
50	0.5520	0.0310	0.0716	0.0011	0.0559	0.0032	446	7
51	0.5610	0.0260	0.0723	0.0013	0.0565	0.0028	450	8
52	0.5610	0.0290	0.0727	0.0011	0.0559	0.0028	452	7
53	0.5600	0.0350	0.0723	0.0012	0.0563	0.0035	450	7
54	0.5530	0.0230	0.0719	0.0009	0.0555	0.0021	448	5
55	0.5560	0.0300	0.0720	0.0010	0.0559	0.0030	448	6
56	0.5480	0.0280	0.0712	0.0014	0.0554	0.0028	443	9
57	0.5550	0.0250	0.0721	0.0010	0.0558	0.0025	449	6
58	0.5550	0.0250	0.0722	0.0010	0.0556	0.0024	449	6
59	0.5530	0.0230	0.0716	0.0009	0.0559	0.0024	446	5
60	0.5510	0.0270	0.0717	0.0010	0.0557	0.0028	446	6
61	0.5640	0.0240	0.0723	0.0011	0.0564	0.0024	450	6
62	0.5510	0.0250	0.0715	0.0010	0.0557	0.0025	445	6
63	0.5570	0.0200	0.0723	0.0008	0.0557	0.0020	450	5

5. Discussion

Large (mm- to cm-sized) zircon fragments, referred to as macro- or megacrysts, occur in a variety of volcanic rocks, including kimberlites, lamprophyres, and alkali basaltoids [31,38,39]. Apart from their large size, seemingly inconsistent with crystallization from a rapidly ascending melt, these macrocrysts share a number of characteristics [31,38–44]:

1. Complex internal structures involving sector and/or oscillatory zoning (as revealed by high-resolution CL imaging) and arising from multistage growth (±resorption);
2. Low Ti contents (<60 and typically, ≤30 ppm) indicative of relatively low crystallization temperatures (typically, ≤850 °C);
3. Chondrite-normalized patterns steeply sloping up towards the heavy lanthanides and characterized by a positive Ce anomaly (δCe in excess of 100 in some cases);
4. Variable, chondritic to strongly superchondritic (~80) Zr/Hf ratios;
5. Highly variable (over one order of magnitude) Th and U abundances and variable, but consistently subchondritic (0.2–1.7) Th/U ratios;
6. U-Pb ages similar to those of their host igneous rock.

Presently, there is no consensus on the origin of these enigmatic crystals. Three principally different interpretations have been proposed in the previous literature, i.e., that:

I.   They represent older mantle-derived xenocrysts entrained in ascending silica-undersaturated melts [4];

II.   They crystallized in mantle rocks affected by metasomatism, possibly related to kimberlitic or alkali-basaltic magmatism, in their source [31,45];

III.   They are early cognate phenocrysts that recorded evolution of their parental magma [38,42].

Macrocrysts from the Drybones Bay pipe were previously investigated by Heaman et al. [4], who interpreted them to derive from a Mesoproterozoic, deep (~1.0 Ga, >150 km) mantle source, and to have remained open to Pb diffusion prior to their entrainment in kimberlite. The temperatures of zircon crystallization obtained in the present work cluster between 780 and 820 °C (Table 2). These values fall well within the temperature range (608–927 °C) determined for kimberlitic zircon from the Siberian, Kaapvaal, Amazonas, and São Francisco cratons by Page et al. [31]. When projected to the average Slave geotherm (Figure 7), our data plot above the field of mantle garnets from the Drybones Bay pipe analyzed by Carbno and Canil [3]. If the studied macrocrysts were indeed derived from a (metasomatized) mantle source, it had to be much shallower than 160 km (estimated from garnet-clinopyroxene equilibria by Carbno and Canil [3].

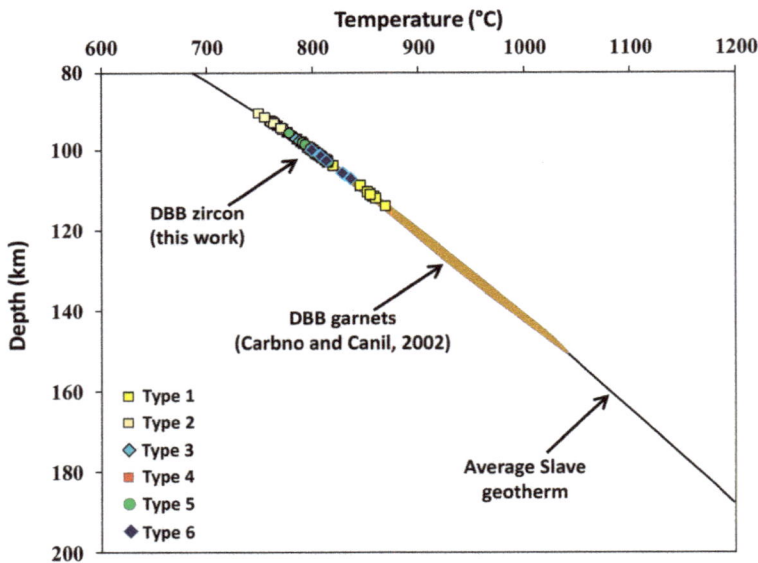

Figure 7. Estimated range of crystallization temperatures and depths for zircons (this work) and garnets [3] from Drybones Bay (DBB) projected on the average Slave Province geotherm.

Notably, the Zr/Hf ratio of our samples is consistently superchondritic ($\geq$43 and, on average, 52 $\pm$ 4), i.e., departs significantly from the mantle values (~36–37: [46,47]). Published data on peridotite-hosted zircon are scarce, but suggest near-chondritic Zr/Hf values for this mineral [45,48,49]. Although it is possible that the trace element budget of the hypothetical mantle source was modified by its interaction with a metasomatizing agent, how likely is the ascending kimberlite to have caused the documented Zr-Hf decoupling? Kimberlites have near-chondritic Zr/Hf ratios [50], which are also observed at Drybones Bay: the average Zr/Hf value, calculated for 21 whole-rock analyses of uncontaminated samples, is 37.5 $\pm$ 1.6 at 390 ppm Zr and 10.4 ppm Hf (authors' unpublished data).

Assuming Rayleigh-style partitioning between a kimberlitic melt and zircon, Zr must partition into the latter ~1.4 times more effectively than Hf to cause the observed decoupling. Note that at very small (but realistic) degrees of fractionation (on the order of 0.01–0.001 wt %), the Zr-Hf budget of kimberlitic melt will not be perceptibly affected and will effectively maintain its near-chondritic Zr/Hf ratio. Unfortunately, Zr-Hf partitioning data are available only for silica-saturated systems, where both $^{Zrn/L}D_{Zr} > {}^{Zrn/L}D_{Hf}$ and $^{Zrn/L}D_{Hf} > {}^{Zrn/L}D_{Zr}$ have been documented (e.g., [51]). At very small degrees of fractionation, preferential uptake of Ce^{4+} by zircon [52] will have little effect on the REE budget of kimberlite because all light lanthanides are strongly incompatible with respect to this mineral. Indeed, the Drybones Bay whole-rock data do not show a Ce anomaly ($\delta Ce = 0.96 \pm 0.02$: authors' unpublished data).

Simonetti and Neal [38] described a trend of increasing Zr/Hf and REE content (39–50 and 20–100 ppm, respectively) in progressively younger zircon megacrysts from Malaita alnöites, Solomon Islands. These authors interpreted the compositional changes in zircon to reflect evolution of its parental alnöitic magma in response to mantle metasomatism. Although our dataset does show a weak correlation between the Zr/Hf and REE values ($R^2 = 0.35$), it is entirely due to overall higher REE and lower Hf abundances in yellow-luminescing areas in comparison with blue-luminescing ones (see above). These variations occur within individual sector-zoned crystals (Figure 3b), i.e., are controlled crystallographically [53]. Hence, we conclude that the data distributions in Figure 4a,e reflect growth phenomena, such as the different rates of REE and Hf uptake by symmetrically non-equivalent faces, rather than magma-evolution processes. Sector zoning, documented in the present work and several earlier studies [31,39,42,43], is indicative of relatively fast crystallization, i.e., not consistent with long residence times in the mantle.

In the present work, we did not observe any evidence of zircon re-equilibration with invading melts or fluids, such as overgrowths or fracture fillings differing from the bulk of the macrocrysts in their trace element budget (e.g., [33]). In fact, the measured REE distributions are remarkably consistent in the Drybones Bay samples irrespective of their CL response (Figure 5). Appreciable intragranular variations are recorded with respect to Nb, Ta, Th, and U contents but, again, correlate with growth sectors or concentric brown bands in dominantly yellow or blue-fluorescing areas (Figure 4b,d). These local fluctuations likely resulted from preferential uptake of these elements by specific crystal faces or changes in their availability due to competitive partitioning between zircon and other minerals.

The U-Pb isotopic data reported in the present work are in good agreement with the previously published "bulk" measurements obtained by thermal ionization mass-spectrometry (441.4 ± 0.8 Ma; [4]). Our data are concordant and do not show any convincing evidence of Pb mobilization by melts or fluids. The recorded variations in Pb within and among the macrocrysts (Figure 4f) can be explained by variations in U content (anchored to Y/Ho values) and do not seem to indicate any Pb loss either. The six types of zircon identified from CL images are considered cogenetic, rather than having been generated by interaction between Precambrian zircon xenocrysts and kimberlitic magma.

On the basis of the mineralogical and geochemical evidence presented above, we conclude that the Drybones Bay zircon macrocrysts most likely crystallized in a shallow (<100 km) mantle source due to its reaction with a transient kimberlitic melt. The conduit–melt interaction was short-lived and produced complexly zoned zircon crystals whose compositional heterogeneity results from interplay between element partitioning and crystallographically controlled uptake of trace elements. We found no evidence for the presence of ancient ($t >> 440$ Ma) zircon in the mantle source(s) of the Drybones Bay kimberlite.

Author Contributions: Conceptualization, A.R.C. and B.E.; methodology, A.R.C., E.P.R., and P.Y.; investigation, A.R.C., E.P.R., and A.R.S.; data curation, A.R.C., E.R., A.R.S., and P.Y.; writing—original draft preparation, A.R.C., E.P.R., and A.R.S.; writing—review and editing, A.R.C. and B.E.; visualization, A.R.C. and E.P.R.; funding acquisition, A.R.C. and B.E.

Funding: This research was funded by the NWT Geological Survey (contract number 63040) and Natural Sciences and Engineering Research Counsel of Canada (NSERC grant number 249596-2013).

Acknowledgments: We would like to thank Guest Editor Sergey Krivovichev for his invitation to contribute to the present issue. We are grateful to two anonymous reviewers who provided constructive comments on its earlier version. The NWT Government and NSERC are acknowledged for their financial support.

Conflicts of Interest: The authors declare no conflict of interest.

References

1. Government of NWT. 2016. Available online: https://www.iti.gov.nt.ca/sites/iti/files/mineraldeposites2-nocrops.pdf (accessed on 24 October 2018).
2. Pell, J.A. Kimberlites in the Slave Craton, Northwest Territories, Canada. *Geosci. Can.* **1997**, *24*, 77–90.
3. Carbno, G.B.; Canil, D. Mantle structure beneath the southwest Slave craton, Canada: Constraints from garnet geochemistry in the Drybones Bay kimberlite. *J. Petrol.* **2002**, *43*, 129–142. [CrossRef]
4. Heaman, L.M.; Kjarsgaard, B.A.; Creaser, R.A. Timing of kimberlite magmatism in North America: Implications for global genesis and diamond exploration. *Lithos* **2003**, *71*, 153–184. [CrossRef]
5. Yang, Y.-H.; Wu, F.-Y.; Wilde, S.A.; Liu, X.-M.; Zhang, Y.-B.; Xie, L.-W.; Yang, J.-H. In situ perovskite Sr–Nd isotopic constraints on the petrogenesis of the Ordovician Mengyin kimberlites in the North China Craton. *Chem. Geol.* **2009**, *264*, 24–42. [CrossRef]
6. Smelov, A.P.; Zaitsev, A.I. The age and localization of kimberlite magmatism in the Yakutian Kimberlite Province: Constraints from isotope geochronology—An overview. In *Proceedings of 10th International Kimberlite Conference*; Pearson, D.G., Grütter, H.S., Harris, J.W., Kjarsgaard, B.A., O'Brien, H., Rao, N.V.C., Sparks, S., Eds.; Springer: New Delhi, India, 2013; pp. 225–234. [CrossRef]
7. Padgham, W.A.; Fyson, W.K. The Slave Province: A distinct craton. *Can. J. Earth Sci.* **1992**, *29*, 2072–2086. [CrossRef]
8. Davis, W.J.; Hegner, E. Neodymium isotopic evidence for the accretionary development of the Late Archean Slave Province. *Contrib. Miner. Petrol.* **1992**, *111*, 493–503. [CrossRef]
9. Thorpe, R.I.; Cumming, G.L.; Mortensen, J.K. A Significant Pb Isotope Boundary in the Slave Province and Its Probable Relation to Ancient Basement in the Western Slave Province. Project Summaries, Canada Northwest Territories Mineral Development Subsidiary Agreement. *Geol. Surv. Can. Open File Rep.* **1992**, *2484*, 279–284. [CrossRef]
10. Bleeker, W.; Ketchum, J.W.F.; Jackson, V.A.; Villeneuve, M.E. The Central Slave Basement Complex: Part I. Its Structural Topology and Autochthonous Cover. *Can. J. Earth Sci.* **1999**, *36*, 1083–1109. [CrossRef]
11. Davis, W.J.; Canil, D.; Mackenzie, J.M.; Carbno, G.B. Petrology and U–Pb geochronology of lower crustal xenoliths and the development of a craton, Slave Province, Canada. *Lithos* **2003**, *71*, 541–573. [CrossRef]
12. Bleeker, W.; Ketchum, J.W.F.; Davis, W.J. The Central Slave Basement Complex: Part II. Age and tectonic significance of high-strain zones along the basement-cover contact. *Can. J. Earth Sci.* **1999**, *36*, 1111–1130. [CrossRef]
13. Davis, W.J.; Bleeker, W. Timing of plutonism, deformation, and metamorphism in the Yellowknife domain, Slave province, Canada. *Can. J. Earth Sci.* **1999**, *36*, 1169–1187. [CrossRef]
14. Davis, W.J.; Kjarsgaard, B.A. A Rb-Sr isochron age for a kimberlite from the recently discovered Lac De Gras Field, Slave Province, Northwest Canada. *J. Geol.* **1997**, *105*, 503–510. [CrossRef]
15. Masun, K.M.; Doyle, B.J.; Ball, S.; Walker, S. The geology and mineralogy of the Anuri kimberlite, Nunavut, Canada. *Lithos* **2004**, *76*, 75–97. [CrossRef]
16. Kretschmar, U. *Drybones Bay Kimberlite: Summary and Exploration Update; Exploration Overview 1996*; Northwest Territories Mining, Exploration and Geological Investigations, NWT Geological Mapping Division: Yellowknife, NT, Canada, 1996; pp. 3.27–23.28.
17. Kretschmar, U. *Drybones Bay Kimberlite: 1995 Spring Drill Report*; Open File 83627; NWT Geological Survey: Yellowknife, NT, Canada, 1995; pp. 1–50.
18. Power, M.A. Seismic signature of the Drybones Bay kimberlite pipe, N.W.T. *CIM Bull.* **1998**, *91*, 66–69.
19. Field, M.; Scott Smith, B.H. Contrasting geology and near-surface emplacement of kimberlite pipes in Southern Africa and Canada. In Proceedings of the 7th International Kimberlite Conference, Cape Town, South Africa, 11–17 April 1998; pp. 217–231.

20. Scott Smith, B.H. Canadian kimberlites: Geological characteristics relevant to emplacement. *J. Volcanol. Geotherm. Res.* **2008**, *174*, 9–19. [CrossRef]

21. Timmins, W.G. Report on the Drybones Bay property. Unpublished report, 2002; pp. 1–24.

22. Bird, H. Yellowknives Dene lose Drybones Bay exploration appeal. *CBC News*, 26 June 2015. Available online: https://www.cbc.ca/news/canada/north/yellowknives-dene-lose-drybones-bay-diamond-exploration-appeal-1.3128489 (accessed on 24 October 2018).

23. Norman, M.D.; Pearson, N.J.; Sharma, A.; Griffin, W.L. Quantitative analysis of trace elements in geological materials by laser ablation ICPMS: Instrumental operating conditions and calibration values of NIST glasses. *J. Geostand. Geoanal. Res.* **1996**, *20*, 247–261. [CrossRef]

24. Jackson, S.E.; Pearson, N.J.; Griffin, W.L.; Belousova, E.A. The application of laser ablation-inductively coupled plasma-mass spectrometry to in situ U–Pb zircon geochronology. *Chem. Geol.* **2004**, *211*, 47–69. [CrossRef]

25. Paces, J.B.; Miller, J.D. Precise U–Pb ages of Duluth Complex and related mafic intrusions, northeastern Minnesota: Geochronological insights into physical, petrogenetic, paleomagnetic and tectonomagmatic processes associated with the 1.1 Ga midcontinent rift system. *J. Geophys. Res.* **1993**, *98*, 13997–14013. [CrossRef]

26. Petrus, J.A.; Kamber, B.S. VizualAge: A novel approach to laser ablation ICPMS U–Pb geochronology data reduction. *J. Geostand. Geoanal. Res.* **2012**, *36*, 247–270. [CrossRef]

27. Woodhead, J.D.; Hellstrom, J.; Hergt, J.; Grieg, A.; Maas, R. Isotopic and elemental imaging of geological materials by laser ablation Inductively Coupled Plasma mass spectrometry. *J. Geostand. Geoanal. Res.* **2007**, *31*, 331–343. [CrossRef]

28. Ludwig, K.R. Isoplot 3.75. A Geochronological Toolkit for Microsoft Excel 2012, Berkeley Geochronology Center Special Publication No. 5. Available online: http://www.bgc.org/isoplot_etc/isoplot/Isoplot3_75-4_15manual.pdf (accessed on 27 September 2018).

29. Lyakhovich, V.V. Zircons in diamond-bearing rocks. *Trans. Russ. Acad. Sci. Earth Sci. Sect.* **1996**, *347*, 179–199.

30. Belousova, E.A.; Griffin, W.L.; Pearson, N.J. Trace element composition and cathodoluminescence properties of southern African kimberlitic zircons. *Min. Mag.* **1998**, *62*, 355–366. [CrossRef]

31. Page, F.Z.; Fu, B.; Kita, N.T.; Fournelle, J.; Spicuzza, M.J.; Schulze, D.J.; Viljoen, F.; Basei, M.A.S.; Valley, J.W. Zircons from kimberlite: New insights from oxygen isotopes, trace elements, and Ti in zircon thermometry. *Geochim. Cosmochim. Acta* **2007**, *71*, 3887–3903. [CrossRef]

32. Belousova, E.A.; Griffin, W.L.; O'Reilly, S.Y.; Fisher, N.I. Igneous zircon: Trace element composition as an indicator of source rock type. *Contrib. Mineral. Petrol.* **2002**, *143*, 602–622. [CrossRef]

33. Kelly, C.J.; Schneider, D.A.; Lajoie, M.-È.; Jackson, S.E.; McFarlane, C.R. U–Pb geochronology and trace element composition of zircon from the Horseshoe Lake greenstone belt, Superior Province, Canada: Implications for the tectonic and metamorphic history. *Can. J. Earth Sci.* **2018**, *55*, 172–187. [CrossRef]

34. Grimes, C.B.; Wooden, J.L.; Cheadle, M.J.; John, B.E. "Fingerprinting" tectono-magmatic provenance using trace elements in igneous zircon. *Contrib. Miner. Petrol.* **2015**, *170*, 46. [CrossRef]

35. Hanchar, J.M.; Van Westrenen, W. Rare Earth Element Behaviour in Zircon-Melt Systems. *Elements* **2007**, *3*, 37–42. [CrossRef]

36. Li, H.-Y.; Chen, R.-X.; Zheng, Y.-F.; Hu, Z. The crust-mantle interaction in continental subduction channels: Zircon evidence from orogenic peridotite in the Sulu orogen. *J. Geophys. Res. Solid Earth* **2016**, *121*, 687–712. [CrossRef]

37. Watson, E.B.; Wark, D.A.; Thomas, J.B. Crystallization thermometers for zircon and rutile. *Contrib. Miner. Petrol.* **2006**, *151*, 413–433. [CrossRef]

38. Simonetti, A.; Neal, C.R. In-situ chemical, U–Pb dating, and Hf isotope investigation of megacrystic zircons, Malaita (Solomon Islands): Evidence for multi-stage alkaline magmatic activity beneath the Ontong Java Plateau. *Earth Planet Sci. Lett.* **2010**, *295*, 251–261. [CrossRef]

39. Sutherland, L.; Graham, I.; Yaxley, G.; Armstrong, R.; Giuliani, G.; Hoskin, P.; Nechaev, V.; Woodhead, J. Major zircon megacryst suites of the Indo-Pacific lithospheric margin (ZIP) and their petrogenetic and regional implications. *Miner. Petrol.* **2016**, *110*, 399–420. [CrossRef]

40. Hinton, R.W.; Upton, B.G.J. The chemistry of zircon: Variations within and between large crystals from syenite and alkali basalt xenoliths. *Geochim. Cosmochim. Acta* **1991**, *55*, 3287–3302. [CrossRef]

41. Berryman, A.K.; Stiefenhoffer, J.; Shee, S.R.; Wyatti, B.A.; Belousova, E.A. The discovery and geology of the Timber Creek kimberlites, Northern Territory, Australia. In Proceedings of the 7th International Kimberlite Conference, Cape Town, South Africa, 11–17 April 1998; Gurney, J.J., Dawson, B., Nixon, P.H., Eds.; Red Roof Design: Cape Town, South Africa, 1999; pp. 30–39.

42. Visonà, D.; Caironi, V.; Carraro, A.; Dallai, L.; Fioretti, A.M.; Fanning, M. Zircon megacrysts from basalts of the Venetian Volcanic Province (NE Italy): U–Pb ages, oxygen isotopes and REE data. *Lithos* **2007**, *94*, 168–180. [CrossRef]

43. Yu, Y.; Xu, X.; Chen, X. Genesis of zircon megacrysts in Cenozoic alkali basalts and the heterogeneity of subcontinental lithospheric mantle, eastern China. *Miner. Petrol.* **2010**, *100*, 75–94. [CrossRef]

44. Cong, F.; Li, S.-Q.; Lin, F.-C.; Shi, M.F.; Zhu, H.-P.; Siebel, W.; Chen, F. Origin of zircon megacrysts from cenozoic basalts in northeastern Cambodia: Evidence from U-Pb age, Hf-O isotopes, and inclusions. *J. Geol.* **2016**, *126*, 221–234. [CrossRef]

45. Dawson, J.B.; Hill, P.G.; Kinny, P.D. Mineral chemistry of a zircon-bearing, composite, veined and metasomatized upper-mantle peridotite xenolith from kimberlite. *Contrib. Miner. Petrol.* **2001**, *140*, 720–733. [CrossRef]

46. McDonough, W.F.; Sun, S.-S. The composition of the Earth. *Chem. Geol.* **1995**, *120*, 223–253. [CrossRef]

47. Palme, H.; O'Neill, H.S.C. Cosmochemical estimates of mantle composition. In *Treatise on Geochemistry*; Carlson, R.W., Holland, H.D., Turekian, K.K., Eds.; Elsevier: Amsterdam, The Netherlands, 2003; Volume 2, pp. 1–38. [CrossRef]

48. Pearson, D.G.; Canil, D.; Shirey, S.B. Mantle Samples Included in Volcanic Rocks: Xenoliths and Diamonds. In *Treatise on Geochemistry*; Carlson, R.W., Holland, H.D., Turekian, K.K., Eds.; Elsevier: Amsterdam, The Netherlands, 2003; Volume 2, pp. 171–275. [CrossRef]

49. Zheng, J.; Griffin, W.L.; O'Reilly, S.Y.; Zhang, M.; Pearson, N. Zircons in mantle xenoliths record the Triassic Yangtze–North China continental collision. *Earth Planet. Sci. Lett.* **2006**, *247*, 130–142. [CrossRef]

50. Chakhmouradian, A.R.; Reguir, E.P.; Kamenetsky, V.S.; Sharygin, V.V.; Golovin, A.V. Trace-element partitioning in perovskite: Implications for the geochemistry of kimberlites and other mantle-derived undersaturated rocks. *Chem. Geol.* **2013**, *253*, 112–131. [CrossRef]

51. Daniela Rubatto, D.; Hermann, J. Experimental zircon/melt and zircon/garnet trace element partitioning and implications for the geochronology of crustal rocks. *Chem. Geol.* **2007**, *241*, 38–61. [CrossRef]

52. Burnham, A.D.; Berry, A.J. An experimental study of trace element partitioning between zircon and melt as a function of oxygen fugacity. *Geochim. Cosmochim. Acta* **2012**, *95*, 196–212. [CrossRef]

53. Hoffman, J.F.; Long, J.V.P. Unusual sector zoning in Lewisian zircons. *Min. Mag.* **1984**, *48*, 513–517. [CrossRef]

![minerals logo] *minerals*

MDPI

Review

Beryllium Mineralogy of the Kola Peninsula, Russia—A Review

Lyudmila M. Lyalina *, Ekaterina A. Selivanova, Dmitry R. Zozulya and Gregory Yu. Ivanyuk

Geological Institute, Kola Science Centre, Russian Academy of Sciences, Fersmana 14, 184209 Apatity, Russia; selivanova@geoksc.apatity.ru (E.A.S.); zozulya@geoksc.apatity.ru (D.R.Z.); ivanyuk@geoksc.apatity.ru (G.Y.I.)
* Correspondence: lialina@geoksc.apatity.ru; Tel.: +7-81555-7-96-66

Received: 1 October 2018; Accepted: 21 December 2018; Published: 25 December 2018

![check for updates]

Abstract: This paper reviews the available information on the beryllium mineralogy of the different type of occurrences in the Kola Peninsula, northwest Russia. Beryllium mineralization in the region is mainly associated with alkaline and felsic rocks, which differ significantly in petrological, geochemical, mineralogical features and age. In total 28 beryllium minerals are established on the Kola Peninsula up today. Beryl is one of the ore minerals in the differentiated granite pegmatites of the Kolmozerskoe lithium deposit. A large diversity of beryllium minerals occur in the pegmatites and hydrothermal veins formed in the late stages of the Lovozero and Khibiny alkaline massifs. Most of these minerals, as leifite, lovdarite, odintsovite, sphaerobertrandite and tugtupite are rare in other environments and have unique properties. These minerals formed under conditions of extreme alkalinity and their formation was favored by abrupt changes in the alkalinity regimes. Some of minerals, as chrysoberyl in xenoliths of hornfels, genthelvite and unique intergrowth of meliphanite and leucophanite formed in contrasting geochemical fronts between felsic/intermediate and mafic rocks.

Keywords: beryllium minerals; chemical composition; mineral data; alkaline rocks; granite; pegmatites; hydrothermal veins; Kola Peninsula

1. Introduction

In 2018, 220 years have passed since the discovery of beryllium by the French chemist N.-L. Vauquelin. He made a report "De l'aigue marine, ou béril; et découverte d'une terre nouvelle dans cette pierre" (About the aquamarine or beryl, and discovery of a new earth in this stone) at the National Institute of France on the 14 February 1798 [1,2]. However, beryl and their colored varieties: emerald, aquamarine, and "chrysoberyl" (golden beryl, not the sensu stricto chrysoberyl), were known from much earlier times.

Many branches of contemporary industry use this chemical element due to the unique properties of the beryllium-bearing compounds. Thus, in metallurgy, beryllium is used effectively in alloys in order to increase their strength, hardness, and corrosion resistance. Such additives increase the service life of parts in several times. Alloys with beryllium are incredibly light and heat-resistant, which determines their use in the aerospace industry. Beryllium is indispensable in the nuclear and electronic industry, in the electro- and radiotechnics and many other high-tech industries.

USA is the main producer of beryllium ore from bertrandite-bearing tuff in the Spor Mountain deposit, Utah. Russia plans to resume the mining of beryllium ore (bertrandite-phenakite) in Ermakovsky deposit, Buryatia in 2019. The Russian production of beryllium is planned to be started in 2020.

History, properties, distribution, analytical methods of determination, economics, fields of application, mineralogy, petrology, geochemistry and crystal chemistry of beryllium can be found in numerous reviews [3–16].

Beryllium is a rare lithophile element; it is concentrated mainly in felsic and alkaline rocks. The beryllium content in rocks is low, and the average concentration of Be in the Earth's upper crust is 2.1 ppm, while in primitive mantle it shows a 30-fold decreasing up to 0.07 ppm Be [17]. Beryllium tends to accumulate in late derivatives of alkaline and felsic rocks, i.e., pegmatites and hydrothermal veins. Due to specific crystal chemistry properties [6,18], there is a significant amount of Be minerals (121 minerals, according to IMA list from March 2018). Discoveries of beryllium minerals are accelerating and 18 new minerals were described since 2010 (IMA list of Minerals, March 2018). Minerals of extremely unusual composition have been defined among them, for example verbierite $BeCr^{3+}_2TiO_6$ [19].

2. Geological Overview of the Kola Peninsula. Deposit and Occurrences of Beryllium Minerals

The Kola Peninsula is the northeastern part of the Fennoscandian Shield, the largest representative of the Early Precambrian crystalline basement of the East-European craton. The major structural units and rocks are given in Figure 1. An essential role of alkaline magmatism in the Kola Peninsula should be emphasized [20]. It took place during the formation of the Fennoscandian Shield from the Archean to Paleozoic. Keivy block is formed by the rocks of Archean peralkaline granite magmatism (2.6–2.7 Ga) [21]. The emplacement of Sakharjok nepheline syenite within the Keivy terrane occurred during the final stage of the peralkaline granite magmatismin (2.6 Ga). Alkaline magmatism activity reached the peak during the Paleozoic (350–400 Ma) and gave the world's largest alkaline plutons, Khibiny and Lovozero, and a number of other alkaline-ultrabasic massifs, all of them containing carbonatites [22].

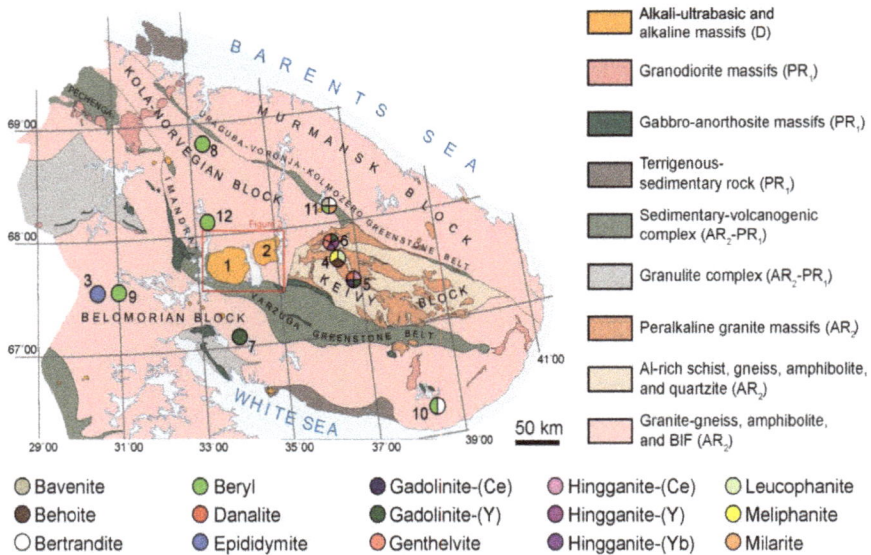

Figure 1. Simplified geological map of the Kola Peninsula (after [22]). The numbers indicate be deposit and occurrences: 1 = Khibiny, 2 = Lovozero, 3 = Kovdor, 4 = Sakharjok, 5 = El'ozero, 6 = Keivy, 7 = Kanozero, 8 = Shongui, 9 = Jona, 10 = Strel'na, 11 = Kolmozero-Voronja, 12 = Olenegorsk. Abbreviations: AR–Archaean, PR–Proterozoic, D–Devonian, BIF–Banded Iron Formation.

The Uraguba-Kolmozero-Voronja greenstone belt (Figure 1) is composed mainly by Neoarchean basic-intermediate-acid metavolcanic and metasedimentary suites metamorphosed at amphibolite facies. Younger igneous events are represented by voluminous intrusions of plagio-microcline granites and small stocks of tourmaline granites. Six pegmatite fields, comprising Polmostundra,

Kolmozero and Vasin-Mylk (Voronja Tundra) ore deposits, with a total ensemble of more than 100 bodies are confined to the Kolmozero-Voronja belt. The pegmatites intrude in amphibolites, metasediments and rarely gabbro-anorthosite and have 50–700 m long and 10–35 m in thickness. Quartz-albite-microcline pegmatite bodies are extremely enriched in rare element minerals: spodumene (18–20 vol %, locally up to 50 vol %), tantalite, microlite, beryl, lithiophillite, holmquistite, pollucite, and lepidolite. The Kolmozero-Voronja pegmatites are of complex type, spodumene subtype with Li, Cs, Be, Ta geochemical specialization and belong to LCT (lithium-cesium-tantalum) family (according to the classification of Černý and Ercit [23]). The overall schematic structure and composition of pegmatites is the following (from contact to central part): (1) aplitic zone of a 1–10 cm width consisting by plagioclase and quartz with minor biotite, tourmaline, apatite, and epidote; (2) granoblastic medium-grained quartz-albite zone of 0.3–5 m in width with minor and accessory muscovite, microcline, and spodumene; (3) medium-, coarse-grained quartz-albite-spodumene zone with minor and accessory muscovite, apatite, spessartine, beryl, and tantalite; (4) coarse-grained quartz-spodumene blocks and lenses within zone 3 of 0.5–2 m size with minor euhedral (up to 10 cm) muscovite, beryl, tantalite, apatite, spessartine, lithiophilite; (5) quartz-muscovite lenses of 0.1–1 m length within zone 3 with minor beryl, tantalite, and lithiophilite. The pegmatites crystallized under relatively high pressure (3–4 kbar) with peraluminous S-type granite as the source magma. The ore contains 0.9–1.25 wt % of Li_2O, 0.37 wt % of Cs_2O, 0.027–0.053 wt % of BeO, 0.004–0.03 wt % of Ta_2O_5 [24]. The grades and estimated resources of Kolmozerskoe lithium deposit within Kolmozero-Voronja belt are comparable to those from giant pegmatite-hosted Li-Ta-Be deposits, i.e., Greenbushes, Wodgina and Mt Cattlin in Australia, Bernic Lake, La Corne and Wekusko in Canada, and North Carolina in USA [25].

The Shongui occurrence in Kolmozero-Voronja belt (Figure 1) is represented by granite pegmatite but published geological data are not available.

The peralkaline granite-syenite complex (named as Keivy alkaline province) is an essential part of the Keivy block (Figure 1). It consists of six large alkali feldspar granite massifs (Western Keivy, Beliye Tundry, Ponoy, Lavrentyevskiy, Pacha and Purnach), some small nepheline syenite massifs (Sakharjok, Kulijok) and numerous late- and postmagmatic bodies (intragranite pegmatite, peralkaline granite pegmatite, amazonite pegmatite, metasomatically altered rocks, etc.). The late- and postmagmatic bodies are different in size, composition, location in exo- as well as endocontact zones of parent granite massifs but often contain rare metal mineralization.

The Sakharjok is a pint-sized massif (1.5–2 × 8 km) located within peralkaline granites of the Keivy block (Figure 1). The main rocks are alkaline syenites, nepheline syenites and genetically related pegmatoid schlierens and veins. The geology and petrology of the Sakharjok massif are described in detail in [26,27]. Large essexite (alkaline gabbro) body outcrops in an area up to 80 × 200 m within nepheline syenite. Essexite is made up by phlogopite-pyroxene-plagioclase rock with minor nepheline and amphibole. Syenite intrudes essexite resulting in the formation of numerous fractures subsequently filled by a pegmatitic material. Pegmatite has a complex internal structure and is enriched in beryllium. The surface of pegmatite body outcrop is approximately 30 m^2. Contact aureole of metasomatically altered rocks contaning essential biotite up to 80–90% (biotite metasomatite) wraps around the pegmatite.

The El'ozero rare-metal occurrence is confined to a linear tectonic zone along the contact of amphibolitized gabbro-anorthosite with the peralkaline granite of the Keivy block (Figure 1). The granite intruded and metasomatically altered the host rocks. As a result different rock types, including mineralized granites, peralkaline granite veins, pegmatites, granite-aplite and locally quartz veins and various metasomatically altered rocks were formed at the contact zone. The numerous Zr-Y-REE-Nb ore occurrences and deposits are associated with these rocks [22,28].

The Kanozero peralkaline granite massif is situated in the south of Kola Peninsula within the Belomorian block (Figure 1). The main rocks are peralkaline granite, granite-aplite and alaskite [29]. Numerous amazonite pegmatites occur in exo- and endocontact zones respect to the parental granites of the Kanozero massif. The exocontact zone of the massif is made up by a breccia, in which pegmatitic materials with peralkaline granite and granite-aplite filled the space between the fragments of the basic country rocks. These pegmatite bodies most probably have formed by metasomatic reworking and recrystallization of aplite. In the endocontact zone, pegmatitic bodies filled the cracks in parental granite. In some of pegmatitic bodies pink or greenish microcline forms instead of amazonite. This indicates a higher temperature of pegmatitic process.

The Khibiny alkaline massif is located in the center of the Kola Peninsula, at the contact of the Imandra-Varzuga Proterozoic greenstone belt with the Archaean metamorphic complexes of the Kola-Norwegian megablock (Figures 1 and 2). In plan view, the massif is elliptical (45 × 35 km) with concentrically-zoned structure; in vertical section it is cone-like, with its apex pointing downward. The massif consists dominantly of foyaite/khibinite (a nepheline syenite with predominant K-feldspar; about 70% of the outcrop area) and foidolite (8% of the outcrop area) that intruded into the foyaite massif along the Main Ring fault [30]. Highly potassic poikilitic nepheline syenites, "rischorrites" (10% of the outcrop area), commonly occur between the rocks of the Main Ring and the foyaite. The foidolite of the Main Ring accommodates all the apatite deposits and occurrences. Khibiny is somewhat a Mecca for mineralogists and mineral collectors due to numerous pegmatites and hydrothermal veins comprising unique assemblages of rare minerals. Their mineral composition strongly depends on the type of host rock [31]. Processes of local albitization, micatization and astrophyllitization are usually also imposed on the veins, leading to their transformation to albitites and other metasomatically altered rocks, practically indistinguishable from surrounding altered syenite rocks. Rather peculiar veins of quartz-syenite, syenite and nepheline-syenite with high concentrations of corundum, hercynite, topaz, almandine, pyrrhotite, pyrite and scarce accessories such as chrysoberyl, gadolinite-(Ce), chevkinite-(Ce), buergerite, native iron, sulfur, crichtonite, alabandite, and akaganeite are associated with xenoliths of hornfels. Ultimately, there are two different geological settings: veins in alkaline rocks and veins in xenoliths of hornfels.

The Lovozero alkaline massif intruded Archean granite-gneiss and Devonian tuff-basalt strata of the Kola-Norwegian megablock (Figures 1 and 2). It comprises regularly alternating subhorizontal layers of foyaite-malignite (a nepheline syenite containing ~50 vol % aegirine; "lujavrite") and ijolite-urtite. The alkaline rock strata are divided into two complexes—Differentiated (bottom) and Eudyalite (top) that differ in eudyalite content, nepheline-syenite–foidolite proportion and thickness of individual layers. The near-contact zone of fenitized rocks extends for 50–200 m. Fenites are represented by eudialite-microcline and amphibole-oligoclase varieties. Numerous nepheline syenite veins and alkaline pegmatites penetrated the country rocks at the distance of more than 100 m from the contact. Like at the Khibiny, pegmatites and hydrothermal veins are widespread within the massif [32]. They are very diverse in size, geological environment, composition and genesis. There are pegmatites also in xenoliths of country rocks.

Kovdor is an alkaline ultrabasic carbonatite-bearing pluton within Belomorian block (Figure 1) [33–35]. The outer contacts are subvertical or dip steeply to the centre of the massif. The contacts between different rock types inside the massif are also subvertical, thus the Kovdor massif can be represented as a neck. The Kovdor massif has concentric zonation, with axial peridotite; marginal melilitolite and intermediate diopsidite, phlogopitite and skarn-like rocks. Final stages of the massif formation include a pipe of phoscorite and carbonatite in western part of the melilitolite–foidolite ring as well as numerous dykes of carbonatites, nepheline syenites, and foidolites.

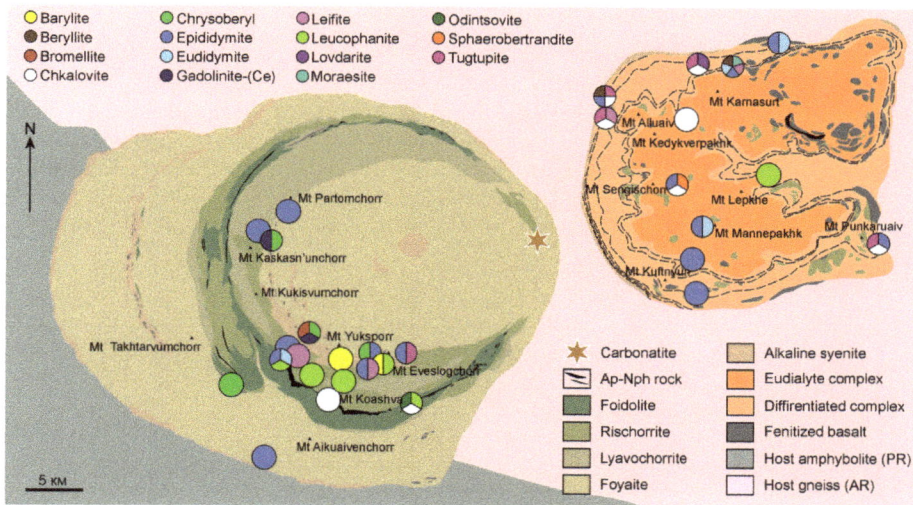

Figure 2. Simplified geological map of the Khibiny-Lovozero alkaline complex (after [22]), with localities of Be minerals.

The Jona occurrence is represented by Proterozoic granite pegmatite bodies within metamorphic rocks—gneisses, migmatites and amphibolites (Figure 1). The lenticular pegmatite bodies extend up to 500 m in depth. These pegmatites are a source of mica and ceramic raw materials [36].

The Strel'na occurrence is represented by Strel'na pegmatite field in South-East of Kola Peninsula (Figure 1). Plagioclase-microcline pegmatites are hosted by muscovite gneisses [37].

3. The Source Data for the Review

The most of the reports and papers on beryllium minerals of the Kola Peninsula were published in the 1950–70s and only in Russian, and therefore are not accessible to a wide audience. That is why, in this review the authors tried to collect all the data in brief form from previous Russian publications. The new original data are also provided. In this paper, beryllium minerals of the Kola Peninsula are described. The authors want to emphasize the following points. (1) Geological settings, the rocks and names of minerals are presented according to the author's text if it is not possible to clarify them. For example, "apatite" or "garnet" means members of these groups without an proper determination of their chemical composition. (2) The authors provide a synthesis of all the available information on mineral paragenesis. According to the source data we present the "mineral paragenesis" as a common occurrence of minerals which formed at the same time, or in a sequence, during a single mineralization process.

The following designations are used in the tables of mineral composition: EMP—electron microprobe and WCA—wet chemistry analyses, correspondingly; $_{calc}$—beryllium content was calculated, $_{wet}$—beryllium content was determined by wet chemistry method. The photos of minerals are selected to show the most typical morphologies and mineral associations.

Sometimes the distinctive features of minerals comparing to type localities are presented. The last ones include variations in composition, unusual morphology, special physical properties and other characteristics. A special attention is paid to the rate of study of individual minerals—for some of them only the presence in the occurrence is indicated, while others have been repeatedly studied. As a result, the description of minerals varies greatly in volume and completeness of information.

4. Beryllium Minerals from the Kola Peninsula

Kola Peninsula is not remarkable for significant resources of beryllium. On a whole the north-western part of Russia possesses 13.9% of all-Russian beryllium reserves (http://vims-geo.ru). Nevertheless, the Kola Peninsula demonstrates the unique diversity of beryllium minerals.

The first Be mineral discovered on the Kola Peninsula was apparently leucophanite from the Khibiny massif and mentioned by Chernik [38] and Kostyleva [39,40]. In total 28 beryllium minerals are established on the Kola Peninsula up today (Table 1). According with the classification of Pekov [5], among them only beryl is widespread, others are less common (as chrysoberyl and bertrandite) and most occur as trace minerals.

Table 1. List of beryllium minerals on Kola Peninsula.

Location/ Geological Era	Geological Setting	Minerals of Beryllium in Location and Geological Setting According Column 2
Kolmozero-Voronja belt/AR	spodumene pegmatite	*Silicates* bavenite, **bertrandite, beryl** *, milarite
Shongui/AR?	granite pegmatite	*Silicates* **beryl**
Keivy alkaline province/AR	1. pegmatite of peralkaline granite 2. quartz-epidote metasomatites 3. amazonite pegmatite 4. veins of peralkaline granite 5. intragranite pegmatite of peralkaline granite 6. nepheline syenite pegmatite 7. biotite metasomatite	*Silicates* danalite (2), **gadolinite-(Ce)** (2), **gadolinite-(Y)** (1,3,4,5), **genthelvite** (1), hingganite-(Ce) (2), hingganite-(Y) (3,5), hingganite-(Yb) (3), **leucophanite** (7), meliphanite (6,7) *Oxides and Hydroxides* behoite (6)
Kanozero/PR	amazonite pegmatite	*Silicates* **gadolinite-(Y)**
Strel'na/PR	granite pegmatite	*Silicates* **beryl, bertrandite**
Jona/PR?	granite pegmatite	*Silicates* **beryl**
Khibiny massif/PZ	1. pegmatite and hydrothermal veins within alkaline rocks 2. veins in xenolithes of hornfels 3. alkaline syenite	*Silicates* **barylite** (1), **beryllite** (1), **chkalovite** (1), **epididymite** (1), **eudidymite** (3), **gadolinite-(Ce)** (2), **leifite** (1), **leucophanite** (1), odintsovite (1), **sphaerobertrandite** (1), **tugtupite** (1) *Oxides and Hydroxides* bromellite (2), chrysoberyl (2)
Lovozero massif/PZ	1. pegmatite and hydrothermal veins within alkaline rocks 2. quartz-albite rock (without nepheline) 3. albitite 4. contact fenite 5. feldspar pegmatite in xenolithes	*Silicates* **barylite** (1), **beryllite** (1), **chkalovite** (1), eirikite (1), **epididymite** (1,2,4,5), **eudidymite** (1,3), **genthelvite** (5), **leifite** (1), **leucophanite** (1), lovdarite (1), **sphaerobertrandite** (1), **tugtupite** (1) *Phosphates* moraesite (1)
Kovdor/PZ	fenite	*Silicates* **epididymite**

* **Bold** for minerals occurring in more than one location.

The agpaitic massifs Khibiny and Lovozero have the highest number of Be minerals—17 species up today (Figure 2). Other occurrences are from perakaline granites and syenites of the Keivy alkaline province (11), spodumene pegmatites of Kolmozero-Voronja belt (4), Kovdor alkaline-ultrabasic massif (1), peralkaline granites of Kanozero (1), and from granite pegmatites of Jona (1), Shongui (1), and Strel'na (2). The distribution of these minerals in the mineral classes is as follows: 24 mineral species for silicates, 3 for oxides and hydroxides, 1 for phosphate. The discoveries of new beryllium minerals are related mainly to Khibiny and Lovozero massifs, including chkalovite [41], beryllite [42], tugtupite (Tugtupite was discovered almost simultaneously at Ilímaussaq (named tugtupite) and Lovozero (named beryllosodalite); the former is considered the type locality) [43], and lovdarite [44]. Hingganite-(Yb) was firstly described from pegmatites related to Keivy peralkaline granite [45].

As was mentioned above, beryllium is accumulated in the late products of felsic and alkaline magmatic rocks. Thereby the abundant beryllium minerals occur almost in pegmatites, hydrothermal veins and metasomatically altered rocks. Similar occurrences are described for Kola Peninsula: the most rich and diverse Be mineralization is found in highly differentiated alkaline pegmatites. The age span for Kola alkaline rocks ranges from Late Archean to Paleozoic suggesting the high perspectivity of them for beryllium minerals.

Barylite BaBe$_2$Si$_2$O$_7$ was found [46] and thoroughly studied [47] in pegmatites intruding in nepheline syenite of Mts Eveslogchorr, Rasvumchorr and Yuksporr, Khibiny massif (Figure 2). All these pegmatites are zonal, with microcline-dominant marginal zones and a natrolite core [20].

The most famous occurrence is in a barylite-bearing lens which is located within gneissose foyaite in Mt Yuksporr [20]. Barylite forms sheaf-like and radiated aggregates up to 3 cm in diameter, composed of lamellar crystals of pale-creamy color (Figure 3). Barylite aggregates occur in interstices amongst blocky feldspar, long prismatic aegirine and tabular catapleiite, associated with apatite group minerals, labuntsovite-Mn, astrophyllite, titanite and strontianite. In other cases, interstices in microcline aggregate are filled by radially-fibrous natrolite with sheaf-like splices of lamellar barylite crystals. Composition of Yuksporr barylite is given in Table 2. Barylite from Khibiny massif is of two rhombic polytypes: MDO$_1$ polytype, space group *Pmn2$_1$*; and MDO$_2$ polytype, space group *Pmnb* [20,48,49]. Barylite from the Yuksporr pegmatite has a rhombic cell *Pmn2$_1$* [50]. Now clinobarylite is discredited and is referred to MDO$_1$ polytype of barylite.

Table 2. Chemical composition of barylite.

Sample *	1, EMP
Component	wt %
SiO$_2$	36.74
CaO	0.03
BeO	15.50$_{wet}$
BaO	47.02
Total	99.29

* Sample: 1—Yuksporr, Khibiny [47].

Another occurrence of barylite is known from the natrolite core of a pegmatite at Mt Kuamdespahk, Lovozero massif [49]. The mineral composition and crystal structure are similar to the described in the Yuksporr occurrence. Barylite forms lamellar crystals and is associated with microcline, aegirine, natrolite, fluorapatite, titanite, catapleiite, labuntsovite, astrophyllite and strontianite.

There is a personal communication of Voloshin [49] about the barylite finding in the Vuorijarvi carbonatite.

Figure 3. Sheaf-like aggregate of barylite (**1**) on microcline (**2**), with natrolite (**3**), from a microcline-natrolite vein in foyaite, Mt Yuksporr, Khibiny.

Bavenite Ca$_4$Be$_2$Al$_2$Si$_9$O$_{26}$(OH)$_2$ was found in spodumene pegmatites of the Kolmozero-Voronja belt (Figure 1) and confirmed by X-ray study [51]. The mineral is rare and occurs in two associations: (a) in association with late suggary (granular) albite as radially-fibrous aggregates of 0.1–1 cm (Figure 4) or as tiny veins of 0.01–0.1 mm thick, and (b) in cavities produced by leaching of spodumene, where it is a late mineral, covering quartz, microcline, albite, and stilbite and mantled on its turn by later chabazite-(Ca) and calcite.

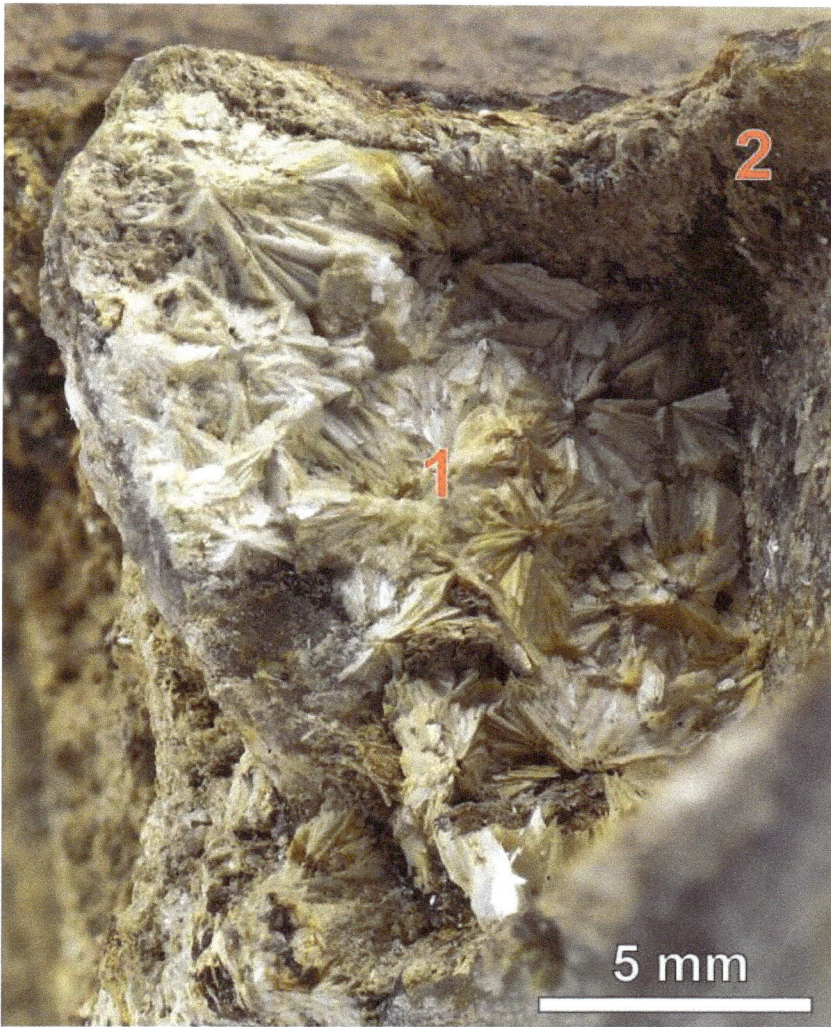

Figure 4. Radiated bavenite aggregates (**1**) in albite (**2**), from a spodumene pegmatite, Kolmozero-Voronja.

Comparing to type locality (Baveno, Italy) and other occurrences [51–53], bavenite from Kolmozero-Voronja pegmatites is characterized by moderately high BeO and Na$_2$O and very low CaO (Table 3). The relations of elements in this mineral are determined by the following scheme $^{T(3)}$Be + O(OH) + Na + 2$^{T(4)}$Si $\rightarrow$ $^{T(3)}$Si + 2O + Ca + 2$^{T(4)}$Al [52].

Table 3. Chemical composition of bavenite.

Sample *	1, WCA	2, EMP	3, WCA
Component		wt %	
SiO_2	58.88	61.48	56.93
Al_2O_3	7.83	3.54	12.38
Fe_2O_3	0.46	-	-
CaO	15.74	23.44	24.47
BeO	10.3	8.96	2.67
MgO	0.20	-	0.12
Na_2O	1.46	0.30	0.29
H_2O^+	4.92	3.48	2.49
Total	99.79	101.20	99.72

* Samples: 1—Kolmozero-Voronja [51]; 2,3—Baveno, Italy [52,53].

Behoite Be(OH)$_2$ is described from nepheline syenite pegmatite in Sakharjok massif (Figure 1), Keivy alkaline province [54]. Behoite occurs in leaching cavities up to 1.5–2 mm in size between euhedral aegirine-augite crystals (Figure 5). The mineral forms brown powder-like aggregates and closely associates with prehnite. Identification of behoite is based on coincidence of its X-ray powder diffraction pattern with that of behoite from Texas [55] and absence of characteristic lines of any elements in EDS spectrum. Behoite is a product of hydrothermal alteration of meliphanite.

Figure 5. Powder-like aggregate of behoite (**1**) with prehnite (**2**) in leaching cavities of aegirine-augite (**3**) in nepheline-syenite pegmatite, Sakharjok.

Bertrandite Be$_4$Si$_2$O$_7$(OH)$_2$ is an extremely rare mineral in Kola Peninsula, despite the wide range of conditions suitable for its formation. It was found [51] in clayish aggregate from a cavity in beryl crystal in spodumene pegmatite, Kolmozero-Vornja belt (Figure 1). The mineral forms very thin plates up to 0.2 mm in size. Bertrandite is apparently a secondary mineral. Its mineral identification is confirmed by X-ray study, but compositional data are not available.

The bertrandite formed along with the clay minerals at the expense of beryl in pegmatites of the Strel'na field [37]. It is characterized by tabular or elongated flattened crystals set amidst the mica unit. Only qualitative spectral analysis of this mineral is given.

The finds of bertrandite are mentioned in some publications on the Lovozero massif [32,56], but most likely the authors mean sphaerobertrandite [57] which was considered a variety of bertrandite.

Beryl Be₃Al₂Si₆O₁₈ is a common mineral in quartz-albite-microcline pegmatite bodies from Kolmozero-Voronja, Jona, Strel'na, Shongui and Olenegorsk occurrences (Figure 1). Beryl is one of the exploited minerals in Kolmozerskoe deposit.

Beryl from the Kolmozero-Voronja belt was formed during several stages [51]. The earliest beryl (I) occurs as long-prismatic crystals (up to 20 cm long and 5 cm in diameter) amongst coarse-grained quartz-albite-microcline aggregates of the albite-microcline pegmatite (Figure 6a).

(a)   (b)

Figure 6. Beryl (**1**): (**a**) section of long-prismatic beryl crystal from granite pegmatite of the Olenegorsk deposit; (**b**) crystal with quartz (**2**) from quartz-albite-microcline pegmatite, Kolmozero-Voronja.

The second stage beryl (II) occurs as euhedral short-prismatic crystals (up to 5 cm long) within medium-grained quartz-albite-spodumene aggregates (zone 3). The mineral includes the grains of quartz, spodumene and albite and is cut by muscovite laths.

The third stage beryl (III) occurs within the coarse-grained quartz-spodumene blocks (zone 4), where equidimensional mineral grains reach 0.5 m. It crystallizes after spodumene and microcline but before albite and muscovite. Beryl from quartz-muscovite lenses (zone 5) also belongs to the third stage. Here mineral forms coarse agglomerations.

The latest beryl (IV) was formed during recrystallization of the earlier beryl and its replacement by suggary albite. This mineral occurs as euhedral crystals of 0.2–10 mm size. During spodumene leaching and zeolite formation beryl decomposes to bertrandite and bavenite.

In ceramic pegmatites cutting rocks of the Olenegorsk Banded Iron-Formation, there are well shaped long-prismatic beryl crystals up to 50 cm long and 10 cm in diameter (Figure 6b) associated with phlogopite, muscovite, members of the dravite-shorl tourmaline series, spessartine, fluorapatite, ilmenite and ferrocolumbite.

Several dozens of beryl chemical analyses are available [51,58]. From the content of major constituents (BeO, Al₂O₃, SiO₂) beryl from Kolmozero-Voronja pegmatites are fully consistent to stoichometric formula (Table 4). BeO varies from 11.36 to 13.90 wt %, whereas zone 3 beryl contains an average BeO 12.25 wt %, zone 4 beryl—11.9 wt %, zone 5 beryl—12.6 wt %. The latest fine-grained beryl from albitization zones contains the lowest BeO—11.3 wt %. Beryl from Kolmozero-Voronja is characterized by elevated contents of alkaline elements: Li_2O ranges from 0.3 to 1 wt %, Na_2O 1.1–2.3 wt %, Rb_2O 0.01–0.06 wt %, Cs_2O 0.12–0.87 wt %. At a whole, the total content of alkalis in beryl increases from marginal to central part of pegmatites. Sosedko [59] mentions the 7.23 wt % of total alkalis (from which Cs_2O content is 4.13 wt %) in beryl from central zone 4. Beryl decomposition led to crystallization of other Be minerals, as bertrandite and bavenite.

Table 4. Chemical composition of beryl.

Sample *	1, WCA	2, WCA	3, WCA	4, WCA	5, EPM
Component			wt %		
SiO_2	64.84	63.30	63.78	64.90	67.00
TiO_2	0.04	-	-	0.02	-
Al_2O_3	16.10	17.97	18.23	17.49	18.63
Fe_2O_3	1.00	0.21	0.28	-	0.44
FeO	-	0.26	0.13	0.72	-
MnO	-	-	0.12	-	-
BeO	12.90	12.37	11.60	12.72	13.66_{calc}
MgO	0.78	0.02	0.15	0.23	-
CaO	0.19	0.25	0.42	0.12	-
Na_2O	1.00	1.67	1.71	1.20	0.27
K_2O	0.55	-	-	-	-
Li_2O	0.04	0.74	0.74	0.41	-
Cs_2O	-	0.20	-	-	-
Rb_2O	-	0.04	-	-	-
H_2O^+	2.45	2.75	2.63	2.52	-
Total	100.07	100.18	99.83	100.33	100.00

* Sample: 1—1 stage, 2—2 stage, 3—3 stage, Kolmozero-Voronja [51]; 4—Strel'na [37]; 5—Olenegorsk (analyst Ya.A. Pakhomovsky, KSC RAS).

Beryl is found as an accessory mineral in undifferentiated plagioclase-microcline pegmatites occurring among the mica gneisses in pegmatites of Strel'na field [37]. Beryl is related to the plagioclase areas. Beryl is represented by well-faceted prismatic crystals up to 15 cm in length. Those crystals are often opaque, covered by a dense network of cracks intersecting in different directions.

Samples of beryl collected from granite pegmatites of Jona and Shongui are stored in the Museum of GI KSC RAS.

Beryllite $Be_3SiO_4(OH)_2 \cdot H_2O$ was firstly described from the concentrically zoned natrolitized alkaline pegmatite Natrolite Stock, hosted by sodalite-nepheline syenite of Mt Karnasurt in the Lovozero massif (Figure 2) [42]. The mineral overgrowths epididymite forming white to brown spherulites up to 2–3 mm and waxy crusts of these spherulites (up to 2 mm thick) in leaching cavities of chkalovite (Figure 7). Beryllite is associated with fine-grained albite and epididymite which it overgrows. Composition of beryllite is given in Table 5. Another beryllite locality is large (above 2 m in diameter) ussingite pegmatite in foidolite–lujavrite of Mt Alluaiv (Figure 2), where similar berrylite crystals (up to 1 mm thick) together with epididymite, hydroxycancrinite, and K-Ca zeolites incrust intersices in the marginal aegirine-microcline zone.

Figure 7. Beryllite (**1**) pseudomorph after chkalovite, in albite (**2**), from an aegirine-feldspar lens in sodalite-nepheline syenite, Mt Karnasurt, Lovozero. (**3**)—Natrolite, (**4**)—microcline.

Table 5. Chemical composition of beryllite.

Sample *	1, WCA
Component	wt %
SiO_2	34.10
Al_2O_3	1.63
Fe_2O_3	0.12
BeO	40.00
CaO	0.50
Na_2O	2.42
H_2O^+	18.95
Total	97.22

* Sample: 1—Lovozero [42].

Other beryllite findings in the Kola Peninsula are unreliable. The mineral is mentioned from Mt Mannepahk in the Lovozero massif [60], as fibrous aggregates overgrowing chkalovite from Mt Punkuruaiv in the Lovozero massif and from natrolite vein in Mt Kuelpor, Khibiny massif [61].

Kola Peninsula is the type locality for beryllite. Crystal structure of mineral has not yet been solved.

Bromellite BeO was found in a sodalite-nepheline-orthoclase vein, along with corundum, siderophyllite, ilmenite and fluorapatite, cutting hornfels xenoliths at Mt Kukisvumchorr, Khibiny massif [20,46]. Bromellite forms colorless hemimorphic crystals up to 3 mm as well as irregularly-shaped grains between the grains of nepheline and potassium feldspar. The mineral identification has been confirmed by powder X-ray study.

Chkalovite Na₂BeSi₂O₆ is the first beryllium mineral discovered on the Kola Peninsula. It was found by Gerasimovsky [41] in natrolitized ussingite pegmatites hosted by eudialyte-rich foyaite of Mt Malyi Punkuruaiv, Lovozero massif (Figure 2, Table 6). The mineral forms large isometric crystals, up to 10 cm, white to semitransparent and partially replaced by snow-white epididymite and crimson tugtupite. Characteristic associated minerals are sodalite, analcime, murmanite, epistolite, manganoeudyalite, manganoneptunite, tainiolite, manganonordite-(Ce)–ferronordite-(Ce), steenstrupine-(Ce), belovite-(Ce), rhabdophane-(Ce), sphalerite and galena. The crystal structure of chkalovite has been solved later [62,63].

Table 6. Chemical composition of chkalovite.

Sample *	1, WCA	2, WCA	3, EMP
Component		wt %	
SiO_2	56.81	57.08	56.74
Fe_2O_3	0.30	-	-
FeO	0.12	-	-
CaO	0.37	-	-
BeO	12.67	12.78	13.44
Na_2O	28.93	29.49	29.07
K_2O	0.13	-	-
H_2O^+	0.23	0.38	-
SO_3	0.22	-	-
Total	99.78	99.73	99.25

* Sample: 1—M. Punkuruaiv, Lovozero [41], 2—Sengischorr, Lovozero [61], 3—Koashva, Khibiny [47].

The following numerous findings of chkalovite indicate that this mineral is typical of ultra-agpaitic ussingite pegmatites of the Lovozero massif: Mts Sengischorr, Punkuruaiv, Karnasurt, Lepkhe-Nelm, Kedykverpahk, Alluaiv [32,64–66]. Chkalovite is the earliest high temperature beryllium mineral from the pegmatites and it is a precursor for most Be mineral phases: tugtupite, beryllite, epididymite, lovdarite and sphaerobertrandite [32]. For example, the ussingite pegmatite "Palitra" in rocks of the Lovozero Differentiated complex at Mt Kedykvyrpakh contains transparent tabular chkalovite crystals (up to 5 cm in diameter) included in dark violet ussingite together with sodalite, natrolite, villiaumite, natrosilite, manaksite (Figure 8a), serandite, kapustinite, lomonosovite, vuonnemite, manganoneptunite, and other rare minerals [64]. In natrolitized protoussingite pegmatite "Yubileinaya", there are large (up to 15 cm in diameter) colorless chkalovite crystals partially or completely replaced by lovdarite and tugtupite, and associated with analcime, lorenzenite, lomonosovite, eudialyte, (mangano)nordite-(Ce), manganoneptunite, umbozerite, steenstrupine-(Ce) and products of their alteration (bornemanite, terskite, vitusite-(Ce), narsarsukite, zorite, raite, etc.) [32].

(a)   **(b)**

Figure 8. Chkalovite crystals (**1**): (**a**) in manaksite (**2**), from the "Palitra" pegmatite, Mt Kedykvyrpakhk, Lovozero; (**b**) in villiaumite (**3**), from a sodalite-microcline-aegirine vein in urtite, Mt Koashva, Khibiny.

In the Khibiny massif (Figure 2) chkalovite was firstly found in 20–40 cm thick microcline-nepheline-sodalite-katophorite veins, hosted by urtite in Mt Rasvumchorr [67]. Pectolite-villiaumite-chkalovite or monomineralic chkalovite aggregates fill the cracks in the veins; in many cases, these minerals fill out the veins completely.

The mineral was found also in a large sodalite-microcline-aegirine pegmatite in apatite-bearing urtite of Mt Koashva, Khibiny massif [20,47]. Chkalovite forms anhedral grains up to 4 × 5 cm and rarely aggregates of euhedral 0.5–3 cm crystals. The mineral sets amongst microcline crystals and natrolite-pectolite aggregates and is associated with lomonosovite and lamprophyllite. Vitusite-(Ce) and nacaphite crystallize in leaching cavities in chkalovite. Other remarkable occurrence of chkalovite is the microcline-pectolite-sodalite-aegirine ultraagpaitic vein of Mt Koashva, Khibiny massif [20], where large (up to 10 cm) euhedral colorless chkalovite crystals are included in dark-red villiaumite (Figure 8b) in close association with natrophosphate, lomonosovite, shcherbakovite, nefedovite, fluorcaphite, vitusite-(Ce), chlorbartonite, djerfisherite and cobaltite.

Chrysoberyl BeAl$_2$O$_4$ was found in drill core near Maly Vud'iavr Lake, Khibiny massif (Figure 2). The host rocks are fenitized basalt and tuff (so-called "hornfels") found as xenoliths in foyaite situated near the Main foidolite ring [68]. The mineral forms thick tabular crystals (up to 5 mm in diameter), twins and penetration trillings of pseudohexagonal habit. Composition of chrysoberyl is given in Table 7.

Contact of alkaline magma with xenolites of basalt and its tuff leads to the formation of a peraluminic assemblage of minerals: anorthoclase, cordierite–sekaninaite, phlogopite, muscovite-paragonite, corundum, hercynite, topaz, etc. This process takes place in a highly alkaline medium enriched with fluorine and carbon dioxide. The dispersed beryllium is leached from the primary rocks to form easily mobile complex compounds. Decomposition of these complexes and the formation of chrysoberyl occur in areas of fenites schistosity on the contact with the later biotite-orthoclase veins due to the lower alkalinity of solutions.

Table 7. Chemical composition of chrysoberyl.

Sample *	1, WCA	2, EMP
Component	wt %	
V_2O_5	-	0.32
SiO_2	0.80	-
TiO_2	-	0.41
Al_2O_3	75.90	79.43
Fe_2O_3	3.10	-
FeO	-	0.61
CaO	1.50	-
BeO	19.50	19.43$_{wet}$
Total	100.80	100.20

* Sample: 1—M.Vud'iavr, Khibiny [68], 2—Kaskasn'unchorr, Khibiny [47].

This mineral was further described in similar xenoliths from the foyaites of Mts Eveslogchorr, Kukisvumchorr and Kaskasn'unchorr, Khibiny massif [20,47]. The mineral occurs in veins of nepheline syenite and nepheline-sodalite syenite emplaced into biotite-feldspar hornfels. Chrysoberyl here forms tabular penetration trillings of pseudohexagonal habits and 1.5 × 5 mm in size (Figure 9). The mineral is included into nepheline and sodalite. Chrysoberyl is rarely represented by short prismatic crystals of 1.5 × 3 cm in size. Associated beryllium minerals are epididymite (Eveslogchorr) and gadolinite-(Ce) (Kaskasn'unchorr).

Figure 9. Crysoberyl twin (**1**) in a biotite-nepheline-anorthoclase pegmatite in fenite, Mt Kukisvumchorr, Khibiny. (**2**)—Nepheline, (**3**)—biotite.

Danalite Be₃Fe²⁺₄(SiO₄)₃S occurs in quartz-feldspar vein emplaced in amphibolized metagabbro-anorthosites of the El'ozero rare-metal ore occurrence in the Keivy alkaline province [69,70]. The mineral is associated with quartz, allanite-(Ce) and thorite. Danalite forms tetrahedral crystals and, on rare occasions, tiny equidimensional grains. The length of the edge in the largest crystals reaches 2 cm. Danalite includes numerous grains of quartz, albite, zircon and euxenite-(Y). The mineral identification was performed by XRD and wet chemical analysis. Danalite composition from El'ozero occurrence (Table 8) may be represented as a combination of the end members: 46.3% danalite (Be₃Fe²⁺₄(SiO₄)₃S), 36.5% genthelvite (Be₃Zn₄(SiO₄)₃S) and 17.2% helvine (Be₃Mn²⁺₄(SiO₄)₃S). Thus, danalite is characterized by broad isomorphic schemes, both isovalent ($Fe^{2+} \rightleftarrows Zn^{2+}$; $Fe^{2+} \rightleftarrows Mn^{2+}$) and heterovalent ($Be^{2+} \rightleftarrows Al^{3+}$).

Table 8. Chemical composition of danalite.

Sample *	1, WCA
Component	wt %
SiO_2	31.92
TiO_2	0.07
ZrO_2	0.14
Al_2O_3	1.44
FeO	22.38
ZnO	19.99
MnO	8.15
MgO	0.10
CaO	0.50
BeO	12.44
H_2O^-	0.17
H_2O^+	0.69
S	5.31
−O = S	2.66
Total	100.30

* Sample: 1—El'ozero [69].

Eirikite KNa₆Be₂(Si₁₅Al₃)O₃₉F₂. There are no published data on eirikite from Kola Peninsula. Nevertheless, Pekov [71] mentioned "potassium leifite" from Mt Alluaiv in Lovozero massif (Figure 1). However, the chemical composition of this mineral and FTIR spectra are similar to those of eirikite. In addition, leifite from Khibiny massif contains 1.48 wt % of K₂O (0.45 *apfu*) [47]. The total content of K₂O, Cs₂O and Rb₂O is about 0.5 *apfu* and can exceed this value, and then the formula of the mineral according to [72] can be recalculated as eirikite.

Epididymite Na₂Be₂Si₆O₁₅·H₂O was established simultaneously in the Khibiny and Lovozero massifs (Figure 2). Those were the first findings in the former Soviet Union [73]. In Lovozero, this mineral was found in a number of agpaitic pegmatites at Mts Karnasurt and Kuivchorr, while in Khibiny it was found in nepheline syenite pegmatites at Mt Yuksporr (Table 9). In fact, epididymite is a common mineral in nepheline syenite pegmatites worldwide. In addition to nepheline-bearing pegmatites, Lovozero epididymite is found also in nepheline-free albite veins with quartz and narsarsukite, which is a similar context to that found in the Narsarsuk pegmatites in Greenland, the type locality for epididymite. The mineral forms as individual euhedral platy and tabular crystals and pseudohexagonal tabular complex twins up to 2 cm size (Figure 10a).

Also, epididymite can form large (up to 20–30 cm) aggregates of different morphology: spherules, fibrous, fine-grained, chalky-like, clayey and dense porcelaine-like (Mts Karnasurt, Mannepahk, Sengischorr, Kuftnyun, etc.) and represents the most abundant beryllium mineral at the Lovozero massif [32,61,74,75]. Usually, epididymite grows after chkalovite (Figure 10b) and is replaced by a mineral sequence comprising first beryllite, then bertrandite and finally moraesite.

Table 9. Chemical composition of epididymite.

Sample *	1, WCA	2, WCA	3, EMP	4, EMP
Component	wt %			
SiO_2	72.60	70.70	71.82	74.27
Al_2O_3	0.20	0.60	-	0.13
Fe_2O_3	0.13	-	-	-
MgO	0.07	0.10	-	-
CaO	0.26	0.25	-	-
BeO	10.42	12.15	10.51_{wet}	-
Na_2O	12.63	11.81	11.72	11.91
K_2O	0.20	-	0.05	0.18
H_2O^+	4.00	4.50	3.81_{wet}	-
Total	100.51	100.16	97.91	86.49

* Sample: 1—Lovozero [73], 2—Khibiny [73], 3—Eveslogchorr, Khibiny [47], 4—Kovdor [33].

(a)   (b)

Figure 10. Epididymite (**1**): (**a**) skeletal trilling on albite (**2**), from an aegirine-feldspar lens in foyaite, Mt Kuftnyun, Lovozero; (**b**) complete pseudomorph after chkaklovite, in natrolite (**3**), from a microcline-natrolite lens in sodalite-nepheline syenite, Mt Karnasurt, Lovozero.

After the first occurrence in Mt Yuksporr, Khibiny massif, epididymite was also found in alkaline pegmatites of Mts Kukisvumchorr, Aikuaivenchorr and Partomchorr [20,47,76,77]. In the Aikuaivenchorr occurrence, epididymite forms complex tabular twins up to 1.2 cm in diameter and occurs in the albite zone together with catapleiite, aegirine, polylithionite and chabazite-Sr. Epididymite found in a central part of aegirine-feldspar pegmatite in gneissose foyaite of Mt Eveslogchorr is associated with leifite. Here, epididymite grows as large aggregates up to 20 cm and it is represented by two morphological types: (a) as fine-grained aggregates and (b) as druses of flattened crystals. The mineral can also be a product of alteration of tugtupite, as was established in another natrolite-feldspar vein in Mt Eveslogchorr [20,47].

Epididymite is the unique beryllium mineral known from Kovdor alkaline-ultrabasic massif [33]. It was found in the cavities in pyroxenised fenite near the contact with phoscorites. Crystals of epididymite grow on albite and aegirine-augite along with saponite, labuntsovite and catapleiite.

The existence of a dimorphism between epididymite and eudidymite was established in 1972 [78] and the crystal-chemical formula of H_2O-containing epididymite was proposed. However, many authors continued to use the formulae with OH-groups until 2008 [79] due to the unusually high dehydration temperature (810–830 °C according [73]).

Eudidymite $Na_2Be_2Si_6O_{15} \cdot H_2O$ was found in albitites of Mt Flora, Lovozero massif (Figure 2), in association with epididymite and narsarsukite [73]. Later, eudidymite was found in the central part of the "Natrolite Stock" pegmatite in sodalite-nepheline syenite of Mt Karnasurt, where where it forms characteristic twins up to 5 mm in diameter (Figure 11) in association with epididymite and leifite [80]. Similar twins of eudidymite associated with epididymite occur in and aegirine-feldspar pegmatite in eudialyte lujavrite of Mt Mannepakh.

Figure 11. Eudydimite twin (**1**) on albite (**2**), from an aegirine-feldspar lens in sodalite-nepheline syenite, Mt Karnasurt, Lovozero.

"Eudidymite" was mentioned from pegmatite of Mt Partomchorr, Khibiny massif [79], but later was discredited and identified as epididymite [47]. Thus, the occurrences of eudidymite in Khibiny massif include the feldspar rock in Mt Restinjun and natrolitized aegirine-nepheline-microcline pegmatite in urtite of Mt Kukisvumchorr [20], where it forms colorless flattened-prismatic crystals and sheaf-like aggregates up to 5 mm long in close association with epididymite.

Gadolinite-(Ce) $Ce_2Fe^{2+}Be_2O_2(SiO_4)_2$ is found in sodalite-biotite-orthoclase veinlets within hornfels at Mts Kukisvumchorr, Rischorr and Kaskasn'unchorr, the Khibiny massif (Figure 2) as dark green irregularly-shaped grains up to 2 mm in diameter in the interstices of nepheline and potassium feldspar [20,46]. Fayalite, zirconolite, fluorcalciobritholite, zircon, molybdenite, monazite-(Ce), löllingite, graphite, and chrysoberyl are the characteristic associated minerals. Gadolinite-(Ce) is metamict and yields the distinct X-ray powder pattern after heating up to 900 °C.

Also, gadolinite-(Ce) is a relatively common mineral in quartz-epidote metasomatites of the El'ozero rare-metal occurrence (Figure 1), Keivy alkaline province [28]. Gadolinite-(Ce) forms anhedral elongated grains up to 100 μm rimmed by hingganite-(Ce) (Figure 12a). The grains are clearly emanating from chevkinite-(Ce). Areas between the grains are composed of ferriallanite-(Ce), ilmenite and titanite. Gadolinite-(Ce) from El'ozero was diagnosed only on the results of microprobe analyses without beryllium determination (Table 10). It is shown that the measured and calculated composition of gadolinite-(Ce) is cation deficient in the positions of Ce and Fe. Occurrence has not been confirmed by XRD due to gadolinite-(Ce) metamict state. The mineral crystallizes during the late stages of fluid-induced alteration of chevkinite-(Ce).

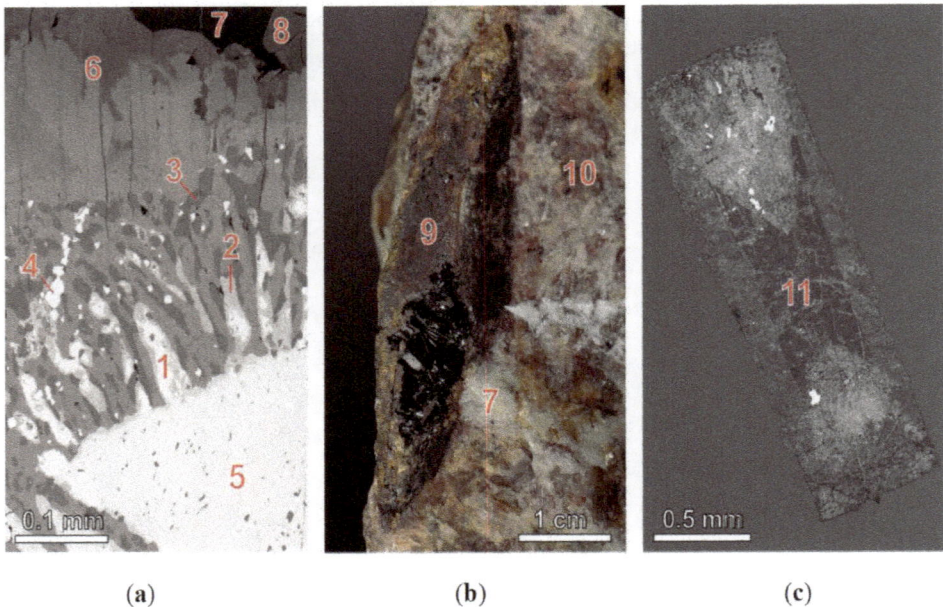

Figure 12. Gadolinite-(Ce), gadolinite-(Y) and hingganite-(Y): (**a**) corona-like aggregates of gadolinite-(Ce) (**1**), hingganite-(Ce) (**2**), ilmenite (**3**) and thorite (**4**) around chevkinite-(Ce) inclusion (**5**) in ferriallanite-(Ce) (**6**), with quartz (**7**) and titanite (**8**), from quartz–epidote metasomatite of the El'ozero deposit, Keivy; (**b**) crystal of gadolinite-(Y) (**9**) in quartz (**7**) and microcline (**10**), from Mt Ploskaya, Keivy; (**c**) sectoral crystal of hingganite-(Y) (**11**), from alkaline granite pegmatite of the Belye Tundry massif, Keivy.

Table 10. Chemical composition of gadolinite-(Ce).

Sample *	1, EMP	2, EMP
Component	wt %	
SiO_2	21.14	21.47
TiO_2	-	0.29
ThO_2	4.75	4.12
UO_2	-	0.44
Y_2O_3	0.84	5.44
La_2O_3	18.41	5.80
Ce_2O_3	25.36	19.37
Pr_2O_3	1.19	2.73
Nd_2O_3	3.14	11.39
Sm_2O_3	0.40	3.31
Gd_2O_3	-	2.36
Fe_2O_3	12.59	-
BeO	9.99_{wet}	8.94_{calc}
CaO	1.82	0.83
MnO	0.65	-
FeO	-	11.68 **
Total	100.28	98.17

* Sample: 1—Kaskasn'unchorr, Khibiny [20], 2—El'ozero [28]. ** all Fe as Fe^{2+}. BeO_{calc} from Be = 2 *apfu*.

Gadolinite-(Y) $Y_2Fe^{2+}Be_2O_2(SiO_4)_2$. The first findings of gadolinite-(Y) on the Kola Peninsula were made by Belkov [81] in amazonite pegmatites related to peralkaline granites of Western Keivy massif of Keivy alkaline province and of Kanozero massif (Figure 1). Morphology and abundance of the mineral was noted to depend on pegmatite zones. As a whole, numerous prismatic or ellipsoidal gadolinite-(Y) crystals up to 5–8 cm in length always coated by alteration crusts, composing from hydroxides of Fe and REE (Figure 12b), are usual for border zone (selvage) of pegmatites. Crystal accumulations or, on the contrary, sporadic crystals are usual for albitized pegmatite zones. It is assumed that gadolinite-(Y) in the selvages has crystallized at an earlier high-temperature stage, and in the albitized zones it crystallized at a later hydrothermal stage [81]. Another origin of gadolinite-(Y), namely metasomatic, is assumed by Lunz [82]. Chemical composition of gadolinite-(Y) is close to stoichiometric formulae with insignificant deficiency in Fe position, especially for Kanozero samples [81] (Table 11).

Gadolinite-(Y) was found also in peralkaline granite veins of the El'ozero occurrence, Keivy alkaline province [84]. The chemical composition of minerals stands out with high content of Ca and Al, and lower content of Be.

At last, gadolinite-(Y) is found in intragranite pegmatites of peralkaline granite massifs of Keivy alkaline province. The pegmatites are assumed as the earliest postmagmatic product of peralkaline granites [85]. The mineral is represented by thick tabular, prismatic crystals and anhedral grains up to 5 mm in size. Gadolinite-(Y) is associated with hingganite-(Y), fergusonite-(Y), tengerite-(Y), britholite group minerals, and astrophyllite. Gadolinite-(Y) from intragranite pegmatites differs from many other occurrences by its crystalline state.

Table 11. Chemical composition of gadolinite-(Y).

Sample *	1, EMP	2, WCA	3, EMP
Component		wt %	
U_3O_8	-	0.07	-
ThO_2	0.36	0.36	0.39
Al_2O_3	-	2.00	0.00
Fe_2O_3	-	1.27	-
Y_2O_3	27.52	-	16.46
La_2O_3	1.65		2.75
Ce_2O_3	5.52		11.42
Pr_2O_3	0.65		1.14
Nd_2O_3	1.83		5.76
Sm_2O_3	0.47		2.19
Eu_2O_3	0.12		-
Gd_2O_3	0.60	$REE_2O_3 = 48.34$	3.69
Tb_2O_3	0.15		0.56
Dy_2O_3	2.54		4.05
Ho_2O_3	0.75		0.65
Er_2O_3	3.31		1.61
Tm_2O_3	0.55		0.00
Yb_2O_3	4.02		0.37
Lu_2O_3	0.41		-
SiO_2	23.96	24.81	23.09
TiO_2	-	0.22	-
BeO	9.97_{calc}	9.13	9.61_{calc}
PbO	-	0.12	-
FeO	10.15 *	9.79	9.83 **
ZnO	-	-	0.07
MnO	0.28	0.29	0.36
MgO	-	0.25	-
CaO	0.26	2.64	0.37
H_2O^+	-	0.71	-
Total	95.07	100.09	94.37

* Sample: 1—Kanozero [83], 2—El'ozero [84], 3—Keivy [85]. ** all Fe as Fe^{2+}. In sample 3: 7.11 wt % FeO, 2.90 wt % Fe_2O_3 (wet chemistry analysis, analyst L.I. Konstantinova, KSC RAS); 9.80 wt % BeO, 0.20 wt % B_2O_3 (ICP-MS, analyst S.V. Drogobuzhskaya, ICT KSC RAS). BeO_{calc} calculated from the atomic ratio Si:Be = 1:1.

Gadolinite-(Y) is one of the main Y and *REE* carriers in pegmatites of peralkaline granite of Kola Peninsula [83]. There are differences in concentration of individual REE in gadolinite-(Y) from peralkaline granite veins, amazonite pegmatites and intragranite pegmatites [85–87]. Based on the REE distribution, it is assumed that in intragranite pegmatites the mineral crystallizes during the late magmatic/pegmatitic stage, while in amazonite pegmatites it crystallizes during the hydrothermal stage [85].

Genthelvite $Be_3Zn_4(SiO_4)_3S$. The mineral was found [88] in the feldspar pegmatites located in a small outcrop of augite-porphyrites on the roof of nepheline syenite intrusion, Mt Flora, Lovozero massif (Figure 2). It was the first finding of genthelvite within the former Soviet Union. The mineral forms anhedral 1×2 mm grains, rarely up to 5×10 mm. Its color changes from green to colorless. High MnO contents (up to 10.21 wt %) are determined in this mineral (Table 12). Genthelvite is associated with Mn-bearing ilmenite and zircon. Its formation is related to input of Fe, Mn, Mg from hosting basic rocks into alkaline pegmatitic melt/solution. In that case genthelvite can form instead of Be silicates of sodium and calcium (leucophanite, meliphanite, epididymite, eudidymite, chkalovite), typical for alkaline pegmatites.

Table 12. Chemical composition of genthelvite.

Sample *	1, WCA	2, WCA
Component	wt %	
SiO_2	27.35	32.08
Fe_2O_3	6.70	0.78
Y_2O_3	-	
Yb_2O_3		0.32
Ce_2O_3	-	
FeO	-	13.71
MnO	10.21	5.21
ZnO	40.00	33.54
CaO	-	
SrO	-	0.28
BeO	12.00	11.71
Na_2O	-	
K_2O	-	0.12
H_2O^+	-	0.12
S	5.74	5.49
$-O = S$	2.87	2.78
Total	99.13	100.58

* Sample: 1—Mt Flora, Lovozero [88], 2—Keivy alkaline province [89].

Genthelvite occurs also in amazonite pegmatites of the Keivy alkaline province [89,90], where it usually forms anhedral 2–3 mm grains and their aggregates. Vasiliev [89] observed these anhedral grains within beryl crystals. In amazonite pegmatite of Mt Rovgora, genthelvite forms well shaped crystals (Figure 13) and its debris reaches up to 8 × 12 cm [90]. Their chemical analysis returns on significant variations of Fe and Mn, which can be related to influence of pegmatite' host rocks [70]. It is suggested that genthelvite is of metasomatic origin and formed during albitization of pegmatites [89].

Figure 13. Genthelvite crystal (**1**), with albite (**2**), from the amazonite pegmatite, Mt Rovgora, Keivy.

Hingganite-(Ce) BeCe(SiO₄)(OH) is found [28] in quartz-epidote metasomatic rocks of the El'ozero rare-metal occurrence in the Keivy alkaline province (Figure 1). The mineral replaces gadolinite-(Ce) (Figure 12a). Chemical composition of hingganite-(Ce) from this locality (Table 13) is similar to those from other occurrences [85,91], with the difference of higher CaO values (up to 4.4 wt %). The mineral has not been confirmed by X-ray studies.

Table 13. Chemical composition of hingganite-(Ce).

Sample *	1, EMP
Component	wt %
P_2O_5	0.05
Ta_2O_5	0.06
SiO_2	25.30
TiO_2	0.32
ThO_2	3.38
UO_2	0.09
Al_2O_3	0.80
Y_2O_3	5.07
La_2O_3	5.27
Ce_2O_3	17.13
Pr_2O_3	2.38
Nd_2O_3	10.31
Sm_2O_3	2.68
Gd_2O_3	2.02
Tb_2O_3	0.24
Dy_2O_3	0.86
Yb_2O_3	0.25
BeO	10.53_{calc}
CaO	3.41
MnO	0.23
FeO	6.07 **
PbO	0.26
Total	96.60

* Sample: 1—El'ozero [28]. ** all Fe as Fe^{2+}. BeO_{calc} from Be = 2 *apfu*.

Hingganite-(Y) BeY(SiO₄)(OH). The first description of hingganite-(Y) in Kola was made [83,87] in the amazonite pegmatite of Mt Ploskaya, the Keivy alkaline province (Figure 1), where it forms radiated aggregates of acicular crystals. Hingganite-(Y) is colorless or has white color.

The intragranite pegmatites in the Keivy alkaline province were the second occurrence of this mineral [85]. Hingganite-(Y) here is represented by sectoral prismatic (Figure 12c) and pseudo-rhombohedral crystals and by anhedral grains. Hingganite-(Y) from intragranite pegmatite has significantly higher Fe content comparing to that from amazonite pegmatites (Table 14). Its high content of CaO (up to 5 wt %) allows suggest the possible transition to hypothetic "calciohingganite", by analogy with the series "gadolinite"—synthetic calciogadolinite. *REE* distribution in hingganite-(Y) from intragranite pegmatites is different from those in amazonite pegmatites and characterized by higher LREE, lower HREE, and weak positive slope of chondrite-normalized values ($LREE_n/HREE_n$ = 0.7). Also, hingganite-(Y) from intragranite pegmatites shows the larger cell parameters (Å): a = 10.05, b = 7.72, c = 4.76. However, the link between cell parameters and content of Y, REE, Ca, Fe is not well established [85].

Table 14. Chemical composition of hingganite-(Y).

Sample *	1, EMP	2, EMP
Component	wt %	
ThO_2	0.60	0.48
Al_2O_3	0.16	-
Y_2O_3	14.97	26.91
La_2O_3	0.97	-
Ce_2O_3	4.75	-
Pr_2O_3	0.59	-
Nd_2O_3	3.78	-
Sm_2O_3	1.58	-
Gd_2O_3	2.89	0.53
Tb_2O_3	0.51	0.33
Dy_2O_3	4.04	2.39
Ho_2O_3	0.74	0.21
Er_2O_3	1.83	4.42
Tm_2O_3	0.34	1.38
Yb_2O_3	0.73	15.31
Lu_2O_3	-	1.69
SiO_2	26.79	27.59
BeO	11.15_{calc}	11.48_{calc}
FeO	5.14 **	0.73
MnO	0.22	-
CaO	4.98	2.42
Na_2O	0.21	-
Total	86.97	95.87

* Sample: 1—intragranite pegmatite, Keivy [85], 2—amazonite pegmatite, Keivy [83]. ** all Fe as Fe^{2+}. BeO_{calc} from the atomic ratio Si:Be = 1:1.

Hingganite-(Yb) BeYb(SiO₄)(OH) was described [45] as a new mineral in amazonite pegmatites of the Keivy alkaline province (Figure 1). The mineral forms radiated aggregates made up by tiny needles of 0.1–0.2 mm in width, covering on "plumbomicrolite" crystals or growing in interstices of fluorite. The hingganite-(Yb) has lower Ca content (Table 15) and nearly complete absence of Fe comparatively to composition of hingganite-(Y) from amazonite pegmatites. Both minerals are characterized by different *REE* patterns. It is suggested that hingganite-(Yb) is one of the latest minerals in amazonite pegmatite and crystallized due to the hydrothermal/metasomatical alteration of gadolinite-(Y) combined with selective enrichment of solutions by ytterbium [45]. A possible new mineral species, an ytterbium sorosilicate with thortveitite-like structure is associated with hingganite-(Yb).

Leifite Na₇Be₂(Si₁₅Al₃)O₃₉(F,OH)₂ was firstly noted at the Kola Peninsula in [92], although it was misidentified as a new mineral and described under the name "karpinskiite". The mineral occurs in natrolite-feldspar lens close to the pegmatite "Natrolite Stock", Mt Karnasurt, Lovozero massif (Figure 2). The lens consists mainly of coarse natrolite, albite, potassium feldspar and chabazite. eifite forms white or colorless, thin prismatic to fibrous crystals, assembled in radially-fibrous rosettes (Figure 14). Leifite occurs in cavities and cracks cutting granular albite, and rarely natrolite. Micheelsen and Petersen re-investigated the material and established that "X-ray data on karpinskiite show that it is a mixture of leifite and a zinc-bearing clay of the montmorillonite group". The redefinition of leifite and the discrediting of karpinskiite were approved before publication by the commission on New Minerals and Mineral Names, IMA [93].

Table 15. Chemical composition of hingganite-(Yb).

Sample *	1, EMP
Component	wt %
Y_2O_3	8.56
Gd_2O_3	0.11
Tb_2O_3	0.05
Dy_2O_3	2.47
Ho_2O_3	1.03
Er_2O_3	8.22
Tm_2O_3	3.10
Yb_2O_3	34.07
Lu_2O_3	4.50
SiO_2	22.11
BeO	10.90
CaO	1.14
Total	96.26

* Sample: 1—amazonite pegmatite, Keivy [45].

Figure 14. Radiated aggregate of leifite (**1**) on albite (**2**), with aegirine (**3**), from an aegirine-feldspar lens in sodalite-nepheline syenite, Mt Karnasurt, Lovozero.

Leifite was subsequently discovered in other Lovozero pegmatites, in particular [94] and in the "Shomiokitovoe" and "Nastrofitovoe" pegmatites [32].

Leifite was found in the Khibiny massif (Figure 2) in pegmatite veins, up to 2 m thick, cutting gneissose foyaite in the Mt Eveslogchorr [20,31]. Detailed studies demonstrated that the mineral occurs in the cavities of rose albite, forming white prismatic crystals (1–2 mm long) and spherules, up to 6 mm in diameter, associated with epididymite [47]. Also, flat white spherulites of leifite (up to 4 mm in diameter) are found in analcime-microcline veinlet in urtite of Mt Kukisvumchorr [20].

The mineral contains high tenors of Cs_2O, 0.88 wt %, and Rb_2O, 1.77 wt % (Table 16), and can be regarded as the main concentrator of these elements [71].

Table 16. Chemical composition of leifite.

Sample *	1, EMP	2, WCA	3, WCA	4, EMP
Component	wt %			
SiO_2	69.65	56.68	62.00	64.13
Al_2O_3	8.05	16.40	16.53	10.32
Fe_2O_3	-	0.06	-	-
ZnO	0.71	3.26	0.07	-
MgO	-	0.78	-	-
MnO	-	-	0.41	-
BeO	3.71_{calc}	2.58	-	3.72_{wet}
Na_2O	14.78	9.18	10.50	13.32
K_2O	0.71	1.55	-	1.48
Cs_2O	0.04	-	-	0.88
Rb_2O	0.31	-	-	1.77
H_2O^+	-	5.00	5.50	-
H_2O^-	-	2.50	-	-
F	2.99	-	3.75	3.77_{wet}
$-O = F$	1.26	-	-	-
Total	99.69	97.99 **	100.12	99.39

* Sample: 1–3—Karnasurt, Lovozero [61,72,92], 4—Eveslogchorr, Khibiny [47]. BeO_{calc} from Be = 2 *apfu*. ** in sample 2 [92] the misleading total 99.99.

Leucophanite $NaCaBeSi_2O_6F$ was firstly mentioned by Chernik [38] in Khibiny massif (Figure 2), but it was not properly studied. Full description and study of the mineral from the eudialyte-nepheline-aegirine-microcline vein in gneissose rischorrite at Mt Eveslogchorr, Khibiny massif, was provided by Men'shikov [31]. Leucophanite forms individual platy pale-creamy transparent crystals up to 10 cm in diameter and 5 mm wide (Figure 15a) and the color may be green because of abundant inclusions of aegirine or brown because the occurrence of astrophyllite inclusions. In the central part of vein the leucophanite crystals are among the mass of natrolite, which also contains wadeite, pectolite, rinkite, barylite, shcherbakovite, perlialite, thorite, fluorapatite, galena, löllingiteand safflorite [20].

Later, leucophanite was found as a common mineral in many hydrothermal veins, located in nepheline syenites and foidolites of Khibiny massif [47]. The mineral from pectolite-natrolite veins in ijolite-urtite and foyaite of Mts Kukisvumchorr and Yuksporr occurs in cavities of central part of vein and forms individual tabular crystals of milky-white to light-gray color up to 7 × 8 × 1 cm in size. It is associated with fluorite, strontianite, calcite and sphalerite. Leucophanite crystals become more abundant in the vein margins, but their grain size is smaller.

Leucophanite from the central part of a feldspar-natrolite vein (in the rischorrite of Mt Rasvumchorr) forms platy milky-white crystals up to 6 cm in diameter among columnar aggregates of natrolite, in association with titanite, cafetite, lorenzenite, mosandrite-(Ce), astrophyllite, priderite and shcherbakovite [20].

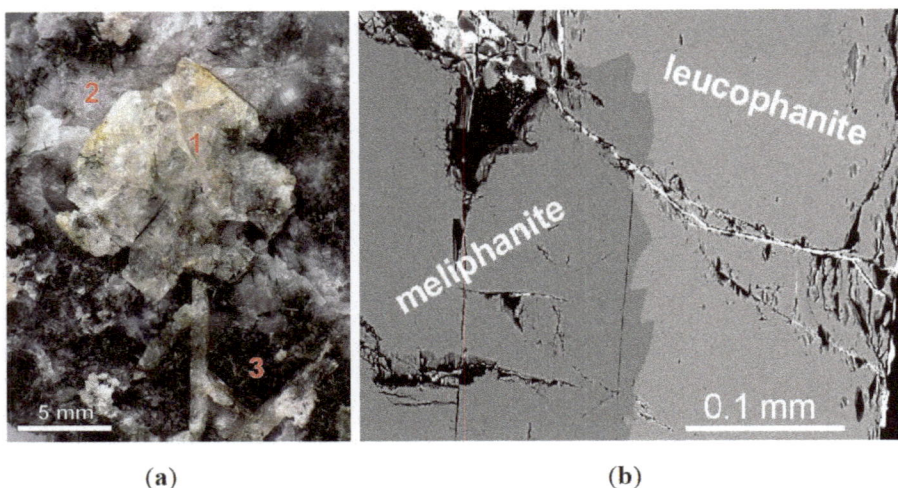

Figure 15. Leucophanite and meliphanite: (**a**) leucophanite crystals (**1**) in natrolite (**2**), with aegirine (**3**), from eudialyte-nepheline-aegirine-microcline vein in rischorrite, Mt Eveslogchorr; (**b**) fragment of porphyroblastic meliphanite crystal with leucophanite zone from biotite metasomatite, Sakharjok. SEM, BSE image.

Leucophanite is also found in differentiated pegmatites with agpaitic mineralization in foidolites of Mt Koashva, Khibiny massif [95]. The largest crystals are of 5 cm and 0.5 cm thickness. The mineral forms tabular semitransparent colorless and light-beige crystals among blocky villiaumite, in association with thermonatrite, aegirine, microcline, lamprophyllite, astrophyllite and lorenzenite. Crystals of loparite-(Ce) are embedded in leucophanite.

Semenov [96] described leucophanite from the natrolite core of large natrolite-feldspar pegmatite in sodalite-nepheline syenite of Mt Lepkhe-Nelm, Lovozero massif (Figure 2). Spherules up to 1 cm of this mineral occur in cavities and cracks in natrolite aggregate together with fluorapatite, fluorite, polylithionite, manganoneptunite and kupletskite. Pink-red UV luminescence is indicated for the mineral.

Leucophanite is established also in biotite metasomatite rimming the large nepheline-syenite pegmatite in Sakharjok massif (Figure 1), Keivy alkaline province [97]. It forms outer zones of poikilitic meliphanite crystals (Figure 15b) with uneven boundary between minerals.

Composition of leucophanite is given in Table 17.

Lovdarite K$_2$Na$_6$Be$_4$Si$_{14}$O$_{36}$·9H$_2$O, a very rare late hydrothermal mineral, was found by Men'shikov [44] in "Yubileinaya" pegmatite, Mt Karnasurt, Lovozero massif (Figure 2). Lovdarite occurs as white and yellowish white rims up to 1–5 cm width on chkalovite crystals and sometimes is replacing them completely. Lovdarite aggregates are massive with no distinguishing features or radial. Druses of small colorless prismatic crystals of lovdarite (1–2 mm in length) are present in the cavities in the aggregates of lovdarite and chkalovite (Figure 16). Ilmajokite, leucosphenite, zorite, mountainite, natrolite, raite, serandite, chkalovite, and some other associate with lovdarite. Composition of lovdarite is given in Table 18.

Table 17. Chemical composition of leucophanite.

Sample *	1, WCA	2, EMP	3, WCA	4, EMP
Component		wt %		
SiO_2	49.69	50.55	45.98	47.85
Al_2O_3	0.55	-	2.32	-
FeO	-	0.04	0.22	-
MgO	0.10	-	0.30	-
MnO	-	-	-	0.03
CaO	20.90	22.26	23.65	22.43
BeO	10.77	10.39_{calc}	11.52	9.96_{calc}
SrO	-	0.39	-	-
Na_2O	13.14	12.40	10.79	12.58
K_2O	0.02	0.01	0.70	-
H_2O	1.29	-	0.83	-
F	5.71	4.04	7.04	7.08
$-O = F$	2.39	1.70	2.96	2.98
Total	99.78	96.53	100.40	96.95

* Sample: 1—Eveslogchorr, Khibiny [31], 2—Koashva, Khibiny [95], 3—Lepkhe-Nelm, Lovozero [96], 4—Sakharjok [97]. BeO_{calc} from the atomic ratio Si:Be = 2:1.

Figure 16. Lovdarite crystals (**1**) in porous pseudomorph after chkalovite, from the aegirine-feldspar-natrolite vein "Yubileinaya", the Karnasurt mine, Lovozero.

Table 18. Chemical composition of lovdarite.

Sample *	1, WCA
Component	wt %
P_2O_5	0.05
SiO_2	56.13
TiO_2	0.15
Al_2O_3	1.77
Fe_2O_3	0.18
MgO	0.06
CaO	0.49
BaO	0.20
BeO	6.90
Na_2O	14.95
K_2O	6.28
H_2O^+	10.85
H_2O^-	1.44
F	0.07
$-O = F$	0.02
Total	99.50

* Sample: 1—Karnasurt, Lovozero [44].

Lovdarite was found also at the dumps of the Karnasurt mine [32] as radial and fibrous aggregates of acicular crystals up to 2 mm in length in the cavities of the selvage of an unknown large pegmatite. The study of the crystal structure [98] showed that lovdarite is some sort of berylosilicate zeolite, and allowed to write the current structural formula of this mineral.

Meliphanite $Ca_4(Na,Ca)_4Be_4AlSi_7O_{24}(F,O)_4$ presence is reliable only in alkaline rocks of Sakharjok massif, Keivy alkaline province (Figure 1). The mineral existence in the West Keivy peralkaline granite massif (Mt Rovgora) and Khibiny massif (https://www.mindat.org) has not been confirmed by publications.

First data on meliphanite (morphology, macroscopic and optical properties, X-ray powder data and chemical composition) from Sakharjok were published in [99]. Meliphanite is widespread in the nepheline syenite pegmatite as platy or tabular crystals up to 2–3 cm in size and up to 0.5 cm in thickness (Figure 17a), also as radiating and randomly oriented aggregates [99,100]. Meliphanite from biotite metasomatite occurs as porphyroblastic crystals and grains up to 5 cm in size (Figure 17b). X-ray investigation of Sakharjok meliphanite by Belkov and Denisov allowed at first time determine the correct type of the unit cell of the mineral as body-centered, not primitive. Recent investigation including infrared and Raman spectroscopy, chemical and electron microprobe analyses (Table 19), thermogravimetric analysis, and the single crystal X-ray diffraction had revealed the presence of OH-groups in meliphanite from Sakharjok and had refined the empirical formula of this mineral, particularly in its anionic part [100]. Meliphanite is the main beryllium mineral in the rocks of Sakharjok alkaline massif. A noteworthy fact is that leucophanite and meliphanite occur as intergrowths [101]. Meliphanite is "parent" for behoite crystallization.

(**a**)   (**b**)

Figure 17. Meliphanite (**1**): (**a**) platy crystals with nepheline (**2**), natrolite (**3**), aegirine (**4**) in alkaline pegmatite, Sakharjok; (**b**) porphyroblastic crystals with natrolite (**3**) and biotite (**5**) in biotite metasomatite, Sakharjok.

Table 19. Chemical composition of meliphanite.

Sample *	1, EMP
Component	wt %
SiO_2	43.30
Al_2O_3	4.95
FeO	0.05
CaO	28.83
BeO	10.23_{calc}
Na_2O	8.55
F	5.41
$-O = F$	2.28
H_2O^+	0.47_{wet}
Total	99.51

* Sample: 1—Sakharjok [100]. BeO_{calc} from the atomic ratio (Si + Al):Be = 2:1.

Milarite $KCa_2(Be_2AlSi_{12})O_{30} \cdot H_2O$ was found in a quartz-albite-microcline pegmatite in the Kolmozero-Voronja belt [102,103]. Milarite together with bavenite and chabazite grow in cavities and fractures on the microcline and albite crystals. Co-crystallization of these minerals indicate that milarite is a late low-temperature hydrothermal mineral. Milarite from its type locality [104] and the Kolmozero-Voronja (Table 20) differ by Be:Al *apfu* ratio—2:1 and 1:1, respectively. The substitution of Be^{2+} by Al^{3+} needs charge balance, for example through the replacement of O^{2-} anions by OH^--groups [103].

Table 20. Chemical composition of milarite.

Sample *	1, WCA
Component	wt %
SiO_2	71.12
Al_2O_3	7.70
CaO	11.55
BeO	3.57
K_2O	4.80
Na_2O	0.30
H_2O^-	0.14
H_2O^+	1.25
Total	100.43

* Sample: 1—Kolmozero-Voronja [103].

Moraesite $Be_2(PO_4)(OH) \cdot 4H_2O$ is the only beryllium phosphate which was found by Pekov [105] in the "Natrolite Stock" pegmatite at Mt Karnasurt, Lovozero massif (Figure 2). Moraesite forms spherical or hemispherical aggregates up to 1 mm in diameter made of radiating acicular crystals up to 0.05 mm in size. Moraesite aggregates grow on the tabular complex twins of epididymite or in cavities in it (Figure 18). Mineral association of moraesite with bertrandite is noted. Other beryllium minerals as leifite and beryllite are also present in this pegmatite. The mineral was detected by XRD and FTIR, and the presence of beryllium is confirmed by spectral analysis.

Figure 18. Moraesite aggregates (**1**) growing on epididymite (**2**) and albite (**3**), from the "Natrolite Stock" pegmatite, Mt Karnasurt, Lovozero.

Odintsovite $K_2Na_4Ca_3Ti_2Be_4Si_{12}O_{38}$ was found in the pegmatite vein in a drilling core at Mt Koashva, Khibiny massif (Figure 2). Detailed investigation of this mineral was carried out by Khomyakov and Pekov [95]. Odintsovite has grayish-pink color and occurs as granular aggregates up to 4 cm in size in which it is associated with K-feldspar, sodalite, aegirine, fluorapatite and lamprophyllite. This is the third odintsovite location in the world after Murun alkaline massif in East Siberia (type locality) [106] and Ilimaussaq massif, South West Greenland [107]. The mineral from Khibiny (Table 21) is similar to odintsovite from Murun massif in composition, X-ray and IR-spectroscopy data, but different by blue-white color fluorescence in short-wave UV light.

Table 21. Chemical composition of odintsovite.

Sample *	1, EMP
Component	wt %
SiO_2	52.49
TiO_2	11.51
Fe_2O_3	0.48
CaO	11.91
BeO	7.28_{calc}
Na_2O	8.35
K_2O	7.36
Total	99.38

* Sample: 1—Koashva, Khibiny [95].

Sphaerobertrandite $Be_3(SiO_4)(OH)_2$. This mineral has a long intricate history. Sphaerobertrandite was first described as a new mineral by Semenov [108], who found the mineral in a few pegmatites at Mts Mannepakhk, Kuftnyun, and Sengischorr in Lovozero massif (Figure 2). In all these pegmatites, sphaerobertrandite occurs as spherulites in cavities of epididymite druses (Figure 19). The formula $Be_5Si_2O_7(OH)_4$ was proposed for sphaerobertrandite. The absence of single crystal X-ray data and/or structural analogies with known phases did not allow the determination of the symmetry or the unit cell dimensions of sphaerobertrandite. Sphaerobertrandite was therefore not included in the system of valid mineral species by the time of foundation of the Commission on New Minerals and Mineral Names of the IMA in 1959. On the other hand, the mineral has never been formally discredited. Later Semenov [61] described one more occurrence of sphaerobertrandite at Mt Lepkhe-Nelm, and gave new chemical analysis of sphaerobertrandite from Mt Kuftnyun. The unique crystal structure of sphaerobertrandite was studied on a crystal from Mt Sengischorr (Table 22), Lovozero massif only in 2003 [57] and an ideal structural formula, $Be_3SiO_4(OH)_2$, was suggested for sphaerobertrandite based on structural data. The mineral is now confirmed and takes its valid place in the mineralogical nomenclature.

"Gelbertrandite" was described simultaneously with sphaerobertrandite [108]. Upon further decrease of the temperature, "gelbertrandite", $Be_4Si_2O_7(OH)_2 \cdot nH_2O(?)$ appears in these pegmatites. The latter mineral is an insufficiently described phase, which has an X-ray powder pattern very similar to bertrandite but with broad and weak lines [108]. It is probably a poorly crystalline, hydrated variety of bertrandite.

Sphaerobertrandite was noted also at Mt Yuksporr, Khibiny massif, but without further description [108].

Figure 19. Sphaerobertrandite spherulites (**1**) on/in epididymite trillings (**2**), from an aegirine-feldspar vein in sodalite-nepheline syenite, Mt Sengischorr, Lovozero.

Table 22. Chemical composition of sphaerobertrandite.

Sample *	1, WCA	2, EMP
Component	wt %	
Al_2O_3	1.40	-
Fe_2O_3	0.07	-
SiO_2	41.03	38.46
BeO	45.20	45.88_{wet}
H_2O^+	12.00	12.54
Total	99.70	96.88

* Sample: 1—Mannepakhk, Lovozero [108], 2—Sengischorr, Lovozero [57].

Tugtupite $Na_4AlBeSi_4O_{12}Cl$ is found in pegmatite veins of Mts Sengischorr and Punkaruaiv, Lovozero massif (Figure 2). Previously the mineral was considered as a new mineral species and published as "beryllosodalite" [43]. The mineral forms rounded and irregular grains up to 0.3 cm colored in blue or pale green, but it tends to be pink. "Beryllosodalite" occurs as a hydrothermal alteration product of chkalovite. Simultaneously with Semenov and Bykova [43] other authors [109] described unnamed beryllium mineral in a pegmatite from the nepheline syenite of Ilimaussaq, South West Greenland. According to Fleischer [110] both papers describe the same mineral. Two years later, analysis of mineral from Ilimaussaq was published and it is very close to that of beryllosodalite of Semenov and Bykova [111]. Nevertheless, the priority of type locality and the name of the mineral are assigned to the mineral from Ilimaussaq, Greenland.

Later the mineral was found in the ussingite zone of the pegmatite body "Shkatulka" at Mt Alluaiv, Lovozero massif. Different morphological types were discovered here: (a) 3–5 mm thick crimson-colored borders around large aggregates of chkalovite; (b) pink round aggregates among ussingite and small spherules in the cavities [47]. Transparent tugtupite with shades of green and blue grows into eudialyte, along with ussingite, terskite, manganoneptunite, lomonosovite, bornemanite, etc. in the pegmatite vein "Sirenevaya" at Mt Alluaiv. In natrolitized proto-ussingite pegmatite "Shkatulka", there are complete pseudomorphs of tugtupite after well-shaped chkalovite crystals (Figure 20a).

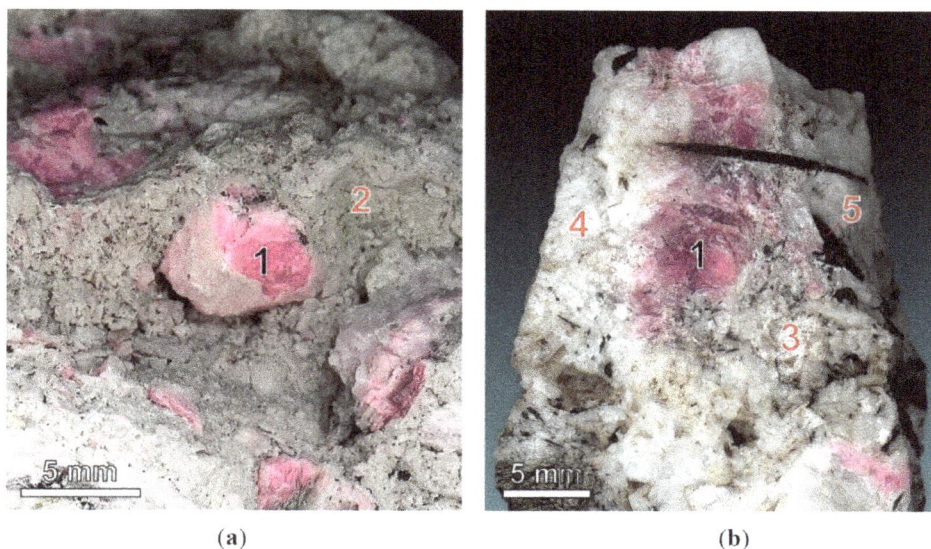

(a)   (b)

Figure 20. Tugtupite (**1**): (**a**) pseudomorphs after chkalovite, on natrolite (**2**), from the aegirine-feldspar-natrolite vein "Yubileinaya", Lovozero; (**b**) irregularly shaped segregation partially replaced by epididymite (**3**) in microcline (**4**), with aegirine (**5**), from a microcline-natrolite vein in foyaite, Mt Eveslogchorr, Khibiny.

A peculiarity of tugtupite from the Lovozero massif is its high content of gallium (up to 0.04 wt % Ga_2O_3 [43]) (Table 23).

Table 23. Chemical composition of tugtupite.

Sample *	1, WCA	2, EMP
Component	wt %	
SiO_2	50.45	51.86
Al_2O_3	12.56	10.31
Ga_2O_3	0.04	-
CaO	0.50	-
BeO	5.30	5.36_{wet}
Na_2O	23.26	25.22
K_2O	0.40	0.09
H_2O^+	1.50	-
H_2O^-	1.51	-
Cl	6.04	7.07
$-O = Cl$	1.40	-
Total	100.26	99.91

* Sample: 1—Sengischorr, Lovozero [43], 2—Eveslogchorr, Khibiny [47].

Tugtupite is found and described in a natrolite-feldspar vein intruding nepheline syenite at Mt Eveslogchorr, Khibiny massif [20,31]. Crimson-colored anhedral crystals of tugtupite up to 1 cm in size are located in feldspar aggregate and associated with lamprophyllite, manganoneptunite, murmanite, belovite-(La), and safflorite. Tugtupite grains are usually partially replaced by epididymite (Figure 20b).

5. Discussion

The distribution of beryllium mineralization on the Kola Peninsula serves a confirmation of chemical and crystal chemical patterns in beryllium behavior [4,6,8,17]: (a) beryllium species are more diverse in younger rocks in comparison with ancient ones; (b) beryllium tends to be accumulated in the differentiated products of alkaline magmatism, saturated or undersaturated; (c) alteration of primary Be minerals may produce complex secondary paragenesis during hydrothermal or weathering stages.

Beryllium is present in ore mainly in the form of its own minerals, and, less frequently, in the form of isomorphic impurity in rock-forming minerals. Thus, about 98% of beryllium from Kolmozerskoe deposit is concentrated in beryl and about 1% is incorporated in mica as isomorphic impurity [51].

It can be seen from the Table 1 that in the Kola Peninsula some minerals were formed in different environments. For example, epididymite occurs in different rocks of alkaline and alkaline- ultramafic complexes, gadolinite-(Y) is connected to different late products of peralkaline granite magmatism, leucophanite, genthelvite and gadolinite-(Ce) are connected with both alkaline and peralkaline granite complexes.

Beryllium mineralization of the Khibiny massif is associated with two different types of alkaline pegmatites: (1) in nepheline syenite; (2) in xenoliths of fenitized basic volcanites in foyaite. Geological environment of beryllium minerals in Lovozero massif is more diverse: (1) pegmatites within alkaline rocks, (2) albitites; (3) contact fenites, (4) feldspar pegmatites in xenoliths of host trapps and gneisses. However, the vast majority of beryllium minerals are related to alkaline pegmatites and related hydrothermalites. In the Keivy alkaline province Be-minerals are present in different late- and postmagmatic products of peralkaline granite magmatism and also in metasomatically altered host rocks associated with the intrusion of peralkaline granites. Beryllium minerals of Kolmozero-Voronja belt are identified only in rare-metal pegmatites.

Kolmozerskoe deposit of Li, Be, Nb and Ta contain beryl as quite common ore mineral of spodumene LCT-type pegmatites like worldwide (e.g., Greenbushes, Wodgina and Mt Cattlin in Australia, Bernic Lake, La Corne and Wekusko in Canada, and North Carolina in USA [25]).

Like a beryl, the Kola meliphanite, leucophanite, chkalovite and epididymite are also found in different types of alkaline rocks in significant amounts.

Meliphanite and leucophanite have very similar crystal structures and both of them are postmagmatic primary minerals, although their associated locations are different, particularly in alkalinity: more structural-ordered and high-calcium meliphanite is associated with alkaline metasomatically altered rock of Sakharjok massif and less ordered leucophanite is associated with alkaline pegmatites of Khibiny and Lovozero. It was considered impossible for them to coexist [101] but unique overgrowths have been discovered [100].

Chkalovite is a characteristic hyper-alkaline mineral which forms closed intergrowths with villiaumite and occurs in the border or core zones of pegmatites in the Khibiny and Lovozero massifs. Large aggregates and grains of chkalovite remain unaltered in relatively "dry" ussingite pegmatites. However, chkalovite fully or partially alters into epididimite, tugtupite or sometimes later minerals in those pegmatites showing a distinctive hydrothermal process.

During the development of pegmatite-hydrothermal process in high-alkaline environment, a change of parageneses is being controlled by the downward trend of a mineral framework density [71]. For example, lovdarite with the highest alkali content and porous zeolite-like crystal structure has been crystallized in ultraagpaitic environment in the late hydrothermal stage of the "Yubileinaya" pegmatite formation.

Gradual decreasing alkalinity during the pegmatite formation leads to densification of framework and polymerization of beryllium tetrahedrons in the crystal structures of sequently crystallizing minerals. On this reason, leifite (Na_2O content of about 13 wt %) or epididymite (Na_2O content of about 12 wt %) were crystallized after chkalovite (Na_2O content of about 29–30 wt %). A sharp decline in alkalinity at the late stage of the hydrothermal process leads to polymerization of beryllium tetrahedrons and Be shifting to the cationic part of a crystal structure. As a result such minerals as sphaerobertrandite, beryllite and moraesite have been formed in many Lovozero pegmatites (i.e., in "Natrolite Stock" pegmatite).

In case of peralkaline granite pegmatite in the Keivy province, calcium, yttrium and lanthanides, become to be involved in beryllium mineral formation. So, gadolinite subgroup minerals are common in peralkaline granite pegmatites, amazonite pegmatites and peralkaline granite veins, especially in exocontact zones of the granite massifs. Conversely, in the Khibiny massif, the minerals of gadolinite subgroup can only be found in pegmatites that cut the mafic rock xenoliths, which is similar in broad sense to crystallization environment of Keivy granites. For the similar reason, chrysoberyl can be found in contact zones of high-Al hornfels with host alkaline rocks.

The unique properties of a number of Be minerals have the potential use in material science. For example, intensive fluorescence and piezoelectric effect in meliphanite allow using it in different fields of ferroelectrics. Leifite and eirikite crystal structures are able to trap the significant amount of large alkali cations, cesium and rubidium. Lovdarite is a unique berylosilicate with zeolite structure.

Author Contributions: Conceptualization, L.M.L.; Writing—Original Draft Preparation, L.M.L., E.A.S., D.R.Z.; Writing—Review & Editing, L.M.L., E.A.S.; Visualization, L.M.L., G.Y.I.

Funding: This research was funded by the Russian Foundation for Basic Research (grant No. 16-05-00427) and Russian Government (grants Nos. 0226-2019-0051, 0226-2019-0053).

Acknowledgments: We are grateful to A. Chernyavsky, P. Goryainov, N. Frishman, N. Zhikhareva and V. Yakovenchuk for permission to use photos of minerals from their collections.

Conflicts of Interest: The authors declare no conflict of interest.

References

1. Vauquelin, N.-L. De l'aigue marine, ou béril; et découverte d'une terre nouvelle dans cette pierre. *Ann. Chim. Phys.* **1798**, *26*, 155–177, (For a Partial Translation, see Anonymous 1930).
2. Anonymous. Discovering the sweet element "A classic of science". *Sci. News Lett.* **1930**, *18*, 346–347. (A Translation of the Essential Parts of Vauquelin 1798).
3. Kogan, B.I.; Kapustinskaya, K.A.; Topunova, G.A. *Berillium*; Nauka: Moscow, Russia, 1975; p. 371. (In Russian)

4. Grew, E.S. Mineralogy, Petrology and Geochemistry of Beryllium: An Introduction and List of Beryllium Minerals. *Rev. Mineral. Geochem.* **2002**, *50*, 1–76. [CrossRef]
5. Pekov, I.V. Remarkable finds of minerals of beryllium: From the Kola Peninsula to Primorie. *World Stones* **1994**, *4*, 10–26.
6. Pekov, I.V. Genetic crystal chemistry of beryllium in derivates of alkaline complexes. In *Proceedings of XXI Workshop "Earth Alkaline Magmatism"*; Arzamastsev, A.A., Ed.; KSC RAS: Apatity, Russia, 2003; pp. 122–123. (In Russian)
7. Aurisicchio, C.; Fioravanti, G.; Grubessi, O.; Zanazzi, P.F. Reappraisal of the crystal chemistry of beryl. *Am. Mineral.* **1988**, *73*, 826–837.
8. Beus, A.A. *Geochemistry of Beryllium and Genetic Types of Beryllium Deposits*; Freeman: San Francisco, CA, USA, 1966; p. 401.
9. Burt, D.M. Minerals of Beryllium. In *Granitic Pegmatites in Science and Industry*; Černý, P., Ed.; Short Course Handbook 8; Mineralogical Association of Canada: Québec, QC, Canada, 1982; pp. 135–148.
10. Everest, D.A. 9. Beryllium. In *Comprehensive Inorganic Chemistry*; Bailar, J.C., Eméléus, H.J., Nyholm, R., Trotman-Dickenson, A.F., Eds.; Pergamon: Oxford, UK; New York, NY, USA, 1973; Volume 1, pp. 531–590.
11. Gaines, R.V. Beryl–A review. *Mineral. Rec.* **1976**, *7*, 211–223.
12. Ginzburg, A.I. *The Genetic Types of Hydrothermal Beryllium Deposits*; Nedra: Moscow, Russia, 1975; p. 247. (In Russian)
13. Ginzburg, A.I.; Kupriyanova, I.I.; Novikova, M.I.; Shazkaya, V.T.; Shpanov, E.P.; Zabolotnaya, N.P.; Vasilkova, N.N.; Zhurkova, Z.A.; Zubkov, L.B.; Leviush, I.T.; et al. *The Mineralogy of Hydrothermal Beryllium Deposits*; Ginzburg, A.I., Ed.; Nedra: Moscow, Russia, 1976; p. 199. (In Russian)
14. Griffith, R.F. Historical Note on Sources and Uses of Beryllium. In *The Metal Beryllium*; White, D.W., Jr., Burke, J.E., Eds.; The American Society for Metals: Cleveland, OH, USA, 1955; pp. 5–13.
15. Hörmann, P.K. Beryllium. In *Handbook of Geochemistry II/1 Elements H (1)–Al (13)*; Wedepohl, K.H., Ed.; Springer: Berlin, Germnay, 1978; pp. 1–6, ISBN 3540065784.
16. Kupriyanova, I.I.; Shpanov, Y.P.; Novikova, M.I.; Zhurkova, Z.A. *Beryllium of Russia: Situation, Problems of Exploration and of Industrial Development of a Raw-Materials Base*; Komitet Rossiyskoy Federatsii po Geologii i Ispol'zovaniyu Nedr, Geologiya, Metody Poiskov, Razvedki i Otsenki Mestorozhdeniy Tverdykh Poleznykh Iskopayemykh; Obzor, AOZT "Geoinformmark": Moscow, Russia, 1996; p. 40. (In Russian)
17. Grew, E.S.; Hazen, R.M. Beryllium mineral evolution. *Am. Miner.* **2014**, *99*, 999–1021. [CrossRef]
18. Ross, M. *Crystal Chemistry of Beryllium*; US Geol Survey Prof Paper 468; U.S. Geological Survey: Washington, DC, USA, 1964; p. 30.
19. Meisser, N.; Widmer, R.; Armbruster, T.; May, E.; Bussy, F.; Ulianov, A.; Michellod, P.M. Verbierite, IMA 2015-089. CNMNC Newsletter No. 30. *Mineral. Mag.* **2016**, *80*, 408.
20. Yakovenchuk, V.N.; Ivanyuk, G.Y.; Pakhomovsky, Y.A.; Men'shikov, Y.P. *Khibiny*; Laplandia Minerals: Apatity, Russia, 2005; p. 468.
21. Zozulya, D.R.; Bayanova, T.B.; Eby, G.N. Geology and Age of the Late Archean Keivy Alkaline Province, Northeastern Baltic Shield. *Geology* **2005**, *113*, 601–608. [CrossRef]
22. Kalashnikov, A.O.; Konopleva, N.G.; Pakhomovsky, Y.A.; Ivanyuk, G.Y. Rare Earth Deposits of the Murmansk Region, Russia—A Review. *Econ. Geol.* **2016**, *111*, 1529–1559. [CrossRef]
23. Černý, P.; Ercit, T.S. The classification of granitic pegmatites revisited. *Can. Mineral.* **2005**, *43*, 2005–2026. [CrossRef]
24. Afanas'ev, B.V.; Bichuk, N.I.; Dain, A.D.; Zhabin, S.V.; Kamenev, E.A. Minerals and raw material base of Murmansk region. *Miner. Resour. Russia* **1997**, *4*, 12–17. (In Russian)
25. Mohr, S.H.; Mudd, G.M.; Giurco, D. Lithium Resources and Production: Critical Assessment and Global Projections. *Minerals* **2012**, *2*, 65–84. [CrossRef]
26. Batieva, I.D.; Belkov, I.V. *Sakharjokskii Peralkaline Massif, Component Rocks and Minerals*; Kol'skii Filial Akademii Nauk SSSR: Apatity, Russia, 1984; p. 133. (In Russian)
27. Zozulya, D.R.; Lyalina, L.M.; Eby, N.; Savchenko, Ye.E. Ore Geochemistry, Zircon Mineralogy and Genesis of the Sakharjok Y–Zr Deposit, Kola Peninsula, Russia. *Geol. Ore Dépos.* **2012**, *54*, 81–98. [CrossRef]
28. Baginski, B.; Zozulya, D.; Macdonald, R.; Kartashov, P.; Dzierzanowski, P. Low-temperature hydrothermal alteration of a rare-metal rich quartz–epidote metasomatite from the El'ozero deposit, Kola Peninsula, Russia. *Eur. J. Mineral.* **2016**, *28*, 789–810. [CrossRef]

29. Batieva, I.D. Alkaline Granite of Kanozero–Kolvitskoe Ozero District. In *Alkaline Granite of Kola Peninsula*; Izdatel'stvo akademii Nauk SSSR: Moscow, Russia, 1958; pp. 146–179. (In Russian)

30. Ivanyuk, G.; Yakovenchuk, V.; Pakhomovsky, Y.; Kalashnikov, A.; Mikhailova, J.; Goryainov, P. Self-Organization of the Khibiny Alkaline Massif (Kola Peninsula, Russia). In *Earth Sciences*; Dar, I.A., Ed.; InTech: Rijeka, Croatia, 2012; pp. 131–156.

31. Kostyleva-Labuntzova, E.E.; Borutzkii, B.E.; Sokolova, M.N.; Shlyukova, Z.V.; Dorfman, M.D; Dudkin, O.B.; Kosyreva, L.V. *Mineralogy of Khibiny Massif (Minerals) V. 2*; Nauka: Moscow, Russia, 1978; p. 586. (In Russian)

32. Pekov, I.V. *Lovozero Massif: History, Pegmatites, Minerals*; Ocean Press: Moscow, Russia, 2001; ISBN 978-5900395272.

33. Ivanyuk, G.Y.; Yakovenchuk, V.N.; Pakhomovsky, Y.A. *Kovdor*; Laplandia Minerals: Apatity, Russia, 2002; p. 326, ISBN 5900395413.

34. Kukharenko, A.A.; Orlova, M.P.; Bulakh, A.G.; Bagdasarov, E.A.; Rimskaya-Korsakova, O.M.; Nefedov, E.I.; Ilyinsky, G.A.; Sergeev, A.S.; Abakumova, N.B. *Caledonian Complex of Ultrabasic, Alkaline Rocks and Carbonatites of the Kola Peninsula and Northern Karelia*; Nedra: Moscow, Russia, 1965; p. 772. (In Russian)

35. Mikhailova, J.A.; Kalashnikov, A.O.; Sokharev, V.A.; Pakhomovsky, Y.A.; Konopleva, N.G.; Yakovenchuk, V.N.; Bazai, A.V.; Goryainov, P.M.; Ivanyuk, G.Y. 3D mineralogical mapping of the Kovdor phoscorite–carbonatite complex (Russia). *Miner. Depos.* **2016**, *51*, 131–149. [CrossRef]

36. Pozhilenko, V.I.; Gavrilenko, B.V.; Zhirov, D.V.; Zhabin, S.V. *Geology of Mineral Areas of the Murmansk Region*; Kola Science Centre RAS: Apatity, Russia, 2002; p. 359. (In Russian)

37. Aver'yanova, I.M. On the Mineralogical Processes of Beryl Change. In *Materialy po Mineralogii Kolskogo Poluostrova. Vyp.2*; Kola Branch of USSR Academy of Sciences: Apatity, Russia, 1962; pp. 140–142. (In Russian)

38. Chernik, G.P. Results of the analysis of some minerals of the Khibiny laccolite, Kola Peninsula. *Gornyi Zhurnal* **1927**, *12*, 740–753. (In Russian)

39. Kostyleva, E.E. Catapleiite from the Khibiny tundra. *Izvestiya Akademii Nauk SSSR* **1932**, *8*, 1109–1125. (In Russian)

40. Kostyleva, E.E. Leucophan, Melinophan. In *Minerals of Khibiny and Lovozero Tundras*; Izdatel'stvo Akademii Nauk SSSR: Moscow, Russia, 1937; p. 243. (In Russian)

41. Gerasimovsky, V.I. Chkalovite. *Dokl. Akad. Nauk SSSR* **1939**, *22*, 263–267. (In Russian)

42. Kuzmenko, M.V. Beryllite–new mineral. *Dokl. Akad. Nauk SSSR* **1954**, *99*, 3–451. (In Russian)

43. Semenov, E.I.; Bykova, A.V. Beryllosodalite. *Dokl. Akad. Nauk SSSR* **1960**, *133*, 64–66. (In Russian)

44. Men'shikov, Yu.P.; Denisov, A.P.; Uspenskaya, E.I.; Lipatova, E.A. Lovdarite–new hydrous beryllosilicate of alkalis. *Dokl. Akad. Nauk SSSR* **1973**, *213*, 429–432. (In Russian)

45. Voloshin, A.V.; Pakhomovsky, Ya.A.; Men'shikov, Yu.P.; Povarennykh, A.S.; Matvienko, E.N.; Yakubovich, O.V. Hingganite-(Yb), a new mineral in amazonitic pegmatites of the Kola Peninsula. *Dokl. Akad. Nauk SSSR* **1983**, *270*, 1188–1192. (In Russian)

46. Men'shikov, Yu.P. Mineral Assemblages of Postmagmatic veins in Eveslogchorr Tectonic Zone of the Khibiny Massif. In *Mineral Assemblages and Minerals of Magmatic Complexes of the Kola Peninsula: Apatity, Russia*; Kola Branch of USSR Academy of Sciences: Apatity, Russia, 1987; pp. 49–54. (In Russian)

47. Men'shikov, Yu.P.; Pakhomovsky, Y.A.; Yakovenchuk, V.N. Beryllium mineralization within veins in Khibiny massif. *Zap. Vseross. Mineral. Obs.* **1999**, *128*, 3–14. (In Russian)

48. Merlino, S.; Biagioni, C.; Bonaccorsi, E.; Chukanov, N.V.; Pekov, I.V.; Krivovichev, S.V.; Armbruster, T. 'Clinobarylite'-barylite: OD relationships and nomenclature. *Mineral. Mag.* **2015**, *79*, 145–155. [CrossRef]

49. Chukanov, N.V.; Pekov, I.V.; Rastsvetaeva, R.K.; Chilov, G.V.; Zadov, A.E. Clinobarylite BaBe$_2$Si$_2$O$_7$–new mineral from Khibiny massif, Kola Peninsula. *Zap. Vser. Miner.Obshch.* **2003**, *132*, 29–37. (In Russian)

50. Krivovichev, S.V.; Yakovenchuk, V.N.; Armbruster, T.; Mikhailova, Y.A.; Pakhomovsky, Y.A. Clinobarylite, BaBe$_2$Si$_2$O$_7$: Structure refinement, and revision of symmetry and physical properties. *Neues Jb Miner. Monat.* **2004**, *8*, 373–384. [CrossRef]

51. Gordienko, V.V. *Mineralogy, Geochemistry and Genesis of Spodumene Pegmatites*; Nedra: Leningrad, Russia, 1970; p. 240. (In Russian)

52. Lussier, A.J.; Hawthorne, F.C. Short-range constraints on chemical and structural variations in bavenite. *Mineral. Mag.* **2011**, *75*, 213–239. [CrossRef]

53. Schaller, W.T.; Fairchild, J.G. Bavenite, a beryllium mineral, pseudomorphous after beryl, from California. *Am. Mineral.* **1932**, *17*, 409–422.

54. Lyalina, L.M.; Savchenko, Ye.E.; Selivanova, E.A.; Zozulya, D.R. Behoite and mimetite from the Saharjok alkaline pluton, Kola Peninsula. *Geol. Ore Depos.* **2010**, *52*, 641–645. [CrossRef]
55. Ehlmann, A.J.; Mitchell, R.S. Behoite, Beta-Be(OH)$_2$, from the Rode Ranch Pegmatite, Llano County, Texas. *Am. Miner.* **1970**, *55*, 1–9.
56. Arzamastsev, A.; Yakovenchuk, V.; Pakhomovsky, Y.; Ivanyuk, G. The Khibina and Lovozero alkaline massifs: Geology and unique mineralization. In Proceedings of the Guidbook for 33rd International Geological Congress Excursion, Apatity, Russia, 22 July–2 August 2008; p. 58.
57. Pekov, I.V.; Chukanov, N.V.; Larsen, A.O.; Merlino, S.; Pasero, M.; Pushcharovsky, D.Yu.; Ivaldi, G.; Zadov, A.E.; Grishin, V.G.; Asheim, A.; et al. Sphaerobertrandite, Be$_3$SiO$_4$(OH)$_2$: New data, crystal structure and genesis. *Eur. J. Miner.* **2003**, *15*, 157–166. [CrossRef]
58. *Chemical Analyzes of Minerals*; Kola Branch of USSR Academy of Sciences: Apatity, Russia, 1970; p. 508. (In Russian)
59. Sosedko, A.F. *Materials on Mineralogy and Geochemistry of Granite Pegmatites*; Gosgeotekhizdat: Moscow, Russia, 1961; 154p. (In Russian)
60. Vlasov, K.A.; Kuzmenko, M.V.; Es'kova, E.M. *Lovozero Alkaline Massif*; Izdatelstvo AN SSSR: Moscow, Russia, 1959; p. 632. (In Russian)
61. Semenov, E.I. *Mineralogy of Lovozero Alkaline Massif*; Nauka: Moscow, Russia, 1972; p. 307. (In Russian)
62. Pyatenko, Yu.A.; Bokii, G.B.; Belov, N.V. X-ray analysis of the crystal structure of chkalovite. *Dokl. Akad. Nauk SSSR* **1956**, *108*, 1077–1080. (In Russian)
63. Simonov, M.A.; Egorov, Y.K.; Belov, N.V. Refined crystal structure of chkalovite Na$_2$Be(Si$_2$O$_6$). *Dokl. Akad. Nauk SSSR* **1975**, *225*, 1319–1322. (In Russian)
64. Pekov, I.V. The Palitra pegmatite. *Mineral. Rec.* **2005**, *36*, 397–416.
65. Bussen, I.V.; Es'kova, E.M.; Men'shikov, Yu.P.; Mer'kov, A.N.; Sakharov, A.S.; Semenov, E.I.; Khomyakov, A.P. The Main Features of High Alkaline Pegmatites and Hydrothermalites. In *Materialy po Mineralogii I Geokhimii Schelochnyx Kompleksov Porod Kol'skogo Poluostrova*; Materials on Mineralogy and Geochemistry of Alkaline Rocks of the Kola Peninsula: Apatity, Russia, 1975; pp. 102–117. (In Russian)
66. Pakhomovsky, Y.A.; Yakovenchuk, V.N.; Ivanyuk, G.Yu.; Men'shikov, Yu.P. New Data on Mineralogy of "Sirenevaya" Pegmatite, Alluaiv Mnt.; Lovozero Masiif. In *Proceedings of II Fersman Scientific Session of Russian Mineralogical Society*; Woytekhovsky, Yu.L., Voloshin, A.V., Dudkin, O.B., Eds.; K & M: Apatity, Russia, 2005; pp. 72–75. (In Russian)
67. Khomyakov, A.P.; Stepanov, V.I. The first finding of the chkalovite in Khibiny and its paragenesis. *Dokl. Akad. Nauk SSSR* **1979**, *248*, 727–730. (In Russian)
68. Tikhonenkova, R.P. Accessory chrysoberyl and monazite in the Khibiny massif. *Dokl. Akad. Nauk SSSR* **1982**, *266*, 1236–1239. (In Russian)
69. Belolipetsky, A.P.; Denisov, A.P.; Kulchitzkaya, E.A. Danalite Finding on the Kola Peninsula. In *Materialy po Mineralogii Kolskogo Poluostrova. Vyp.4*; Nauka: Leningrad, Russia, 1965; pp. 190–194. (In Russian)
70. Kalita, A.P. *Pegmatites sand Hydrothermalites of Alkaline Granites of the Kola Peninsula*; Nedra: Moscow, Russia, 1974; p. 140. (In Russian)
71. Pekov, I.V. Genetic Mineralogy and Crystal Chemistry of Rare Elements in Highly Alkaline Postmagmatic Systems. Ph.D. Thesis, Moscow State University, Moscow, Russia, 2005. (In Russian)
72. Sokolova, E.; Huminicki, D.M.C.; Hawthorne, F.C.; Agakhanov, A.A.; Pautov, L.A.; Grew, E.S. The crystal chemistry of telyushenkoite and leifite, ANa$_6$[Be$_2$Al$_3$Si$_{15}$O$_{39}$F$_2$], A = Cs, Na. *Can. Mineral.* **2002**, *40*, 183–192. [CrossRef]
73. Shilin, L.L.; Semenov, E.I. Beryllium minerals epididymite and eudidymite in alkali pegmatites of the Kola Peninsula. *Dokl. Akad. Nauk SSSR* **1957**, *112*, 325–328. (In Russian)
74. Sorokhtina, N.V.; Voloshin, A.V.; Pakhomovsky, Y.A.; Selivanova, E.A. The First Find of Daqingshanite-(Ce) on Kola Peninsula. In *Proceedings of I Fersman Scientific Session of Russian Mineralogical Society*; K & M: Apatity, Russia, 2004; pp. 34–36. (In Russian)
75. Yakovenchuk, V.N.; Ivanyuk, G.Y.; Pakhomovsky, Y.A.; Selivanova, E.A.; Men'shikov, Yu.P.; Korchak, J.A.; Krivovichev, S.V.; Spiridonova, D.V.; Zalkind, O.A. Punkaruaivite, LiTi$_2$[Si$_4$O$_{11}$(OH)](OH)$_2$·H$_2$O, a new mineral species from hydrothermal assemblages, Khibiny and Lovozero Alkaline Massifs, Kola Peninsula, Russia. *Can. Mineral.* **2010**, *48*, 41–50. [CrossRef]
76. Khomyakov, A.P. *The Mineralogy of the Ultra-Agpaitic Alkaline Rocks*; Nauka: Moscow, Russia, 1990; p. 196.

77. Tikhonenkov, I.P. *Nepheline Syenites and Pegmatites of the Khibiny Massif and the Role of Postmagmatic Phenomena in Their Formation*; Izdatelstvo AN SSSR: Moscow, Russia, 1963; 247p. (In Russian)

78. Fang, J.H.; Robinson, P.D.; Ohya, Y. Redetermination of the crystal structure of eudidymite and its dimorphic relationship to epididymite. *Am. Mineral.* **1972**, *57*, 1345–1354.

79. Gatta, D.G.; Rotiroti, N.; McIntyre, G.J.; Guastoni, A.; Nestola, F. New insights into the crystal chemistry of epididymite and eudidymite from Malosa, Malawi: A single-crystal neutron diffraction study. *Am. Mineral.* **2008**, *93*, 1158–1165. [CrossRef]

80. Pekov, I.V.; Chukanov, N.V.; Yamnova, N.A.; Zadov, A.E.; Tarassoff, P. Gjerdingenite-Na and gjerdingenite-Ca, two new mineral species of the labuntsovite group. *Can. Mineral.* **2007**, *45*, 529–539. [CrossRef]

81. Belkov, I.V. Yttrium Mineralization of Amazonite Pegmatites of Alkaline Granites of the Kola Peninsula. In *Voprosy Geologii I Mineralogii Kolskogo Poluostrova. Geology and Mineralogy of the Kola Peninsula. Vyp. 1*; Izdatelstvo AN SSSR: Moscow, Russia, 1958; pp. 126–139. (In Russian)

82. Lunz, A.Y. About gadolinite nature from amazonite pegmatite. *Zap. Vseross. Mineral. Obs.* **1961**, *90*, 704–709. (In Russian)

83. Voloshin, A.V.; Pakhomovsky, Ya.A.; Sorokhtina, N.V. Study of the composition of gadolinite group minerals from amazonite randpegmatites of the Kola Peninsula. *Vestnik MGTU* **2002**, *5*, 61–70. (In Russian)

84. Belolipetsky, A.P.; Pletneva, N.I.; Denisov, A.P.; Kulchitzkaya, E.A. Accessory Gadolinite from Pegmatites and Granitic Viens on the Kola Peninsula. In *Materialy po Mineralogii Kolskogo Poluostrova. Materials on Mineralogy of the Kola Peninsula. Vyp. 6*; Nauka: Leningrad, Russia, 1968; pp. 162–173. (In Russian)

85. Lyalina, L.M.; Selivanova, E.A.; Savchenko, Y.E.; Zozulya, D.R.; Kadyrova, G.I. Minerals of the gadolinite-(Y)–hingganite-(Y) series in the alkali granite pegmatites of the Kola Peninsula. *Geol. Ore Depos.* **2014**, *56*, 675–684. [CrossRef]

86. Semenov, E.I. *Mineralogy of Rare Earths*; Izdatelstvo AN SSSR: Moscow, Russia, 1963; p. 412. (In Russian)

87. Voloshin, A.V.; Pakhomovsky, Ya.A. *Minerals and Evolution of Mineral Formation in Amazonite Pegmatites of the Kola Peninsula*; Nauka: Leningrad, Russia, 1986; p. 168. (In Russian)

88. Es'kova, E.M. Genthelvite from alkaline pegmatites. *Dokl. Akad. Nauk SSSR* **1957**, *116*, 481–483. (In Russian)

89. Vasil'ev, V.A. About genthelvite. *Zap. Vseross. Mineral. Obs.* **1961**, *90*, 571–578. (In Russian)

90. Lunz, A.Ya.; Sal'dau, E.P. Genthelvite from pegmatite on the Kola Peninsula. *Zap. Vseross. Mineral. Obs.* **1963**, *92*, 81–84. (In Russian)

91. Demartin, F.; Minaglia, A.; Gramaccioli, C.M. Characterization of gadolinite-group minerals using crystallographic data only; the case of higganite-(Y) from Cuasso Al Monte, Italy. *Can. Miner.* **2001**, *39*, 1105–1114. [CrossRef]

92. Shilin, L.L. Karpinskiite, a new mineral. *Dokl. Akad. Nauk SSSR* **1956**, *107*, 737–739. (In Russian)

93. Fleischer, M. New mineral names. *Am. Miner.* **1972**, *57*, 1003–1006.

94. Khomyakov, A.P.; Bykova, A.V.; Kaptsov, V.V. New Data on Lovozero Leifite. In *New Data on Mineralogy of Alkali Formation Deposits*; Trudy IMGRE: Moscow, Russia, 1979; pp. 12–15. (In Russian)

95. Pekov, I.V.; Nikolaev, A.P. Minerals of the pegmatites and hydrothermal assemblages of the Koashva deposit (Khibiny, Kola Peninsula, Russia). *Mineral. Alm.* **2013**, *18*, 7–65.

96. Semenov, E.I. Leucophan in alkali pegmatites of the Kola Peninsula. *Trudy IMGRE* **1957**, *1*, 60–63. (In Russian)

97. Lyalina, L.; Zozulya, D.; Selivanova, E.; Savchenko, Y. Rare beryllium silicates–meliphanite and leucophanite–from nepheline-feldspar pegmatite, Sakharjok massif, Kola Peninsula. In Proceedings of the XXXII International Conference Alkaline Magmatism of the Earth and Related Strategic Metal Deposits, Moscow, Russia, 7–14 August 2015; Kogarko, L.N., Ed.; GEOKHI RAS: Moscow, Russia, 2015; pp. 70–71.

98. Merlino, S. Lovdarite: Structural features and OD character. *Eur. J. Mineral.* **1990**, *2*, 809–817. [CrossRef]

99. Belkov, I.V.; Denisov, A.P. Melinofane from Sakharjok Alkaline Massif. In *Materialy po Mineralogii Kolskogo Poluostrova, Materials on Mineralogy of the Kola Peninsula. Vyp. 6*; Nauka: Leningrad, Russia, 1968; pp. 221–224. (In Russian)

100. Lyalina, L.M.; Kadyrova, G.I.; Selivanova, E.A; Zolotarev, A.A., Jr.; Savchenko, Y.E.; Panikorovskii, T.L. About composition of meliphanite of the nepheline syenite pegmatite of the Saharjok massif (Kola Peninsula). *Zap. Vseross. Mineral. Obs.* **2018**, *2*, 79–91. (In Russian) [CrossRef]

101. Grice, J.D.; Hawthorne, F.C. New data on meliphanite, $Ca_4(Na,Ca)_4Be_4AlSi_7O_{24}(F,O)_4$. *Can. Mineral.* **2002**, *40*, 971–980. [CrossRef]

102. Sosedko, T.A. The find of milarite on the Kola Peninsula. *Dokl. Akad. Nauk SSSR* **1960**, *131*, 643–647. (In Russian)

103. Sosedko, T.A.; Telesheva, R.L. To the chemical composition of milarite. *Dokl. Akad. Nauk SSSR* **1962**, *146*, 437–439. (In Russian)

104. Palache, C. On the presence of beryllium in milarite. *Am. Miner.* **1931**, *16*, 469–470.

105. Pekov, I.V. Moraesite from alkaline pegmatite in Lovozero massif. *Ural'skii Mineral. Sbornik* **1995**, *5*, 256–260. (In Russian)

106. Konev, A.A.; Vorob'ev, E.I.; Sapozhnikov, A.N.; Piskunova, L.F.; Uschapovskaya, Z.F. Odintsovite–K$_2$Na$_4$Ca$_3$Ti$_2$Be$_4$Si$_{12}$O$_{38}$–new mineral (Murun massif). *Zap. Vseross. Mineral. Obs.* **1995**, *5*, 92–96. (In Russian)

107. Petersen, O.V.; Gault, R.A.; Balic-Zunic, T. Odintsovite from the Ilimaussaq alkaline complex, South Greenland. *Neues Jahrb. Mineral. Monatshefte* **2001**, *5*, 235–240.

108. Semenov, E.I. New hydrous beryllium silicates–gelbertrandite and sphaerobertrandite. *Trudy IMGRE* **1957**, *1*, 64–69. (In Russian)

109. Sørensen, H. Beryllium minerals in a pegmatite in the nepheline syenites of Ilimaussaq, South West Greenland. Internatl. Geol. Congress, Rept. 21st Session, Norden. *Am Mineral.* **1960**, *17*, 31–35.

110. Fleischer, M. New Mineral Names. *Am. Miner.* **1961**, *46*, 241–244.

111. Fleischer, M. New Mineral Names. *Am. Miner.* **1963**, *48*, 1178–1184.

minerals

MDPI

Article

Compositional and Textural Variations in Hainite-(Y) and Batievaite-(Y), Two Rinkite-Group Minerals from the Sakharjok Massif, Keivy Alkaline Province, NW Russia

Ekaterina A. Selivanova *, Lyudmila M. Lyalina * and Yevgeny E. Savchenko

Geological Institute, Kola Science Centre, Russian Academy of Sciences, 14 Fersman Str., 184209 Apatity, Russia; evsav@geoksc.apatity.ru
* Correspondence: selivanova@geoksc.apatity.ru (E.A.S.); lialina@geoksc.apatity.ru (L.M.L.);
  Tel.: +7-81555-7-93-33 (E.A.S.)

Received: 10 September 2018; Accepted: 12 October 2018; Published: 16 October 2018

check for updates

Abstract: Compositional and textural variations in the rinkite group, seidozerite supergroup minerals, batievaite-(Y), hainite-(Y) and close to them titanosilicates from the Sakharjok massif were studied. Statistical analysis allowed for defining two major substitution schemes leading to batievaite-(Y) and cation-deficient titanosilicates forming: $Ca^{2+} + Na^+ + F^- \leftrightarrow \square + Y^{3+} + (OH)^-$ and $Ca^{2+} + Na^+ \leftrightarrow \square + REE^{3+}$. Batievaite-(Y) and other cation-deficient titanosilicates are the earlier minerals formed by solid state transformation of the primary full-cation phase. Hainite-(Y) is a later mineral. It forms rims around earlier titanosilicates, or, less often, separate crystals.

Keywords: batievaite-(Y); hainite-(Y); titanosilicate; rinkite group minerals; Kola Peninsula; Sakharjok massif; Keivy alkaline province; transformation mineral species

1. Introduction

Some titanosilicates, such as the rinkite group seidozerite supergroup minerals, are known to concentrate *REE*. Exploring the geochemical and crystal chemical behavior of Y and *REE*, we expand our notions on rare elements and their compounds for material science.

Currently, there are data on two titanosilicates of the rinkite group, the seidozerite supergroup of the Sakharjok massif, hainite-(Y) [1], and batievaite-(Y) [2]. Batievaite-(Y) $Y_2Ca_2Ti[Si_2O_7]_2(OH)_2(H_2O)_4$ can be considered as a Na-deficient Y-analogue of hainite-(Y) [3].

Three-layer *HOH*-blocks are common structural elements of these minerals. They have an inner *O*-layer formed by Ti, Na, Ca, Mn, and other cations (M^O sites, Table 1), with an octahedral coordination, as a rule. There are also two outer heteropolyhedral *H*-layers that host 6- (7-) coordinated cations of Ca, *REE*, Y, Zr (M^H sites, Table 1) linked to Si_2O_7 groups. The Ti (+Nb + Zr):Si_2O_7 = 1:2 is stoichiometric ratio for the rinkite-group minerals [4].

Table 1. Ideal structural formulae for rinkite group minerals.

Mineral	Ideal Structural Formula						
	$2A^P$	$2M^H$	$4M^O$		$(Si_2O_7)_n$	$2(X^O_M)$	$2(X^O_A)$
Mosandrite-(Ce)	Ca_2	$(CaREE)$	$(H_2O)_2Ca_{0.5}\square_{0.5}$	Ti	$(Si_2O_7)_2$	$(OH)_2$	$(H_2O)_2$
Rinkite-(Ce)	Ca_2	$(CaREE)$	$Na(NaCa)$	Ti	$(Si_2O_7)_2$	(OF)	F_2
Rinkite-(Y)	Ca_2	(CaY)	$Na(NaCa)$	Ti	$(Si_2O_7)_2$	(OF)	F_2
Nacareniobsite-(Ce)	$(Ca, REE)_2$	$(Ca, REE)_2$	Na_3	Nb	$(Si_2O_7)_2$	(OF)	F_2
Seidozerite	Na_2	Zr_2	Na_2Mn	Ti	$(Si_2O_7)_2$	O_2	F_2
Grenmarite	Na_2	Zr_2	Na_2Mn	Zr	$(Si_2O_7)_2$	O_2	F_2
Rosenbuschite	Ca_4	Ca_2Zr_2	Na_2Na_4	TiZr	$(Si_2O_7)_4$	O_2F_2	F_4
Kochite	Ca_2	$MnZr$	Na_3	Ti	$(Si_2O_7)_2$	OF	F_2
Gotzenite	Ca_2	Ca_2	$NaCa_2$	Ti	$(Si_2O_7)_2$	(OF)	F_2
Hainite-(Y)	Ca_2	(CaY)	$Na(NaCa)$	Ti	$(Si_2O_7)_2$	(OF)	F_2
Batievaite-(Y)	Ca_2	Y_2	$(H_2O)_2\square$	Ti	$(Si_2O_7)_2$	$(OH)_2$	$(H_2O)_2$
Fogoite-(Y)	Ca_2	Y_2	Na_3	Ti	$(Si_2O_7)_2$	(OF)	F_2

* Ideal structural formulae are from [5], except for rinkite-(Y) [6].

A number of authors, who studied isomorphism in rinkite group minerals, found a linear negative correlation between Na and Ca. The researchers concluded that the basic isomorphic schemes are related with cations in M^H-A^P sites and M^O sites:

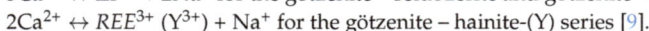

$3Ca^{2+} \leftrightarrow Zr^{4+} + 2Na^+$ for the götzenite – seidozerite and götzenite - kochite series [7,8]

$2Ca^{2+} \leftrightarrow REE^{3+}$ (Y^{3+}) + Na^+ for the götzenite – hainite-(Y) series [9].

Other types of cation substitutions have also been proposed:

(1) $Ca^{2+} + 2F^- \leftrightarrow (Ti,Zr)^{4+} + 2O^{2-}$ as main scheme for götzenite-rosenbuschite from Pian di Celli [10];

(2) $Ca^{2+} + Ti^{4+} \leftrightarrow Na^+ + Nb^{5+}$ for the rinkite-(Ce)–nacareniobsite-(Ce) series from the Ilimaussaq complex [9];

(3) $^M[(H_2O)_2 + \square_{0.5}] + {}^X[(OH)^-_2 + (H_2O)_2] \leftrightarrow {}^M[Na^+_2 + Ca^{2+}_{0.5}] + {}^X[(OF)^{3-} + (F_2)^{2-}]$ the scheme that follows from the ideal mosandrite-(Ce)-rinkite-(Ce) formula from Norway [11];

(4) $Ca^{2+} + Na^+ \leftrightarrow REE^{3+} + \square$ according to factor analysis for rinkite-(Ce)-altered rinkite-(Ce) from the Khibiny massif [12];

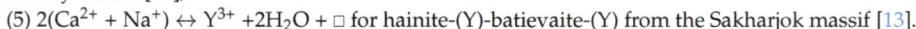

(5) $2(Ca^{2+} + Na^+) \leftrightarrow Y^{3+} + 2H_2O + \square$ for hainite-(Y)-batievaite-(Y) from the Sakharjok massif [13]. In schemes 3–5, the correlation between Na and Ca is positive, unlike the basic pattern.

Reconstructing relations between mineral species becomes a challenge because of the opposite behavior of Na in hainite-(Y) and its involvement in substitutions of the anion part (the full scheme for the hainite-(Y)-batievaite-(Y) series: $Y^{3+} + 4(H_2O) + \square + 2(OH)^- \leftrightarrow 2Ca^{2+} + 2Na^+ + O^{2-} + 3F^-$, according to Table 1).

Since relations between structurally similar hainite-(Y) and batievaite-(Y) are still unclear, the task to analyze compositions of titanosilicates rises. Numerous individuals of those minerals co-existing show that batievaite-(Y), cation-deficient, morphologically, and chemically heterogeneous mineral, occurs in the central part of an individual. Its rims are represented by transparent well-crystallized hainite-(Y). The calcite and analcime aggregates exist of along the boundary between batievaite-(Y) and hainite-(Y) [14] indicates a break in the crystallization of these phases. Also, the paragenesis of analcime and calcite indicates the CO_2 saturation of pegmatitic fluid due to temperature dropping down to 250–100 °C [14].

There are also individuals, where hainite-(Y) rims surround phases with an intermediate chemical composition, which is more or less close to batievaite-(Y) [14]. These compositions were not published in [1] and [2]. According to geochemical indicators—(La/Nd)n and Y/Dy ratios, the batievaite-(Y) crystallized from CO_2-rich fluid with high Ca activity prior to hainite-(Y) that formed from relatively F-rich fluid [14].

But, the question is whether batievaite-(Y) could be formed in the same form as we observe? It looks more like a transformational mineral species that was initially formed as a full-cation phase with the same structural elements, and then it was leached by a hydrothermal fluid, according to the

A.P. Khomyakov principle [15]. We did not identify any sign of precursor mineral to batievaite-(Y). In this paper, we try to reconstruct the titanosilicates origin based on compositional and textural variations in the minerals.

Notably, there is the same problem with the rinkite-(Ce)-mosandrite-(Ce) series. Sokolova and Hawthorne give a cautious remark [11]: "It seems likely that mosandrite is a product of alteration of rinkite, in accord with Slepnev [16], but the rarity of mosandrite and lack of textural context do not allow elucidation of any detail of this process". Truly, though the minerals are quite common for the Khibiny and Lovozero alkaline massifs, they do not coexist. At least, such intergrowths are not thoroughly studied, but just mentioned, which is certainly not enough. It is difficult to reconstruct genetic relations for the rinkite-(Ce)-mosandrite-(Ce) series, since rinkite-(Ce) is poorly crystalline and mosandrite-(Ce) is commonly amorphous.

Some titanosilicates of the seidozerite supergroup, e.g., murmanite and lomonosovite, have similar elements (*HOH*-blocks) in their structures. These minerals have cation-exchange properties [17,18] and they can be easily leached. According to structural analysis, during leaching, cations remove from *O*-layer easier, and Na does so almost completely.

The purpose of this work is to define isomorphic schemes for rinkite group minerals from Sakharjok massif, as well as factors that are responsible for compositional variations and relations between the hainite-(Y) and batievaite-(Y). For that, the authors statistically processed both new and published data on compositions of hainite-(Y) and batievaite-(Y) from the Sakharjok, using Principal Components Analysis (PCA).

2. Materials and Methods

2.1. Occurrence

The Sakharjok is a pint-sized massif (1.5–2 × 8 km) located in the central part of the Kola Peninsula, NW Russia. The main rocks are alkaline syenites, nepheline syenites, and genetically related pegmatoid schlierens and veins. The geology and petrology of the Sakharjok massif are described in details in [19,20]. Alkaline gabbro form minor bodies within the massif. Nepheline syenite affects the alkaline gabbro with the formation of fractures and veins that are filled by a pegmatite material. Pegmatite bodies on the contacts of such contrasting rocks have unusual mineral associations and a complex internal structure with development of aggregates of leucocratic and melanocratic minerals. Three types of mineralization are revealed: rare-earth, beryllium, and chalcophile [7,8,21]. The rinkite group minerals occur in leucocratic (major zeolitic) aggregates.

Minerals of the hainite-(Y)-batievaite-(Y) series occur in nepheline syenite pegmatite, Sakharjok massif as euhedral, subhedral or anhedral separate crystals and touching crystals. The size of individuals is up to 2 mm. In hand specimens, the minerals have a milky-white, brownish, pinkish, creamy color, semi-, and non-transparent. They are colorless or transparent in thin chips. SEM analyses allow identifying heterogeneous structure and intergrowths of different mineral phases in one individual. The study shows that titanosilicates with intermediate composition or/and batievaite-(Y) compose central parts in all mineral intergrowths. Outer rims of intergrowths are formed by hainite-(Y). The contact between batievaite-(Y) and hainite-(Y) is often marked by calcite and analcime aggregate. We divided the titanosilicates into six types with the following morphological features (Figures 1 and 2):

1. *Hainite-(Y)* forms separate prismatic crystals or anhedral grains [1], but more often rims around batievaite-(Y) and/or titanosilicates with intermediate compositions (Figure 1a,c). Hainite-(Y) refers to points 1 on PCA plots (Figure 3).

2. *"Eye-like"* titanosilicates are small isolated areas of rounded, ellipsoidal, or irregular shapes with clear boundaries, can have a porous structure near the boundaries "Eye-like" titanosilicates can be surrounded by hainite-(Y) (Figure 1a) or by a "layered" titanosilicate (Figure 2a,b). Within "eye", there are "heavy" phases (lighter in backscattered electron images, Figure 2a,b). "Eye-like" titanosilicates refer to points 2 on PCA plots (Figure 3).

3. *"Homogenous"* titanosilicates are rounded areas ("coins") with smooth surface that occur as individuals or merge into larger areas. They occur among "loose" or "porous" titanosilicates and look like relics (Figure 2c,d). "Homogenous" titanosilicates refer to points 3 on PCA plots (Figure 3).

4. Mica-like *"layered"* titanosilicates are formed by numerous thin parallel layers (Figure 2f). "Layered" titanosilicates refer to points 4 on PCA plots (Figure 3).

5. *"Loose"* titanosilicates occur as a fractured fragile body, which is difficult to polish (Figure 1c). "Loose" titanosilicates refer to points 5 on PCA plots (Figure 3).

6. *"Porous"* titanosilicates are a heterogenous material with numerous point defects and cracks. The latter can be filled by a "heavy" material marked light in BSE images (Figures 1b and 2e). "Porous" titanosilicates refer to points 6 on PCA plots (Figure 3).

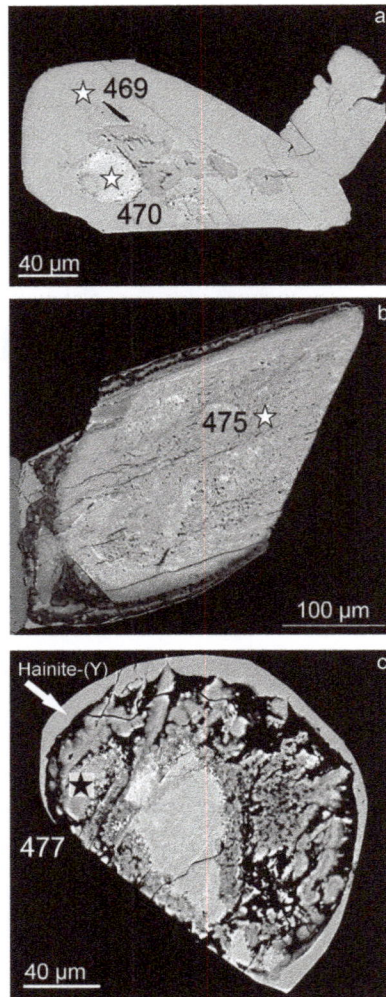

Figure 1. Morphology of titanosilicates from the nepheline syenite pegmatite, Sakharjok massif, Kola Peninsula. Backscattered electron images. Asterisks indicate point of microprobe analyses; analyses numbers correspond to Table 2. (**a**)—"eye-like"titanosilicate (an. 470), hainite-(Y) (an. 469), (**b**)—"porous" (an. 475), (**c**)—"loose" titanosilicate (an. 477).

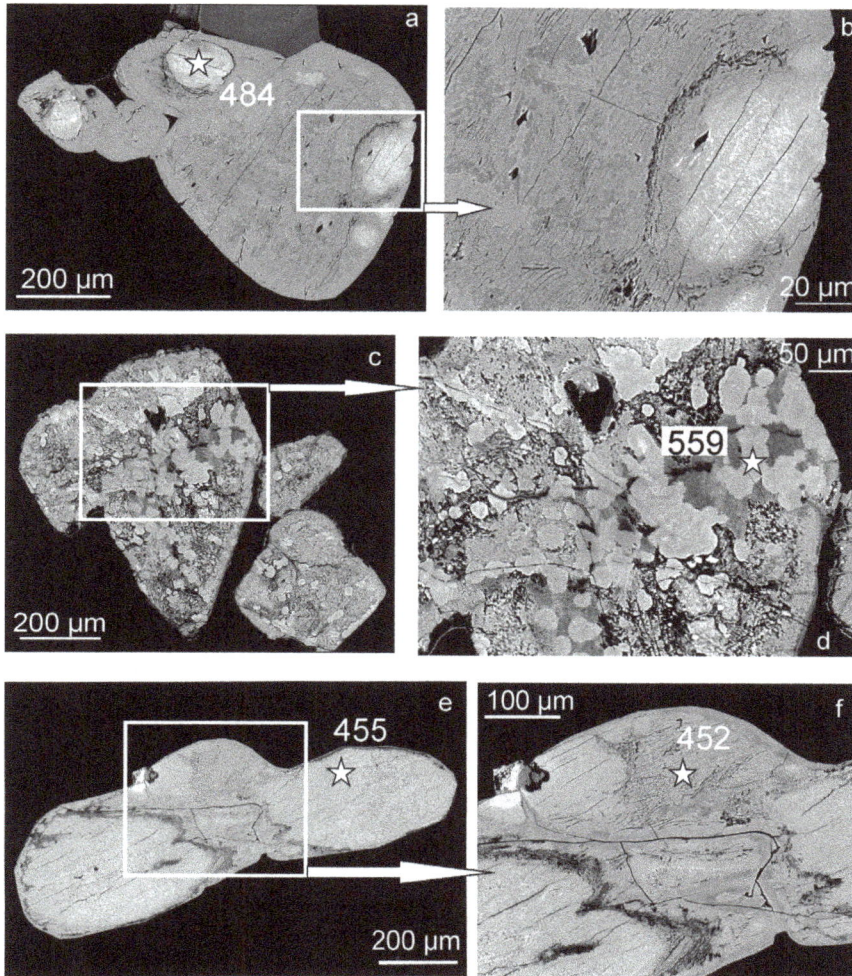

Figure 2. Morphology of titanosilicates from the nepheline syenite pegmatite, Sakharjok massif, Kola Peninsula. SEM, BSE images. Arrows to the right indicate enlarged images of the respective squared areas. Asterisks indicate point of microprobe analyses, analyses numbers correspond to Table 2. (**a**,**b**)—"eye-like" (an. 484), (**c**,**d**)—homogenous" (an. 559), (**e**)—"porous" (an. 455), and (**f**)—"layered" (an. 452), titanosilicates.

Table 2. Chemical compositions (wt %) and formula coefficients (apfu) for different morphological groups of titanosilicates presented in Figures 1 and 2. Variations correspond to 79 analyses of the rinkite group minerals from the nepheline syenite pegmatite, Sakharjok massif, Kola Peninsula.

An. No.*	452	484	455	475	559	470	477	469	Variation
SiO_2	32.39	32.97	28.63	29.46	31.62	30.96	30.54	30.70	24.60–33.03
Al_2O_3	1.13	0.91	2.06	0.90	0.91	0.73	1.98	0.43	0.13–2.30
TiO_2	8.66	9.11	12.43	9.55	5.74	8.13	11.83	7.45	3.80–13.92
ZrO_2	1.48	1.92	1.61	3.04	1.09	2.41	3.60	1.91	0.94–5.18
Nb_2O_5	1.63	1.84	1.60	2.37	1.02	2.18	3.03	1.71	0.35–4.78
P_2O_5	0.00	0.00	0.00	0.00	0.19	0.00	0.00	0.00	0–0.50

Table 2. *Cont.*

An. No.*	452	484	455	475	559	470	477	469	Variation
MnO	0.27	0.33	2.34	1.72	1.19	0.58	0.63	0.15	0.14–8.92
MgO	0.11	0.00	0.27	0.00	0.45	0.00	0.00	0.06	0–0.45
Fe_2O_3	0.19	0.14	0.28	0.08	0.22	0.15	0.21	0.22	0–0.66
CaO	24.64	24.20	21.86	22.72	16.29	26.16	14.67	27.95	12.28–34.73
Na_2O	2.50	1.64	0.74	0.56	0.18	3.19	0.50	6.16	0.16–7.62
K_2O	0.07	0.00	0.04	0.00	0.03	0.02	0.07	0.12	0–0.16
Y_2O_3	10.87	11.07	11.23	12.71	15.35	10.96	10.41	9.46	7.61–18.51
La_2O_3	0.32	0.44	0.60	0.31	1.01	0.17	0.59	0.00	0–1.58
Ce_2O_3	0.28	0.23	0.56	0.47	2.77	0.24	0.64	0.00	0–3.18
Pr_2O_3	0.00	0.00	0.00	0.00	0.21	0.00	0.21	0.00	0–0.57
Nd_2O_3	0.00	0.00	0.12	0.00	0.54	0.14	0.29	0.00	0–1.04
Sm_2O_3	0.00	0.00	0.00	0.00	0.11	0.00	0.00	0.00	0–0.26
Gd_2O_3	0.00	0.00	0.20	0.00	0.39	0.00	0.00	0.00	0–0.40
Tb_2O_3	0.00	0.00	0.00	0.00	0.00	0.00	0.00	0.00	0–0.16
Dy_2O_3	0.57	0.44	0.50	0.38	0.68	0.53	0.49	0.43	0.23–0.93
Ho_2O_3	0.00	0.00	0.00	0.00	0.41	0.00	0.00	0.00	0–0.46
Er_2O_3	0.71	0.90	0.94	0.94	1.27	1.10	1.37	0.94	0.71–1.61
Tm_2O_3	0.11	0.36	0.19	0.34	0.22	0.24	0.35	0.25	0.11–0.46
Yb_2O_3	2.38	2.10	2.00	2.48	2.14	2.94	3.35	2.70	1.83–4.02
Lu_2O_3	0.37	0.52	0.36	0.61	0.27	0.44	0.62	0.39	0.21–0.80
F	3.27	3.46	1.81	1.85	0.00	3.31	1.10	5.95	0–6.80
Cl	0.05	0.00	0.22	0.15	0.00	0.10	0.00	0.00	0–0.35
Sum	92.00	92.58	90.57	90.65	84.27	94.72	86.50	96.90	
apfu based on Si + Al = 4									
Si	3.84	3.87	3.69	3.86	3.87	3.89	3.72	3.94	3.67–3.98
Al	0.16	0.13	0.31	0.14	0.13	0.11	0.28	0.06	0.02–0.33
Ti	0.77	0.80	1.20	0.94	0.53	0.77	1.08	0.72	0.34–1.58
Zr	0.09	0.11	0.10	0.19	0.07	0.15	0.21	0.12	0.06–0.37
Nb	0.09	0.10	0.09	0.14	0.06	0.12	0.17	0.10	0.02–0.31
P	0.00	0.00	0.00	0.00	0.02	0.00	0.00	0.00	0–0.05
Mn	0.03	0.03	0.25	0.19	0.12	0.06	0.06	0.02	0.01–1.07
Mg	0.02	0.00	0.05	0.00	0.08	0.00	0.00	0.01	0–0.08
Fe	0.02	0.01	0.03	0.01	0.02	0.01	0.02	0.02	0–0.07
Ca	3.13	3.05	3.02	3.19	2.13	3.52	1.91	3.84	1.54–4.75
Na	0.57	0.37	0.18	0.14	0.04	0.78	0.12	1.53	0.04–1.86
K	0.01	0.00	0.01	0.00	0.00	0.01	0.02	0.01	0–0.02
Y	0.69	0.69	0.77	0.89	1.00	0.73	0.67	0.65	0.57–1.21
La	0.01	0.03	0.03	0.02	0.05	0.01	0.03	0.00	0–0.08
Ce	0.01	0.03	0.03	0.02	0.12	0.01	0.03	0.00	0–0.15
Pr	0.00	0.00	0.00	0.00	0.01	0.00	0.01	0.00	0–0.03
Nd	0.00	0.01	0.01	0.00	0.02	0.01	0.01	0.00	0–0.05
Sm	0.00	0.00	0.00	0.00	0.00	0.00	0.00	0.00	0–0.01
Gd	0.00	0.01	0.01	0.00	0.02	0.00	0.00	0.00	0–0.02
Tb	0.00	0.00	0.00	0.00	0.00	0.00	0.00	0.00	0–0.01
Dy	0.02	0.02	0.02	0.02	0.03	0.02	0.02	0.02	0.01–0.04
Ho	0.00	0.00	0.00	0.00	0.02	0.00	0.00	0.00	0–0.02
Er	0.03	0.04	0.04	0.04	0.05	0.04	0.05	0.04	0.03–0.07
Tm	0.00	0.01	0.01	0.01	0.01	0.01	0.01	0.01	0.00–0.02
Yb	0.09	0.08	0.08	0.10	0.08	0.11	0.12	0.11	0.07–0.19
Lu	0.01	0.01	0.01	0.02	0.01	0.02	0.02	0.02	0.01–0.03
Sum cations	9.59	9.40	9.94	9.92	8.49	10.38	8.56	11.22	
F	1.23	1.29	0.74	0.77	0.00	1.32	0.42	2.41	0–2.73
Cl	0.01	0.00	0.05	0.03	0.00	0.02	0.00	0.00	0–0.08

Note. Fe_{total} as Fe_2O_3. * **No.** an. Refers to Figures 1 and 2: 452, 484—"layered", 455, 475—"porous", 559—"homogenous", 470—"eye-like", 477—"loose" titanosilicates, 469—hainite-(Y).

2.2. Chemical Composition

Data on 79 analyses of rinkite group minerals were studied. Chemical analyses were carried out by means of a Cameca MS-46 electron microprobe (WDS mode, 22 kV, 20–30 nA, 5–20 μm beam diameter, Cameca, Paris, France). The following standards (and analytical lines) were used: wollastonite (Si$K\alpha$, Ca$K\alpha$), $Y_3Al_5O_{12}$ (Al$K\alpha$, Y$L\alpha$), lorenzenite (Ti$K\alpha$, Na$K\alpha$), $ZrSiO_4$ (Zr$L\alpha$), Nb (Nb$K\alpha$), $MnCO_3$ (Mn$K\alpha$), forsterite (Mg$K\alpha$), hematite (Fe$K\alpha$), wadeite (K$K\alpha$), $LaCeS_2$ (La$L\alpha$), CeS (Ce$L\alpha$), $LiNd(MoO_4)_2$ (Nd$L\alpha$), GdS (Gd$L\alpha$), $Dy_3Al_5O_{12}$ (Dy$L\alpha$), $ErPO_4$ (Er$L\alpha$), $Tm_3Al_5O_{12}$ (Tm$L\alpha$), $Yb_3Al_5O_{12}$ (Yb$L\alpha$), $Y_{2.8}Lu_{0.2}Al_5O_{12}$ (Lu$L\alpha$), atacamite (Cl$K\alpha$), and fluorapatite (P$K\alpha$). The fluorine content was determined using a SEM LEO-1450 (Carl Zeiss AG, Oberkochen, Germany) equipped with an EDS XFlash-5010 Bruker Nano GmbH (Bruker Nano GmbH, Berlin, Germany). The electron microscope was operated at an acceleration voltage of 20 kV, and beam current of 0.5 nA, for an accumulation time of 200 s. Standard-free analysis with the P/B–ZAF method of the QuanTax system was used to analyze F.

Table 2 provides chemical composition of the different textural types of titanosilicates shown on the Figures 1 and 2, variations of components and their formula coefficients (apfu) calculated based on Si + Al = 4. These analyzes should not be considered representative, although they have been selected in accordance with Figures 1 and 2. They do not fully characterize the texture type, the full pool of analyses for the type is given in Supplementary Materials Table S1. Published data on batievaite-(Y) [2] correspond to some layered and porous varieties.

According to Table 2, the hainite-(Y) has the highest oxide and cations sums in comparison with other textural types of titanosilicates. Also there are significant differences in the component contents for mineral phases with a structure similar to batievaite-(Y). Numerous components and their significant and multidirectional content variations do not allow to find correlation between the texture group and the chemical composition using the "standard methods", in other words, these correlations are latent. Therefore, the PCA was chosen to detect any correlations.

2.3. PCA

To identify the most significant factors, and, as a consequence, the factor structure, it is most (entirely) justified to use the principal components analysis (PCA). PCA was applied using the STATISTICA 12.0 program for data processing. Formula coefficients were taken as variables. The procedure of orthogonal rotation "varimax" was applied, the Kaiser's rule became the criterion for estimating the number of principal components (factors). Supplementary Materials Table S1 provides the full set of 79 samples used for PCA.

3. Results

Seventy-nine mineral samples of the hainite-(Y)-batievaite-(Y) series were divided into six material types, according to their morphology. Factor Analysis (PCA) of coefficients in their formulae provided the following results. According to the Kaiser criterion, any factor that displays dispersion with an eigenvalue less than 1.00 ($\lambda j < 1.0$) is viewed as trivial and is not retained. These four factors satisfy this criterion (Table 3).

Table 3. Principal Components Analysis (PCA) results of the mineral composition.

Variable	Factor Loadings			
	Factor 1	**Factor 2**	**Factor 3**	**Factor 4**
REE	0.756320	0.476891	0.041060	−0.082863
Y	0.754054	−0.058515	0.394748	0.054088
Na	−0.624196	−0.337260	−0.420135	−0.412386
K	0.029323	0.141516	0.164027	0.802367
Mg	0.822040	−0.288034	0.016398	0.091280
Ca	−0.734975	−0.183786	0.079822	−0.440946
Ti	0.038583	0.595748	0.154522	0.348134

Table 3. *Cont.*

Variable	Factor Loadings			
	Factor 1	Factor 2	Factor 3	Factor 4
Zr	0.007376	0.643992	0.115776	0.387322
P	0.709521	−0.378585	−0.111755	−0.071313
Nb	−0.074589	0.843153	0.008262	−0.076845
Mn	0.145517	0.123806	0.734688	0.044299
Fe	−0.150331	−0.055727	0.037116	−0.700456
F	−0.715492	−0.339113	−0.316716	−0.439336
Cl	−0.012788	0.038252	0.883334	0.069680
Expl.Var	3.813428	2.240130	1.839141	1.999090
Prp.Totl	0.272388	0.160009	0.131367	0.142792

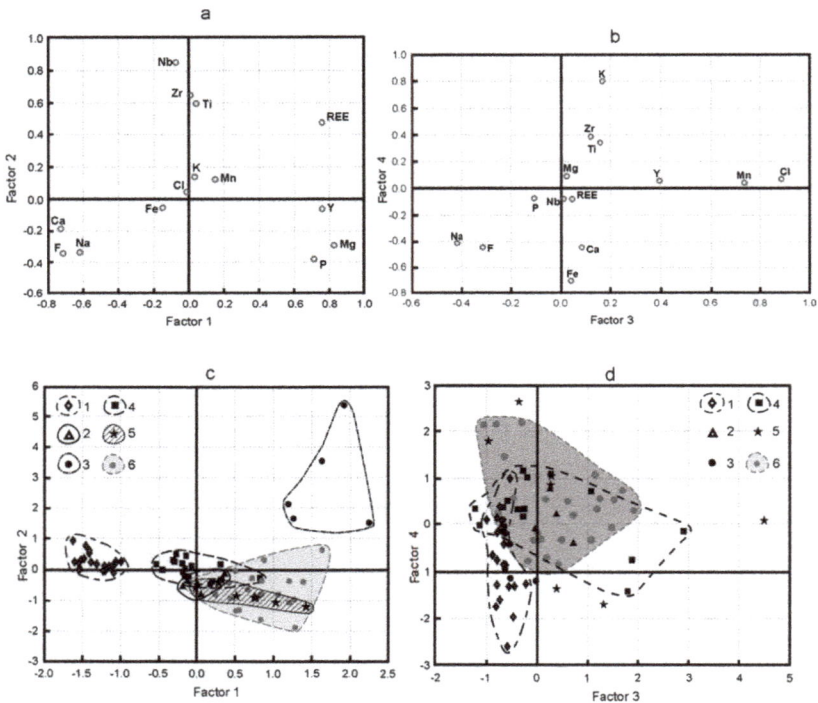

Figure 3. Loadings: Factor 1—Factor 2 (**a**) and Factor 3—Factor 4 (**b**). Scores: Factor 1—Factor 2 (**c**) and Factor 3—Factor 4 (**d**) with marked material types: 1—hainite-(Y), 2—"eye-like", 3—"homogenous", 4—"layered", 5—"loose", 6—"porous".

3.1. Factor F1

Analysis of the factor loading matrix (Table 3, Figure 3a) shows that the first component (factor F1, 27% of dispersion) binds a group of variables (Ca, Na, F) with a close positive correlation. According to scores plot (Figure 3c), F1 clearly differentiates between compositions of hainite-(Y) (points 1) and decationized, cation-deficient samples (points 2–6). In the right part of the plot (Figure 3a, III quadrant), there is the (Y, Mg, P) group that has a negative correlation with the first group of variables. It reflects a relationship between the decationization on the one hand and the increased Y content and admixture elements (Mg, P) present here on the other hand:

(1) $Ca^{2+} + Na^+ + F^- \leftrightarrow \square + Y^{3+} + (OH)^-$ for the main components and $Ca^{2+} + Na^+ + Si^{4+} \leftrightarrow \square + Mg^{2+} + P^{5+}$ or more strictly *(2a)* $Ca^{2+} + 2Na^+ \leftrightarrow \square + 2Mg^{2+}$ or *(2b)* $Ca^{2+} + Na^+ + F^- \leftrightarrow \square + Mg^{2+} + H_2O$ for the admixture elements (Mg).

This trend is best-observed in samples 5 that have loose structure, but neither visible layering, nor porosity (Figure 3c, points 5). "Layered» (points 4) and "porous" (points 6) are divided less clearly, but still distinctly. Thus, the higher the Y, Mg, and P content in a material, the better its destruction is marked.

3.2. Factor F2

Variable *REE* has great loadings both on F1, and on F2 (Figure 3a, II quadrant). It is the opposite of the group (Ca, Na, F) and forms diagonal relationship between the objects. The "homogenous" group is associated with this element on scores plot (Figure 3c, points 3). PCA of a dataset on rinkite-(Ce) (a rare-earth analogue of hainite-(Y)) from the Khibiny massif showed a similar correlation [12]. *REE* are included in the crystal structure according to scheme *(3)*: $Ca^{2+} + Na^+ \leftrightarrow \square + REE^{3+}$. The authors believe that this scheme reflects the processes, partly at least, when rinkite was decationized, as it was altered. Namely, areas with the maximum *REE* content being subject to alteration, just like areas with the high Y content. However, they are not destructed, but preserved as "homogenous" relics. The difference in the way the material changes, i.e., with destruction or without it, is attributed to admixture elements (Mg, P). When the admixture content is high, the mineral is decationized on both schemes *(1)* and *(2a)*, *(2b)* and the material is destructed more intensively. When the admixture content is low, the process follows scheme *(3)* only, with "homogenous" areas being preserved.

Nb associated with Zr and Ti has the greatest loading on Factor F2 (Table 3, Figure 3a). This reflects similar chrystallochemical features of the elements, in the absence of competition in Ti-dominant M^O site (=$M^O(1)$) of the structure. In case of hainite-(Y), the Ti-dominant M^O site can be incompletely filled by these elements. In case of graph points with the high Y and particularly *REE* content, the sum of Ti, Nb and Zr *apfu* can reach 1.5 (non-stoichiometric composition).

3.3. Factor F3

It is difficult to interpret Factor F3 (Table 3, Figure 3b) with great loadings of Mn and Cl. We may see that Mn and Cl are associated with some decationized samples from the groups (points 4–6).

Unlike the (Ca, Na, F) group with a load on F1, (Na, F) group has loads both on F3 and F4 (Figure 3b, IV quadrant), which corresponds to the hainite-(Y) composition on scores plot (Figure 3d). It means that we may clearly associate elements to certain positions in the structure: (Na, F) corresponds with positions in O-layer, and (Ca, Na, F) correspond with all cation sites summarily. Hence, group (Zr, Ti) in II quadrant that is the opposite of (Na, F) explains the location of excessive Ti and Zr. They occur in other sites of O-layer, $M^O(2)$, $M^O(3)$, according to scheme *(4)*: $4Na^+ \leftrightarrow 3\square + Ti^{4+}$. The greater the number of such mixed positions, the more intensively the original material (protophases) alters.

3.4. Factor F4

Factor F4 (Table 3, Figure 3b) provides 14% of dispersion. It follows scheme *(5)* of isomorphic substitutions: $Ca^{2+} + Fe^{3+} \leftrightarrow K^+ + Ti^{4+}(Zr^{4+})$. The scheme works both for hainite-(Y) (points 1 along F4), and for "homogenous" *REE* titanosilicates (points 3) (Figure 3d). It is not confined to leaching.

4. Discussion

Statistic analysis of composition of structurally close minerals enables us to detect any correlations between components, both obvious and latent. These correlations reflected the behavior of main and admixture elements and helped us to reconstruct the mineral genesis.

According to the PCA, the main isomorphism with a negative relation of Na and Ca (see "Introduction") suggested for the rinkite group minerals is insignificant for minerals of the hainite-(Y)-batievaite-(Y) series from Sakharjok. Contents of Y and/or *REE* increase not with the

343

growth of sodium, but, on the contrary, due to its reduction in conjunction with calcium and the appearance (formation) of vacancies according to schemes $Ca^{2+} + Na^+ + F^- \leftrightarrow \square + Y^{3+} + (OH)^-$ and $Ca^{2+} + Na^+ \leftrightarrow \square + REE^{3+}$ (schemes (1) and (3) in Section 3).

Figure 4 shows the temporal relationships between minerals and mineral phases, in addition to the spatial relationships shown in Figures 1 and 2. Statistics data are consistent with geochemical indicators in showing batievaite-(Y) and hainite-(Y) to be minerals of different generations (Figure 4) and batievaite-(Y) to be prior to hainite-(Y).

Batievaite-(Y) and titanosilicates of intermediate compositions: "eye-like", "layered", "porous" or "loose" titanosilicates are likely to be products of postcrystallization alterations of an early full-cation phase (=protophase). The protophase crystallized in conditions of highly active CO_2, while contents Ca, Na, and Y were high enough to produce a full-cation phase and a content of F was low.

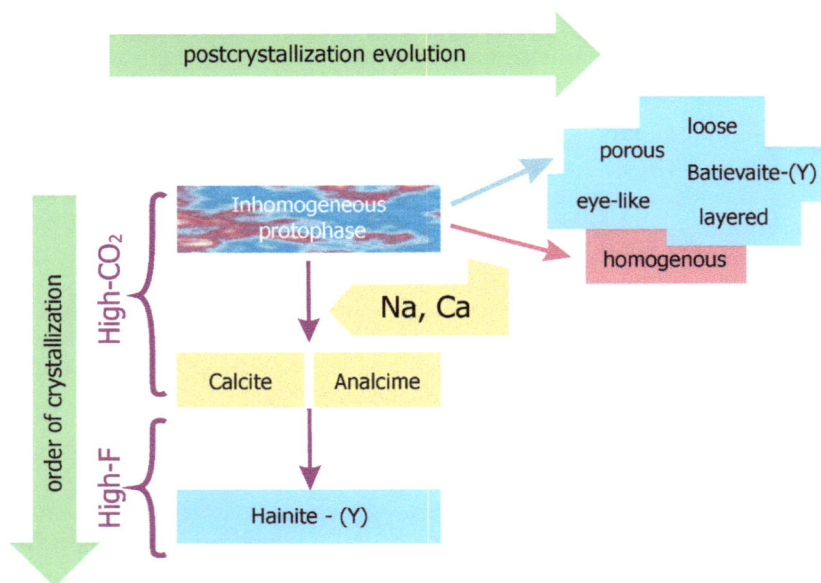

Figure 4. Relationship scheme of the hainite-(Y)-batievaite-(Y) series.

The protophase composition was highly inhomogeneous. It was rich in *REE*, Y, Ti, and Nb that sometimes increased stoichiometric values (1 apfu). For hainite-(Y) as well for batievaite-(Y) apfu Ti + Nb must be 1 in Ti-dominant M^O site (=$M^O(1)$) of the structure. But, for some studied samples, this value is more than 1 (Table 2, an. 455, 477). There were admixture elements (Mg, P) as well. Schemes (2a) and (2b) (Section 3) indicate that magnesium could be present in the cationic positions of the calcium-sodium in protophase in larger content than in hainite-(Y).

It is possible that the inhomogeneity of the protophase could make it unstable to decationization when fluid regimes change. As for Y, we may recall that, e.g., hainite-(Y) from the Sakharjok massif differs from hainite-(Y) from other deposits, not only by its increased Y content, but also by a deficient analysis sum [13].

Postcrystallization alterations in the protophase with various compositional deviations from stoichiometry are related to the different morphology of studied variations of titanosilicates. There are two distinctive main types of alterations; we call them destructive and non-destructive. The leaching of high yttrium areas should be considered as the first type. The products of such destructive alteration are "eye-like", "layered", "porous", or "loose" textural variations of titanosilicates. Due to protophase inhomogeneity different textural variations showed within same mineral grain. PCA of the chemical composition does not allow for distinguishing among these textural variations themselves. Obviously the differences are related to other external, perhaps physical factors. By contrast, *REE*-enriched areas with the high Ti content, "homogenous", retain integrity (non-destructive alteration type). We believe that collective presence of "homogenous" with "loose", as shown in Figure 2c,d also with "porous" or "layered" titanosilicates indicates their transformation during the same postcrystallization events but from different areas of highly inhomogeneous protophase. Excessive Ca and CO_2, which occurred in the crystallization medium of the protophase, as well as Ca and Na, which were separated by postcrystallization, deposited as rims of calcite and analcime (Figure 4) and marked the break in the crystallization of titanosilicates (see "Introduction").

When the fluid regime changes (F content increases, CO_2 content decreases), and, probably, Y (and *REE*) content decreases, the next titanosilicate generation occurs, i.e., hainite-(Y). The mineral overgrowths altered (up to batievaite-(Y)) individuals of the protophase as outer rims. Less often it produces individual crystals.

There is no correlation between the composition and the quantitative ratio of cation-deficient batievaite-like phases to hainite-(Y) in a crystal. The former can be represented either by minor areas ("eye-like"), or by completely altered crystals with a thin hainite-(Y) rim.

Statistic analysis showed that K and Fe admixtures can be present in any minerals of the series. Their content is not related to the fluid regime change and postcrystallization processes. However, only two data sets indicate its position in the structure: hainite-(Y) data and "homogeneous" phase data (Figure 3d). For other titanosilicates, the presence of potassium does not detect any regularity.

Supplementary Materials: The following are available online at www.mdpi.com/2075-163X/8/10/458/s1, Table S1. Chemical composition (wt %), formula coefficients based on Si + Al = 4 for titanosilicates of the rinkite group from nepheline-feldspar syenite of the Sakharjok massif, Kola Peninsula.

Author Contributions: Conceptualization, L.M.L.; Methodology, E.A.S.; Formal Analysis, E.A.S.; Investigation, Y.E.S.; Writing-Original Draft Preparation, L.M.L., E.A.S.; Writing-Review & Editing, L.M.L., E.A.S.; Visualization, L.M.L.

Funding: This research was funded by the Russian Foundation for Basic Research (grant 16-05-00427).

Acknowledgments: The chemical analytical studies were conducted in the Geological Institute, Kola Science Centre, Russian Academy of Sciences. We are grateful to Zozulya D.R. for fruitful discussion and valuable comments.

Conflicts of Interest: The authors declare no conflict of interest.

References

1. Lyalina, L.M.; Zolotarev, A.A., Jr.; Selivanova, E.A.; Savchenko, Y.E.; Zozulya, D.R.; Krivovichev, S.V.; Mikhailova, Y.A. Structural characterization and composition of Y-rich hainite from Sakharjok nepheline syenite pegmatite (Kola Peninsula, Russia). *Miner. Pet.* **2015**, *109*, 443–451. [CrossRef]

2. Lyalina, L.M.; Zolotarev, A.A., Jr.; Selivanova, E.A.; Savchenko, Y.E.; Krivovichev, S.V.; Mikhailova, Y.A.; Kadyrova, G.I.; Zozulya, D.R. Batievaite-(Y), $Y_2Ca_2Ti[Si_2O_7]_2(OH)_2(H_2O)_4$, a new mineral from nepheline syenite pegmatite in the Sakharjok massif, Kola Peninsula, Russia. *Miner. Pet.* **2016**, *110*, 895–904. [CrossRef]

3. Zolotarev, A.; Krivovichev, S.; Lyalina, L.; Selivanova, E. Crystal structure and chemistry of Na-deficient Y-dominant analogue of hainite/götzenite. *Miner. Pet.* **2016**, *329*, 895–904.

4. Sokolova, E. From structure topology to chemical composition: I. Structural hierarchy and stereochemistry in titanium disilicate minerals. *Can. Miner.* **2006**, *44*, 1273–1330. [CrossRef]

5. Sokolova, E.; Camara, F.C. The seidozerite supergroup of TS-block minerals: Nomenclature and classification, with change of the following names: Rinkite to rinkite-(Ce), mosandrite to mosandrite-(Ce), hainite to hainite-(Y) and innelite-1T to innelite-1A. *Miner. Mag.* **2017**, *81*, 1457–1484. [CrossRef]

6. Pautov, L.A.; Agakhanov, A.A.; Karpenko, V.Y.; Uvarova, Y.A.; Sokolova, E.; Hawthorne, F.C. Rinkite-(Y), IMA 2017-043. CNMNC Newsletter No. 39, October 2017, page 1280. *Miner. Mag.* **2017**, *81*, 1279–1286.

7. Cannillo, E.; Mazzi, F.; Rossi, G. Crystal structure of götzenite. *Sov. Phys. Crystallogr.* **1972**, *16*, 1026–1030.

8. Christiansen, C.C.; Johnsen, O.; Makovicky, E. Crystal chemistry of the rosenbuschite group. *Can. Miner.* **2003**, *41*, 1203–1224. [CrossRef]

9. Rønsbo, J.G.; Sørensen, H.; Roda-Robles, E.; Fontan, F.; Monchoux, P. Rinkite-nacareniobsite-(Ce) solid solution series and hainite from the Ilímaussaq alkaline complex: Occurrence and compositional variation. *Bull. Geol. Soc. Den.* **2014**, *62*, 1–15.

10. Sharygin, V.V.; Stoppa, F.; Kolesov, B.A. Zr-Ti disilicates from the Pian di Celle volcano, Umbria, Italy. *Eur. J. Miner.* **1996**, *8*, 1199–1212. [CrossRef]

11. Sokolova, E.; Hawthorne, F.C. From structure topology to chemical composition. XIV. Titaniumsilicates: Refinement of the crystal structure and revision of the chemical formula of mosandrite, $(Ca_3REE)[(H_2O)_2Ca_{0.5\square 0.5}]Ti(Si_2O_7)_2(OH)_2(H_2O)_2$, a Group-I mineral from the Saga mine, Morje, Porsgrunn, Norway. *Miner. Mag.* **2013**, *77*, 2753–2771.

12. Konopleva, N.G.; Ivanyuk, G.Y.; Pakhomovsky, Y.A.; Yakovenchuk, V.N.; Mikhailova, Y.A.; Selivanova, E.A. Typochemistry of Rinkite and Products of Its Alteration in the Khibiny Alkaline Pluton, Kola Peninsula. *Geol. Ore Depos.* **2015**, *57*, 614–625. [CrossRef]

13. Selivanova, E.A.; Lyalina, L.M. *Minerals of rinkite and rosenbuschite groups from Kola Peninsula. Regionalnaya Geologiya, Mineralogiya I Poleznyie Iskopaemyie Kolskogo Poluostrova; Trudy XIII Vserossiyskaya Fersmanovskaya nauchnaya sessiya, Apatity, Russia, 4–5 April 2016;* Voytekhovsky, Y.L., Ed.; K & M: Apatity, Russia, 2016; pp. 97–101. (In Russian)

14. Lyalina, L.M.; Zozulya, D.; Selivanova, E.; Savchenko, Y. Genetic relationship between batievaite-(Y) and hainite-(Y) from Sakharjok nepheline syenite pegmatite, Keivy alkaline province, NW Russia. In Proceedings of the Conference on Accessory Minerals—2017, Vienna, Austria, 13–17 September 2017; pp. 71–72.

15. Khomyakov, A.P.; Yushkin, N.P. The principle of inheritance in crystallogenesis. *Dokl. Akad. Nauk SSSR* **1981**, *256*, 1229–1233.

16. Slepnev, Y.S. The minerals of the rinkite group. *Izv. Akad. Nauk. USSR Ser. Geol.* **1957**, *3*, 63–75.

17. Lykova, I.S.; Pekov, I.V.; Zubkova, N.V.; Chukanov, N.V.; Yapaskurt, V.O.; Chervonnaya, N.A.; Zolotarev, A.A. Crystal chemistry of cation-exchanged forms of epistolite-group minerals, Part, I. Ag- and Cu-exchanged lomonosovite and Ag-exchanged murmanite. *Eur. J. Miner.* **2015**, *27*, 535–549. [CrossRef]

18. Lykova, I.S.; Pekov, I.V.; Zubkova, N.V.; Yapaskurt, V.O.; Chervonnaya, N.A.; Zolotarev, A.A.; Giester, G. Crystal chemistry of cation-exchanged forms of epistolite-group minerals. Part II. Vigrishinite and Zn-exchanged murmanite. *Eur. J. Miner.* **2015**, *27*, 669–682. [CrossRef]

19. Batieva, I.D.; Belkov, I.V. *Sakharjokskii Peralkaline Massif, Component Rocks and Minerals;* Kola Branch of USSR Academy of Sciences: Apatity, Russia, 1984; p. 133. (In Russian)

20. Zozulya, D.R.; Lyalina, L.M.; Eby, N.; Savchenko, Y.E. Ore Geochemistry, Zircon Mineralogy, and Genesis of the Sakharjok Y–Zr Deposit, Kola Peninsula, Russia. *Geol. Ore Depos.* **2012**, *54*, 81–98. [CrossRef]

21. Lyalina, L.M.; Savchenko, Y.E.; Selivanova, E.A.; Zozulya, D.R. Behoite and Mimetite from the Saharjok Alkaline Pluton, Kola Peninsula. *Geol. Ore Depos.* **2010**, *52*, 641–645. [CrossRef]

![minerals logo]

Article

Shkatulkalite, a Rare Mineral from the Lovozero Massif, Kola Peninsula: A Re-Investigation

Andrey A. Zolotarev Jr. [1], Ekaterina A. Selivanova [2], Sergey V. Krivovichev [1,3,*],
Yevgeny E. Savchenko [2], Taras L. Panikorovskii [1,3], Lyudmila M. Lyalina [2], Leonid A. Pautov [4]
and Victor N. Yakovenchuk [2,3]

[1] Department of Crystallography, Institute of Earth Sciences, Saint Petersburg State University,
    University Emb. 7/9, 199034 St. Petersburg, Russia; aazolotarev@gmail.com (A.A.Z.J.);
    rotor_vlg@list.ru (T.L.P.)
[2] Geological Institute, Kola Science Centre, Russian Academy of Sciences, Fersmana 14, 184209 Apatity,
    Russia; selivanova@geoksc.apatity.ru (E.A.S.); evsav@geoksc.apatity.ru (Y.E.S.);
    lyalinalyudmila71@mail.ru (L.M.L.); yakovenchuk@geoksc.apatity.ru (V.N.Y.)
[3] Nanomaterials Research Centre, Kola Science Centre, Russian Academy of Sciences, Fersmana 14,
    184209 Apatity, Russia
[4] Fersman Mineralogical Museum, Russian Academy of Sciences, Leninskiy Av. 18/2, 115162 Moscow, Russia;
    pla58@mail.ru
*   Correspondence: s.krivovichev@spbu.ru or krivovichev@admksc.apatity.ru; Tel.: +7-815-557-53-50

Received: 18 May 2018; Accepted: 12 July 2018; Published: 18 July 2018

check for updates

Abstract: The crystal structure of shkatulkalite has been solved from the crystal from the Lovozero alkaline massif, Kola Peninsula, Russia. The mineral is monoclinic, $P2/m$, $a = 5.4638(19)$, $b = 7.161(3)$, $c = 15.573(6)$ Å, $\beta = 95.750(9)°$, $V = 606.3(4)$ Å^3, $R_1 = 0.080$ for 1551 unique observed reflections. The crystal structure is based upon the HOH blocks consisting of one octahedral (O) sheet sandwiched between two heteropolyhedral (H) sheets. The blocks are parallel to the (001) plane and are separated from each other by the interlayer space occupied by $Na1$ atoms and H$_2$O groups. The $Na2$, $Na3$, and Ti sites are located within the O sheet. The general formula of shkatulkalite can be written as Na$_5$(Nb$_{1-x}$Ti$_x$)$_2$(Ti$_{1-y}$Mn$^{2+}_y$)[Si$_2$O$_7$]$_2$O$_2$(OH)$_2 \cdot n$H$_2$O, where x + y = 0.5 and x ≈ y ≈ 0.25 for the sample studied. Shkatulkalite belongs to the seidozerite supergroup and is a member of the lamprophyllite group. The species most closely related to shkatulkalite are vuonnemite and epistolite. The close structural relations and the reported observations of pseudomorphs of shkatulkalite after vuonnemite suggest that, at least in some environments, shkatulkalite may form as a transformation mineral species.

Keywords: shkatulkalite; titanosilicate; crystal structure; Kola Peninsula; Lovozero alkaline massif; transformation mineral species; vuonnemite; titanium; niobium

---

1. Introduction

Titanosilicates constitute an important group of minerals that have found many applications as materials, including ion-exchange, sorption, catalysis, optics, biocide technologies, etc. [1–5]. Of particular interest are layered titanosilicates of the seidozerite supergroup that currently contains more than forty-five mineral species [6] with several new minerals described very recently [7–13]. These species occur mainly in alkaline massifs such as those in Kola Peninsula, Russia, above the Polar circle. Belov and Organova [14] were the first who considered the crystal chemistry of several minerals of the current murmanite group of seidozerite supergroup [6]. Belov [15] and Pyatenko et al. [16] made further generalizations and called minerals with a "seidozerite block" (=TS block, [17]) and astrophyllite-group minerals titanosilicate analogues of micas. The modular approach to these minerals has been developed by Egorov-Tismenko and Sokolova [18,19], who described a homologous series of Ti-analogues of micas, and Ferraris [20–22], who named those minerals heterophyllosilicates and described them as members of a single polysomatic series. Sokolova [17] quantitatively divided TS-block minerals into four groups based on the content of Ti, structural topology and stereochemistry of the TS block.

Shkatulkalite, $Na_{10}MnTi_3Nb_3(Si_2O_7)_6(OH)_2F \cdot 12H_2O$, was described by Menshikov et al. [23] from the pegmatite "Shkatulka" of the Lovozero alkaline massif, Kola Peninsula, Russia. The mineral was considered to be a Ti–Nb–sorosilicate of the "epistolite group" (now considered a part of the lamprophyllite group [24]). In their review on the seidozerite-supergroup minerals, Sokolova and Cámara [6] pointed out that shkatulkalite is a potential member of the supergroup, but the final assignment of the mineral to a particular group remained unclear, due to the fact that its crystal structure was unknown until now. Menshikov et al. [23] established that the mineral is monoclinic, $a = 5.468(9)$, $b = 7.18(1)$, $c = 31.1(1)$ Å, $\beta = 94.0(2)°$, $V = 1218(8)$ Å^3, $Z = 1$, and commented on the proximity of the a and b parameters of shkatulkalite to those typical for other known Ti and Nb sorosilicates. On the basis of systematic absences, the space groups Pm, $P2$, $P2/m$ were proposed as possible for the mineral. However, due to the poor quality of single-crystal X-ray diffraction data, the structure of the mineral could not be solved at the time. Németh et al. [25] investigated syntactic intergrowths of epistolite, murmanite and shkatulkalite using transmission electron microscopy (TEM) and selected-area electron diffraction (SAED), and reported on the absence of the $l = 2n + 1$ reflections for the latter mineral, pointing out that its c parameter is halved with respect to the value of 31.1 Å reported by Menshikov et al. [23]. Later, shkatulkalite was described in nepheline syenites of the alkaline sill of St. Amable, Quebec, Canada [26], as forming prismatic crystals and radial intergrowth of crystals in small miarolic voids as well as in a hydrothermal cavity in the southeastern part of the Demix quarry. The authors [26] noted that the shkatulkalite is visually indistinguishable from vuonnemite and epistolite found in the same voids. At the same time, Menshikov et al. [23] pointed out that shkatulkalite sometimes forms pseudomorphs after vuonnemite and thus can be considered as a transformation mineral species [27,28], that is, mineral species that forms as a result of a secondary transformation of a primary proto-phase. However, this hypothesis could not be confirmed until the crystal structure of the mineral is solved.

The aim of the present paper is to report the results of crystal-structure determination of shkatulkaite and to re-consider its status as both seidozerite-supergroup mineral and transformation mineral species.

2. Materials and Methods

2.1. Occurrence

The ultra-agpaitic pegmatite body "Shkatulka" located in the western part of the Alluaiv mountain of the Lovozero alkaline massif was discovered by underground excavations in 1990 [29]. Shkatulkalite was found in the marginal zone of the ussingite core of the pegmatite and in the adjacent aegirine zone [23]. The mineral was represented by three morphological varieties: (1) rectangular plates and tabular crystals; (2) aggregates of nacreous mica-like flakes; (3) partial pseudomorphs after vuonnemite. The first variety has the best quality of the material and was used as a holotype sample. In the present work we used material provided by the Museum of the Geological Institute of the Kola Science Center of the Russian Academy of Sciences, sample No. GIM 5968/2-1. It turned out that, in addition to the three types of shkatulkalite identified previously by Menshikov et al. [23], there is also fourth, which is not visually different from vuonnemite and only sometimes has a slightly lighter tone. Some areas of matte from the white or slightly yellowish to cream-colored large vuonnemite plates turned out to be shkatulkalite, and this material proved to be suitable for the single-crystal X-ray diffraction studies. The main difficulty of studying this material was in the preparation of samples for microprobe analysis, due to the poor polishability of its crystals.

2.2. Chemical Composition

The chemical composition of shkatulkalite was studied in three independent laboratories. Initially, the analyzes were performed at the Resource Center "Geomodel" of St. Petersburg State University using the AzTec Energy 350 energy dispersive attachment to the Hitachi S-3400N scanning electron microscope (Hitachi, Tokyo, Japan), operating at 20 kV, 20–30 nA, with a 5 μm beam diameter. The standards used were: lorenzenite (Na), periclase (Mg), diopside (Ca), quartz (Si), microcline (K), barite (Ba), rutile (Ti), rhodonite (Mn), corundum (Al), celestine (Sr), hematite (Fe), and Nb (Nb). Alternatively the chemical composition of shkatulkalite also was determined by the wavelength-dispersive spectrometry on a Cameca MS-46 electron microprobe (Geological Institute, Kola Science Centre, Russian Academy of Sciences, Apatity) operating at 22 kV (for Sr at 30 kV), 20–40 nA. Due to the easy dehydration of the mineral in a vaccum, and its deterioration under an electron beam, the measurements were carried out with a defocused beam up to 20 μm, while manually moving the sample. The standards used were: lorenzenite (NaKα, TiKα), wollastonite (CaKα, SiKα), orthoclase (KKα), synthetic $MnCO_3$ (MnKα), $Y_3Al_5O_{12}$ (AlKα), hematite (FeKα), apatite (PKα) and Nb (NbLα). The fluorine was determined using an Xflash-5010 Bruker Nano Gmbh energy dispersive X-ray spectrometer (Bruker, Bremen, Germany) mounted on a scanning electron microscope LEO-1450 at 20 kV and 0.5 nA. A non-standard procedure for the P/B-ZAF method of Quantax-200 was used. Table 1 provides the summary of analytical results.

Table 1. Chemical composition of shkatulkalite [1].

Component	1		2		3		4	
	Mean	Range	Mean	Range	Mean	Range	Mean	Range
Chemical Composition in wt %								
Na_2O	13.90	13.25–14.30	14.69	13.75–15.39	16.14	15.60–16.84	16.15	14.70–17.57
CaO	0.43	0.28–0.61	0.50	0.43–0.56	0.44	0.40–0.52	0.55	0.49–0.62
SrO	2.07	1.59–2.51	1.03	0.98–1.07	0.46	0.26–0.60	1.01	0.65–1.85
MnO	1.58	1.14–1.90	1.86	1.79–1.91	1.70	1.54–1.81	1.92	1.70–2.28
K_2O	0.27	0.23–0.33	0.18	0.17–0.22	-	-	0.25	0.20–0.41
BaO	1.08	0.82–1.38	n.d.	n.d.	-	-	0.06	0.01–0.29
FeO	n.d.	n.d.	0.08	0.06–0.10	-	-	0.12	0.02–0.19
Fe_2O_3	-	-	-	-	0.07	0.05–0.08	-	-
Al_2O_3	n.d.	n.d.	0.13	0.11–0.16	0.24	0.21–0.28	0.16	0.08–0.62
SiO_2	31.52	31.24–32.07	32.24	31.59–32.60	35.70	35.58–35.84	32.71	31.65–33.74
TiO_2	10.68	10.04–11.41	11.59	11.51–11.64	11.12	11.03–11.23	10.87	10.60–11.19
Nb_2O_5	22.72	21.49–23.92	22.55	21.89–23.05	21.93	21.72–22.22	22.65	21.30–24.14
P_2O_5	n.d.	n.d.	0.32	0.26–0.38	-	-	0.43	0.21–0.98
F	not measured		0.96	0.77–1.17	0.94	-	1.51	0.77–1.83
H_2O	15.75 [2]	-	14.67 [2]	-	11.66 [2]	-	12.25	-
$-O=F_2$	-	-	0.40	-	0.40	-	0.64	-
Total	100.00	-	100.00	-	100.00	-	100.00	-
Chemical Composition in Atoms per Formula Unit (*apfu*)								
Basis	Si = 4		Si + Al = 4		Si + Al + Fe = 4		Si + Al + Fe = 4	
Na	3.42	-	3.52	-	3.47	-	3.80	-
Ca	0.06	-	0.07	-	0.05	-	0.07	-
Sr	0.15	-	0.07	-	0.03	-	0.07	-
Mn	0.17	-	0.19	-	0.16	-	0.20	-
K	0.06	-	0.03	-	-	-	0.04	-
Ba	0.05	-	-	-	-	-	-	-
Fe	-	-	0.01	-	0.01	-	0.01	-
Al	-	-	0.02	-	0.03	-	0.02	-
Si	4.00	-	3.98	-	3.96	-	3.97	-
Ti	1.02	-	1.08	-	0.93	-	0.99	-
Nb	1.30	-	1.26	-	1.10	-	1.24	-
P	-	-	0.03	-	-	-	0.04	-
F	-	-	0.37	-	0.33	-	0.58	-
H	13.34	-	12.09	-	8.64	-	9.91	-

[1] 1–3—this work (1—Hitachi S-3400N; 2—Cameca MS-46, 3—JEOL 733); 4—Menshikov et al. [23]; [2] calculated by difference to 100%.

The chemical composition of shkatulkalite also was studied by the wavelength-dispersive spectrometry on a JEOL Superprobe 733 electron microprobe (Fersman Geological Museum, Russian Academy of Sciences, Moscow) operating at 15 kV and 15 nA. Due to the easy dehydration of the mineral in a vaccum and its deterioration under an electron beam, the measurements were carried out with a defocused beam up to 20 μm. The standards used were: chkalovite (NaKα), diopside (CaKα), microcline (KKα), tephroite (MnKα), almandine (FeKα, SiKα, AlKα), synthetic $AlPO_4$ (PKα), synthetic $BaSO_4$ (BaKα), synthetic $SrTiO_3$ (SrKα, TiKα), fluorphlogopite (FKα), and synthetic $Cs_2Nb_4O_{11}$ for Nb (NbLα). It was observed that shkatulkalite contains rutile inclusions oriented parallel to the (001) plane (Figure 1a). Shkatulkalite shows significant variations of different elements (Figure 1b–f). No elements other than those mentioned above were detected.

The resulting empirical formulae obtained in our study can be written as $(Na_{3.42}Sr_{0.15}Ca_{0.06}K_{0.06}Ba_{0.05})_{3.84}(Nb_{1.30}Ti_{1.02}Mn^{2+}_{0.17})_{2.49}(Si_2O_7)_2O_{0.47}(OH)_2 \cdot 5.67H_2O$ (for sample 1), $(Na_{3.52}Sr_{0.07}Ca_{0.07}K_{0.03})_{3.69}(Nb_{1.26}Ti_{1.08}Mn^{2+}_{0.19})_{2.53}(Si_{1.99}Al_{0.01}O_7)_2[(OH)_{1.63}F_{0.37}]O_{0.44} \cdot 5.23H_2O$ (for sample 2) and $(Na_{3.47}Sr_{0.03}Ca_{0.05})_{3.69}(Nb_{1.10}Ti_{0.93}Mn^{2+}_{0.16})_{2.53}(Si_{1.98}Al_{0.02}O_7)_2[(OH)_{0.80}F_{0.33}]_{1.13} \cdot 3.49H_2O$ (for sample 3). These results are in very general agreement with the results of the crystal-structure study discussed below, taken into account the high instability of the mineral under electron beam.

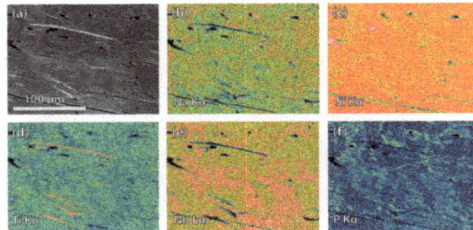

Figure 1. Backscattered electron image of shkatulkalite crystal (**a**) and X-ray distribution maps for Na Kα (**b**); SiKα (**c**); TiKα (**d**); NbLα (**e**); PKα (**f**) radiation.

2.3. Single-Crystal X-ray Diffraction

Single-crystal X-ray diffraction study of shkatulkalite was performed at the Resource Center "X-ray Diffraction Methods" of St. Petersburg State University using Bruker Kappa APEX DUO diffractometer operated at 45 kV and 0.6 mA and equipped with a CCD (charge-coupled device) area detector. The study was done by means of a monochromatic MoKα X-radiation (λ = 0.71073 Å), frame widths of 0.5° in ω and 30 s counting time for each frame. The intensity data were reduced and corrected for Lorentz, polarization and background effects using the Bruker software APEX2 [30]. A semiempirical absorption-correction based upon the intensities of equivalent reflections was applied (SADABS [31]). The unit-cell parameters were refined by least square techniques using 4463 reflections. In general, the unit-cell parameters obtained in this study are in agreement with those reported by Menshikov et al. [23], but with the *c* parameter halved (15.573 instead of 31.1 Å), thus confirming the observations by Németh et al. [25] (see above). The structure was solved and refined in space group *P*2/*m* to $R_1 = 0.080$ ($wR_2 = 0.195$) for 1378 unique observed reflections using ShelX program package [32] within the Olex2 shell [33]. Crystal data, data collection and structure refinement details are given in Table 2; atom coordinates, occupancies and displacement parameters in Tables 3 and 4, selected interatomic distances in Table 5. Occupancies of the cation sites were calculated from the experimental site-scattering factors taking into account empirical formulae. Table 6 provides the results of bond-valence analysis with bond-valence parameters taken from [34].

Table 2. Crystal data and structure refinement for shkatulkalite.

Crystal System	Monoclinic
Space group	$P2/m$
a, Å	5.4638(19)
b, Å	7.161(3)
c, Å	15.573(6)
β, °	95.750(9)
V, Å^3	606.3(4)
Z	1
ρ_{calc}, g/cm^3	2.370
μ, mm^{-1}	1.988
Crystal dimensions, mm	$0.13 \times 0.10 \times 0.05$
$F(000)$	418.0
Radiation	MoKα ($\lambda = 0.71073$)
2Θ range, deg.	2.63–56.00
Index ranges	$-7 \leq h \leq 7, -8 \leq k \leq 9, -17 \leq l \leq 20$
Reflections collected	4463
Independent reflections	1551 [$R_{int} = 0.0550, R_{sigma} = 0.0652$]
Data/restraints/parameters	1551/0/126
GOF (goodness-of-fit)	1.202
Final R indexes [$I \geq 2\sigma(I)$]	$R_1 = 0.0801, wR_2 = 0.1947$
Final R indexes [all data]	$R_1 = 0.0892, wR_2 = 0.1997$
Largest diff. peak/hole/e^-Å^{-3}	1.88/-1.16

Table 3. Atomic coordinates, experimental and calculated site-scattering factors (SSF$_{exp}$ and SSF$_{calc}$, respectively) and isotropic displacement parameters (Å^2) for shkatulkalite.

Atom	x	y	z	Occupancy	SSF$_{exp}$ [e^-]	SSF$_{calc}$ [e^-]	U_{eq}
Nb	0.5901(2)	0	0.30252(7)	Nb$_{0.75}$Ti$_{0.25}$	36.3(7)	36.25	0.0102(4)
Ti	0	1/2	1/2	Ti$_{0.77}$Mn$_{0.20}\square_{0.03}$	22.0(4)	22.00	0.027(1)
Si	0.0947(3)	0.7115(3)	0.3227(1)	Si	14.0(2)	14.00	0.0117(6)
Na1	0.5803(8)	1/2	0.2572(7)	Na$_{0.60}\square_{0.25}$Sr$_{0.07}$Ba$_{0.06}$K$_{0.02}$	13.0(3)	13.00	0.079(4)
Na2	0.5	0.7437(6)	1/2	Na$_{0.87}$Mn$_{0.07}$Ca$_{0.06}$	12.5(2)	12.52	0.019(2)
Na3	0	0	1/2	Na	11.0(3)	11.00	0.018(1)
O1	0.331(1)	0.8064(9)	0.2873(4)	O	8.0(1)	8.00	0.024(1)
O2	0.085(2)	1/2	0.2832(6)	O	8.0(1)	8.00	0.021(2)
O3	0.841(1)	0.8087(9)	0.2841(4)	O	8.0(1)	8.00	0.023(1)
O4	0.1206(9)	0.7024(7)	0.4268(3)	O	8.0(1)	8.00	0.013(1)
O5	0.627(1)	0	0.4167(5)	O	8.0(1)	8.00	0.017(2)
X6	0.684(1)	1/2	0.4282(5)	(OH)$_{0.82}$F$_{0.18}$	8.0(1)	8.18	0.018(2)
O$_w$1	0.542(2)	0	0.1551(7)	(H$_2$O)$_{0.94}$	7.5(3)	7.52	0.040(3)
O$_w$2	0.536(7)	0.410(5)	0.134(2)	(H$_2$O)$_{0.37}$	2.7(2)	2.96	0.09(1)
O$_w$3	0.044(3)	0	0.158(1)	(H$_2$O)$_{0.92}$	7.4(4)	7.36	0.085(8)
O$_w$4	1/2	1/2	0	(H$_2$O)$_{0.19}$	1.5(5)	1.52	0.04(3)
O$_w$5	0.771(8)	1/2	0.001(3)	(H$_2$O)$_{0.24}$	1.9(3)	1.92	0.04(2)
O$_w$6	0.019(5)	0.380(4)	0.058(2)	(H$_2$O)$_{0.14}$	1.1(1)	1.12	0.003(9)
O$_w$7	0	0.259(7)	0	(H$_2$O)$_{0.31}$	2.5(4)	2.48	0.06(2)
O$_w$8	0.65(1)	0	0.001(5)	(H$_2$O)$_{0.25}$	2.0(4)	2.00	0.09(3)

Table 4. Anisotropic displacement atom parameters for shkatulkalite (Å^2).

Atom	U^{11}	U^{22}	U^{33}	U^{23}	U^{13}	U^{12}
Nb	0.0033(5)	0.0060(6)	0.0213(7)	0	0.0004(3)	0
Ti	0.044(2)	0.005(2)	0.038(2)	0	0.029(2)	0
Si	0.0078(9)	0.006(1)	0.021(1)	−0.0003(7)	0.0007(7)	0.0008(7)
*Na*1	0.012(2)	0.006(3)	0.22(1)	0	0.027(3)	0
*Na*2	0.009(2)	0.016(2)	0.032(3)	0	−0.004(1)	0
*Na*3	0.010(3)	0.019(3)	0.024(3)	0	−0.006(2)	0
O1	0.018(3)	0.025(3)	0.032(3)	−0.004(3)	0.008(2)	−0.014(2)
O2	0.029(4)	0.005(4)	0.029(5)	0	0.001(3)	0
O3	0.020(3)	0.022(3)	0.028(3)	0	0.001(2)	0.012(2)
O4	0.012(2)	0.009(3)	0.020(3)	−0.001(2)	0.002(2)	−0.001(2)
O5	0.011(3)	0.016(4)	0.024(4)	0	0.001(3)	0
X6	0.016(4)	0.014(4)	0.025(4)	0	0.006(3)	0
O_w1	0.039(6)	0.048(7)	0.033(6)	0	0.007(4)	0
O_w2	0.11(3)	0.09(3)	0.05(2)	−0.03(2)	0.01(2)	0
O_w3	0.041(9)	0.07(1)	0.15(2)	0	0.02(1)	0

Table 5. Selected bond lengths (Å) in the crystal structure of shkatulkalite.

Nb–O1 [4,18]	1.980(6)	Si1–O1	1.603(6)	Na1–O_w2 [9,18]	2.01(3)
Nb–O_w1	2.29(1)	Si1–O2	1.634(4)	Na1–O3 [9,18]	2.641(7)
Nb–O3 [2,5]	1.979(6)	Si1–O3	1.612(6)	Na1–O1 [7,10]	2.650(7)
Nb–O5	1.770(8)	Si1–O4	1.615(6)	Na1–X6	2.67(1)
<Nb–O>	1.995	<Si–O>	1.616	Na1–O2 [7,18]	2.774(9)
				<Na1–X,O>	2.534
Ti–O4 [8,9,12,18]	1.996(5)	Na2–O4 [6,18]	2.282(5)		
Ti–X6 [4,18]	1.959(8)	Na2–X6 [2,8]	2.353(6)	Na3–O5 [13,18]	2.305(8)
<Ti–O,X>	1.984	Na2–O5 [1,18]	2.389(6)	Na3–O4 [1,2]	2.535(5)
		<Na2–O,X>	2.342	Na3–O4 [5,6]	2.535(5)
				<Na3–O>	2.458

[1] $1 - x, 2 - y, 1 - z;$ [2] $1 + x, +y, +z;$ [3] $1 + x, 1 + y, +z;$ [4] $+x, 2 - y, +z;$ [5] $1 + x, 2 - y, +z;$ [6] $1 - x, +y, 1 - z;$ [7] $-1 + x, +y, +z;$ [8] $-x, 1 - y, 1 - z;$ [9] $+x, 1 - y, +z;$ [10] $-1 + x, 1 - y, +z;$ [11] $-1 + x, -1 + y, +z;$ [12] $-x, +y, 1 - z;$ [13] $2 - x, 2 - y, 1 - z;$ [14] $-1 - x, 1 - y, -z;$ [15] $-x, 1 - y, -z;$ [16] $-x, +y, -z;$ [17] $-1 - x, -y, -z;$ [18] $x, y, z.$

Table 6. Bond-valence analysis (v.u. = valence units) for shkatulkalite.

Atom	Nb	Ti	Si	Na1	Na2	Na3	Total
O1	0.83↓ × 2		1.06	0.10			1.99
O2			0.97→ × 2	0.07, 0.08			2.09
O3	0.83↓ × 2		1.03	0.10↓ × 2			1.96
O4		0.61↓ × 4	1.03		0.27↓ × 2	0.14↓ × 4	2.05
O5	1.47				0.20↓→ × 2	0.26↓ × 2	2.13
X6		0.68↓ × 2		0.10	0.22↓→ × 2		1.22
O1w	0.36						0.36
O2w				0.56↓ × 2			0.56
Total	5.15	3.81	4.09	1.77	1.40	1.06	

3. Results

The crystal structure of shkatulkalite is based upon the *HOH* blocks consisting of one octahedral (*O*) sheet sandwiched between two heteropolyhedral (*H*) sheets (Figure 2a). According to Sokolova [17], the *HOH* blocks are of the type III with one Ti *apfu* present in the *O* sheet (Figure 2b). The blocks are parallel to (001) and are separated from each other (Figure 3a) with interlayer space occupied by *Na*1 atoms and H_2O groups. The *Na*2, *Na*3, and *Ti* sites are located within the *O* sheet. The *H* sheets are formed by NbO_6 octahedra and Si_2O_7 groups sharing common O atoms.

The *Na1* site is coordinated by eight anions (taking into account the disorder observed for the O_w2 site with adjacent O_w2 sites located at 1.318 Å) and is located approximately in the center of a six-membered ring in the *H* layer (Figure 4). The relatively high coordination number (8) and the long average <*Na1*–O> bond length of 2.534 Å indicate the capability of this site to accumulate large cations such as K^+, Sr^{2+} and Ba^{2+} present in shkatulkalite. The structure refinement indicated the presence of rather short $Na1$–O_w2 contact of 2.01 Å, which we explain by the static disorder observed for both $Na1$ and O_w2 sites (note also the high values of their atomic displacement parameters). The *Na2* and *Na3* sites are octahedrally coordinated by six anions each. The $Na2O_6$ octahedron is more compact, with the mean <*Na2*–O> bond length of 2.342 Å, the bond-valence sum (BVS) of 1.40 valence units (v.u.) and the site-scattering factor (SSF) of 12.54 e^- all pointing out that, in addition to Na^+, this site also incorporates Ca^{2+} and Mn^{2+} cations (Table 3). The refinement of the SSF of the *Na3* site is consistent with its full occupancy by Na^+ cations, which is also confirmed by its BVS (1.06 v.u.) and the <*Na3*–O> bond length of 2.458 Å.

The *Nb* and *Ti* sites are both octahedrally coordinated by O atoms. The *Nb* site contains significant amount of Ti (its refined SSF corresponds to the composition $Nb_{0.75}Ti_{0.25}$). The NbO_6 octahedron is essentially distorted with one short (1.770 Å) *Nb*–O5 bond opposite to one long *Nb*–O_w1 (2.29 Å) bond, and four intermediate (~1.98 Å) *Nb*–O bonds. This kind of octahedral distortion is typical for NbO_6 octahedra in *H* sheets and was observed, for instance, in vuonnemite, $Na_{11}TiNb_2(Si_2O_7)_2(PO_4)_2O_3F$ [35,36], and epistolite, $Na_4TiNb_2(Si_2O_7)_2O_2(OH)_2(H_2O)_4$ [37], two minerals most closely related to shkatulkalite (see below). The refined SSF of the *Ti* site is close to 22 e^-, which would account for the full occupancy of this site by Ti. However, this would contradict the predominance of Nb over Ti observed in the chemical analyses (Table 1), which prompted us to suggest that the *Ti* site also accommodates Mn^{2+} cations present in shkatulkalite. It is rather common for divalent cations and, in particular, Mn^{2+} to substitute for Ti in the octahedral sites of the *O* sheet in heterophyllosilicates. Such a substitution was reported, for instance, for sobolevite, $Na_{13}Ca_2Mn_2Ti_3(Si_2O_7)_2(PO_4)_4O_3F_3$ [38,39].

The BVS for the *X6* site in shkatulkalite is 1.22 v.u., which is compatible with its occupancy by $(OH)^-$ or F^- anions. The $(OH)_{0.82}F_{0.18}$ assigned to this site in Table 3 was calculated to conform with the results of the chemical analyses that demonstrate the presence in shkatulkalite of 0.36 F *apfu*.

The interlayer between the adjacent *HOH* blocks in shkatulkalite is occupied by a number of partially occupied O_w sites that belong to H_2O molecules. The O_w1 atom is bonded to the *Nb* site, forming a long apical Nb-H_2O bond in the NbO_6 octahedra. The O_w2 site is linked to the *Na1* site and is split into two sites, with the total occupancy of 68%. The O_w3–O_w8 sites are purely interlayer with the occupancies in the range from 19% to 92%. However, there are several H_2O positions that are mutually excluding (O_w4–O_w5, O_w6–O_w7, etc.). The maximum amount of H_2O in the formula considering all incompatible sites is 10 H_2O per formula unit. Therefore, the amount of H_2O molecules in the ideal formula can be written as *n*, where $n \leq 10$. It seems that the cohesion among different TS-blocks is ensured through hydrogen bonding between the H_2O molecule of the O_w1 site (apical anion of NbO_6 octahedra) and the H_2O molecule of the O_w8 site. However, no precise picture of the hydrogen bonding in shkatulkalite can be derived, owing to the high degree of disorder observed for the interlayer sites.

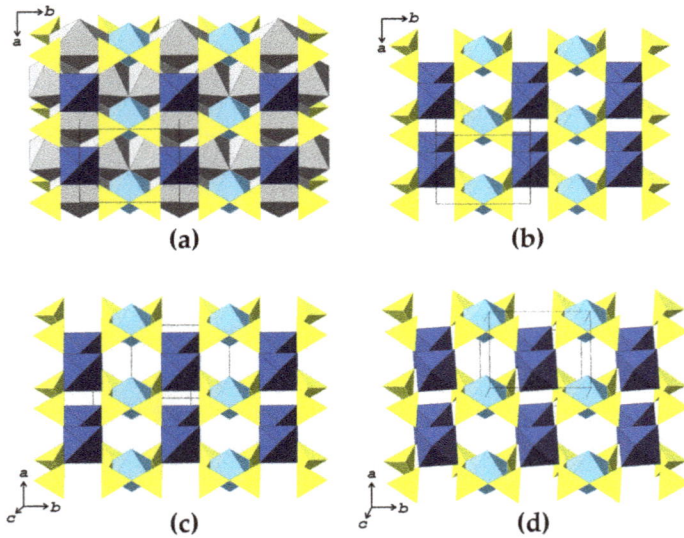

Figure 2. The *HOH* block in shkatulkalite (**a**); and the *HOH* blocks in shkatulkalite (**b**); epistolite (**c**) and vuonnemite (**d**) with the NaO$_6$ octahedra of the *O* sheets omitted for clarity. Legend: Nb, Ti, Na, and Si polyhedra are shown in dark-blue, light-blue, gray, and yellow colors, respectively.

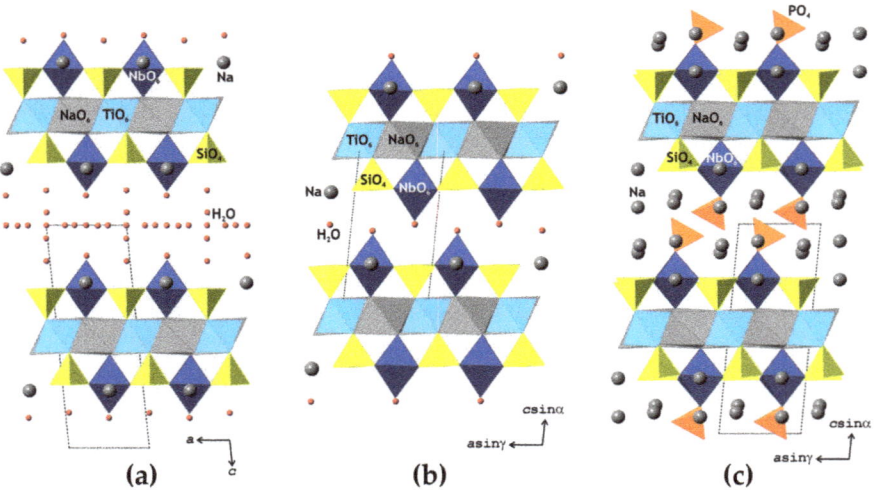

Figure 3. The crystal structures of shkatulkalite (**a**); epistolite (**b**) and vuonnemite (**c**) projected along the *b* axes. Legend: Nb, Ti, P, Na, and Si polyhedra are shown in dark-blue, light-blue, orange, gray and yellow colors, respectively. Interlayer Na and H$_2$O sites are shown as gray and red spheres, respectively.

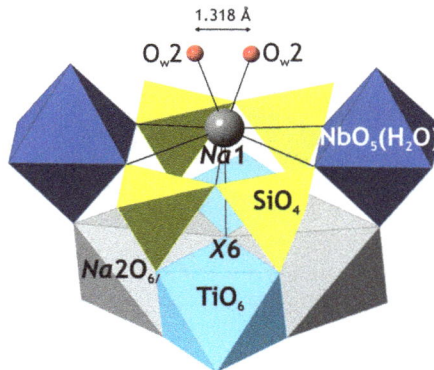

Figure 4. The local coordination environment of the *Na*1 site in the crystal structure of shkatulkalite. Note that the O_w2 sites are partially occupied (site occupation factor = 0.37). Legend as in Figure 3.

According to the crystal-structure refinement, the crystal-chemical formula of shkatulkalite can be written as $[Na_{3.94}\square_{0.50}Sr_{0.14}Mn_{0.14}Ba_{0.12}Ca_{0.12}K_{0.04}][Nb_{1.50}Ti_{1.27}Mn^{2+}_{0.20}\square_{0.03}][Si_2O_7]_2O_2$ $[(OH)_{0.82}F_{0.18}]_2\cdot7.43H_2O$, which is in very general agreement with the empirical chemical formula given above, taking into account the difficulties associated with the study of the mineral by the electron-microprobe analysis. It is noteworthy that, according to the chemical analyses, the H_2O content in shkatulkalite ranges from 4.32 to 6.67 apfu. Thus, by analogy to selivanovaite and murmanite [11,25] the interlayer H_2O content in shkatulkalite is variable and may be defined as nH_2O. Taking into account the cations prevalent at different atomic sites in shkatulkalite, its ideal chemical formula should be written as $Na_5Nb_2Ti[Si_2O_7]_2O_2(OH)_2]\cdot nH_2O$, where $n \leq 10$, which is not electroneutral and contains one extra positive charge. Therefore, we assume that the incorporations of Ti into *Nb* and Mn into *Ti* sites are important for charge-balance considerations, and suggest the general formula should be written as $Na_5(Nb_{1-x}Ti_x)_2(Ti_{1-y}Mn^{2+}_y)[Si_2O_7]_2O_2(OH)_2\cdot nH_2O$, where $x + y = 0.5$, $n \leq 10$. For the sample of shkatulkalite under investigation $x \approx y \approx 0.25$. For the two critical cases of $x = 0.5$ or $y = 0.5$, the end-member formulae would be $Na_5(NbTi)Ti[Si_2O_7]_2O_2(OH)_2\cdot nH_2O$ and $Na_5Nb_2(Ti_{0.5}Mn^{2+}_{0.5})[Si_2O_7]_2O_2(OH)_2\cdot nH_2O$ (or $Na_{10}Nb_4TiMn^{2+}[Si_2O_7]_4O_4(OH)_4\cdot nH_2O$), $n \leq 10$, respectively.

4. Discussion

The structure of shkatulkalite represents the basic structure type B5(GIII), according to Sokolova and Cámara [40] that was inferred by inverse prediction. In fact, Sokolova and Cámara [40] introduced the concept of basic and derivative structures for TS-block minerals and stated that a derivative structure is related to two or more basic structures of the same group. Hence a derivative structure can be built by adding basic structures *via* sharing the central O sheet of the TS blocks of adjacent structural fragments. The inverse prediction of the structure of shkatulkalite to B5(GIII) is now completely confirmed with the present results. Incidentally, the ideal formula of B5(GII) predicted by Sokolova and Cámara [40] is $\square_2Nb_2Na_2M^{2+}Ti(Si_2O_7)_2O_2(OH)_2(H_2O)_8$ (with M^{2+} = Mn, Ca), while the proposed ideal formula by us for shkatulkalite is $Na_5(Nb_{1-x}Ti_x)_2(Ti_{1-y}Mn^{2+}_y)[Si_2O_7]_2O_2(OH)_2\cdot nH_2O$, where x + y = 0.5 and n ≤ 10, which is a remarkable agreement. It is worth noting that Cámara et al. [41] have already confirmed the right prediction of B7(GIV) structure type with the crystal structure of kolskyite.

In agreement with the suggestions by Németh et al. [25] and Sokolova and Cámara [6], shkatulkalite belongs to the seidozerite supergroup, since its structure is based upon the *HOH* [22] or TS [17] blocks. More precisely, shatulkalite is a member of the lamprophyllite group, having Ti + (Nb + Mn) = 3 *apfu*, from which the *H* and O sheets has 2 and 1 (Ti + Nb + Mn) *apfu*, respectively. The most closely related species to shkatulkalite are vuonnemite, $Na_{11}TiNb_2(Si_2O_7)_2(PO_4)_2O_3F$ [35,36], and epistolite, $Na_4TiNb_2(Si_2O_7)_2O_2(OH)_2(H_2O)_4$ [37,38], crystal structures of which are shown in Figure 3b,c, respectively. The three minerals are based upon the same topological type of the *HOH* blocks (Figure 2b–d) with different layer symmetries ($p2/m$ in shkatulkalite, and $p-1$ in vuonnemite and epistolite). In all three minerals, the *HOH* blocks are parallel to (001), but the *c* parameters are different and equal to 15.573, 14.450 and 12.041 Å for shkatulkalite, vuonnemite and epistolite, respectively. In fact, both shkatulkalite and epistolite can be considered derivatives of vuonnemite and can be obtained from the latter at least through the *gedanken* experiment by removing the some Na^+ and all $(PO_4)^{3-}$ ions and subsequent hydration of the interlayer space. The hypothesis that shkatulkalite is a transformation mineral species that forms at the expense of vuonnemite (or some vuonnemite-related proto-mineral) seems quite reasonable, taking into account its high hydration state, which results in the very open packing of adjacent *HOH* blocks that is manifested in the large value of the *c* parameter and can be clearly seen in Figure 3a. In contrast, in the crystal structure of epistolite (Figure 3b), the packing of the *HOH* blocks is quite dense, resulting in the shrinkage of the *c* parameter. The syntactic intergrowths of shkatulkalite and epistolite reported by Németh et al. [25] may point out to the following growth scenarios: (1) both minerals are primary phases that crystallize simultaneously; (2) both minerals are transformation species that form according to the sequence "vuonnemite (or vuonnemite-related proto-mineral) → shkatulkalite → epistolite"; or (3) both minerals are transformation species that form along two different pathways, "vuonnemite → shkatulkalite" and "vuonnemite → epistolite". The reported observations of pseudomorphs of shkatulkalite after vuonnemite suggest that, at least in some environments, shkatulkalite is indeed a transformation mineral species that inherits basic structural features from vuonnemite in accordance with Khomyakov's structural inheritance principle [42]. The secondary nature and the status of shkatulkalite as a transformation mineral species may also account for the difficulties encountered when investigating the mineral by means of electron microprobe and crystal-structure analysis. The absence of precise agreement between the chemical and structural studies (the obvious cation deficiency observed in three series of independent chemical analyses and in the original report [23]) remains an issue that we cannot resolve at the present time.

Author Contributions: Conceptualization, A.A.Z., E.A.S. and S.V.K.; Methodology, A.A.Z. and T.L.P.; Investigation, A.A.Z., E.A.S., Y.E.S., T.L.P., L.M.L., L.A.P., and V.N.Y.; Writing-Original Draft Preparation, A.A.Z.; Writing-Review & Editing, S.V.K.; Visualization, S.V.K.

Funding: This research was funded by the President of Russian Federation grant for leading scientific schools (project NSh-3079.2018.5) and Russian Foundation for Basic Research (grant 16-05-00427).

Acknowledgments: The X-ray diffraction studies were performed in the X-ray Diffraction Resource Centre of St. Petersburg State University. The chemical analytical studies were done in "Geomodel" Resource Centre of St. Petersburg State University and Geological Institute, Kola Science Centre, Russian Academy of Sciences.

Conflicts of Interest: The authors declare no conflict of interest.

References

1. Rocha, J.; Anderson, M.W. Microporous titanosilicates and other novel mixed octahedral–tetrahedral framework oxides. *Eur. J. Inorg. Chem.* **2000**, *2000*, 801–818. [CrossRef]
2. Noh, Y.D.; Komarneni, S.; Mackenzie, K.J.D. Titanosilicates: Giant exchange capacity and selectivity for Sr and Ba. *Sep. Purif. Technol.* **2012**, *95*, 222–226. [CrossRef]
3. Oleksiienko, O.; Wolkersdorfer, C.; Sillanpaa, M. Titanosilicates in cation adsorption and cation exchange—A review. *Chem. Eng. J.* **2017**, *317*, 570–585. [CrossRef]
4. Hu, J.; Ma, Z.; Sa, R.; Zhang, Y.; Wu, K. Theoretical perspectives on the structure, electronic, and optical properties of titanosilicates $Li_2M_4[(TiO)Si_4O_{12}]$ ($M = K^+$, Rb^+). *Phys. Chem. Chem. Phys.* **2017**, *19*, 15120–15128. [CrossRef] [PubMed]
5. Prech, J. Catalytic performance of advanced titanosilicate selective oxidation catalysts—A review. *Catal. Rev. Sci. Eng.* **2018**, *60*, 71–131. [CrossRef]

6. Sokolova, E.; Cámara, F. The seidozerite supergroup of TS-block minerals: Nomenclature and classification, with change of the following names: Rinkite to rinkite-(Ce), mosandrite to mosandrite-(Ce), hainite to hainite-(Y) and innelite-1T to innelite-1A. *Mineral. Mag.* **2017**, *81*, 1457–1484. [CrossRef]

7. Sokolova, E.; Cámara, F.; Abdu, Y.A.; Hawthorne, F.C.; Horváth, P.; Pfenninger-Horváth, E. Bobshannonite, $Na_2KBa(Mn,Na)_8(Nb,Ti)_4(Si_2O_7)_4O_4(OH)_4(O,F)_2$, a new TS-block mineral from Mont Saint-Hilaire, Québec, Canada: Description and crystal structure. *Mineral. Mag.* **2015**, *79*, 1791–1811. [CrossRef]

8. Lykova, I.S.; Pekov, I.V.; Chukanov, N.V.; Belakovskiy, D.I.; Yapaskurt, V.O.; Zubkova, N.V.; Britvin, S.N.; Giester, G. Calciomurmanite, $(Na,vac)_2Ca(Ti,Mg,Nb)_4[Si_2O_7]_2O_2(OH,O)_2(H_2O)_4$, a new mineral from the Lovozero and Khibiny alkaline complexes, Kola Peninsula, Russia. *Eur. J. Mineral.* **2016**, *28*, 835–845. [CrossRef]

9. Lyalina, L.M.; Zolotarev, A.A., Jr.; Selivanova, E.A.; Savchenko, Y.E.; Krivovichev, S.V.; Mikhailova, Y.A.; Kadyrova, G.I.; Zozulya, D.R. Batievaite-(Y), $Y_2Ca_2Ti[Si_2O_7]_2(OH)_2(H_2O)_4$, a new mineral from nepheline syenite pegmatite in the Sakharjok massif, Kola Peninsula, Russia. *Mineral. Petrol.* **2016**, *110*, 895–904. [CrossRef]

10. Sokolova, E.; Cámara, F.; Hawthorne, F.C.; Semenov, E.I.; Ciriotti, M.E. Lobanovite, $K_2Na(Fe^{2+}_4Mg_2Na)Ti_2(Si_4O_{12})_2O_2(OH)_4$, a new mineral of the astrophyllite supergroup and its relation to magnesioastrophyllite. *Mineral. Mag.* **2017**, *81*, 175–181. [CrossRef]

11. Pakhomovsky, Y.A.; Panikorovskii, T.L.; Yakovenchuk, V.N.; Ivanyuk, G.Y.; Mikhailova, J.A.; Krivovichev, S.V.; Bocharov, V.N.; Kalashnikov, A.O. Selivanovaite, $NaTi_3(Ti,Na,Fe,Mn)_4[(Si_2O_7)_2O_4(OH,H_2O)_4]·nH_2O$, a new rock-forming mineral from the eudialyte-rich malignite of the Lovozero alkaline massif (Kola Peninsula, Russia). *Eur. J. Mineral.* **2018**, *30*. [CrossRef]

12. Cámara, F.; Sokolova, E.; Abdu, Y.A.; Hawthorne, F.C.; Charrier, T.; Dorcet, V.; Carpentier, J.-F. Fogoite-(Y), $Na_3Ca_2Y_2Ti(Si_2O_7)_2OF_3$, a Group I TS-block mineral from the Lagoa do Fogo, the Fogo volcano, São Miguel Island, the Azores: Description and crystal structure. *Mineral. Mag.* **2017**, *81*, 369–381. [CrossRef]

13. Andrade, M.B.; Yang, H.; Downs, R.T.; Färber, G.; Contreira Filho, R.R.; Evans, S.H.; Loehn, C.W.; Schumer, B.N. Fluorlamprophyllite, $Na_3(SrNa)Ti_3(Si_2O_7)_2O_2F_2$, a new mineral from Poços de Caldas alkaline massif, Morro do Serrote, Minas Gerais, Brazil. *Mineral. Mag.* **2018**, *82*, 121–131. [CrossRef]

14. Belov, N.V.; Organova, N.I. Crystal chemistry and mineralogy of the lomonosovite group in the light of the crystal structure of lomonosovite. *Geokhimiya* **1962**, *1*, 4–13.

15. Belov, N.V. *Essays on Structural Mineralogy*; Nedra: Moscow, Russia, 1976; 344p. (In Russian)

16. Pyatenko, Y.A.; Voronkov, A.A.; Pudovkina, Z.V. *Mineralogical Crystal Chemistry of Titanium*; Nauka: Moscow, Russia, 1976; 155p. (In Russian)

17. Sokolova, E. From structure topology to chemical composition. I. Structural hierarchy and stereochemistry in titanium disilicate minerals. *Can. Mineral.* **2006**, *44*, 1273–1330. [CrossRef]

18. Egorov-Tismenko, Y.K.; Sokolova, E.V. Comparative crystal chemistry of a group of titanium silicate analogues of mica. In *Comparative Crystal Chemistry*; Moscow State University: Moscow, Russia, 1987; pp. 96–106. (In Russian)

19. Egorov-Tismenko, Y.K.; Sokolova, E.V. Homologous series seidozerite-nacaphite. *Mineral. ZH* **1990**, *12*, 40–49. (In Russian)

20. Ferraris, G. Polysomatism as a tool for correlating properties and structure. In *Modular Aspects of Minerals*; Merlino, S., Ed.; Eötvös University Press: Budapest, Hungary, 1997; pp. 275–295.

21. Ferraris, G. Modular structures—The paradigmatic case of heterophyllosilicates. *Z. Kristallogr.* **2008**, *223*, 76–84. [CrossRef]

22. Ferraris, G.; Ivaldi, G.; Khomyakov, A.P.; Soboleva, S.V.; Belluso, E.; Pavese, A. Nafertisite, a layer titanosilicate member of a polysomatic series including mica. *Eur. J. Mineral.* **1996**, *8*, 241–249. [CrossRef]

23. Menshikov, Y.P.; Khomyakov, A.P.; Polezhaeva, L.I.; Rastsvetaeva, R.K. Shkatulkalite–$Na_{10}MnTi_3Nb_3(Si_2O_7)_6(OH)_2F$·$12H_2O$ a new mineral. *Zap. Vseross. Mineral. Obs.* **1996**, *125*, 120–126. (In Russian)

24. Sokolova, E.; Cámara, F. From structure topology to chemical composition. XXI. Understanding the crystal chemistry of barium in TS-block minerals. *Can. Mineral.* **2016**, *54*, 79–95. [CrossRef]

25. Németh, P.; Ferraris, G.; Radnóczi, G.; Ageeva, O.A. TEM and X-ray study of syntactic intergrowths of epistolite, murmanite and shkatulkalite. *Can. Mineral.* **2005**, *43*, 973–987. [CrossRef]

26. Horvath, L.; Pfenniger-Horvath, E.; Gault, R.A.; Tarasoff, P. Mineralogy of the Saint-Amable Sill, Varennes and Saint-Amable, Quebec. *Miner. Rec.* **1998**, *29*, 83–118.

27. Khomyakov, A.P. Transformation mineral species and their use in paleomineralogical reconstructions. In Proceedings of the 30th International Geological Congress, Beijing, China, 4–14 August 1996; p. 450.

28. Khomyakov, A.P. The largest source of minerals with unique structure and properties. In *Minerals as Advanced Materials I*; Krivovichev, S.V., Ed.; Springer: Heidelberg/Berlin, Germany, 2008; pp. 71–77, ISBN 978-3-540-77123-4.

29. Pekov, I.V. *Lovozero Massif: History, Pegmatites, Minerals*; Ocean Press: Moscow, Russia, 2000; ISBN 978-5900395272.

30. Bruker-AXS. *APEX2*, Version 2014.11-0; Bruker-AXS: Madison, WI, USA, 2014.

31. Sheldrick, G.M. *SADABS*, University of Goettingen: Goettingen, Germany, 2007.

32. Sheldrick, G.M. Crystal structure refinement with *SHELXL*. *Acta Crystallogr.* **2015**, *C71*, 3–8.

33. Dolomanov, O.V.; Bourhis, L.J.; Gildea, R.J.; Howard, J.A.K.; Puschmann, H. Olex2: A complete structure solution, refinement and analysis program. *J. Appl. Crystallogr.* **2009**, *42*, 339–341. [CrossRef]

34. Brese, N.E.; O'Keeffe, M. Bond-valence parameters for solids. *Acta Crystallogr.* **1991**, *B47*, 192–197. [CrossRef]

35. Bussen, I.V.; Denisov, A.P.; Zabavnikova, N.I.; Kozyreva, L.V.; Menshikov, Y.P.; Lipatova, E.A. Vuonnemite, a new mineral. *Zap. Vseross. Mineral. Obs.* **1973**, *102*, 423–426. (In Russian) [CrossRef]

36. Ercit, T.S.; Cooper, M.A.; Hawthorne, F.C. The crystal structure of vuonnemite, $Na_{11}Ti^{4+}Nb_2(Si_2O_7)_2(PO_4)_2$ $O_3(F,OH)$, a phosphate-bearing sorosilicate of the lomonosovite group. *Can. Mineral.* **1998**, *37*, 1311–1320.

37. Sokolova, E.; Hawthorne, F.C. The crystal chemistry of epistolite. *Can. Mineral.* **2004**, *42*, 797–806. [CrossRef]

38. Khomyakov, A.P.; Kurova, T.A.; Chistyakova, N.I. Sobolevite $Na_{14}Ca_2MnTi_3P_4Si_4O_{34}$—A new mineral. *Zap. Vseross. Mineral. Obs.* **1983**, *112*, 456–461. (In Russian)

39. Sokolova, E.; Hawthorne, F.C.; Khomyakov, A.P. Polyphite and sobolevite: Revision of their crystal structures. *Can. Mineral.* **2005**, *43*, 1527–1544. [CrossRef]

40. Sokolova, E.; Cámara, F. From structure topology to chemical composition. XVI. New developments in the crystal chemistry and prediction of new structure topologies for titanium disilicate minerals with the TS block. *Can. Mineral.* **2013**, *51*, 861–891. [CrossRef]

41. Cámara, F.; Sokolova, E.; Abdu, Y.A.; Hawthorne, F.C.; Khomyakov, A.P. Kolskyite, $(Ca\square)Na_2Ti_4(Si_2O_7)_2O_4$ $(H_2O)_7$, a Group-IV Ti-disilicate mineral from the Khibiny alkaline massif, Kola Peninsula, Russia: Description and crystal structure. *Can. Mineral.* **2013**, *51*, 921. [CrossRef]

42. Khomyakov, A.P.; Yushkin, N.P. The principle of inheritance in crystallogenesis. *Dokl. Akad. Nauk SSSR* **1981**, *256*, 1229–1233.

minerals

MDPI

Review

Advanced Techniques of Saponite Recovery from Diamond Processing Plant Water and Areas of Saponite Application

Valentine A. Chanturiya [1], Vladimir G. Minenko [1], Dmitriy V. Makarov [2,*], Olga V. Suvorova [3] and Ekaterina A. Selivanova [4]

[1] Institute of Comprehensive Exploitation of Mineral Resources of the Russian Academy of Sciences, Kryukovsky Tupik, 4, 111020 Moscow, Russia; vchan@mail.ru (V.A.C.); vladi200@mail.ru (V.G.M.)
[2] Institute of North Industrial Ecology Problems, Kola Science Centre of the Russian Academy of Sciences, Fersman St., 14a, 184209 Apatity, Russia
[3] I.V. Tananaev Institute of Chemistry and Technology of Rare Elements and Mineral Raw Materials, Kola Science Centre of the Russian Academy of Sciences, Fersman St., 26a, 184209 Apatity, Russia; suvorova@chemy.kolasc.net.ru
[4] Geological Institute, Kola Science Centre of the Russian Academy of Sciences, Fersman St., 14, 184209 Apatity, Russia; selivanova_e_a@mail.ru
* Correspondence: mdv_2008@mail.ru; Tel.: +7-81555-79-3-37

Received: 22 October 2018; Accepted: 22 November 2018; Published: 26 November 2018

check for updates

Abstract: Methods of cleaning and processing of saponite-containing water from diamond processing plants in the Arkhangelsk region, Russia, are discussed. The advantages of electrochemical separation of saponite from process water enabling to change its structural-texture, physico-chemical and mechanical properties are demonstrated. Possible areas of saponite and modified-saponite products application are considered.

Keywords: saponite-containing waters; diamond processing plants; cryogenic treatment; electrochemical separation; saponite product applications

---

1. Introduction

The Lomonosov diamond deposit (Archangelsk province, Russia) currently contains 10 kimberlite pipes. Before launching quarrying operations, it is necessary to extract about 300 million tons of diamond-bearing ore and barren rock [1,2]. Almost all rock in the deposit pipes is represented by clay minerals, mostly saponite, the share of which reaches 90% in the vent facies [1].

Saponite belongs to the smectite group and is characterized by high physico-chemical activity and low density in aqueous media due to its tendency to hydrate. When in an aqueous medium, saponite disperses, forming a suspension, hindering the managing of both the tailing ponds and water circulation at the processing plants (Figure 1).

Figure 1. Tailing pond of processing plant, Lomonosov GOK, Severalmaz JSC. Reproduced with permission from the author, V.I. Bogachev.

Therefore, searching, validating and developing effective techniques for recovery of saponite-containing water at diamond processing plants is a current issue. The immediate objective is to develop an effective water rotating system, which will boost diamond recovery, reduce the ecological stress, and open ways for manufacturing of target-oriented saponite-containing products for multi-sector applications.

2. Structure and Properties of Saponite

The structure and properties of saponite, which are highly attractive for the industry, are described in detail in textbooks, reference books, and professional reports. According to the International Mineralogical Association (IMA) list [3], the saponite mineral consists of the formula $(Ca,Na)_{0.3}(Mg,Fe)_3(Si,Al)_4O_{10}(OH)_2 \cdot 4H_2O$; it is one of the three most common members in the smectite group, along with montmorillonite and nontronite [4].

Like all smectites, saponite has unique physical and chemical properties that arouse scientific interest, namely, high cation exchange capacity, swelling and rheological properties, hydration and dehydration, high plasticity, bonding capacity, and the ability to react with inorganic and organic reagents [5]. These properties are the results of:

-the layered nature of the crystal structure containing weakly bound cations;
-a wide range of chemical composition variations;
-extremely small particle size, flat shape and, accordingly, great surface area.

2.1. Crystal Structure

The basic structure of smectite group minerals is well known and has been illustrated in many publications (one of them is shown in Figure 2). Smectites are three-layer minerals. This three-layer package has two silica tetrahedral sheets joined to a central octahedral sheet. On this basis, smectite structures can be classified as 2:1 phyllosilicates. Due to substitutions in the tetrahedral or octahedral sheets, or the existence of vacancies in the octahedral sheet, the surface of the three-layer package has a negative charge, which creates a charge imbalance.

According to the Nomenclature Committee of the Association Internationale pour l'Etude des Argiles (AIPEA) [6] the layer charge of smectites varies from 0.2 to 0.6 electrons per half unit cell (e/h.u.c.). The layer charge is balanced by the interlayer cations Na, Ca, K, Mg, Fe, which are weakly bonded and exchangeable. The interlayer space contains water molecules. The number of molecules is not constant and may increase depending on the grade of the interlayer cation, causing the mineral particle to swell. The interlayer space also contains water molecules in varying quantities, which may increase depending on the interlayer cation kind and cause the layers to distend and the mineral particles to swell. Indeed, it has been shown that the layer charge is related to the colloidal properties of smectites such as swelling. Charge heterogeneity, which includes both charge magnitude and charge localization, is also related to these properties [7].

Saponite belongs to the trioctahedral series of minerals of the smectite group. Notwithstanding both the wide occurrence of saponite in nature and extensive previous study, the saponite structure has not been determined yet. The authors are of the opinion that the reason for it is the absence of material applicable for single-crystal investigations.

2.2. Chemical Composition and Properties

The chemical composition of naturally occurring saponite is highly variable due to common Fe^{2+}, Fe^{3+} and Al^{3+} substitutions for Mg^{2+} in the octahedral sheet, which are accompanied by partial Al^{3+} and Fe^{3+} substitutions for Si^{4+} in the tetrahedral sheet [8,9]. Thus, saponite from the Arkhangelsk kimberlite province contains 5.09% Al_2O_3, 3.14% Fe_2O_3 and 2.56% FeO [10]; crystal-chemical formula, calculated on 22 charges, is the following: $(Ca_{0.1},Na_{0.1},K_{0.1})_{0.3}(Mg_{2.6},Fe^{2+}_{0.1},Fe^{3+}_{0.2})_{2.9}(Si_{3.6},Al_{0.4})_4O_{10}(OH)$ $2.8H_2O$.

Saponite is different from other smectites because a part of the negative tetrahedral charge is balanced by substitution of octahedral Mg^{2+} by trivalent cations Al^{3+} or Fe^{3+}, i.e., the octahedral sheet bears a positive charge. However, the tetrahedral charge, due to substitution of Si^{4+} by Al^{3+}, is much greater and outbalances any possible positive octahedral charge [11].

These substitutions, their quantity and cation kind essentially affect the mineral's properties. Moreover, the mineral's properties are further changed when the iron, present in the mineral as an isomorphous impurity, varies its oxidation degrees in response to certain conditions.

Fairly often, the composition of natural saponite is heterogeneous and, like all smectites and other layered silicates, it may contain fragments of other layered silicates, forming sometimes mixed-layer structures. Saponite is commonly found in association with montmorillonite or talc [12–14]. According to the authors of [14], mixed-layered aggregates of this type have a greater surface area (up to 283 m^2/g) and a high concentration of mesopores comparable with quality sorbents.

In order to obtain a chemically- and phase-wise homogeneous composition, saponite is synthesized to achieve controllable properties, the surface charge distribution in the first place [15–17]. One recent example is the use of synthetic saponites as hydro processing catalyst components, described in a patent of Chevron Corp [18]. The growing interest in synthesis methods over the past decade has been generated by the competitive advantages of synthetic saponite, namely, its mesoporosity, controllable acidic and basic properties, and stability [19,20].

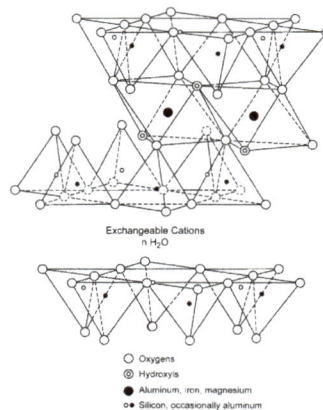

Exchangeable Cations
n H_2O

○ Oxygens
◎ Hydroxyls
● Aluminum, iron, magnesium
◦● Silicon, occasionally aluminum

Figure 2. Schematic diagram of smectite structure [21].

Being weakly bonded with the structure packages, the interlayer cations Na, Ca, K, Mg can be easily removed from it, or substituted for other cations, without destroying the three-layer package.

Traditionally, the cations are removed via acid treatment. Researching in the area of acidic modification of saponite is also provoked by the possibility of altering the coordination of octahedral cations to the extent of obtaining a separate silicate layer with a super-high surface area [22–25].

In the case of trioctahedral saponite, the conditions providing an almost complete removal of the octahedral layer with resulting surface increase of up to 300 m^2/g and microporosity are rather mild [26].

There is no evidence of studies of the sorption mechanism of natural saponite, but some results have been reported on the sorption of saponite-containing products [27,28].

Regarding thermal properties, the Arkhangelsk saponite is close to other smectites [10]. Differences can be observed in the area of the second endothermic effect, which is attributable to cation substitutions in the octahedral layer.

There is a unique property of smectite crystals that has been widely researched [29]. It consists in increasing, or decreasing, of the number of water molecules between the layers from 0 to 4, causing the basal distance to increase from 10 to 22 Å. Swelling occurs in polar organic solvents as well. The layer charge effect, produced on the swelling, has been fairly well researched [7], but no studies on saponite swelling have been reported yet.

Investigating of smectite-water suspensions is by far more difficult. Usually, smectites form fine crystals of size 0.5 μm. When in small concentrations, the suspensions have Newtonian properties created by hydrodynamic forces, whereas at increasing concentrations, the inter-particle interactions trigger non-Newtonian behavior. The variety of interactions caused by electrostatic and electrokinetic characteristics is of great current interest (see, for instance, [5]). Saponite is the least studied among the smectites, even though it is also known to possess high and ambiguous electrokinetic potential values in aqueous suspensions [30].

3. Techniques of Circulating Water Clarification and Obtaining of Thickened Saponite Product

Nowadays the rate of saponite particles precipitation and sediment compaction is increased by using reagent and cryogenic treatment, bubbling with carbon dioxide, and exposure to acoustic, electric and magnetic fields [1,31–34].

K.M. Asonchik, A.V. Utin and coauthors have proposed carbonizing to accelerate clearing of the slurry tailings and circulating water at the diamond processing plant of the Lomonosov GOK, Severalmaz JSC (DPP LGOK) [31,32]. The method incorporates the interaction of pure carbon dioxide with saponite-containing water to produce carbonic acid, which further interacts with the saponite calcium and magnesium compounds and forms water-soluble hydrocarbonates $Mg(HCO_3)_2$ and $Ca(HCO_3)_2$. According to the authors, the passing of calcium and magnesium ions to the solution promotes destruction of the colloid structure, thus purifying the water.

During a pilot testing of the technology at DPP LGOK, the water was clarified to less than 0.5 mg/dm^3 of solids. It should be noted, however, that the initial solid content in the experiments did not exceed 110 g/dm^3 and became 158 g/dm^3 in the thickened slurry after four days of settling [31,32], whereas the natural settling of dumped water samples with initial solids contents of 50–100 g/dm^3 takes seven or more days. Gravity separation yields clarified water and a precipitate containing less than 5 g/dm^3 and up to 200 g/dm^3 of solids, respectively [7,8].

In 2013, S.A. Bakharev conducted at DPP LGOK a commercial testing of a unit producing a complex acoustic impact on circulating water (CAIP-CW). The geometric size of the pond was 307 × 37 × 3.7 m [33]. Prior to commercial testing, the circulating water was highly turbid, both in the lower pond horizon, representing the settling area, and at the outlet of discharge pipes. In the middle horizon (0.4 M), the slime particles concentration was about 3.31 g/dm^3; in the upper horizon (0.2 M) it was 1.26 g/dm^3. The CAIP-CW testing showed a much greater diminishing of water turbidity compared to gravity clarification: the slime particle concentrations in the middle and upper horizons became, respectively, 0.32 g/dm^3, and 0.16 g/dm^3.

The results of the acoustic method of thickening of saponite-containing precipitate at CAIP-CW are of practical importance because it makes possible the extraction and recycling of saponite. Using the reagent-free (acoustic) method of thickening of saponite-containing precipitate helped increase the particles' concentration in the precipitate from 89 to 743–790 g/dm^3. The precipitate structure was similar to modeling clay.

However, the report [33] does not describe the experimental conditions of the saponite-containing precipitate thickening method. In addition, the high solids values in the resulting saponite-containing products (743–790 g/dm^3 at a power consumption of 0.5 W/m^3 of the slurry) are untypical of hygroscopic saponite, which may indicate the presence of high quartz, dolomite and other mineral contents in the samples. Unfortunately, the author does not disclose the results of mineralogical analyses of the precipitates. Besides, the data on sieve composition of the initial and resulting products, presented by the author [33], seem questionable due to the small size of pure saponite particles (less than 7 μm) and, which is worse, of the obtained precipitates, because when the content of the highly hygroscopic saponite is 400 g/dm^3, the suspension turns viscous and non-flowing. Therefore, the sieve analysis of precipitates with saponite contents of 743–790 g/dm^3 is likely to have been achieved only after repeat diluting with water and vigorous stirring, which must have resulted in destruction of the floccules believed to be forming by the author.

The authors of works [1,34] have tested some unconventional techniques, including cryogenic treatment, applicable in clarification of circulating water at diamond treatment plants. Cryogenic treatment incorporates freezing and defrosting of a saponite suspension. The diffusion layer of the particles, which prevents particle convergence, is destroyed, initiating the saponite sediment genesis. The key factor to be observed is the frosting-defrosting regime. The precipitate is the densest at a slow rate, providing the draining of separating moisture. The precipitate carcass density achieves 0.74 g/dm^3, which is 4-fold higher than that of the initial suspension. Furthermore, the resulting precipitate, as believed by the author of work [1], acquires the ability to further diagenetic alterations, because the increase of gravity loads promotes the growth of the number and area of contacts between the particles, which strengthens the precipitate structure.

Figure 3 presents the general layout of a tailing pond with cryogenic treatment [1,35]. It can be seen that beyond the main pond and the protecting dam is located a pond of clarified water. After saponite sedimentation, the water flows through a drainage installation in the bottom part of the main tailing pond to the clarified water pond, where it accumulates for reverse water supply. The tailing pond is divided into two sections by a dam. One section receives the pulp and accumulates the saponite suspension; the other is used for freezing of suspension in winter and draining of thawed out water during the summers. The freezing-defrosting process promotes the sedimentation of saponite, which is extracted from the pond, pumped, and stored in a specially arranged storage representing depressions in natural relief. After being emptied, this section of the tailing pond is ready to be filled with the pulp again, while the accumulated suspension will be freezing in the other section. The cycle is repeated so that every year the pond sections change places.

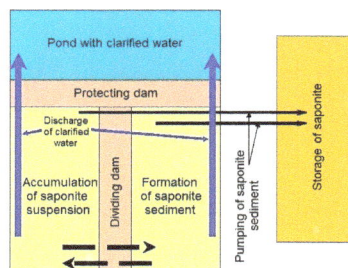

Figure 3. Layout of a tailing pond with cryogenic treatment. Reproduced with permission from F.S. Karpenko [1].

The efficiency of reagent application in saponite sediment thickening was experimentally tested by F.S. Karpenko [1], who used 18 various flocculants and inorganic coagulating agents $AlCl_3$ and Al_2SO_4 and also examined the effect of electric and magnetic fields on saponite precipitation and thickening. As a result of reagent use, the saponite settling rate was increased several hundred-fold, but this neither increased the density nor reduced the porosity of the precipitate so that it differed (slightly) from natural sediment. What is more, the precipitate in the experiments was unstable, forming a water suspension if stirred. It should also be remembered that the quantity of flocculating and coagulating agents consumed was considerable (up to 180 g/t and to 150 g/t, respectively). The most effective of the tested reagents were anionic flocculants of the series Praestol 2540 and Magnafloc 156, which provided a precipitation rate of 30 cm/h and a maximum carcass density of 0.24 g/cm^3.

Using the electric and magnetic fields for saponite precipitate settling and thickening proved to be ineffective [1].

The above-discussed process questions the efficiency of both reagent treatment, bubbling with carbon dioxide and acoustic impact at high content (over 50–100 g/dm^3) of fine slimes. What is more, none of the above-mentioned methods allow varying of the precipitate's mineral composition.

In works [36–40], a reagent-free electrochemical method and implementation for saponite recovery from diamond-treatment plant process water is reported (Figure 4). The designed electrochemical separators accommodate the processes of electrophoretic extraction of saponite-containing product at the anode and osmic evolution of water at the cathode. The obtained concentrate (electrochemically modified saponite) is characterized by high contents of solids (up to 620 g/dm^3 of suspension) and saponite and montmorillonite (more than 74.5%) and low quartz and dolomite contents (less than 12% and 5%, respectively), compared to the initial saponite-containing product (60–68%, 14–20%, and 6–10%, respectively). Owing to its chemical composition (SiO_2, Al_2O_3, CaO, Fe_2O_3, FeO, TiO_2, Na_2O, K_2O, SO_3 etc. as main components), denser packing, high content of smectite group minerals and the presence of exchangeable cations, the modified saponite-containing product can be used in the manufacturing of quality building materials and sorbents [41].

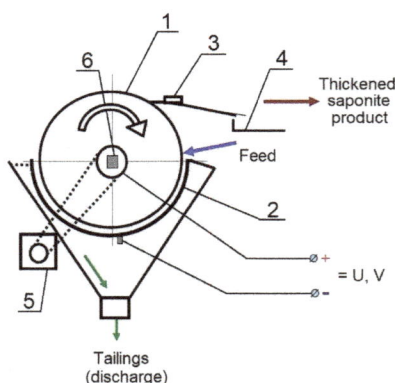

Figure 4. Schematic diagram of the electrochemical separator. 1—drum; 2—separating bath; 3—scraper; 4—chute; 5—engine; 6—current-collecting device.

It is evident that the most effective techniques for obtaining and thickening of saponite precipitate are cryogenic treatment and electrochemical separation. However, the former has the drawbacks of seasonal use and likely dispersion of defrosting of saponite-containing product by melting water.

4. Range of Saponite Product Applications

Saponite has been actively researched in recent years [42–94] as a valuable product with unique properties and a wide range of applications, including the chemical, food and consumer goods industries, agriculture, medicine and pharmacology, foundry practices, metallurgy, and construction.

4.1. Application of Saponite in Agriculture

Much of the research has been devoted to utilizing saponite in agriculture, livestock husbandry, and veterinary as an active or suspending agent and mineral additive to fodder [42–48].

The authors of [44,45,48] have developed a compound for pre-sowing treatment of winter wheat incorporating the raxil pesticide—a 2% wettable powder and saponite, i.e., saponite-based thixotropic water suspensions for plant protection, and a method of improving the agrochemical performance of ammonium saltpeter, whereby saponite as a mineral adsorbent is added in 2% per mass of the ready product.

Inventions reported in [46,47] present the processes of manufacturing of cattle fodder admixed with saponite, and also a KANIR-3 amide-concentration mineral additive based on grain offal and carbamide admixed with saponite, potato starch, and sodium sulfate.

Thus, the Ukraine standards [42,43] establish the general technical requirements for the quality of saponite flour as an integral-action ameliorant—a magnesium-containing fertilizer produced from saponite clay with high magnesium content (up to 12%) [41], and also to the quality of saponite-containing polymineral preparations used as additives to broiler-chick fodder [43].

4.2. Application of Saponite in Cosmetics Industry

Saponite is now used as an ingredient in the manufacturing of cosmetics and preparations, hygiene, deterging, and bleaching materials [49–57].

Research works [49,50] have been proposed some cosmetic composites containing clay mineral powders (talc, kaolin, saponite, mica, etc.), treated by fluorine, and an almost water-free oily component. The share of the clay minerals in the cosmetic product varies between 0.5 to 50%.

The authors of [51–55] have developed the following preparations: facial cosmetics containing an ultraviolet adsorbent with 0.001–0.005% mol per 10 g of a clay mineral, such as saponite; a cosmetic composition for skin and hair care based on water-soluble derivatives of chitin, clay minerals and intercalated clay; cosmetic deodorants containing up to 40 wt % of clay minerals (saponite, montmorillonite, beidellite, kaolin, etc.). These cosmetic hair, scalp and/or skin detergents (shampoos and shower gels) contain up to 50% of detergents (surface-active materials), 0.001 to 5% of insoluble conditioning agents and up to 15% of clay [55].

The authors of [56,57] report a water-softening reagent for household washing and dishwashing machines based on fine-crystal zeolite (50–70%), clay minerals (2–10%) such as saponite, montmorillonite and hectorite, a sodium salt of (co) polymeric carboxylic acid (5–15%), sodium sulfate (1–10%), an organic surface-active material (0–3%), and water. A method for the production of a bleaching agent consisting of a bleaching activator; an inorganic bonding substance (montmorillonite, saponite or hectorite with an ion-exchange capacity of 50–100 meq/100 g) has also been proposed.

4.3. Application of Saponite in Pharmaceutical Industry

Works [58–60] demonstrate the possibility of medicine-related applications of saponite, such as in the manufacturing of drugs and medications.

The authors of [58] have proposed a method for the manufacturing of α-tocopherol derivatives used as antisterile vitamins, anticholesterol agents boosting blood flow, antioxidants, etc. Tocopherol derivatives can be commercially manufactured using saponite, bentonite or montmorillonite as catalysts, in which the mobile cations are substituted for one of the following metals: scandium, yttrium, aluminum, iron, tin, copper, titanium, zinc, nickel, gallium, or zirconium.

Work [59] describes a method for the production of a clay mineral containing an IB-group metal (Au, Ag, Cu) intercalated into it by the contact of a clay aqueous dispersion (0.5–6.0%) with a cation-exchange resin stoichiometrically associated with an IB-group metal ion. The method provides substituting of over 30% of exchange cations in the clay mineral for an IB-group metal. For instance, an Ag-containing clay material is used as a basic component or thickener for coating materials with antifungal properties.

The authors of work [60] have proposed a manufacturing method of antimicrobial nano-size clay inhibiting the growth and proliferation of microorganisms by substituting the clay interlayer cations for an alkylamine. Clay of this kind can prevent the adverse health impact of microorganisms such as malignant bacteria and fungi. Manufacturing of the antimicrobial nano-size clay includes the following stages: Dispersing of nanoparticles of clay mineral, such as montmorillonite, saponite, hectorite, etc. in organic acid solutions, obtained by adding an organic solvent (ethanol, methanol, isopropyl alcohol, acetone, dimethylformamide, dimethylsulfoxide, or N-methylpyrrolidone) to a solution containing distilled water and chlorohydric acid; adding of alkylamine; filtering and drying of the obtained product.

4.4. Application of Saponite in Various Technology Processes

Saponite is known [61–74] to be used in various technological processes as an ingredient of sorbents, catalysts, carrying agents, thickeners and pigments, either in natural or activated form.

M.K. Uddin presented a survey of research of clay materials as heavy-metal adsorbents, carried out mostly in 2006–2016 [61]. The work describes the structure, classification and chemical composition of different clay minerals and analysis of their adsorption behavior. Although analysis of previous studies has confirmed the effectiveness of both natural and modified clay minerals in water treatment from heavy metal ions, the author considers that further research is needed on the modification and synthesis of new clay materials for adsorbing of dissimilar pollutants from the environment.

The Cu^{2+} ions removal from aqueous solutions by using natural and acid-activated clays is reported in work [62]. The adsorption isotherms, process kinetics and thermodynamics have been investigated. Interacting of the Cu^{2+} ions with clays and their acid-activated species was studied in an equilibrium batch process. The experiments were conducted at varied pH, interaction time, Cu^{2+} ion concentrations, clay quantities, and temperatures. The clays had a satisfactory adsorption ability of Langmuir monolayer of 9.2–10.1 mg/g; the acid-activated clays—31.8–32.3 mg/g. The interactions were endothermic, promoting the entropy increment and decreasing of the Gibbs energy. Adsorption of the Cu^{2+} ions on clay surface is unlikely to seriously affect the surface configuration, but the noticeable entropy change may be linked to increasing of randomness state, with Cu^{2+} ions taking up positions on the solid surface. Major structural changes and adjustments in the adsorbent surface structure may be ruled out. However, the surface structure might have been affected by the release of ions such as H^+ and K^+ from the clay surface to solution and also by partial desolvation of the metal ions. Thus, although the interactions were endothermic, they were driven by a positive entropy change [62].

Unlike untreated clay minerals, the acid-activated ones acquire a greater adsorption capacity due to developed surface area and greater pore volume. The Cu^{2+} ions release is affected by solution pH; as pH increases, the quantity of adsorbed matter increases to the point when the ions begin to precipitate as insoluble hydroxides at pH higher than 6.0. Adsorption of Cu^{2+} ions is precipitous at the onset of the substance-adsorbent interaction. The process kinetics is highly complicated and, although the authors have applied different kinetic models, they failed to draw a definite conclusion about the process rate mechanisms. It is highly probable, however, that the adsorption follows a second-order process. The validity of Langmuir isotherm in relation to the Cu^{2+}-clay interactions suggests that the Cu^{2+} ions are largely retained at the clay surface owing to chemical reactions.

Work [63] demonstrates the performance of mechanically and chemically activated saponite as Cu^{2+} and Ni^{2+} ions adsorbent from aqueous solutions. Saponite was activated via high-energy grinding in a planetary ball mill. Its structure was profoundly changed by the mechanical stresses of high-energy grinding, creating an active Mg–OH surface with a high acid neutralization. Activated saponite is a better adsorbent of Cu^{2+} ions as $Cu(OH)_2$ than of Ni^{2+}. As a result of activation, the maximum removal capacity of saponite in relation to Cu^{2+} ions increased from 33.2 to 287 mg/g, whereas for Ni^{2+} ions it increased from 46.0 to 124 mg/g. It has been established that the adsorbed metal ions form insoluble hydroxides on the saponite surface and are therefore inaccessible to further ion exchange. The formation of insoluble $Cu(OH)_2$ was confirmed by XRD and SEM analysis, and the formation of $Ni(OH)_2$—by SEM analysis. This research has illustrated the feasibility of smectites modification and usability in purification of heavy metal-contaminated process water.

As discovered by V.G. Minenko et al. [64], the saponite-containing product, obtained via electrochemical separation of process water, is characterized by high sorption capacity in relation to Ni^{2+} and Cu^{2+} ions (40 and 90 mg/g, respectively). After calcining at 750 °C, the sorption capacity increases to 189 and 224 mg/g, respectively. The metal-containing phases were diagnosed and the metals' sorption mechanism was validated using the XRD and SEM with EDS. The product thermally activated at 750 °C was successfully tested in nickel and copper solutions (50 and 250 mg/dm^3, respectively) with resulting solution concentrations of 0.01 and 0.001 mg/dm^3, respectively, meeting the maximum admissible concentration requirements to fishery water.

The method proposed in patent [65] describes the ion adsorption of Cr^{6+} on natural ferro-saponite followed by Cr^{6+} reduction and is aimed at removal of Cr^{6+} from water. The method incorporates the interaction of 0.04 M potassium dichromate solutions with natural ferro-saponite clays at 50 to 200 °C during 1 to 3.0 h.

Work [66] has demonstrated that the saponite-based sorption technique can be applied in purification of uranium-contaminated water (over 90%). Equilibrium in the U^{6+}–saponite system was found to be established within 8 h. The saponite–pH relationship has a clear maximum at pH 5–7 (Figure 5). The isotherm of U^{6+} sorption on saponite is described by the Langmuir empirical equation. The sorption process was shown to be affected by complexing reagents (ethylenediaminetetraacetic acid–EDTA, carbonate ions and fulvic acids).

Intercalating of Eu^{3+} ions into the interlayer space of a layered silicate (magadiite) occurred via the ion-exchange reactions between magadiite and europium chloride [67]. As indicated by both the X-ray diffraction and elemental analyses, the Eu^{3+} cations were intercalated into the magadiite interlayer space. The ion exchange between Eu^{3+} and Na^+ occurred preferentially so that the adsorbed Eu^{3+} amounts were controlled quantitatively. Due to intercalated Eu^{3+}, the resulting compounds were photo-luminescent. The luminescence intensity was directly dependent on the amount of Eu^{3+} absorbed. The intensity was also affected by heat treatment, corresponding to the changing environment of adsorbed Eu^{3+} caused by elimination of adsorbed water molecules and hydroxyl groups.

The authors of [68] report the results of quantitative determination and comparison of reactive properties of bentonite clays FEBEX and MX-80 and saponite in creating geochemical barriers and europium immobilization. (Hydrothermal treatment was performed using $Eu(NO_3)_3$ ((^{151}Eu and ^{153}Eu, with 52.2% ^{153}Eu)) and radioactive ^{152}Eu for quantitative assessment of reactions.) Saponite proved to be a better barrier material than bentonite. The calculations have shown that with the use of saponite, FEBEX and MX-80 at 200 °C $\leq$ T $\leq$ 350 °C europium will be immobilized within 8.5 months and at 80 °C $\leq$ T $\leq$ 200 °C—within several years. The reaction rate was not affected by the clay type, but the immobilizing ability of bentonite was lower than that of saponite.

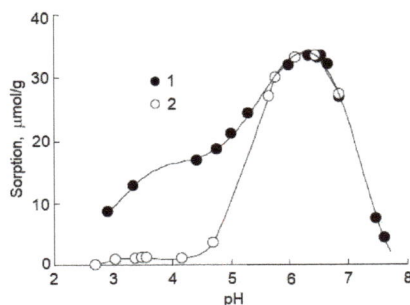

Figure 5. U^{6+} sorption on saponite vs the solution pH (ionic force (NaClO$_4$) 0.01(1); 0.1 (2); $C(U^{6+}) = 10^{-4}$ M; m = 0.1 g; V = 50 cm^3) Reproduced with permission from G.N. Pshinko et al. [66].

Technology using a composite and composite-based catalyst is proposed in work [69]. The composite consists of laminated clay homogeneously dispersed in an inorganic oxide matrix. The clayish dispersion is arranged so that to totally enclose the clay layers by the matrix. The inorganic oxide was selected from a group consisting of Al$_2$O$_3$, TiO$_2$, SiO$_2$, ZrO$_2$, P$_2$O$_5$ and their mixtures. The clay share is 5 wt % to 80 wt % of the composite. The clays used in the composite were saponite, montmorillonite, etc. The product can be applied as a catalyst in alkylation or hydrocracking.

The patent in [70] presents a process of glycerol polymers with the saponite catalyst. In broader terms, this invention refers to mineral polymerization techniques, including glycerol and its derivatives. The process incorporates the heating of a mineral in the presence of at least one magnesium saponite clay catalyst in the "H-form" to achieve polycondensation until glycerol polymers or its derivatives are formed. The "H-form" saponite catalyst is obtained in the ion-exchange process via substituting the Na$^+$ ions for H$^+$ and NH$_4^+$ ions.

The authors of work [71] have proposed an aqueous composition for surface friction reduction containing a solid lubricant and a modified, or synthetic, clay mineral. The clay minerals are smectites such as saponite, montmorillonite, hectorite etc. The share of solid lubricant is usually 1 wt % of the aqueous composite, more rarely 3% to 40%. The clay mineral is modified via pre-mixing with a water-soluble organic polymer in water, followed by spray drying. If necessary, the resulting dried mixture is ground to the size of 1 μm to 20 μm. The content of water-soluble polymer in the mixture is 0.1 wt % to 40 wt %.

The invention described in [72] proposes the process of a granular adsorbent characterized by high hardness, water-resistance, and adsorbing performance. The granular adsorbent is obtained by mixing 0.1 to 300 weight parts of a layered silicate mineral with 100 weight parts of an alkali metal hydroxide, which is followed by pelletizing and calcining. The hydroxides of magnesium, calcium, etc. are used. The mineral layered silicate is a natural mineral such as saponite, montmorillonite, beidellite, kaolinite, bentonite, etc. If needed, inorganic fibers, a pigment or an antibacterial agent can be added. The recommended adsorbent size is 0.5 to 10 mm.

The authors of work [73] propose an effective composition for eliminating metal pollution based on a composite containing a substrate, an organic ion, and a metal-binding agent. The substrate contains saponite, montmorillonite, natural and synthetic zeolite, polymer resin, lignite, kaolinite or a combination thereof. The organic ion includes quaternary amines, imidazolium salts, phosphonium salts, tetra alkyl ammonium, bis-(hydrogenated tallow)-dimethyl-ammonium chloride, bis-(hydrogenated tallow)-benzyl-methyl-ammonium chloride, 4,5-dihydro-1-methyl-2-nortallow-alkyl-1-(2-tallow-amidoethyl)-imidazolium methyl sulfate, 1-ethyl-4,5-dihydro-3-(2-hydroxyethil)-2-(8-heptadecenyl); -imidazolium ethyl sulfate, or combinations thereof. The metal-binding agent comprises mercaptan, carboxylic acid, chelating agents, amines, esters, carboxylic acids, alcohols, ethers, aldehydes, ketones, alkenes, mercaptans, thiols, tert-dodecanethiol, nonanethiol, octanethiol, n-stearic acid, palmitic acid, or combinations thereof.

4.5. Saponite in Nanocomposite, Textile and Paper Manufacture

Saponite is well known in the manufacturing of polymeric and polymer-ceramic composites, nano-composites, textile and paper [74–86].

Nityashree and coauthors [74] have synthesized some nanocomposites, where the layers of anionic clay are intercalated between the layers of cationic clay, by mixing an aqueous colloid dispersion of laminated copper hydroxide layers, or α-cobalt hydroxide, with an aqueous colloid dispersion of saponite. The thermal decomposition behavior of nano-composites, resembling a chlorite-like mineral with metal hydroxide layers intercalated between saponite layers, and of original layered solid substances was different. Thermal cracking of the chlorite-like nanocomposites caused the emergence of metal oxide nano-particles (CuO/Co_3O_4) uniformly distributed throughout the saponite matrix. The obtained oxide nano-particles can be of different sizes by varying decomposition time and temperature.

W. Wang and coauthors [75] have proposed and researched some nanocomposites consisting of quaternary fulvic acid (QFA) and saponite. The fulvic acid (FA) was obtained from sodium humate and nitric acid and further synthesized into QFA. The QFA-intercalated saponite (QFA-saponite) was prepared by using ultrasonic radiation. In order to improve the thermal and mechanical characteristics of polylactic acid (PLA), the QFA-saponite/PLA nano-composites were obtained from QFA-saponite and PLA via the solution intercalation. Anti-bacterial properties of QFA-saponite/PLA nano-composites have been investigated. Due to nano-size of QFA-saponite dispersion in the PLA matrix, the composites had better thermal and mechanical properties compared to pure PLA. The thermal and mechanical properties of the composites were optimal when the QFA-saponite content was 1 wt %. Moreover, the QFA-saponite/PLA nano-complexes were found to possess high bacteriostatic activity.

The authors of [76] have proposed a technique yielding a nanocomposite material from a composite consisting of 5 wt% to 90 wt% of synthetic thermoplastic polymer (polyopheline), a nanosize filler obtained from natural or synthetic phyllosilicates or laminated silicate clay (saponite, montmorillonite, etc.) (5 wt% to 80 wt%), and a deflocculating agent obtained by controlled free-radical polymerization (5 wt% to 50 wt%).

Work [77] also proposes a polyolephynic nanocomposite process incorporating stirring in a melt at 120–290 °C of a mixture of polyophylline, a filler (1 to 15% in terms of polyolephyne) represented by a laminated silicate clay (saponite, montmorillonite, etc.), a laminated hydroxycarbonate or phyllosilicate and a non-ionogenic surface-active material (0.1 to 7.5% in terms of polyolephyne mass). The non-ionogenic surface-active material is a block- or graft-copolymer with hydrophilic (ethylene oxide block) and hydrophobic (polyolephyne, fluorocarbon, siloxane or low-molecular methacrylate) segments. The surface-active non-ionogenic materials can be sorbitan ester, dimethylsiloxane-ethylenoxide copolymer block, or poly(methylmetalkrylate)-poly(oxoethylen) copolymer block.

Nanocomposite materials can be admixed [76,77] with phenol antioxidants, light stabilizers, solvents, pigments, coloring and plasticizing agents, admixtures enhancing impact resistance, thixotropic agents, acid acceptors and/or metal deactivators.

These nanocomposites are used in the manufacturing of foam plastic, fibers, various building, hygienic, packing, insulating and textile materials, strand or molded articles, storage basins, footware, printing forms, image carriers and circuit boards, optic and magnetic materials, furniture, playthings, sports and household items, etc.

The authors of [78] proposed a composite material obtainable at the temperature of 190–220 °C from thermoplastic (80.0–99.5%) and a filler (0.5–20%). The filler is a laminated silicate (saponite, montmorillonite, palygorskite, kaolin, bentonite, etc.) modified by QAS (a quaternary ammonium salt) at a silicate to QAS ratio of (2–200):1. This material is used in the manufacturing of technical parts such as plain bearings.

The invention described in work [79] proposes a method for preparation of high-porous composite materials from aluminum oxide and water-swellable clay (saponite, montmorillonite, etc.) homogeneously dispersed in the aluminum oxide component. The average pore diameter of the composite material is 1 μm to 150 μm. Calcining at 537.8 °C for 2 h causes the surface area to increase to about 200 m^2/g; the average pore diameter and pore volume are 60 Å to 400 Å and 0.5 cm^3/g to 2.0 cm^3/g, respectively.

The patent [80] has proposed a polymeric nanocomposite process based on clay with dispersed olefin and polyolefin resin. The nanocomposite is characterized by high stretching and thermal resistance, without a detrimental effect on transparency. The process incorporates immersing of the catalyst and co-catalyst into the clay, followed by olefin immersion, and polymerization thereof. The clay material is montmorillonite, hectorite, saponite, vermiculite, etc. The catalyst of olefin polymerization is an organic complex bonded with one of the following metals: Zr, Ti, Ni and Pd. The co-catalyst represents an alumoorganic compounds, such as $(C_2H_5)_3Al$, $(C_2H_5)_2AlCl$, $(C_2H_5)AlCl_2$, $(t-C_4H_9)_3Al$ and $(iso-C_4H_9)_3Al$.

The invention [81] proposes a method for preparation of a composite clay material with improved properties whereby the clays are treated with an agent containing 10–80 mol % of hydroxoorganic onium ions and vinyl alcohol polymers. The organic onium ions were obtained from an acrylic oligomer with an average molecular mass of 1000–15,000. The feed clay represents laminated clay minerals such as saponite, montmorillonite, etc.

Patent [82] proposes a nanocomposite process based on polymeric clay (montmorillonite, saponite, bentonite, hectorite, etc.) modified with an agent including two or more hydroxyl groups, silanol or alkyldiisocyanate with carbon atom values of 2–10.

P. O'Connor and S. Daamen proposed a technique yielding stable biomass suspensions containing inorganic particles [83]. Liquid suspensions containing a suspending medium, fine biomass particles and fine inorganic particles have been developed. The suspensions are stabilized due to the presence of fine (3 mm to 50 μm) inorganic material (saponite, alumina, transition metals hydroxides, sepiolite, etc.), which makes it possible to avoid sludging during the pipeline or cistern transportation. The suspensions are used in biofuel manufacture. The biomass contains polymeric materials such as cellulose and lignocellulose. The liquid suspension medium may also contain either water or alcohol, or carbonic acid. The process of biomass suspension manufacture can be accelerated by varying the temperature, pH and evaporation of the liquid suspension medium, individually or in combination.

The authors of [84] have demonstrated the advantages of microwave synthesis yielding mesoporous acid saponites characterized by a surface area of 603 m^2/g and lamellae crystallite size of about 4 nm by using a quaternary ammonium salt, surfactants, or polymer, as template, and researched the effect of the pH, temperature and H_2O/Si molecular ratio on the process. Required acidity was obtained by calcination of NH_4-form of saponite.

Work [85] proposes a method for the preparation and use of laminar phyllosilicate particles (silicate clay particles), with or without surface modification, with controllable size of 0.05 to 15 μm. Powders of this type are admixed to plastic or ceramic matrices. Pre-grinding to the size of 5–100 μm is performed in flushing mills in either dry or wet regimes. This is followed by controlled extraction of the particles sized 0.05 to 15 μm. Extraction is either dry, in dynamic classifiers, or wet, in centrifuges. The surface of phyllosilicate particles is modified using acrylic bioactive materials and acetate, alcohol solutions, or silver, copper, iron, nickel or cobalt salts.

The authors of [86] proposed a method for the production of fire-proof paper, incorporated in fire-proof film laminate for use in heat- and acoustic-insulating systems in civil aviation and other areas. The fire-proof paper contains inorganic biodegradable fibers, organic reinforcing fibers, organic and inorganic fibers or binder and, although not obligatorily, fire-proof ceramic fibers. The inorganic binding consists of clay (saponite, bentonite, montmorillonite, etc.) and ground inorganic or ceramic fibers, molten silicon oxide, etc.

4.6. Saponite in Construction Materials Manufacture

The recovery of saponite from kimberlite ore concentration waste in the manufacturing of construction materials such as binders, plasters, ceramics, etc., has been widely researched.

Thus, the authors of works [2,87,88] propose a method for the pelletizing of iron-ore concentrates and producing high-quality building materials.

In works [89–91], it is proposed to utilize clays with saponite composition as binders in the manufacturing of steam-cured silicate materials.

Kim Dae Hee developed a method for the production of construction plastering mortar from natural minerals, including clays [92]. Compared to conventional cement mortars, the resulting mixture is characterized by low fissuring, better acoustic absorption, adhesion and initial strengths.

Work [93] proposes a method for controlling the water-cement ratio in mortar curing by adding a fine saponite-containing product obtained from kimberlite ore concentration waste. As a result of a 7% addition of saponite-containing material, the concrete strength increased 2-fold and the frost-resistance quality was improved to F150.

It has been proved that thickened saponite-containing material can be recovered in ceramic brick manufacture [94]. The maximum compressive strength of the samples was 13 MPa. The authors also obtained a 250 quality cement clinker (75% of lime and 25% of thickened saponite-containing product) usable in the manufacturing of Portland cement.

Modified saponite-containing product has been used to produce high-quality ceramic bricks (Figure 6), with compressive and bending strengths of 800–1000 °C of 61.3 ± 2.8–80.9 ± 3.6 MPa and 11.9 ± 1.7–26.7 ± 2.1 MPa, respectively, within the sintering temperature range [30,95]. The bulk density, water absorption and fire shrinkage of ceramic materials at various sintering temperatures are presented in Figure 7. As expected, sample density increased with increase of sintering temperature: 1.9 ± 0.1 g/cm^3 (800 °C)–2.2 ± 0.2 g/cm^3 (1000 °C). Water absorption diminished from $13.6 \pm 0.4\%$ (800 °C) to $11 \pm 0.3\%$ (1000 °C). The fire shrinkage values in the examined sintering temperature range were: $1.15 \pm 0.22\%$ (800 °C)–$3.2 \pm 0.3\%$ (1000 °C).

It is of interest to continue researching the applicability of saponite-containing concentrates in pellets manufacture and using the electrochemically obtained concentrates as drilling mud components due to small dimensions (less than 7 µm) of quartz particles in the concentrate.

Figure 6. Ceramic samples obtained from modified saponite product.

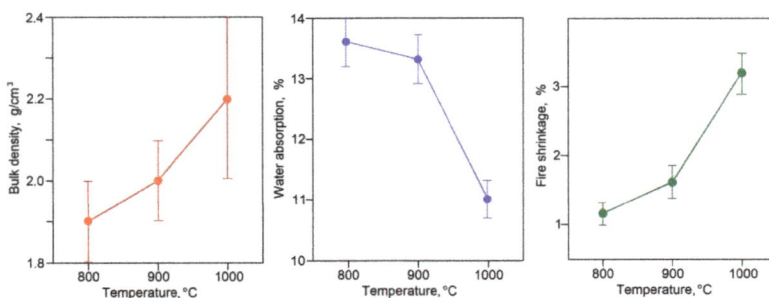

Figure 7. Dependences of bulk density, water absorption and fire shrinkage of ceramic materials obtained from modified saponite product on sintering temperature.

5. Conclusions

Analyses of the capacity, mining and concentration techniques, and saponite-containing waste recycling, as well as the research of structural and mineral composition of saponite-containing products carried out at the Lomonosov kimberlite deposit, has led us to conclusions on the preferability of methods of cryogenic treatment and electrochemical separation in the purification of recycling water at Lomonosov GOK plants.

Electrochemical separation helps obtain modified saponite-containing products with high smectite-group minerals concentrations, lower mineral particles size, more compact structure, and greater surface area. These characteristics open possibilities for the manufacture of high-quality ceramics and heavy-metal sorbents from saponite-containing products.

The application area of saponite-containing products should be extended by utilizing them in iron-ore pellet production and as drilling fluid components.

It was also shown that pure saponite is a valuable material for different industries, especially in the manufacturing of composites and other advanced materials. Therefore, the new methods for thickening and purification of saponite concentrate from Lomonosov deposit and producing of pure saponite represent a considerable scientific and practical interest. The properties exhibited by electrochemically modified saponite are also intriguing and potentially useful.

Author Contributions: Conceptualization, V.A.C., V.G.M., D.V.M., O.V.S. and E.A.S.; Methodology, V.A.C., V.G.M., D.V.M., O.V.S. and E.A.S.; Investigation, V.A.C., V.G.M., D.V.M., O.V.S. and E.A.S.; Writing-Original Draft Preparation, V.A.C., V.G.M., D.V.M., O.V.S. and E.A.S.; Writing-Review & Editing, V.A.C., V.G.M., D.V.M., O.V.S. and E.A.S.; Visualization, V.A.C., V.G.M., D.V.M., O.V.S. and E.A.S.

Funding: This research was funded by Russian Government grant 01201374315 (0138-2014-0002).

Acknowledgments: We are grateful to anonymous referees whose comments improve the manuscript greatly.

Conflicts of Interest: The authors declare no conflict of interest.

References

1. Karpenko, F.S. Saponite-Containing Precipitation Accumulation Conditions and Thickening in Tailing Ponds of Lomonosov Diamond Deposit. Ph.D. Thesis, Sergeev Institute of Environmental Geoscience RAS, Moscow, Russia, 2009. (In Russian)
2. Bezborodov, S.M.; Verzhak, V.V.; Verzhak, D.V.; Garanin, V.K.; Garanin, K.V.; Zuev, V.M.; Kudryavtseva, G.P.; Pylaev, N.F. Way of Recycling Diamond Industry Wastes. RU Patent 2,206,534, 20 June 2003.
3. The official IMA-CNMNC List of Mineral Updated List of IMA-Approved Minerals (November 2018). Available online: http://ima-cnmnc.nrm.se/IMA_Master_List_%282018-11%29.pdf (accessed on 26 November 2018).
4. Mindat.org. Smectite Group. Available online: https://www.mindat.org/min-11119.html (accessed on 26 November 2018).

5. Odom, I.E. Smectite clay minerals: Properties and uses. *Philos. Trans. R. Soc. A* **1984**, *311*, 391–409. [CrossRef]

6. Guggenheim, S.; Adams, J.M.; Bain, D.C.; Bergaya, F.; Brigatti, M.F.; Drits, V.A.; Formoso, M.L.L.; Galaʹn, E.; Kogure, T.; Stanjek, H. Summary of recommendations of nomenclature committees relevant to clay mineralogy: Report of the Association Internationale pour l'Etude des Argiles (AIPEA) Nomenclature Committee for 2006. *Clay Miner.* **2006**, *41*, 863–877. [CrossRef]

7. Laird, D.A. Influence of layer charge on swelling of smectites. *Appl. Clay Sci.* **2006**, *34*, 74–87. [CrossRef]

8. Decarreau, A.; Bonnin, D. Synthesis and crystallogenesis at low temperature of Fe(III)-smectites by evolution of coprecipitated gels: Experiments in partially reducing conditions. *Clay Miner.* **1986**, *21*, 861–877. [CrossRef]

9. Jasmund, K.; Lagaly, G. *Tonminerale und Tone. Struktur, Eigenschaften, Anwendung und Einsatz in Industrie und Umwelt*; Steinkopff Verlag: Darmstadt, Germany, 1993; p. 490.

10. Ogorodova, L.P.; Kiseleva, I.A.; Mel'chakova, L.V.; Vigasina, M.F.; Krupskaya, V.V.; Sud'in, V.V. Calorimetric determination of the enthalpy of formation of natural saponite. *Geochem. Int.* **2015**, *53*, 617–623. [CrossRef]

11. Christidis, G.E. The concept of layer charge of smectites and its implications for important smectite-water properties. *EMU Notes Mineral.* **2011**, *11*, Chapter 6. 239–260.

12. Alietti, A.; Mejsner, J. Structure of a talc/saponite mixed-layer mineral. *Clay Clay Min.* **1980**, *28*, 388–390. [CrossRef]

13. Eberl, D.D.; Jones, G.; Khoury, H.N. Mixed-layer kerolite/stevensite from the Amargosa Desert, Nevada. *Clay Clay Min.* **1982**, *57*, 115–133. [CrossRef]

14. Steudel, A.; Friedrich, F.; Schuhmann, R.; Ruf, F.; Sohling, U.; Emmerich, K. Characterization of a fine-grained interstratification of turbostratic talc and saponite. *Minerals* **2017**, *7*, 5. [CrossRef]

15. Breukelaar, J.; van Santen, R.A.; De Winter, A.W. Synthetic Saponite-Derivatives, a Method for Preparing Such Saponites and Their Use in Catalytic (Hydro) Conversions. US Patent 5,089,458, 18 February 1992.

16. Farmer, V.C.; McHardy, W.J.; Elsass, F.; Robert, M. hk-ordering in aluminous nontronite and saponite synthesized near 90 °C: Effects of synthesis conditions on nontronite composition and ordering. *Clay Clay Min.* **1994**, *42*, 180–186. [CrossRef]

17. Grauby, O.; Petit, S.; Decarreau, A.; Baronnet, A. The nontronite-saponite series: An experimental approach. *Eur. J. Mineral.* **1994**, *6*, 99–112. [CrossRef]

18. Kuperman, A.E.; Maesen, T.; Dykstra, D.; Uckung, I.J. Magnesium Aluminosilicate Clays-Synthesis and Catalysis. US Patent 20,100,087,313, 8 April 2010.

19. Tkachenko, O.P.; Kustov, L.M.; Kapustin, G.I.; Mishina, I.V.; Kuperman, A. Synthesis and acid-base properties of Mg-saponite. *Mendeleev Commun.* **2017**, *27*, 407–409. [CrossRef]

20. Baldermann, A.; Dohrmann, R.; Kaufhold, S.; Nickel, C. The Fe-Mg-saponite solid solution series–A hydrothermal synthesis study. *Clay Miner.* **2014**, *49*, 391–415. [CrossRef]

21. Murray, H.H. *Applied Clay Mineralogy. Occurrences, Processing and Application of Kaolins, Bentonites, Palygorskite-Sepiolite, and Common Clays*; Elsevier B.V.: Oxford, UK, 2007; p. 188.

22. Krupskaya, V.V.; Zakusin, S.V.; Tyupina, E.A.; Dorzhieva, O.V.; Zhukhlistov, A.P.; Belousov, P.E.; Timofeeva, M.N. Experimental study of montmorillonite structure and transformation of its properties under treatment with inorganic acid solutions. *Minerals* **2017**, *7*, 49. [CrossRef]

23. Komadel, P. Structure and chemical characteristics of modified clays. In *Natural Microporous Materials in Environmental Technology*; Misealides, P., Macasek, F., Pinnavaia, T.J., Colella, C., Eds.; Kluwer: Alphen aan den Rijn, The Netherlands, 1999; pp. 3–18.

24. Tkac, I.; Komadel, P.; Muller, D. Acid-treated montmorillonites—A study by ^{29}Si and ^{27}Al MAS NMR. *Clay Miner.* **1994**, *29*, 11–19. [CrossRef]

25. Kumar, P.; Jasra, R.V.; Bhat, T.S.G. Evolution of Porosity and Surface Acidity in Montmorillonite Clay on Acid Activation. *Ind. Eng. Chem. Res.* **1995**, *34*, 1440–1448. [CrossRef]

26. Vicente, M.A.; Suarez Barrios, M.; Lopez Gonzalez, J.D.; Banares Munoz, M.A. Characterization, surface area, and porosity analyses of the solids obtained by acid leaching of a saponite. *Langmuir* **1996**, *12*, 566–572. [CrossRef]

27. Morozova, M.V.; Frolova, M.A.; Makhova, T.A. Sorption-desorption properties of saponite-containing material. *J. Phys. Conf. Ser.* **2017**, *929*, 012111. [CrossRef]

28. Minenko, V.G.; Samusev, A.L.; Selivanova, E.A.; Bajurova, J.L.; Silikova, A.R.; Makarov, D.V. Study of copper ions sorption with electrochemically modified saponite. In Proceedings of the Mineralogy of

Technogenesis–2017, Institute of Mineralogy, Ural Branch of RAS, Miass, Russia, 22–25 June 2017; pp. 190–199. (In Russian)

29. Norrish, K. The swelling of montmorrilonite. *Discuss. Faraday Soc.* **1954**, *18*, 120–134. [CrossRef]

30. Chanturiya, V.; Minenko, V.; Suvorova, O.; Pletneva, V.; Makarov, D. Electrochemical modification of saponite for manufacture of ceramic building materials. *Appl. Clay Sci.* **2017**, *135*, 199–205. [CrossRef]

31. Asonchik, K.M.; Utin, A.V.; Kovkova, T.M.; Kostrov, A.M. Tailings slurry carbonization plant pilot-scale testing at the Lomonosovsky mining and concentration complex. *Obogashchenie Rud (Miner. Process.)* **2016**, *1*, 47–53. [CrossRef]

32. Utin, A.V. Method of Thickening Saponite Suspension. RU Patent 2,448,052, 20 April 2012.

33. Bakharev, S.A. The return water acoustical treatment at a diamonds recovery plant tailings storage facility cell. *Obogashchenie Rud (Miner. Process.)* **2014**, *6*, 3–7.

34. Dvoychenkova, G.P.; Minenko, V.G.; Kaplin, A.I.; Kobelev, D.A.; Bychkova, G.M. Experimental substantiation of the use of non-standard methods for recycled water clarification at the processing plants AK ALROSA. In Proceedings of the International Conference "The Plaksin's Readings—2007", Kola Science Centre of RAS, Apatity, Russia, 1–7 October 2007; pp. 332–336. (In Russian)

35. Osipov, V.I.; Karpenko, F.S. Method of Compacting Saponite-Bearing Sediments of Tailings Dumps. RU Patent 2,475,454, 20 February 2013.

36. Chanturiya, V.A.; Minenko, V.G.; Timofeev, A.S.; Dvoychenkova, G.P.; Samofalov, Y.L. Electrochemical method of the extraction of montmorillonite group minerals from the tailing dump waters. *Gornyi Zhurnal (Min. J.)* **2012**, *12*, 83–87.

37. Chanturiya, V.A.; Minenko, V.G.; Samusev, A.L.; Timofeev, A.S.; Ostrovskaya, G.K. Electrochemical separation of OAO «Severalmaz» facilities saponite-containing tailings pulp. *Obogashchenie Rud (Miner. Process.)* **2014**, *1*, 49–52.

38. Minenko, V.G. Justification and design of electrochemical recovery of saponite from recycled water. *J. Min. Sci.* **2014**, *50*, 595–600. [CrossRef]

39. Chanturiya, V.A.; Minenko, V.G.; Samusev, A.L.; Dvoychenkova, G.P.; Kur'janov, M.B.; Timofeev, A.S. Method of Deslimation of Circulating Saponite-Containing Waters and Device Its Implementation. RU Patent 2,529,220, 27 September 2014.

40. Chanturiya, V.A.; Trofimova, E.A.; Bogachev, V.I.; Minenko, V.G.; Dvoychenkova, G.P.; Kur'janov, M.B.; Timofeev, A.S. Method of Extraction of Saponite-Containing Substances from Return Water and Device for Its Implementation. RU Patent 2,535,048, 10 December 2014.

41. Osipov, V.I.; Sokolov, V.N. *Clays and Their Properties. Composition, Structure and Formation of Properties*; GEOS: Moscow, Russia, 2013; p. 576. (In Russian)

42. National Standard of Ukraine DSTU 7110:2009. Saponite Meal. Ameliorant of the Combined Action. General Specifications. Available online: http://document.ua/boroshno-saponitove.-meliorant-kompleksnoyi-diyi.-zagalni-te-std3600.html (accessed on 26 November 2018).

43. National Standard of Ukraine DSTU 4906:2008. Food for Animals. Polimineral Substances Based Saponitis and Glauconitis for Broiler Chickens. Specifications. Available online: http://document.ua/kormi-dlja-tvarin_-preparati-polimineralni-na-osnovi-saponit-std36156.html (accessed on 26 November 2018).

44. Derecha, O.A.; Klyuchevich, M.M. Composition for Presowing Treatment of Winter Wheat Seeds. UA Patent 54,892, 17 March 2003.

45. Frisch, G.; Maier, T. Thixotropic Aqueous Plant Protection Agent Suspensions. CA Patent 2,158,711, 22 March 1996.

46. Kulik, M.F.; Velichko, I.N.; Ovsienko, A.I.; Khimich, V.V.; Gricyk, V.E.; Vasilenko, S.V.; Gerasimchuk, A.P. Method of Obtaining Food for Pigs. SU Patent 1,748,780, 23 July 1992.

47. Karunskyi, O.Y.; Nikil'bursky, M.I.; Riznichuk, I.F. Amidoconcentrate Mineral Additive Kanir-3. UA Patent 48,445, 15 August 2002.

48. Roik, M.V.; Hurskyi, D.S.; Barshtein, L.A.; Heiko, V.D.; Musich, V.I.; Metalidi, V.S.; Zaryshniak, A.S.; Boiko, V.S.; Cherednychok, I.I.; Yanov, V.P. Method for Improving the Agrochemical Properties of Ammonium Nitrate. UA Patent 46,004, 15 May 2002.

49. Tomoko, S.; Kanemaru, T.; Matsuzaki, F.; Yanaki, T. Powdery Composition. Patent EP 1,402,875, 31 March 2004.

50. Tokubo, K.; Yamaguchi, M.; Suzuki, J.; Yoshioka, T.; Kanda, F.; Fukuda, M.; Ikeda, T.; Kawaura, T.; Yagita, Y. Spherical Clay Mineral Powder, Process for Production Thereof and Composition Containing the Same. US Patent 5,165,915, 24 November 1992.

51. Takuo, S.; Kenji, S. Ultraviolet Absorbent Composition and Cosmetic Containing the Same. JP Patent 60,081,124, 9 May 1985.

52. Takeshi, Y.; Tomiyuki, N. Composition for Skin and Hair. JP Patent 63,275,507, 14 November 1988.

53. Herve, D.; Jocelyne, B.; Maguy, J.; Lucia, L.A. Procede D'exfoliation D'argiles Intercalees. FR Patent 2,882,997, 15 September 2006.

54. Klein, W.; Kaden, W.; Röckl, M. Desodorierende Kosmetische Mittel (Deodorizing Cosmetic Products). DE Patent 4,009,347, 26 September 1991.

55. Decoster, S.; Beauquey, B.; Cotteret, J. Composition Cosmetiques Detergents et Utilisation (Cosmetic Detergent Composition and Utilization). FR Patent 2,722,091, 12 January 1996.

56. Upadek, H.; Schwadtke, K.; Seiter, W.; Pioch, L. Granulares, Phosphatfreies Wasserenthartungsmittel (Granular, Phosphate-Free Water Softener). DE Patent 3,931,871, 4 April 1991.

57. Hoeghst, A.G. Granulierte Blaichaktivatoren und Ihre Herstellung (Granulated Bleach Activators and Their Preparation). DE Patent 4,439,039, 9 May 1996.

58. Matsui, M.; Yamamoto, H. Process for the Preparation of Tocopherol Derivatives and Catalyst. US Patent 5,536,852, 16 July 1996.

59. Koga, S.; Sugiyama, H.; Suzuki, K. Preparation of Clay Mineral Containing Metal. JP Patent 02,116,611, 1 May 1990.

60. Hong, S.I.; Park, H.W.; Cho, Y.J.; Rhim, J.W. Antimicrobial Nano-Particle Clay and Manufacturing Method. Thereof. Patent KR 20,080,075,813, 19 August 2008.

61. Uddin, M.K. A review on the adsorption of heavy metals by clay minerals, with special focus on the past decade. *Chem. Eng. J.* **2017**, *308*, 438–462. [CrossRef]

62. Bhattacharyya, K.G.; Gupta, S.S. Removal of Cu(II) by natural and acid-activated clays: An insight of adsorption isotherm, kinetic and thermodynamics. *Desalination* **2011**, *272*, 66–75. [CrossRef]

63. Petra, L.; Billik, P.; Melichová, Z.; Komadel, P. Mechanochemically activated saponite as materials for Cu^{2+} and Ni^{2+} removal from aqueous solutions. *Appl. Clay Sci.* **2017**, *143*, 22–28. [CrossRef]

64. Minenko, V.G.; Makarov, D.V.; Samusev, A.L.; Suvorova, O.V.; Selivanova, E.A. New efficient techniques of saponite recovery from process water of diamond treatment plants yielding high-quality marketable products (abstract). In Proceedings of the XXIX International Mineral Processing Congress, Moscow, Russia, 17–21 September 2018; pp. 187–188.

65. Parthasarathy, G.; Sreedhar, B.; Boyapati, M.C. Method for Adsorption and Reduction of Hexavalent Chromium by Using Ferrous-Saponite. US Patent 2,006,016,757, 26 January 2006.

66. Pshinko, G.N.; Kobets, S.A.; Bogolepov, A.A.; Goncharuk, V.V. Treatment of waters containing uranium with saponite clay. *J. Water Chem. Technol.* **2010**, *32*, 10–16. [CrossRef]

67. Mizukami, N.; Tsujimura, M.; Kuroda, K.; Ogawa, M. Preparation and characterization of Eu- magadiite intercalation compounds. *Clay Clay Miner.* **2002**, *50*, 799–806. [CrossRef]

68. Villa-Alfagemea, M.; Hurtado, S.; Castro, M.; Mrabet, S.; Orta, M.; Pazosc, M.; Alba, M. Quantification and comparison of the reaction properties of FEBEX and MX-80 clays with saponite: Europium immobilisers under subcritical conditions. *Appl. Clay Sci.* **2014**, *101*, 10–15. [CrossRef]

69. Holmgren, J.S.; Schoonover, M.W.; Gembicki, S.A.; Kocal, J.A. Catalysts Containing Homogeneous Layered Clay/Inorganic Oxide. Patent EP 0,568,741, 10 November 1993.

70. Kraft, A. Method for Preparing Polymers of Glycerol with a Saponite Catalyst. US Patent 20,030,105,274, 5 June 2003.

71. Nozoe, T.; Tsuji, Y.; Black-Wood, W.; Kojima, K.; Ozaki, M.; Hori, S. Friction Reducing Coatings. Patent WO 2,011,082,137, 27 October 2010.

72. Ota, S.; Kurosaki, K. Production of Granular Adsorbent. Patent JP 10,137,581, 26 May 1998.

73. Angeles-Boza, A.M.; Landis, C.R.; Shumway, W.W. Composition and Method for Removing Metal Contaminants. US Patent 2,012,261,609, 18 October 2012.

74. Nityashree, N.; Gautam, U.K.; Rajamathi, M. Synthesis and thermal decomposition of metal hydroxide intercalated saponite. *Appl. Clay Sci.* **2014**, *87*, 163–169. [CrossRef]

75. Wang, W.; Zhen, W.; Bian, S.; Xi, X. Structure and properties of quaternary fulvic acid–intercalated saponite/poly (lactic acid) nanocomposites. *Appl. Clay Sci.* **2015**, *109–110*, 136–142. [CrossRef]

76. Moad, G.; Sajmon, D.F.; Din, K.M.; Li, G.; Mejjadann, R.T.A.; Vermter, K.; Pfehndner, R. Dispersants in Nanocomposites. RU Patent 2,404,208, 20 November 2010.

77. Moad, G.; Sajmon, D.F.; Din, K.M.; Li, G.; Mejjadann, R.T.A.; Pfehndner, R.; Vermter, K.; Shnajder, A. Method of Preparing Polyolefin Nanocomposites. RU Patent 2,360,933, 10 July 2009.

78. Burmistr, M.V.; Sukhyi, K.M.; Ovcharov, V.I.A. Composition Material. Patent UA 77,823, 15 January 2007.

79. Ljuss'er, R.Z.; Plesha, S.; Vehar, C.S.; Uiterbi, G.D. Hydrothermally Stable High-Porous Composite Materials of the Type Alumina/Swelled Clay and Methods for Preparation and Use Thereof. RU Patent 2,264,254, 20 November 2005.

80. Jin, Y.H.; Kim, J.A.; Kim, J.G.; Kwak, S.J.; Park, H.J.; Park, M. Preparation Method of Clay-Dispersed Olefin-Based Polymer Nanocomposite. Patent KR 20,030,025,308, 29 March 2003.

81. Usuki, A.; Hiruta, O.; Okada, A. Composite Clay Material. Patent JP 10,158,459, 16 June 1998.

82. Hwang, S.Y.; ImSeung, S. Modified clay, A Treating Method Thereof, Clay-Polymer Nanocomposite and a Manufacturing Method Thereof. Patent KR 20,100,068,823, 24 June 2010.

83. O'Connor, P.; Daamen, S. Stable Suspensions of Biomass Comprising Inorganic Particulates. Patent WO 2,008,020,046, 21 February 2008.

84. Gebretsadik, F.; Mance, D.; Baldus, M.; Salagre, P.; Cesteros, Y. Microwave synthesis of delaminated acid saponites using quaternary ammonium salt or polymer as template. Study of pH influence. *Appl. Clay Sci.* **2015**, *114*, 20–30. [CrossRef]

85. Lagaron, C. Method for Obtaining Laminar Phyllosilicate Particles Having Controlled Size and Products Obtained Using Said Method. Patent WO 2,011,101,508, 25 August 2011.

86. Garvi Chad, E. Fire-Proof Film Laminate. RU Patent 2,448,841, 27 April 2012.

87. Apollonov, V.N.; Verzhak, V.V.; Garanin, K.V.; Garanin, V.K.; Kudrjavtzeva, G.P.; Shlykov, V.G. Saponite from the Lomonosov diamond deposit. *Izv. Vuzov. Geol. Razved.* **2003**, *3*, 20–37. (In Russian)

88. Posukhova, T.V.; Dorofeev, S.A.; Garanin, K.V.; Siaoin, G. Diamond industry wastes: Mineral composition and recycling. *Mosc. Univ. Geol. Bull.* **2013**, *68*, 96–107. [CrossRef]

89. Volodchenko, A.N. Cementing based magnesial clays for autoclave silicate material. In Proceedings of the International Conference: Scientific researches and their practical application. Modern state and ways of development, Odessa, Ukraine, 2–12 October 2012; pp. 38–42.

90. Volodchenko, A.N.; Zhukov, R.V.; Lesovik, V.S. Silicate-based materials overburden Arkhangelsk diamond province-ray. *Univ. N. North Cauc. Reg. Tech. Sci. Ser.* **2006**, *3*, 67–70. (In Russian)

91. Volodchenko, A.N. Effect of mechanical activation of lime-saponite binding on the properties of silicate materials autoclave. *Bull. BSTU Named V.G. Shukhov* **2011**, *3*, 12–16. (In Russian)

92. Kim, D.H. Manufacturing Method of Construction Plastering Mortar Using Natural Minerals. Patent KR 20,020,026,897, 12 April 2012.

93. Morozova, M.V.; Ayzenstadt, A.M.; Makhova, T.A. The use of saponite-containing material for producing frost-resistant concretes. *Ind. Civ. Constr.* **2015**, *1*, 28–31. (In Russian)

94. Oblitcov, A.Y.; Rogalev, V.A. Prospective ways of diamondiferous rock enrichment wastes utilization at M.V. Lomonosov diamond deposit. *Proc. Min. Inst.* **2012**, *195*, 163–167. (In Russian)

95. Chanturiya, V.A.; Minenko, V.G.; Samusev, A.L.; Masloboev, V.A.; Makarov, D.V.; Suvorova, O.V. Method of Manufacturing Wall Products and Tiles. RU Patent 2,640,437, 9 January 2018.

Article

Hydrometallurgical Processing of Low-Grade Sulfide Ore and Mine Waste in the Arctic Regions: Perspectives and Challenges

Vladimir A. Masloboev [1], Sergey G. Seleznev [2], Anton V. Svetlov [1] and Dmitriy V. Makarov [1,*]

[1]   Institute of North Industrial Ecology Problems, Kola Science Centre of the Russian Academy of Sciences, 184209 Apatity, Russia; masloboev@mail.ru (V.A.M.); antonsvetlov@mail.ru (A.V.S.)
[2]   Ural State Mining University, 620144 Yekaterinburg, Russia; seleznev.s.ek@mail.ru
*   Correspondence: mdv_2008@mail.ru; Tel.: +7-81555-79-5-94

Received: 1 August 2018; Accepted: 1 October 2018; Published: 7 October 2018

Abstract: The authors describe the opportunities of low-grade sulfide ores and mine waste processing with heap and bacterial leaching methods. By the example of gold and silver ores, we analyzed specific issues and processing technologies for heap leaching intensification in severe climatic conditions. The paper presents perspectives for heap leaching of sulfide and mixed ores from the Udokan (Russia) and Talvivaara (Finland) deposits, as well as technogenic waste dumps, namely, the Allarechensky Deposit Dumps (Russia). The paper also shows the laboratory results of non-ferrous metals leaching from low-grade copper-nickel ores of the Monchepluton area, and from tailings of JSC Kola Mining and Metallurgical Company.

Keywords: heap leaching; bacterial leaching; cryomineralogenesis; low-grade copper-nickel ore; raw materials

1. Introduction

On the one hand, off-balance sulfide ores from abandoned and producing deposits, overburden rocks, mill tailings, and non-ferrous slag are considered to be one of the largest sources for non-ferrous metals production. On the other hand, they are well-known environmental hazards. Therefore, the use of mine dumps and tailings, and the ores left in the ground as raw products accompanied by environmental impact decrease is considered to be a critical objective in terms of the environment and economy [1].

Obviously, methods of heap and bacterial leaching are potentially productive for processing such low-grade ore and mine waste containing non-ferrous metals [2–6]. Introduction of this technology will facilitate step-by-step transition to a circular economy during the implementation of the closed procedures for solution circulation and waste material disposal in construction industry [7–9].

Previously, based on a number of sulfide-containing technogenic deposits that were located in Murmansk region, Russia, it was proved that, not only fine, but also coarse disperse wastes represent environmental hazards. These are the Allarechensky Deposit Dumps, as well as copper-nickel ore processing tailings from the Pechenga deposits. These tailings are characterized by low sulfide content and high chemical activity of nonmetallic minerals [10].

It should also be noted that sulfide oxidation begins even at early stages of waste storage. This has resulted in degradation of ore minerals processing properties. Valuable components also undergo dilution, and start moving to lower levels due to supergene processes [10,11]. It could be seen the redistribution of the ratios of silicate and sulfide nickel forms in tailings. [10,12]. Due to a long-term storage, technogenic deposits lose their value and they become a constant source of adverse impact on the environment [9,10,13].

A heap leaching process involves the following requirements. Solution containing sulfuric acid, oxidizing agent (oxygen, iron (III) ions, etc.), as well as microorganisms (for example, *Acidithiobacillus ferrooxidans, Acidithiobacillus thiooxidans*), are fed to the surface of a heap (ore stack) or inside. The solution is distributed evenly over the surface and dump heap mass by means of reservoirs, drainage channels, network of perforated tubes, or by spraying. Pregnant solution coming from under the heap is enriched in non-ferrous metals. It is collected in channels or tubes, and directed to further processing [2–6].

In comparison with heap leaching, biological leaching is a very young method of a valuable component recovery in the world processing industry. In the former USSR (Union of Soviet Socialist Republics), pilot tests were carried out in the 60 s of the past century, but they did not become a frequent practice in the industry [2,14,15].

The world's hydrometallurgical experience proves perceptiveness of the heap leaching method mainly for recovery of gold, copper, and uranium, from low-grade ores, mining and processing waste [2–6,16–19]. The twenty percent of the world's copper production is accomplished by heap leaching of ores from mines and processing plants as well as waste ores from dumps and further processing by solvent extraction and electrowinning (SX/EW) [16].

Biological leaching was proved to be efficient for sulfides of Co, Ga, Mo, Ni, Zn, and Pb [16]. Sulfide minerals, including metals of the platinum group (Pt, Rh, Ru, Pd, Os, and Ir) can be also subjected to preprocessing with microorganisms [20]. In recent decade, pilot tests on heap leaching of low-grade copper and nickel sulfide ores were launched (the mines involved are Radio Hill in the Western Australia; Talvivaara, Sotkamo in Finland; Khami, Sintszyan in China) [21–30].

The world industry has accumulated scientific and production experience that is related to assessment of ore amenability to heap leaching, engineering, and design of heap leach pads, solution irrigation systems, and collection of pregnant solutions, heap leaching schemes optimization, and chemicals regeneration [2,3,31]. Mathematical modeling techniques are being developed [32–35].

When using heap leaching, ore grades can be significantly lower than in the conventional metallurgical technologies. For example, for the heap leaching process used by the Finnish Company "Talvivaara Mining Company Plc" at its nickel mine situated in the subarctic zone of the north-east of Finland, average nickel grade amounts to 0.27%, copper 0.14%, cobalt 0.02%, and zinc 0.56% [36]. In cases of technogenic waste processing, the standard ore grade can be even lower. This can be explained by the fact that ore mining costs have been already spent, and transport expenses are allocated to the remediation budget [15].

The Russian Federation has all necessary conditions for a wide use of the heap leaching method for metal recovery. Undoubtedly, certain complications may result from unfavorable climate in the mining and processing regions of Russia. Most of the foreign mines are situated in the regions with warm climate, where the lowest temperature is above zero even in cold seasons. Therefore, it should be considered the specifics of the heap leaching process implementation in the regions with arctic and subarctic climate. The paper addresses these specifics by the example of gold and silver ores, as well as several case studies.

2. Specifics of Metal Heap Leaching in Severe Climatic Conditions

Currently, there are more than 300 heap leaching facilities in the world. About 70 of them (about 25%) are operated in the regions with arctic and subarctic climate with subzero average annual temperatures. These mines are located in North Europe, Asia, North America, as well as in the Andean regions of South America. Leading companies in the Arctic area are gold mines. They include Kinross Fort Knox Gold mine (Fairbanks, Alaska), Eagle and Coffee-Gold (Yukon Territory, Canada), as well as Casino Project (Yukon Territory, Canada) processing sulfide and oxidized copper ores bearing gold [37].

Low temperature and permafrost complicate implementation and operation of heap leaching pads and increase their costs. The following measures are taken to keep the leaching heaps warm [17,38,39]:

- construction of ore stacks in cells;
- use of drip emitters irrigation system consisting of pressure emitter trickles of labyrinth type;
- use of snow cover as a natural thermal insulation layer over a dump;
- freezing of "ice glaze" on a heap surface;
- covering a heap with thermally insulating materials during a cold season (polymer geo-textiles, polyethlene films with heated air supply underneath a covering mining material, with a layer thickness of up to 1 m);
- heating of process solutions.

In addition to these measures, a heap leach pad should be located in a place ensuring the maximum possible use of solar energy and minimum wind impact. A pad should not be built on permafrost soil, and frozen ore should not be placed on a dump as well.

Russian specialists have developed the method for precious metals leaching from mineral raw materials at low temperature (down to $-40\,°C$), without any need for solvent percolation through ore lumps [17]. This method was called a "passive leaching". In brief, the method includes the following steps: the ore is crushed, mixed with alkaline cyanide solution to complete the impregnation of the ore mass and held for some time, until all free gold transfers into a water-soluble state. On completing the process, the ore mass is subjected to water leaching by one of the two methods depending on technical and economic calculations: directly in heaps and dumps in-situ; or, based on a counterflow scheme in any flushing apparatus. The main feature of the passive leaching method is its applicability in severe cold conditions for gold and silver raw material processing. This is important for the Far North regions.

Quite a wide range of specialists used to think that geotechnologies, i.e., leaching in heaps, dumps, sludge storage facilities is unreasonable and unrealistic in cold climates with subzero average annual temperatures, permafrost soils, and short warm period (three months a year). However, yet, in the eighties of the 20th century, processing parameters and indicators of percolation (heap) and passive leaching were studied and determined, based on the ores from 12 deposits and the material of two gold-processing plants in Yakutia and Magadan region, Russia. Those parameters included: water retention capacity of gold- and silver-bearing materials, absorption of chemical agents, optimum solvent concentration, temperature impact, and metal recovery [17]. For the first time, the experiments proved that gold and silver minerals can be transferred into a water-soluble form by addition of cyanides under the temperature range from 0 down to $-40\,°C$.

After many years of cryomineralogenesis studies in mining and technogenic massifs, Ptitsyn developed a geochemical basis for metal geotechnology in the permafrost environment. Analysis of copper ore samples from the Udokan deposit experimentally showed conditions of copper solutions formation and migration under subzero temperatures [40,41]. Corrosive solution originates at the contact between sulfide ores and atmosphere. By reacting with ores during gravity migration, it forms an oxidation zone. The solutions penetrate into ice intergranular space, and between ice and minerals contacts. Cryogenic zones should be considered as a common feature in the areas of permafrost development [41,42].

The most important factors of cryomineralogenesis in mining and technogenic massifs are the following [43]:

(1) wide development of film water;
(2) exothermic effect of sulfides oxidation;
(3) significant impact of psychrophilic bacteria, including understudied silicate bacterium and other simple organisms;
(4) cryogenic solutions formation and migration conditions;
(5) low freezing temperature of high-concentrated percolated solutions, mainly sulfate ones;
(6) multiple geochemical barriers causing and ensuring mineralization;

(7) long-term existence of tectonic activity zones, which are conductors of both active endogenous fluid streams and exogenous supergenesis agents;

(8) frequency of changes in thermal conditions with various duration and intensity.

Ptitsyn's monograph describes an essential opportunity for using geotechnology for recovery of non-ferrous, rare, and precious metals in permafrost areas with cold climate [40]. Metals transition into a liquid form is achieved by combined treatment of ores with film-type solutions and solutions produced by cryogenic concentration. An additional factor intensifying a heap leaching process is frost cracking, which increases the contact surface between ore and solution. Due to gravity migration through rock conglomerated by ice, a concentrated freeze-resistant metal-containing solution can be accumulated on a waterproof foundation in the lower part of a massif. It ensures a possibility for the metal extraction and processing. The author pointed out that silting of massif could be reduced considerably during leaching at subzero temperatures.

Ptitsyn recommends using geotechnology methods at subzero temperatures for leaching gold, silver, copper, beryllium, magnesium, mercury, tin, bismuth, molybdenum, tungsten, manganese, nickel, as well as a number of rare metals found in significant quantity in highly-mineralized solutions [40]. The author assumed that lead, zinc, cobalt and platinum group metals can be recovered with the same method. Depending on ore composition, both acid (sulfate, chloride) and alkali (carbonate, ammonium, etc.) solutions can be used for leaching.

Nitrogen oxygen compounds (NO, N_2O_3, NO_2, N_2O_4, HNO_2, NO_2^-, HNO_3, NO_3^-, NO^+, NO_2^+, etc.) are likely to be oxidizing substances for oxidized leach cap, but they need further studies [42,44–47]. For example, nitrous acid activates various oxidizing processes, including leaching of galena (PbS), pyrites (FeS_2), pyrrhotite ($Fe_{1-x}S$), sphalerite (ZnS), and chalcocite (Cu_2S) in sulfuric acid solutions [42]. The activating effect of HNO_2 is intensified with its concentration increase and it is limited only by the disproportion of HNO_2 at the positive temperature range and low pH values. The experiments proved that nitrous acid stability under cryogenic conditions increases significantly even in highly acidic medium. Frost weathering modeling on the samples of copper-sulfide ores from the Udokan deposits (Russia) showed that copper-sulfide oxidation accelerated considerably with nitrous acid addition, and it has practically the same intensity throughout the analyzed range of parameters. The experiment involved the leaching of fraction 0.063–0.2 mm with 0.5 M sulfuric acid solution, using 0.1 M of nitrous acid and without it, at the temperatures of −20 and +20 °C, at S:L ratio of 1:5. The activating effect of HNO_2 appears to a greater degree during freezing. Copper recovery from solid phase increases by 2.7 times at a room temperature, and by 4.7 times under cryogenic conditions [45].

Passive leaching of gold and silver raw materials, as well as the justification of metals transition into a liquid phase with new oxidizing agents under subzero temperatures need generalization and experimental proofs for using it for sulfide ores and mine wastes processing.

In the next sections, the authors describe case studies and perspective of non-ferrous metals recovery from sulfide and mixed ores with heap leaching in the northern regions of the Russian Federation and abroad.

3. Geotechnological Methods of Ore Processing at the Udokan Mine

The Udokan deposit (in the North of Zabaykalsky Krai, Russia) is one of the largest copper deposits in the world, which is located 30 km to the south from the Novaya Chara railway station, on the mountain range Udokan. The territory of the deposit is associated with the Russian Far North. Continuous permafrost area extends with thickness from 65 m under water courses to 950 m under the watershed divide. Permafrost rock temperature is from −7 °C to −8 °C, the active layer thickness is about 1 m. By 2020, it is planned to build a mining and metallurgical complex with the capacity of 474,000 tons of copper per year. In addition to that, they plan to extract 277 t of silver per year, and to process additionally off-balance ores using the heap leaching method [46,47].

According to Khalezov, high-grade and amenable ores of the Udokan deposit should be processed with a traditional processing method, i.e., melting process [48]. Oxidized, sulfide-oxidized, and low-grade mixed ores should be processed using geotechnologies, i.e., heap leaching. Such ores should be dumped on specially arranged pads during mining. For this purpose, the corresponding ore classification process should be used. This process has already been used for a long time by a number of mining companies in Russia. Pyrometallurgical processing will be a source for sulfuric acid and heat to warm heap leaching solutions. Consequently, pyro- and hydrometallurgical methods are suggested for ore processing. This will ensure a technical and economic effect, and it enables increasing raw material usage. Copper minerals content is as follows: 68.5% of malachite ($Cu_2CO_3(OH)_2$) and brochantite ($Cu_4SO_4(OH)_6$); 29.6% of chalcocite (Cu_2S), covellite (CuS), bornite (Cu_5FeS_4); and, 1.9% of chalcopyrite ($CuFeS_2$). The ore has a vein, finely disseminated mineralization. Rock-forming minerals include quartz, feldspar, sericite, they amount up to 90%, and resistant to acid. This fact determines the low consumption of sulfuric acid. If leaching at Udokan is carried out only during warm season (140 days per year), leaching of ore with below 400 mm grain size will take 5–7 years. In case of a year-around operation with solution heating, the leaching will take 2–3 years. The SX/EW process is considered to be the most suitable method for copper extraction. The technology with closed water circulation enables recovering accompanying elements from ores, including rare and noble metals [48]. Solid residues of ore processing can be used for land reclamation and in construction industry.

4. Heap Bioleaching of Polymetallic Nickel Ore at the Talvivaara Deposit in Sotkamo, Finland

The Talvivaara deposit of polymetallic ores in Sotkamo, Finland, has become the most striking instance of heap bioleaching implementation for sulfide ores processing under the Far North conditions [49–53]. This deposit is considered to be one of the largest sulfide deposits in Europe situated in the geographical environment that closely resembles the Murmansk region in Russia.

Talvivaara black schist ore contains pyrrhotite ($Fe_{1-x}S$), pyrite (FeS_2), sphalerite (ZnS), pentlandite ($(Fe,Ni)_9S_8$), violarite ($(Fe,Ni)_3S_4$), chalcopyrite ($CuFeS_2$), and graphite (C). Nickel is distributed in the different sulfide minerals as follows: pentlandite 71%, pyrrhotite 21% and pyrite 8%. Cobalt is distributed in the following mineralogical phases: pentlandite 11%, pyrrhotite 26%, and pyrite 63%. The total copper is concentrated in chalcopyrite and the total zinc in sphalerite. The main silica-containing phases are quartz, mica, anorthite, and microcline [52]. Ore bioleaching in Talvivaara was studied during two decades: beginning from laboratory experiments in leaching columns, and up to construction of a pilot heap [49–53].

The selected mining method is a large-scale open pit mining. During processing, ore is crushed and screened in four stages to 80% passing 8 mm (Figure 1). After primary crushing, ore is conveyed to a secondary crushing plant, where it is crushed and screened in three stages. All of the material below 10 mm reports to agglomeration in a rotating drum, where pregnant leach solution (PLS) is added to the ore in order to consolidate fine particles with coarser particles. This preconditioning step makes ore permeable to air and water for heap leaching. After agglomeration, ore is conveyed and stacked in piles from eight to ten meters hight on the primary heap pad. After 13–14 months of bioleaching on the primary pad, the leached ore is reclaimed, conveyed, and re-stacked onto the secondary heap pad, where it is leached further in order to recover metals from those parts of the primary heaps, which had poor contact with the leaching solution. During the recovery process metals are precipitated from the PLS while using gaseous hydrogen sulfide and pH adjustment [53].

The primary heap is irrigated at a rate of 5 L/(m^2·h). The irrigation solution pH is adjusted to 1.7–2.0, and the solution is distributed evenly over the heap surface. PLS is collected with a drainage system and is discharged to PLS collection ponds. About 10–20% of the solution is pumped to a metals recovery plant and the rest is recirculated to the heap (Figure 1).

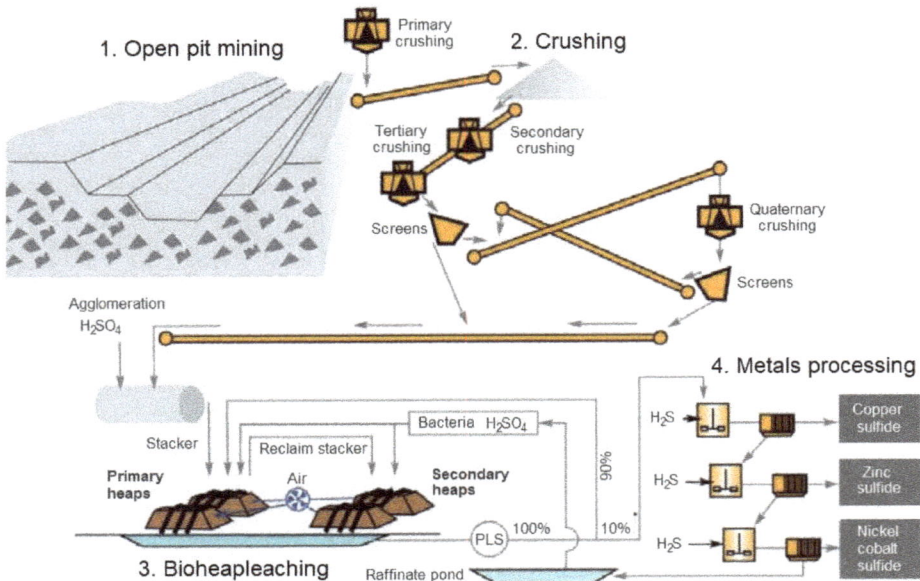

Figure 1. Process flow diagram of the "Talvivaara Mining Company Plc" [52].

The temperatures within the heap varied between 30–90 °C, despite the climate, and the PLS temperatures were 40–50 °C [49]. The microbial community in the heap (monitored by leachate analyses) varied during the first few months, becoming dominated by *At. ferrooxidans* and *Desulfotomaculum geothermicum*; other species identified were *Thiomonas arsenivorans*, *Alicyclobacillus tolerans*, and *Ferromicrobium acidophilum* [51]. Analyses of ore samples from the heap revealed *At. ferrooxidans*, *At. caldus*, and *F. acidophilum*. An uncultured bacterium clone H70 was found in both leachate and ore samples. Cell numbers varied between 10^5–10^8 cells/mL in leachate and between 10^5–10^7 cells/g in ore. In 500 days, recoveries were Ni 92%, Zn 82%, Co 14%, and Cu 2%. The low copper recovery was explained by the minerals' electrochemical properties [21].

Talvivaara became the first commercial nickel sulfide heap leaching operation. Nickel production was launched in October 2008. It was originally planned to reach the design capacity 33,000 t of nickel, 1200 t of cobalt, 60,000 t of zinc, and 10,000 t of copper by 2010 [21]. However, the peak performance was reached only by 2011, and the parameters amounted only to the half of the planned figures (16,087 t of nickel) [54].

The project engineering and development at commercial scale occurred during the period, when non-ferrous metal prices were rather high [54]. In 2007, average annual price of refined nickel at London Metal Exchange increased by 52.5% USD/t in comparison with 2006, i.e., from 24,416 to 37,230. The perspectives of successful operation at commercial scale were attractive. However, during the following years, prices for non-ferrous metals (including nickel) dropped dramatically.

The tense situation in the world market led to the bankruptcy of Talvavaara Mining Company Plc [54]. Regardless of the negative economic results, the company left a number of useful experiments, proved by experience, and having strategic value for further research and pilot tests; the microorganisms used in bioleaching process still exist and develop in the feed ore. They are endemic and well-adapted to the environment, what results in processing efficiency; high quality of materials recovered from solutions at commercial scale was demonstrated.

5. Copper-Nickel Ores and Technogenic Waste in Murmansk Region

Early studies of hydrometallurgical methods for processing of copper-nickel ores from the Murmansk region deposits were launched in the 70 s of the previous century [55]. For referencing purposes, the main types of disseminated ores from the deposits Zhdanovsky, Kaula, Sopchinsky Plast, and Lovnoozersky were selected. The metal content varied: 0.46–1.49% Ni and 0.01–0.62% Cu. In the first experiments, the solution containing 150 g/L of H_2SO_4 and 135 g/L of NaCl was used as a leaching reagent. The laboratory and further large-scale tests proved the possibility of underground metal leaching from the Lovnoozersky deposit ores.

The first data on a number of thionic bacteria in mine water of copper-nickel deposits in the Murmansk region are presented in the works of Lyalikova and Karavaiko et al. [56,57]. The possibility of their use for non-ferrous leaching was studied by Golovko et al. [55].

Currently, the research on justification of non-ferrous metal leaching from sulfide-containing natural and technogenic deposits of the Murmansk region continues [58–63].

5.1. Perspectives for Biological Leaching of Sulfide Copper-Nickel Ores from the Allarechensky Deposit Dumps

The technogenic deposit, the Allarechensky Deposit Dumps, is situated in the north-western part of the Murmansk region, 45 km to the south from the village Nickel, the Pechenga district. The deposit is a dump of mine waste that formed after open-pit mining of the Allarechensky sulfide copper-nickel ore deposit, which was completed in 1971 [13].

The ores of the deposit are represented by two morphological types: massive ore containing the following valuable components: 5–18% Ni, 0.15–8% Cu, up to 0.3% Co; and, disseminated ore containing 0.2–7.9% Ni, 0.12–4.9% Cu, up to 0.12% Co; with grades ranging: from 7.9%, 4.9%, and 0.12% (for high-grade ores) and up to 0.2%, 0.12%, and 0.008% (for low-grade ores), respectively. Ore studies enabled to define two key properties, which could be useful for ore processing: gravity and magnetic contrast [13,58,59].

Industrial tests showed that magnetic separation enables to efficiently process both raw and high-grade ones within the size range of 5–60 mm, and to produce a high-grade concentrate with cumulative content of Ni 2.0–3.7%, Cu 1.5–2.2%, and Co 0.03–0.08% [13,58,59].

For the processing of a fine fraction constituting 10–15% of all the deposit volume, it was studied whether non-ferrous metal recovery is possible by using biotechnology. The deposit ores are amenable to bioleaching due to their structural and textural properties. The chain of sulfide leaching was determined: pyrrhotite ($Fe_{1-x}S$) $\rightarrow$ pentlandite ($(Fe,Ni)_9S_8$) $\rightarrow$ chalcopyrite ($CuFeS_2$), and pyrite (FeS_2). The crystal structure, fracturing, substitution by violarite ($(Fe,Ni)_3S_4$) and bravoite ($(Fe,Ni,Co)S_2$), contributing to faster mineral destructurization, encourage pentlandite ($(Fe,Ni)_9S_8$) leaching. A large portion of minerals has excessive sorption capacity and it is characterized by ore mineral binding. These factors are referred to adversity [13,58,59].

JSC Irgiredmet has performed large-scale laboratory tests and calculation of technical and economic parameters on applicability of non-ferrous bacteria heap leaching technology for magnetic separation products in the dumps. One of the analyzed process flow diagrams is presented in Figure 2 [13,59].

Regardless of the fact that the resources of the deposit are not large, and, according to various estimates, amount to about one million tons of ore containing on the average 0.54% Ni and 0.47% Cu, the development of only explored reserves is considered to be a commercially viable investment project with low risks [59].

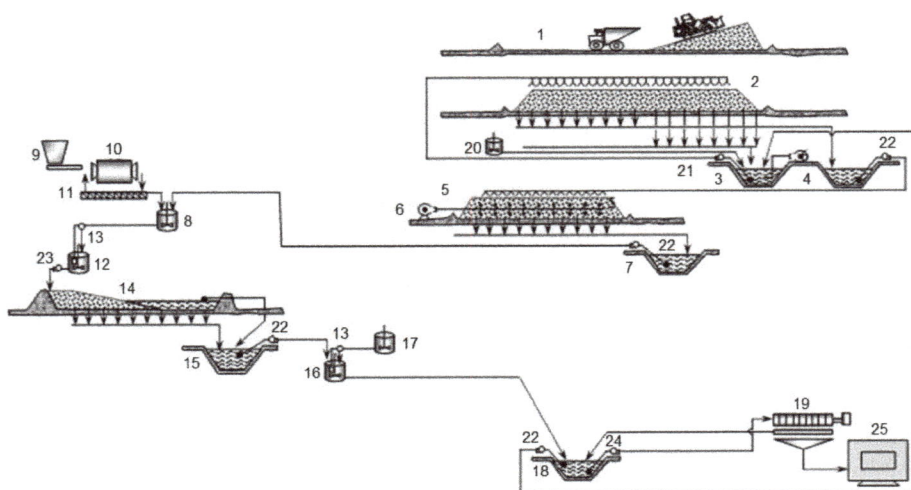

Figure 2. Process flow diagram of the ore heap leaching process at the Allarechensky Deposit Dumps [59]. 1—heap at the forming stage; 2—heap at the irrigation stage; 3—raffinate pond; 4—pregnant leach solution (PLS) pond; 5—heap for iron oxidation; 6—aeration system; 7—oxidized PLS pond; 8—neutralizing tank; 9—crushed lime bunker; 10—ball mill; 11—spiral classifier; 12—setting tank; 13—automatic titration system; 14—sludge storage; 15—storage pond for neutralized PLS; 16—precipitating tank; 17—reagent tank; 18—settling pond; 19—filter-press; 20—bacteria cultivation tank; 21–24 —pumps; 25—furnace.

5.2. Copper-Nickel Ore Tailings

Copper-nickel ore processing tailings from the Processing Plant no.1 of the Pechenganickel Complex of JSC Kola Mining and Metallurgical Company in the town of Zapolyarny, the Murmansk region, are thought to be one of the largest technogenic deposits in Russia. These ores are characterized by the predominance of grain size below 0.1 mm. In many cases, up to 50% of grains are below 0.044 mm. Serpentines (about 60%) prevail in the mineral composition of tailings [10]. Piroxenes, amphiboles, talcum, chlorites, quartz, and feldspar are abundant, as well. The main ore minerals are magnetite (Fe_3O_4), pyrrhotite ($Fe_{1-x}S$), pentlandite (($Fe,Ni)_9S_8$), and chalcopyrite ($CuFeS_2$). Total content of sulfide minerals is 1–3%. The losses of nickel with tailings during the enrichment of disseminated copper-nickel ores is up to 30%. Processing of finely-dispersed technogenic products or natural raw materials with high content of laminal hydrosilicates (clay minerals) involve colmatation problems, which decrease dump permeability for leaching solutions, and they lead to the process shutdown. Such effects were observed during the storage of copper-nickel ores processing tailings. Presence of chlorites, and mixed layer buildups in mature tailings results in formation of artificial clay-like ground, and decrease of filtration coefficient by more than 100 times [12].

One of the ways to solve this problem is material agglomeration with binders. Taking into account the excess amount of sulfuric acid produced by operations of the Kola Mining and Metallurgical Company (Zapolyarny City, Russia) and problems with it is sales; it is believed that a sulfatisation process involving pelletizing in agglomerators would be a promising technology. H_2SO_4 is used as a binding material in this process [60]. Pellets were experimentally produced at the ratio S:L = 5–3:1. H_2SO_4 solution with concentration of 10% was used as a binder. The pellet diameter for the tests was 0.8–1 cm. The highest pellet hardness under compression amounted to 2.8–3 MPa.

The tailings contained: Ni 0.17%, Cu 0.07%, Co 0.01%. Percolation leaching was fulfilled using 1% sulfuric acid in the columns with 45 mm diameter during 110 days.

About 60% of nikel contained in tailings transfers into solution during 110 days. Copper demonstrates a lower leaching rate (about 44%) as the metal occurs in the form of chalcopyrite ($CuFeS_2$) [2,21]. Relatively low leaching rate for cobalt (about 41%) are likely to be explained by the metal existence in the form of isomorphic contaminant in magnetite (Fe_3O_4). The leaching rate of metals per day amounted to: 0.55% of nickel, 0.37% of cobalt, and 0.4% of copper.

5.3. Low-Grade Copper-Nickel Ores of the Monchepluton Deposits

Previous mineralogical and processing tests that were performed on the off-balance copper-nickel ores from the Monchepluton deposits demonstrated their general amenability to hydrometallurgical processing [60,61]. It was necessary then to find process design solutions, which could ensure the intensification of sulfide mineral dissolution.

The samples of low-grade copper-nickel ores from three deposits were used for studies: Lake Moroshkovoye, Nyud Terrasa, and Nittis-Kumuzhya-Travyanaya (NKT). Nickel and copper content in the samples is presented in the Table 1 below.

To intensify heap leaching, we carried out ore grinding with further sulfuric agglomeration. The ore was ground to 0.05–1 mm.

Table 1. Nickel and copper content in the ore samples.

Content, %					
Nickel			Copper		
Lake Moroshkovoye	NKT	Nyud Terrasa	Lake Moroshkovoye	NKT	Nyud Terrasa
0.547	0.567	0.465	0.036	0.363	0.044

The ore from the Lake Moroshkovoye deposit demonstrated the most intensive nickel leaching. After 32 days, the recovery was more than 60%. Herewith, about 20% of nickel that was transferred to solution at the stage of water leaching during one day (Figure 3a).

Nickel from the Nyud Terrasa deposit showed much worse leaching performance. The Ni recovery was about 10% during the same period. Obviously, it can be explained by fine sulfide impregnations prevailing in this type of ore. Therefore, after dissolving coarser minerals at the water leaching stage, the next growth of nickel recovery amounted to less than 2% up to the end of the test (Figure 3a).

As anticipated, copper leached much slower in comparison with nickel, and that was due to its presence in the form of chalcopyrite [2,21]. The best copper recoveries were achieved during the tests with the NKT deposit ore, they amounted to 8% (Figure 3b). The lowest Cu recovery, as in the case with Ni, was observed for the Nyud Terrasa deposit ore, it amounted to 1.95%. Leaching intensities of the non-ferrous metals per day are presented in Table 2.

Table 2. Leaching intensities of non-ferrous metals per day.

Nickel			Copper		
Lake Moroshkovoye	NKT	Nyud Terrasa	Lake Moroshkovoye	NKT	Nyud Terrasa
1.87%	0.97%	0.32%	0.13%	0.24%	0.06%

Consequently, sulfuric acid agglomeration of finely ground ores leads to considerable improvement of leaching performance. For example, during leaching of the Lake Moroshkovoye deposit ore crushed to 2–3 mm, the rate of recovery into solution per day amounted to 0.48% for nickel, 0.08% for copper, which is lower than the values that were demonstrated by ore in pellets by 3.9 and 1.6 times, respectively.

An optimization of the agglomeration mode and parameters is necessary to allow for the usage of increased grain sizes for pellets production. An increase of acid concentration for pelletizing enables further water leaching with solution circulation [62].

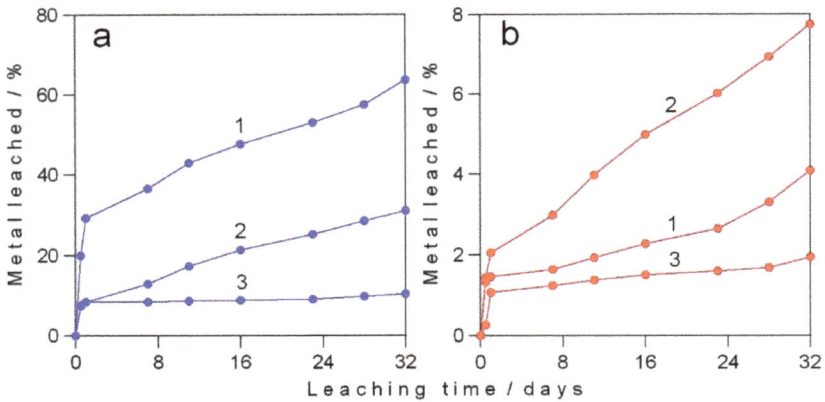

Figure 3. Kinetics of nickel (**a**) and copper (**b**) leaching from the ore samples. Deposits: Lake Moroshkovoye (1), Nittis-Kumuzhya-Travyanaya (NKT) (2), and Nyud Terrasa (3).

5.4. Percolation Bioleaching of Non-Ferrous Metals from Low-Grade Copper-Nickel Ore

Thionic bacteria facilitating bioleaching of sulfide ores were produced from sulfides and runoffs from the Allarechensky Deposit Dumps copper-nickel ores. For further work in a bioreactor at 27 °C and steady aeration bacterial biomass with a cell number of 10^9 cells/mL was produced. A scale-up process was carried out during 10–12 days on the mineral medium containing ferrous iron.

Test leaching of ores from the Nyud Terrasa deposit was conducted. The content of Ni and Cu in the ore amounted to 0.42% and 0.15%, respectively. The ore was crushed to 1–3 mm, and loaded to percolators. Bacterial leaching was carried out with circulation of a certain solution portion. The ore was preliminary saturated with water. Solution was added once in two days in the amount of 25 mL.

Figure 4 demonstrates changes of solutions pH and Eh after filtration in a percolator during 76 days of the test. The first data points indicate ore water saturation.

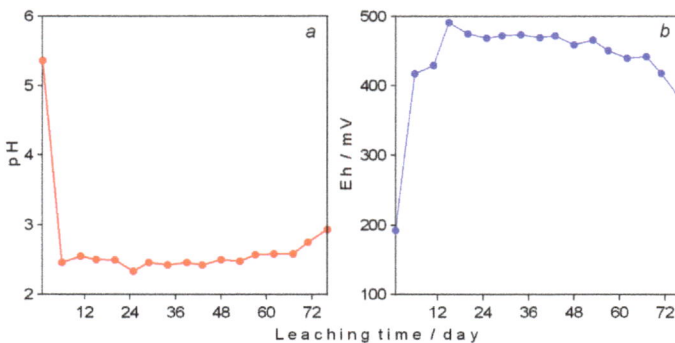

Figure 4. Changes of pH (**a**) and redox potential (**b**) in the solutions after filtration.

Correlation of ferrous and ferric iron concentrations in the solutions after filtration in a percolator indicates the oxidation of Fe^{2+} (Figure 5a).

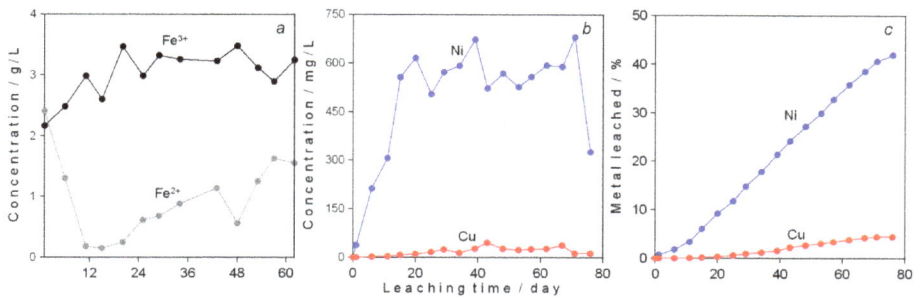

Figure 5. Ferrous and ferric ion concentrations (**a**), non-ferrous metal concentrations in the leach solutions (**b**), and their extraction (**c**).

Figure 5b,c present non-ferrous metal concentrations in pregnant solutions. As can be seen, nickel concentrations are sufficiently high. Copper concentrations are expectedly much lower. In general, the solutions are acceptable for further processing. Figure 5c shows the curves of nickel and copper recovery vs. process duration. Obviously, bioleaching enabled achieving acceptable values of non-ferrous metal recovery [63].

6. Conclusions

Experience of precious metals heap leaching in the regions with arctic and subarctic climate, and subzero average annual temperatures proves the applicability of these methods for geotechnological processing of sulfide ores containing non-ferrous metals.

Cryomineralogenesis studies in natural and technogenic sulfide-containing massifs, development of passive leaching methods for gold- and silver-bearing raw materials, as well as metals transfer into liquid phase under subzero temperature with new oxidizing agents deserve consideration and comprehensive experiments that are based on sulfide copper-nickel ores.

When processing the Udokan deposit copper ores, it is economically reasonable to combine pyro- and hydrometallurgical methods to increase the use of raw materials. Oxidized, oxide-sulfide, and low-grade complex ores should be processed with heap leaching.

The authors analyzed the experience of Talvivaara Mining Company Plc (a Finnish company), related to the nickel mine operation at the Talvivaara polymetallic deposit situated in the subarctic area, in the north-east of Finland.

Using the technogenic deposit, Allarechensky Deposit Dumps, as an example, we determined optimal combination of magnetic separation and biological leaching for sulfide copper-nickel ores processing ensuring the operation profitability and investment attractiveness of such technogenic deposits.

At the level of laboratory tests, we showed solutions and directions for more intensive methods of non-ferrous metals leaching from low-grade ores and waste of the mining and processing complex in the Murmansk region. The results will form a scientific basis for development of optimal conditions and parameters for ore and mine waste hydrometallurgical processing. This will enable justifying the involvement of low-grade copper-nickel ores and processing tailings containing strategic non-ferrous metals into commercial processing.

Author Contributions: Conceptualization, V.A.M., S.G.S., A.V.S. and D.V.M.; Methodology, V.A.M., S.G.S., A.V.S. and D.V.M.; Investigation, V.A.M., S.G.S., A.V.S. and D.V.M.; Writing-Original Draft Preparation, V.A.M., S.G.S., A.V.S. and D.V.M.; Writing-Review & Editing, V.A.M., S.G.S., A.V.S. and D.V.M.; Visualization, V.A.M., S.G.S., A.V.S. and D.V.M.

Funding: The work has been performed with support of RFBR (project No. 18-05-60142 Arctic).

Conflicts of Interest: The authors declare no conflict of interest.

References

1. Chanturiya, V.A.; Kozlov, A.P. Development of physical-chemical basis and working out of innovation technologies of deep processing of anthropogenic mineral raw materials. *Gorn. Zhurnal (Min. J.)* **2014**, *7*, 79–84. (In Russian)

2. Halezov, B.D. *Copper and Copper–Zinc Ore Heap Leaching*; Ural Branch of Russian Academy of Sciences: Ekaterinburg, Russia, 2013; p. 360. (In Russian)

3. Petersen, J. Heap leaching as a key technology for recovery of values from low-grade ores—A brief overview. *Hydrometallurgy* **2016**, *165*, 206–212. [CrossRef]

4. Karavaiko, G.I.; Rossi, D.; Agate, A.; Grudev, S.; Avakyan, Z.A. *Biogeotechnology of Metals: Practical Guide*; Tsentr Mezhdunarodnykh Proektov GKNT: Moscow, Russia, 1989; p. 375. (In Russian)

5. Kondrat'eva, T.F.; Bulaev, A.G.; Muravyov, M.I. *Microorganisms in Biotechnologies of Sulfide Ores Processing*; Nauka: Moscow, Russia, 2015; p. 212. (In Russian)

6. Watling, H.R. Review of biohydrometallurgical metals extraction from polymetallic mineral resources. *Minerals* **2015**, *5*, 1–60. [CrossRef]

7. McDonough, W.; Braungart, M.; Anastas, P.T.; Zimmerman, J.B. Applying the principles of GREEN engineering to cradle-to-cradle design. *Environ. Sci. Technol.* **2003**, *37*, 434A–441A. [CrossRef] [PubMed]

8. Pakhomova, N.V.; Richter, K.K.; Vetrova, M.A. Transition to circular economy and closed-loop supply chains as driver of sustainable development. *St Petersbg. Univ. J. Econ. Stud.* **2017**, *33*, 244–268. [CrossRef]

9. Lèbre, É.; Corder, G.D.; Golev, A. Sustainable practices in the management of mining waste: A focus on the mineral resource. *Miner. Eng.* **2017**, *107*, 34–42. [CrossRef]

10. Masloboev, V.A.; Seleznev, S.G.; Makarov, D.V.; Svetlov, A.V. Assessment of eco-hazard of copper-nickel ore mining and processing waste. *J. Min. Sci.* **2014**, *50*, 559–572. [CrossRef]

11. Chanturiya, V.A.; Makarov, V.N.; Makarov, D.V.; Vasil'eva, T.N.; Pavlov, V.V.; Trofimenko, T.A. Influence exerted by storage conditions on the change in properties of copper-nickel technologies products. *J. Min. Sci.* **2002**, *38*, 612–617. [CrossRef]

12. Chanturiya, V.A.; Makarov, V.N.; Makarov, D.V.; Vasil'eva, T.N. Forms of nickel in storage tailings of copper–nickel ores. *Dokl. Earth Sci.* **2004**, *399*, 1150–1152.

13. Seleznev, S.G.; Boltyrov, V.B. Ecology of Mining-Generated Object "Allarechensky Deposit Dumps" (Pechenga District, Murmansk Region). *News High. Inst. Min. J.* **2013**, *7*, 73–79. (In Russian)

14. Adamov, E.V.; Panin, V.V. *Biotechnology of Metals: Course of Lectures*; MISiS: Moscow, Russia, 2008; p. 153. (In Russian)

15. Kashuba, S.G.; Leskov, M.I. Heap leaching in the Russian practice—An overview of the experience and analysis of prospects. *Zoloto I Tehnologii* **2014**, *1*, 10–14. (In Russian)

16. Watling, H.R. The bioleaching of sulphide minerals with emphasis on copper sulphides—A review. *Hydrometallurgy* **2006**, *84*, 81–100. [CrossRef]

17. Dement'ev, V.E.; Druzhinina, G.E.; Gudkov, S.S. *Heap Leaching of Gold and Silver*; Irgiredmet: Irkutsk, Russia, 2004; p. 252. (In Russian)

18. Sovmen, V.K.; Gus'kov, V.N.; Belyi, A.V.; Kuzina, Z.P.; Drozdov, S.V.; Savushkina, S.I.; Maiorov, A.M.; Zakraevskii, M.P. *Processing of Gold-Bearing Ores with the Use of Bacterial Oxidation in the Conditions of Far North*; Nauka: Novosibirsk, Russia, 2007; p. 144. (In Russian)

19. Golik, V.I.; Zaalishvili, V.B.; Razorenov, Yu.I. Uranium leaching experience. *Min. Inf. Anal. Bull.* **2014**, *7*, 97–103.

20. Anjum, F.; Shahid, M.; Akcil, A. Biohydrometallurgy techniques of low grade ores: A review on black shale. *Hydrometallurgy* **2012**, *117–118*, 1–12. [CrossRef]

21. Watling, H.R. The bioleaching of nickel sulphides. *Hydrometallurgy* **2008**, *91*, 70–88. [CrossRef]

22. Watling, H.R.; Elliot, A.D.; Maley, M.; van Bronswijk, W.; Hunter, C. Leaching of a low-grade, copper-nickel sulfide ore. 1. Key parameters impacting on Cu recovery during column bioleaching. *Hydrometallurgy* **2009**, *97*, 204–212. [CrossRef]

23. Maley, M.; van Bronswijk, W.; Watling, H.R. Leaching of a low-grade, copper-nickel sulfide ore 2. Impact of aeration and pH on Cu recovery during abiotic leaching. *Hydrometallurgy* **2009**, *98*, 66–72. [CrossRef]

24. Maley, M.; van Bronswijk, W.; Watling, H.R. Leaching of a low-grade, copper-nickel sulfide ore 3. Interactions of Cu with selected sulfide minerals. *Hydrometallurgy* **2009**, *98*, 73–80. [CrossRef]

25. Halinen, A.-K.; Rahunen, N.; Kaksonen, A.H.; Puhakka, J.A. Heap bioleaching of a complex sulfide ore: Part I. Effect of temperature on base metal extraction and bacterial compositions. *Hydrometallurgy* **2009**, *98*, 92–100. [CrossRef]

26. Halinen, A.-K.; Rahunen, N.; Kaksonen, A.H.; Puhakka, J.A. Heap bioleaching of a complex sulfide ore: Part II. Effect of temperature on base metal extraction and bacterial compositions. *Hydrometallurgy* **2009**, *98*, 101–107. [CrossRef]

27. Qin, W.; Zhen, S.; Yan, Z.; Campbell, M.; Wang, J.; Liu, K.; Zhang, Y. Heap bioleaching of a low-grade nickel-bearing sulfide ore containing high levels of magnesium as olivine, chlorite and antigorite. *Hydrometallurgy* **2009**, *98*, 58–65. [CrossRef]

28. Zhen, S.; Yan, Z.; Zhang, Y.; Wang, J.; Campbell, M.; Qin, W. Column bioleaching of a low grade nickel-bearing sulfide ore containing high magnesium as olivine, chlorite and antigorite. *Hydrometallurgy* **2009**, *96*, 337–341. [CrossRef]

29. Yang, C.; Qin, W.; Lai, S.; Wang, J.; Zhang, Y.; Jiao, F.; Ren, L.; Zhuang, T.; Chang, Z. Bioleaching of a low grade nickel-copper-cobalt sulfide ore. *Hydrometallurgy* **2011**, *106*, 32–37. [CrossRef]

30. Bhatti, T.M.; Bigham, J.M.; Vuorinen, A.; Tuovinen, O.H. Chemical and bacterial leaching of metals from black schist sulfide minerals in shake flasks. *Int. J. Miner. Process.* **2012**, *110–111*, 25–29. [CrossRef]

31. Mandziak, T.; Pattinson, D. Experience-based approach to successful heap leach pad design. *Min. World* **2015**, *12*, 28–35.

32. Petersen, J.; Dixon, D.G. Modelling zinc heap bioleaching. *Hydrometallurgy* **2007**, *85*, 127–143. [CrossRef]

33. Mellado, M.E.; Galvez, E.D.; Cisternas, L.A.; Ordonez, J. A posteriori analysis of analytical models for heap leaching. *Miner. Metall. Process.* **2012**, *29*, 103–112.

34. Ding, D.; Song, J.; Ye, Y.; Li, G.; Fu, H.; Hu, N.; Wang, Y. A kinetic model for heap leaching of uranium ore considering variation of model parameters with depth of heap. *J. Radioanal. Nucl. Chem.* **2013**, *298*, 1477–1482. [CrossRef]

35. McBride, D.; Gebhardt, J.E.; Croft, T.N.; Cross, M. Modeling the hydrodynamics of heap leaching in sub-zero temperatures. *Miner. Eng.* **2016**, *90*, 77–88. [CrossRef]

36. Lodeyschikov, V.V. Processing of nickel ores by heap leaching bacteria. The experience of the Finnish company Talvivaara. *Zolotodobyicha.* **2009**, *132*, 12–14. (In Russian)

37. Sinha, K.P.; Smith, M.E. Cold climate heap leaching. In Proceedings of the 3rd International Conference on Heap Leach Solutions, Reno, NV, USA, 12–16 September 2015.

38. Smith, K.E. Cold weather gold heap leaching operational methods. *JOM* **1997**, *49*, 20–23. [CrossRef]

39. Shesternev, D.M.; Myazin, V.P.; Bayanov, A.E. Heap gold leaching in permafrost zone in Russia. *Gornyi Zhurnal (Min. J.)* **2015**, *1*, 49–54. (In Russian)

40. Ptitsyn, A.B. *Geochemical Fundamentals of Metal Geotechnology in Permafrost Conditions*; Nauka: Novosibirsk, Russia, 1992; p. 120. (In Russian)

41. Ptitsyn, A.B.; Sysoeva, E.I. Cryogenic mechanism of the Udokan oxidizing area formation. *Russ. Geol. Geophy.* **1995**, *36*, 90–97.

42. Abramova, V.A.; Ptitsyn, A.B.; Markovich, T.I.; Pavlyukova, V.A.; Epova, E.S. *Geochemistry of Oxidation in Permafrost Zones*; Nauka: Novosibirsk, Russia, 2009; p. 88. (In Russian)

43. Yurgenson, G.A. The cryomineralogenesis of minerals in the technological massifs. In *Proceedings of Mineralogy of technogenesis—2009*; Institute of Mineralogy, Ural Branch of RAS: Miass, Russia, 2009; pp. 61–75. (In Russian)

44. Markovich, T.I. Processes of Heavy Metal Sulphides Oxidation with Nitrous Acid. Ph.D. Thesis, Trofimuk Institute of Petroleum Geology and Geophysics, Siberian Branch of the Russian Academy of Sciences (IPGG SB RAS), Novosibirsk, Russia, February 2000. (In Russian)

45. Ptitsyn, A.B.; Markovich, T.I.; Pavlyukova, V.A.; Epova, E.S. Modeling cryogeochemical processes in the oxidation zone of sulfide deposits with the participation of oxygen-bearing nitrogen compounds. *Geochem. Int.* **2007**, *45*, 726–731. [CrossRef]

46. Abramova, V.A.; Parshin, A.V.; Budyak, A.E. Physical and chemical modeling of the influence of nitrogen compounds on the course of geochemical processes in the cryolithozone. *Kriosfera Zemli.* **2015**, *9*, 32–37. (In Russian)

47. Abramova, V.A.; Parshin, A.V.; Budyak, A.E.; Ptitsyn, A.B. Geoinformation modeling of sulfide frost weathering in the area of Udokan deposit. *J. Min. Sci.* **2017**, *53*, 501–597. [CrossRef]

48. Khalezov, B.D. Problems of Udokansky field ores processing. *Min. Inf. Anal. Bull.* **2014**, *8*, 103–108. (In Russian)

49. Riekkola-Vanhanen, M. Talvivaara black schist bioheapleaching demonstration plant. *Adv. Mater. Res.* **2007**, *20–21*, 30–33. [CrossRef]

50. Puhakka, J.A.; Kaksonen, A.H.; Riekkola-Vanhanen, M. Heap leaching of black schist. In *Biomining*; Rawlings, D.E., Johnson, D.B., Eds.; Springer-Verlag: Berlin, Germany, 2007; pp. 139–151.

51. Halinen, A.K.; Rahunen, N.; Määttä, K.; Kaksonen, A.H.; Riekkola-Vanhanen, M.; Puhakka, J. Microbial community of Talvivaara demonstration bioheap. *Adv. Mater. Res.* **2007**, *20–21*, 579. [CrossRef]

52. Riekkola-Vanhanen, M. Talvivaara Sotkamo mine—Bioleaching of a polymetallic nickel ore in subarctic climate. *Nova Biotechnol.* **2010**, *1011*, 7–14.

53. Riekkola-Vanhanen, M.; Palmu, L. Talvivaara Nickel Mine—From a project to a mine and beyond. In *Proceedings of Symposium Ni-Co 2013*; Battle, T., Moats, M., Cocalia, V., Oosterhof, H., Alam, S., Allanore, A., Jones, R., Stubina, N., Anderson, C., Wang, S., Eds.; Springer International Publishers: Cham, Switzerland, 2016; pp. 269–278.

54. *Annual Report Talvivaara 2013*; Talvivaara Sotkamo Ltd.: Tuhkakylä, Finland, 2013.

55. Golovko, E.A.; Rozental, A.K.; Sedel'nikov, V.A.; Suhodrev, V.M. *Chemical and Bacterial Leaching of Copper-Nickel ores*; Nauka: Leningrad, Russia, 1978; p. 199. (In Russian)

56. Lyalikova, N.N. Bacteria role in the sulfide ores oxidizing of the copper-nickel deposits on the Kola Peninsula. *Microbiology* **1961**, *30*, 135–139. (In Russian)

57. Karavaiko, G.I.; Kuznetsov, S.I.; Golomzik, A.I. *Role of Microorganisms in Leaching Metals from Ores*; Nauka: Moscow, Russia, 1972; p. 248. (In Russian)

58. Seleznev, S.G. Unconventional effective ways to enrich the sulfide copper-nickel ores on the example of Allarechenskiy technogenic deposit. *News High. Inst. Min. J.* **2011**, *8*, 118–125. (In Russian)

59. Seleznev, S.G. The Specificity and Development Problems for the Dumps of Sulfide Copper-Nickel Ores Allarechensky Deposit. Ph.D. Thesis, Ural State Mining University, Yekaterinburg, Russia, December 2013. (In Russian)

60. Svetlov, A.; Kravchenko, E.; Selivanova, E.; Seleznev, S.; Nesterov, D.; Makarov, D.; Masloboev, V. Perspectives for heap leaching of non-ferrous metals (Murmansk Region, Russia). *J. Pol. Min. Eng. Soc. (Inzynieria Mineralna)* **2015**, *36*, 231–236.

61. Svetlov, A.; Seleznev, S.; Makarov, D.; Selivanova, E.; Masloboev, V.; Nesterov, D. Heap leaching and perspectives of bioleaching technology for the processing of low-grade copper-nickel sulfide ores in the Murmansk region, Russia. *J. Pol. Min. Eng. Soc. (Inzynieria Mineralna)* **2017**, *39*, 51–59.

62. Svetlov, A.V.; Makarov, D.V.; Goryachev, A.A. Directions for intensification of leaching of non-ferrous metals on the example of low-grade copper-nickel ore deposits in the Murmansk region. In *Proceedings of Mineralogy of Technogenesis—2017*; Institute of Mineralogy, Ural Branch of RAS: Miass, Russia, 2017; pp. 154–162. (In Russian)

63. Fokina, N.V.; Yanishevskaya, E.S.; Svetlov, A.V.; Goryachev, A.A. Functional activity of microorganisms in mining and processing of copper-nickel ores in the Murmansk Region. *Vestnik of MSTU (Murmansk State Technical University)* **2018**, *21*, 109–116. (In Russian)

![minerals logo] *minerals*

MDPI

Article

Alkali-Activated Binder Based on Milled Antigorite

Elena V. Kalinkina, Basya I. Gurevich and Alexander M. Kalinkin *

Tananaev Institute of Chemistry-Subdivision of the Federal Research Centre, Kola Science Centre of the Russian Academy of Sciences, Akademgorodok 26a, 184209 Apatity, Murmansk Region, Russia; kalinkina@chemy.kolasc.net.ru (E.V.K.); gurevich1931@yandex.ru (B.I.G.)
* Correspondence: kalinkin@chemy.kolasc.net.ru; Tel.: +7-81555-79523

Received: 29 September 2018; Accepted: 2 November 2018; Published: 4 November 2018

check for updates

Abstract: Antigorite is a very common rock-forming mineral and it is often present in mining wastes. Utilization of these wastes is a very important issue from the environmental point of view. A potential use for mining wastes is for the production of building materials. This study investigated the alkali activation of antigorite and antigorite-containing ore dressing tailings (AT) milled in a planetary ball mill in an air or CO_2 atmosphere. The specific surface area, amorphisation, and dehydroxylation of milled antigorite and AT were examined, and their effect on the cementitious properties was investigated. Binders were prepared by mixing the milled antigorite or AT with liquid glass and curing at 20 ± 2 °C in dry (relative humidity of $65 \pm 5\%$) or humid (relative humidity of $95 \pm 5\%$) conditions for up to 28 days. Curing at dry conditions was found to produce binders with increased strengths. The compressive strength of the alkali-activated binder also increased with increased milling time. For AT milled in air for 4 min and cured in dry conditions for 28 days, the compressive strength was 49 MPa. The milling atmosphere (air or CO_2) influenced the cementitious properties of the alkali activated binder to a small extent.

Keywords: antigorite; ore dressing tailings; mechanical activation; alkali-activated binder

1. Introduction

Antigorite, along with lizardite and chrysotile, is one of the major polymorphs of the serpentine group with ideal composition of $Mg_3Si_2O_5(OH)_4$. These hydrated, 1:1 layered magnesium-rich silicates are very common rock-forming minerals. Their crystal structures—which consist of alternating tetrahedral (silica-like) and octahedral (brucite-like) sheets—are essentially the same, differing only in the curvature of the lattice planes [1].

Large amounts of serpentine minerals are often present in mining wastes (overburden rocks, ore dressing tailings and so on), including waste from copper and nickel mines [2,3]. In fact, the ore dressing tailings of the Pechenganickel works (JSC Kola Mining and Metallurgical Company, Murmansk Region, Russia) typically contain an average of 60% serpentine (mainly in the form of antigorite) [4]. Considering that the total volume of the accumulated tailings is around 250 Mt [5], the amount of serpentine in them can be estimated to be 150 Mt. Utilization of the ore dressing tailings of the Pechenganickel works is a very important issue from an environmental point of view [4,5].

A potential use for mining waste is for the production of building materials. It has long been known that thermal treatment of inert serpentinite confers cementitious properties to the ground rock [6]. Heating serpentinite in specific conditions (500–700 °C) results in dehydroxylation, amorphisation and the formation of reactive metastable phases. However, overheating leads to crystallisation of forsterite and enstatite, as well as loss of hydraulic activity. It should be noted that cements based on thermally-treated serpentine have not yet found a wide practical application. This is mainly due to the fact that achieving the desired binding properties requires the calcination of large volumes of raw materials within a narrow temperature range (depending on serpentinite composition), with strictly fixed residence. Recently, the increased reactivity of dehydroxylated serpentine minerals has been extensively studied with respect to their aqueous dissolution and carbonation (reviewed in [7,8]). This research has been driven by current interests in the extraction of metals from serpentine, and the reduction of anthropogenic CO_2 emissions. Mechanical activation of serpentine and other kinds of pre-treatment can have similar effects as thermal treatment with regards to dehydroxylation and reactivity [7–12].

A promising pathway for the preparation of binders from industrial waste is alkali-activation technology. During the last two decades, this technology has played an increasingly important part in the development of novel, eco-friendly alternatives to Portland cement. Alkali-activated materials are binder systems that can be prepared by the reaction of ground silicate (aluminosilicate) with an alkali agent (for example, sodium hydroxide solution or soluble sodium silicate). Geopolymers—such as ones based on class F fly ashes and calcined clays—are regarded as a subclass of alkali-activated materials, and are synthesised almost exclusively from aluminosilicate powder with low contents of calcium and other divalent and trivalent metals such as magnesium and iron [13–15]. A large body of research has shown that alkali-activated binders based on various industrial wastes often display comparable or even higher physical-mechanical and other properties—as well as a notably lower CO_2 footprint—than conventional Portland cement [16–24].

To our knowledge, no previous research has investigated alkali-activated materials based on serpentine minerals. In the present study, we investigated the cementitious properties of antigorite as a component of alkali-activated binder. Because of the low hydraulic activity of serpentine, we employed a pre-treatment step of mechanical activation. The efficiency of mechanical activation for increasing the reactivity of solids is well established [25–28], and is a particularly effective technique for improving the mechanical properties of alkali-activated materials [29–35]. The term "mechanical activation" (MA) in this paper refers to the structural-chemical changes on the surface and in the bulk of minerals caused by milling.

Previous studies have revealed that Ca- and Mg-containing silicates—including serpentine—chemisorb considerable amounts of atmospheric CO_2 (up to ~10–12 wt. %) in the form of carbonate ions during prolonged grinding in air in a mortar grinder [36,37]. The carbonisation of silicates is notably increased if MA is carried out in a pure CO_2 atmosphere [38–40]. It has also been shown that MA of magnesia ferriferous Cu-Ni slag in CO_2 resulted in an alkali-activated slag binder with higher compressive strength compared with that produced using MA in air [41]. The aim of this study was to investigate the influence of MA on the alkaline-activation behaviour of antigorite and antigorite-containing ore dressing tailings. The MA was carried out in both air and in CO_2 atmospheres in order to elucidate the influence of carbonisation on the binding properties of antigorite.

2. Materials and Methods

2.1. Mineral Samples and Preparation

Monomineral samples of antigorite (>98%) were obtained from the Pilgujärvi massif of the Kola Peninsula, and copper-nickel ore dressing tailings were obtained from the Pechenganickel works (JSC Kola Mining and Metallurgical Company, Murmansk Region, Russia). Antigorite was crushed and hand-picked to remove identifiable impurities and analysed using X-ray diffraction (XRD). It was then ground and sieved with a 250 μm mesh. This sample is referred to as initial antigorite. The antigorite-containing ore dressing tailings, hereafter referred as AT, contained ~70 wt. % serpentine (mainly in the form of antigorite); 10–15 wt. % talc; 10–15 wt. % olivine, pyroxenes, amphiboles and magnetite and minor amounts of carbonate minerals and sulphides (pyrrhotite, pentlandite and chalcopyrite). The Brunauer–Emmett–Teller (BET) surface areas of initial antigorite and AT were 1.63 and 1.44 m²/g, respectively. The chemical compositions of antigorite and AT are given in Table 1.

Table 1. Chemical composition of antigorite and antigorite-containing ore dressing tailings (AT) (wt. %).

Mineral Component	SiO_2	MgO	Fe_2O_3	FeO	CaO	Al_2O_3	NiO	Na_2O	K_2O	MnO	CO_2	LOI
Antigorite	38.74	33.82	3.99	8.83	0.66	1.29	0.04	0.06	0.05	0.19	0.13	11.4
AT	35.64	25.2	6.42	12.63	2.16	3.29	0.19	0.39	0.6	0.21	0.50	9.84

2.2. Mechanical Activation

Mechanical activation was carried out in an AGO-2 laboratory planetary mill (Novic, Novosibirsk, Russia) [26] at a centrifugal force of 40 g in an air or CO_2 atmosphere. Steel balls 8 mm in diameter were used as milling bodies. For the majority of experiments, the ratio of the mass of steels balls to the mass of mineral sample was six (240 g of steels balls and 40 g of mineral sample). These conditions are referred to as moderate conditions of MA. In order to study the BET surface area of antigorite as a function of milling time under more intensive mechanical treatment, several experiments were carried out changing the ratio of the mass of steel balls to the mass of mineral to 20 (referred to as hard conditions of MA). All other experiments employed moderate milling conditions. MA was carried out batchwise for different periods of time, ranging from 30 s to 600 s.

Prior to MA in a CO_2 atmosphere, the grinding vial containing the balls and sample was purged with a slow jet of CO_2 to displace all air. The moisture content of CO_2 was less than 0.05 vol. %. Every 60 s, the mill was stopped and any powder sticking to the vial was scraped off and mixed thoroughly with a spatula. Milling was resumed after refilling the vial with CO_2.

2.3. Powder Characterisation

Analysis by XRD was performed using a XRD 6000 diffractometer (Shimadzu Corporation, Tokyo, Japan) with Cu-Kα radiation; 2-theta/step size: 0.02°; dwell time: 1 s. Fourier transform infrared (FT-IR) transmission spectra were recorded with a Nicolet 6700 FTIR spectrometer (Thermo Electron Scientific Instruments Corporation, Madison, WI, USA) using potassium bromide tablets. The carbon contents of the slag samples were determined using an ELTRA CS-2000 analyser (ELTRA GmbH, Haan, Germany). The specific surface area was measured using the standard Blaine test and nitrogen BET method with a Flow-Sorb II 2300 instrument (Micromeritics, Norcross, GA, USA). The thermal analyses of differential thermal analysis (DTA) and thermogravity (TG) measurements were performed on a STA 409 PC/PG Luxx® instrument (NETZSCH, Selb, Germany), heating the samples at a rate of 10 °C/min in an alumina crucible under argon.

2.4. Leaching Analysis

Leaching analyses were performed by mixing mechanically activated antigorite or AT with distilled water or 2 M and 5 M NaOH solution at a mineral:liquid ratio of 1:40, and then stirring continuously for 5 h at 22 ± 2 °C. After filtration, the concentrations of elements in solution were determined with an Optima 8300 ICP-OES Spectrometer (PerkinElmer, Inc., Waltham, MA, USA).

2.5. Preparation and Characterisation of Alkali-Activated Binders

Milled antigorite or AT was mixed with sodium silicate solution ("liquid glass") and distilled water by hand for three minutes, resulting in a paste with plastic consistency. The amount of liquid glass added was expressed by amount of Na_2O in liquid glass relative to the amount of mineral component, wt. %. The water content was adjusted to give the same workability of the pastes; therefore, the water content varies. Water to solid ratio (w/s) was defined without considering water present in the liquid glass. Characteristics of sodium silicate solutions are as follows: (1) solid content of 45.2% with $SiO_2 = 31.0\%$, $Na_2O = 14.2\%$ and a silica modulus (molar ratio of SiO_2/Na_2O in liquid glass) of 2.25; (2) solid content of 40.7% with $SiO_2 = 30.7\%$, $Na_2O = 10.0\%$ and a silica modulus of 3.18. Examples of mixture compositions are given in Table 2.

Table 2. Examples of mixture compositions.

Mineral Component	Milling Time, s	Liquid Glass		w/s	Paste Composition, wt. %		
		% Na$_2$O *	Silica Modulus		Mineral Component	Liquid Glass	Added Water
Antigorite	600	3.0	2.25	0.26	67.97	14.36	17.67
AT**	600	3.0	2.25	0.22	69.87	14.76	15.37
AT	240	4.0	3.18	0.25	60.53	24.33	15.14

* Amount of Na$_2$O in liquid glass relative to the amount of mineral component, wt. %. ** Antigorite-containing ore dressing tailings.

The pastes were cast into $1.41 \times 1.41 \times 1.41$ cm^3 cubic molds. Prepared specimens were cured in a relative humidity of $65 \pm 5\%$ (referred to as dry conditions) or $95 \pm 5\%$ (referred to as humid conditions) at $22 \pm 2\,^{\circ}$C for 24 h. After demolding, the specimens were further cured to testing time in the same condition as applied in the first 24 h. Compressive strength data were obtained from an average of 3 samples in 1, 7 and 28 days.

3. Results and Discussion

3.1. Effect of Mechanical Activation on Antigorite and AT

3.1.1. BET Specific Surface Area

The effects of milling time and conditions on the BET specific surface area (S_{sp}) of antigorite and AT are shown in Figure 1. When moderate MA was carried out in air, the S_{sp} continuously increased, reaching 30 m^2/g and 23 m^2/g by 600 s of milling for antigorite and AT, respectively. The lower S_{sp} of AT is likely to be related to the difference in the mineral composition.

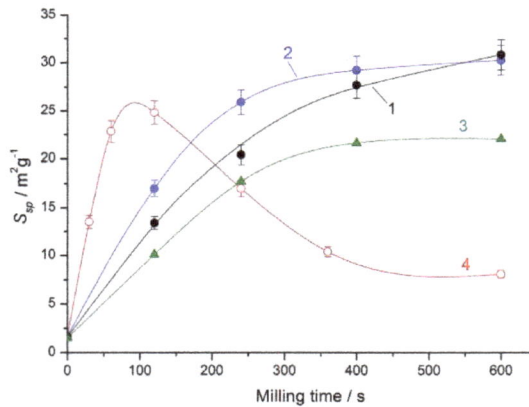

Figure 1. Specific surface areas of antigorite and antigorite-containing ore dressing tailings (AT) as a function of milling time. Curves **1** and **2**: moderate mechanical activation (MA) of antigorite in air and CO$_2$, respectively; **3**: moderate MA of AT in air; **4**: hard MA of antigorite in air.

The S_{sp} of antigorite was higher in the first few minutes of MA when the mineral was milled using moderate conditions in CO$_2$ than when milling was carried out in air (Figure 1). The S_{sp} values converged with increased milling time, to be almost equivalent by 10 min. Therefore, the S_{sp} of antigorite has some dependence on the milling atmosphere. The measured carbon contents of antigorite samples milled for 600 s in air and in CO$_2$ were 0.17 and 0.81%, respectively.

It is well established that the particle size decrease that occurs during intensive MA results in the following effects: (i) irreversible plastic deformations of the particles' surface layers leading to the formation of structural defects; (ii) strengthening of the particles due to unfavourable conditions for crack formation; (iii) reduction of the energy of milling bodies due to increased inter-particle contacts and initiation of viscous behaviour of a fine powder; and (iv) aggregation of particles. As a result, with increased milling time, the rate of aggregation becomes equal to the rate of disintegration, at which point S_{sp} reaches the highest value obtainable in experimental conditions [25,26,42]. It can be seen from Figure 1 that, during moderate conditions of MA (curves 1, 2 and 3), the S_{sp} values approached their maximum values within 8–10 min of milling. However, when hard MA of antigorite was carried out in air, the S_{sp} increased sharply to reach ~25 m^2/g in the first 2 min, then decreased gradually to reach ~8 m^2/g by 10 min (Figure 1, curve 4). This can be explained by acceleration of the above-mentioned processes leading to particle aggregation outweighing disintegration after 2 min of MA. A similar dependence of the S_{sp} has been reported for serpentine milled in air for 240 min in a Fritsch Pulverissette-7 planetary mill (3 g of the sample were milled with 7 agate balls of 15 mm diameter; the rotational speed of the mill was 700 rpm) [12].

3.1.2. X-Ray Diffraction

The XRD data revealed that MA of antigorite in air resulted in disorder of the mineral lattice predominantly along the c axis, as the largest decrease of intensity was observed for the (001) reflection at 2θ = 12.20° (Figure 2). This is in line with the results of Drief and Nieto, who studied structural changes of antigorite following MA in a vibrational mill [10].

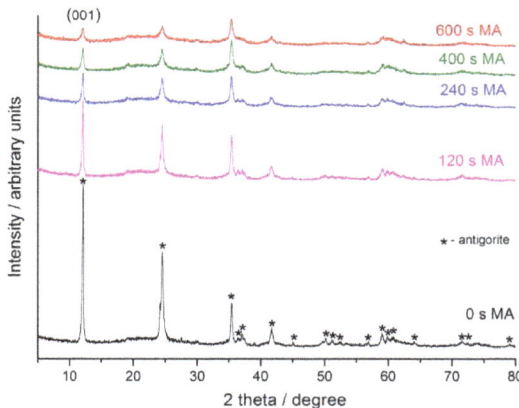

Figure 2. The X-ray diffraction patterns of initial antigorite (0 s MA) and antigorite that had been milled in air in moderate conditions.

The degree of amorphisation of antigorite (D_{am}) was calculated from the peak areas corresponding to (001) reflection of initial antigorite (A_0) and mechanically activated antigorite (A_{MA}) according to Equation (1):

$$D_{am} = \left(1 - \frac{A_{MA}}{A_0}\right) \cdot 100\%. \tag{1}$$

From Figure 3, it can be seen that D_{am} increased rapidly to ~70% during the first 200 s of MA. The D_{am} continued to increase smoothly, reaching ~85% by 600 s.

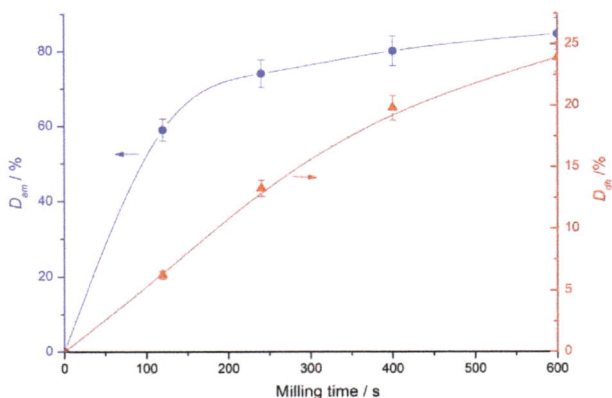

Figure 3. Degree of amorphisation (D_{am}) and degree of dehydroxylation (D_{dh}) of antigorite as a function of milling time (moderate mechanical activation in air).

3.1.3. Thermal Analysis

The DTA and TG curves of initial and antigorite subjected to moderate MA in air are presented in Figures 4 and 5, respectively. The endothermic peaks at ~80 °C correspond to the removal of adsorbed water, and their position was slightly influenced by milling time (Figure 4). The endothermic peak at 763 °C in the DTA spectrum of initial antigorite is due to dehydroxylation of the mineral (the removal of structural water). It was found that MA caused this peak to shift to a lower temperature region. The exothermic peak at 830 °C corresponds to crystallisation of forsterite and/or enstatite [7,11,12]. Its position did not change with increased milling time.

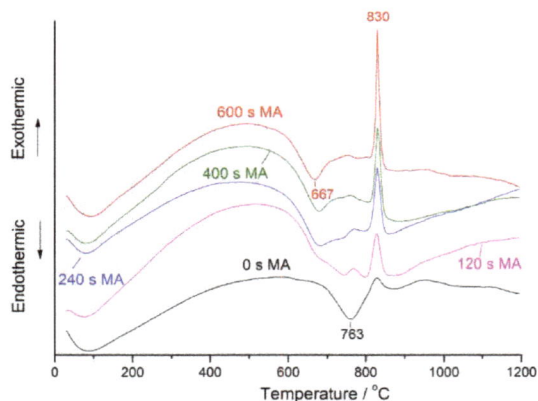

Figure 4. Differential thermal analysis curves for initial antigorite (0 s MA) and mechanically-activated (MA) antigorite (moderate mechanical activation in air).

The overall water loss increased slightly with increased milling time due to mechanosorption of atmospheric moisture and CO_2 (Figure 5). The TG data confirmed that disorder of the crystal lattice (Figure 2) led to recombination of structural hydroxyls and their transformation into adsorbed water. As a result, the dehydration reaction accelerated considerably [10,12]. The FT-IR spectra support this, showing decreased intensity of the sharp OH stretching vibration band at 3675 cm^{-1} with increased milling time. In contrast, the intensity of a broad band in the 3500–3350 cm^{-1} region—corresponding to adsorbed water—increased with increased milling times (Figure 6). It is thus conceivable that mechanically-induced dehydroxylation could increase the reactivity of milled antigorite in the synthesis of an alkali-activated binder.

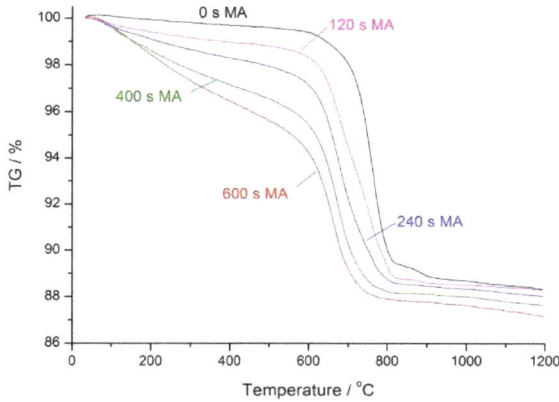

Figure 5. Thermogravity curves for initial and mechanically-activated antigorite (moderate mechanical activation in air).

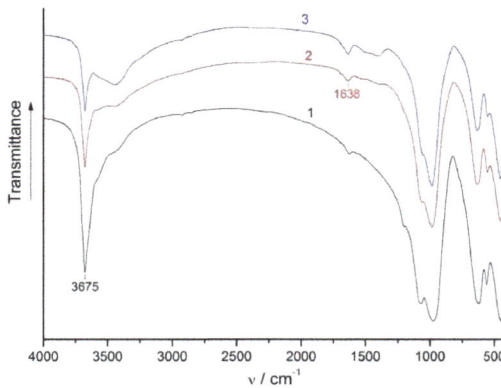

Figure 6. Fourier transform infrared spectra of **1**: initial antigorite, **2**: antigorite milled in moderate conditions for 600 s in air, **3**: alkali-activated binder prepared using antigorite milled for 600 s in moderate conditions in air and liquid glass with a modulus of 2.25 (the amount of liquid glass added corresponded to 3% Na_2O, curing time in dry conditions was 28 days).

The degree of dehydroxylation (D_{dh}) of antigorite as a function of milling time was estimated from the TG data (Figure 5) using Equation (2):

$$D_{dh} = \left(1 - \frac{W_{MA}}{W_0}\right) \cdot 100\%, \tag{2}$$

where W_0 and W_{MA} are the weight losses relative to the temperature range from the onset of the dehydroxylation endotherm (Figure 4) to 1200 °C for initial and MA antigorite, respectively. From Figure 3, it can be seen that D_{dh} increased almost linearly with increased milling time, reaching ~24% after 600 s of milling.

The D_{dh} values as functions of D_{am} were derived from the data shown in Figure 3, and are illustrated in Figure 7. It should be noted that, for vibratory milled antigorite, the peak area of

(001) reflection and loss of structural water have been determined as functions of milling time in previous publications [10]. We used these data to evaluate D_{am} and D_{dh} values as described above; the corresponding curve is also shown in Figure 7. Although the MA conditions (type of mill, milling bodies to sample ratio, milling time and so on) are quite different for the two data sets, the resulting curves are very similar. This confirms the close relationship between the disordering of the crystal structure of antigorite and dehydroxylation, regardless of the type of mill used for MA.

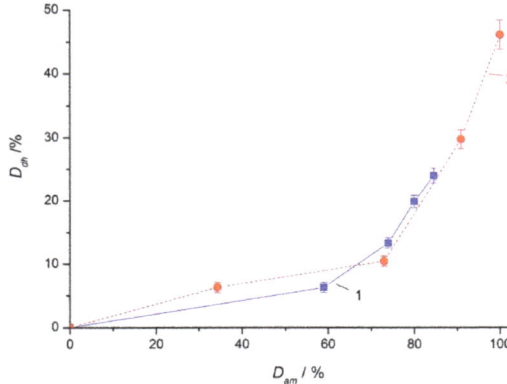

Figure 7. Degree of dehydroxylation as a function of the degree of amorphisation of antigorite from **1**: this study (mechanical activation in moderate conditions in air in the planetary mill); and **2**: vibratory milling in air [10].

3.2. Binding Properties of Milled Antigorite and Antigorite-Containing Ore Dressing Tailings

To estimate the hydraulic activity of antigorite, leaching analyses were performed. The products of complete decomposition of serpentine by NaOH solution are brucite and soluble sodium silicates [43]. In this study, we carried out alkaline leaching of both antigorite milled in moderate conditions and AT, using 2 or 5 M NaOH. In addition, antigorite milled in air was leached with water. The degree of leaching of Si (D_{Si}) was defined by the ratio of silicon in solution to the initial silicon content in antigorite or AT (Figure 8).

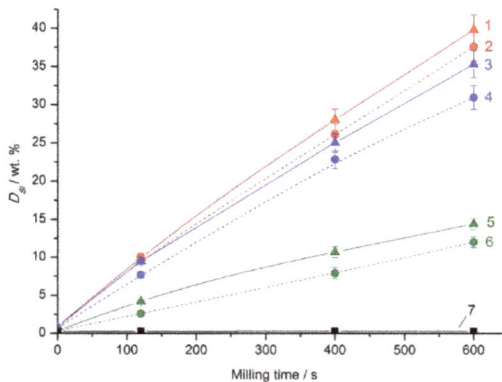

Figure 8. Leaching of silicon (D_{Si}) from antigorite milled in moderate conditions in air (1, 2) and in CO_2 (3, 4) and from AT milled in air (5, 6) as a function of milling time. For curves **1–6**, solid and dashed lines designate leaching with 5 M and 2 M NaOH, respectively. For comparison, curve **7** represents leaching of silicon from antigorite milled in moderate conditions in air with water.

From Figure 8, it can be seen that MA notably enhanced the alkaline leaching of silicon. Following milling in air for 10 min, the D_{Si} values of antigorite samples leached with 2 M and 5 M NaOH were 37.5% and 40.0%, respectively. The D_{Si} values of initial antigorite were less than 1% for all conditions. Leaching of silicon is only slightly affected by the milling atmosphere (air or CO_2). The leaching of Si from AT (Figure 8, curves 5 and 6) was about 2.5 times lower than that from antigorite in all cases. This is consistent with the S_{sp} values of milled antigorite and AT (Figure 1). This could be due also to the reactivity of talc—which is present in considerable amounts in AT—with NaOH solution, which is considerably lower than that of antigorite [43].

Another key factor that is known to affect the degree of leaching is structural disorder and the appearance of lattice defects. In the present study, the D_{Si} increased almost linearly with time for antigorite milled in either air or CO_2 (Figure 8, curves 1–4), while S_{sp} (Figure 1) and D_{am} (Figure 3) both exhibited an initial rapid increase but began to plateau during the last 200 s of milling. Because D_{am} of antigorite was determined from the peak area corresponding to (001) reflection, it is reasonable to assume that the consistent increase of D_{Si} between the 400 and 600 s time points (Figure 8) is due to the formation of structural defects in directions other than that of the c axis. The identification of such defects and discussion of their impact on leaching is beyond the scope of this paper. It is pertinent to note that D_{dh} and D_{Si} continuously increased with milling time (Figures 3 and 8). The degree of alkaline leaching of silicon from milled antigorite is therefore more strongly associated with D_{dh} of the mineral than with D_{am} or S_{sp}.

Leaching of silicon with water from antigorite milled in air was low, with D_{Si} found to be no more than 0.3% at all time points (Figure 8, curve 7). However, binders produced using antigorite or AT milled for 600 s in moderate conditions exhibited a certain cementitious properties even when prepared with only water and alkali activation was not carried out (Table 3). After curing for 28 days in dry conditions, the compressive strengths of the binders based on antigorite or AT milled in air were 3.4 MPa and 7.1 MPa, respectively. Due to the different mineral compositions, more water is required to create an antigorite-based paste with suitable workability compared with an AT-based paste. This explains the lower compressive strength of the binder prepared with antigorite.

Table 3. Compressive strengths (R) of binders prepared using water and antigorite or antigorite-containing ore dressing tailings (AT) milled in moderate conditions for 600 s. (w/s: water to solid ratio).

Mineral Component	Hardening Conditions	Milling Atmosphere	w/s	R, MPa		
				1 Day	7 Days	28 Days
Antigorite	dry	air	0.35	1.8	3.4	3.4
Antigorite	dry	CO_2	0.35	1.8	1.6	1.4
Antigorite	humid	air	0.35	1.8	2.1	2.2
Antigorite	humid	CO_2	0.35	1.4	2.8	2.4
AT	dry	air	0.26	4.0	5.1	7.1
AT	dry	CO_2	0.27	4.4	6.0	8.8
AT	humid	air	0.26	2.5	3.5	4.7
AT	humid	CO_2	0.27	2.1	2.8	3.0

It has been established that humid conditions are preferable to dry conditions for curing of thermally-activated serpentine with respect to the mechanical properties of the binder [6]. However, our results indicate that the reverse is true in the case of MA serpentine (Table 3), suggesting that the mechanism of hardening differs between MA and thermally-activated serpentine. The compressive strength was not significantly affected by the milling atmosphere.

The synthesis of alkali-activated binders was investigated by varying the following parameters: milling time, milling atmosphere, amount of liquid glass added and curing conditions (dry or humid). The compressive strength of the alkali-activated binder based on AT milled in air and cured in dry conditions was significantly greater than that of the same binder cured in humid conditions (Figure 9). A similar trend was observed for the binder prepared without alkali activation (Table 3). Compressive strength was seen to increase with increased milling time for all samples. The compressive strengths of binders based on initial AT (0 s MA) are 1 MPa and 19 MPa for binders cured for 28 days in humid and dry conditions, respectively. The corresponding values for the binders prepared from AT milled for 240 s in air are 15 MPa and 49 MPa, respectively (Figure 9). Using liquid glass with a modulus equal to 3.18, the best results for compressive strength were achieved when binders were prepared with compositions corresponding to 4% Na_2O (Figure 10).

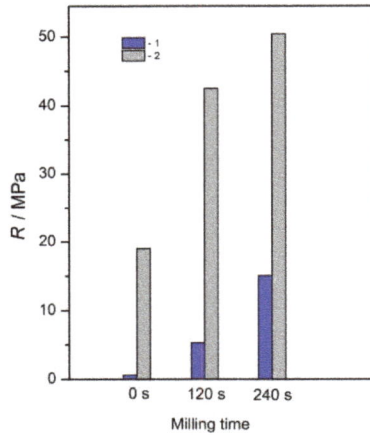

Figure 9. Compressive strength (*R*) of alkali-activated binders prepared using antigorite-containing ore dressing tailings (AT) milled in air in moderate conditions and liquid glass with a modulus of 3.18. The amount of liquid glass added corresponded to 4% Na_2O. Curing time was 28 days. Hardening conditions were **1**: humid, or **2**: dry.

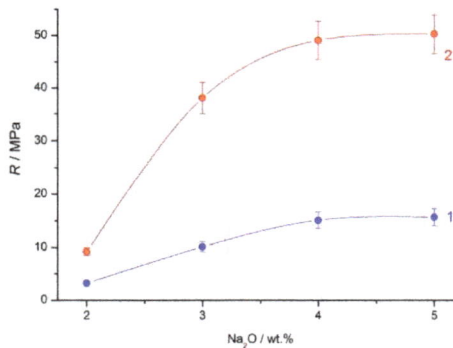

Figure 10. Compressive strengths (*R*) of alkali-activated binders prepared using antigorite-containing ore dressing tailings (AT) milled in air in moderate conditions for 240 s as a function of amount of added liquid glass. The modulus of liquid glass was 3.18. Curing time was 28 days. Hardening conditions were **1**: humid, or **2**: dry.

Table 4 presents the compressive strengths of alkali-activated binders prepared using liquid glass with a modulus of 2.25 (the amount of liquid glass added corresponded to 3 % Na_2O) and antigorite or AT milled in air or in CO_2 for 600 s and cured in dry or humid conditions. After curing for 28 days in dry conditions, the compressive strength of the binders based on antigorite and AT milled in air were 40.8 MPa and 30.8 MPa, respectively. The corresponding compressive strengths of binders based on antigorite and AT milled in CO_2 were 29.1 MPa and 23.0 MPa, respectively. The increased compressive strength of the binders based on antigorite cured in dry conditions compared with those based on AT is consistent with the alkali-leaching experiments (Figure 8).

Table 4. Compressive strengths (*R*) of alkali-activated binders prepared using antigorite or antigorite-containing ore dressing tailings (AT) milled for 600 s in moderate conditions and liquid glass with a modulus of 2.25. The specific surface area ($S_{sp}{}^{(B)}$) was measured using the Blaine method. The amount of liquid glass added corresponded to 3% Na_2O. (w/s: water to solid ratio).

Mineral Component	$S_{sp}{}^{(B)}$, m²/kg	Hardening Conditions	Milling Atmosphere	w/s	R, MPa		
					1 Day	7 Days	28 Days
Antigorite	1259	dry	air	0.26	12.2	20.6	40.8
Antigorite	1255	dry	CO_2	0.25	15.6	28.0	29.1
Antigorite	1259	humid	air	0.26	3.5	4.1	5.3
Antigorite	1255	humid	CO_2	0.25	5.0	6.7	6.8
AT	1280	dry	air	0.22	10.8	22.6	30.8
AT	1288	dry	CO_2	0.25	10.2	21.8	23.0
AT	1280	humid	air	0.22	6.8	8.7	7.8
AT	1288	humid	CO_2	0.25	1.3	1.9	2.0

From Table 4, it can be seen that MA in CO_2 is not preferable to air to achieve high compressive strengths, as has been described for magnesia ferriferous slag [41]. However, the compressive strengths of binders prepared following MA in CO_2—particularly in the case of AT cured in dry conditions—were similar to those prepared from mineral component that had been milled in air. This suggests that serpentinite tailings milled in CO_2, which is carried out for CO_2 capture, could be used to produce alkali-activated binders.

The XRD data revealed no newly formed crystalline compounds in the binder prepared using liquid glass and antigorite milled in air and cured in dry conditions (Figure 11). Consequently, the cementitious phase is likely amorphous in nature. FT-IR spectrum of this binder is shown in Figure 6. When this FT-IR spectrum is compared with that of the milled antigorite (Figure 6), it is apparent that the intensities of the OH stretching vibration band in the 3500–3350 cm^{-1} region and the OH bending vibration band at 1638 cm^{-1} increased, suggesting that the notable hydration of antigorite occurred. The FT-IR spectrum of the binder shows increased intensity of the carbonate group stretching vibration band in the 1550–1350 cm^{-1} region due to chemisorption of the atmospheric carbon dioxide.

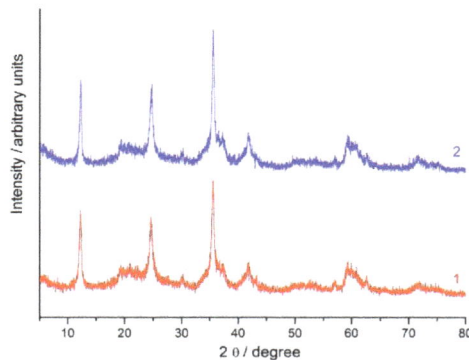

Figure 11. The X-ray diffraction patterns of **1**: antigorite milled in moderate conditions for 600 s in air, **2**: alkali-activated binder prepared using antigorite milled for 600 s in moderate conditions in air and liquid glass with a modulus of 2.25 (the amount of liquid glass added corresponded to 3% Na_2O, curing time in dry conditions was 28 days).

4. Conclusions

From our results, we can draw the following conclusions:

1. The S_{sp} of antigorite milled in a planetary mill in air is 30–40% greater than that of AT milled in equivalent conditions. This can be explained by the difference in the mineral compositions of antigorite and AT. The atmosphere of MA (air or CO_2) affects the S_{sp} of antigorite, although to a lesser extent than the choice of mineral component.

2. Milling of antigorite in a planetary mill results in rapid amorphisation. The D_{am} along the c axis reached 85% after MA for 600 s in air in moderate conditions.

3. According to thermal analysis and FT-IR data, milling of antigorite leads to dehydroxylation. After 600 s of milling in air, the D_{dh} of antigorite was 24%. Comparison of our results with literature data confirms that dehydroxylation of MA antigorite is closely related to its amorphisation, irrespective of the type of mill that is used.

4. Mechanical activation of antigorite increases the alkaline leaching of Si into solution. The degree of silicon leaching for antigorite milled in air for 600 s and leached with 5 M NaOH was 40%. The corresponding value for initial antigorite was less than 1%. Water leaching of silicon from antigorite milled in air for 600 s resulted in leaching of only 0.3% of silicon. The degree of alkaline leaching of silicon from MA antigorite is more significantly associated with the D_{dh} of the mineral than with D_{am} or S_{sp}.

5. The reduced alkali leaching of Si from milled AT compared with that from milled antigorite can be explained by the differences in S_{sp} of the solid components, and by the lower reactivity of talc, as one of the main admixtures in AT, with a NaOH solution compared with that of antigorite.

6. Binders prepared using antigorite or AT that have been milled for 600 s in moderate conditions in air and in CO_2 show cementitious properties even if the paste is prepared without the use of alkali. The compressive strengths of the binders based on antigorite or AT milled in air for 600 s and cured for 28 days in dry conditions were found to be ~3 MPa and ~7 MPa, respectively. The lower compressive strength of the binder based on milled antigorite was concluded to be related to the higher water demand of the mineral.

7. The cementitious properties of alkali-activated binder increased with increasing milling time of the antigorite. Alkali-activated binder that was prepared using liquid glass and cured at dry conditions produced a preferable product compared with that cured in humid conditions. The compressive strength of the alkali-activated binder based on AT milled in air for 240 s and cured for 28 days in dry or humid conditions was 49 MPa and 15 MPa, respectively. The corresponding values for binders prepared using initial AT were 19 MPa and 1 MPa, respectively.

8. In line with the mineral compositions of antigorite and AT and the results of leaching experiments, alkali-activated binders based on milled antigorite and cured at dry conditions show higher compressive strengths than those based on AT. According to XRD data, the cementitious phase is amorphous in nature. The milling atmosphere (air or CO_2) has a relatively small effect on the cementitious properties of alkali-activated binders.

Author Contributions: Designing the study, E.V.K., B.I.G. and A.M.K.; pre-treatment of mineral components, preparation and characterisation of binders, B.I.G.; Thermal analysis and leaching experiments, E.V.K.; FT-IR and XRD studies, A.M.K.; Writing the initial draft of the paper, E.V.K. and B.I.G.; Writing the final paper, A.M.K.

Funding: This research received no external funding.

Acknowledgments: The authors are thankful to E.S. Serova and A.G. Ivanova for help with experimental work.

Conflicts of Interest: The authors declare no conflict of interest.

References

1. Wicks, F.J.; O'Hanley, D.S. Serpentine minerals: structures and petrology. *Rev. Miner. Geochem.* **1988**, *19*, 91–167.
2. Barnes, S.-J.; Melezhik, V.A.; Sokolov, S.V. The composition and mode of formation of the Pechenga nickel deposits, Kola Peninsula, North Western Russia. *Can. Mineral.* **2001**, *39*, 447–471. [CrossRef]
3. Hitch, M.; Ballantyne, S.M.; Hindle, S.R. Revaluing mine waste rock for carbon capture and storage. *Int. J. Min. Reclam. Environ.* **2010**, *24*, 64–79. [CrossRef]
4. Svetlov, A.; Makarov, D.; Vigdergauz, V. Nonoxidative leaching of non-ferrous metals from wastes of Kola Mining by sulphuric acid. *Int. J. Min. Sci.* **2015**, *1*, 17–24.
5. Masloboev, V.A.; Makarov, D.V.; Baklanov, A.A.; Amosov, P.V.; Seleznev, S.G. Methods to reduce the environmental hazards of mining and processing of minerals in the Arctic regions. In Proceedings of the International Mineral Processing Congress 2016 XXVIII, Quebec City, QC, Canada, 11–15 September 2016.
6. Budnikov, P.P.; Mchedlov-Petrosyan, O.P. Manifestation of hydraulic binding properties by dehydrated serpentinite. *Dokl. Akad. Nauk SSSR* **1950**, *73*, 539–540. (In Russian)
7. Dlugogorski, B.Z.; Balucan, R.D. Dehydroxylation of serpentine minerals: Implications for mineral carbonation. *Renu. Sust. Energy Rev.* **2014**, *31*, 353–367. [CrossRef]
8. Li, J.; Hitch, M.; Power, I.M.; Pan, Y. Integrated mineral carbonation of ultramafic mine deposits—A review. *Minerals* **2018**, *8*, 147–164. [CrossRef]
9. Li, J.; Hitch, M. Ultra-fine grinding and mechanical activation of mine waste rock using a high-speed stirred mill for mineral carbonation. *Int. J. Min. Met. Mater.* **2015**, *22*, 1005–1016. [CrossRef]
10. Drief, A.; Nieto, F. The effect of dry grinding on antigorite from Mulhacen, Spain. *Clay Clay Min.* **1999**, *47*, 417–424. [CrossRef]
11. Kim, D.J.; Sohn, J.S.; Ahn, J.G.; Chung, H.S. Extraction of metals from mechanically milled serpentine. *Geosyst. Eng.* **2008**, *11*, 25–28. [CrossRef]
12. Zhang, Q.; Sugiyama, K.; Saito, F. Enhancement of acid extraction of magnesium and silicon from serpentine by mechanochemical treatment. *Hydrometallurgy* **1997**, *45*, 323–331. [CrossRef]
13. Brice, D.G.; Lesley, S.-L.; Provis, J.L.; van Deventer, J.S.J. Conclusions and the future of alkali activation technology. In *Alkali-Activated Materials: State of the Art Report of RILEM TC 224-AAM*; Provis, J.L., van Deventer, J.S.J., Eds.; Springer Science & Business Media: Dordrecht, Holland, 2014; pp. 381–388.
14. Shi, C.; Krivenko, P.V.; Roy, D.M. *Alkali-Activated Cements and Concretes*; CRC Press: Boca Raton, FL, USA, 2006; p. 376.
15. Provis, J.L.; Palomo, A.; Shi, C. Advances in understanding alkali-activated materials. *Cem. Concr. Res.* **2015**, *78A*, 110–125. [CrossRef]
16. Part, W.K.; Ramli, M.; Cheah, C.B. An overview on the influence of various factors on the properties of geopolymer concrete derived from industrial by-products. *Constr. Build. Mater.* **2015**, *77*, 370–395. [CrossRef]
17. Provis, J.L.; Bernal, S.A. Geopolymers and related alkali-activated materials. *Annu. Rev. Mater. Res.* **2014**, *44*, 299–327. [CrossRef]
18. Ke, X.; Criado, M.; John, L.; Provis, J.L.; Bernal, S.A. Slag-Based Cements That Resist Damage Induced by Carbon Dioxide. *ACS Sustain. Chem. Eng.* **2018**, *6*, 5067–5075. [CrossRef]
19. Mucsi, G.; Kumar, S.; Csőke, B.; Kumar, R.; Molnár, Z.; Rácz, Á.; Mádai, F.; Debreczeni, Á. Control of geopolymer properties by grinding of land filled fly ash. *Int. J. Min. Proc.* **2015**, *143*, 50–58. [CrossRef]
20. Nath, S.K.; Maitra, S.; Mukherjee, S.; Kumar, S. Microstructural and morphological evolution of fly ash based geopolymers. *Constr. Build. Mater.* **2016**, *111*, 758–765. [CrossRef]
21. Ma, C.-K.; Awang, A.Z.; Omar, W. Structural and material performance of geopolymer concrete: A review. *Constr. Build. Mater.* **2018**, *186*, 90–102. [CrossRef]
22. Arbi, K.; Nedeljković, M.; Zuo, Y.; Ye, G. A Review on the Durability of Alkali-Activated Fly Ash/Slag Systems: Advances, Issues, and Perspectives. *Eng. Chem. Res.* **2016**, *55*, 5439–5453. [CrossRef]
23. Ding, Y.; Dai, J.-G.; Shi, C.-J. Mechanical properties of alkali-activated concrete: A state-of-the-art review. *Constr. Build. Mater.* **2016**, *127*, 68–79. [CrossRef]
24. Luukkonen, T.; Abdollahnejad, Z.; Juho Yliniemi, J.; Kinnunen, P.; Illikainen, M. One-part alkali-activated materials: A review. *Cem. Concr. Res.* **2018**, *103*, 21–34. [CrossRef]
25. Heinicke, G. *Tribochemistry*; Akademie-Verlag: Berlin, Germany, 1984; p. 495.

26. Avvakumov, E.G. *Mechanical Methods of Activation of Chemical Processes*; Nauka: Novosibirsk, Russia, 1986; p. 306. (In Russian)
27. Boldyrev, V.V. Mechanochemistry and mechanical activation of solids. *Russ. Chem. Rev.* **2006**, *75*, 177–189. [CrossRef]
28. Baláž, P. *Mechanochemistry in Nanoscience and Minerals Engineering*; Springer: Heidelberg, Germany, 2008; p. 413.
29. Tchadjie, L.N.; Ekolu, S.O. Enhancing the reactivity of aluminosilicate materials toward geopolymer synthesis. *J. Mater. Sci.* **2018**, *53*, 4709–4733. [CrossRef]
30. Kumar, S.; Kumar, R. Mechanical activation of fly ash: effect on reaction, structure and properties of resulting geopolymer. *Ceram. Int.* **2011**, *37*, 533–541. [CrossRef]
31. Djobo, J.N.Y.; Elimbi, A.; Tchakoute, H.K.; Kumar, S. Mechanical activation of volcanic ash for geopolymer synthesis: Effect on reaction kinetics, gel characteristics, physical and mechanical properties. *RSC Adv.* **2016**, *6*, 39106–39117. [CrossRef]
32. Wei, B.; Zhang, Y.; Bao, S. Preparation of geopolymers from vanadium tailings by mechanical activation. *Constr. Build. Mater.* **2017**, *145*, 236–242. [CrossRef]
33. Mucsi, G. Mechanical activation of power station fly ash by grinding: a review. *J. Silic. Based Compos. Mater.* **2016**, *68*, 56–61.
34. Marjanovic, N.; Komljenovic, M.; Bascarevic, Z.; Nikolic, V. Improving reactivity of fly ash and properties of ensuing geopolymers through mechanical activation. *Constr. Build. Mater.* **2014**, *57*, 151–162. [CrossRef]
35. Temuujin, J.; Williams, R.P.A.; van Riessen, A. Effect of mechanical activation of fly ash on the properties of geopolymer cured at ambient temperature. *J. Mater. Process. Technol.* **2009**, *209*, 5276–5280. [CrossRef]
36. Kalinkina, E.V.; Kalinkin, A.M.; Forsling, W.; Makarov, V.N. Sorption of atmospheric carbon dioxide and structural changes of Ca and Mg silicate minerals during grinding. I. Diopside. *Int. J. Miner. Process.* **2001**, *61*, 273–288. [CrossRef]
37. Kalinkina, E.V.; Kalinkin, A.M.; Vasil'eva, T.N.; Mazukhina, S.I.; Belyaevskii, A.T. Study of the mechanically activated serpentine's sorption properties in respect to copper (II) cations. *Geoekologiya* **2012**, *3*, 229–236. (In Russian)
38. Kalinkin, A.M.; Kalinkina, E.V.; Politov, A.A.; Makarov, V.N.; Boldyrev, V.V. Mechanical interaction of Ca silicate and aluminosilicate minerals with CO_2. *J. Mater. Sci.* **2004**, *39*, 5393–5398. [CrossRef]
39. Turianicová, E.; Obut, A.; Tuček, L.; Zorkovská, A. Interaction of natural and thermally processed vermiculites with gaseous carbon dioxide during mechanical activation. *Appl. Clay Sci.* **2014**, *88–89*, 86–91. [CrossRef]
40. Turianicová, E. CO_2 utilization for mechanochemical carbonation of celestine. *GeoSci. Eng.* **2015**, *6*, 20–23. [CrossRef]
41. Kalinkin, A.M.; Kumar, S.; Gurevich, B.I.; Alex, T.C.; Kalinkina, E.V.; Tyukavkina, V.V.; Kalinnikov, V.T.; Kumar, R. Geopolymerization behavior of Cu–Ni slag mechanically activated in air and in CO_2 atmosphere. *Int. J. Miner. Process.* **2012**, *112–113*, 101–106. [CrossRef]
42. Khodakov, G.S. *Physics of Grinding*; Nauka: Moscow, Russia, 1972; p. 307. (In Russian)
43. Vlasov, V.V.; Remiznikova, V.I. On interaction of clay minerals and some sheet silicates with alkaline solutions. In *XRD Studies on Mineral Raw Materials*; Sidorenko, G.A., Ed.; Nedra: Moscow, Russia, 1967; p. 140. (In Russian)

![minerals logo] *minerals*

MDPI

Article

Titanite Ores of the Khibiny Apatite-Nepheline-Deposits: Selective Mining, Processing and Application for Titanosilicate Synthesis

Lidia G. Gerasimova [1,2], **Anatoly I. Nikolaev** [1,2,*], **Marina V. Maslova** [1,2], **Ekaterina S. Shchukina** [1,2], **Gleb O. Samburov** [2], **Victor N. Yakovenchuk** [1] and **Gregory Yu. Ivanyuk** [1]

[1] Nanomaterials Research Centre of Kola Science Centre, Russian Academy of Sciences, 14 Fersman Street, Apatity 184209, Russia; gerasimova@chemy.kolasc.net.ru (L.G.G.); maslova@chemy.kolasc.net.ru (M.V.M.); shuki_es@chemy.kolasc.net.ru (E.S.S.); yakovenchuk@geoksc.apatity.ru (V.N.Y.); ivanyuk@admksc.apatity.ru (G.Y.I.)
[2] Tananaev Institute of Chemistry of Kola Science Centre, Russian Academy of Sciences, 26a Fersman Street, Apatity 184209, Russia; samgleb@yandex.ru
*   Correspondence: nikol_ai@chemy.kolasc.net.ru; Tel.: +7-815-557-9231

Received: 4 September 2018; Accepted: 10 October 2018; Published: 12 October 2018

✓ check for updates

Abstract: Geological setting and mineral composition of (apatite)-nepheline-titanite ore from the Khibiny massif enable selective mining of titanite ore, and its processing with sulfuric-acid method, without preliminary concentration in flotation cells. In this process flow diagram, titanite losses are reduced by an order of magnitude in comparison with a conventional flotation technology. Further, dissolution of titanite in concentrated sulfuric acid produces titanyl sulfate, which, in turn, is a precursor for titanosilicate synthesis. In particular, synthetic analogues of the ivanyukite group minerals, SIV, was synthesized with hydrothermal method from the composition based on titanyl-sulfate, and assayed as a selective cation-exchanger for Cs and Sr.

Keywords: apatite-nepheline-titanite ore; sulfuric-acidic decomposition; titanyl sulfate; hydrothermal synthesis; ivanyukite

1. Introduction

The world largest apatite-nepheline deposits of the Khibiny massif (NW Russia) host complex ores containing five economic minerals are: Fluorapatite, nepheline, titanite, aegirine and titanomagnetite. However, at present, only two of them (fluorapatite and nepheline) are economically attractive, while the rest are accumulated in tailings ponds [1]. Titanite, $CaTiSiO_5$, is the most important mineral among them because titanium and its compounds are widely used in different industrial technologies.

In addition to titanite and titanomagnetite, the apatite-nepheline deposits contain numerous pegmatites with rare titanium minerals that have important functional properties, for example, ferroelectric loparite-(Ce) [2], molecular sieve chivruaiite [3,4] (Ca-analog of synthetic microporous titanosilicate ETS-4 [5,6]), cation exchangers kukisvumite and punkaruaivite [7–10] (respectively, Zn and Li analogues of synthetic titanosilicate AM-4 [11,12]), cation exchangers sitinakite [13] (natural analogue of synthetic titanosilicate Ionsiv IE-911 [14–16]), and minerals of the ivanyukite group [17–20]. Besides, these and other synthetic titanium-based compounds (ETS-2, ETS-10, JDF-L1, LHT-9, MIL-125, etc.) are utilized as agents for cation relocation, hydrogen storage and heat transformation, dehydrating agents, drug nanocarriers, adsorbents and membranes [20–25].

In particular, ivanyukite-Na and its synthetic analogue SIV described herein have pharmacosiderite-related crystal structure consisting of Ti_4O_{24}-clusters coupled into a framework by single SiO_4 tetrahedra

(Figure 1). In this framework, there is a system of wide channels ($\approx$6.1 Å), with extra-framework cations of K and Na and water molecules inside that can be easily exchanged with Rb^+, Cs^+, Tl^+, Ag^+, Sr^{2+}, Ba^{2+}, Cu^{2+}, Ni^{2+}, Co^{2+}, La^{3+}, Ce^{3+}, Eu^{3+}, Th^{4+}, etc. [18–20,26]. Moreover, (^{137}Cs, ^{90}Sr, ^{154}Eu)-exchanged forms of ivanyukite can be transformed by heating into stable SYNROC-like titanate ceramics for long-time immobilization of these radionuclides [20].

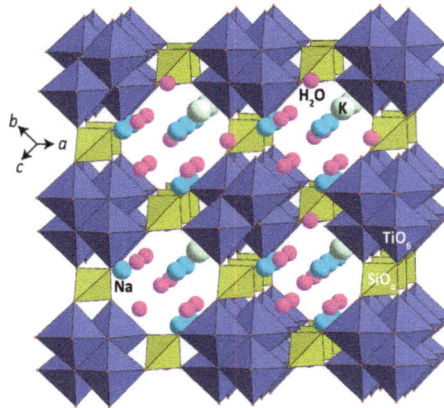

Figure 1. Crystal structure of ivanyukite-Na-*T* [17]: cubano-like clusters Ti_4O_{24} consisting of four edge-shared TiO_6 octahedra (dark-blue) are connected by vertexes SiO_4 tetrahedra (yellow) to form the microporous framework, with Na (cyan), K (pale green), and H_2O (magenta) in channels.

There are a lot of studies related to titanite concentrate chemical processing [27–29]. Among the known technologies, sulfuric-acidic ones are the most perspective because they enable to obtain a wide spectrum of materials including titanium pigments, fillers, adhesives, tanning agents, sorbents and molecular sieves. High titanite content is a typical property of apatite-titanite and nepheline-titanite rocks, which are currently stockpiled in dumps due to difficulties with titanite separation from apatite and nepheline by flotation. We would like to show herein new perspectives for production of synthetic ivanyukite-type compounds and other materials mentioned above with sulfuric-acidic cleaning followed by hydrometallurgical processing of ground apatite-titanite and nepheline-titanite ores. This approach features better productiveness in relation to a conventional technologies of titanite extraction by flotation and titanosilicate synthesis from compositions based on titanium chlorides.

2. Geological Setting

The world's largest Khibiny alkaline massif (35 × 45 km) is situated in the West of the Kola Peninsula (Murmansk Region, Russia), at the contact between low-metamorphosed volcano-sedimentary rocks of the Proterozoic Imandra-Varzuga greenstone belt and the Archaean ultra-metamorphic rocks of the Kola-Norwegian megablock (Figure 2). Its age is 380–360 million years [30]. The main rock of the massif is coarse-grained nepheline syenite (foyaite) that forms funnel-shaped homogeneous body divided into two parts by the Main Ring intrusion of foidolite (melteigite–urtite rock series), with related poikilitic (kalsilite)-nepheline syenites (rischorrite and lyavochorrite) and fine-grained alkaline syenite with xenolites of feinitized volcano-sedimentary rocks [31,32]. The Main Ring thickness on the day surface varies from 0.1 to 10 km, and rapidly decreases with depth. Within the outer part of the foyaite body, there is another one semi-ring zone of foidolites, alkaline syenite and fenite. In addition, monchiquite–carbonatite veins and explosion pipes occur in the Eastern part of the Main Ring.

The rock-forming minerals of foyaite include nepheline, microcline, orthoclase, albite, aegirine, augite, Na–Ca- and Na-amphiboles, aenigmatite, titanite, eudialyte, lamprophillite, and annite. Rischorrite and lyavochorrite contain rock-forming nepheline, sodalite, orthoclase, aegirine, KNa-amphiboles, annite, titanite, aenigmatite, ilmenite, lamprophillite, astrophyllite, and fluorapatite [31,32]. The main constituents of the foidolites are nepheline, clinopyroxenes of the diopside–aegirine-augite series, KNaCa-amphiboles (potassicrichterite, potassicferrorichterite, etc.), annite, titanite, fluorapatite, titanomagnetite (members of the magnetite–ulvöspinel series), ilmenite, and eudialyte, with wide-ranging relations between them (Figure 3).

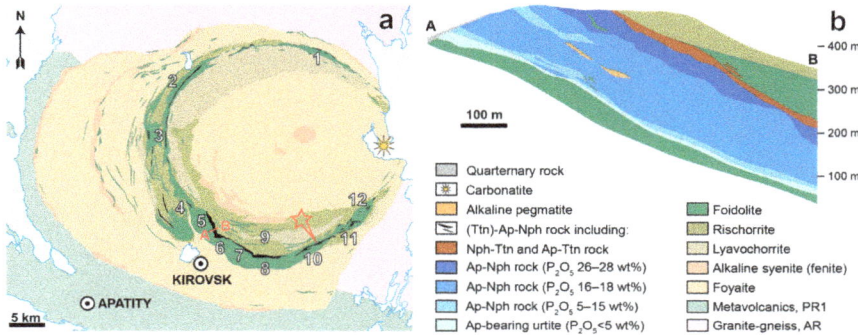

Figure 2. Geological map of the Khibiny massif (**a**, after [32]) and vertical section of the Kukisvumchorr titanite-apatite-nepheline ore body along A–B line (**b**, after F.M. Onokhin). Titanite-apatite-nepheline deposits: 1—Valepakhk, 2—Partomchorr, 3—Kuelporr, 4—Snezhny Circus, 5—Kukisvumchorr, 6—Yuksporr, 7—Apatite Circus, 8—Plato Rasvumchorr, 9—Eveslogchorr, 10—Koashva, 11—Niorkpakhk, and 12—OleniyRuchei. Red star indicates the place of titanite ore sampling.

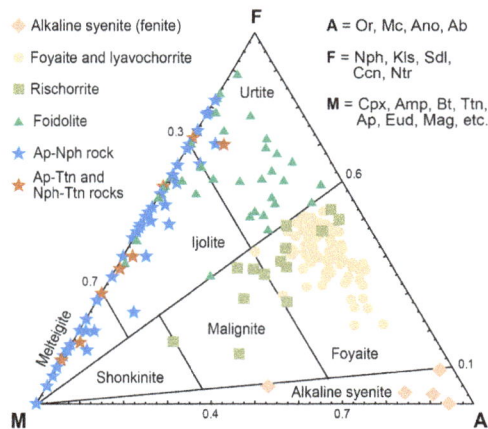

Figure 3. Modal composition of alkaline rocks estimated with grain squares of feldspars (A), feldspathoids (F), and dark colored minerals (M) in polished hand-sized specimens of the Khibiny massif (after [32,33], with additions). Ab—albite, Amp—amphiboles, Ano—anorthoclase, Ap—fluorapatite, Bt—biotite, Ccn—cancrinite, Cpx—clinopyroxenes, Eud—eudialyte, Kls—kalsilite, Mag—magnetite, Mc—microcline, Nph—nepheline, Ntr—natrolite, Or—orthoclase, Sdl—sodalite, and Ttn—titanite.

At apical parts of the Main Ring, there are fractal stockworks of specific apatite-rich foidolite, apatite-nepheline rock (see Figures 2 and 3), which form 12 apatite deposits developed by two mining companies: JSC "Apatit", and JSC "North-Western Phosphorus Company" [32]. Major minerals of apatite-nepheline rock are fluorapatite, nepheline, diopside–aegirine-augite, potassicrichterite, potassicferrorichterite, orthoclase, titanite, magnetite, and ilmenite, contents of that vary in a wide range of values (Figure 3).

Titanite is a typical apatite-nepheline rock mineral of accessory to minor rock-forming types (Table 1). It occurs as separate fine-grained lenses (up to 5 cm in diameter) within fluorapatite bands (Figure 4a). Content and size of titanite lenses increase towards the upper contact of apatite-nepheline rock covered by rischorrite–lyavochorrite that accommodates layers (up to 50 m thick) of titanite-dominant (up to 80 modal %) apatite-titanite and nepheline-titanite rocks (see Figure 2b). In these rocks, the main part of titanite is presented by brown prismatic crystals (up to 1 cm long) with lustrous faces (Figure 4b–d). Average chemical composition of such titanite (Table 2) corresponds to the formula: $(Ca_{0.95}Na_{0.04}Sr_{0.01})_{\Sigma1.00}(Ti_{0.96}Fe^{3+}_{0.02}Nb_{0.01})_{\Sigma0.99}(Si_{0.99}Al_{0.01})_{\Sigma1.00}O_5$ [33].

Table 1. Titanite resources of the Khibiny titanite-apatite-nepheline deposits [33].

Deposit	Measured Ttn Resources, kt	Average Ttn Content in (Ttn)-Ap-Nph ore, wt %	Fraction of Ap-Ttn and Nph-Ttn Ores, vol %
Partomchorr	59,615	6.9	13.6
Kuelporr	995	5.2	4.3
Kukisvumchorr	17,550	4.2	6.9
Yuksporr	23,340	4.4	8.1
Apatite Circus	3840	3.2	1.9
Rasvumchorr	13,395	4.0	2.1
Eveslogchorr	45	8.9	27.7
Koashva	38,780	4.7	8.4
Niorkpakhk	1595	2.4	0.0
OleniyRuchei	12,075	3.1	0.0

Table 2. Chemical composition of titanite from titanite-apatite-nepheline rock, wt % [33].

Constituent	*n*	Mean	Min	Max	SD
Na_2O	15	0.61	b.d.	1.18	0.30
Al_2O_3	15	0.24	b.d.	0.66	0.17
SiO_2	15	30.31	29.61	32.38	0.62
K_2O	15	0.01	b.d.	0.20	0.05
CaO	15	27.16	25.22	28.04	0.95
TiO_2	15	38.93	37.28	42.86	1.30
V_2O_3	15	0.02	b.d.	0.15	0.04
MnO	15	0.01	b.d.	0.06	0.02
FeO	15	0.88	0.33	1.29	0.29
SrO	15	0.32	b.d.	0.53	0.15
ZrO_2	15	0.14	b.d.	0.43	0.15
Nb_2O_5	15	0.38	b.d.	1.35	0.41
La_2O_3	15	0.05	b.d.	0.17	0.07
Ce_2O_3	15	0.30	b.d.	0.68	0.16
Nd_2O_3	15	0.04	b.d.	0.26	0.08

At the apatite deposits, content of (apatite)-nepheline-titanite rock reaches 30 vol % (Table 1) with average titanite content about 20 wt %. Taking into account stable position of this rock at the contact between bedding foidolite (including apatite-nepheline rock) and covering rischorrite–lyavochorrite, as well as significant thickness of these layers (Figure 2b), selective mining of titanite ores gains practical importance. Under the present studies, about 1.5 tons of nepheline-titanite rock were selectively mined from the apical part of the Koashva ore-body, and then processed with the sulfuric-acid method without preliminary concentration.

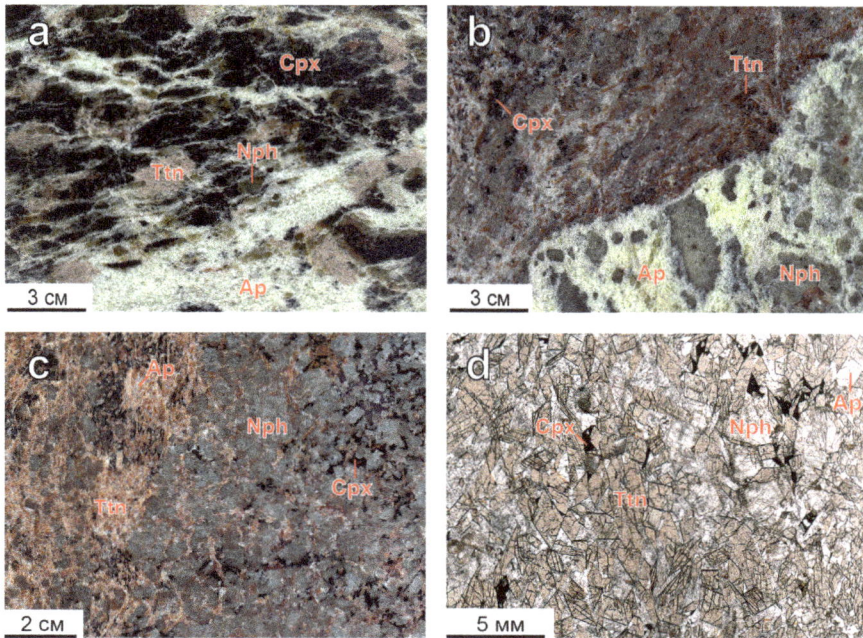

Figure 4. Titanite segregations in apatite-nepheline (**a**), apatite-titanite (**b**) and nepheline-titanite (**c**,**d**) ores of the Koashva apatite deposit, the Khibiny massif. Photos of polished samples (**a**–**c**) and thin section of the sample c in transmitted light (**d**). Ap—fluorapatite, Cpx—clinopyroxene, Nph—nepheline, Ttn—titanite.

3. Materials and Methods

For this study, we used blocks (up to 0.5 m in diameter) of (apatite)-nepheline-titanite ore, mined at the Koashva open pit. The reagents used included reagent-grade sulfuric and hydrochloric acids, sodium and potassium hydroxides (H_2SO_4 93.0%, HCl 37.5%, NaOH 99%, KOH 99%), analytical reagent grade pentahydrate sodium metasilicate, $Na_2SiO_3 \cdot 5H_2O$, and ammonium sulfate, $(NH_4)_2SO_4$, (Reachem, Moscow, Russia), and distilled water (Tananaev Institute of Chemistry, Apatity, Russia).

At first, ore blocks were crushed on a jaw crusher, and then in a 1.5 kW AGO-2 planetary mill (NOVIC, Novosibirsk, Russia) for 6 h, with the rock:balls ratio of 1:5 [34]. Then classification (dry sieving) was carried out (with separation of a powder fraction less than 40 µm). Modal and chemical composition of the crushed sample was determined by means of a FT IR 200 spectrophotometer (Perkin Elmer, Waltham, MA, USA).

The produced powder sample was separated from apatite and nepheline in H_2SO_4 or HCl dilute water solutions (50–100 g/L), with solid/liquid ratio S:L = 1:3–1:4 (Figure 5). The suspension was mixed in a magnetic stirrer at the temperature of 50 ± 5 °C for 2 h. After dissolution of apatite and nepheline, a purified titanite-aegirine concentrate was separated by filtration.

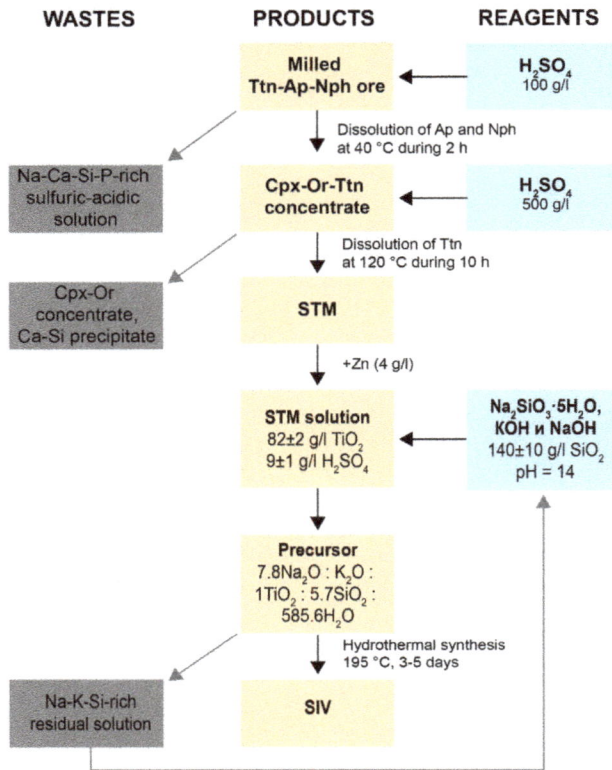

Figure 5. A principal scheme of apatite-nepheline-titanite ore processing for SIV synthesis. Mineral abbreviations see in Figure 3.

Then the concentrate was mixed with sulfuric acid (500 ± 100 g/L H_2SO_4) at the ratio of S:L = 1:3.5, and the mixture was poured into a flask equipped with a stirrer and a condenser [35]. The concentrate interacted with H_2SO_4 at the boiling point of the reaction mass for 9–10 h. The produced Ti-bearing sulfuric acid liquid phase was separated from insoluble aegirine by filtration, and used for crystallization of titanyl sulfates $TiOSO_4 \cdot H_2O$ (STM) and $(NH_4)_2TiO(SO_4)_2 \cdot H_2O$ (STA) [28,36]. The last compound was obtained by means of reaction of STM with ammonium sulfate.

Hydrothermal synthesis of microporous titanosilicates (ETS-4, SIV, AM-4, etc.) was performed by analogy with ETS-10 [37], with STM as Ti source. To prepare a titanium precursor, we used STM solution with concentrations of 82 ± 2 g/L TiO_2 and 9 ± 1 g/L H_2SO_4. Four-valent titanium contained in the solution was partially reduced by addition of zinc powder (4.4 ± 0.1 g/L). Zinc consumption was assumed on the basis of Ti_2O_3 concentration of 5–10 g/L. Solution of sodium silicate $Na_2SiO_3 \cdot 7H_2O$ containing 140 ± 10 g/L of SiO_2 was also used. The reduced titanium solution was gradually dosed into the sodium silicate solution; after that, sodium and potassium alkali solutions were added to adjust pH. Total time for the mixture preparation was 3–3.5 h. The resulting gel-like suspension was placed in the Parr 4666-FH-SS autoclave (Parr Instrument, Moline, IL, USA) and held at 195 ± 5 °C for 3–5 days. The produced precipitate was separated and, after washing, placed in a drying cabinet. Drying time at the temperature of 70 ± 5 °C was 7–10 h.

Modal composition of the synthesized samples was determined by means of a XRD-6000 diffractometer (Shimadzu, Kyoto, Japan), with a Cu X-ray tube operated at 60 kV and 55mA.

Bulk chemical compositions of the powder samples were determined using an X-ray fluorescence spectrometer Spectroscan MAKS-GV (Spectron, St. Petersburg, Russia), operating in WDS mode at 40 kV. Morphology of the produced particles was analyzed with an optical microscope DM-2500P (Leica, Wetzlar, Germany) and a scanning electron microscope XL 30 (Philips, Amsterdam, The Netherlands) equipped with an Oxford INCA EDS analyzer (Oxford Instruments, Oxford, UK).

The BET surface properties and the porosity of the final products were characterized by nitrogen and adsorption/desorption method at 77 K using a surface area analyzer Micromeritics TriStar 3020. The average pore diameter (Dav) was calculated as 4 V/S. Prior to adsorption/desorption measurements the sample was degased at 393 K for about 24 h. Low degassing temperature was chosen to avoid any structural changes of the material.

4. Results

4.1. STA and STM Production

Acidic cleaning of (apatite)-nepheline-titanite ore occurs due to consecutive fluorapatite and nepheline dissolution under the influence of diluted sulfuric or hydrochloric acids:

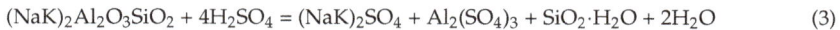

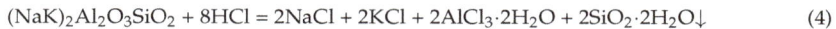

$$Ca_5(PO_4)_3F + H_2SO_4 \rightarrow 5CaSO_4\downarrow + 3H_3PO_4 + HF \tag{1}$$

$$Ca_5(PO_4)_3F + 10HCl \rightarrow 5CaCl_2 + 3H_3PO_4 + HF \tag{2}$$

$$(NaK)_2Al_2O_3SiO_2 + 4H_2SO_4 = (NaK)_2SO_4 + Al_2(SO_4)_3 + SiO_2 \cdot H_2O + 2H_2O \tag{3}$$

$$(NaK)_2Al_2O_3SiO_2 + 8HCl = 2NaCl + 2KCl + 2AlCl_3 \cdot 2H_2O + 2SiO_2 \cdot 2H_2O\downarrow \tag{4}$$

Because of close titanite intergrowth with nepheline and especially fluorapatite, it is necessary to ensure fine grinding of ore (<63 μm). It is obvious that at this stage we should rather use hydrochloric acid, which does not cause formation of silica and insoluble calcium compounds that can inhibit titanite dissolution at the next stage.

We found out that cold acidic cleaning of homogeneous finely ground ore was not very effective. Temperature increase up to 50 °C activates apatite dissolution with the corresponding phosphorus transfer into the liquid phase (Table 3); however, nepheline remains undissolved. Two-stage ore cleaning with cold and then heated sulfuric acid gives satisfactory results for both apatite (P) and nepheline (Al), without Ti losses. Addition of hydrochloric acid is the most effective for apatite separation; however, absence of insoluble secondary products compensates quite well slow dissolution of nepheline.

Table 3. Chemical composition of ground ore cleaned under different conditions.

Experiment	Conditions	Ore Composition, wt %		
		TiO$_2$	Al$_2$O$_3$	P$_2$O$_5$
1	H$_2$SO$_4$, 80 g/L, S:L = 1:3, 4 h, 18 °C	30.5	0.97	4.23
2	H$_2$SO$_4$, 80 g/L, S:L = 1:4, 2h, 50 °C	28.0	2.69	1.00
3	1 stage: H$_2$SO$_4$, 80 g/L, S:L = 1:3, 4 h, 18 °C; 2 stage: H$_2$SO$_4$, 100 g/L, S:L = 1:4, 2 h, 50 °C	32.0	0.27	1.15
4	HCl, 50 g/L, S:L = 1:4, 2 h, 50 °C	31.5	1.62	0.21

Electron-microscope analyses showed that titanite grains cleaned with sulfuric acid were covered with the smallest particles of Ca sulfate and aluminum alum (Figure 6a). After cleaning with hydrochloric acid, they became free from any secondary phases, as we predicted (Figure 6b).

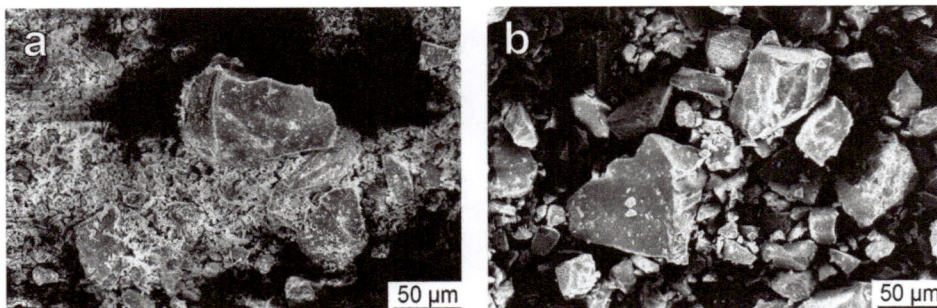

Figure 6. SE-images of titanite grains cleaned by sulfuric acid (**a**) and hydrochloric acid (**b**).

The titanite-aegirine concentrate processing stage included dissolution of titanite in heated concentrated sulfuric acid with formation of insoluble calcium sulfate and amorphous silica:

$$CaSiTiO_5 + 2H_2SO_4 = TiOSO_4 + SiO_2 \cdot H_2O\downarrow + CaSO_4\downarrow + H_2O.$$

This reaction can be divided into two consecutive stages: kinetic stage and diffusion stage. The first stage gradually transforms into the second one due to accumulation of solid reaction products. Both stages, especially the second one, are diffusion limited. The mechanism of titanite dissolution is the same for both stages, but the dissolution rate is different (Figure 7). In practice, proportion of leached titanium (IV) in the liquid phase (i.e., proportion of dissolved titanite) ranges from 35 to 90% [28]. Optimal sulfuric acid concentration of 600 g/L prevents from precipitation of titanyl sulfate that exists in such solution only in a stable molecular-dispersive form.

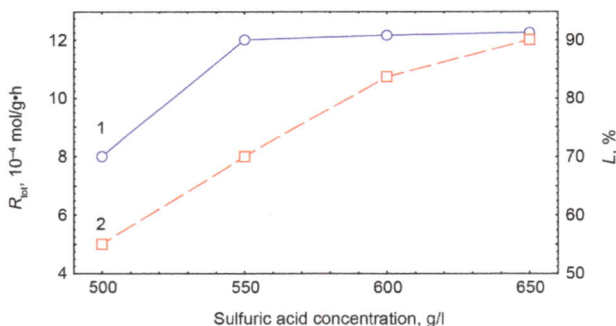

Figure 7. Ratio of Ti(IV) leaching from titanite (R_{tot}, line 1) and titanium recovery into the liquid phase (L, line 2) vs. sulfuric acid concentration at S:L = 1:3.5.

During the third stage of the experiment, the produced titanyl-sulfate solution was used to crystallize STA and STM salts. Their bulk crystallization from the solutions can be initiated by addition of salting-out reagents, in particular, ammonium sulfate [38]:

$$TiOSO_4(L) + (NH_4)_2SO_4(S) + H_2O = (NH_4)_2TiO(SO_4)_2 \cdot H_2O(S).$$

This composition becomes supersaturated when total concentration of ammonium sulfate and sulfuric acid in the solution reaches 550 g/L, with the ammonium sulfate:acid ratio to be 1.2:1. If this ratio decreases to the value of 1:2 due to growth of sulfuric acid concentration (400 g/L H_2SO_4

and 80–85 g/L TiO_2), another solid "acidic" compound $(NH_4)_2TiO(SO_4)_2 \cdot 0.15H_2SO_4 \cdot 0.89H_2O$ will precipitate instead of STA.

To produce STM, we pre-evaporated or diluted titanyl-sulfate solution with sulfuric acid up to 850–900 g/L H_2SO_4. Crystalline STM precipitates during boiling of this solution for 5–7 h.

4.2. Synthetic Ivanyukite-Na (SIV) Synthesis

For hydrothermal synthesis of ivanyukite-Na (SIV), we used STM water solutions pre-reacted with metallic zinc to transform Ti^{4+} to Ti^{3+} [39]. The experimental conditions are shown in Table 4.

Table 4. Experimental conditions of SIV synthesis.

Experiment	Pre-Reaction with Zn	$TiO_2:SiO_2$, Moles	pH	T °C	Time, Day
1	+	1:4	11.5	200	3
2	+	1:5	12.5	190	3
3	+	1:4	12.5	200	5
4	–	1:4	11.5	200	5
5	–	1:5	12.5	200	3

Long-term crystallization of SIV occurs in the basic (pH up to 12.5) system Na_2O-K_2O-TiO_2-H_2SO_4-SiO_2-H_2O supersaturated with Na and Si [39,40]. After a three-day synthesis (experiments 1 and 2), the precipitates consist of two SIV polymorphs: rhombohedral SIV-*T*, $Na_4(TiO)_4(SiO_4)_3 \cdot nH_2O$, and cubic SIV-*C*, $Na_3H(TiO)_4(SiO_4)_3 \cdot nH_2O$ (Table 5). Longer crystallization (experiment 3) causes protonation of SIV-*T* on the schema $Na + TiO_6 + H_2O \rightleftarrows \square + TiO_5(OH) + Na(OH)$, with the corresponding transformation of SIV-*T* into SIV-*C* (Figure 8) by analogy with natural ivanyukites [19]. When Ti^{4+} predominates over Ti^{3+} (experiments 4 and 5), ETS-4, $Na_6Ti_5Si_{12}O_{34}(O,OH)_5 \cdot 11H_2O$, crystallizes instead of SIV.

Table 5. Modal composition and surface properties of titanosilicate precipitates.

Experiment	Modal Composition	Surface Properties		
		S, m^2/g	V, cm^3/g	D_{av}, nm
1	SIV-*T* + SIV-*C*	148.0 ± 0.9	0.68	18.4 ± 0.2 (3.4–68.0)
2	SIV-*T* + SIV-*C*	158 ± 1	0.73	18.6 ± 0.1 (2.9–65.4)
3	SIV-*C*	143.3 ± 0.6	0.75	20.9 ± 0.1 (4.3–75.0)
4	70% ETS-4 + 30%-SIV-*T*	81.0 ± 0.4	0.18	8.9 ± 0.1 (2.1–54.3)
5	65% ETS-4 + 30% SIV-*T* + 5% TiO_2	85.1 ± 0.5	0.22	10.3 ± 0.1 (1.8–58.2)

S—specific surface area, *V*—specific volume of pores, D_{av}—average diameter of pores, mean $\pm$ SD (min–max).

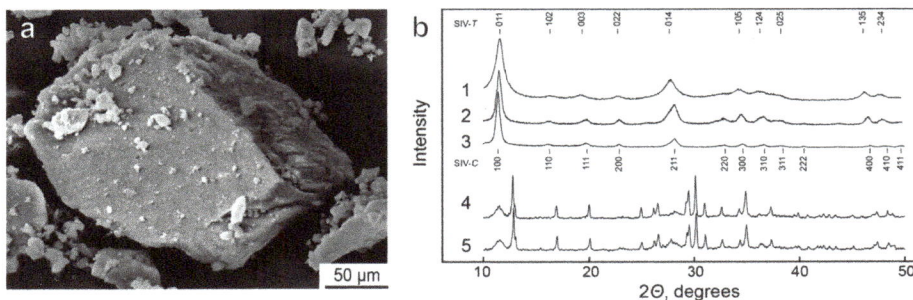

Figure 8. SE-image of SIV-*C* aggregates (experiment 3) (**a**) and powder X-Ray diffraction patterns of products obtained in experiments 1–5 (**b**).

Size of SIV particles varies from 2 to 15 μm. High specific surface area (143–158 m^2/g) and general pore volume (0.68–0.75 cm^3/g) of SIV aggregates as well as presence of extra-framework cations and water molecules in wide channels of SIV crystal structure (see Figure 1) result in impressive SIV cation-exchange properties [20,26].

5. Discussion

Environmental technologies play an increasingly important role in various industrial applications, including production of new materials. Being the main world suppliers of new zircon-, niobo- and titanosilicates, massifs of the Kola Alkaline Province are important sources of information on conditions of mineral formation and synthesis of the corresponding mineral-like functional materials. Again, we should like to emphasize that titanosilicate materials of practical importance, such as ETS-4, IE-911, SL3 and SIV, are complete analogues of zorite, sitinakite, punkaruaivite and ivanyukite-Na discovered in Lovozero and Khibiny massifs. On the other hand, synthesis of the natural minerals analogues allows us to study conditions of natural minerals formation for effective exploration of the minerals—prototypes of functional materials.

The Kola Science Center of the Russian Academy of Sciences has developed technologies for recovery of all listed mineral-like materials from the products of chlorine processing of the Lovozero's loparite, and sulfuric acid processing of the Khibiny's titanite (its resources is about 170 Mt [33]). The latest technology presented in this paper seems to be safer and more profitable, and enables to produce about 1 t of titanium tanning agent/precursor of sorbents, and then about 1 t of SIV from 1 t of titanite concentrate (1.5–2 t of apatite-nepheline-titanite ore), with simultaneous recycling of sulfuric acid produced by the Monchegorsk Copper-Nickel Combine.

The main consumer of SIV is the nuclear power industry (direct processing of liquid radioactive waste into titanate ceramics with reduction in waste volume by 2–3 orders). Precursor of STA/STM is used in the leather industry (due to higher safety of titanium tannic in comparison with chromium ones). Other potential applications of the developed materials are selective or collective extraction of non-ferrous and, especially, precious metals from process solutions and effluents, (photo)catalysis and medicine. In addition, during decomposition of titanite with sulfuric acid, a moist precipitate is produced. It contains calcium, silicon and titanium compounds (wt %): $CaSO_4$ 40–45, SiO_2 15–20, TiO_2 5–10, and sulfuric acid. Neutralization of sediment with lime milk followed by annealing produces a titanium pigment filler, which is widely used in paint-and-varnish industry.

6. Conclusions

(1) Large lenses and layers of (apatite)-nepheline-titanite ore, up to 50 m thick, are widespread in apical parts of the most apatite deposits in the Khibiny massif. This ore can be selectively mined and processed as a new kind of titanium raw materials;

(2) This ore can be processed by acidic cleaning only, without a conventional flotation stage. With a new approach, the yield of Ti-salts becomes 5–6 times higher than in the known process flow diagram of titanite recovery from apatite-nepheline ore with flotation followed by sulfuric-acidic processing of titanite concentrate;

(3) Sulfuric-acidic processing of titanite ore enables to recover up to 90% Ti into the liquid phase that becomes a common precursor for hydrothermal synthesis of functional titanosilicates, in particular, synthetic analogues of the ivanyukite-group minerals. To prevent from ETS-4 crystallization, the rate of titanium cations hydrolysis should be reduced by partial transformation of Ti^{4+} into Ti^{3+};

(4) Perspectives of SIV application include recovering of liquid radioactive wastes into Synroc-type titanate ceramics, selective or collective extraction of non-ferrous and, especially, precious metals from process solutions and effluents.

Author Contributions: A.I.N., L.G.G. and M.V.M. designed the experiments, performed ore processing and SIV synthesis, wrote and reviewed the manuscript. E.S.S. and G.O.S. participated in the experiments. G.Y.I. and V.N.Y. took and analyzed ore samples and wrote the manuscript. All authors discussed the manuscript.

Funding: The research is supported by the Russian Foundation for Fundamental Researches, Grant 18-29-12039, and the Program 35 of the Presidium of the Russian Academy of Sciences.

Acknowledgments: We are grateful to JSC "Apatite" for help with titanite ore sampling. We would like to thank the anonymous reviewers for their suggestions and comments.

Conflicts of Interest: The authors declare no conflict of interest. The founding sponsors had no role in the design of the study; in the collection, analyses, or interpretation of data; in the writing of the manuscript, and in the decision to publish the results.

References

1. Fedorov, S.G.; Nikolaev, A.I.; Brylyakov, Y.E.; Gerasimova, L.G.; Vasilieva, N.Y. *Chemical Processing of Mineral Concentrates of the Kola Peninsula*; The Kola Science Center Press: Apatity, Russia, 2003. (In Russian)
2. Popova, E.A.; Zalessky, V.G.; Yakovenchuk, V.N.; Krivovichev, S.V.; Lushnikov, S.G. Ferroelectric phase transition and relaxer-like behavior of loparite-(Ce). *Ferroelectrics* **2014**, *469*, 130–137. [CrossRef]
3. Men'shikov, Y.P.; Krivovichev, S.V.; Pakhomovsky, Y.A.; Yakovenchuk, V.N.; Ivanyuk, G.Y.; Mikhailova, J.A.; Armbruster, T.; Selivanova, E.A. Chivruaiite, $Ca_4(Ti,Nb)_5[(Si_6O_{17})_2[(OH,O)_5]\cdot13-14H_2O$, a new mineral from hydrothermal veins of Khibiny and Lovozero alkaline massifs. *Am. Mineral.* **2006**, *91*, 922–928. [CrossRef]
4. Yakovenchuk, V.N.; Krivovichev, S.V.; Men'shikov, Y.P.; Pakhomovsky, Y.A.; Ivanyuk, G.Y.; Armbruster, T.; Selivanova, E.A. Chivruaiite, a new mineral with ion-exchange properties. In *Minerals as Advanced Materials I*; Krivovichev, S.V., Ed.; Springer: Berlin/Heidelberg, Germany, 2008; pp. 57–63. ISBN 978-3-540-77122-7.
5. Kuznicki, S.M.; Bell, V.A.; Nair, S.; Hillhouse, H.W.; Jacubinas, R.M.; Braunbarth, C.M.; Toby, B.H.; Tsapatis, M. A titanosilicate molecular sieve with adjustable pores for size-selective adsorption of molecules. *Nature* **2001**, *412*, 720–724. [CrossRef] [PubMed]
6. Braunbarth, C.; Hillhouse, H.W.; Nair, S.; Tsapatis, M.; Burton, A.; Lobo, R.F.; Jacubinas, R.M.; Kuznicki, S.M. Structure of strontium ion-exchanged ETS-4 microporous molecular sieves. *Chem. Mater.* **2000**, *12*, 1857–1865. [CrossRef]
7. Yakovenchuk, V.N.; Pakhomovsky, Y.A.; Bogdanova, A.N. Kukisvumite, a new mineral from alkaline pegmatites of the Khibiny massif (Kola Peninsula). *Mineral. Zhurnal* **1991**, *13*, 63–67. (In Russian)
8. Yakovenchuk, V.N.; Ivanyuk, G.Y.; Pakhomovsky, Y.A.; Selivanova, E.A.; Men'shikov, Y.P.; Korchak, J.A.; Krivovichev, S.V.; Spiridonova, D.V.; Zalkind, O.A. Punkaruaivite, $LiTi_2[Si_4O_{11}(OH)](OH)_2\cdot H_2O$, a new mineral species from hydrothermal assemblages, Khibiny and Lovozero alkaline massifs, Kola Peninsula, Russia. *Can. Mineral.* **2010**, *48*, 41–50. [CrossRef]
9. Yakovenchuk, V.N.; Krivovichev, S.V.; Pakhomovsky, Y.A.; Selivanova, E.A.; Ivanyuk, G.Y. Microporous titanosilicates of the lintisite-kukisvumite group and their transformation in acidic solutions. In *Minerals as Advanced Materials II*; Krivovichev, S.V., Ed.; Springer: Berlin/Heidelberg, Germany, 2012; pp. 229–238, ISBN 978-3-642-20017-5.
10. Kalashnikova, G.O.; Selivanova, E.A.; Pakhomovsky, Y.A.; Zhitova, E.S.; Yakovenchuk, V.N.; Ivanyuk, G.Y.; Nikolaev, A.I. Synthesis of new functional materials by the self-assembly of titanosilicate nanolayers $Ti_2Si_4O_{10}(OH)_4$. *Perspect. Mater.* **2015**, *10*, 64–72. (In Russian)
11. Dadachov, M.S.; Rocha, J.; Ferreira, A.; Lin, Z.; Anderson, M.W. Ab initio structure determination of layered sodium titanium silicate containing edge-sharing titanate chains (AM-4) $Na_3(Na,H)Ti_2O_2[Si_2O_6]_2\cdot2.2H_2O$. *Chem. Commun.* **1997**, *24*, 2371–2372. [CrossRef]
12. Perez-Carvajal, J.; Lalueza, P.; Casado, C.; Téllez, C.; Coronas, J. Layered titanosilicates JDF-L1 and AM–4 for biocide applications. *Appl. Clay Sci.* **2012**, *56*, 30–35. [CrossRef]
13. Men'shikov, Y.P.; Sokolova, E.V.; Egorov-Tismenko, Y.K.; Khomyakov, A.P.; Polezhaeva, L.I. Sitinakote $Na_2KTi_4Si_2O_{13}(OH)\cdot4H_2O$, a new mineral. *Zap. RMO* **1992**, *1*, 94–99. (In Russian)
14. Clearfield, A.; Bortun, L.N.; Bortun, A.I. Alkali metal ion exchange by the framework titanium silicate $M_2Ti_2O_3SiO_4\cdot nH_2O$ (M=H, Na). *React. Funct. Polym.* **2000**, *43*, 85–95. [CrossRef]
15. Mann, N.R.; Todd, T.A. Removal of cesium from acidic radioactive tank waste by using Ionsiv IE-911. *Sep. Sci. Technol.* **2005**, *39*, 2351–2371. [CrossRef]

16. Dyer, A.; Newton, J.; O'Brien, L.; Owens, S. Studies on a synthetic sitinakite-type silicotitanate cation exchanger: Part 1: Measurement of cation exchange diffusion coefficients. *Micropor. Mesopor. Mater.* **2009**, *117*, 304–308. [CrossRef]

17. Yakovenchuk, V.N.; Nikolaev, A.P.; Selivanova, E.A.; Pakhomovsky, Y.A.; Korchak, J.A.; Spiridonova, D.V.; Zalkind, O.A.; Krivovichev, S.V. Ivanyukite-Na-T, ivanyukite-Na-C, ivanyukite-K, and ivanyukite-Cu: New microporous titanosilicates from the Khibiny massif (Kola Peninsula, Russia) and crystal structure of ivanyukite-Na-T. *Am. Mineral.* **2009**, *94*, 1450–1458. [CrossRef]

18. Yakovenchuk, V.N.; Selivanova, E.A.; Ivanyuk, G.Y.; Pakhomovsky, Y.A.; Spiridonova, D.V.; Krivovichev, S.V. First natural pharmacosiderite-related titanosilicates and their ion-exchange properties. In *Minerals as Advanced Materials I*; Krivovichev, S.V., Ed.; Springer: Berlin/Heidelberg, Germany, 2008; pp. 27–35, ISBN 978-3-540-77122-7.

19. Yakovenchuk, V.N.; Selivanova, E.A.; Krivovichev, S.V.; Pakhomovsky, Y.A.; Spiridonova, D.V.; Kasikov, A.G.; Ivanyuk, G.Y. Ivanyukite-group minerals: Crystal structure and cation-exchange properties. In *Minerals as Advanced Materials II*; Krivovichev, S.V., Ed.; Springer: Berlin/Heidelberg, Germany, 2012; pp. 205–211, ISBN 978-3-642-20017-5.

20. Britvin, S.N.; Gerasimova, L.G.; Ivanyuk, G.Y.; Kalashnikova, G.O.; Krzhizhanovskaya, M.G.; Krivovivhev, S.V.; Mararitsa, V.F.; Nikolaev, A.I.; Oginova, O.A.; Panteleev, V.N.; et al. Application of titanium-containing sorbents for treating liquid radioactive waste with the subsequent conservation of radionuclides in Synroc-type titanate ceramics. *Theor. Found. Chem. Eng.* **2016**, *50*, 598–606. [CrossRef]

21. Britvin, S.N.; Lotnyk, A.; Kienle, L.; Krivovichev, S.V.; Depmeier, W. Layered hydrazinium titanate: Advanced reductive adsorbent and chemical toolkit for design of titanium dioxide nanomaterials. *J. Am. Chem. Soc.* **2011**, *133*, 9516–9525. [CrossRef] [PubMed]

22. Perera, A.S.; Coppens, M.O. Titano-silicates: Highlights on development, evolution and application in oxidative catalysis. In *Catalysis 28*; Spivey, J., Dooley, K.M., Han, Y.F., Eds.; Royal Society of Chemistry: London, UK, 2016; pp. 119–143, ISBN 1782626859.

23. Du, H.; Fang, M.; Chen, J.; Pang, W. Synthesis and characterization of a novel layered titanium silicate JDF-L1. *J. Mater. Chem.* **1996**, *6*, 1827–1830. [CrossRef]

24. Kim, S.N.; Kim, J.; Kim, H.Y.; Cho, H.Y.; Ahn, W.S. Adsorption/catalytic properties of MIL-125 and NH₂-MIL-125. *Catal. Today* **2013**, *204*, 85–93. [CrossRef]

25. Vilela, S.; Salcedo-Abraira, P.; Colinet, I.; Salles, F.; de Koning, M.; Joosen, M.; Serre, C.; Horcajada, P. Nanometric MIL-125-NH₂ Metal–Organic Framework as a Potential Nerve Agent Antidote Carrier. *Nanomaterials* **2017**, *7*, 321. [CrossRef] [PubMed]

26. Milyutin, V.V.; Nekrasova, N.A.; Yanicheva, N.Y.; Kalashnikova, G.O.; Ganicheva, Y.Y. Sorption of cesium and strontium radionuclides onto crystalline alkali metal titanosilicates. *Radiochemistry* **2017**, *59*, 65–69. [CrossRef]

27. Nikolaev, A.I.; Larichkin, F.D.; Gerasimova, L.G.; Glushchenko, Y.G.; Novoseltseva, V.D.; Maslova, M.V.; Nikolaeva, O.A. *Titanium and Titanium Compounds: Resources, Production, and Markets—Current State and Trends*; The Kola Science Center Press: Apatity, Russia, 2011. (In Russian)

28. Gerasimova, L.G.; Maslova, M.V.; Nikolaev, A.I. *Investigation of Non-Equilibrium Chemical Processes in Raw-Material Technologies*; LKM Press: Moscow, Russia, 2014. (In Russian)

29. Maslova, M.V.; Gerasimova, L.G.; Nikolaev, A.I. Treatment of apatite nepheline ore enrichment waste. *Mod. App. Sci.* **2015**, *9*, 81–92.

30. Arzamastsev, A.A.; Arzamastseva, L.V.; Travin, A.V.; Belyatsky, B.V.; Shamatrina, A.M.; Antonov, A.V.; Larionov, A.N.; Rodionov, N.V.; Sergeev, S.A. Duration of formation of magmatic system of polyphase Paleozoic alkaline complexes of the central Kola: U-Pb, Rb-Sr, Ar-Ar data. *Dokl. Earth Sci.* **2007**, *413*, 432–436. [CrossRef]

31. Yakovenchuk, V.N.; Ivanyuk, G.Y.; Pakhomovsky, Y.A.; Men'shikov, Y.P. *Khibiny*; Wall, F., Ed.; Laplandia Minerals: Apatity, Russia, 2005; ISBN 5900395480.

32. Ivanyuk, G.; Yakovenchuk, V.; Pakhomovsky, Y.; Kalashnikov, A.; Mikhailova, J.; Goryainov, P. Self-Organization of the Khibiny Alkaline Massif (Kola Peninsula, Russia). In *Earth Sciences*; Ahmad Dar, I., Ed.; INTECH: London, UK, 2012; pp. 131–156, ISBN 978-953-307-861-8.

33. Ivanyuk, G.Y.; Konopleva, N.G.; Pakhomovsky, Y.A.; Yakovenchuk, V.N.; Mikhailova, J.A.; Bazai, A.V. Titanite of the Khibiny Alkaline Massif (Kola Peninsula). *Zap. RMO* **2016**, *3*, 36–55. (In Russian)

34. Kalinkin, A.M. Physicochemical processes in mechanical activation of titanium- and calcium-containing minerals. *Russ. J. Appl. Chem.* **2007**, *80*, 1613–1620. [CrossRef]

35. Lazareva, I.V.; Gerasimova, L.G.; Maslova, M.V.; Okhrimenko, R.F. Interaction of titanite with sulfuric acid solution. *Russ. J. Appl. Chem.* **2006**, *79*, 18–21. [CrossRef]

36. Melikhov, I.V. Ways of using crystallization to produce solid products with specified properties. *Chem. Ind.* **1997**, *7*, 488–500. (In Russian)

37. Ji, Z.; Yilmaz, B.; Warzywoda, J.; Sacco, A., Jr. Hydrothermal synthesis of titanosilicate ETS-10 using $Ti(SO_4)_2$. *Micropor. Mezpor. Mat.* **2005**, *81*, 1–10. [CrossRef]

38. Motov, D.L. *Physico-Chemistry and Sulphate Technology of Titanium-Rare Metal Raw Materials. Part 2*; The Kola Science Center Press: Apatity, Russia, 2002. (In Russian)

39. Gerasimova, L.G.; Nikolaev, A.I.; Kuz'mich, J.V.; Ivanyuk, G.Y.; Yakovenchuk, V.N.; Schukina, E.S. Patent RU-2539303: Method of Titanium-Silicon Sodium-Containing Composition Obtaining. U.S. Patent RU2539303C1, 20 January 2015.

40. Gerasimova, L.G.; Maslova, M.V.; Nikolaev, A.I. Synthesis of new nano-porous titanosilicates using ammonium oxysulphotitanite. *J. Glass Phys. Chem.* **2013**, *39*, 846–855. [CrossRef]

MDPI

St. Alban-Anlage 66

4052 Basel

Switzerland

Tel. +41 61 683 77 34

Fax +41 61 302 89 18

www.mdpi.com

Minerals Editorial Office

E-mail: minerals@mdpi.com

www.mdpi.com/journal/minerals

www.ingramcontent.com/pod-product-compliance
Lightning Source LLC
Chambersburg PA
CBHW051705210326
41597CB00032B/5371